Multiphase Flow Dynamics 3

Nikolay Ivanov Kolev

Multiphase Flow Dynamics 3

Thermal Interactions

Springer

Author

Dr. Nikolay Ivanov Kolev
Möhrendorferstr. 7
91074 Herzogenaurach
Germany
E-mail: Nikolay.Kolev@herzovision.de

ISBN 978-3-662-50117-7 ISBN 978-3-642-21372-4 (eBook)

DOI 10.1007/978-3-642-21372-4

Softcover re-print of the Hardcover 1st edition 2011

Typeset & Cover Design: Scientific Publishing Services Pvt. Ltd., Chennai, India.

Printed on acid-free paper

9 8 7 6 5 4 3 2 1

springer.com

To Iva, Rali and Sonja with love!

Morning, July. 2004, Nikolay Ivanov Kolev, 36×48cm oil on linen

Nikolay Ivanov Kolev, PhD, DrSc
Born 1.8.1951, Gabrowo, Bulgaria

Summary

This monograph contains theory, methods and practical experience for describing complex transient multiphase processes in arbitrary geometrical configurations. It is intended to help applied scientists and practicing engineers to understand better natural and industrial processes containing dynamic evolutions of complex multiphase flows. The book is also intended to be a useful source of information for students in the high semesters and in PhD programs.

This monograph consists of five volumes:

Vol. 1 Fundamentals, 4th edn, (14 Chapters and 2 Appendices), 782 pages;
Vol. 2 Mechanical Interactions, 4th edn, (11 Chapters), 364 pages;
Vol. 3 Thermal Interactions, 4th edn, (16 Chapters), 678 pages;
Vol. 4 Turbulence, Gas Absorption and Release by Liquid, Diesel Fuel Properties, 2nd edn, (13 Chapters), 328 pages;
Vol. 5 Nuclear Thermal Hydraulics, 2nd edn, (17 Chapters), 848 pages.

In Volume 1 the concept of three-fluid modeling is presented in detail "from the origin to the applications". This includes derivation of local volume- and time-averaged equations and their working forms, development of methods for their numerical integration and finally finding a variety of solutions for different problems of practical interest.

Special attention is paid in Volume 1 to the link between the partial differential equations and the constitutive relations called closure laws without providing any information on the closure laws.

Volumes 2 and 3 are devoted to these important constitutive relations for the mathematical description of the mechanical and thermal interactions. The structure of the volumes is in fact a state-of-the-art review and selection of the best available approaches for describing interfacial transfer processes. In many cases the original contribution of the author is incorporated in the overall presentation. The most important aspects of the presentation are that they stem from the author's long years of experience developing computer codes. The emphasis is on the practical use of these relationships: either as stand-alone estimation methods or within a framework of computer codes.

Volume 4 is devoted to the turbulence in multiphase flows.

The nuclear thermal hydraulic is the science providing knowledge about the physical processes occurring during the transferring the fission heat released in structural materials due to nuclear reactions into its environment. Along its way to the environment the thermal energy is organized to provide useful mechanical work or useful heat or both. Volume 5 is devoted to the nuclear thermal hydraulics. In a way this is the most essential application of the multiphase fluid dynamics in analyzing steady and transient processes in nuclear power plants.

In particular, Volume 3 contains information on how to describe the flow patterns and the specific thermal interactions between the velocity fields in flight.

Chapter 1 presents nucleation in liquids. Chapter 2 presents different aspects of the bubble growth in superheated liquid and the connection to computational system models. Condensation of pure steam bubbles is considered in Chapter 3.

Chapter 4 considering the bubble departure mechanism on heated walls is updated with new information. The theory is extended to subcooled liquid.

Chapter 5 presents a solution of the coupled problem of transient bubble growth or collapse and relates it to the state-of-the-art. A comparison with experimental data is given. Finally, a discussion is provided on how to couple such low-scale physics with large-scale physics in multiphase flow computer codes.

Chapter 6 considering nucleate boiling is updated. Additional information coming from boiling of fluids with nanoparticles confirms the sound physical basics of the author's method. Additional information is provided for nucleation site density at high pressure.

An interesting result of the theory presented in these chapters is the prediction of a critical heat flux without an empirical correlation for critical heat flux as a result of the mutual bubble interaction for increasing wall superheating. Applying the theory for the inverted problem of the flashing of superheated water in pipes as given in Chapter 7, surprisingly supported the validity of the new approach.

The state-of-the-art in boiling theory is presented in Chapters 8 to 11 where boiling in subcooled liquid, natural convection film boiling, critical heat flux, forced convection film boiling, and film boiling on vertical plates and spheres is presented. The emphasis is on the elaboration of all coupling terms between the fluids, and between the wall and fluid required for closure of the overall description. Chapter 9 considering boiling of subcooled liquid is updated. Chapter 20 considering forced convection boiling is updated. The 2005 look-up table by *Groeneveld* et al. is demonstrated to be an excellent scaling base for critical heat flux in bundles for variety of geometries and conditions.

Chapter 12 provides information on all heat- and mass-transfer processes across a droplet interface starting with the nucleation theory, and going through the droplet growth, self-condensation stop, heat transfer across a droplet interface without mass transfer, direct contact condensation of pure steam on a subcooled droplet, spontaneous flushing of a superheated droplet, evaporation of a saturated droplet in superheated gas, and droplet evaporation in a gas mixture. A similar approach is applied to the description of the interface processes at a film/gas interface in Chapter 13 with a careful treatment of the influence of the turbulent pulsation on the interfacial heat and mass transfer. A set of updated empirical methods for prediction of condensation on cooled walls with and without a noncondensable is presented in Chapter 14. Test examples are provided for demonstration of the application of the theory.

Chapter 15 provides information on the implementation of the discrete ordinate method for radiation transport in multiphase computer codes. In this chapter the dimensions of the problem, the differences between micro- and macrointeractions and the radiation transport equation are discussed. Then, the finite-volume representation of the radiation transport is given and different aspects of the numerical integration are discussed. The computation of some material properties is discussed. Then three specific radiation transport cases of importance for the melt–water interaction are discussed in detail: a spherical cavity of gas inside a molten material; concentric spheres of water droplets, surrounded by vapor surrounded by molten material; clouds of spherical particles of radiating material surrounded by a layer of vapor surrounded by water. For the last case the useful Lanzenberger's solution is presented and its importance is demonstrated in two practical cases.

The book ends with Chapter 16, which provides information on how to verify multiphase flow models by comparing them with experimental data and analytical solutions. First, Section 16.2 contains a quick look of the IVA computer code with several interesting demonstration of the power of the technology described in this four Volumes work. A list of references is provided in Sections 16.3 and 16.4 documenting the IVA-code development and validation. The complexity of the problems is gradually increased from very simple ones to problems with very complex melt–water interaction multifluid flows with dynamic fragmentation and coalescence and strong thermal and mechanical interactions. In particular the following cases are described and compared with the prediction using the basics presented in different chapters of this book: material relocation – gravitational waves (2D), U-tube benchmarks such us adiabatic oscillations, single-phase natural convection in a uniformly heated vertical part of a U-tube, single-phase natural convection in a uniformly heated inclined part of a U-tube, single-phase natural convection in a U-tube with an inclined part heated by steam condensation, steady-state single-phase nozzle flow, pressure waves in single-phase flow, 2D gas explosion in a space filled previously with gas, 2D gas explosion in space with internals previously filled with liquid, film entrainment in pipe flow, water flashing in nozzle flow, pipe blow-down with flashing, single-pipe transients, complex pipe network transients, boiling in pipes and rod bundles, critical heat flux, post-critical heat flux heat transfer, film boiling, behavior of clouds of cold and very hot spheres in water, experiments with dynamic fragmentation and coalescence like the FARO L14, 20, 24, 28, 31 experiments, PREMIX 13, 15, 17, 18 experiment, RIT and IKE experiment. Chapter 16 also contains some additional experiments and movies documenting the performance of the method for fast pressure wave propagation in 2D geometry and interesting acoustical problems of melt-water interaction. Review of the state-of-the-art of the instability analyses of boiling systems is provided and interesting comparison of the modern IVA predictions with large-scale AREVA- experiments is provided in Section 16.6.3. Powerful demonstration of the methods for analyzing pressure-wave propagations in single- and two-phase systems is given by comparison with the Interatome experiments from 1983 performed on simple and complex pipe networks, Section 16.8.3. Section 16.15 contains comparison with 333 experiments for variety of bundles, flow regimes including dry out, steady state and transients. It clearly demonstrates the power of the method and its well-defined uncertainty to simulate boiling processes in complex geometry. In addition an application of a powerful method for investigation of the propagation of input and model uncertainties on the final results by using the Monte Carlo method and regression analysis is demonstrated for the prediction of nonexplosive melt-water interactions. Benchmarks for testing the 3D capabilities of computer codes are provided for the rigid-body steady-rotation problem, pure radial symmetric flow, radial-azimuthal symmetric flow. Examples of very complex 3D flows are also given such as small break loss of coolant, asymmetric steam–water interaction in a vessel, and melt relocation in a pressure vessel. And finally, Section 16.21 contains the discussion regarding: Is it possible to design a universal multiphase flow analyzer? It contains my personal vision for future development of the multiphase fluid dynamics.

Chapter 16 of this volume together with Chapter 14 of Volume 1 are available from the WEB site of Springer. In addition many animated sequences (movies) are presented there. HTML documents are then executed using any Web browser available on the local computer of the reader.

29.12.2010 Nikolay Ivanov Kolev
Herzogenaurach

Table of Contents

1. Nucleation in liquids

After reviewing the literature for description of the nucleation in superheated liquids the following conclusions and recommendations have been drawn. The maximum superheating in technical systems is a function of the depressurization velocity and of the produced turbulence. The maximum superheating can be predicted by the Algamir and Lienhard and by the Bartak correlations within an error band of 48.5%. Flashing in short pipes and nozzles leads to critical flows driven by the pressure difference equal to the entrance pressure minus the flushing inception pressure. For the prediction of the maximum achievable superheating, which represents the spinoidal line the Skripov correlation is recommended. The wetting angle is an important property of the polished surface characterizing its capability to activate nucleation sites. For the prediction of the activated nucleation sites the correlation obtained by Wang and Dhir is recommended. The establishing of a vapor film around a heated surface having temperature larger than the minimum film boiling temperature takes a finite time. The availability of small bubbles of noncondensing gases reduces the superheating required to initiate evaporation. Evaporation at lower than the saturation temperature is possible.

1.1 Introduction

Liquids having temperature above the saturation temperature corresponding to the local pressure are called superheated liquids. Superheated liquids are unstable and start to disintegrate. This process is in generally called *flashing*. The process of rupturing a *continuous* liquid by decrease in pressure at roughly constant liquid temperature is often called *cavitation* – a word proposed by *Froude*. The process of rupturing a *continuous* liquid by increase the temperature at roughly constant pressure is often called *boiling*. The fluctuation of molecules having energy larger than that characteristic of a stable state causes the formation of clusters of molecules, which after reaching some critical size are called nuclei. The theory of the nucleation provides us with information about the generation of nuclei per unit time and unit volume of the liquid as a function of the local parameter.

1.2 Nucleation energy, equation of *Kelvin* and *Laplace*

Let us abstract from a superheated *nonstable liquid a spherical volume*, having an initial radius R_{10}, a volume $\frac{4}{3}\pi R_{10}^3$, pressure p, temperature T_2 and density $\rho_2 = \rho_2(p, T_2)$. After some time the selected sphere liquid volume increases due to *evaporation* to the radius R_1 (respectively to the volume $\frac{4}{3}\pi R_1^3$) and reaches a pressure $p'(T_2)$. The pressure inside the bubble is assumed to be uniform because of the small bubble size. The density of the evaporated steam inside the sphere is $\rho'' = \rho''[p'(T_2)] = \rho''(T_2)$. The *initial* and the *end spheres have the same mass* per definition, therefore

$$\left(\frac{R_{10}}{R_1}\right)^3 = \frac{\rho''}{\rho_2}. \tag{1.1}$$

Consequently, the initial volume of the sphere is changed by

$$\frac{4}{3}\pi\left(R_1^3 - R_{10}^3\right) = \frac{4}{3}\pi R_1^3\left(1 - \frac{\rho''}{\rho_2}\right). \tag{1.2}$$

During this expansion a *mechanical work*

$$4\pi\int_{R_{10}}^{R_1}\left[p'(T_2) - p\right]r^2 dr \approx \frac{4}{3}\pi\left(R_1^3 - R_{10}^3\right)\left[p'(T_2) - p\right]$$

$$= \frac{4}{3}\pi R_1^3\left(1 - \frac{\rho''}{\rho_2}\right)\left[p'(T_2) - p\right] \tag{1.3}$$

is performed and transferred into total kinetic energy of the surrounding liquid *Skripov* et al. (1980). In other words, this work is introduced into the liquid. For the creation of a sphere with a free surface, additional work

$$\int_0^{R_1}\frac{2\sigma}{r}4\pi r^2 dr = 4\pi R_1^2\sigma \tag{1.4}$$

is needed. The surface tension of water in N/m, σ, in contact with its vapor is given in *Lienhard* (1976) with great accuracy by

$$\sigma = 0.2358\left[1 - \frac{T'(p)}{T_c}\right]^{1.256}\left\{1 - 0.625\left[1 - \frac{T'(p)}{T_c}\right]\right\}, \tag{1.5}$$

where the T_c is the thermodynamic critical temperature (for water T_c = 647.2 K). For the region of 366 to 566 K the above equation can be approximated by

$$\sigma = 0.14783\left(1 - T / T_c\right)^{1.053} \tag{1.6}$$

with an error of $\pm 1\%$. The surface tension is a function of the surface temperature, which can strongly vary in transients. The surface tension is usually measured at macroscopic surfaces. Whether the so-obtained information is valid for the microscopic metastable bubbles is not clear. Next, we assume that this relationship holds also for microscopic surfaces.

Thus, the work necessary to create a single bubble with radius R_1 is

$$\Delta E_1 = 4\pi R_1^2 \sigma - \left[p'(T_2) - p\right]\frac{4}{3}\pi R_1^3 \left(1 - \frac{\rho''}{\rho_2}\right) = 4\pi\sigma\left(R_1^2 - \frac{2}{3}\frac{R_1^3}{R_{1c}}\right),\ p'(T_2) > p \tag{1.7}$$

We see that this work depends on the bubble radius and has a maximum

$$\Delta E_{1c} = \frac{16\pi\sigma^3}{3\left[p'(T_2) - p\right]^2 \left(1 - \frac{\rho''}{\rho_2}\right)^2} = \frac{4}{3}\pi\sigma R_{1c}^2 \approx \frac{4}{3}\pi\sigma\left[\frac{T'(p)}{T_2 - T'(p)}\frac{2\sigma}{\rho''(h'' - h')}\right]^2 \tag{1.8}$$

for

$$R_{1c}^2 = \frac{3\Delta E_{1c}}{\sigma 4\pi} = \left\{\frac{2\sigma}{\left[p'(T_2) - p\right]\left(1 - \rho''/\rho_2\right)}\right\}^2. \tag{1.9}$$

The corresponding bubble volume is then

$$V_{1c} = \frac{4}{3}\pi\left(\frac{3\Delta E_{1c}}{\sigma 4\pi}\right)^{3/2}. \tag{1.10}$$

The equation (1.9) is known as the *Laplace* and *Kelvin* equation. *Gibbs* (1878) noted that the expression for the maximum " ... does not involve any geometrical magnitudes".

1.3 Nucleus capable of growth

The above consideration did not lead to any conclusion whether a bubble with size R_{1c} will further grow or collapse. It says only that at that size the mechanical energy needed to create a bubble at the initial state $R_1 = 0$ and at the final state $R_{1\infty} > R_{1c}$ possesses a maximum at R_{1c} and no more. There are many papers in which this size is taken to represent the bubble size, at which the bubble is further capable to grow, which as we will show below is not true.

Next, we consider the conservation of the liquid mechanical energy in order to describe the bubble growth from the beginning through the critical radius. The mass-conservation equation of the liquid can be approximated by

$$\frac{1}{r^2}\frac{\partial}{\partial r}\left(r^2 u_2\right) \approx 0, \tag{1.11}$$

or integrating with the boundary condition

$$r = R_1, \quad u_2 = \frac{\partial R_1}{\partial \tau} \tag{1.12}$$

$$u_2 = const / r^2 = \frac{\partial R_1}{\partial \tau}\left(R_1 / r\right)^2 . \tag{1.13}$$

This gives the liquid velocity as a function of the radius if the liquid is assumed to be incompressible during the bubble expansion. The total kinetic energy of the liquid environment estimated using the above equation is therefore

$$\int_{R_1}^{\infty} \frac{1}{2}\rho_2 u_2^2 dVol_2 = \frac{1}{2}\rho_2\left(\frac{\partial R_1}{\partial \tau}\right)^2 R_1^4 4\pi \int_{R_1}^{\infty}\frac{dr}{r^2} = 2\pi\rho_2\left(\frac{\partial R_1}{\partial r}\right)^2 R_1^4 \int_{R_1}^{\infty} d\left(-\frac{1}{r}\right)$$

$$= 2\pi\rho_2\left(\frac{\partial R_1}{\partial \tau}\right)^2 R_1^3 . \tag{1.14}$$

During the bubble growth, the work performed by the bubble expansion is transferred in total kinetic energy of the liquid environment, i.e.,

$$2\pi\rho_2\left(\frac{\partial R_1}{\partial \tau}\right)^2 R_1^3 = -4\pi\sigma\left(R_1^2 - \frac{2}{3}\frac{R_1^3}{R_{1c}}\right) \equiv \Delta E_1$$

or

$$\left(\frac{\partial R_1}{\partial \tau}\right)^2 = \frac{2\sigma}{\rho_2}\left(\frac{2}{3}\frac{1}{R_{1c}} - \frac{1}{R_1}\right). \tag{1.15}$$

The constant 2/3 valid for bubble growth in a bulk liquid should be replaced by $\pi/7$ if a spherical bubble grows on a flat surface.

Equation (1.15) is a very important result. We see that real bubble growth is possible *if and only if*

$$R_1 > \frac{3}{2} R_{1c} . \tag{1.16}$$

For the case of $R_1 >> R_{1c}$ Eq. (1.15) transforms into

$$\left(\frac{\partial R_1}{\partial \tau}\right)^2 = \frac{4}{3}\frac{\sigma}{\rho_2 R_{1c}}. \tag{1.17}$$

This mechanism of bubble growth is called *inertially controlled bubble growth.* For low pressure where the assumption $\rho''/\rho_2 = 0$ is reasonable the above equation reduces to one obtained for the first time by *Besand* (1859) and in a more elegant way by *Rayleigh* (1917). This mechanism controls the bubble growth within the first 10^{-8} s of the life of the stable bubble.

It follows from the above consideration that for the creation of a bubble with a critical unstable diameter 3/2 R_{1c}, a surplus of internal energy of the liquid is needed greater than E_{1c}, and for creation of bubble that is capable to grow, a surplus of internal energy of the liquid – greater than (9/4) E_{1c}.

1.4 Some useful forms of the *Clausius–Clapeyron* equation, measures of superheating

Usually, the liquid superheating is expressed by the temperature difference $T_2 - T'(p)$. Sometimes, the pressure difference $\Delta p = p'(T_2) - p$ corresponding to the superheating of the liquid with respect to the saturation temperature $T_2 - T'(p)$ is also used as a measure for the liquid superheating. Using the *Clausius–Clapeyron* equation

$$\frac{dT}{dp} = \frac{v'' - v'}{s'' - s'} = \frac{\rho' - \rho''}{s'' - s'}\frac{1}{\rho'\rho''} = T'(p)\frac{\rho' - \rho''}{h'' - h'}\frac{1}{\rho'\rho''} \tag{1.18}$$

integrated between the initial state $[p, T'(p)]$ and the final state $[p'(T_2), T_2]$ for

$$\frac{\rho' - \rho''}{h'' - h'}\frac{1}{\rho'\rho''} \approx const \tag{1.19}$$

one obtains

$$p'(T_2) - p = (h'' - h')\frac{\rho'\rho''}{\rho' - \rho''}\ln\left[1 + \frac{T_2 - T'(p)}{T'(p)}\right]. \tag{1.20}$$

With this result the critical bubbles size can be approximated by

$$R_{1c} \approx 2\sigma / \left\{(h'' - h')\rho' \ln\left[1 + \frac{T_2 - T'(p)}{T'(p)}\right]\right\}. \tag{1.22}$$

For $T_2 - T'(p) << T'(p)$ we have for the tensile pressure difference

$$p'(T_2) - p = (h'' - h')\frac{\rho'\rho''}{\rho' - \rho''}\frac{T_2 - T'(p)}{T'(p)}, \tag{1.23}$$

and for the critical bubble size

$$R_{1c} \approx \frac{T'(p)}{T_2 - T'(p)}\frac{2\sigma}{\rho''(h'' - h')}. \tag{1.24}$$

Thus, one can use as a measure of liquid superheating either the *tension pressure difference* $p'(T_2) - p$ or the *liquid superheat* $T_2 - T'(p)$.

Equation (1.23) makes it possible to compute the liquid superheating corresponding to the radius R_{1c} from Eq. (1.9)

$$[T_2 - T'(p)]/T'(p) = \frac{2\sigma}{R_{1c}}\frac{(\rho' - \rho'')\rho_2}{(\rho_2 - \rho'')\rho'\rho''}\frac{1}{h'' - h'} \approx 2\frac{\sigma}{R_{1c}}\frac{1}{\rho''(h'' - h')} \tag{1.25}$$

or corresponding to $3\,R_{1kr}/2$

$$[T_2 - T'(p)]/T'(p) = \frac{4}{3}\frac{\sigma}{R_{1c}}\frac{(\rho' - \rho'')\rho_2}{(\rho_2 - \rho'')\rho'\rho''}\frac{1}{h'' - h'} \approx \frac{4}{3}\frac{\sigma}{R_{1c}}\frac{1}{\rho''(h'' - h')}. \tag{1.26}$$

For the atmospheric pressure Eq. (1.23) gives

$$T_2 - T'(p) = 6.4197 \times 10^{-5}/D_{1c}. \tag{1.27}$$

This linearization is valid for very small deviations from the saturation. In general, the participating properties are not a linear function of temperature. A nonlinear expression is obtained also for low pressure for which the density difference between liquid and vapor is very large. In this case Eq. (1.18) simplifies to $dT/dp = v''/(T\Delta h)$. This equation has been known since 1828 in France as the *August* equation. Assuming the vapor behave as a perfect gas results in $dp/p = (\Delta h/R)dT/T^2$, where R is the vapor gas constant for the specific substance. Integrating between an initial state $[p, T'(p)]$ and a final state $[p'(T_2), T_2]$ and rearranging results in useful expression for computing the relation between *tension pressure difference* $p'(T_2) - p$ and the *liquid superheat* $T_2 - T'(p)$ of metastable liquid,

$$\frac{p'(T_2) - p}{p} = \exp\left[\frac{\Delta h}{RT_2}\frac{T_2 - T'(p)}{T'(p)}\right] - 1. \tag{1.28}$$

With this result the *Laplace* and *Kelvin* equation reads

$$R_{1c} \approx \frac{2\sigma}{\left\{\exp\left[\frac{\Delta h}{RT_2}\frac{T_2 - T'(p)}{T'(p)}\right] - 1\right\}p(1 - \rho''/\rho_2)}. \tag{1.29}$$

One should carefully use the approximations of the *Laplace* and *Kelvin* equation. At low pressure the approximations are good as shown in Fig.1.1. In this case at higher superheat Eq. (1.29) is much closer to the accurate equation.

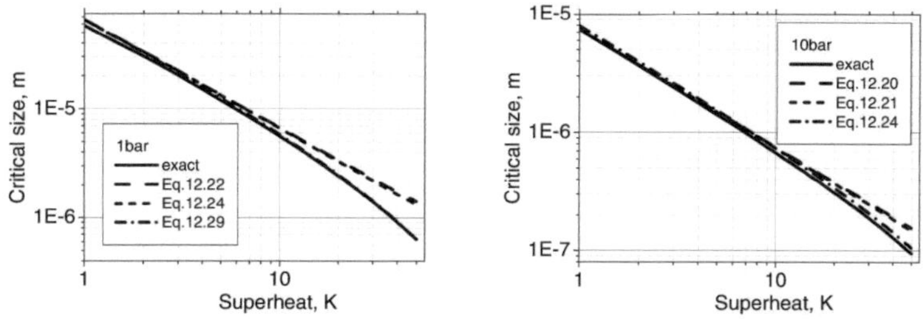

Fig. 1.1 Critical bubbles size at low pressure as a function of the liquid superheat

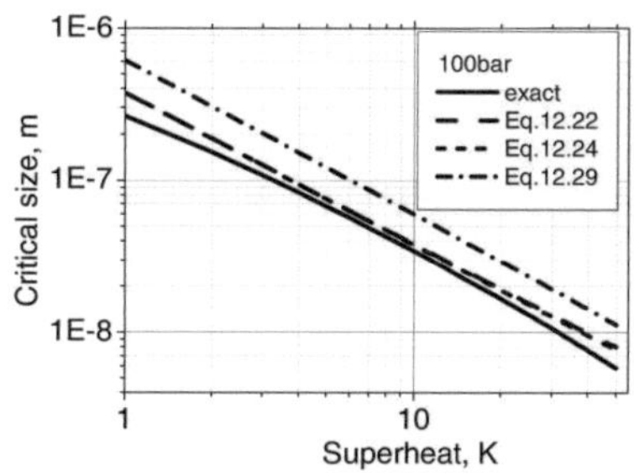

Fig. 1.2 Critical bubbles size at high pressure as a function of the liquid superheat

At high pressure the approximations may deviate substantially from the exact solution as shown in Fig. 1.2.

The dimensionless number constructed as follows

$$Gb_2 = \frac{\Delta E_{1c}}{kT_2} = \frac{\Delta S_{1c}}{k} = \frac{16\pi\sigma^3}{kT_2 3\left[p'(T_2) - p\right]^2 \left(1 - \rho''(T_2)/\rho_2\right)^2} \tag{1.30}$$

is called the *Gibbs* number. Here, $k = 13.805\times10^{-24}$ J/K is the *Boltzmann* constant. The *Gibbs* number is frequently used as a dimensionless measure of superheating. The greater the superheating the smaller the *Gibbs* number. Gb_2 converging to infinity means no superheating. For $T_2 = T_c$, $\sigma = 0$ and $Gb_2 = 0$.

However, it should be noted that neither of the measures discussed above tells us at what ΔT liquid starts boiling or flashing.

1.5 Nucleation kinetics

1.5.1 Homogeneous nucleation

Now we discuss the dependence between liquid superheating and production rates of bubbles with critical size called nuclei. Consider a continuous liquid characterized by average state parameters p and T_2. An observer measures a long time $\Delta\tau$ the time $\Delta\tau_n$ in which a mass Δm_{2c} of the continuum departs from the normal, average, state. The entropy change necessary for this fluctuation is $\Delta m_{2c}\Delta s_{2c} = \Delta S_{2c}$. *Volmer* (1939) p. 81 following *Boltzmann* and *Einstein* assumed that

$$\Delta\tau_n / \Delta\tau \approx const\ \mathrm{e}^{-\Delta S_{2c}/k} = const\ e^{-Gb_2} \tag{1.31}$$

where k is the *Boltzmann* constant. This means that the smaller the needed entropy change for the fluctuation the higher the probability of this fluctuation. This is the leading idea for describing such processes. It did not change with the years. Actually it is known from experimental observations that the greater the superheating the greater the probability of initiation of bubbles with critical size. The probability of initiation of a single bubble is defined as

$$\text{probability of initiation of single bubble} = \frac{nucleation\ \ events}{molecular\ \ colisions}. \tag{1.32}$$

Thus,

$$\frac{nucleation\ \ events}{molecular\ \ colisions} = \exp(-Gb_2) \tag{1.33}$$

Obviously for $T_2 = T_c$ and $\sigma_2 = 0$, the probability of initiation of a single bubble $exp(-Gb_2) = 1$. Nuclei with critical size can originate in the bulk liquid. This phenomenon is called *homogeneous* nucleation. Nuclei with critical size can originate also at the wall. This phenomenon is called *heterogeneous* nucleation. The theories modeling the nucleation in the bulk liquid are called homogeneous nucleation theories. The theories modeling the nucleation at surfaces are called heterogeneous nucleation theories. Next, we give the main result of the homogeneous nucleation theory.

Kaishew and *Stranski* computed in 1934 the number of the created nuclei per unit time in unit volume of the liquid as follows

$$\dot{n}_{1cin} = N_2 \left\{ \frac{6\sigma}{\left[2 + p/p'(T_2)\right]\pi\, m_2} \right\}^{1/2} \exp(-Gb_2)$$

$$= \exp\left\{ \ln\left[\rho_2 \left(\frac{N_A}{m_\mu} \right)^{3/2} \left\{ \frac{6\sigma}{\left[2 + p/p'(T_2)\right]\pi\, m_2} \right\}^{1/2} \right] - Gb_2 \right\}. \tag{1.34}$$

Here, $N_2 = \rho_2 / m_2 = \rho_2 N_A / m_\mu$ ($\approx$ 3.3–10^{28} for water) is the number of the molecules in one cubic meter, $m_2 = m_\mu / N_A$ is the mass of a single molecule, m_μ is the kg-mole mass (18 *kg* for water), $N_A = 6.02\times 10^{26}$ (1/kg-mole) is the number of molecules in one kilogram-mole mass, the *Avogadro* number, and

$$\left\{\frac{6\sigma}{\left[2+p/p'(T_2)\right]\pi m_2}\right\}^{1/2} \tag{1.35}$$

is the frequency with which each single liquid molecule interacts with its neighbors. The logarithmic expression is not very sensitive with respect to p and T_2 and ranges over

$$\ln\left[\rho_2\left(\frac{N_A}{m_\mu}\right)^{3/2}\left\{\frac{6\sigma}{\left[2+p/p'(T_2)\right]\pi m_2}\right\}^{1/2}\right] \approx 80 \textit{ to } 83 \tag{1.36}$$

for water.

1.5.2 Heterogeneous nucleation

1.5.2.1 Characteristics of the surfaces and of the liquid/surface contact

Technical surfaces possess roughness as a result of the manufacturing procedure. The structure of the roughness is an important characteristic of the nucleation processes at the surface. Another important characteristic is the molecular interaction between the surface and the liquid. It is usually characterized by the so-called contact angle. The angle between the tangent to the interface and the wall θ is called the contact angle. Hydrophobic surfaces, $\theta > 0$, cause heterogeneous nucleation at much reduced pressure difference (tensile strength). In this section, we give some examples of the characteristics of surfaces influencing the heterogeneous nucleation kinetics.

At real surfaces there are several cavities. For ordinary machined surfaces the cavities have sizes from 2 to 6 μm with density in the range of

$$n''_w \approx (10 \textit{ to } 250)\times 10^4 m^{-2} \tag{1.37}$$

For mirror finished copper surfaces as described in *Wang and Dhir* (1993) physically existing cavities have been found to have surface densities depending on the equivalent cavity size in the range

$$n''_w \approx 9\times 10^{-9} / D_{cav}^2 \quad \text{for} \quad D_{cav} \ge 5.8\times 10^{-6} m \tag{1.38}$$

$$n''_w \approx 10.3\times 10^4 + 1.5\times 10^{-25.2} / D_{cav}^{5.2} \quad \text{for} \quad 3.5\times 10^{-6} \le D_{cav} \le 5.8\times 10^{-6} m \tag{1.39}$$

$$n''_w \approx 2.2135\times10^7 + 3.981\times10^{-26.4} / D_{cav}^{5.4} \quad \text{for} \quad D_{cav} \le 3.5\times10^{-6} m . \qquad (1.40)$$

These are surface properties, which depend on the surface manufacturing only. The authors discovered that only cavities having a nearly spherical form and mouth angle less than the static contact angle serve as nucleation sites. The mouth angle is the angle between the boiling surface and the internal surface of the cavity forming the mouth. The region of changing the heat flux covers the entire nucleate boiling region.

The following tables give some information on how the contact angle can change for different wall materials and different surface preparation procedures.

Table 1.1 Static contact angles θ for distilled water at polished surfaces.

Steel	$\pi/3.7$	*Siegel* and *Keshock* (1964)
Steel, Nickel	$\pi/4.74$	*Bergles* and *Rohsenow* (1964)
Nickel	$\pi/4.76$ to $\pi/3.83$	*Tolubinsky* and *Ostrovsky* (1966)
Nickel	$\pi/4.74$ to $\pi/3.83$	*Siegel* and *Keshock* (1964)
Chrome-Nickel Steel	$\pi/3.7$	*Arefeva* and *Aladev* (1958)
Silver	$\pi/6$ to $\pi/4.5$	*Labuntsov* (1963) for p =1 to 150 bar
Zinc	$\pi/3.4$	*Arefeva* and *Aladev* (1958)
Bronze	$\pi/3.2$	*Arefeva* and *Aladev* (1958)
Zr-4	$\pi/3.16$	*Basu*, et al. (2002)
Note the contradictory data for copper in the literature		
Copper	$\pi/4$	*Arefeva* and *Aladev* (1958)
Copper	$\pi/3$	*Gaertner* and *Westwater* (1960)
Copper	$\pi/2$	*Wang* and *Dhir* (1993)
Stainless steel 304 (25 °C)	$\theta_{25} \approx \pi/2.25$	*Hirose* et al. (2006)
Zircaloy (25 °C)	$\theta_{25} \approx \pi/2.46$	*Hirose* et al. (2006)
Aluminum (25 °C)	$\theta_{25} \approx \pi/2$	*Hirose* et al. (2006)

Table 1.2 Static contact angle θ for distilled water at thermally or chemically treated polished surfaces

Copper heated to 525 K and exposed to air one hour:	$\pi/5.14$	*Wang* and *Dhir* (1993)
Copper heated to 525 K and exposed to air two hour:	$\pi/10$	*Wang* and *Dhir* (1993)
Chrome-Nickel Steel chemically treated:	$\pi/2.9$	*Arefeva* and *Aladev* (1958)

Hirose et al. (2006) found that the wetting angle is a function of the wall temperature. He correlated measurements within $\pm 12.5\%$ uncertainty band for nitrogen and argon and water–stainless steel 304, water–zircaloy and water–aluminum for pressures 1 to 151 bar and temperatures 20 to 300 °C by the expression

$$\theta = \theta_{25} \tanh\left(2.55 \times T^{*1.72}\right) \text{ for } T^* > 0.3,$$

$$\theta = \theta_{25} \tanh\left(2.55 \times 0.3^{1.72}\right) \text{ for } T^* < 0.3,$$

where $T^* = \dfrac{T_{cr} - T_w}{T_{cr} - (273.15 + 25)}$, T_{cr} is the critical water temperature and T_w is the wall temperature and $\theta_{25} = \theta(25°C)$. Note that there is no zero wettability at the critical temperature. This is important finding. The work by *Hirose* et al. (2006) shows that measurements for wetting angles can be compared to each other if they are made at the same wall temperature.

In the case of nucleation at walls the segment of a bubble attached at the cavity possesses

$$\textit{surface area} = \varphi 4\pi R_{1c}^2, \tag{1.41}$$

which is less than the nucleation surface by a factor of φ. Therefore, less energy is needed for the creation of bubbles at walls

$$\Delta E_{1c}^* = \varphi \Delta E_{1c}, \tag{1.42}$$

than inside the liquid. Consequently, the superheating that can be achieved in technical systems is much smaller than in spontaneous bulk nucleation systems. φ is frequently called the work-reduction factor in the literature.

Tolubinski (1980) computed the work-reduction factor, φ, for an idealized wall surface having *plane geometry* without cavities as follows

$$\varphi = \frac{1}{4}(1+\cos\theta)^2(1-\cos\theta). \tag{1.43}$$

Using the values from Table 1.1 we have φ s in the range of 0.078 to 0.25.

The work-reduction factor for surface with cavities calculated by *Kottowski* (1973) is

$$\varphi = \frac{1}{4}\left[2 - 3\sin(\theta - \Phi) + \sin^3(\theta - \Phi)\right]. \tag{1.44}$$

Here, Φ is the cavity angle if the cavity is idealized as a cone.

1.5.2.2 Nucleation theories and experimental observations

Blander and *Katz* (1975) computed for the heterogeneous nucleation the number of the nucleations per unit liquid volume and unit time in a pipe with wetted diameter D_{hy} as follows

$$\dot{n}_{1cin} = \frac{4}{D_{hy}} N_2^{2/3} \frac{1+\cos\theta}{2} \left\{ \frac{6\sigma}{\left[2+p/p'(T_2)\right]\pi\, m_2 \varphi} \right\}^{1/2} \exp(-\varphi\, Gb_2) =$$

$$\exp\left\{ \ln\left[\frac{4}{D_{hy}} \rho_2^{2/3} \left(\frac{N_A}{m_\mu}\right)^{7/6} \frac{1+\cos\theta}{2} \left\{ \frac{6\sigma}{\left[2+p/p'(T_2)\right]\pi\, m_2\varphi} \right\}^{1/2} \right] - \varphi\, Gb_2 \right\}. \tag{1.45}$$

Instead of N_2 in the *Kaishew* and *Stranski* equation $N^{2/3}$ is used because the nucleation occurs at the wall instead in the bulk.

The problem with the homogeneous and heterogeneous nucleation theories is that they did not describe appropriate nucleation rates in real liquid. *Rohatgi* and *Reshotko* (1975), *Algamir* and *Lienhard* (1981) data comparisons indicated substantial deviation from the above theories in real technical systems.

Rohatgi and *Reshotko* (1975) tried to match *Simoneau's* experimental data (1975) with an expression

$$\dot{n}_{1cin} = (2.236 \textit{ to } 4.472)\times 10^3 \left(\frac{2\sigma}{\pi m_2 \varphi}\right)^{1/2} \exp(-\varphi\, Gb_2), \tag{1.46}$$

where

$$\varphi = 5\times 10^{-6}. \tag{1.47}$$

As will be discussed below *Algamir* and *Lienhard* (1981) found

$$\varphi = 0.055 \textit{ to } 2\times 10^{-7} \tag{1.48}$$

for different depressurization speeds. *Algamir* and *Lienhard* (1981) found that depending on the depressurization speed the nucleation process can be delayed. They use the work-reduction factor φ to correlate experimental data with the following relationship

$$\varphi = 0.1058 (T_2/T_c)^{28,46} \left\{ 1 + 14\left[\left(\gamma_v \frac{\delta p}{\delta \tau} + \mathbf{V}\gamma.grad\ p \right) / 10^{11} \right]^{0.8} \right\} \tag{1.49}$$

taking values as already mentioned of 0.055 to 2×10^{-7} for different depressurization rates. The correlation is valid for decompression speeds within

$$0.004 \le (dp/d\tau)10^{-11} \le 1.803 \tag{1.50}$$

in channels with constant cross-section in the temperature range

$$0.62 \le T_2/T_c \le 0.935 . \tag{1.51}$$

Here, the decompression rate $dp/d\tau$ is measured in Pa/s.

For some practical applications it is useful to compute the time required to cover strongly superheated walls with nuclei. This is in fact the time that is required to establish film boiling at walls heated at temperatures above the film boiling temperature. Next we discuss briefly this problem. If the growth of the nucleus is neglected and the initial conditions are assumed to remain constant the nucleation number density originating during the time interval $\Delta\tau$ is

$$n_1 - \int_0^{\Delta\tau} \dot{n}_{1cin} d\tau = \dot{n}_{1cin}\Delta\tau . \tag{1.52}$$

If the nuclei are spherical in shape and their centers form a triangular array at the wall the maximum number of nucleation sites per unit volume in a pipe is

$$n_{1,\max} = \frac{4}{D_h}\frac{2}{\sqrt{3}}\frac{1}{D_{1c}^2} \tag{1.53}$$

and the time required to reach it is

$$\Delta\tau_{\max} = \frac{4}{D_h}\frac{2}{\sqrt{3}} / \left(D_{1c}^2 \dot{n}_{1cin} \right). \tag{1.54}$$

If the cause for the superheating is depressurization with a $dp/d\tau$ starting from the saturation state the number of the produced particles is

$$n_1 = \int_0^{\Delta\tau} \dot{n}_{1cin} d\tau = n_1 = \frac{1}{dp/d\tau}\int_0^{\Delta p} \dot{n}_{1cin} dp . \tag{1.55}$$

If Δp corresponding to experimentally observed superheating is known and n_1 is set equal to $n_{1,max}$ the critical work-reduction factor can be estimated. This procedure was used by *Algamir* and *Lienhard* (1981). The authors found, as already mentioned

$$\varphi = 0.055 \text{ to } 2\times 10^{-7} \tag{1.56}$$

for different depressurization speeds. Obviously, only the functional form derived from the nucleation theories is useful to fit experimental data but not the theoretical formula itself. One of the reasons why this is so, may be neglecting nucleus growth of the already existing bubble generations during the nucleation.

Next, we improve the computation by allowing each generation to grow. During a time interval $N\ \Delta\tau$, N generation of nuclei, or $n_{1,n=1N}$ nuclei per unit flow volume are produced

$$n_{1,n} = \int_0^{\Delta\tau} \dot{n}_{1cin} d\tau, \qquad n = 1, N\ . \tag{1.57}$$

Each generation has its own bubble growth history. If we assume that the bubble growth is thermally controlled following the following relationship

$$\frac{dD_{1,n}}{d\tau} = \frac{B}{\sqrt{\tau}}, \tag{1.58}$$

where B is a function of a local parameter the bubble size of the nth generation after the time $N\ \Delta\tau$ is

$$D_{1,n} = D_{1c} + \sum_{k=1}^{N-n+1} \int_0^{\Delta\tau} \frac{dD_{1,n}}{d\tau} d\tau = D_{1c} + 2\left(N - n + 1\right) B \Delta\tau^{1/2}. \tag{1.59}$$

Thus, the total number of nuclei and bubbles after N time steps of $\Delta\tau$ is

$$n_1 = \sum_{n=1}^{N} n_{1,n} = N \int_0^{\Delta\tau} \dot{n}_{1cin} d\tau\ . \tag{1.60}$$

The cross-section at the wall surface covered by attached bubbles per unit flow volume is

$$F_1 / Vol = \frac{\pi}{4} \sum_{n=1}^{N} n_{1,n} D_{1,n}^2, \tag{1.61}$$

which for deformable bubble cannot be larger than $4/D_h$, where D_h is the hydraulic diameter of the pipe, i.e.

$$\frac{4}{D_h} \approx \frac{\pi}{4} \sum_{n=1}^{N} n_{1n} D_{1,n}^2\ . \tag{1.62}$$

Analogous conditions can be derived for a hot sphere with diameter D_3 immersed in liquid

$$D_3^2 \approx \frac{1}{4} \sum_{n=1}^{N} D_{1,n}^2\ . \tag{1.63}$$

By known nucleation and bubble growth laws and prescribed time step $\Delta\tau$, the above equation defines the number of bubble generation, N, and therefore the time needed to cover the surface with bubbles and nuclei, $N\ \Delta\tau$. At this time, the maximum possible superheating is reached. At the end of this process the void fraction generated within $N\ \Delta\tau$ is

$$\alpha_1 = \frac{4}{3}\pi \sum_{n=1}^{N} n_{1,n} D_{1,n}^3 \tag{1.64}$$

and the volume-averaged bubble size is then

$$D_1^{*3} = \left(\frac{3}{4\pi} \alpha_1 / n_1 \right)^{1/3} . \tag{1.65}$$

After this moment, the bubble-production frequency is governed by combined thermal interaction and mechanical entrainment phenomena, which will be considered later.

1.6 Maximum superheat

Skripov et al. (1980) p. 136, 143, found that rapid decompressions can bring the water into a state of maximum superheating defined by Gb_2 = 6.8 to 9.6, which means

$$p'(T_2^{spin}) - p = (1.57\ to\ 1.32) \frac{\sigma^{3/2}}{(kT_2)^{1/2} \left[1 - \rho''(T_2)/\rho'(T_2) \right]} . \tag{1.66}$$

Lienhard found in 1976 the following explicit approximation of the homogeneous nucleation temperature that can be taken as a spinoidal temperature

$$T_2^{spin} = T'(p) + \left\{ 0.905 - \frac{T'(p)}{T_c} + 0.095 \left[\frac{T'(p)}{T_c} \right]^8 \right\} T_c . \tag{1.67}$$

Under rapid decompression we understand decompression with

$$dp/d\tau \ \geq 2\times 10^{11}\ \text{Pa/s}. \tag{1.68}$$

In addition, we should mention that *Hutcherson* et al. (1983) classified the boundary between “fast” and “slow” depressurization as about 400×10^6 Pa/s. The line in the *p-T* diagram, defined by Eq. (1.66) or Eq. (1.67) is called the spinoidal line. The database defining the spinoidal line for water is given by *Skripov* (1980) p. 144 Table 6.4. *Lienhard* (1981) proposed for the constant the value 1.349 corresponding to $Gb_2 \approx 9.2$. For n-pentane, n-hexane, n-nepthane, ether and benzene *Eberhard* and *Schnyders* reported in 1973 $Gb_2 \approx 11.5$. The authors observed experimentally negative pressure up to $\approx 0.4 p_c$.

In technical systems the superheating is a function of several parameters not considered in our previous discussion. Next, we discuss some of them.

The question of the maximal achievable superheating in technical systems has attracted the attention of the engineers since 1947. Probably, *Burnell* was the first who investigated in 1947 experimentally the flow of boiling water through orifices. The inlet conditions are corresponding to saturated water at $p \approx (0.34$ to $12.7)\times 10^5$ Pa. *Burnell* found that a superheating of $\Delta p_{Fi}^* = p'(T_2) - p_{Fi}$ is necessary to observe flashing inception at the throat. The flashing inception at the throat reduces dramatically the mass flow rate compared to the prediction of the *Bernoulli* equation for subcooled water. *Burnell* found that the product of the initial saturated pressure multiplied by the critical bubble radius is constant

$$p'(T_2)R_{1c} \approx const \tag{1.69}$$

and therefore

$$\Delta p_{Fi}^* = p'(T_2) - p_{Fi} = 2\sigma_2(T_2)/R_{1c} \approx 2\sigma_2(T_2)p'(T_2)/const \approx f(\sigma_2)p'(T_2) \tag{1.70}$$

and

$$f(\sigma_2) = 0.264\frac{\sigma_2\left[T'(p)\right]}{\sigma_2'\left[T'(12.066\times 10^5 Pa)\right]} \tag{1.71}$$

where the surface tension of the saturation condition was computed as follows

$$\sigma_2'(T_2) = 0.07548\left[1 - 1.85\times 10^{-3}(T_2 - 273.15)\right] \text{ in N/m.} \tag{1.72}$$

Thus, the critical bubble radius in this region is

$$R_{1c} = 2\sigma_2(T_2)/\Delta p_{Fi}^* . \tag{1.73}$$

Algamir and *Lienhard* correlated several experimental data for the reached water superheating with a standard deviation of $\pm 10\%$ with the following value of the dimensionless critical work Gb_2 = 28.2 ± 5.8 that after using Eqs. (1.30) and (1.49) results in the following relationship for flashing inception,

$$\Delta p_{Fi}^* = \left\{\frac{16\pi\sigma^3}{3kT_c\left[1-\rho''(T_2)/\rho'(T_2)\right]^2 \dfrac{T_2}{T_c}\dfrac{Gb_2}{\varphi}}\right\}^{1/2}$$

$$= \Delta p_{Fi}^o\sqrt{1+14\left[\left(\gamma_v\,\partial p/\partial\tau + \mathbf{V}\gamma.grad\ p\right)/10^{11}\right]^{0.8}}\ , \tag{1.74}$$

where

$$\Delta p_{Fi}^{o} = 0.253\frac{\left[\sigma(T_2)\right]^{3/2}(T_2/T_c)^{13.73}}{(kT_c)^{1/2}\left[1-\rho''(T_2)/\rho'(T_2)\right]} \tag{1.75}$$

is the pressure difference corresponding to the maximum possible superheating in an isobar system and T_2 is the initial temperature before the pressurization. For

$$T_2 = 273.15 + 100 \text{ K}, \ \Delta p_{Fi}^{*} > 0.2\times10^{5} \text{ Pa}, \tag{1.76}$$

and for

$$T_2 \approx 273.15 + 300 \text{ K}, \ \Delta p_{Fi}^{*} < 9.5\times10^{5} \text{ Pa} \tag{1.77}$$

For $T_2 > 273.15 + 300\text{K}$ Δp_{Fi}^{o} decreases to zero due to vanishing surface tension as the critical point is approached. In fact, Eq. (1.75) is the modified *Skripov* equation. Instead of the constant the expression $0.253(T_2/T_c)^{13.73}$ is used.

Bartak (1990) correlated the data of several authors including his own data with the following correlation

$$\Delta p_{Fi}^{*} = \left\{\frac{16\pi\sigma^3}{3kT_2\left[1-\rho''(T_2)/\rho'(T_2)\right]^2 \dfrac{T_2}{T_c}\dfrac{Gb_2}{\varphi}}\right\}^{1/2}, \tag{1.78}$$

where

$$\frac{Gb_2}{\varphi} = f_1 f_2, \tag{1.79}$$

$$\log f_1 = 11 - 0.0274(T_2 - 273.15), \tag{1.80}$$

$$f_2 = 36/\left[\left(\gamma_v\, \partial p/\partial\tau + \mathbf{V}\gamma.grad\ p\right)/10^6\right]^{0.37}, \tag{1.81}$$

with a mean relative error of 16%. This is valid for water in the temperature range from 373.15 to 583.15 K and depressurization rates of 4×10^8 to 2×10^{11} Pa/s. *Bartak* reported that the *Algamir* and *Lienhard* correlation 65 approximates his own data with 21% mean relative error. In fact, *Baratak's* correlation predicts more then 30% higher values than the *Algamir* and *Lienhard* correlation. Consequently, it can be concluded that the achievable prediction accuracy is within the error band of about 48.5%.

If the maximum possible superheating under given conditions is reached, the maximum nucleation density is reached too. Note that the reduction factor, Eq. (1.48) obtained by *Algamir* and *Lienhard* can be used to compute the heterogeneous nucleation densities using the *Blander* and *Katz* equation (1.45) only together with the assumption of no bubble growth during the nucleation process.

Jones found in 1979 that the turbulization of the flow could reduce the maximum superheating

$$\Delta p_{Fi} = \Delta p_{Fi}^{*} - 27\tfrac{1}{2}\rho_2 V_2'^2 , \tag{1.82}$$

where the turbulent pulsation velocity in a straight pipe is 7.2% from the local flow velocity

$$\sqrt{V'^2} = 0.072 V_2 \; or \; V_2'^2 = 5.184 x 10^{-3} V_2^2 . \tag{1.83}$$

For a converging nozzle the local fluctuation velocity can be expressed as a function of the entrance velocity V_{2o} using the continuity equation

$$V_2'^2 = 5.184 x 10^{-3} V_{2o}^2 \left(F_o / F\right)^2 . \tag{1.84}$$

Comparing with experimental data *Abuaf* et al. (1983) found that the turbulent pulsation velocity varies in accordance with

$$V_2'^2 = 5.184 \times 10^{-3} V_{2o}^2 \left(F_o / F\right)^n , \tag{1.85}$$

where

$$n = 1.75 \; for \; F / F_o \ge 1/6 , \tag{1.86}$$

and for stronger contraction

$$n = 1.4 \; for \; F / F_o < 1/6 . \tag{1.87}$$

Jones (1982) shows that for $G > 15\ 000$ kg/(m^{2}s) no superheating is observed. To what extent the superheating reduction explained by *Jones* (1979) with turbulization is separable from the convective transport of nuclei by high liquid velocities is not clear.

1.7 Critical mass flow rate in short pipes, orifices and nozzles

The most important outcome of the heterogeneous nucleation theory for the engineering design practice is the possibility to compute in a simple manner the critical mass flow rate in short pipes, orifices and nozzles for saturated and subcooled water. Probably *Burnell* was the first approximating critical flow in nozzles with saturated and subcooled inlet condition modifying the *Bernoulli* equation as follows

$$G^* = \sqrt{2\rho_2(p_o - p_{Fi})}\,, \tag{1.88}$$

where the critical pressure is set to the flashing inception pressure

$$p_{Fi} = p'(T_2) - \Delta p_{Fi}\,. \tag{1.89}$$

Jones (1982) described successfully critical mass flow rate in nozzles modifying Eq. (1.88) to

$$G^* = (0.93 \pm 0.04)\sqrt{2\rho_2\left(p_o - p_{Fi}\right)} \tag{1.90}$$

The error of this approach was reported to be $\pm 5\%$ for inlet conditions of $p = (28$ to $170) \times 10^5$ Pa and $T_2 = 203$ to 288 °C and use of the *Algamir* and *Lienhard* correlation. *Fincke* (1984) found for the discharge coefficient instead of 0.93 the value 0.96 and reported that this coefficient does not depend on the *Reynolds* number.

1.8 Nucleation in the presence of noncondensable gases

In the liquids usually used in technology there are *dissolved inert gases* and *microbubbles*. It is known that at $p = 10^5$ Pa, $T_{2o} = 298.15$ K the amount of dissolved gases and microbubbles in the coolant is $\alpha_{1o} \cong 0.005$, for boiling-water reactors, and $\alpha_{1o} \cong 0.001$, for pressurized-water reactor *Malnes* and *Solberg* (1973). *Brennen* (1995) p. 20 reported that it takes weeks of deaeration to reduce the concentration of air in the water tunnel below 3 ppm (saturation at atmospheric pressure is about 15 ppm).

This amount is dissolved in a form of

$$n_{1o} = \alpha_{1o} / (\pi D_{1o}^3 / 6) \tag{1.91}$$

bubbles per unit volume, so that before starting the nucleation n_{1o} nucleation sites already exist. Here, the initial bubble diameter is in any case less than the bubble diameter computed after equating the buoyancy force and the surface force

$$D_{1o} < \sqrt{6}\lambda_{RT}\,. \tag{1.92}$$

Note that the so-called free-stream nuclei number density is subject to distribution depending on the nucleation size, e.g., $n_{10} \sim R_{10}^{-4}$, $R_{10} > 5 \times 10^{-6}$ as reported by *Brennen* (1995).

Let us suppose that at p_o, T_{2o} we have no water steam in the single bubble *Besand* (1859) and therefore the mass of the single bubble is

$$m_{\sum n1o} = p_{\sum n1} \frac{\frac{4}{3}\pi R_{1o}^3}{R_{\sum n1} T_{2o}} = (p + \frac{2\sigma}{R_{1\sigma}}) \frac{\frac{4}{3}\pi R_{1o}^3}{R_{\sum n1} T_{2o}} = (p + \frac{2\sigma}{R_{1\sigma}}) \frac{\alpha_{1o} / n_{1o}}{R_{\sum n1} T_{2o}}. \quad (1.93)$$

Further, let us assume that the steam evaporating from the surface into the bubble is saturated. The initial internal bubble pressure $\frac{3}{4}\frac{m_{\sum n1o} R_{\sum n1} T'}{\pi R_1^3}$ increases with $p'(T_2)$ and equals the external pressure $p + \frac{2\sigma}{R_1}$

$$p'(T_2) - p = \frac{2\sigma}{R_1} - \frac{3}{4}\frac{m_{\sum n1o} R_{\sum n1} T'}{\pi R_1^3}, \qquad T_2 > T'(p). \quad (1.94)$$

With increasing liquid superheating the bubble radius increases to its critical size

$$R_{1c} = 3\sqrt{\frac{m_{\sum n1o} R_{\sum n1} T'(p)}{8\pi\sigma}}, \quad (1.95$$

Blake (1949), *Neppiras* and *Noltingk* (1951), corresponding to the superheating

$$T_2 - T'(p) = \frac{2^{5/2}}{3\sqrt{\pi}} \sigma^{3/2} \sqrt{T'(p)} \frac{(\rho' - \rho'')\rho_2}{(\rho_2 - \rho'')\rho'\rho''} \frac{1}{h'' - h'} \frac{1}{\sqrt{m_{\sum n1o} R_{\sum n1}}}. \quad (1.96)$$

This is the maximum possible superheating. Thereafter, the bubble loses its stability and starts to increase in size. We see that the presence of dissolved gases in the liquid decreases the superheating necessary for starting the intensive evaporation. A considerable amount of dissolved gases can lead to flashing initiation at temperatures *lower than the saturation temperature.*

For the first time steps, where thc nucleation occurs, the instantaneous source terms are

$$\mu_{21} = \rho'' V_{1c} \dot{n}_{1cin}, \quad (1.97)$$

$$\dot{q}_2^{\prime\prime\prime 1\sigma} = -\mu_{21} (h'' - h_{M2}). \quad (1.98)$$

1.9 Activated nucleation-site density – state-of-the-art

Not all cavities at the wall serve as a nucleation sites. It is experimentally observed that the increase of the wall superheating activates an increasing number of nucleation sites, Fig. 1.3. Many authors tried to represent this observation by setting

$$n''_{1w} \approx const / D_{1c}^n, \quad (1.99)$$

where the exponent n varies from 2 to 6. If nucleation sites form a triangular array and are touching each other their maximum number per unit surface corresponding to the local superheating is

$$n''_{1w,\max} = 2/\left(\sqrt{3}D_{1c}^2\right). \tag{1.100}$$

Note that at atmospheric pressure this expression reduces to

$$n''_{1w,\max} = 0.20\times10^9 \Delta T^2 . \tag{1.101}$$

Labuntsov assumed in 1963 that the active nucleation sites density is a part of this value

$$n''_{1w,\max} = const\ 2/\left(\sqrt{3}D_{1c}^2\right). \tag{1.102}$$

Later in 1977 this equation was used by *Avdeev* et al. (1977) for modeling of flashing flows

$$n''_{1w,\max} = 20\times10^{-5}/D_{1c}^2 , \tag{1.103}$$

which for atmospheric pressure gives an activated site density four order of magnitude lower than the thinkable maximum. The constant $20–10^{-5}$ was estimated by comparison of the critical mass flow rates with those experimentally measured. For atmospheric pressure Eq. (1.92) results in

$$n''_{1w,\max} = 4.85\times10^4 \Delta T^2 . \tag{1.104}$$

Johov (1969) plotted data for water and nonwater liquids for the nucleation-site density as a function of the critical diameter. The data have been in the region $400 \le n''_{1w} \le 4\times10^8$. The data correlated with

$$n''_{1w} = 4\times10^{-12}/D_{1c}^3 , \tag{1.105}$$

$\left[= 15.12\Delta T^3\right]$ for p = 0.1 MPa, with considerable data spreading within two orders of magnitude.

Mikic and *Rohsenow* (1969) proposed the expression

$$n''_{1w} = c\left(D_1^*/D_{1c}\right)^{3.5} , \tag{1.106}$$

where D_1^* is the diameter for which n''_{1w} would be 1 per unit area, and c is a dimensional constant ($1/m^2$) depending on the cavity size distribution. Here, n''_{1w} is interpreted as the number of the active bubble sites, per unit area, having sizes greater than D_1^*. For a triangular array

$$D_1^* = \left(2/\sqrt{3}\right)^{1/2}. \tag{1.107}$$

Hutcherson et al. (1983) used instead of the critical nucleation size the bubble departure size and obtained good agreement with experimental data for the initial depressurization of a vessel blow-down assuming

$$n''_{1w} D_{1d}^2 \approx 0.01 \times 2/\sqrt{3} \tag{1.108}$$

using the *Fritz* equation for the bubble departure diameter D_{1d}.

$$D_{1d} \equiv D_{1d,Fritz} = 1.2\theta \left\{ \frac{\sigma}{g\left[\rho_2 - \rho''(p)\right]} \right\}^{1/2}. \tag{1.109}$$

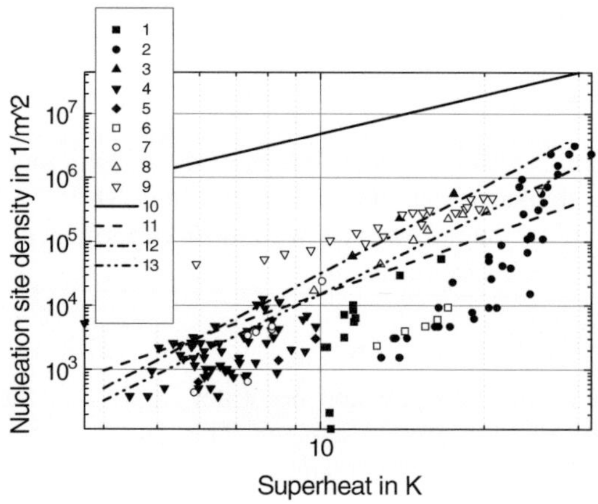

Fig. 1.3 Active nucleation-site density as a function of superheat. Saturated water at 0.1 MPa. Data: 1) *Gaertner* 1965, 4/0 polished copper, 2) *Gaertner* and *Westwater* 1960, 4/0 polished copper, 20 p.c. nickel salt-water solution, 3) *Sultan* and *Judd* 1978, diamond grid 600 polished copper, 4) *Yamagata* et al. 1955, fine polished brass, 5) *Jakob* and *Linke* 1933, polished steel, 6) *Cornwell* and *Brown* 1978, 4/0 polished copper, 7) *Kurihara* and *Myers* 1960, 4/0 polished copper, 8) *Rallis* and *Jawurek* 1964, nickel wire, 9) *Faggiani* et al. [13] 1981, polished 316 steel horizontal cylinder. Prediction with correlations proposed by 10) *Avdeev* et al. (1977), 11) *Johov* (1969), 12) *Cornwell* and *Brown* (1978) and 13) *Kocamustafaogullari* and *Ishii* (1983)

For the case of nucleate boiling at the wall, *Kocamustafaogullari* and *Ishii* (1983) proposed the following correlation to predict the active nucleation sites density

$$n''_{1w} D_{1d}^2 = \left[\left(D_{1d}/D_{1c}\right)^{4.4} 2.157 \times 10^{-7} \left(1 + 0.0049\bar{\rho}\right)^{4.13} / \bar{\rho}^{3.2} \right], \tag{1.110}$$

where

$$\bar{\rho} = (\rho' - \rho'')/\rho'', \tag{1.111}$$

$$D_{1d} = 0.0012\bar{\rho}^{0.9} D_{1d,Fritz} \tag{1.112}$$

and the size of the activated sites, D_{1c}, is computed using Eq. (1.9) with a temperature difference $S\ [T_w - T'(p)]$ and properties taken at the bulk temperature. Here, S is the so-called *Chen* superposition factor for taking into account the effective superheating of the wall boundary layer. $S = 1$ for free convection pool boiling and ≤ 1 for forced-convection boiling. The authors reported good agreement with data for pressures $p = 0.1$ to 19.8 MPa and for both subcooled and pool boiling.

Jones (1992) found

$$n''_{1w} D_{1d}^2 = 10^{-7} \left(D_{1d} / D_{1c} \right)^4 \tag{1.113}$$

for critical flashing flow in a nozzle using the equilibrium between drag and surface forces in the viscous boundary layer to compute departure diameter. *Riznic* and *Ishii* (1989) demonstrated the applicability of this correlation for flashing adiabatic flow using D_{1c} computed with the equation by *Kocamustafaogullari* and *Ishii* for the temperature difference $T_2 - T'(p)$.

Cornwell and *Brown* (1978) reported experimental data for pool boiling of saturated water at atmospheric pressure on a copper surface correlated by

$$n''_{1w} \approx \Delta T^{4.5} \approx 1.36 \times 10^{-19} / D_{1c}^{4.5}. \tag{1.114}$$

Plotting the available data for nucleation-site density as a function of superheating for boiling water at atmospheric pressure, Fig. 1.4, gives an average trend that can be represented by

$$n''_{1w} \approx \Delta T^{4.18} \approx 2.988 \times 10^{-18} / D_{1c}^{4.18}. \tag{1.115}$$

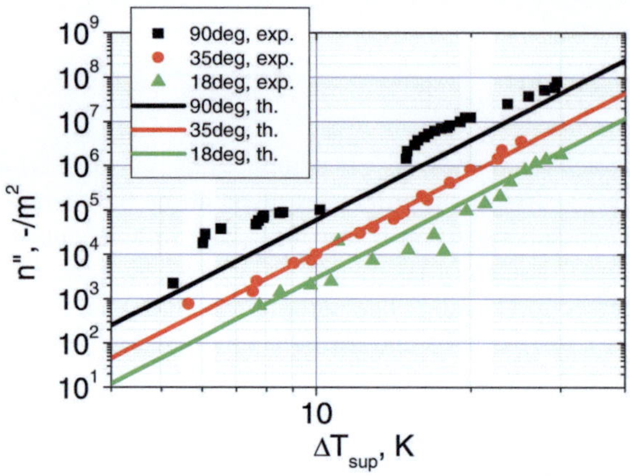

a)

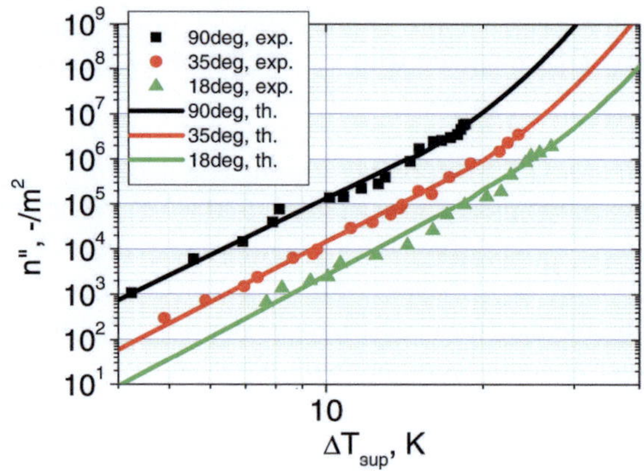

b)

Fig. 1.4 Active nucleation-site density as a function of superheating. Saturated water at 0.1 MPa. *Wang* and *Dhir* (1993) data for three different static contact angles 1) 90°, 2) 35° and 3) 18°. Prediction of the same data with their correlation 4), 5), and 6), respectively. Larger static contact angle results of larger active nucleation-site density by the same superheating. a)

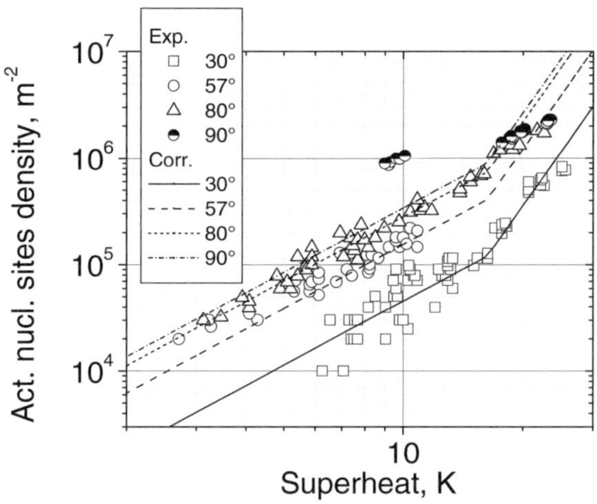

Fig. 1.5. Active nucleation-site density as a function of superheating. Saturated water at 0.1 MPa. *Basu*, et al. (2002) data for different static contact angles: 30°,57°,80°, 90°. $124 < G < 886 kg/(m^2 s)$, $6.6 < \Delta T_{sub} < 52.5K$, $25 < \dot{q}''_{w2} < 960 kW/m^2$. Prediction of the same data with their correlation, respectively

The minimum of the nucleation-site density can be represented by

$$n''_{1w} \approx \left(\Delta T / 3.729\right)^{2.593} \text{for } \Delta T \le 21.1, \tag{1.116}$$

and

$$n''_{1w} \approx 100\left(\Delta T / 16.135\right)^{16.53} \text{ for } 21.19 \le \Delta T \le 40. \tag{1.117}$$

Comparing the predictions for n''_{1w} by the different above mentioned methods with each other shows discrepancies of several orders of magnitude in different regions that leads to the conclusion that the knowledge in this field is unsatisfactory for practical application.

An important step in this field was made by *Wang* and *Dhir* (1993). *Wang* and *Dhir* (1993) succeeded to quantify the effect of the static contact angle θ, and correlated their data for active nucleation sites density as a function of the wall superheating as follows

$$n''_{1w} = 5\times10^{-27}(1-\cos\theta)/D_{1c}^6, \tag{1.118}$$

where the constant has a dimension m^4. The uncertainty in measuring the static contact angle was $\pm 1/60$ rad; the uncertainty of the measured nucleation-site density was estimated to $\pm 20\%$. The data are collected for atmospheric pool boiling of water with static contact angle of π 90/180, π 35/180 and π 18/180 rad. No instruction is given under which temperature the properties should be computed. If the properties are taken at the bulk temperature the agreement with the data is shown in Fig. 1.4a. Because I will use these data later for checking my nucleate boiling theory I approximate them as accurately as possible: in the lower-temperature region linear in log–log representation and in the upper-temperature region exponential in the normal coordinate system. The result is

$$n''_{1w}(18°) = 0.001758368925\Delta T_2^{6.17928} \text{ for } \Delta T_w < 20 \text{ and}$$
$$n''_{1w}(18°) = 411.05386 e^{\Delta T_w/3.18706} \text{ for } \Delta T_w \geq 20,$$

$$n''_{1w}(35°) = 0.014362498\Delta T_2^{6.00464} \text{ for } \Delta T_w < 20 \text{ and}$$
$$n''_{1w}(35°) = 524.77739 e^{\Delta T_w/2.66626} \text{ for } \Delta T_w \geq 20,$$

$$n''_{1w}(90°) = 0.283302234\Delta T_2^{5.67893} \text{ for } \Delta T_w < 15 \text{ and}$$
$$n''_{1w}(90°) = 1687.11274 e^{\Delta T_w/2.27907} \text{ for } \Delta T_w \geq 15.$$

Figure 1.4b shows the quality of this approximation. Note that the character of such approximation corresponds also to the measurements by *Basu* et al. (2002) presented in Fig. 1.5.

Note that Eq. (1.118) should be used together with the experimentally observed relationship between the averaged nearest-neighbor distance and the nucleation-site density,

$$\ell \approx 0.84 / n''_{1w}. \tag{1.119}$$

In 2002 *Basu*, et al. (2002) proposed a replacement for Eq. (1.118)

$$n''_{1w} = 3400(1-\cos\theta)\Delta T^2 = 2.183\times10^{-5}(1-\cos\theta)/D_{1c}^2 \tag{1.120}$$

for $\Delta T_{inb} < \Delta T < 16.298K$, and

$$n''_{1w} = 0.34(1-\cos\theta)\Delta T^{5.3} = 2.048\times10^{-23}(1-\cos\theta)/D_{1c}^{5.3} \tag{1.121}$$

for $16.298K \leq \Delta T$. Here, ΔT_{inb} is the superheat required for initiation of the nucleate boiling. The correlation reproduces data given Fig. 1.5 for $\theta = \pi/2$ and $\theta = \pi/6$ within $\pm 40\%$. The correlation is valid in the range $124 < G < 886 kg/(m^2 s)$, inlet subcooling $6.6 < \Delta T_{sub} < 52.5K$, $25 < \dot{q}''_{w2} <$ 960 kW/m^2, and $\pi/6 < \theta < \pi/4$.

Shoji et al. performed experiments in 2005 with conical, cylindrical and spherical form. From the wall-temperature fluctuation at the cavity they found that the conical cavities are not holding bubble nucleus as effectively as the cylindrical or spherical cavity do, which in fact conformation of the *Wang* and *Dhir* finding. At the same heat flux the conical cavity showed much larger magnitude of the temperature fluctuations compared to the other forms and waiting-time reduction with increasing heat flux.

Benjamin and *Balakrishnan* reported in 1997 the only correlation taking into account the thermal properties of the wall and its roughness as follows

$$n''_{1w} = 218.8 \frac{\Pr_2^{1.63}}{\gamma \delta_w^{*0.4}} \Delta T^3 . \tag{1.122}$$

The liquid *Prandtl* number, the dimensionless roughness and the dimensionless factor defining the transient interface temperature are defined as follows

$$\Pr_2 = c_{p2}\eta_2/\lambda_2 , \tag{1.123}$$

$$\delta_w^* = 14.5 - 4.5\frac{\delta_w p}{\sigma_2} + \left(\frac{\delta_w p}{\sigma_2}\right)^{0.4} , \tag{1.124}$$

$$\gamma = \left[\lambda_w \rho_w c_{pw} / \left(\lambda_2 \rho_2 c_{p2}\right)\right]^{1/2} . \tag{1.125}$$

δ_w is the arithmetic averaged deviation of the roughness surface counter from the averaged line. The correlation is valid within $1.7 < \Pr_2 < 5$, $4.7 < \gamma < 93$, $0.02 < \delta_w < 1.17 mm$, $5 < \Delta T < 25K$, $13\times10^{-3} < \sigma < 59\times10^{-3} N/m$, $2.2 < \delta_w^* <$ 14. Table 1.3 provides useful information about the roughness of differently polished materials.

Table 1.3 Roughness of differently polished materials

Material	Finish	Roughness in μm	Reference
Cooper	3/0 emery paper	0.14	*Griffith* and *Wallis* (1960)
Cooper	4/0 emery paper	0.07	*Kurihara* and *Myers* (1960)
Cooper	4/0 emery paper	0.07	*Gaertner* and *Westwater* (1960)
Cooper	4/0 emery paper	0.07	*Gaertner* (1965)
Cooper	Mirror finish	<0.02	*Wang* and *Dhir* (1993)
Nickel	4/0 emery paper	0.045	*Zuber* (1963)
Stainless steel	1/0 emery paper	0.2	*Benjamin* and Balakrishnan (1997)
Aluminum	2/0 emery paper	1.17	*Benjamin* and Balakrishnan (1997)

Aluminum	3/0 emery paper	0.89	*Benjamin* and Balakrishnan (1997)
Aluminum	4/0 emery paper	0.52	*Benjamin* and Balakrishnan (1997)

Many authors reported that the wall nucleation is not dependent on the liquid subcooling but only on the wall superheating with respect to the saturation line, e.g., *Judd* and *Hwang* (1976).

There is no attempt to describe theoretically the activated nucleation sites per unit surface as a function of the local superheating.

1.10 Conclusions and recommendations

1. The maximum superheating in technical systems is a function of the depressurization velocity and of the produced turbulence.
2. The maximum superheating can be predicted by the *Algamir* and *Lienhard* and by the *Bartak* correlations within an error band of 48.5%.
3. Flashing in short pipes and nozzles leads to critical flows driven by the pressure difference equal to the entrance pressure minus the flashing inception pressure.
4. For the prediction of the maximum achievable superheating that represents the spinoidal line the *Skripov* correlation is recommended.
5. The wetting angle is an important property of the polished surface characterizing its capability to activate nucleation sites.
6. For the prediction of the activated nucleation sites the *data* obtained by *Wang* and *Dhir* is recommended.
7. The establishment of a vapor film around a heated surface having temperature larger than the minimum film boiling temperature takes a finite time.
8. Availability of small bubbles of noncondensing gases reduces the superheating required to initiate evaporation. Evaporation at lower than the saturation temperature is possible.

Nomenclature

B thermally controlled bubble growth parameter
D_{cav} equivalent cavity size, m
D_h hydraulic diameter, m
$D_{1,n}$ bubble diameter belonging to *n*th bubble generation, m

D_{1c}	critical bubble diameter, m
D_{1d}	bubble departure diameter, m
d	differential, dimensionless
F_1 /Vol	cross-section at the wall surface covered by attached bubbles per unit flow volume, $1/m$
G^*	critical mass flow rate, $kg/(m^2 s)$
Gb_2	$= \dfrac{\Delta E_{1kr}}{kT_2}$, *Gibbs* number, dimensionless
h	specific enthalpy, J/kg
k	$= 13.805\mathrm{x}10^{-24}$, *Boltzmann* constant, J/K
m_2	$= m_\mu/N_A$, mass of single molecule, kg
m_μ	*kg-mole* mass (18 *kg* for water), kg/mole
N_2	$= \rho_2/m_2 = \rho_2 N_A/m_\mu$ ($\approx 3.3\times10^{28}$ for water), number of the molecules per unit volume, $1/m^3$
N_A	$= 6.02\times10^{26}$, *Avogadro* number, number of molecules in one kilogram-mole mass, 1/kg-mole
$\dot{n}_{1cin}$	number of the created nuclei per unit time in unit volume of the liquid, $1/(m^3 s)$
n''_w	cavity per unit surface, $1/m^2$
n''_{1w}	bubble generating (active) cavity per unit surface, $1/m^2$
n_1	bubble number density, $1/m^3$
$n_{1,n}$	bubble number density belonging to *n*th bubble generation, $1/m^3$
$n_{1,\max}$	maximum number of nucleation sites per unit volume in a pipe, $1/m^3$
p	pressure, Pa
$\dot{q}_2'''^{1\sigma}$	thermal energy consumed from the liquid per unit time and unit flow volume, W/m^3
R	gas constant, J/(kg K)
R_1	bubble radius, m
R_{10}	initial bubble radius, m
R_{1c}	critical bubble radius, m
$R_{1\infty}$	final radius of the bubble, m
r	radius, m
s	specific entropy, J/(kg K)
T	temperature, K
u	radial velocity, m/s
$\mathbf{V}$	velocity vector, m/s
V_{1c}	volume of the bubble with critical size
v	specific volume, m^3/kg
Vol	volume, m^3

Greek

α_1	bubble volumetric fraction, m³/m³
γ_v	volume occupied by the flow divided by the volume of the system, dimensionless
γ	cross-section flowed by the flow divided by the total cross-section of the system, dimensionless
Δ	finite difference
ΔE_1	work necessary to create a single bubble with radius R_1, J
ΔE_{1c}	work necessary to create a single bubble with radius R_{1c}, J
ΔE_{1c}^*	$=\varphi \Delta E_{1c}$, energy for the creation of a single bubble at a wall, J
Δp_{Fi}^*	$=p'(T_2)-p_{Fi}$, flashing inception pressure difference, Pa
$\Delta\tau$	time interval, s
θ	static contact angle, rad
λ_{RT}	*Rayleigh–Taylor* wavelength, m
μ_{21}	evaporating mass per unit time and unit flow volume, kg/(m³ s)
ρ	density, kg/m³
σ	surface tension, N/m
τ	time, s
Φ	cavity angle if the cavity is idealized as a cone, rad
φ	work-reduction factor, dimensionless

Superscripts

$'$	saturated liquid
$''$	saturated vapor
spin	spinoidal line

Subscripts

1	vapor, gas
2	liquid
c	critical state
s	at constant entropy
Fi	flashing inception
$\sum_{n1o}$	sum of all noncondensing gases
M	evaporating chemical component

References

Abuaf, N., Jones Jr., O.C., Wu, B.J.C.: Critical flashing flows in nozzles with subcooled inlet conditions. J. Heat Transfer 105, 379–383 (1983)

Algamir, M., Lienhard, J.H.: Correlation of pressure undershoot during hot-water depressurisation. J. Heat Transfer 103(1), 60 (1981)

Arefeva, E.I., Aladev, I.T.: O wlijanii smatchivaemosti na teploobmen pri kipenii. Injenerno - Fizitcheskij Jurnal 1(7), 11–17 (1958) (in Russian)

Avdeev, A.A., Maidanik, V.N., Selesnev, L.I., Shanin, V.K.: Calculation of the critical flow rate with saturated and subcooled water flashing through a cylindrical duct. Teploenergetika 24(4), 36–38 (1977)

Bartak, J.: A study of the rapid depressurization of hot water and the dynamics of vapor bubble generation in superheated water. Int. J. Multiphase Flow 16(5), 789–798 (1990)

Basu, N., Warrier, G.R., Dhir, V.K.: Onset of nucleate boiling and active nucleation site density during subcooled flow boiling. J. Heat Transfer 124, 717–728 (2002)

Benjamin, R.J., Balakrishnan, A.R.: Nucleation site density in pool boiling of saturated pure liquids: effect of surface micro roughness and surface and liquid physical properties. Exp. Thermal Fluid Sci. 15, 32–42 (1997a)

Benjamin, R.J., Balakrishnan, A.R.: Nucleation site density in pool boiling of binary mixtures: effect of surface micro roughness and surface and liquid physical properties. Can. J. Chem. Eng. 75, 1080–1089 (1997b)

Bergles, A.E., Rohsenow, W.M.: The determination of forced convection surface-boiling heat transfer. ASME J. Heat Transfer 1, 365–372 (1964)

Besand, W.H.: Hydrostatics and hydrodynamics, p. 170. Deighton Bell, Cambridge (1859)

Blake, F.G.: The onset of cavitation in liquids. I. Acoustic Res. Lab., Harvard Univ., Tech. Memo. no 12 (1949)

Blander, M., Katz, J.L.: Bubble nucleation in liquids. AIChE 21, 833–838 (1975)

Brennen, C.E.: Cavitation and bubble dynamics. Oxford University Press, Oxford (1995)

Burnell, J.G.: Flow of boiling water through nozzles, orifices and pipes. Engineering, 572–576 (December 12, 1947)

Cornwell, K., Brown, R.D.: Boiling surface topology. In: Proc. 6th Int. Heat Transfer Conf. Heat Transfer, Toronto, vol. 1, pp. 157–161 (1978)

Eberhard, J.G., Schnyders, M.C.: Application of the mechanical stability condition to the prediction of the limit of superheat for normal alkanes, ether and water. J. Phys. Chem. 77(23), 2730–2736 (1973)

Faggiani, S., Galbiati, P., Grassi, W.: Active site density, bubble frequency and departure on chemically etched surfaces. La Termotechnica 29(10), 511–519 (1981)

Fincke, J.R.: Critical flashing flow of subcooled fluids in nozzles with contour discontinuities, Basic Aspects of Two Phase Flow and Heat Transfer. In: 22nd Nat. Heat Transfer Conference and Exhibition, Niagara Falls, New York, August 5-8. HTD, vol. 34, pp. 85–93 (1984)

Gaertner, R.F.: Photographic study of nucleate pool boiling on a horizontal surface. Transaction of the ASME, J. Heat Transfer 87, 17–29 (1965)

Gaertner, R.F., Westwater, J.W.: Population of active sites in nucleate boiling heat transfer. In: Chem. Eng. Progr. Symp. Ser., vol. 30(30), pp. 39–48 (1960)

Gibbs, J.W.: Thermodynamische Studien (Leipzig 1892). Amer. J. Sci. Arts 16, 454–455 (1878)

Griffith, P., Wallis, G.B.: The role of the surface conditions in nucleate boiling. In: Chem. Eng. Prog. Symp., vol. 56, pp. 49–63 (1960)

Hirose, Y., Hayashi, T., Hazuku, T., Takamasa, T.: Experimental study on contact angle of water droplet in high temperature conditions. In: Proc. of ICONE14, Int. Conf. on Nucl. Engineering, Miami, Florida, USA, July 17-20 (2006)

Hutcherson, M.N., Henry, R.E., Wollersheim, D.E.: Two-phase vessel blowdown of an initially saturated liquid - Part 2: Analytical. Trans. ASME, J. Heat Transfer 105, 694–699 (1983)

Jakob, M., Linke, W.: Der Wärmeübergang von einer waagerechten Platte an siedendes Wasser. Forsch. Ing. Wes. 4, 75–81 (1933)

Johov, K.A.: Nucleations number during steam production, Aerodynamics and Heat Transfer in the Working Elements of the Power Facilities, Leningrad. In: Proc. CKTI, vol. 91, pp. 131–135 (1969) (in Russian)

Jones Jr., O.C.: Flashing inception in flowing liquids, non-equilibrium two-phase flow. In: Chan, J.C., Bankoff, S.G. (eds.) Non-Equilibrium Two- Phase Flow, pp. 29–34. ASME, New York (1979)

Jones Jr., O.C.: Towards a unified approach for thermal non-equilibrium in gas-liquid systems. Nucl. Eng. Design 69, 57–73 (1982)

Jones, O.C.: Non-equilibrium phase change 1. Flashing inception, critical flow, and void development in ducts. In: Lahey Jr., R.T. (ed.) Boiling Heat Transfer, pp. 189–234. Elsevier Science Publishers BV, New York (1992a)

Jones, O.C.: Nonequilibrium phase change 2. Relaxation models, general applications, and post heat transfer. In: Lahey Jr., R.T. (ed.) Boiling Heat Transfer, pp. 447–482. Elsevier Science Publishers BV, New York (1992b)

Judd, R.L., Hwang: A comprehensive model for nucleate pool boiling heat transfer including microlayer evaporation. Trans. ASME, J. Heat Transfer, 623–629 (November 1976)

Kaishew, R., Stranski, I.N.: Z. Phys. Chem. 26, 317 (1934)

Kocamustafaogullari, G., Ishii, M.: Interfacial area and nucleation site density in boiling systems. Int. J. Heat Mass Transfer 26, 1377–1389 (1983)

Kottowski, H.M.: Nucleation and superheating effects on activation energy of nucleation. In: Dwyer, O.E. (ed.) Progress in Heat and Mass Transfer, vol. 7, pp. 299–324. Pergamon, New York (1973)

Kurihara, H.M., Myers, J.E.: The effect of superheat and surface roughness on boiling coefficients. AIChE Journal 6(1), 83–91 (1960)

Labuntsov, D.A.: Approximate theory of heat transfer by developed nucleate boiling (Russ.), Izvestiya AN SSSR, Energetika i transport, no 1 (1963)

Lienhard, J.H.: Correlation for the limiting liquid superheat. Chem. Eng. Sci. 31, 847–849 (1976)

Lienhard, J.H.: Homogeneous nucleation and the spinoidal line. J. Heat Transfer 103, 72 (1981)

Lienhard, J.H.: A Heat Transfer Textbook. Phlogiston Press, Cambridge (2005)

Malnes, D., Solberg, K.: A fundamental solution to the critical two-phase flow problems, applicable to loss of coolant accident analysis. SD-119, Kjeller Inst., Norway (May 1973)

Mikic, B.B., Rohsenow, W.M.: A new correlation of pool-boiling data including the effect of heating surface characteristics. Trans. ASME, J. Heat Transfer 91, 245–250 (1969)

Neppiras, E.A., Noltingk, B.E.: Cavitation by ultrasonic: theoretical conditions for onset of cavitation. Proc. Phys. Soc. 64B, 1032–1038 (1951)

Rallis, C.J., Jawurek, H.H.: Latent heat transport in saturated nucleate boiling. Int. J. Heat Transfer 7, 1051–1068 (1964)

Rayleigh, L.: On the pressure developed in a liquid during the collapse of spherical cavity. Phil. Mag. 34, 94 (1917)

Riznic, J., Ishii, M.: Bubble number density and vapor generation in flashing flow. Int. J. Heat Mass Transfer 32, 1821–1833 (1989)

Rohatgi, U.S., Reshotko, E.: Non-equilibrium one dimensional flow in variable area channels. In: Lahey, R.T., Wallis, G.B. (eds.) Non-Equilibrium Two Phase Flows. ASME, New York (1975)

Shoji, M., Zhang, L., Chatpun, S.: Nucleation site interaction in pool nucleate boiling. In: 6th World Conference on Experimental Heat Transfer, Fluid Mechanics, and Thermodynamics, Matsushima, Miyagi, Japan, April 17-21 (2005)

Siegel, R., Keshock, E.G.: Effects of reduced gravity on nucleate boiling bubble dynamics in saturated water. AIChE Journal 10(4), 509–517 (1964)

Simoneau, R.J.: Pressure distribution in converging diverging nozzle during two phase choked flow of subcooled nitrogen. In: Lahey, R.T., Wallis, G.B. (eds.) Non-Equilibrium Two Phase Flows. ASME, New York (1975)

Skripov, V.P., et al.: Thermophysical properties of liquids in metastable state, Moskva - Atomisdat (1980) (in Russian)

Sultan, M., Judd, R.L.: Spatial distribution of active sites and bubble flux density. Trans. ASME, J. Heat Transfer 100, 56–62 (1978)

Tolubinski, I.S.: Boiling heat transfer. Kiev, Naukova Dumka (1980) (in Russian)

Tolubinsky VI, Ostrovsky, J.N.: On the mechanism of boiling heat transfer (vapor bubbles growth rate in the process of boiling in liquids, solutions, and binary mixtures). Int. J. Heat Mass Transfer 9, 1463–1470 (1966)

Volmer, M.: Kinetik der Phasenbildung, Dresden und Leipzig, Verlag von Theodor Steinkopf (1939)

Wang, C.H., Dhir, V.K.: Effect of surface wettability on active nucleation site density during pool boiling of water on a vertical surface. ASME J. Heat Transfer 115, 659–669 (1993)

Yamagata, K., Hirano, F., Nishikawa, K., Matsuoka, H.: Nucleate boiling of water on the horizontal heating surface. Mem. Fac. Engng; Kyushu Univ. 15, 97 (1955)

Zuber, N.: Nucleate boiling: The region of isolated bubbles and the similarity with natural convection. Int. J. Heat Mass Transfer 6, 53–79 (1963)

2. Bubble growth in superheated liquid

This chapter presents the achievements of the theory of bubble growth in superheated liquids. The thermally controlled bubble growth solutions are summarized and the derivation of the Mikic equation is discussed in some detail. Then the link between the solutions for the bubble growth and the mass source terms for the averaged conservation equations for two-phase flow are presented. The way to derive nonaveraged mass source terms and time-averaged mass source terms is given. The effect of the steam superheating is discussed. A brief discussion of diffusion-controlled bubble growth initially containing noncondensable gases is also given.

2.1 Introduction

The *spontaneous flashing of superheated liquid* is associated with the *bubble growth* in the liquid. The mechanism of the *bubble growth* is described already by *Jakob* in 1932 in the following way:

> "...it can be imagined that during the small explosion which starts the growth of a bubble, the interface temperature, because of the consumed heat for vaporization, drops immediately from the superheat temperature to the saturation temperature, for example from 110 °C to 100 °C. As a consequence of the heat transfer from the liquid to the vapor bubble the liquid envelope is being cooled progressively from the inside toward the outer boundary; a temperature boundary layer is created with a constantly decreasing temperature drop. This thermal boundary layer increases in thickness until the thermal wave, which advances from the vapor bubble interface into the liquid, has reached the outer limit of the hydrodynamics boundary layer. The decrease in thickness of the hydrodynamics boundary layer because of the evaporation at the interface is, initially, a small fraction of the total thickness...".

The rigorous mathematical description of the processes in a system of spherical coordinates with the origin equivalent with the center of the mass of the nucleation site, needs the solution of a system of mass-, momentum-, and energy-conservation equations for the corresponding initial and boundary conditions, see *Nigmatulin* (1978), *Beylich* (1991) among others. An important result of such a solution is the *bubble radius as a function of time* and of the local thermodynamic parameters describing the nonequilibrium of the mixture. A general analytical solution of these problems is not known to the author. For practical applications there are several approximate solutions obtained under different simplifying assumptions. The most successful compromise between strict formulation and sound simplifying assumption was made by *Mikic* et al. (1970). In what follows we present the arguments used by *Mikic*.

2.2 The thermally controlled bubble growth

Consider an infinite liquid with temperature T_2 that is superheated with respect the saturation temperature by the system pressure p, i.e.,

$$T_2 > T'(p). \tag{2.1}$$

The liquid-side bubble interface temperature is $T_2^{1\sigma}$,

$$T_2^{1\sigma} < T_2, \tag{2.2}$$

Due to the intensive steam production from the bubble surface and the fact that the vapor sound velocity is higher than the bubble interface velocity, it can be assumed that the pressure inside the bubble is uniform and that the *vapor temperature in the bubble is equal to the surface temperature*

$$T_1 = T_2^{1\sigma}. \tag{2.3}$$

The pressure inside the bubble can be assumed to be the saturation pressure at the vapor temperature and therefore

$$p'(T_1) = p'\left(T_2^{1\sigma}\right). \tag{2.4}$$

The change of the bubble mass with the time is equal to the mass evaporating from the surface per unit time, i.e.,

$$\frac{d}{d\tau}\left(\frac{4}{3}\pi R_1^3 \rho_1\right) = 4\pi R_1^2 (\rho w)_{21}, \tag{2.5}$$

see *Bosnjakovic* (1930). This is the vapor mass-conservation equation. Here, $(\rho w)_{21}$ is the mass flow rate from the surface into the bubble. This equation is used in the literature in different forms, e.g.,

$$\frac{1}{3}\frac{d}{d\tau}\left[\left(R_1 \rho_1^{1/3}\right)^3\right] = R_1^2 (\rho w)_{21} \tag{2.6}$$

or

$$\frac{d}{d\tau}\left[\left(R_1 \rho_1^{1/3}\right)\right] = (\rho w)_{21} / \rho_1^{2/3} \tag{2.7}$$

or in integrated form

$$R_1 = \left(\rho_{10}/\rho_1\right)^{1/3} R_{10}\left\{1+\frac{1}{R_{10}\rho_{10}}\int_0^{\Delta\tau}\left[\left(\rho w\right)_{21}/\rho_1^{2/3}\right]d\tau\right\}, \tag{2.8}$$

which is appropriate to take into account the density change with the time, or

$$\frac{dR_1}{d\tau} = \left(\rho w\right)_{21}/\rho_1 - \frac{R_1}{3\rho_1}\frac{d\rho_1}{d\tau}. \tag{2.9}$$

The enthalpy of the liquid experiencing evaporation before crossing the interface is $h'(T_2^{1\sigma})$ and after crossing the interface $h''(T_2^{1\sigma})$. Thus, the energy-jump condition at the interface, assuming no heat transfer between the bubble and the interface, $\dot{q}_1''^{2\sigma} = 0$, is

$$\left(\rho w\right)_{21}\left[h''(T_2^{1\sigma}) - h'(T_2^{1\sigma})\right] = -\dot{q}_2''^{1\sigma}. \tag{2.10}$$

Solving with respect to the evaporation mass flow rate and substituting into the mass-conservation equation (2.9) we obtain

$$\frac{dR_1}{d\tau} = -\dot{q}_2''^{1\sigma}/\left\{\rho_1\left[h''(T_2^{1\sigma}) - h'(T_2^{1\sigma})\right]\right\} - \frac{R_1}{3\rho_1}\frac{d\rho_1}{d\tau}. \tag{2.11}$$

The heat transferred from the superheated bulk liquid to the surface is controlled by the *Fourier* equation, in spherical coordinates (see *Fourier*, (1822))

$$\frac{\partial T}{\partial \tau} = -a_2\frac{1}{r^2}\frac{\partial}{\partial r}\left(r^2\frac{\partial T}{\partial r}\right) \tag{2.12}$$

solved for the initial and boundary conditions

$$\tau = 0,\ r = R_1,\ T = T_2^{1\sigma}, \tag{2.13}$$

$$\tau = 0,\ r > R_1,\ T = T_2, \tag{2.14}$$

$$r = R_1,\ T = T_2^{1\sigma}, \tag{2.15}$$

$$r = \infty,\ T = T_2\ . \tag{2.16}$$

The text book solution for a *thin thermal boundary layer, Carlsaw and Jaeger* (1959), is used to compute the temperature gradient at the bubble surface and to compute the resulting heat flux as a function of time

$$\dot{q}_{i2}'' = -\lambda_2\frac{\partial T}{\partial r}\bigg|_{r=R_1} = -\rho_2 c_{p2}\left(T_2 - T_2^{1\sigma}\right)\sqrt{\frac{3a_2}{\pi\tau}}$$
$$= -\rho_2 c_{p2}\left[T_2 - T'(p)\right]\sqrt{\frac{3a_2}{\pi\tau}}\left[1 - \frac{T_2^{1\sigma} - T'(p)}{T_2 - T'(p)}\right]. \tag{2.17}$$

Usually, the so-obtained heat flux is substituted in Eq. (2.11) and the resulting equation is rewritten in the form

$$\frac{dR_1}{d\tau} = \frac{1}{2}\frac{B}{\sqrt{\tau}}\left[1 - \frac{T_2^{1\sigma} - T'(p)}{T_2 - T'(p)}\right] - \frac{R_1}{3\rho_1}\frac{d\rho_1}{d\tau} \tag{2.18}$$

where

$$B = Ja\sqrt{\frac{12a_2}{\pi}}. \tag{2.19}$$

The group

$$Ja = \frac{\rho_2 c_{p2}\left[T_2 - T'(p)\right]}{\rho_{10}\left[h''(T_2^{1\sigma}) - h'(T_2^{1\sigma})\right]} \tag{2.20}$$

is named the *Jakob* number in honor of the German scientist who considered for the first time this problem in *Jakob* (1932). *Fritz* and *Ende* obtained in 1936 Eq. (2.18) for $\rho_1 \approx const$ and plane geometry. For processes with dramatic pressure change the vapor compressibility is important and $d\rho_1/d\tau$ cannot be neglected. Equation (2.18) controls the bubble growth if the surface temperature $T_2^{1\sigma}$ is known. If the surface temperature is assumed to be saturation temperature at the system pressure

$$T_2^{1\sigma} = T'(p), \tag{2.21}$$

the obtained solution for $\rho_1 \approx const$ is the so-called thermally controlled bubble growth solution obtained by *Plesset* and *Zwick* in 1954

$$\frac{dR_1}{d\tau} = \frac{1}{2}\frac{B}{\sqrt{\tau}} \text{ or } R_1 = B\sqrt{\tau}. \tag{2.22}$$

Golovin, et al. (1963), *Labuntsov* et al. (1964b) confirmed experimentally the dependence $R_1/\sqrt{a_2\tau} = f(Ja)$, see Fig. 2.1. *Miheev* and *Miheeva* (1973) reproduced their data by $R_1/\sqrt{a_2\tau} = 3.5\sqrt{Ja}$ for $Ja < 20$.

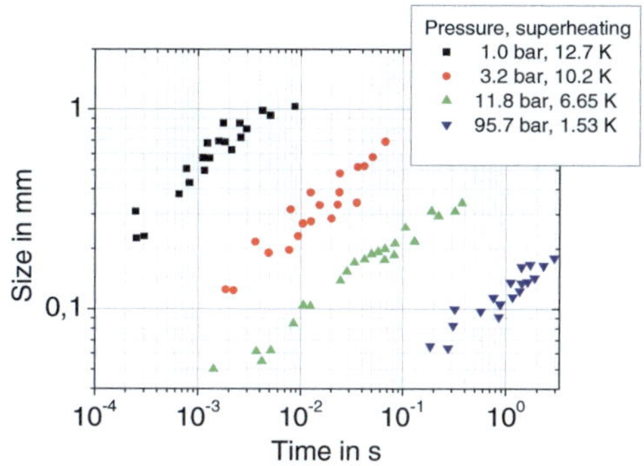

Fig. 2.1 Bubble diameter as a function of time for water boiling at horizontal silver plate, *Golovin*, et al. (1963), *Labuntsov* et al. (1964b)

There are several other authors contributing to this topic. Their results are summarized in Appendix 2.1.

2.3 The *Mikic* solution

Note that in accordance with the assumption that the pressure inside the bubble is equal to the saturation pressure as a function of the interface temperature, there is no pressure difference between the bubble and the environment, which is not true. Am improved solution can be obtained if one allows a surface temperature different from the saturation temperature at the system pressure. In this case, one needs one more equation to close the description. *Mikic* uses the following method: The mass-conservation equation of the liquid can be approximated by

$$\frac{1}{r^2}\frac{\partial}{\partial r}(r^2 u_2) \approx 0, \tag{2.23}$$

Or, integrating with the boundary condition

$$r = R_1,\ u_2 = \frac{\partial R_1}{\partial \tau}, \tag{2.24}$$

$$u_2 = const / r^2 = \frac{\partial R_1}{\partial \tau}\left(R_1/r\right)^2. \tag{2.25}$$

It gives the liquid velocity as a function of the radius if the liquid were incompressible during the bubble expansion. During the bubble growth the work performed by the bubble expansion

$$4\pi \int_0^{R_1} (p_1 - p) r^2 dr \approx \frac{4}{3}\pi R_1^3 \left(1 - \frac{\rho''}{\rho_2}\right)(p_1 - p), \tag{2.26}$$

is transferred in total kinetic energy of the liquid environment

$$\int_{R_1}^{\infty} \frac{1}{2}\rho_2 u_2^2 dVol_2 = \frac{1}{2}\rho_2 \left(\frac{dR_1}{d\tau}\right)^2 R_1^4 4\pi \int_{R_1}^{\infty} \frac{dr}{r^2} = 2\pi\rho_2 \left(\frac{dR_1}{d\tau}\right)^2 R_1^4 \int_{R_1}^{\infty} d(-\frac{1}{r})$$

$$= 2\pi\rho_2 \left(\frac{dR_1}{d\tau}\right)^2 R_1^3,$$

i.e.,

$$\left(\frac{dR_1}{d\tau}\right)^2 = \frac{2}{3}\left(1 - \frac{\rho''}{\rho_2}\right)(p_1 - p)/\rho_2 = \frac{2}{3}\left(1 - \frac{\rho''}{\rho_2}\right)\left(p'(T_2^{1\sigma}) - p\right)/\rho_2 = \frac{2}{3}\frac{2\sigma}{R_{1c}}/\rho_2, \tag{2.27}$$

which is the second needed equation. The constant 2/3 valid for bubble growth in a bulk liquid should be replaced by π/7 if a spherical bubble growing on a flat surface is considered. In fact, this is the mechanical energy-conservation equation. If one assumes that the interface temperature is equal to the liquid temperature

$$T_2^{1\sigma} = T_2, \tag{2.28}$$

and therefore the bubble pressure is the saturation pressure at the liquid temperature the above equation describes only the bubble growth. This is the limiting case giving the fastest bubble growth with the liquid superheating. This solution is named the inertial-controlled solution and was obtained for the first time by *Besand* (1859) and in the above-described more elegant way by *Rayleigh* (1917) for $\rho''/\rho_2 = 0$. The solution is valid within the first 10^{-8} s of the life of the bubble.

In reality, the bubble in a superheated liquid behaves between the two above-described limiting mechanisms, so the bubble growth is governed by the equations (2.22) and (2.27) simultaneously. To transfer the pressure difference $p'(T_2^{1\sigma}) - p$ in the last equation into a temperature difference $T_2^{1\sigma} - T'(p)$ one can use the linearized equation of *Clausius* and *Clapeyron*

$$\frac{p'(T_2^{1\sigma}) - p}{T_2^{1\sigma} - T'(p)} = \left(\frac{dp}{dT}\right)_{sat}, \tag{2.29}$$

where

$$\left(\frac{dp}{dT}\right)_{sat} = \frac{1}{T'(p)} \frac{h''(p) - h'(p)}{v''(p) - v'(p)}, \tag{2.30}$$

or

$$\left[p'(T_2^{1\sigma})-p\right]/\rho_2=\left[T_2^{1\sigma}-T'(p)\right]\frac{1}{\rho_2}\left(\frac{dp}{dT}\right)_{sat}. \tag{2.31}$$

Note that this dependence is in general nonlinear. The result is

$$\left(\frac{dR_1}{d\tau}\right)^2=\frac{2}{3}\left(1-\frac{\rho''}{\rho_2}\right)\left[T_2^{1\sigma}-T'(p)\right]\frac{1}{\rho_2}\left(\frac{dp}{dT}\right)_{sat}$$
$$=\frac{2}{3}\frac{1}{\rho_2}\left(\frac{dp}{dT}\right)_{sat}\left(1-\frac{\rho''}{\rho_2}\right)\left[T_2-T'(p)\right]\frac{T_2^{1\sigma}-T'(p)}{T_2-T'(p)} \tag{2.32}$$

or

$$\left(\frac{dR_1}{d\tau}\right)^2=A^2\frac{T_2^{1\sigma}-T'(p)}{T_2-T'(p)}, \tag{2.33}$$

where

$$A^2=\frac{2}{3}\frac{1}{\rho_2}\left(\frac{dp}{dT}\right)_{sat}\left(1-\frac{\rho''}{\rho_2}\right)\left[T_2-T'(p)\right]. \tag{2.34}$$

Eliminating the temperature difference from the both equations, Eqs. (2.22) and (2.33), *Mikic* et al. obtained the following final form of the equation describing best the available experimental data

$$\frac{1}{A^2}\left(\frac{dR_1}{d\tau}\right)^2+\frac{2\sqrt{\tau}}{B}\frac{dR_1}{d\tau}-1+\frac{2\sqrt{\tau}}{B}\frac{R_1}{3\rho_1}\frac{d\rho_1}{d\tau}=0, \tag{2.35}$$

or after solving for $dR_1/d\tau$ and substituting $\left(dR_1/d\tau\right)/A=dR_1^+/d\tau^+$

$$\frac{dR_1^+}{d\tau^+}=\left(\tau^++1-\frac{2}{3}R_1^+\frac{\tau_1^{+1/2}}{\rho_1^+}\frac{d\rho_1^+}{d\tau^+}\right)^{1/2}-\left(\tau^+\right)^{1/2}, \tag{2.36}$$

where

$$R^+=AR_1/B^2, \tag{2.37}$$

$$\tau^+=A^2\tau/B^2. \tag{2.38}$$

Substituting the time derivative of the bubble radius from Eq. (2.36) into Eq. (2.33) we compute the surface temperature

$$T_2^{1\sigma}=T'(p)+\left[T_2-T'(p)\right]\left[\left(\tau^++1-\frac{2}{3}R_1^+\frac{\tau_1^{+1/2}}{\rho_1^+}\frac{d\rho_1^+}{d\tau^+}\right)^{1/2}-(\tau^+)^{1/2}\right]^2 \tag{2.39}$$

which is necessary to compute the enthalpies of the transferred mass before and after crossing the interface.

Integrating the equation (2.35) for $R_1^+ = R_{1o}$ at $\tau^+ = 0$, and $\rho_1 = const$, *Mikic* et al. obtained the general bubble-growth relation

$$R_1^+ - R_{1o}^+ = \frac{2}{3}\left[\left(\tau^+ + 1\right)^{3/2} - (\tau^+)^{3/2} - 1\right], \tag{2.40}$$

which for $\tau^+ << 1$ simplifies to the *Rayleigh* solution, and for $\tau^+ >> 1$ to the thermally controlled bubble growth solution.

The validity of the Eq. (2.40) is confirmed experimentally by *Lien* and *Griffith* (1969) for superheated water over the pressure range 0.0012 to 0.038 MPa, superheating range 8 to 15 K, and 58 $\le Ja \le$ 2690 and by *Lee* and *Merte* (1996) p. 14 for 0.0012 to 0.1 MPa, 0.001 $\le Ja \le$ 10000. The comparison with experimental data made by *Tolubinskii* (1980) shows that in a earlier phase of the bubble growth $\tau^+ < 10^{-3} (\tau \le 10^{-8} s)$ the bubble growth is inertial dominated. For $\tau^+ > 10^{-3}$ the *Mikic* et al. solution is the best one. For the time $\tau^+ > 10$ all of the heat diffusion-controlled solutions tend to the asymptotic solution, the so-called thermally controlled bubble growth.

Olek et al. (1990) obtained a general counterpart solution to that of *Mikic* et al. (1970) using the hyperbolic heat conduction equation, instead of the classical *Fourier* (parabolic) equation. The new solution is important for bubble growth in fluids like helium II.

The experimental data from the pioneer work by *Plesset* and *Zwick* (1954) presents the bubble size as a function of time for subatmospheric pressure. The bubbles are generated on heated surface at subatmospheric pressure and wall superheat given in Table 2.1.

Table 2.1 Bubble-growth experiments on heated plate performed by Plesset and Zwick 1954

p in Pa	T' in K H_2O/Na	$T_w - T'$ in K	Ja H_2O/Na
101 325	373.12/1154.6	3.1	9.3/2.78
101 325	373.12/1154.6	4.5	13.48/4.04
101 325	373.12/1154.6	5.3	15.9/4.76

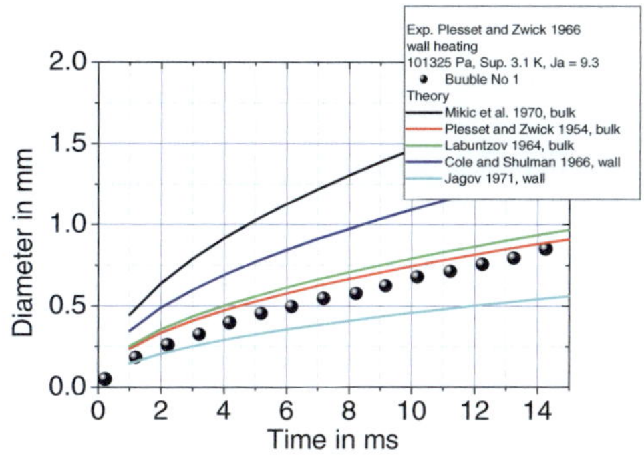

Fig. 2.2 Bubble diameter as function of time. Bubble generated at heated wall. Experimental data for water by *Plesset* and *Zwick* (1954). Comparison with five theoretical models

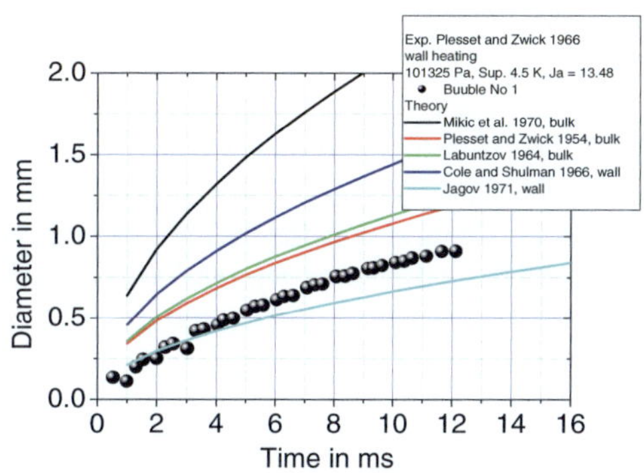

Fig. 2.3 Bubble diameter as function of time. Bubble generated at heated wall. Experimental data for water by *Plesset* and *Zwick* (1954). Comparison with five theoretical models

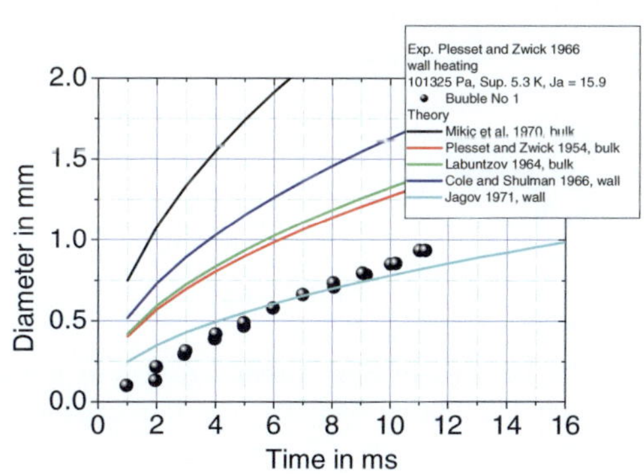

Fig. 2.4 Bubble diameter as function of time. Bubble generated at heated wall. Experimental data for water by *Plesset* and *Zwick* (1954). Comparison with five theoretical models

The predictions for water by fife of the models from Appendix 2.1 are presented at Figs. 2.2, 2.3 and 2.4. The experimental data are also entered in these figures.

The data in the paper by *Cole* and *Shulman* (1966) presents also bubble size as a function of time at atmospheric pressure. The bubbles are generated also on a heated surface at subatmospheric pressure and wall superheat given in Table 2.2.

Table 2.2 Bubble-growth experiments on heated plate performed by Cole and Shulman (1966). Comparison with five theoretical models

p in Pa	T' in K H_2O/Na	$T_w - T'$ in K	Ja H_2O/Na	$\dot{q}_2^{''w\sigma}$ in kW/m²
47 996	353.45/1078.39	15	89/26	62.78/?
25 998	338.99/1023.36	18.3	191/57	47.79/?
13 065	324.29/968	15	297/89	37.38/?
6666	311.24/919.5	20.56	760/229	67.03/?

The predictions for water by the five methods given in Appendix 2.1 are presented at Figs. 2.5 through 2.9. The experimental data are also entered in these figures.

From this comparison we realize that the bubble growth at heated surfaces is slower then the bubble growth in the bulk. We realize also that uncertainties of the existing state-of-the-art for mathematical description of such processes still exist.

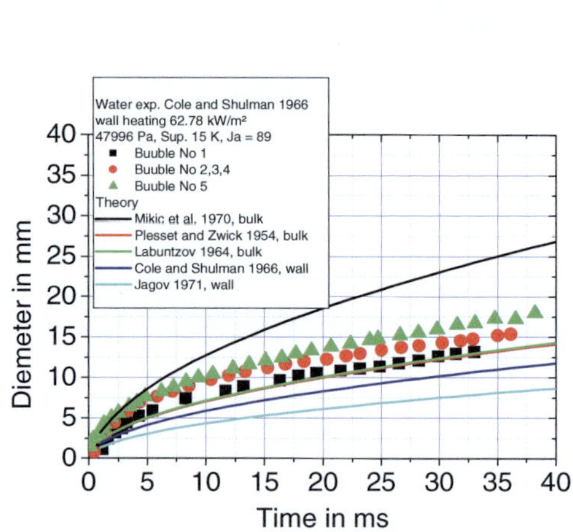

Fig. 2.5 Bubble diameter as function of time. Bubble-growth experiments on heated plate performed by *Cole* and *Shulman* (1966). Comparison with five theoretical models

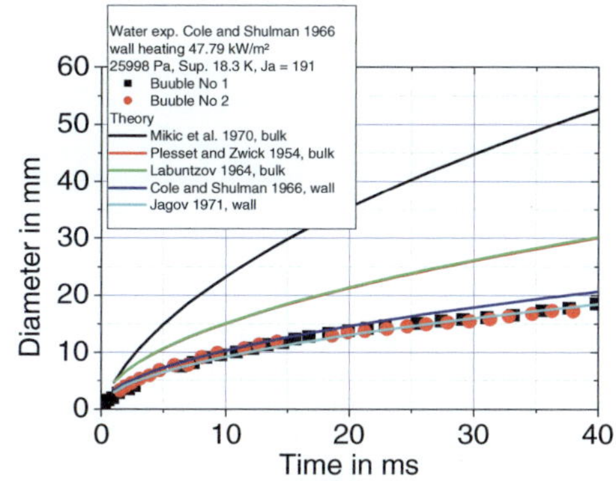

Fig. 2.6 Bubble diameter as function of time. Bubble-growth experiments on heated plate performed by *Cole* and *Shulman* (1966). Comparison with five theoretical models

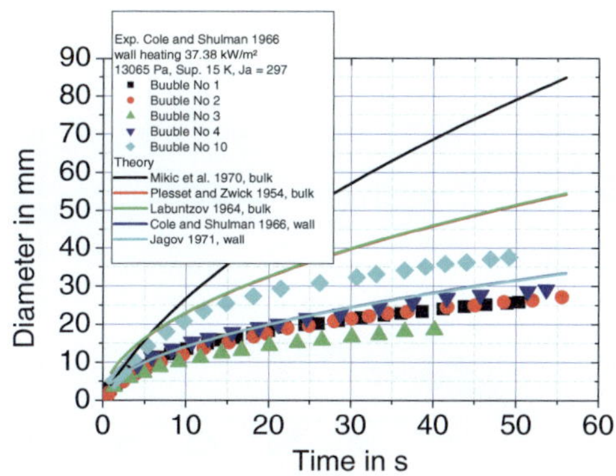

Fig. 2.7 Bubble diameter as function of time. Bubble-growth experiments on heated plate performed by *Cole* and *Shulman* (1966). Comparison with five theoretical models

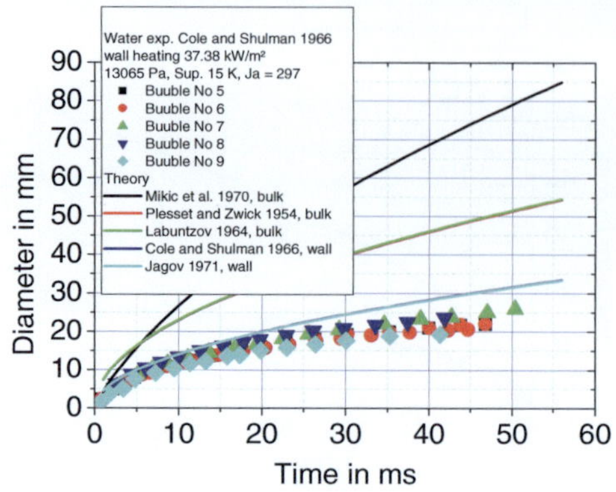

Fig. 2.8 Bubble diameter as function of time. Bubble-growth experiments on heated plate performed by *Cole* and *Shulman* (1966). Comparison with five theoretical models

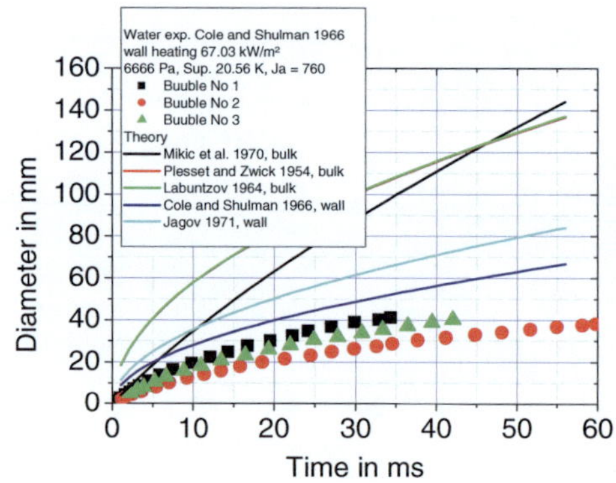

Fig. 2.9 Bubble diameter as function of time. Bubble-growth experiments on heated plate performed by *Cole* and *Shulman* (1966). Comparison with five theoretical models

Labuntsov et al. (1964a) provided the data for high-pressure bubble growth given Table 2.3 proving the dependence $R_1/\sqrt{\tau} = const$ also for high pressure. Obviously, the higher the pressure the slower the bubble growth at the same superheating.

Table 2.3 Experimental data for the averaged characteristics $R_1/\sqrt{\tau}$

P in bar	$T_w - T'$ in K	$R_1/\sqrt{\tau}$ in m/s$^{1/2}$
1.33	9.1	0.009
1.33	9.8	0.009
3.14	8.14	0.006
3.14	6.7	0.005
11.6	2.6	0.00125
11.6	3.0	0.0015
20.5	1.4	0.0007
20.5	1.8	0.0007
0.98	12.7	0.022
3.2	9.74	0.0043
3.2	10.2	0.0046
11.8	6.65	0.0017
11.8	5.1	0.00145
31.4	3.0	0.0009
47.7	2.15	0.00062
77.5	1.55	0.0004
77.5	1.67	0.00067
95.7	1.57	0.00034
95.7	1.53	0.00032

2.4 How to compute the mass source terms for the averaged conservation equations?

2.4.1 Nonaveraged mass source terms

In what follows we show how to link the results obtained in the previous sections with the macroscopic flow description from Volume 1 of this monograph.

The evaporated mass per unit mixture volume and unit time is equal to the product of the number of the bubbles per unit mixture volume n_1 and the mass change of a single bubble $\rho''\left(T_2^{1\sigma}\right)dV_1/d\tau$. Therefore

$$\mu_{21} = \rho''(T_2^{1\sigma})n_1 dV_1 / d\tau = \rho''(T_2^{1\sigma})n_1 4\pi R_1^2 dR_1 / d\tau, \tag{2.41}$$

or replacing

$$R_1 = \left(\frac{3}{4\pi}\frac{\alpha_1}{n_1}\right)^{1/3}, \tag{2.42}$$

$$\mu_{21} = 3^{1/3}(4\pi)^{2/3}\rho''(T_2^{1\sigma})n_1^{2/3}\alpha_1^{1/3}R_1 dR_1 / d\tau. \tag{2.43}$$

So, if the bubble growth function is $\frac{dR_1}{d\tau}=\frac{1}{2}\frac{B}{\sqrt{\tau}}$ with solution $R_1-R_{1c}=B\sqrt{\tau}$ where the critical bubble size is almost zero $R_{1c}\approx 0$, the bubble acceleration

$$R_1\frac{dR_1}{d\tau}=\frac{1}{2}B^2$$

is time independent. Replacing $R_1 dR_1/d\tau$ with one of the expressions from Appendix 2.1 we have for the nonaveraged mass source term for the thermally controlled bubble growth

$$\mu_{21}=C\rho''(T_2^{1\sigma})n_1^{2/3}\alpha_1^{1/3}B^2=C\rho''(T_2^{1\sigma})n_1^{2/3}\alpha_1^{1/3}\left(\frac{B}{Ja}\right)^2\left[\frac{\rho_2 c_{p2}}{\rho''(h''-h')}\right]^2[T_2-T'(p)]^2, \tag{2.44}$$

with constant C ranging from 2.2 to 14.89. It is very interesting to note that the evaporation source term depends on the void fraction, on the bubble number density, on the liquid temperature, and on the system pressure. All of these values change during a single time interval.

The heat taken away from the liquid and used for the evaporation per unit mixture volume and unit time is

$$\dot{q}_2^{\prime\prime\prime 1\sigma}=-\mu_{21}\left[h''(T_2^{1\sigma})-h'(T_2^{1\sigma})\right], \tag{2.45}$$

or for the thermally controlled bubble growth

$$\dot{q}_2^{\prime\prime\prime 1\sigma}=-const\ n_1^{2/3}\alpha_1^{1/3}\frac{\rho_2 c_{p2}\lambda_2}{\rho''(h''-h')}[T_2-T'(p)]^2. \tag{2.46}$$

We see that it is nearly proportional to the square root of the difference between the liquid temperature and the saturation temperature. This is the nonaveraged energy source term. A bit more complex is the use of the *Mikic* et al. solution for constructing the source term. The interface acceleration is then

$$R_1\frac{dR_1}{d\tau}=AR_{10}\left[\left(\tau_0^+ +1\right)^{1/2}-\left(\tau_0^+\right)^{1/2}\right],$$

where the dimensionless time τ_0^+ at which the bubble reached the local dimensionless size R_{10}^+ is the solution of the equation

$$R_{10}^+=\frac{2}{3}\left[\left(\tau_0^+ +1\right)^{3/2}-(\tau_0^+)^{3/2}-1\right].$$

Because the equation is transcendental a *Newton* method is applied for bringing the function

$$f = \left(\tau_0^+ + 1\right)^{3/2} - (\tau_0^+)^{3/2} - 1 - \frac{3}{2} R_{10}^+$$

by iteration to zero with

$$\tau_0^{+new} = \tau_0^{+old} - f^{old} \Big/ \left(df \big/ d\tau_0^+ \right).$$

where

$$\frac{df}{d\tau_0^+} = \frac{3}{2}\left[\left(\tau_0^+ + 1\right)^{1/2} - (\tau_0^+)^{1/2} \right].$$

An initial value of $\tau_0^+ = 0.001$ is a good one to reach convergence after two three cycles.

The use of the nonaveraged source terms is justified for time steps considerably smaller then the characteristic time interval needed to reach the equilibrium. Otherwise, the averaged source terms have to be used. Further, we discuss how to compute the averaged source terms.

2.4.2 The averaged mass source terms

For a constant number of bubbles per unit volume n_{10} during the time step, the *volume difference* between the end and the beginning of the time step, *multiplied* by the *steam density* and the *bubble number per unit mixture volume*, gives the *integral mass evaporating* during the considered time step per unit mixture volume. Dividing this mass by the time step we obtain the averaged mass source term

$$\mu_{21} = \rho'' n_{10} V_1 / \Delta\tau \, , \; \alpha_1 = 0 \, , \tag{2.47}$$

for the first integration step when the bubble growth just starts, and

$$\mu_{21} = n_{10} \frac{\rho_1 V_1 - \rho_{10} V_{10}}{\Delta\tau} = \frac{\rho_{10} n_{10} V_{10}}{\Delta\tau} \left(\frac{\rho_1 V_1}{\rho_{10} V_{10}} - 1 \right) = \frac{\rho_{10} \alpha_{10}}{\Delta\tau} \left[\frac{\rho_1}{\rho_{10}} \left(\frac{R_1}{R_{10}} \right)^3 - 1 \right], \; \alpha_1 > 0 \tag{2.48}$$

for the next time steps.

If thermally controlled bubble growth is assumed, the bubble growth function is $R_1 - R_{1c} = B\sqrt{\tau}$, where the critical bubble size is almost zero $R_{1c} \approx 0$, the effective time required to reach the bubbles size R_{10} is

$$\Delta\tau_0 = \left(\frac{R_{10}}{B}\right)^2 = \frac{1}{B^2}\left(\frac{3}{4\pi}\frac{\alpha_{10}}{n_{10}}\right)^{2/3}. \tag{2.49}$$

R_{10} defines at which state the bubble growth is if it started at the actual local conditions. Then, the bubble size R_1 is $R_1 = B\sqrt{\Delta\tau_0 + \Delta\tau}$,

$$\left(\frac{R_1}{R_{10}}\right)^3 = \left[1+\left(\frac{B}{R_{10}}\right)^2 \Delta\tau\right]^{3/2} = \left(1+\frac{\Delta\tau}{\Delta\tau_0}\right)^{3/2},$$

and finally

$$\mu_{21} = \frac{\rho_{10}\alpha_{10}}{\Delta\tau}\left[\frac{\rho_1}{\rho_{10}}\left(1+\frac{\Delta\tau}{\Delta\tau_0}\right)^{3/2} - 1\right]. \tag{2.50}$$

Surprisingly, the larger the time interval over which we average the larger the source term because the term $\Delta\tau^{3/2}/\Delta\tau$.

If the *Mikic* solution is to be used,

$$R_1^+ - R_{1c}^+ = \frac{2}{3}\left[\left(\tau^+ + 1\right)^{3/2} - (\tau^+)^{3/2} - 1\right],$$

where the critical bubble size is almost zero $R_{1c} \approx 0$, the source term algorithm is more complicated. The effective time required to reach the bubbles size R_{10} has to satisfy the equation

$$R_{10}^+ = \frac{2}{3}\left[\left(\tau_0^+ + 1\right)^{3/2} - (\tau_0^+)^{3/2} - 1\right].$$

Then, the dimensionless radius of the bubble at the end of time step is

$$R_1^+ = \frac{2}{3}\left[\left(\tau_0^+ + \Delta\tau^+ + 1\right)^{3/2} - \left(\tau_0^+ + \Delta\tau^+\right)^{3/2} - 1\right],$$

and correspondingly the radius ratio

$$\left(\frac{R_1}{R_{10}}\right)^3 = \left(\frac{R_1^+}{R_{10}^+}\right)^3 = \left\{\frac{2}{3}\frac{1}{R_{10}^+}\left[\left(\tau_0^+ + \Delta\tau^+ + 1\right)^{3/2} - \left(\tau_0^+ + \Delta\tau^+\right)^{3/2} - 1\right]\right\}^3. \tag{2.51}$$

The final form of the source term is then

$$\mu_{21} = \frac{\rho_{1o}\alpha_{1o}}{\Delta\tau}\left[\frac{\rho_1}{\rho_{1o}}\left\{\frac{2}{3}\frac{1}{R_{10}^+}\left[\left(\tau_0^+ + \Delta\tau^+ + 1\right)^{3/2} - \left(\tau_0^+ + \Delta\tau^+\right)^{3/2} - 1\right]\right\}^3 - 1\right]. \tag{2.52}$$

Using Eq. (2.40) we obtain for the very beginning of the bubble growth $\alpha_{1a} = 0$

$$\mu_{21} = \frac{\rho'' n_{1o}}{\Delta\tau}\left(\frac{4}{3}\pi R_1^3\right) = \frac{\rho'' n_{1o}}{\Delta\tau}\frac{4}{3}\pi\left\{\frac{B^2}{A}\frac{2}{3}\left[\left(\Delta\tau^+ + 1\right)^{3/2} - \left(\Delta\tau^+\right)^{3/2} - 1\right]\right\}^3. \tag{2.53}$$

Having in mind that for the spontaneous evaporation $\mu_{12} = 0$ and $\dot{q}_1^{\prime\prime 2\sigma} = 0$, we obtain the corresponding energy source term from the energy-jump condition on the bubble surface, i.e., Eq. (2.45).

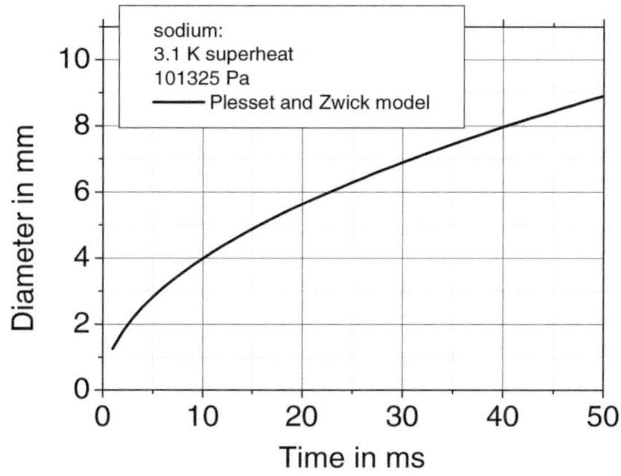

a)

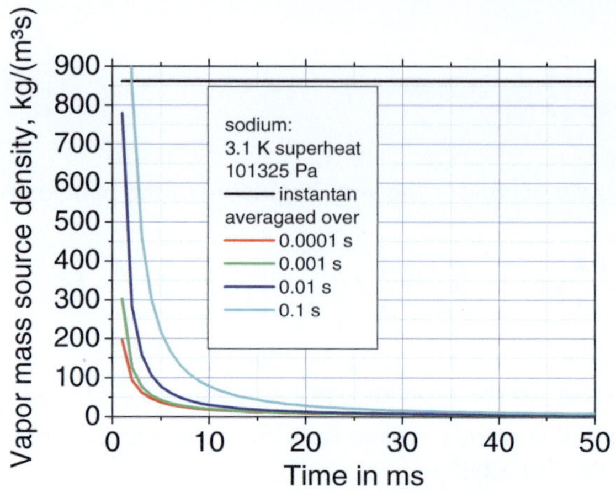

b)

Fig. 2.10 a) Bubble diameter as a function of time; b) The difference between the instant and time-averaged vapor mass source term as a function of the time since the beginning of the bubble growth

To understand the difference between the instant and time-averaged vapor mass source term let as consider Fig. 2.10. Sodium at atmospheric pressure and 3.1 K superheating contains voids with volumetric fraction of 0.1 distributed in 10^9 bubbles per cubic meter. The bubble size as a function of time in accordance with the *Plesset* and *Zwick* model is presented on Fig. 2.10a. The instantaneous vapor mass source term is a constant over the growing period of the bubble – see Fig. 2.10 b). But if we compute the generated steam per cubic meter over a specified time step and then divide to this step we in fact obtain the time-averaged source term. Obviously, at a later stage of the bubble growth it is smaller than in the preceding stages. We see also that the period of time over which we project the bubble growth in the future influences the time average.

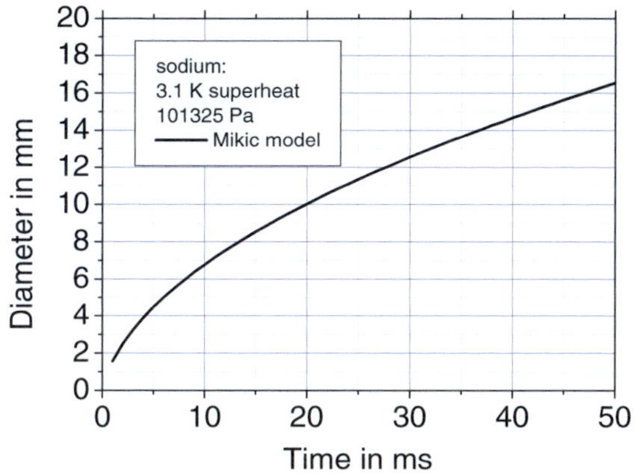

a)

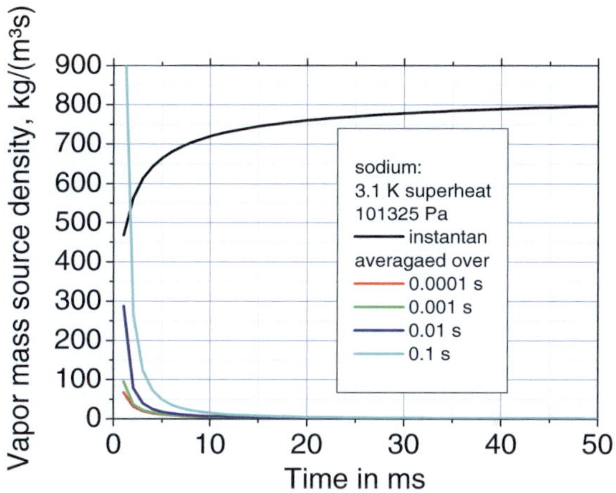

b)

Fig. 2.11 a) Bubble diameter as a function of time; b) The difference between the instant and time-averaged vapor mass source term as a function of the time since the beginning of the bubble growth

The bubble size as a function of time in accordance with the *Mikic* model is presented on Fig. 2.11a. Unlike in the *Plesset* and *Zwick* model the instantaneous vapor mass source term is a constant over the growing period of the bubble – see Fig. 2.11b. We see also for this model that the period of time over which we project the bubble growth in the future influences the time average.

2.5 Superheated steam

The above-derived formalism is strictly valid if the bubble surface temperature is equal to the gas temperature T_1,

$$T_2^{1\sigma} = T_1 . \tag{2.54}$$

In the following, we denote the evaporation rate for this case with μ_{21}'. In general, it is possible to have superheated or subcooled steam with respect to the surface temperature due to the history of the gas–liquid mixing process. In such cases there is a small amount of heat transferred from the bubble to the surface if $T_1 > T_2^{1\sigma}$ or from the surface to the bubble, if $T_1 < T_2^{1\sigma}$, due to a mechanism similar to natural convection inside the cavity. In this case, the above theory should be revised. An approximate correction of the theory can be introduced as follows:

$$\dot{q}_2^{\prime\prime\prime 1\sigma} = -\mu_{21}'\left[h''(T_2^{1\sigma}) - h'(T_2^{1\sigma})\right], \tag{2.55}$$

$$\mu_{12} = 0 , \tag{2.56}$$

$$\dot{q}_1^{\prime\prime\prime 2\sigma} = a_{21} h_{NC}\left(T_2^{1\sigma} - T_1\right) \tag{2.57}$$

$$\mu_{21} = \mu_{21}' - \dot{q}_1^{\prime\prime\prime 2\sigma} / \left[h''(T_2^{1\sigma}) - h'(T_2^{1\sigma})\right], \tag{2.58}$$

where the heat-transfer coefficient due to natural, e.g., convection in a cavity is

$$h_{NC} = const \frac{\lambda_1}{D_1}(Gr_1 \Pr_1)^m , \tag{2.59}$$

$$Gr_1 = g\left|\rho_1 - \rho_1''(T_2^{1\sigma})\right| \rho_1 D_1^3 / \eta_1^2 , \tag{2.60}$$

$$\Pr_1 = \eta_1 c_{p1} / \lambda_1 , \tag{2.61}$$

$$const \approx 0.59 \div 0.9 , \tag{2.62}$$

$$m \approx 1/4 . \tag{2.63}$$

The second term in Eq. (2.58) is usually neglected as very small compared to the first one. This term becomes important only for mass-transfer processes on the bubble surface in saturated or nearly saturated liquid.

2.6 Diffusion-controlled evaporation into mixture of gases inside the bubble

Consider a bubble consisting of a mixture of vapor and noncondensable gases. The partial pressure of the vapor is p_{M1}. If the interface pressure $p'(T_2)$ is larger than the partial vapor pressure there is an evaporation mass flux from the interface into the bubble. The emitted steam molecules from the surface enter the boundary layer of the bubble by diffusion. The nonaveraged instantaneous mass source term is

$$\mu_{21} = a_{21}(\rho w)_{21} = -a_{12}\beta \frac{C_{M1}\rho_1}{p_{M1}}\left[p_{M1} - p'(T_2)\right]$$

$$= -a_{21}\alpha_c \frac{M_{M1}}{M_1} \ln \frac{p - p'(T_2)}{\sum p_{n1}} / (c_{p1} Le^{2/3}) \tag{2.64}$$

see in *VDI-Wärmeatlas* (1984), where

$$\beta = \frac{p_{M1}}{C_{M1}\rho_1\left[p'(T_2) - p_{M1}\right]} \alpha_c \frac{M_{M1}}{M_1} \ln \frac{p - p'(T_2)}{\sum p_{n1}} / (c_{p1} Le_1^{2/3}), \tag{2.65}$$

$$p_{M1} < p'(T_2) < p, \tag{2.66}$$

$$Le_1 = \lambda_1 / (\rho_1 c_{p1} D_{M\to\sum n}) \tag{2.67}$$

$$M_1 = \frac{\sum p_{n1}}{p}\sum M_{n1} + \frac{p - \sum p_{n1}}{p} M_{M1} \tag{2.68}$$

$$(\textstyle\sum M_{n1} = 28.96\ kg \text{ for air, } M_{M1} = 18.96\ kg \text{ for water steam}) \tag{2.69}$$

p' is the partial pressure of the steam in the boundary layer, where the steam is supposed to be saturated at a temperature nearly equal to the liquid temperature

$$p' = p'(T_2). \tag{2.70}$$

During the evaporation the liquid is cooled by

$$\dot{q}_2^{\prime\prime\prime 1\sigma} = -\mu_{21}(h'' - h_{M2}). \tag{2.71}$$

Because this process is relatively slow in most of the practical cases averaging in the time steps of the source terms is not necessary.

2.7 Conclusions

1. After the initiation of bubbles with stable bubble size the evaporation of the superheated liquids is controlled by the bubble growth. The bubble growth is over a very short time controlled by the inertia and is well described by the *Besand* and *Rayleigh* solution. For the time after the inertia-controlled period the thermally controlled bubble growth controls the evaporation. From nine known models for this region the model by *Mikic* is recommended. The *Mikic* model interpolates properly between inertia-controlled and thermally controlled regions.

2. Thermally controlled bubble growth at heated walls is described by modifying the solutions for thermally controlled bubble growth in the bulk.

3. Evaporation processes are the driving forces for pressure excursions. The pressure change has an important feedback on the bubble growth by changing the driving temperature difference controlling the energy transfer between the bulk and the interface. For pressure change with semi-constant pressure change velocity over a given time step the bubble growth is best approximated by the solution of *Jones* and *Zuber*.

Appendix 2.1 Radius of a single bubble in a superheated liquid as a function of time

Thermally controlled bubble growth: $\frac{dR_1}{d\tau} = \frac{1}{2}\frac{B}{\sqrt{\tau}}$, $R_1 = B\sqrt{\tau}$, $R_1 \frac{dR_1}{d\tau} = B^2/2$

where $Ja = \frac{\rho_2 c_{p2}(T_2 - T')}{\rho''(h'' - h')}$, $a_2 = \frac{\lambda_2}{\rho_2 c_{p2}}$ and B is given below.

Bosnjakovic (1930), *Fritz* and *Ende* (1936) for plane geometry:

$$B = Ja\sqrt{\frac{4a_2}{\pi}}.$$

Plesset and *Zwick* (1954) for $Ja < 100$:

$$B = Ja\sqrt{\frac{12a_2}{\pi}}.$$

Foster and *Zuber* (1954), *Dergarabedijan* see in *Tolubinskii* (1980) for $Ja < 100$:

$$B = Ja\sqrt{\pi a_2}.$$

Labuntsov et al. (1964a) approximate the analytical solution obtained by *Scriven* (1959) for bubble growth in uniformly heated liquid with an error of 2%

$$B = Ja\sqrt{\frac{12a_2}{\pi}}\left[1+\frac{1}{2}\left(\frac{\pi}{6Ja}\right)^{2/3}+\frac{\pi}{6Ja}\right]^{1/2}.$$

The *Labunzov* approximation gives close results to the *Foster* and *Zuber* (1954) solution.

Olek et al. [19] (1990):

$$B = \sqrt{\frac{a_2}{\pi}}Ja\left[1+\left(1+\frac{2\pi}{Ja}\right)^{1/2}\right].$$

Cole and *Schulman* (1966):

$$B = Ja\sqrt{\frac{\pi a_2}{4}}.$$

Cooper - see in *Wang* (1989) for H_2O:

$$B = 1.57\,Ja\sqrt{\frac{a_2}{\Pr_2}}.$$

Mikic and *Rohsenow* (1969)

$$B = 0.83\,Ja\sqrt{\pi a_2}.$$

Prisnjakov (1970) for $Ja < 500$:

$$B = Ja\sqrt{\frac{16a_2}{9\pi}}.$$

Avdeev and *Zudin* (2002) for $0 \le Ja \le \infty$, $0 \le \rho''/\rho' \le 1$:

$$\frac{B}{\sqrt{a_2\tau}} = \sqrt{\frac{3}{\pi}}Ja\psi + \sqrt{\frac{3}{\pi}(Ja\psi)^2 + 2Ja}$$

$$\psi = 1+\sqrt{\frac{\pi}{2}}\left(\frac{1}{\sqrt{1-N}}-1\right),\quad N = \frac{\rho''}{\rho'}Ja.$$

$\pm 2.4\%$ deviation from the *Scriven* solution, 5.2% maximum deviation at $\rho''/\rho' = 1$.

Bubble growth at heated wall: no complete analytical description, empirical correlations recommended for practical use.

Jagov (1971)

$$B=\left(\gamma Ja+\sqrt{\gamma^2 Ja^2+2\beta\, a_2 Ja}\right)a_2^{1/2},$$

$\gamma=0.3$, $\beta=6$, valid for $10^{-1}<Ja<1000$. For $Ja<<1$, high pressure, the *Jagov*'s expressions is equivalent with the *Labuntsov* et al. result from (1964), validated for water boiling at silver surface at p = 1 to 100 bar.

$$B=\sqrt{2\beta\, a_2 Ja}\ ,$$

where

$$\beta=\frac{2(\cos\frac{\theta}{2})\ln(\delta_v/\delta_m)}{(1+\cos\theta)^2(2-\cos\theta)}\approx\frac{25.3\cos(\theta/2)}{(1+\cos\theta)^2(2-\cos\theta)},$$

$\theta\approx\pi 35/180$, $\delta_m\approx 10^{-9}$ to $10^{-10} m$, is the averaged distance between the liquid molecules, and $\delta_v\approx 10^{-4} m$ is the thickness of viscous boundary layer of the boiling surface. $\beta\approx 5.6$ to 6.7, $\beta=6$ recommended. For $Ja>>1$, high pressure, the *Jagov*'s expression tends to $B=2\gamma Ja a_2^{1/2}$ characteristic for bubble growth in bulk liquid.

Piening (1971) for $8x10^{-2}<Ja<500$:

$$B=\sqrt{a_2 Ja}\ .$$

Kutateladze (1982) p.120 for bubble growth at heated wall:

$$B\approx const\sqrt{a_2 Ja}\ .$$

Hutcherson et al. (1983):

$$B=\sqrt{12 a_2 Ja/\pi}\ .$$

Wang (1989) for $p\neq const$:

$$R_1=(\rho_{10}/\rho_1)^{1/3}R_{10}\left[1+\frac{1}{R_{10}\rho_{10}}\int_0^{\Delta\tau}\left[(\rho w)_{21}/\rho_1^{2/3}\right]d\tau\right],$$

where

$$(\rho w)_{21,wall}=\beta(\rho w)_{21,\inf\ init\ liquid\ see},$$

$$\beta = \frac{2\left(1+\cos\theta\right)+\sin^2\theta}{2+\cos\theta(2+\sin^2\theta)}, \qquad \theta = 1.05 \ to \ 1.31 \ rad .$$

Cole and *Shulman* (1966):

$$B = \frac{5}{2} Ja^{3/4}\sqrt{a_2} .$$

Galin and *Kirilov* (1987)

$$B = \sqrt{2\beta a_2 Ja\varphi\left(Ja\right)\psi\left(\sqrt{\frac{\lambda_w \rho_w c_{pw}}{\lambda_2 \rho_2 c_{p2}}}\right)}, \ \varphi\left(Ja\right) = 1 + \frac{\gamma^2}{\beta} Ja + \frac{\gamma}{\beta}\sqrt{2\beta Ja + \gamma Ja^2} ,$$

$\beta = 6$, $\gamma = 0.3$, ψ unknown.

Avdeev and *Zudin* (2002) for $0 \le Ja \le \infty$, $0 \le \rho''/\rho' \le 1$:

$$\frac{B}{\sqrt{a_2 \tau}} = 0.3Ja\psi + \sqrt{\left(0.3Ja\psi\right)^2 + 12Ja}$$

$$\psi = 1 + \sqrt{\frac{\pi}{2}}\left(\frac{1}{\sqrt{1-N}} - 1\right), \ N = \frac{\rho''}{\rho'} Ja .$$

Jones and *Zuber* (1978), *Kroshilin* et al. (1986) for $p \neq$ const in uniformly heated liquid.

The heat flux at the bubble surface $\dot{q}_2^{\prime\prime\prime 1\sigma}$ is computed using the *Carlsaw* and *Jaeger* [4] (1959) solution of the *Fourier* equation for heat slab with variable temperature difference at the one boundary

$$T_2 = const , \ \Delta T(\tau) = T_2 - T_2^{1\sigma}(\tau), \ \ \Delta T(0) = T_2 - T'\left(p_0\right),$$

$$\dot{q}_2^{\prime\prime\prime 1\sigma} = -k_s \frac{\lambda_2}{\sqrt{\pi a_2}}\left[\frac{\Delta T(0)}{\sqrt{\tau}} + \int_0^\tau \frac{dT_2^{1\sigma}(\tau)/d\tau}{\sqrt{\tau - \eta}} d\eta\right]$$

k_s is the sphericity correction = $\pi/2$ for *Foster–Zuber*, $\sqrt{3}$ for *Plesset–Zwick*. *Winters* and *Merte* (1979) replaced the integral in the above equation with its finite difference sum representation

$$\dot{q}_2^{\prime\prime\prime 1\sigma} = -k_s \frac{\lambda_2}{\sqrt{\pi a_2}}\left[\frac{\Delta T(0)}{\sqrt{\tau}} + \sum_{m=1}^{n} \frac{T_2^{1\sigma}\left(\tau_m\right) - T_2^{1\sigma}\left(\tau_{m-1}\right)}{\left(\tau - \tau_m\right)^{1/2}}\right],$$

in their blow-down analysis. For linear pressure decrease during the time step $\Delta\tau$,

$$\frac{dT_2^{1\sigma}(\tau)}{d\tau} = \frac{dT'(p_1)}{d\tau} = \frac{dT'(p_1)}{dp_1}\frac{dp_1}{d\tau} = \frac{dT'(p_1)}{dp_1}\frac{p_1 - p_{10}}{\Delta\tau}$$

$$\approx \frac{dT'(p_1)}{dp_1}\frac{p - p_0}{\Delta\tau} = b = const$$

the heat flux is

$$\dot{q}_2^{\prime\prime 1\sigma} = -k_s \frac{\lambda_2}{\sqrt{\pi a_2}}\left(\frac{\Delta T(0)}{\sqrt{\tau}} + b\int_0^\tau \frac{d\eta}{\sqrt{\tau - \eta}}\right) = -\frac{k_s}{\sqrt{\pi}}\sqrt{\lambda_2 \rho_2 c_{p2}}\left(\frac{\Delta T(0)}{\sqrt{\tau}} + 2b\sqrt{\tau}\right).$$

Neglecting the expression containing the density change and substituting into the mass-transport equations with

$$(\rho w)_{21} = -\dot{q}_2^{\prime\prime 1\sigma} / (h'' - h'),$$

we obtain

$$\frac{R_1}{R_{10}} = 1 + \frac{k_s}{\sqrt{\pi}}\frac{1}{\rho_1 (h'' - h') R_{10}}\sqrt{\lambda_2 \rho_2 c_{p2}} \int_0^{\Delta\tau}\left[\frac{\Delta T(0)}{\sqrt{\tau}} + 2b\tau^{1/2}\right]d\tau .$$

After the integration the result is

$$\frac{R_1}{R_{10}} = 1 + \frac{2k_s}{\sqrt{\pi}}\frac{1}{\rho_1 (h'' - h') R_{10}}\sqrt{\lambda_2 \rho_2 c_{p2}}\left(\Delta T(0)\Delta\tau^{1/2} + \frac{2}{3} b \Delta\tau^{3/2}\right).$$

The equation rewritten in dimensionless form is

$$\frac{R_1}{R_{10}} = 1 + \frac{2k_s}{\sqrt{\pi}}\left(Ja_T Fo^{1/2} + \frac{2}{3} Ja_p Fo^{3/2}\right),$$

where

$$Ja_T = \frac{\rho_2 c_{p2}}{\rho_{10}(h'' - h')}\left[T_2 - T'(p_0)\right],$$

$$Ja_p = b\frac{\rho_2 c_{p2}}{\rho_{10}(h'' - h')}\frac{R_{10}^2}{a_2},$$

$$Fo = a_2 \tau / R_{10}^2 .$$

This result was obtained by *Jones* and *Zuber*. The time-averaged evaporation mass per unit time and unit mixture volume is therefore

$$\mu_{21} = \frac{\rho_{1o}\alpha_{1o}}{\Delta\tau}\left\{\left\{1+\frac{2k_s}{\sqrt{\pi}}\left[Ja_T Fo^{1/2}+\frac{2}{3}Ja_p Fo^{3/2}\right]\right\}^3 -1\right\} \quad \text{for} \quad \alpha_1 > 0 .$$

Inertia controlled bubble growth, $\Delta\tau \le 10^{-8}$ s, *Besand* (1859), *Rayleigh* (1917).

$$\frac{dR_1}{d\tau} = \left\{\frac{2}{3}\left[p'(T_2)-p\right]/\rho_2\right\}^{1/2} = \left\{\frac{2}{3}\frac{1}{\rho_2}\left(\frac{dp}{dT}\right)_{sat}\left[T_2 - T'(p)\right]\right\}^{1/2}$$

$$R_1 = \left\{\frac{2}{3}\frac{1}{\rho_2}\left(\frac{dp}{dT}\right)_{sat}\left[T_2 - T'(p)\right]\right\}^{1/2}\tau .$$

Inertia and thermal controlled bubble growth, *Mikic* et al. (1970):

$$R_1^+ - R_{1o}^+ = \frac{2}{3}\left[\left(\tau^+ + 1\right)^{3/2} - \tau^{+3/2} - 1\right],$$

where

$$R_1^+ = R_1\,(B^2/A), \quad \tau^+ = \frac{\tau}{B^2/A^2}$$

$$Ja = \frac{\rho_2 c_{p2}\left[T_2 - T'(p)\right]}{\rho_1\left[h''(T_2^{1\sigma}) - h'(T_2^{1\sigma})\right]}, \quad B = \left(\frac{12}{\pi}a_2\right)^{1/2} Ja$$

$$A^2 = c\frac{1}{\rho_2}\left(\frac{dp}{dT}\right)_{sat}\left[T_2 - T'(p)\right],$$

$$\left(\frac{dp}{dT}\right)_{sat} = \frac{1}{T'(p)}\frac{h''(p)-h'(p)}{v''(p)-v'(p)},$$

$c = 2/3$ for bubble growth in infinite continuum,

$c = \pi/7$ for bubble growth on a wall surface.

Instantaneous mass transfer per unit time and unit mixture volume:

$$\mu_{21} = 3^{1/3}(4\pi)^{2/3}\rho''(T_2^{1\sigma})n_1^{2/3}\alpha_1^{1/3}R_1 dR_1/d\tau$$

with $R_1\frac{dR_1}{d\tau} = AR_{10}\left[\left(\tau_0^+ + 1\right)^{1/2} - \left(\tau_0^+\right)^{1/2}\right]$, where the dimensionless time τ_0^+ at which the bubble reached the local dimensionless size R_{10}^+ is the solution of the equation

$$R_{10}^+ = \frac{2}{3}\left[\left(\tau_0^+ + 1\right)^{3/2} - (\tau_0^+)^{3/2} - 1\right].$$

Nomenclature

Latin

a	liquid thermal diffusivity, m^2/s
a_{21}	interfacial area density between the liquid 2 and the vapor 1, 1/m
C	mass concentration, kg/kg
c_p	specific heat at constant pressure, J/(kg K)
$D_{M\to\sum n}$	diffusion constant for the species M into the mixture $\sum_n$, 1/m
D_{cav}	equivalent cavity size, m
D_h	hydraulic diameter, m
$D_{1,n}$	bubble diameter belonging to nth bubble generation, m
D_{1c}	critical bubble diameter, m
D_{1d}	bubble departure diameter, m
d	differential, dimensionless
F_1/Vol	cross-section at the wall surface covered by attached bubbles per unit flow volume, 1/m
g	gravitational acceleration, m/s²
Ja	$= \dfrac{\rho_2 c_{p2}\left[T_2 - T'(p)\right]}{\rho_{10}\left[h''(T_2^{1\sigma}) - h'(T_2^{1\sigma})\right]}$, *Jakob* number, dimensionless
G^*	critical mass flow rate, kg/(m²s)
Gb_2	$= \dfrac{\Delta E_{1kr}}{kT_2}$, *Gibbs* number, dimensionless
Gr_1	$= g\left\|\rho_1 - \rho_1''(T_2^{1\sigma})\right\|\rho_1 D_1^3/\eta_1^2$, bubble *Grashoff* number, dimensionless
h	specific enthalpy, J/kg
h_{NC}	heat-transfer coefficient by natural circulation, W/(m²K)
k	$= 13.805\times 10^{-24}$, *Boltzmann* constant, J/K
Le_1	$= \lambda_1/(\rho_1 c_{p1} D_{M\to\sum n})$, gas *Lewis* number, dimensionless
m_2	$= m_\mu/N_A$, mass of single molecule, kg
m_μ	*kg-mol* mass (18 *kg* for water), kg/mole
N_2	$= \rho_2/m_2 = \rho_2 N_A/m_\mu$ ($\approx 3.3\times 10^{28}$ for water), number of the molecules per unit volume, 1/m³

N_A = 6.02×10^{26}, *Avogadro* number, number of molecules in one kilogram-mole mass, 1/kg-mole
$\dot{n}_{1cin}$ number of the created nuclei per unit time in unit volume of the liquid, $1/(m^3\ s)$
n''_w cavity per unit surface, $1/m^2$
n''_{1w} bubble generating (active) cavity per unit surface, $1/m^2$
n_1 bubble number density, $1/m^3$
$n_{1,n}$ bubble number density belonging to *n*th bubble generation, $1/m^3$
$n_{1,\max}$ maximum number of nucleation sites per unit volume in a pipe, $1/m^3$
Pr_1 $= \eta_1 c_{p1} / \lambda_1$, gas *Prandtl* number, dimensionless
p pressure, Pa
$\dot{q}_2'''^{1\sigma}$ thermal energy flow density coming from the vapor interface and introduced into the bulk liquid, W/m^3
$\dot{q}_1''^{2\sigma}$ thermal energy flux coming from the liquid interface and introduced into the bulk vapor, W/m^2
$\dot{q}_2''^{1\sigma}$ thermal energy flux coming from the vapor interface and introduced into the bulk liquid, W/m^2
R gas constant, J/(kg K)
R_1 bubble radius, m
R_{10} initial bubble radius, m
R_{1c} critical bubble radius, m
$R_{1\infty}$ final radius of the bubble, m
r radius, m
s specific entropy, J/(kg K)
T temperature, K
u radial velocity, m/s
$\mathbf{V}$ velocity vector, m/s
V_{1c} volume of the bubble with critical size, m^3
v specific volume, m^3/kg
Vol volume, m^3

Greek

α_1 bubble volumetric fraction, m^3/m^3
β diffusion coefficient
γ_v volume occupied by the flow divided by the volume of the system, dimensionless

γ cross-section flowed by the flow divided by the total cross-section of the system, dimensionless
Δ finite difference
ΔE_1 work necessary to create a single bubble with radius R_1, J
ΔE_{1c} work necessary to create a single bubble with radius R_{1c}, J
ΔE_{1c}^* $= \varphi \Delta E_{1c}$, energy for the creation of a single bubble at a wall, J
Δp_{Fi}^* $= p'(T_2) - p_{Fi}$, flashing inception pressure difference, Pa
$\Delta \tau$ time interval, s
η dynamic viscosity of liquid, kg/(m s)
θ static contact angle, rad
λ_{RT} *Rayleigh-Tailor* wavelength, m
λ thermal conductivity, W/(m K)
μ_{21} evaporating mass per unit time and unit flow volume, kg/(m³ s)
μ_{12} condensing mass per unit time and unit flow volume, kg/(m³ s)
ρ density, kg/m³
$(\rho w)_{21}$ evaporation mass flow rate, kg/(m² s)
σ surface tension, N/m
M_{M1} kg-mole mass of the condensing gases, kg/mol
$\sum M_{n1}$ kg-mole mass of the non condensing gases, kg/mol
M_1 kg-mole mass of the gas mixture, kg/mol
τ time, *s*
Φ cavity angle if the cavity is idealized as a cone, rad
φ work-reduction factor, dimensionless

Superscripts

$'$ saturated liquid
$''$ saturated vapor
spin spinoidal line
1σ interface with vapor
2σ interface with liquid

Subscripts

1 vapor, gas
2 liquid
c critical state
s at constant entropy
sat saturated
Fi flashing inception

$\sum_{n1o}$ sum of all noncondensing gases
M evaporating or condensing chemical component
n inert (non condensing) chemical component

References

Avdeev, A.A., Zudin, Y.B.: Thermal energy scheme of vapor bubble growth (Universal approach solution). High Temperature 40(2), 264–271 (2002)

Besand, W.H.: Hydrostatics and hydrodynamics, p. 170. Deighton Bell, Cambridge (1859)

Beylich, A.E.: Dynamics and thermodynamics of spherical vapor bubbles. Int. Chem. Eng. 31(1), 1–28 (1991)

Bosnjakovic, F.: Techn. Mech. Thermodyn. 1, 358 (1930)

Carlsaw, H.S., Jaeger, J.C.: Conduction of heat in solids, Oxford (1959)

Cole, R., Shulman, H.L.: Bubble growth rates of high Jacob number. Int. J. Heat Mass Transfer 9, 1377–1390 (1966)

Golovin, V.S., Kalchugin, B.A., Labuntsov, D.A.: Injenerno fizicheskiy jurnal 6(2) (1963)

Foster, H.K., Zuber, N.: J. Appl. Phys. 25, 474–478 (1954)

Fourier, J.: Theory Analytique de la Chaleur (1822)

Fritz, W., Ende, W.: Ueber den Verdampfungsvorgang nach kinematographischen Aufnahmen an Dampfblasen. Phys. Z. 37, 391–401 (1936)

Galin, H.M., Kirilov, P.L.: Teplomassoobmen, Moskva, Energoatomisdat (1987)

Hutcherson, M.N., Henry, R.E., Wollersheim, D.E.: Two-phase vessel blow down of an initially saturated liquid - Part 2: Analytical. Trans. ASME 105, 694–699 (1983)

Jagov, V.V.: Issledovanie kipenija jidkostej v oblasti niskih davlenij. PhD Thesis, Moskva (1971)

Jakob, M.: Z. d. Ver. Dtsch. Ing. 76, 1161 (1932)

Jones Jr., O.C., Zuber, N.: Bubble growth in variable pressure fields. J. Heat Transfer 100, 453–458 (1978)

Kroshilin, A.E., Kroshilin, V.E., Nigmatulin, B.I.: Growth of a vapor bubble in a superheated liquid volume for various laws governing the pressure variation in the liquid. Teplofizika Vysokih Temperatur 24(3), 533–538 (1986)

Kutateladze, S.S.: Analiz podobija v teplofisike (Similarity analysis in thermal physics), Nauka (1982) (in Russian)

Labuntsov, D.A., Kol'chugin, V.A., Golovin, V.S., et al.: Teplofiz. Vys. Temp. 3, 446–453 (1964a)

Labuntsov, D.A., Kol'chugin, V.A., Golovin, V.S., Saharova, E.A., Vladimirova, L.H.: Teplofiz. Vys. Temp. 2(3) (1964b)

Lee, H.S., Merte Jr., H.: Spherical vapor bubble growth in uniformly superheated liquids. Int. J. Heat Transfer 39(12), 2427–2447 (1996)

Lien, Y., Griffith, P.: Bubble growth in reduced pressure. Ph. D. Thesis, Massachusetts Institute of Technology, Cambridge (1969)

Miheev, M.A., Miheeva, I.M.: Osnovy teploperedachi, Energiya, Moskva (1973)

Mikic, B.B., Rohsenow, W.M.: Bubble growth rates in non uniform temperature fields. Progr. Heat Mass Transfer 2, 283–293 (1969)

Mikic, B.B., Rohsenhow, W.M., Griffith, P.: On bubble growth rates. Int. J. Heat Mass Transfer 13, 657–666 (1970)

Nigmatulin, R.I.: Basics of the mechanics of the heterogeneous fluids, Moskva, Nauka (1978) (in Russian)
Olek, S., Zvirin, Z., Elias, E.: Bubble growth prediction by the hyperbolic and parabolic heat conduction equations. Waerme- Stoffuebertragung 25, 17–26 (1990)
Piening, J.: Der Wärmeübergang an eine an der Heizwand wachsende Dampfblase beim Sieden. Dissertation, Berlin TU (1971)
Plesset, M.S., Zwick, S.: The growth of bubbles in superheated liquids. J. Appl. Phys. 25(4), 493–500 (1954)
Prisnjakov, V.F.: Bubble growth in liquids. J. Eng. Phys. 18(1), 584–588 (1970)
Rayleigh, L.: On the pressure developed in a liquid during the collapse of spherical cavity. Phil. Mag. 34, 94 (1917)
Scriven, L.E.: On the dynamics of phase growth. Chem. Eng. Sci. 10(1), 113 (1959)
Tolubinskii, I.S.: Boiling heat transfer. Kiev, Naukova Dumka (1980) (in Russian)
VDI-Waermeatlas, 4. Auflage, VDI-Verlag (1984)
Wang, Z.: Transient level swell and liquid carryover phenomena of a vapor-liquid pool. PhD Dissertation, Northwestern University, Evanston, Illinois, USA (December 1989)
Winters Jr., W.S., Merte Jr., H.: Experiments and nonequilibrium analysis of pipe blow down. Nucl. Sci. Eng. 69, 411–429 (1979)

3. Condensation of a pure steam bubble in a subcooled liquid

This is a review chapter about the different methods for description of bubble collapse due to vapor condensation in subcooled liquids. First, we consider the case of stagnant bubbles. The solutions for moving bubbles are then presented. The available experimental data are then discussed. Then, the link between the solutions for the bubble growth and the mass source terms for the averaged conservation equations for two-phase flow is presented. The way to derive nonaveraged mass source terms and time-averaged mass source terms is given. The very interesting case of disappearance of the bubbles from the size distribution spectrum due to condensation is discussed in some detail. The influence of the liquid turbulence on the bubble condensation in pipes is discussed too. Then, a brief description of the vapor condensation from gas mixtures inside bubbles surrounded by subcooled liquid is given for the two limiting cases a) thermally controlled collapse and b) diffusion-controlled collapse.

3.1 Introduction

Injecting steam in subcooled water causes condensation. If a pressure wave crosses a water–steam system in thermodynamic equilibrium condensation also happens. One of the basic condensation mechanisms in multiphase systems is the condensation of a single bubble in subcooled water. Several authors study condensation of a saturated bubble in a subcooled liquid. As for the evaporation, the theoretical results of the approximate integration of the process governing a system of PDEs describe the radius as a function of time. Unlike the evaporation, the condensation is associated with a given *initial bubble size*. There are two classes of solutions: a) for a stagnant bubble and b) for a bubble in motion with respect to the surrounding liquid. The motion increases the heat transfer between the bubble interface and the bulk liquid and therefore increases the condensation.

3.2 Stagnant bubble

Table 3.1 contains some of the results known from the literature, obtained with the assumption that there is no relative motion of the bubble with respect to the surrounding liquid.

Table 3.1 Size of a bubble condensing in a subcooled liquid as a function of time by $\Delta V = 0$

$$\tau_H = \frac{4}{\pi} Ja^2 Fo, \quad Ja = \frac{\rho_2}{\rho''} \frac{c_{p2}\left[T'(p) - T_2\right]}{h'' - h'}, \quad Fo = \frac{a_2 \tau}{R_{1o}^2}$$

Forschuetz and *Chao* (1965) for $\tau \le \dfrac{\pi R_{1o}^2}{4 Ja^2 a_2}$

$$\frac{R_1}{R_{1o}} = 1 - \sqrt{\tau_H} \, .$$

Plesset and *Zwick* (1954):

$$\tau_H = \frac{2}{3}\left(\frac{R_{1o}}{R_1}\right) + \frac{1}{3}\left(\frac{R_1}{R_{1o}}\right)^2 - 1, \; \tau_{99\%} = 65 \frac{\pi R_{1o}^2}{4 Ja^2 a_2} \, .$$

Prisnjakov (1970) for $\tau \le \dfrac{1}{4\varepsilon^2} \dfrac{\pi R_{1o}^2}{4 Ja^2 a_2}$:

$$\frac{R_1}{R_{1o}} = 1 - 2\varepsilon \sqrt{\tau_H} \, , \qquad \varepsilon = 1 / \left[1 - \frac{\rho''}{\rho_2} + 2 \frac{h_2}{h'' - h'} \right] .$$

Zuber (1961) for $\tau \le 4\tau_m$:

$$\frac{R_1}{R_{1m}} = \left(\frac{\tau}{\tau_m}\right)^{1/2} \left[2 - \left(\frac{\tau}{\tau_m}\right)^{1/2} \right] .$$

τ_m time when the bubble reaches its maximum radius R_{1m}. For bubble growing and collapsing on a heated wall in subcooled water

$$\tau_m = \frac{1}{\pi a_2} \left(\lambda_2 \frac{T_w - T'}{\dot{q}_w''} \right)^2 , \; R_m = \frac{b}{\pi} Ja \sqrt{\pi a_2 \tau_m} \, , \; b \approx 1 \div \sqrt{3} \, ,$$

$$Ja = \frac{(T_w - T') c_{p2} \rho_2}{(h'' - h') \rho''} \, .$$

Bucher and *Nordman* (1978) show that the best fit to the experimental data is obtained with *Zuber's* equation from Table 3.1.

3.3 Moving bubble

The influence of the relative velocity on the temperature profile around the bubble and therefore on the condensation process can be taken into account by using the results of several authors describing the heat transfer around a solid sphere, as summarized in Table 3.2.

Table 3.2 Averaged heat-transfer coefficient on the surface of solid sphere moving in a liquid

$$Pe_2 = \frac{D_1 \Delta V_{12}}{a_2},\quad a_2 = \frac{\lambda_2}{\rho_2 c_{p2}},\quad Pr_2 = \frac{\eta_2}{\rho_2 a_2},\quad Re_1 = \frac{D_1 \rho_2 \Delta V_{12}}{\eta_2},\quad Nu_1 = \frac{D_1 h_k}{\lambda_2},$$

$$Fo = \frac{a_2 \tau}{R_{1o}^2},\quad Re_{1o} = \frac{D_{1o} \rho_2 \Delta V_{12}}{\eta_2},\quad Fo \le 1/(Nu_1 Ja)$$

Potential flow $\left(\frac{\partial R_1}{\partial \tau} << \Delta V\right)$, $Re_1 << 1$:

Soo (1969) for $Pe_2 << 1$:

$$Nu_1 = 2 + \frac{9}{16} Pe_2 + \frac{9}{64} Pe_2^2 + ...$$

Nigmatulin (1978) for $Pe_2 >> 1$:

$$Nu_1 = \frac{0.65}{\left(1 + \frac{\eta_1}{\eta_2}\right)^{1/2}} Pe_2^{1/2},\quad \eta_1 / \eta_2 \le 1.$$

Hunt (1970) for a single-component system:

$$Nu_1 = \frac{2}{\pi^{1/2}} (c Pe_2)^{1/2},$$

where $c = 1$.

For a two-component system *Isenberg* and *Sideman* (1970) take into account viscous effects in the *Hunt* equation using $c = 0.25\ Pr_2^{-1/3}$ resulting in

$$Nu_1 = \frac{1}{\pi^{1/2}} \mathrm{Re}_{12}^{1/2}\, Pr_2^{1/3}.$$

Kendouch (1976) for bubbles in a swarm:

$$Nu_1 = \frac{2}{\pi^{1/2}} Pe_2^{1/2} (1-\alpha_1)^{-1/2}.$$

Wilson (1965) for $Re_1 < 1$:

$$Nu_1 = 2 + 0.37 Re_1^{3/5} Pr_2^{1/3}.$$

Nigmatulin (1978) for $Re_1 < 1$ or $Pe_2 < 10^3$ (e.g., for $D_{1H2O} < 0.1\ mm$):

$$Nu_1 = 2 + \frac{0.65 Pe_2^{1.7}}{1+Pe_2^{1.3}}, \quad \eta_1/\eta_2 \cong 0.$$

Brauer et al. (1976) for bubbles without internal circulation:

$$Nu_1 = 2 + \frac{0.65 Pe_2^{1.7}}{\left[1+(0.84 Pe_2^{1.6})^3\right]^{1/3} (1+Pe_2^{1.2})}.$$

Soo (1969) for $1 < Re_1 < 7\times 10^4$, $0.6 < Pr_2 < 400$:

$$Nu_1 = 2 + (0.55 \text{ to } 0.7) Re_1^{1/2} Pr_2^{1/3}.$$

Akiyama (1973): $Nu_1 = 0.37 Re_1^{0.6} Pr_2^{1/3}$.

Zeitoun et al. (1995): $Nu_1 = 2.04 Re_1^{0.61} Ja^{-0.308} \alpha_1^{0.328}$.

There are a limited number of correlations, approximating directly experimental data for a *condensing moving bubble in a subcooled liquid.* Some of them are summarized in Table 3.3. Next, we first show the analytical way to derive such an expression that can be successfully modified to correlate the data. Then, to illustrate the idea how the approximate form of the correlations in Table 3.3 is obtained we apply the theory using the heat-transfer mechanism described by the correlation by *Hunt* and *Isenberg*.

The energy-conservation equation with neglected convection and diffusion from the surrounding computational cells gives

$$\rho'' \frac{dR_1}{d\tau} = -h_c \frac{T'-T_2}{h''-h_2}, \tag{3.1}$$

or

$$\frac{dR_1}{d\tau} = -\frac{h_c}{\rho''}\frac{T'-T_2}{h''-h_2} = -\frac{Nu_1 \lambda_2 (T'-T_2)}{R_1 2\rho''(h''-h_2)} = -c/R_1 = -\frac{a_2 Nu_1 Ja}{2R_1}, \tag{3.2}$$

or

$$\frac{d}{dFo}\left(\frac{R_1}{R_{1o}}\right)^2 = -Nu_1 Ja\,. \qquad (3.3)$$

Integrating the above equation for $Nu_1 = const$ we obtain

$$\frac{R_1}{R_{1o}} = \left(1 - \frac{2c\Delta\tau}{R_{1o}^2}\right)^{1/2} = \sqrt{1 - \Delta\tau/\Delta\tau_o} = (1 - Nu_1 JaFo)^{1/2} = (1 - Fo/Fo_o)^{1/2}.$$

$$(3.4)$$

The last form of the above notation contains the important information for the dimensionless time, Fo_o, within which the bubble collapses completely if the mixture still does not reached the saturation. The bubble collapses entirely

$$R_1 = 0, \qquad (3.5)$$

within

$$\Delta\tau_o = R_{1o}^2/(2c), \quad \left[Fo_o = 1/(Nu_1 Ja)\right]. \qquad (3.6)$$

After this time interval Eq. (3.4) is no longer valid. For the third power of the radii ratio we have

$$\left(\frac{R_1}{R_{1o}}\right)^3 = \left(1 - \Delta\tau/\Delta\tau_o\right)^{3/2} = (1 - Fo/Fo_o)^{3/2} \qquad (3.7)$$

If we drop the assumption $Nu_1 = const$ and use the *Hunt* and *Isenberg*'s equation from Table 3.2 we obtain

$$\left(\frac{R_1}{R_{1o}}\right)^3 = (1 - Fo/Fo_o)^2, \qquad (3.8)$$

where

$$Fo_o = 1/\left(\frac{3}{2}\frac{1}{2\sqrt{\pi}}\mathrm{Re}_{1o}^{1/2}\,\mathrm{Pr}_2^{1/3}\,Ja\right), \qquad (3.9)$$

see the first equation in Table 3.3. It differs from the *Moalem* and *Sideman* (1973) correlation only in the constant.

As seen from Table 3.3 all the correlations are based on the above form modifying the exponents except the correlation by *Dushkin* and *Kolomenzev* (1989).

Table 3.3 Bubble size of saturated steam condensing in a subcooled liquid $\Delta V \neq 0$. Approximation of experimental data. Fo_o is the dimensionless lifetime of condensing bubble

Theory

$Nu_1 \neq f(r_1)$, and Eq. (3.3)

$$\frac{R_1}{R_{1O}} = (1 - Nu_1 Ja Fo)^{1/2} = (1 - Fo / Fo_o)^{1/2}, Fo = 1/(Nu_1 Ja)$$

Nu_1 = *Hunt* and *Isenberg* Table 3.2, and Eq. (3.3) + $\Delta V \neq f(R_1)$

$$\frac{R_1}{R_{1O}} = (1 - \frac{3}{2}\frac{1}{2\sqrt{\pi}} Re_{1o}^{1/2} Pr_2^{1/3} Ja Fo)^{2/3}$$

$$= (1 - 0.423 Re_{1o}^{1/2} Pr_2^{1/3} Ja Fo)^{2/3} = (1 - Fo / Fo_o)^{2/3},$$

$$Fo_o = 1/(0.423 Re_{1o}^{1/2} Pr_2^{1/3} Ja).$$

Experiment

Mayinger and *Nordmann* (1979) for $4 < Ja < 300$, p = 0.2 to 5 bar:

$$\frac{R_1}{R_{1O}} = \left[1 - \frac{5}{3}\frac{10^3}{2}\left(\frac{\rho_1}{\rho_2}\right)^{0.79} Re_{1o}^{1/3} Pr_2^{2/3} Ja^{1.27} Fo\right]^{3/5}$$

$$= \left[1 - 0.833\times 10^3 \left(\frac{\rho_1}{\rho_2}\right)^{0.79} Re_{1o}^{1/3} Pr_2^{2/3} Ja^{1.27} Fo\right]^{3/5} = (1 - Fo/Fo_o)^{3/5},$$

$$Fo_o = 1/\left[0.833\times 10^3 \left(\frac{\rho_1}{\rho_2}\right)^{0.79} Re_{1o}^{1/3} Pr_2^{2/3} Ja^{1.27}\right].$$

Condensation of bubbles in a pool behind nozzles

Akiyama (1973) for $\Delta V_{12} = 1.18\left[\frac{\sigma g(\rho_2 - \rho'')}{\rho_2^2}\right]^{1/4}$

$$\frac{R_1}{R_{1o}} = \left(1 - 0.259 Re_{1o}^{0.6} Pr_2^{1/3} Ja Fo\right)^{1/1.4} = \left(1 - Fo/Fo_o\right)^{1/1.4},$$

$$Fo_o = 1/\left(0.259 Re_{1o}^{0.6} Pr_2^{1/3} Ja\right).$$

Brucker and *Sparrow* (1977) for $\Delta T = 15$ *to* $100K$, $p = (10.3$ *to* $62.1)\times 10^5 Pa$, $Pe_2 = 2000$ *to* 3000, $Nu_1 \pm 50\%$:

$$\frac{R_1}{R_{1o}} = \left(1 - 0.259 Re_{1o}^{0.6} Ja Fo\right)^{1/1.4} = \left(1 - Fo / Fo_o\right)^{1/1.4},$$

$$Fo_o = 1/\left(0.259 Re_{1o}^{0.6} Ja\right).$$

Moalem and *Sideman* (1973):

$$\frac{R_1}{R_{1O}} = \left(1 - \frac{2}{3}\frac{1}{\sqrt{\pi}} Re_{1o}^{1/2} Pr_2^{1/2} Ja Fo\right)^{2/3}$$

$$= \left(1 - 0.846 Re_{1o}^{1/2} Pr_2^{1/2} Ja Fo\right)^{2/3} = \left(1 - Fo / Fo_o\right)^{2/3},$$

$$Fo_o = 1/\left(0.846 Re_{1o}^{1/2} Pr_2^{1/2} Ja\right).$$

Dushkin and *Kolomenzev* (1989) for $0.8\times10^6 \le p \le 2\times10^6$ Pa, $15 \le T' - T_2 \le 25$ K, $2\times10^{-3} \le D_{1o} \le 4\times10^{-3}$ *m*, or $0.1 < Ja < 20$, $0 < Fo < 0.2$, $10^3 < Pe_2 < 10^5$, $0.1 \le \frac{R_1}{R_{1o}} \le 1$, $\alpha_1 = 0.2$ to 0.7, $Re_1 = 10^4$ to 10^6, $566 \le \alpha_1\rho_1 w_1 \le 1\,698$ kg/(m^2 s), $11\,317 \le \alpha_2\rho_2 w_2 \le 56\,588$ kg/(m^2 s) approximate experimental data of several authors for a vertical upwards flow, and their own data for downward flow of Freon-22 in a tube with $D_h = 0.015$ with

$$\frac{R_1}{R_{1o}} = \exp\left[-\frac{4}{3}\left(Pe_2^{0.43} + 1\right) Ja Fo\right].$$

Chen and *Maynger* (1989), experiment $2 < Pr_2 < 15$, $1 < Ja < 120$.

$$\frac{R_1}{R_{1o}} = \left(1 - 0.56 Re_{1o}^{0.7} Pr_2^{1/2} Ja Fo\right)^{0.9} = \left(1 - Fo / Fo_o\right)^{0.9},$$

$$Fo_o = 1/\left(0.56 Re_{1o}^{0.7} Pr_2^{1/2} Ja\right).$$

At *Jacob* numbers above 70 inertia-controlled phenomena start to dominate.

Bucher and *Nordman* (1978) compared experimental data with the predictions of the equations from Table 3.3 and found that the *Akiyama* equation shows the best results up to 1978.

3.4 Nonaveraged source terms

The nonaveraged source terms required for microscopic flow description are given below. First, the evaporation mass per unit volume and unit time is equal to zero,

$$\mu_{21} = 0 . \tag{3.10}$$

There is no heat transfer between the bubble interface and the vapor

$$\dot{q}_1'''^{2\sigma} = 0 . \tag{3.11}$$

Only the heat transfer between the interface and the bulk liquid

$$\dot{q}_2'''^{1\sigma} = a_{12} h_c \left[T'(p) - T_2 \right], \tag{3.12}$$

controls the condensed mass per unit mixture volume and unit time

$$\mu_{12} = \dot{q}_2'''^{1\sigma} / \left[h_{M1} - h'(p) \right] . \tag{3.13}$$

Here the instantaneous heat-transfer coefficient between the interface and the bulk liquid is

$$h_c = \frac{\lambda_2}{D_1} Nu_1 , \tag{3.14}$$

where the *Nusselt* number is a function of the local parameter of the flow.

In what follows we show how to link the empirically obtained results for the change of the radius of the condensing bubble with the macroscopic flow description. The condensed mass per unit mixture volume and unit time is equal to the product of the number of the bubbles per unit mixture volume and the mass change of a single bubble.

$$\mu_{12} = -\rho''(p) n_1 dV_1 / d\tau = -\rho''(p) n_1 4\pi R_1^2 dR_1 / d\tau \tag{3.15}$$

or replacing

$$R_1 = \left(\frac{3}{4\pi} \frac{\alpha_1}{n_1} \right)^{1/3} , \tag{3.16}$$

$$\mu_{12} = -3^{2/3} (4\pi)^{1/3} \rho''(p) n_1^{1/3} \alpha_1^{2/3} \frac{dR_1}{d\tau} = 4.84 \rho''(p) n_1^{1/3} \alpha_1^{2/3} \frac{dR_1}{d\tau} , \tag{3.17}$$

where for $dR_1 / d\tau$ one can use one of the expressions from Table 3.2 or 3.3. The heat absorbed by the liquid during the condensation per unit mixture volume and unit time is

$$\dot{q}_2'''^{1\sigma} = \mu_{12} \left[h''(p) - h'(p) \right]. \tag{3.18}$$

3.5 Averaged source terms

In order to calculate the averaged mass source term we proceed as follows.

The integral expression for the averaged condensation within the time interval $\Delta\tau$, where $\Delta\tau < \Delta\tau_o$, is

$$\mu_{12} = \rho'' n_1 \frac{V_{1o} - V_1}{\Delta\tau} = \frac{\rho'' n_1 V_{1o}}{\Delta\tau}\left(1 - \frac{V_1}{V_{1o}}\right) = \frac{\rho'' \alpha_1}{\Delta\tau}\left(1 - \left(\frac{R_1}{R_{1o}}\right)^3\right). \tag{3.19}$$

If the time interval is selected larger than the lifetime of the bubble, $\Delta\tau \ge \Delta\tau_o$, we have

$$\mu_{12} = \rho'' \alpha_1 / \Delta\tau . \tag{3.20}$$

The last relationship is practically the condition for the entire condensation of the available steam within the time step $\Delta\tau$. Having in mind that during the condensation

$$\mu_{21} = 0 \tag{3.21}$$

$$\dot{q}_1^{\prime\prime\prime 2\sigma} = 0, \tag{3.22}$$

we obtain from the energy-jump condition on the bubble surface the averaged energy source term

$$\dot{q}_2^{\prime\prime\prime 1\sigma} = \mu_{12}(h_{M1} - h'). \tag{3.23}$$

The above relationship is strictly valid if the bubble surface temperature

$$T_2^{1\sigma} = T'(p_{M1}) \tag{3.24}$$

is equal to the gas temperature T_1. We note the condensation rate in this case with μ_{12}'. In general, there is a small amount of heat transfer from the bubble to the surface if $T_1 > T'(p_{M1})$ or from the surface to the bubble if $T_1 < T'(p_{M1})$ due to a mechanism similar to natural convection inside the cavity. In this case we have

$$\dot{q}_2^{\prime\prime\prime 1\sigma} = \mu_{12}'\left[h_{M1} - h'(p)\right]. \tag{3.25}$$

$$\mu_{21} = 0, \tag{3.26}$$

$$\dot{q}_1^{\prime\prime\prime 2\sigma} = a_{21} h_{NC}\left[T'(p_{M1}) - T_1\right]. \tag{3.27}$$

$$\mu_{12} = \mu_{12}' + \dot{q}_1^{\prime\prime\prime 2\sigma} / \left[h_{M1} - h'(p)\right], \tag{3.28}$$

where h_{NC} is the heat-transfer coefficient due to natural convection in a cavity. The second term in Eq. (3.28) is usually neglected as very small compared to the first one. This term becomes important only for mass-transfer processes on the bubble surface in saturated or nearly saturated liquid.

3.6 Change of the bubble number density due to condensation

Now let as answer the question what happens to the particle number density during the condensation of bubbles? Suppose the bubble-size distribution obeys the *Nukiama–Tanasawa* law (1938)

$$P(D_1) = 4\left(\frac{D_1}{D_1'}\right)^2 e^{-2(D_1/D_1')} . \tag{3.29}$$

Here, $P(D_1)$ is the probability that a bubble has its size between D_1 and $D_1 + \delta D_1$. D_1' is the most probable particle size, i.e., the size where the probability function has its maximum value. The particle sizes may take values between zero and a maximum value

$$0 < D_1 < D_{1max}. \tag{3.30}$$

Thus, if we know D_1' and D_{1max} the particle distribution is uniquely characterized. *Mac Vean*, see in *Wallis* (1969), found that a great deal of data could be correlated by assuming that

$$D_1' = D_1/2 \tag{3.31}$$

where D_d is the volume averaged particle size. The relationship between D_d and the maximum particle size, D_{1max}, is reported as

$$D_{1max} \approx (2.04 \text{ to } 3.13)\, D_1, \tag{3.32}$$

see *Pilch* et al. (1981), *Kataoka* et al. (1983), among others. Suppose that the distribution remains unchanged during the time step $\Delta\tau$. For that time the volume-averaged diameter changes as

$$\Delta D_1 = D_{10} - D_1 = D_{10}(1 - R_1/R_{10}) = D_{10}\left[1 - \left(1 - \frac{\mu_{12}\Delta\tau}{\rho_1\alpha_1}\right)^{1/3}\right] \tag{3.33}$$

and all particles having sizes

$$0 \le D_1 \le 2\Delta D_1 \tag{3.34}$$

namely

$$n_1 \int_0^{2\Delta D_1} P(D_1) dD_1 = 2n_1 \left\{ 1 - e^{-4\Delta D_1 / D_1'} \left[1 + 4\frac{\Delta D_1}{D_1'} \left(1 + 2\frac{\Delta D_1}{D_1'} \right) \right] \right\} \quad (3.35)$$

disappear. The averaged particle sink per unit time and unit mixture volume is consequently

$$\dot{n}_1 \approx -(n_1 / \Delta\tau) 2 \left\{ 1 - e^{-4\Delta D_1 / D_1'} \left[1 + 4\frac{\Delta D_1}{D_1'} \left(1 + 2\frac{\Delta D_1}{D_1'} \right) \right] \right\}, \quad (3.36)$$

for $\Delta D_1 \le D_1'$. The latter condition means that approximately within the interval $2\,\Delta D_1$ the averaged bubble size is ΔD_1.

3.7 Pure steam bubble drifting in turbulent continuous liquid

In a bubble flow or in a churn-turbulent flow with considerable turbulence, the bubbles are moving practically with the *same* velocity as the liquid. The mechanism governing the condensation is quite different compared to the mechanism described in the previous sections. There are several authors assuming that the heat released during the condensation is transported *by exchange of turbulent eddies* between the boundary layer and the bulk liquid. The characteristic time of one cycle, estimated by dimensional analysis of the turbulent characteristics of the continuous velocity field, is of the order of

$$\Delta\tau = const\ (l_{e2}^2 / \varepsilon_2)^{1/2} \quad (3.37)$$

for high-frequency pulsations in accordance with the *statistical theory of turbulence*. Replacing the characteristic size of the turbulent eddies in the liquid with

$$l_{e2} = const\ (\nu_2^3 / \varepsilon_2)^{1/4} \quad (3.38)$$

we obtain for the time constant

$$\Delta\tau = const(\nu_2 / \varepsilon_2)^{1/2}. \quad (3.39)$$

After replacing the dissipated specific kinetic energy of the turbulent pulsations with

$$\alpha_2 \rho_2 \varepsilon_2 = \frac{1}{2}\frac{\rho^2}{\rho_2}\left[u^3 \Phi_{2ou}^2 \left(\frac{\lambda_{Ru}}{D_{hu}} + \frac{\zeta_u}{\Delta r} \right) + v^3 \Phi_{2ov}^2 \left(\frac{\lambda_{Rv}}{D_{hv}} + \frac{\zeta_\theta}{r^\kappa \Delta\theta} \right) + w^3 \Phi_{2ow}^2 \left(\frac{\lambda_{Rw}}{D_{hw}} + \frac{\zeta_w}{\Delta z} \right) \right] \quad (3.40)$$

we obtain

$$\Delta\tau = const\frac{\rho_2}{\rho}\left(2\alpha_2\nu_2 \Bigg/ \left[\begin{array}{l} u^3\Phi_{2ou}^2\left(\frac{\lambda_{Ru}}{D_{hu}}+\frac{\zeta_u}{\Delta r}\right)+v^3\Phi_{2ov}^2\left(\frac{\lambda_{Rv}}{D_{hv}}+\frac{\zeta_\theta}{r^\kappa\Delta\theta}\right) \\ +w^3\Phi_{2ow}^2\left(\frac{\lambda_{Rw}}{D_{hw}}+\frac{\zeta_w}{\Delta z}\right)\end{array}\right]\right)^{1/2}. \tag{3.41}$$

The heat flux on the bubble surface can be determined to the accuracy of a constant as

$$\dot{q}_2^{\prime\prime I\sigma} = const(T'-T_2)\sqrt{\frac{\lambda_2\rho_2 c_{p2}}{\Delta\tau}}, \quad \left(const=\frac{2}{\sqrt{\pi}}\ \text{plane}\right), \tag{3.42}$$

where $\Delta\tau$ is the time interval in which the high-frequency eddy is in contact with the bubble surface. During this time, the heat is transported from the surface to the eddy by molecular diffusion. Thereafter, the eddy is transported into the bulk flow again, and its place on the surface is occupied by another one. In this way, the heat released during the condensation is transported from the bubble to the turbulent bulk liquid. We substitute $\Delta\tau$ from Eq. (3.41) into Eq. (3.42) and obtain

$$\dot{q}_2^{\prime\prime I\sigma} = const(T'-T_2)\lambda_2 Pr_2^{1/2}\sqrt{\frac{\rho}{\eta_2\nu_2^{1/2}\alpha_2^{1/2}}}$$
$$\times\left[\begin{array}{l} u^3\Phi_{2ou}^2\left(\frac{\lambda_{Ru}}{D_{hu}}+\frac{\zeta_u}{\Delta r}\right)+v^3\Phi_{2ov}^2\left(\frac{\lambda_{Rv}}{D_{hv}}+\frac{\zeta_\theta}{r^\kappa\Delta\theta}\right) \\ +w^3\Phi_{2ow}^2\left(\frac{\lambda_{Rw}}{D_{hw}}+\frac{\zeta_w}{\Delta z}\right)\end{array}\right]^{1/4}. \tag{3.43}$$

The constant can be determined by comparison with a result obtained by *Avdeev* in (1986) for one-dimensional flow without local resistance ($\xi = 0$). *Avdeev* used the known relationship for the *friction coefficient of turbulent flow*

$$\lambda_{Rw} = 0.184 Re_2^{-0.2}, \tag{3.44}$$

where

$$Re_2 = \frac{D_{hw}w_2}{\nu_2}, \tag{3.45}$$

compared the so-obtained equation with experimental data, and estimated the constant in

$$\dot{q}_2^{\prime\prime 1\sigma} = const\ 0.184^{1/4}(T' - T_2)\frac{\lambda_2}{D_{hw}} Pr_2^{1/2} Re_2^{0.7} (\Phi_{2ow}^2 / \alpha_2)^{1/4} \tag{3.46}$$

as

$$\text{const } 0.184^{1/4} = 0.228 \tag{3.47}$$

or

$$\text{const} = 0.348. \tag{3.48}$$

The final relationship recommended by *Avdeev* for one-dimensional flow

$$Nu = 0.228 Pr_2^{1/2} Re_2^{0.7} (\Phi_{2ow}^2 / \alpha_2)^{1/4} \tag{3.49}$$

describes his own data within $\pm$ 30% error band for $D_1 / D_{hw} > 80 / Re_2^{0.7}$. Note that in the *Avdeev* equation ρ/ρ_2 is set to one. Thus, the so-estimated constant can be successfully applied to three-dimensional flows. For comparison let us write the relationships obtained by *Hancox* and *Nikol* (see in *Hughes* et al. (1981)

$$Nu = 0.4 Pr_2 Re_2^{2/3} \tag{3.50}$$

and *Labunzov* (see in *Labunzov* (1974))

$$Nu = \frac{\lambda_{Rw}/8}{1 - 12\sqrt{\dfrac{\lambda_{Rw}}{8}}} Pr_2 Re_2 \frac{D_1}{D_h} \cong \frac{0.023}{1 - 1.82 Re_2^{-0.1}} \frac{D_1}{D_h} Pr_2 Re_2^{0.8}. \tag{3.51}$$

We see that in the three equations obtained independently from each other the dependence on Re_2 is $Re_2^{0.7 \text{ to } 0.8}$, and the dependence on Pr_2 is $Pr_2^{0.5 \text{ to } 1}$. In the case of $Re_2 \approx 0$ the energy dissipated behind the bubbles should be taken into account.

3.8 Condensation from a gas mixture in bubbles surrounded by subcooled liquid

In the *presence of a noncondensing gas*, the temperature of the bubble surface T_k' is not equal to the saturation temperature T' at the system pressure *p*. During the condensation the partial pressure of the inert components increases with the decreasing bubble size due to the simultaneously decreasing partial steam pressure until T'_k becomes equal to the liquid temperature and the condensation ceases. Note that in this case the *bubble remains with a stable equilibrium volume*. Because in accordance with *Dalton*'s law the noncondensing mixture occupies the whole bubble volume at the beginning of the condensation as well as at its end the ratio of the end volume to the initial volume is inversely proportional to the densities of the noncondensing mixture in both states, respectively

$$\left(\frac{R_{1\infty}}{R_1}\right)^3 = \frac{\sum \rho_{n1o}}{\sum \rho_{n1\infty}} = \frac{1}{\sum \rho_{n1\infty}} \rho_{1o}(1 - C_{M1o}) = \frac{R_{n1}^* T_{1o}}{p - p'(T_2)} \rho_{1o}(1 - C_{M1o}) \approx \frac{\sum p_{n1o}}{p - p'(T_2)}. \tag{3.52}$$

This expression is necessary for the estimation of the *maximum duration of the bubble collapse*.

There are different approaches to estimate the condensing mass per unit mixture volume and unit time.

3.8.1 Thermally controlled collapse

One of the approaches assumes that the noncondensing components are *uniformly* distributed through the whole bubble volume and the condensation is controlled, as in one-component bubble condensation, by the heat transport from the bubble surface to the bulk liquid. The only difference is that the surface temperature, assumed in the single-component case to be equal to the saturation temperature at the steam partial pressure, is not equal to the saturation temperature at the system pressure. Thus, the equation describing the change of the bubble radius for a one-component bubble can simply be corrected as follows

$$\frac{dR_1}{d\tau} = -\frac{h_c}{\rho''} \frac{T_c' - T_2}{h'' - h_2} = -\frac{Nu_1 \lambda_2 (T_c' - T_2)}{R_1 2\rho''(h'' - h_2)} = -c_T \frac{Nu_1 \lambda_2 (T' - T_2)}{R_1 2\rho''(h'' - h_2)}$$
$$= -\frac{a_2 Nu_1 Ja}{2R_1} = -c_T c / R_1, \tag{3.53}$$

where

$$c_T = \frac{T_c' - T_2}{T' - T_2} \tag{3.54}$$

is the dimensionless surface temperature. Thus, the derivation of the averaged source terms is similar to the derivation shown in Section 3.4 for single-component bubble condensation.

3.8.2 Diffusion-controlled collapse

Another approach considers the *concentration profile* in the neighborhood of the bubble surface and models the condensation as a diffusion process. The nonaveraged mass condensing per unit mixture volume and unit time is

$$\mu_{12} = a_{12}(\rho w)_{12} = -a_{12}\beta \frac{C_{M1}\rho_1}{p_{M1}} \left[p_{M1} - p'(T_2)\right]$$

$$= -a_{21}\alpha_c \frac{M_{M1}}{M_1} \ln \frac{[p - p'(T_2)]}{\sum p_{n1}} / \left(c_{p1} Le^{2/3}\right), \text{ see in } \textit{VDI-Wärmeatlas}. \text{ (1984)} \tag{3.55}$$

where

$$\beta = \frac{p_{M1}}{C_{M1}\rho_1 \left[p'(T_2) - p_{M1}\right]} \alpha_c \frac{M_{M1}}{M_1} \ln \frac{p - p'(T_2)}{\sum p_{n1}} / (c_{p1} Le^{2/3}), \tag{3.56}$$

$$p'(T_2) < p_{M1}, \tag{3.57}$$

$$Le = \lambda_1 / (\rho_1 c_{p1} D_{M \to \sum n}), \tag{3.58}$$

$$M_1 = \frac{\sum p_{n1}}{p} \sum M_{n1} + \frac{p - \sum p_{n1}}{p} \sum M_{M1}, \tag{3.59}$$

p' is the partial steam pressure in the boundary layer, where the steam is supposed to be saturated, having a temperature nearly equal to the liquid temperature

$$p' = p'(T_2). \tag{3.60}$$

The released heat during the condensation is transported into the liquid

$$\dot{q}_2^{\prime\prime\prime 1\sigma} = \mu_{12} \left[h_{M1} - h'(p)\right]. \tag{3.61}$$

Nomenclature

a	thermal diffusivity, m²/s
a_2	$= \dfrac{\lambda_2}{\rho_2 c_{p2}}$, liquid thermal diffusivity, m²/s
a_{21}	interfacial area density between the liquid *2* and the vapor 1, 1/m
C	mass concentration, kg/kg
c_p	specific heat at constant pressure, J/(kg K)
$D_{M \to \sum n}$	diffusion constant for the species M into the mixture $\sum n$, 1/m
D_{cav}	equivalent cavity size, m
D_h	hydraulic diameter, m
$D_{1,n}$	bubble diameter belonging to *n*th bubble generation, m
D_{1c}	critical bubble diameter, m
D_{1d}	bubble departure diameter, m
D_1	bubble diameter, m

$D_{1\max}$ maximum bubble diameter, m
D_1' most probable bubble size, m
d differential, dimensionless
F_1 / Vol cross-section at the wall surface covered by attached bubbles per unit flow volume, 1/m
Fo $= \dfrac{a_2 \tau}{R_{1o}^2}$, *Fourier* number – condensation time scale, dimensionless
Fo_o time within which the bubble collapses completely, dimensionless
g gravitational acceleration, m/s²
Ja $= \dfrac{\rho_2 c_{p2} \left[T_2 - T'(p)\right]}{\rho_{10} \left[h''(T_2^{1\sigma}) - h'(T_2^{1\sigma})\right]}$, evaporation *Jacob* number, dimensionless
Ja $= \dfrac{\rho_2}{\rho''} \dfrac{c_{p2} \left[T'(p) - T_2\right]}{h'' - h'}$, condensation *Jacob* number, dimensionless
G^* critical mass flow rate, kg/(m² s)
Gb_2 $= \dfrac{\Delta E_{1kr}}{kT_2}$, *Gibbs* number, dimensionless
Gr_1 $= g\left|\rho_1 - \rho_1''(T_2^{1\sigma})\right| \rho_1 D_1^3 / \eta_1^2$, bubble *Grashoff* number, dimensionless
h specific enthalpy, J/kg
h_{NC} heat-transfer coefficient by natural circulation, W/(m² K)
h_c heat-transfer coefficient by forced convection, W/(m² K)
k $= 13.805 \times 10^{-24}$, *Boltzmann* constant, J/K
Le_1 $= \lambda_1 / (\rho_1 c_{p1} D_{M \to \sum n})$, gas *Lewis* number, dimensionless
l_{e2} $= const\ (\nu_2^3 / \varepsilon_2)^{1/4}$, characteristic size of the turbulent eddies in the liquid, m
m_2 $= m_\mu / N_A$, mass of single molecule, kg
m_μ kg-mol mass (18 *kg* for water), kg/mole
Nu_1 $= \dfrac{D_1 h_k}{\lambda_2}$, bubble *Nusselt* number, dimensionless
N_2 $= \rho_2 / m_2 = \rho_2 N_A / m_\mu$ ($\approx 3.3 \times 10^{28}$ for water), number of the molecules per unit volume, 1/m³
N_A $= 6.02 \times 10^{26}$, *Avogadro* number, number of molecules in one kilogram mole mass, 1/kg-mole
$\dot{n}_{1cin}$ number of the created nuclei per unit time in unit volume of the liquid, 1/(m³s)
n_w'' cavity per unit surface, 1/m²
n_{1w}'' bubble generating (active) cavity per unit surface, 1/m²

n_1	bubble number density, 1/m³
$\dot{n}_1$	bubble number density change per unit time, 1/(m³ s)
$n_{1,n}$	bubble number density belonging to *n*th bubble generation, $1/m^3$
$n_{1,\max}$	maximum number of nucleation sites per unit volume in a pipe, $1/m^3$
$P(D_1)$	probability that a bubble has its size between D_1 and $D_1 + \delta D_1$, dimensionless
Pr_1	$=\eta_1 c_{p1}/\lambda_1$, gas *Prandtl* number, dimensionless
Pr_2	$=\dfrac{\eta_2}{\rho_2 a_2}$, liquid *Prandtl* number, dimensionless
Pe_2	$=\dfrac{D_1 \Delta V_{12}}{a_2}$, bubble *Peclet* number, dimensionless
p	pressure, Pa
Re_1	$=\dfrac{D_1 \rho_2 \Delta V_{12}}{\eta_2}$, bubble *Reynolds* number, dimensionless
Re_{1o}	$=\dfrac{D_{1o} \rho_2 \Delta V_{12}}{\eta_2}$, bubble *Reynolds* number based on the initial bubbles size, dimensionless
$\dot{q}_2'''^{1\sigma}$	thermal energy flow density coming from the vapor interface and introduced into the bulk liquid, W/m³
$\dot{q}_1'''^{2\sigma}$	thermal energy flow density coming from the liquid interface and introduced into the bulk vapor, W/m³
$\dot{q}_1''^{2\sigma}$	thermal energy flux coming from the liquid interface and introduced into the bulk vapor, W/m²
$\dot{q}_2''^{1\sigma}$	thermal energy flux coming from the vapor interface and introduced into the bulk liquid, W/m²
R	gas constant, J/(kg K)
R_1	bubble radius, m
R_{10}	initial bubble radius, m
R_{1c}	critical bubble radius, m
$R_{1\infty}$	final radius of the bubble, m
r	radius, m
s	specific entropy, J/(kg K)
T	temperature, K
u	radial velocity, m/s
$\mathbf{V}$	velocity vector, m/s
V_{1c}	volume of the bubble with critical size, m³
V_1	volume of the bubble, m³

V_{10} initial volume of the bubble, m^3
v specific volume, m^3/kg
Vol volume, m^3

Greek

α_1 bubble volumetric fraction, m^3/m^3
α_2 liquid volumetric fraction, m^3/m^3
β diffusion coefficient, $kg/(m^2\ s)$
γ_v volume occupied by the flow divided by the volume of the system, dimensionless
γ cross-section flowed by the flow divided by the total cross-section of the system, dimensionless
Δ finite difference
ΔE_1 work necessary to create a single bubble with radius R_1, J
ΔE_{1c} work necessary to create a single bubble with radius R_{1c}, J
ΔE_{1c}^* $= \varphi \Delta E_{1c}$, energy for the creation of a single bubble at a wall, J
Δp_{Fi}^* $= p'(T_2) - p_{Fi}$, flashing inception pressure difference, Pa
$\Delta \tau$ time interval, s
$\Delta \tau_o$ time within which the bubble collapses completely, s
ε dissipation rate for kinetic energy from turbulent pulsations, irreversibly dissipated power by the viscous forces due to turbulent pulsations, W/kg
ξ local friction coefficient
η dynamic viscosity of liquid, $kg/(m\ s)$
θ static contact angle, rad
λ_{RT} *Rayleigh-Taylor* wavelength, m
λ thermal conductivity, $W/(m\ K)$
λ_R *Darsy* friction coefficient, dimensionless
μ_{21} evaporating mass per unit time and unit flow volume, $kg/(m^3\ s)$
μ_{12} condensing mass per unit time and unit flow volume, $kg/(m^3\ s)$
ρ density, kg/m^3
$(\rho w)_{21}$ evaporation mass flow rate, $kg/(m^2\ s)$
σ surface tension, N/m
M_{M1} kg-mole mass of the condensing gases, kg/mole
$\sum M_{n1}$ kg-mole mass of the noncondensing gases, kg/mole
M_1 kg-mole mass of the gas mixture, kg/mole
τ time, s

τ_H	$=\dfrac{4}{\pi}Ja^2Fo$, condensation time scale , dimensionless
Φ	cavity angle if the cavity is idealized as a cone, rad
Φ_{2o}^2	two-phase friction multiplier, dimensionless
φ	work-reduction factor, dimensionless

Superscripts

$'$	saturated liquid
$''$	saturated vapor
spin	spinoidal line
1σ	interface with vapor
2σ	interface with liquid

Subscripts

1	vapor, gas
2	liquid
c	critical state
s	at constant entropy
sat	saturated
Fi	flashing inception
$\sum_{n1o}$	sum of all noncondensing gases
M	evaporating or condensing chemical component
n	inert (non condensing) chemical component
u,v,w	in the positive direction of the velocity components u, v and w
R	friction

References

Akiyama, A.: Bubble collapse in subcooled boiling. Bull. JSME 16(93), 570–575 (1973)

Avdeev, A.A.: Growth and condensation velocity of steam bubbles in turbulent flow. Teploenergetika 1, 53–55 (1986) (in Russian)

Brauer, H., et al.: Chem. Ing. Tech. 48, 737–741 (1976)

Brucker, G.G., Sparrow, E.M.: Direct contact condensation of steam bubbles in water of high pressure. Int. J. Heat Mass Transfer 20, 371–381 (1977)

Bucher, B., Nordman, D.: Investigations of subcooled boiling problems in two-phase transport and reactor safety. In: Veziroglu, R.N., Kakac, S. (eds.) vol. 1, pp. 31–49. Hemisphere, Washington (1978)

Chen, Y.M., Maynger, F.: Measurement of heat transfer at the phase interface of condensing bubbles. In: ANS Proceedings Nat. Heat Transfer Conference, Philadelphia, Pennsylvania, August 6-9, vol. 4, pp. 147–152 (1989)

Dushkin, A.L., Kolomenzev, A.I.: Steam bubble condensation in subcooled liquid. High Temperature Physics 27(1), 116–121 (1989) (in Russian)

Forschuetz, L., Chao, B.T.: On the mechanics of vapor bubble collapse. Trans. ASME, J. Heat Transfer, 209–220 (1965)

Hughes, E.D., Paulsen, M.P., Agee, L.J.: A drift-flux model of two-phase flow for RETRAN. Nuclear Technology 54, 410–420 (1981)

Hunt, D.L.: The effect of delayed bubble growth on the depressurization of vessels Containing high temperature water. UKAEA Report AHSB(S) R 189 (1970)

Isenberg, J., Sideman, S.: Direct contact heat transfer with change of phase: Bubble condensation in immiscible liquids. Int. J. Heat Mass Transfer 13, 997–1011 (1970)

Kataoka, I., Ishii, M., Mishima, K.: Trans. ASME 105, 230–238 (1983)

Kendouch, A.A.: Theoretical and experimental investigations into the problem of transient two-phase flow and its application to reactor safety. Ph.D. Thesis, Department of Thermodynamics and Fluid Mechanics, University of Strathclyde, U.K. (1976)

Labunzov, D.A.: State of the art of the nuclide boiling mechanism of liquids. Heat Transfer and Physical Hydrodynamics, 98–115 (1974) (in Russian)

Mayinger, F., Nordmann, D.: Temperature, pressure and heat transfer near condensing bubbles. In: Proc. of the Heat and Mass Transfer, Dubrovnik, vol. 1. Hemisphere Publ. Corp., New York (1979)

Moalem, D., Sideman, S.: The effect of motion on bubble collapse. Int. J. Heat Mass Transfer 16, 2321–2329 (1973)

Nigmatulin, R.I.: Basics of the mechanics of the heterogeneous fluids. Nauka, Moskva (1978) (in Russian)

Nukiama, S., Tanasawa, Y.: Trans. Soc. Mech. Engs. (Japan) 4(14), 86 (1938)

Pilch, M., Erdman, C.A., Reynolds, A.B.: Acceleration induced fragmentation of liquid drops. Department of Nucl. Eng., University of Virginia, Charlottesville, NUREG/CR-2247 (August 1981)

Plesset, M.S., Zwick, S.A.: The growth of bubbles in superheated liquids. J. Appl. Phys. 25(4), 493–500 (1954)

Prisnjakov, V.F.: Bubble growth in liquids. J. Eng. Phys. 18(1), 584–588 (1970)

Soo, S.L.: Fluid dynamics of multiphase systems. Woltham, Massachusetts (1969)

Wilson, J.F.: Primary separation of steam from water by natural separation. US/EURATOM Report ACNP-65002 (1965)

VDI-Waermeatlas, 4. Auflage, VDI-Verlag (1984)

Wallis, G.B.: One-dimensional two-phase flow. McGraw-Hill, New York (1969)

Zeitoun, O., Shoukri, M., Chatoorgoon, V.: Interfacial heat transfer between steam bubbles and subcooled water in vertical upward flow. J. Heat Transfer 117, 402–407 (1995)

Zuber, N.: The dynamics of vapor bubbles in non uniform temperature fields. Int. J. Heat Mass Transfer 2, 83–98 (1961)

4. Bubble departure diameter

The tangential shear due to the volume- and time-averaged pulsation velocity caused by the thermally controlled bubble growth and successive cyclic departure is found to be responsible for the natural switch from the regime of isolated bubble growth and departure into the regime of mutual interaction bubble growth and departure as the temperature difference between the bubble interface and the surrounding liquid increases. The proposed theoretical model, based only on the first principles, agrees well with the experimental data for boiling water to which it was compared. The new model successfully predicts the isolated and the mutual interaction bubble departure size, and the quantitative description of the natural transition between the two regimes is the essential new feature of this work. This chapter is a slightly abbreviated version of the work published primarily in Kolev (1994).

4.1 How accurately can we predict bubble departure diameter for boiling?

After 60 years of worldwide research on boiling there are still "classical" problems of practical importance that have been not satisfactorily resolved. Such a problem is the analytical description of the bubble departure diameter during boiling or flashing. We demonstrate this problem as follows. Consider saturated water pool boiling at atmospheric pressure on a heated horizontal surface. The experimental data from the literature *Gaertner* and *Westewater* (1960), *Gaertner* (1965), *Tolubinsky* and *Ostrovsky* (1966), *Siegel* and *Keshock* (1964), *van Stralon* et al. (1975), *Roll* and *Mayers* (1964). are depicted in Fig. 4.1 as a function of the wall superheating. Note that data for bubble sizes measured away from the surface and sometimes reported as bubble departure diameters are excluded. Now we try to predict the data behavior. For the prediction of the bubble departure diameter we use some of the frequently cited theories from *Fritz* (1935), *van Krevelen* and *Hoftijzer* (1950), *Kocamustafaogullari* and *Ishii* (1983), *Cole* and *Rohsenow* (1969), *Moalem* et al. (1977), *Klausner* et al. (1993). The results presented in Figs. 4.1 and 4.2 are disappointing. None of these theories predicts even an appropriate data trend. In particular, none of the existing theories gives an explanation of why the bubble departure diameter starts to decrease after a certain superheating is exceeded. The interested reader will find reviews on modeling by *Hsu* and *Graham* (1976)up to 1975, by *van Stralen* and *Cole* (1979) up to 1979, and by *Klausner* et al. (1993) *Zeng* et al. (1993), Zeng et al. (1993). for more recent work. These works and the references given there present the state-of-the-art in this field and will not be repeated here.

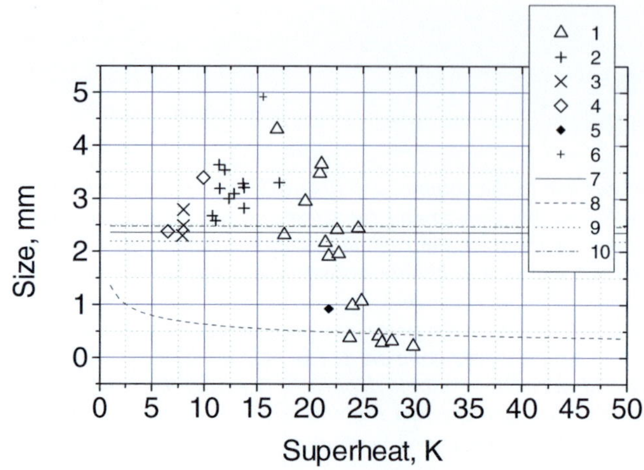

Fig. 4.1 Bubble departure diameter as a function of superheating. Saturated water pool boiling at 0.1 MPa pressure. Data: 1 *Gaertner* and *Westwater* (1960), 2 *Gaertner* (1965), 3 *Tolubinsky* and *Ostrovsky* (1966), 4 *Siegel* and *Keshock* (1964), 5 *van Stralen* et al (1975), 6 *Roll* and *Mayers* (1964). Theories: 7 *Fritz* (1935), 8 *van Krevelen* and *Hoftijzer* (1950), 9 *Kocamustafaogullari* and *Ishii* (1983), 10 *Cole* and *Rohsenow* (1969)

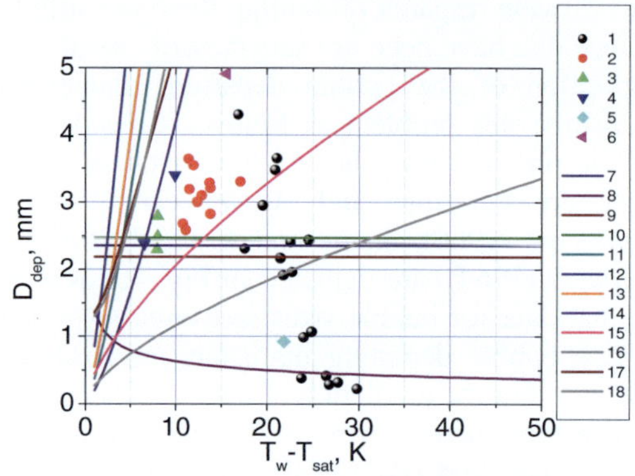

Fig. 4.2 Bubble departure diameter as a function of superheating. Saturated water pool boiling at 0.1 MPa pressure. Data: 1 *Gaertner* and *Westwater* (1960), 2 *Gaertner* (1965), 3 *Tolubinsky* and *Ostrovsky* (1966), 4 *Siegel* and *Keshock* (1964), 5 *van Stralen* et al. (1975), 6 *Roll* and *Mayers* (1964). Theories: 7 *Fritz* (1935), 8 *van Krevelen* and *Hoftijzer* (1950), 9 *Kocamustafaogullari* and *Ishii* (1983), 10 *Cole* and *Rohsenow* (1969), 11 *Moalem* et al. (1977), 12 *Klausner* et al. $n = 1/2$ (1993), 13 *Klausner* et al. $n = 1/3$ (1993), 14 *Ruckenstein* (1961), 15 *Voloshko* and *Vurgaft* (1970), 16 *Golorin* et al (1978), 17 *Kutateladze* and *Gogonin* (1979), 18 *Jensen* and *Memmnel* (1986)

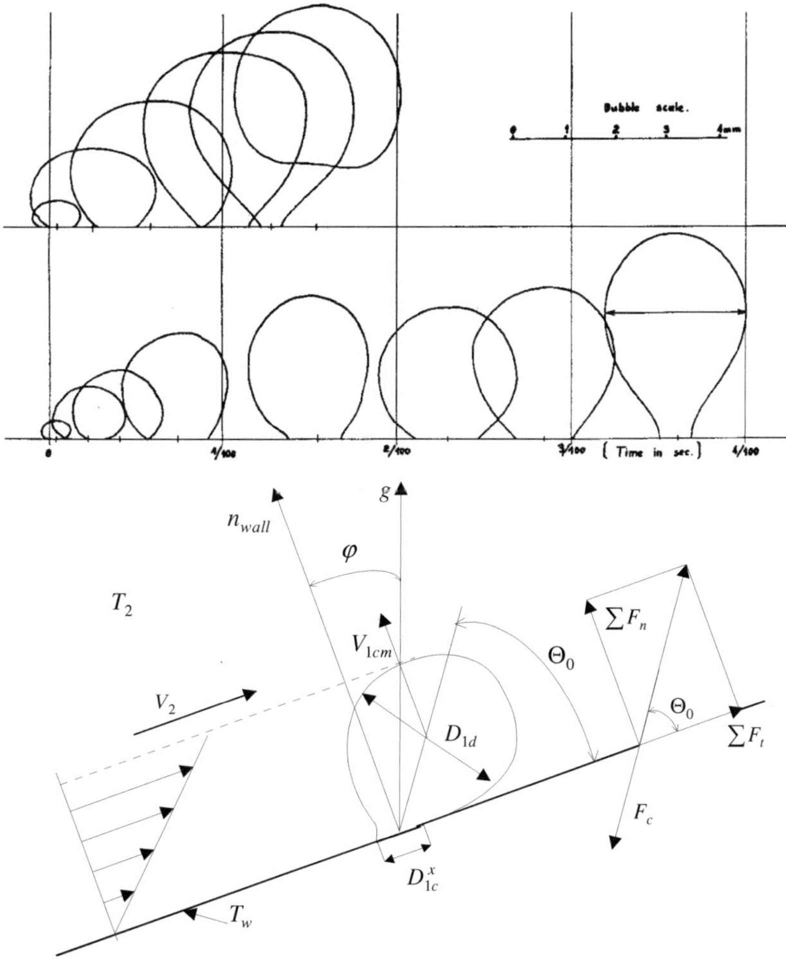

Fig. 4.3 a) *Morin's* (1964) observations of water bubble detachment in pool at heated horizontal copper wall: a) 5°C subcooling; b) no subcooling; c) Model of a bubble detachment in heated wall in flow boiling used in this work

We present in this section a new model for the prediction of the bubble departure diameter as a function of the local flow parameter, which agrees well with data for pool and flow boiling at atmospheric and elevated pressures. We show that the mutual interaction during bubble growth and departure introduces enough shear stress to cut off the adjacent bubbles at an earlier stage of their growth.

4.2 Model development

Consider bubble growth at a heated surface as shown in Fig. 4.3. The surface vector and the upwards-directed vertical form an angle φ. There is a flow from left to right with an average velocity V_2. The derived expressions rely on the following assumptions:

1) The bubble is capable of growing after reaching the size given by Eq. (1.16)

$$D_{1c}^* = \frac{3}{2} D_{1c} \tag{4.1}$$

where D_{1c} is the critical bubble size.

2) The bubble growth is thermally controlled. The expression governing the bubble growth is of the type given in Appendix 13.1

$$D_1 = 2B\tau^n \tag{4.2}$$

or

$$dD_1 / d\tau = 2nB\tau^{n-1}. \tag{4.3}$$

It is generally agreed that the exponent n is of order of 1/2 for saturated liquids. For the purpose of data comparison we use in this work *Labuntsov*'s approximation of the *Scriven* solution, *Labuntsov* (1974), $n = 1/2$ and

$$B = c\ Ja\ a_2^{1/2}, \tag{4.4}$$

where

$$c = (12/\pi)^{1/2} \left[1 + \frac{1}{2}\left(\frac{\pi}{6Ja}\right)^{2/3} + \frac{\pi}{6Ja} \right]^{1/2}. \tag{4.5}$$

3) The base bubble diameter at the wall is equal to D_{1c}^* for high pressure. For low pressure, the base diameter initially grows and thereafter collapses, so that at the moment of detachment it again approximates D_{1c}^*.

4) At the moment of detachment the bubble is deformed such that its diameter measured parallel to the surface is larger than the volume-averaged bubble departure diameter D_{1d}.

5) The presence of the wall increases the drag coefficient of the bubble in the tangential drag force compared to the case of a bubble in an infinite medium.

6) The bubble axis inclination, θ_0, coincides with the main force inclination with respect to the wall.

7) At the moment of departure, the elongation of the bubble base to a cylindrical form with the axis coincident with the bubble axis is the reason for the tangential surface tension force component.

Some useful consequences of assumption 2 that will be used later are given below. At the moment of detachment, $\Delta\tau_{1d}$, the bubble radius is

$$R_{1d} = B\,\Delta\tau_{1d}^{n} \tag{4.6}$$

the bubble center of mass (c.m.) velocity normal to the wall is

$$V_{1cm} = dR_{1d}/d\tau = nB\Delta\tau_{1d}^{n-1} \tag{4.7}$$

and the c. m. temporal acceleration normal to the wall is

$$dV_{1cm}/d\tau = n(n-1)B\Delta\tau_{1d}^{n-2} \tag{4.8}$$

Consequently the total acceleration is

$$\left(\frac{dV_1}{d\tau}+V_1\frac{dV_1}{dy}\right)_{cm} = dV_{1cm}/d\tau + V_{1cm}^2/R_{1d} = n(n-1)B\Delta\tau_{1d}^{n-2} + n^2 B\Delta\tau_{1d}^{n-2}$$

$$= nB\Delta\tau_{1d}^{n-2}(2n-1)\,. \tag{4.9}$$

For n = 1/2, which, as already mentioned, is the case for bubble growth in saturated liquid, the total bubble acceleration is zero, and therefore the bubble does not exert any dynamic forces like inertia and virtual mass force. This is a very surprising result. It simply means that at the moment of departure only the static mechanical equilibrium governs the bubble departure.

Before writing the momentum equation we compute the force components in normal and tangential directions. The buoyancy force components are

$$F_{b,n} = \frac{\pi}{6}D_{1d}^3(\rho_2-\rho_1)g\sin\varphi \tag{4.10}$$

and

$$F_{b,t} = \frac{\pi}{6}D_{1d}^3(\rho_2-\rho_1)g\cos\varphi\,. \tag{4.11}$$

For computation of the normal drag force resisting the bubble growth, we modify the Eq. (2.20) derived by *Ishii* and *Zuber* (1978) for bubbles in infinite liquid

$$F_{d,n} = 3\pi\eta_2 c_{form} D_{1d} V_{1cm}\left[1+0.1\left(c_{form}D_{1d}V_{1cm}\rho_2/\eta_2\right)\right]^{3/4}. \tag{4.12}$$

The product

$$D_{1d}V_{1cm} = 2nB^2\Delta\tau_{1d}^{2n-1}, \tag{4.13}$$

which for $n = 1/2$

$$D_{1d}V_{1cm} = B^2 \tag{4.14}$$

is obviously a function of superheating only. Consequently,

$$F_{d,n} = 3\pi\eta_2 c_{form}B^2\left[1+0.1\left(c_{form}B^2\rho_2/\eta_2\right)\right]^{3/4}. \tag{4.15}$$

The form coefficient, c_{form}, takes into account the ratio of the bubble size parallel to the wall to the volume-averaged bubble size as well the departure of the c.m. bubble velocity normal to the wall due to complicated bubble growth and deformation at the moment of detachment. It was found through the data comparison for pool boiling only that

$$c_{form} \approx 2. \tag{4.16}$$

The drag force parallel to the wall is

$$F_{d,t} = 0.3\pi\rho_2\left(c_{wall}D_{1d}V_{21d}\right)^2. \tag{4.17}$$

Here, V_{21d} is the boundary layer velocity at distance R_{1d} from the wall. We compute this velocity using the *Reichardt* (1951) universal velocity profile for $y << D_{1d}/2$

$$\frac{V_{21d}}{w*} = 2.5\ln\left(1+0.4y^+\right) + 7.8\left[1-\exp\left(-y^+/11\right)-\frac{y^+}{11}\exp\left(-0.33y^+\right)\right],$$

in exactly the same manner as described in *Klausner* et al. (1993). Here, $y^+ = yw*/\nu$ is thc dimensionless distance from the wall and $w*$ is the friction velocity. The deviation from the idealized profile in the real flow, caused by the bubbles themselves, the bubble deformation, and the wall influence are pooled into the wall correction coefficient, which was found to be

$$c_{wall} = 2 \tag{4.18}$$

through comparison with flow boiling data for negligible mutual interaction. The lift force caused by the rotation of the nonuniform velocity field at the wall is

$$F_{l,n} = \pi\frac{1}{2}\rho_2\left(D_{1d}V_{21d}\right)^2 c_{lift}, \tag{4.19}$$

where

$$c_{lift} = 3.877\beta^{1/2}\left(0.014\beta^2 + \mathrm{Re}_{12}^{-2}\right)^{1/4},$$

$$\mathrm{Re}_{12} = V_{21d} D_1 / \nu_2 ,$$

$$\beta = \frac{D_1/2}{V_{21d}} \left| \frac{dV_{21d}}{dy} \right| ,$$

$$\frac{dV_{21d}}{dy} = \frac{w^{*2}}{\nu} \left\{ \frac{1}{1+0.4y^+} + \frac{7.8}{11} \left[\exp\left(-y^+/11\right) + \left(0.33y^+ - 1\right) \exp\left(-0.33y^+\right) \right] \right\}$$

is computed in accordance with *Klausner* et al. (1993).

Finally, we consider a force never before considered in the literature in the momentum equation describing bubble departure – the drag force caused by the growth of the neighboring bubbles

$$F'_{d,t} = 0.3\pi\rho_2 (D_{1d} \bar{\bar{V}}'_2)^2 . \tag{4.20}$$

For an incompressible liquid denoted with subscript 2 the mass-conservation equation for the liquid velocity gives $u_2 = (dR_1/d\tau)(R_1/r)^2$ in the region between R_1 and $R_{2,inf}$, where $R_{2,inf}$ is half of the average center-to-center spacing. *Zuber* (1963) obtained the volume-average liquid velocity as follows

$$\bar{V}'_2 = \frac{1}{R_{2,\inf} - R_1} \int_{R_1}^{R_{2,\inf}} R_1^2 \frac{dR_1}{d\tau} \frac{dr}{r^2} = \frac{R_1}{R_{2,\inf}} \frac{dR_1}{d\tau} = 2R_1 \frac{dR_1}{d\tau} / D_{2,\inf} = B^2 / D_{2,\inf} , \tag{4.21}$$

where

$$B^2 = 2R_1 \frac{dR_1}{d\tau} . \tag{4.22}$$

for n = 1/2. For the idealized case of a triangular array at a plane surface, the influence circle has a diameter $D_{2,\inf} = \left(\frac{2}{\sqrt{3}} \frac{1}{n''_{1w}} \right)^{1/2} \approx 1.074 / n_{1w}^{1/2}$, but the experimentally observed distribution of the active nucleation sites is a *Poisson* distribution *Gaertner* (1963), *Sultan* and *Judd* (1978), *Wang* and *Dhir* (1993). It gives an average distance between two neighboring bubble sites of

$$\left(D_{2,\inf}^2 \right)^{1/2} = \left(\frac{1}{\pi n''_{1w}} \right)^{1/2} \approx 0.56 / n_{1w}^{1/2} . \tag{4.23}$$

More recent measurements by *Wang* and *Dhir* (1993) give $\left(D_{2,\inf}^2 \right)^{1/2} \approx 0.84 / n_{1w}^{1/2}$. For the purpose of the data comparisons we use in this work Eq. (4.23). The *time-averaged fluctuation velocity* (called sometimes microconvection velocity) during $\Delta\tau_w + \Delta\tau_d$ is therefore

$$\overline{\overline{V}}_2' = \left(B^2/D_{2,\inf}\right)\frac{\Delta\tau_d}{\Delta\tau_w+\Delta\tau_d} = B^2\frac{\Delta\tau_d}{\Delta\tau_w+\Delta\tau_d}\left(\pi n_{1w}''\right)^{1/2}, \quad (4.24)$$

which is generally applicable to different situations of bubble production at walls such as boiling, flashing, and gas injection through perforated plates. Note that nucleation at inner and outer walls of a cylinder with diameter D_h is characterized by

$$D_{2,\inf} = \left(\frac{1}{\pi n_{1w}''}\right)^{1/2}\left(1 \pm D_{1d}/D_h\right), \quad (4.23a)$$

where the minus sign is valid for the inner surface. Consequently,

$$\overline{\overline{V}}_2' = \left(B^2/D_{2,\inf}\right)\frac{\Delta\tau_d}{\Delta\tau_w+\Delta\tau_d} = B^2\frac{\Delta\tau_d}{\Delta\tau_w+\Delta\tau_d}\left(\pi n_{1w}''\right)^{1/2}\Big/\left(1 \pm D_{1d}/D_h\right)$$

$$= D_{1d}^2 f_{1w}\left(\pi n_{1w}''\right)^{1/2}\Big/\left(1 \pm D_{1d}/D_h\right). \quad (4.24a)$$

Consequently, with increased wall superheating, bubbles in a pipe interact more strongly than bubbles at a plane, and bubbles at an outer cylinder wall interact less intensively than bubbles at a plane. The higher the value of D_{1d}/D_h the stronger is the influence of the geometry. For large values of superheating, $\Delta\tau_{1d}/\left(\Delta\tau_{1w}+\Delta\tau_{1d}\right)$ approximates 1. For small values, this force is not important. Thus, the approximation of the exact equation

$$\overline{\overline{V}}' = B^2\left(\pi n_{1w}''\right)^{1/2} \quad (4.25)$$

can be used to compute the bubble departure diameter. Now we compute the bubble inclination angle using assumption 6

$$\theta_0 = \arctan\left[\left(F_{b,n} - F_{d,n} + F_{1,n}\right)\Big/\left(F_{b,t} + F_{d,t} + F_{d,t}'\right)\right], \quad (4.26)$$

and write the static force balance along the axis of the bubble divided by π

$$\frac{1}{6}D_{1d}^3\left(\rho_2-\rho_1\right)g\sin\left(\varphi+\theta_0\right) + \frac{1}{2}\rho_2\left(D_{1d}V_{21d}\right)^2 c_{lift}\sin\theta_0$$

$$+0.3\rho_2\left(c_{wall}D_{1d}V_{21d}\right)^2\cos\theta_0 + 0.3\rho_2\left(D_{1d}\overline{\overline{V}}_2'\right)^2\cos\theta_0$$

$$= D_{1c}^*\sigma + 3\eta_2 c_{form}B^2\left[1+0.1\left(c_{form}B^2\rho_2/\eta_2\right)\right]^{3/4}\sin\theta_0$$

or simply

$$\boxed{\left(D_{1d}/D_{1d,nc}\right)^3 + \left(D_{1d}/D_{1d,fc}\right)^2 = 1} \quad (4.27)$$

where

$$D_{1d,nc} = \left\{ 6A / \left[\left(\rho_2 - \rho_1 \right) g \sin\left(\varphi + \theta_0 \right) \right] \right\}^{1/3}, \tag{4.28}$$

$$D_{1d,fc} = \left\{ A \Big/ \left[\rho_2 V_{21d}^2 \left(\frac{1}{2} c_{lift} \sin\theta_0 + 0.3 c_{wall}^2 \cos\theta_0 \right) + 0.3 \rho_2 \overline{\overline{V}}_2'^2 \cos\theta_0 \right] \right\}^{1/2}, \tag{4.29}$$

and

$$A = D_{1c}^* \sigma + 3\eta_2 c_{form} B^2 \left[1 + 0.1 \left(c_{form} B^2 \rho_2 / \eta_2 \right) \right]^{3/4} \sin\theta_0 . \tag{4.30}$$

Note that the inclination angle for forced convection is a real one. For pool boiling it is the cone angle within which the fluctuation of the spatial inclination occurs. The angular position within this angle changes chaotically. That is why one experimentally sees fluctuation but never inclination of the vapor column. It explains the well-known oscillating of the bubble columns. Note that the bubbles at a surface never start and grow simultaneously – such an idealized picture should give zero interaction force, which is never the real case.

Obviously, we have two transcendental equations with two unknowns D_{1d} and θ_0 which are solved by iteration. It is important to note that after each iteration-step the new inclination is computed as an average value between the previous and the new computed values. This ensures a stable algorithm that converges to the desired accuracy after a maximum of seven iterations. For low superheating and low flow velocities, D_{1d} approximates $D_{1d,nc}$, and for large superheating and large flow velocities, D_{1d} approximates $D_{1d,fc}$. It is important to note that the surface tension and the latent heat of vaporization are functions of the wall temperature for boiling and of the liquid temperature for flashing.

Note that the contribution of the macroconvection in boiling heat transfer is generally accepted. Consequently we have to take it into account also in the momentum balance.

4.3 Comparison with experimental data

We have confined our attention to water data only because this model is operating in computer code IVA in all of its variants, which has water as one of the three flow components, Kolev (1993). For further information on this code see the references given in Kolev (1993). The prediction of bubble departure diameter is a part of the analytical modeling of transient sources and sinks of bubbles and droplets in addition to the information presented in Kolev (1993), which is used in IVA3. Figure 4.4 presents a comparison of the prediction of the new theory with the already-discussed data for saturated water pool boiling at a horizontal surface. Without considering the mutual bubble interaction the trend of the curve for superheating is lower than 15 K continuous to an asymptotic value of about 5 mm as shown in Fig. 4 in Kolev (1994). The line on Fig. 4.4 represents the new model considering the mutual bubble interaction.

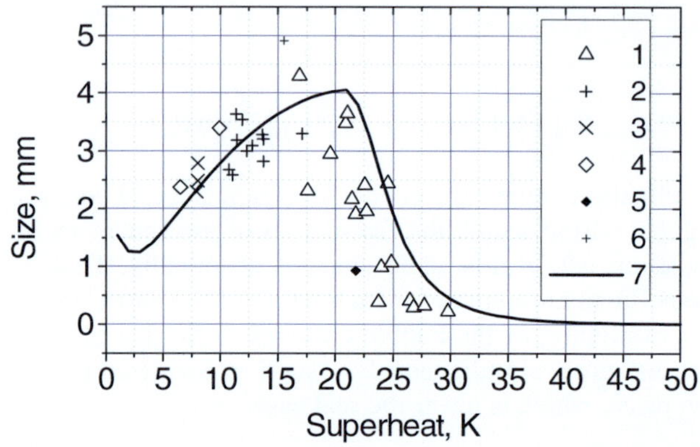

Fig. 4.4 Bubble departure diameter as a function of superheating. Saturated water pool boiling at 0.1 MPa pressure. Data 1: *Gaertner* and *Westwater* (1960), 2 *Gaertner* (1965), 3 *Tolubinski* and *Ostrovsky* (1966), 4 *Siegel* and *Keshock* (1964), 5 *van Stralen* et al. (1975), 6 *Roll* and *Mayers* (1964), 7 IVA5 model with bubble interactions

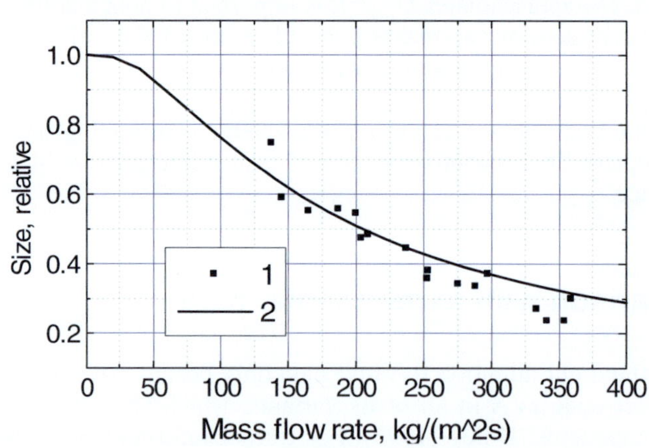

Fig. 4.5 Bubble departure diameter as function of mass flow rate. Saturated water flow boiling at 0.1 MPa pressure. $D_{hy} = 0.019\ m$, $T_w - T_2 = 15$ K. 1 Data of *Koumoutsos* et al. (1968), 2 IVA5 model without bubble interaction

The particle number density for the *Gaertner* and *Westwater* (1960) data, presented in Fig. 1.3 and approximated by Eqs. (1.110) and (1.112) is used for this comparison. In the original publication, *Gaetner* and *Westwater* (1960) called the bubble departure diameter "lift off diameter", and in a later work by *Gaetner* (1963) it is called "bubble column stem" diameter. However, the proportionality constant between both diameters is around one. Note that the *Gaetner* and *Westwater* data for particle number density and bubble departure diameter as a function of the wall superheating are confirmed also by the later measurements by *Iida* and *Kobayasi* (1970), who measured particle number density and the thickness of the thermal boundary layer for pool boiling. Figure 12.1 presents the nucleation-site density as a function of superheating also by other authors *Gaertner* (1965), *Labuntsov* (1974), *Yamagata* et al. (1955), *Jakob* (1932), *Cornwell* and *Brown* (1978), *van Stralen* and *Cole* (1979), *Rallis* and *Jawurek* (1964). This is done to classify the *Gaertner* and *Westwater* data as the data giving the lowest possible nucleation-site density for saturated pool boiling of water. The consideration of the mutual bubble interaction has a striking effect that can not be overlooked. This interaction is responsible for the decrease of the bubble diameter when the nucleation-site density exceeds some value, that is, when the wall superheating exceeds some value. This regime causes a change of the character of the *Nukiama* diagram containing data for saturated water pool boiling at atmospheric pressure on polished surfaces *Gaertner* (1965), *van Stralen* and *Cole* (1979), *Yamagata* et al. (1955), *Sultan* and *Judd* (1978), *Wiebe* (1970), *Gaertner* and *Westwater* (1960), *Borishanskii* et al. (1961), *Jakob* and *Fritz* (1931), *Cornwell* and *Brown* (1978), *Vachon* et al. (1968), *Nishikawa* et al. (1984). As shown in Fig. 4.7 and is in fact quite different from boiling with negligible interaction, which in the boiling literature is called the regime of isolated bubbles. Our theory predicts a smooth transition between the two regimes, which is the expected behavior in nature.

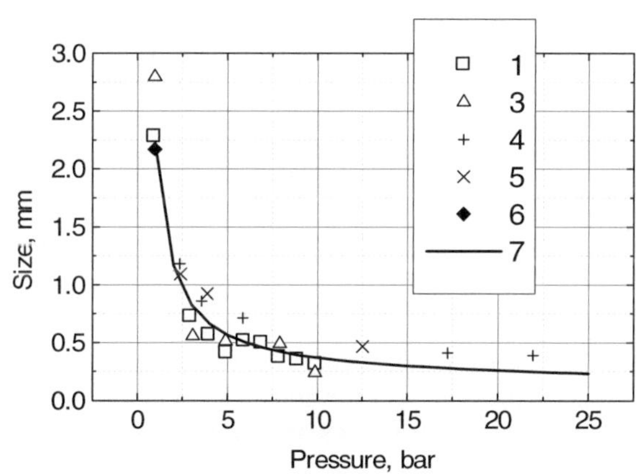

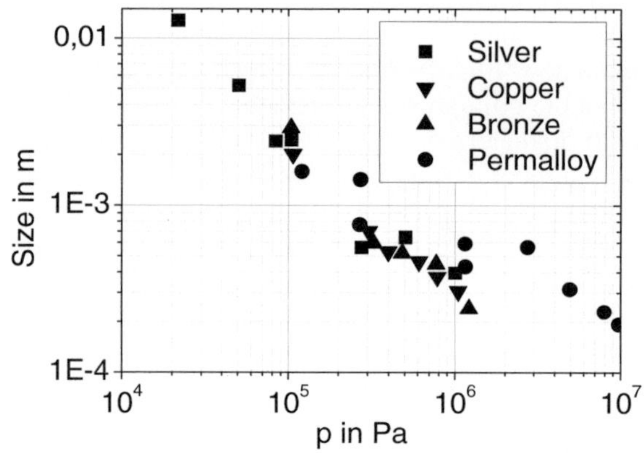

Fig. 4.6. Bubble departure diameter as a function of pressure. a) Saturated water pool boiling, superheat = 7.2 K. Data of *Tolubinsky* and *Ostrovsky*: 1 Permalloy, 2 brass, 3 copper. Data of *Semeria*: 4 wire D = 0.8 mm, 5 plate. 6 Model IVA5 with bubble interaction; b) Water at horizontal surface, *Labuntzov* et al. (1966), *Tolubinsky* and *Ostrovsky* (1966)

Shoji et al. confirmed again in 2005 by carefully performing experiments that decreasing of the spacing between two neighboring nucleation sites decreases the bubble departure diameter definitely and vice versa: the authors found that for $D_{2,\inf} / D_{1d} > 3$ the influence is almost negligible.

In Fig. 4.5 we compare the prediction of our theory with *flow boiling* data given in *Koumoutsos* et al. (1968). On the abscissa is entered the bubble departure diameter divided by the bubble departure diameter for pool boiling. The trend of the data is reasonably reproduced. Figure 8 in *Kolev* (1994) shows the same data compared also with predictions using theories of other authors *Levy* (1967), *Yang* and *Weisman* (1991), *Jones* (1992a), *Jones* (1992b) and with prediction of our theory neglecting the normal force components. We find from this comparison that neglecting the normal force component is allowed for mass flow rates larger than 150 kg/m^2 s. The theory of *Jones* (1992) gives unbounded values for small mass flow rate, and the *Yang* and *Weisman* modification *Yang* and *Weisman* (1991) of *Levy*'s theory *Levy* (1967) underpredicts the data as shown in Fig. 8 in *Kolev* (1994).

Figure 4.6 gives a comparison with data at elevated pressures reported in Refs. *Tolubinsky* and *Ostrovsky* (1966), *Semeria* (1962). In this case also, the trend of the data is properly reproduced by our theory.

Shoji et al. (2005) demonstrated that for cylindrical cavity serving as nucleation sites the cavity diameter (5 to 100μm) does not influence the bubble departure sizes. But the decrease of the depth of the cavity (80, 40, 20 μm) decreases the wall superheat at constant power and reduces the bubble departure diameter: 8.5 K 1.85 mm, 7.5 K, 1.5 mm, 5.8 K, 1.3 mm. 4 W distributed on a circle with 12 mm was the heat source in this experiment. This slope corresponds to those obtained by other authors in this temperature region, as shown in Fig. 4.4.

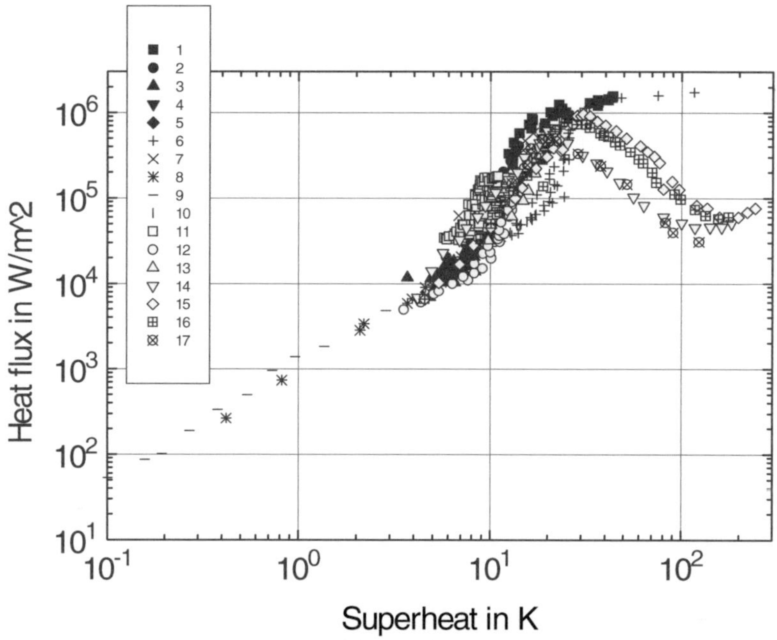

Fig. 4.7 Heat flux as a function of superheat. Saturated water at 0.1 MPa. 1) *Gaertner* 1965, 4/0 polished copper, 2) *Kurihara* and *Myers* 1960, 4/0 polished copper, 3) *Yamagata* et al. 1955, fine polished brass, 4) *Sultan* and *Judd* 1978, diamond grid 600 polished copper, 5) *Wiebe* 1970, diamond 600 polished copper, 6) *Gaertner* and *Westwater* 1960, 4/0 polished copper, 20 p.c. nickel salt-water solution, 7) *Borishanskii* et al. 1961, steel, 5 to 8 p.c. error, 8) *Fritz* 1935, polished steel, 9) *Jakob* and *Linke* 1933, polished steel, 10) *Cornwell* and *Brown* 1978, 4/0 polished cooper, 11) *Vachon* et al. 1968, emery grid 600 polished 304 stainless steel, 12) *Nishikawa* et al. 1984, emery grid No.0/10 polished copper, 13) *Rallis* and *Jawurek* 1964, nickel wire, 14) *Wang* and *Dhir* (1993) nuclide and transition boiling. Contact angle 90°, 15) *Wang* and *Dhir* (1993) nuclide and transition boiling. Contact angle 18°, 16) *Wang* and *Dhir* (1993) nuclide and transition boiling. Contact angle 35°, 17) *Wang* and *Dhir* (1993) nuclide and transition boiling. Contact angle 35°. Liaw data

From the above discussion the following conclusions are in order:

1. The mutual interaction of the growing and departing bubbles causes significant shear at the surface and forces the bubbles to detach at an earlier stage of their growth.

2. The proposed theory successfully describes quantitatively the natural transition between the isolated bubble boiling and boiling with mutual bubble interaction with changing temperature difference.

4.4 Significance

The proposed method can be used in various fields of engineering practice, a few of them being

1. Flashing in channel flow due to depressurization and critical flow – Chapter 17.
2. Boiling in pools and channels – Chapter 16.
3. Modeling of multiphase flow in system computer codes.
4. Reexamination of all approaches for analytical description of boiling heat transfer using the bubble departure diameter as an input variable.
5. Reexamination of all approaches for analytical description of bubble diameter produced by gas injection into a liquid through perforated plates – Section 10.7.

4.5 Summary and conclusions

The *tangential shear* due to the volume- and time-averaged *pulsation velocity* caused by the thermally controlled bubble growth and successive cyclic departure is found to be responsible for the natural switching from the regime of *isolated* bubble growth and departure into the regime of *mutually interacting* bubble growth and departure as the temperature difference between the bubble interface and the surrounding liquid increases. The proposed theoretical model, based only on the first principles, agrees well with the experimental data for boiling water to which it was compared. The new model, which successfully predicts the isolated and the mutually interacting bubble departure sizes, and the quantitative description of the natural transition between the two regimes are the essentially new features of the work presented in this chapter.

4.6 Extension of the theory to subcooled liquids

The theory presented above successfully describes the bubble departure in saturated liquid. In this section we will extend the theory to subcooled boiling following the already introduced method of combined thermal analysis with force equilibrium.

Ünal proposed in 1976 to consider the subcooled boiling bubble growth as superposition of evaporation feed by the conduction from the wall and partial condensation at the bubble top. Using the imagination of *Zuber* (1961) that the superheated microlayer is created before the explosion of the nucleus and neglecting the nucleation diameter as a small compared to any later bubble size after the initiation we can write the following equation for the growth of the bubble mass

$$\frac{d}{d\tau}\left(\rho''\frac{\pi}{6}D_1^3\right)=\frac{\pi D_1^2}{\Delta h}\left[-\dot{q}_2''^{1\sigma}-c_1\left(h_{2,w=0}^{1\sigma}+h_{2,w\neq 0}^{1\sigma}\right)\left(T'-T_2\right)\right], \tag{4.31}$$

which for constant vapor density results in

$$\frac{dR_1}{d\tau}=-\frac{\dot{q}_2''^{1\sigma}}{\rho''\Delta h}-c_1\left(h_{2,w=0}^{1\sigma}+h_{2,w\neq 0}^{1\sigma}\right)\frac{\left(T'-T_2\right)}{\rho''\Delta h}. \tag{4.32}$$

For bubble growth at a wall the expression obtained by *Labuntsov* et al. (1964) in 1964 can be used

$$-\frac{\dot{q}_2''^{1\sigma}}{\rho''\Delta h}=\frac{B_{\sup}}{2\sqrt{\tau}}, \tag{4.33}$$

where

$$B_{\sup}=2.5\sqrt{2a_2\beta Ja_{2w}}\,, \tag{4.34}$$

$Ja_{2w}=\dfrac{\rho_2 c_{c2}\left(T_w-T'\right)}{\rho''\Delta h}$, and $\beta\approx 6$. We assume that the disc with a cross-section $\pi D_1^2/4$ receive the evaporation mass flow rate corresponding to $\dot{q}_2''^{1\sigma}$. The constant c_1 reflects the part of the bubble surface πD_1^2 contacting the subcooled liquid e.g., $c_1 = 1/2$. The condensation is assumed to be driven by a superposition of molecular heat conduction

$$h_{2,w=0}^{1\sigma}\frac{\left(T'-T_2\right)}{\rho''\Delta h}=Ja_{2,sub}\sqrt{\frac{1}{\pi}\frac{a_2}{\tau}}, \tag{4.35}$$

$$Ja_{2,sub}=\frac{\rho_2 c_{p2}\left(T'-T_2\right)}{\rho''\Delta h}, \tag{4.36}$$

Forschluetz and *Chao* (1965), and eddies renewal due to turbulence

$$\frac{h_{2,w\neq 0}^{1\sigma}D_h}{\lambda_2}=0.228\,\mathrm{Re}_2^{0.8}\,\mathrm{Pr}_2^{1/2}\left(\Phi_{20}^2/\alpha_2\right)^{1/4}, \tag{4.37}$$

Avdeev (1986). Replacing the heat flux components in the mass-conservation equation results in

$$\frac{dR_1}{d\tau}=\frac{B}{2\sqrt{\tau}}-A, \tag{4.38}$$

where

$$B = B_{\sup} - \sqrt{\frac{a_2}{\pi}} Ja_{2,sub} , \tag{4.39}$$

$$A = 0.228 \frac{1}{2} \frac{a_2}{D_h} Ja_{2,sub} \, \mathrm{Re}_2^{0.8} \, \mathrm{Pr}_2^{1/2} \left(\Phi_{20}^2 / \alpha_2 \right)^{1/4} . \tag{4.40}$$

Note that there is a driving mechanism for a bubble growth in subcooled liquid only if

$$2.5\sqrt{2\beta\pi Ja_{2w}} > Ja_{2,sub} . \tag{4.41}$$

The radius of the bubble as a function of time is then

$$R_1 \approx R_{1c} + B\sqrt{\tau} - A\tau . \tag{4.42}$$

The radius possesses a maximum

$$R_{1m} \approx R_{1c} + \frac{1}{4} \frac{B^2}{A} \tag{4.43}$$

at

$$\tau_{1m} = \frac{1}{4} \left(\frac{B}{A} \right)^2 . \tag{4.44}$$

If the maximum of the diameter is smaller than the bubble departure diameter the bubble will stay at the surface, else the bubble will depart with the bubble departure diameter. The bubble growth equation can be rewritten in dimensionless form as proposed by *Zuber* in 1961

$$\frac{R_1 - R_{1c}}{R_{1m} - R_{1c}} = 2\sqrt{\frac{\tau}{\tau_{1m}}} - \frac{\tau}{\tau_{1m}} . \tag{4.45}$$

Note that the form of the equation is as obtained by *Zuber* but not the contents of the different terms. As expected for a given radius there are two solutions for the time

$$\frac{\tau}{\tau_{1m}} = \left(1 \pm \sqrt{1 - \frac{R_1 - R_{1c}}{R_{1m} - R_{1c}}} \right)^2 . \tag{4.46}$$

The center-of-mass velocity corresponding to this bubble growth mechanism is then

$$V_{1cm} = \frac{dR_1}{d\tau} = \frac{R_{1m} - R_{1c}}{\tau_{1m}} \left[\left(\frac{\tau}{\tau_{1m}} \right)^{-1/2} - 1 \right] . \tag{4.47}$$

The temporal bubble acceleration is

$$\frac{\partial V_{1cm}}{\partial \tau} = -\frac{1}{2}\frac{R_{1m} - R_{1c}}{\tau_{1m}^2}\left(\frac{\tau}{\tau_{1m}}\right)^{-3/2}. \tag{4.48}$$

The total bubble acceleration is

$$\frac{\partial V_{1cm}}{\partial \tau} + V_{1cm}\frac{V_{1cm}}{R_1 - R_{1c}} = \frac{R_{1m} - R_{1c}}{\tau_{1m}^2}\left\{\frac{R_{1m} - R_{1c}}{R_1 - R_{1c}}\left[\left(\frac{\tau}{\tau_{1m}}\right)^{-1/2} - 1\right]^2 - \frac{1}{2}\left(\frac{\tau}{\tau_{1m}}\right)^{-3/2}\right\}. \tag{4.49}$$

Neglecting the virtual mass force we apply now the equations (4.26) and (4.27) by using the already-described mechanism of the bubble growth and collapse. The result for horizontal plane at 1 bar pressure and varying subcooling is presented in Fig. 4.8.

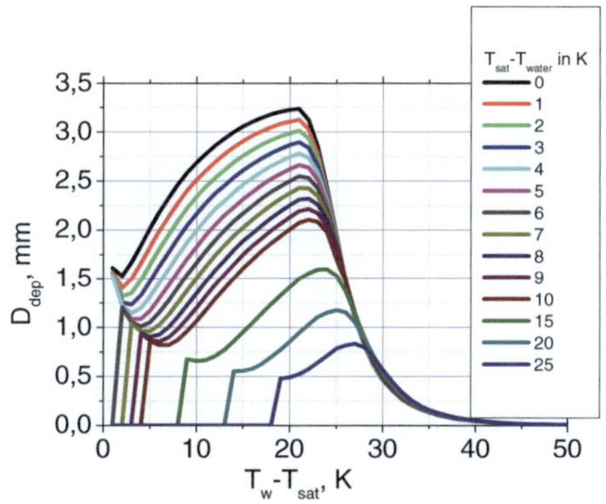

Fig. 4.8 Bubble departure diameter at horizontal surface without flow as function of the wall superheating at 1 bar. Parameter – water subcooling. Stagnant liquid

For zero subcooling we reproduce well the experimental data as already discussed in the pervious section. For increasing subcooling the bubble departure diameter decreases. The zero values are artificially set to indicate in the graph that in the particular combination of superheating and subcooling there are no conditions for bubble departure at all.

4.7 Influence of the wall material

Morin (1964) measuring the temperature in the wall just below the growing bubble found that there is a cyclic decrease and increase corresponding to the growing cycle.

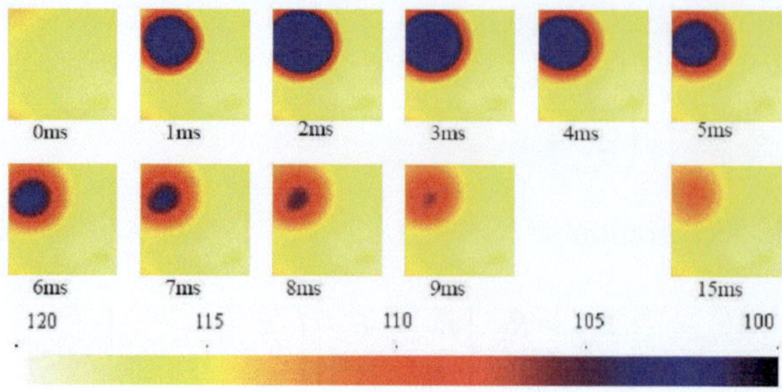

Fig. 4.9 Wall-temperature distribution under the growing bubble at 100 kW/m², *Golobic* et al. (2006)

Better heat-conduction materials react faster on the changing heat exhausting conditions due to evaporation of the liquid around the bubble contact line with the wall. 42 years later *Golobic* et al. (2006) reported two-dimensional temperature distribution around the nucleation center as given in Fig. 4.9. Therefore, the theory reported above can farther be improved by considering this phenomenon.

Nomenclature

Latin

a_2	liquid thermal diffusivity, m^2/s
B^2	$= 2R_1 dR_1 / d\tau$, m^2/s
D_1	bubble diameter, m
D_{1c}	critical diameter, m
D_{1d}	bubble departure diameter, m
$D_{2,inf}$	$= 2R_{2,\mathrm{inf}}$, m
$D_{1d,fc}$	bubble departure diameter for strongly predominant forced convection, $V_{21d} >> \overline{\overline{V_2'}}$, m
$D_{1d,nc}$	bubble departure diameter for natural circulation, $V_{21d} << \overline{\overline{V_2'}}$, m
D	total differential, dimensionless
G	mass flow rate, kg/(m^2 s)
g	gravitational acceleration, m/s^2
h'	saturated liquid specific enthalpy, J/(kg K)
h''	saturated steam specific enthalpy, J/(kg K)

Ja	$= \dfrac{[T_2 - T'(p)]c_{p2}}{h''(p) - h'(p)} \dfrac{\rho'}{\rho''}$, *Jacob* number, dimensionless
n''_{1w}	active nucleation-site density, $1/m^2$
p	pressure, Pa
p_c	critical pressure, Pa
R_1	bubble radius, m
$R_{2,inf}$	half of the average center-to-center spacing, m
$T'(p)$	saturation temperature at system pressure p, K
T_2	liquid temperature, K
T_w	wall temperature, K
V_{1cm}	center-of-mass bubble velocity at the moment of detachment, m/s
V_{21d}	tangential velocity in the boundary layer of thickness D_{1d}, m/s
u_2	liquid velocity in the region between R_1 and $R_{2,inf}$ m/s
$\bar{V}'_2$	volume-averaged fluctuation velocity, m/s
$\bar{\bar{V}}'_2$	time- and volume-averaged fluctuation velocity over $\Delta\tau_w + \Delta\tau_d$, m/s

Greek

∂	partial differential, dimensionless
$\Delta\rho_{21}$	$= \rho_2 - \rho_1$, kg/m³
$\Delta\tau_d$	time need from the initiation of bubble with critical size to the bubble departure from the wall, s
$\Delta\tau_w$	delay time, s
η_2	dynamic viscosity of liquid, kg/(m s)
φ	angle between the flow direction and the upwards-directed vertical, – $\mathbf{n}_g$, rad
ρ_1	gas density, kg/m³
ρ_2	liquid density, kg/m³
ρ''	saturated steam density, kg/m³
ρ'	saturated liquid density, kg/m³
$\bar{\rho}$	$= (\rho' - \rho'')/\rho$, dimensionless
σ	surface tension, N/m
τ_{2w}	$= \dfrac{\xi}{8} G^2/\rho$ shear stress, N/m²
τ	time, s
θ_0	angle between the bubble axis and the wall, rad

References

Avdeev, A.A.: Growth and condensation velocity of steam bubbles in turbulent flow. Teploenergetika 1, 53–55 (1986) (in Russian)

Borishanskii, V., Bobrovich, G., Minchenko, F.: Heat transfer from a tube to water and to ethanol in nucleate pool boiling. In: Kutateladze, S.S. (ed.) Symposium of Heat Transfer and Hydraulics in Two-Phase Media, Gosenergoizdat, Moscow (1961)

Cole, R., Rohsenow, W.M.: Correlation of bubble departure diameters for boiling of saturated liquids. In: Chem. Eng. Prog. Symp. Ser., vol. 65(92), pp. 211–213 (1969)

Cornwell, K., Brown, R.D.: Boiling surface topology. In: Proc. 6th Int. Heat Transfer Conf., Toronto, vol. 1, pp. 157–161 (1978)

Forschuetz, L., Chao, B.T.: On the mechanics of vapor bubble collapse. Trans. ASME, J. Heat Transfer, 209–220 (1965)

Fritz, W.: Berechnung des maximalen Volumens von Dampfblasen. Phys. Z. 36(11), 379–384 (1935)

Gaertner, R.F.: Distribution of active sites in the nucleate boiling of liquids. In: Chem. Eng. Prog. Symp. Series, vol. 59(41), pp. 52–61 (1963)

Gaertner, R.F.: Photographic study of nucleate pool boiling on a horizontal surface. Transaction of the ASME, J. Heat Transfer, 17–29 (February 1965)

Gaertner, R.F., Westwater, J.W.: Population of active sites in nucleate boiling heat transfer. In: Chem. Eng. Prog. Symp. Ser., vol. 30, pp. 39–48 (1960)

Golobic, I., Petkovsek, J., Baselj, M., Papez, A., Kenning, D.B.R.: Experimental determination of the transient wall temperature distribution close to growing vapor bubbles. In: ECI International Conference on Boiling Heat Transfer Spoleto, May 7-12 (2006)

Golorin, V.S., Kol'chugin, B.A., Zakharova, E.A.: Investigation of the mechanism of nucleate boiling of ethyl alcohol and benzene by means of high-speed motion-picture photography. Heat Transfer-Sov. Res. 10(4), 79–98 (1978)

Hsu, Y.Y., Graham, R.W.: Transport processes in boiling and two - phase systems. Hemisphere Publishing Corporation, McGraw-Hill Book Company, Washington, New York (1976)

Iida, Y., Kobayasi, K.: An experimental investigation on the mechanism of pool boiling phenomena by a probe method. In: 4th Int. Heat Transfer Conf., Paris – Versailles, vol. 5(3 B13), pp. 1–11 (1970)

Ishii, M., Zuber, N.: Relative motion, interfacial drag coefficients in dispersed two - phase flow of bubbles, drops and particles. In: AIChE 71st Ann. Meeting, Miami, Paper 56a (1978)

Jakob, M.: Kondensation und Verdampfung. Zeitschrift des Vereins deutscher Ingenieure 76(48), 1161–1170 (1932)

Jakob, M., Fritz, W.: Forsch. Ing.-Wes. 2, 435 (1931)

Jakob, M., Linke, W.: Der Waermeuebergang von einer waagerechten Platte an siedendes Wasser. Forsch. Ing. Wes. 4, 75–81 (1933)

Jensen, M.K., Memmnel, G.J.: Evaluation of Bubble Departure Diameter Correlations. In: Proc. Eighth Int. Heat Transfer Conf., vol. 4, pp. 1907–1912 (1986)

Jones, O.C.: Nonequilibrium phase change 1. Flashing inception, critical flow, and void development in ducts. In: Lahey Jr., R.T. (ed.) Boiling Heat Transfer, pp. 189–234. Elsevier Science Publishers B.V., New York (1992a)

Jones, O.C.: Nonequilibrium Phase Change 2. Relaxation models, general applications, and post heat transfer. In: Lahey Jr., R.T. (ed.) Boiling Heat Transfer, pp. 447–482. Elsevier Science Publishers B.V., New York (1992b)

Klausner, J.F., Mei, R., Bernard, D.M., Zeng, L.Z.: Vapor bubble departure in forced convection boiling. Int. J. Heat Mass Transfer 36(3), 651–662 (1993)

Kocamustafaogullari, G., Ishii, M.: Interfacial area and nucleation site density in boiling systems. Int. J. Heat Mass Transfer 26(9), 1377–1389 (1983)

Kolev, N.I.: The code IVA3 for modeling of transient three-phase flows in complicated 3D geometry. Kerntechnik 58(3), 147–156 (1993a)

Kolev, N.I.: Fragmentation and coalescence dynamics in multi-phase flows. In: Experimental Thermal and Fluid Science, vol. 6, pp. 211–251. Elsevier, New York (1993b)

Kolev, N.I.: The influence of mutual bubble interaction on the bubble departure diameter. In: Experimental Thermal and Fluid Science, vol. 8, pp. 167–174. Elsevier, New York (1994)

Kolev, N.I.: IVA3 NW: Computer code for modeling of transient three phase flow in complicated 3D geometry connected with industrial networks. In: Proc. of the Sixth International Topical Meeting on Nuclear Reactor Thermal Hydraulics, Grenoble, France, October 5-8 (1993)

Koumoutsos, N., Moissis, R., Spyridonos, A.: A study of bubble departure in forced-convection boiling. J. Heat Transfer, Trans. ASME, 223–230 (May 1968)

Kurihara, H.M., Myers, J.E.: The effect of superheat and surface roughness on boiling coefficients. AIChE J. 6(1), 83–91 (1960)

Kutateladze, S.S., Gogonin, I.I.: Growth rate and detachment diameter of a vapour bubble in free convection boiling of saturated liquids. High Temperature 17, 667–671 (1979)

Labuntsov, D.A.: State of the art of the nucleate boiling mechanism of liquids. Heat Transfer and Physical Hydrodynamics, 98–115 (1974) (in Russian)

Labuntsov, D.A., Kol'chugin, B.A., Golovin, V.S., et al.: Teplofiz. Vys. Temp. 3, 446–453 (1964a)

Labuntsov, D.A., Kol'chugin, B.A., Golovin, V.S., Saharova, E.A., Vladimirova, L.N.: Teplofiz. Vys. Temp. 2(3) (1964b)

Liaw, S.P.: Experimental and analytical study of nucleate and transition boiling on a vertical surface. Ph. D. dissertation, University of California, Los Angeles, USA (1988)

Levy, S.: Forced convection subcooled boiling prediction of vapor volume fraction. Int. J. Heat Mass Transfer 10, 951–965 (1967)

Moalem, D., Yijl, W., van Stralen, S.J.D.: Nucleate boiling at a liquid-liquid interface. Letters Heat and Mass Transfer 4, 319–329 (1977)

Morin, R.: Wall temperature fluctuations during bubble generation in boiling. In: Borishanskii, V.M., Paleev, I.I. (eds.) Convective Heat Transfer in Two-phase and One-phase Flows, Energiya, Moskva (1964)

Nishikawa, K., Fujita, Y., Uchida, S., Ohta, H.: Effect of surface configuration on nucleate boiling heat transfer. Int. J. Heat Mass Transfer 27(9), 1559–1571 (1984)

Rallis, C.J., Jawurek, H.H.: Latent heat transport in saturated nucleate boiling. Int. J. Heat Transfer 7, 1051–1068 (1964)

Reichardt, H.: Vollständige Darstellung der turbulenten Geschwindigkeiten in glaten Leitungen. Z. angew. Math. Mech., Bd. 31 Nr. 7, S.208–S.219 (1951)

Roll, J.B., Mayers, J.C.: The effect of surface tension on factors in boiling heat transfer. AIChE J. 330–344 (1964)

Ruckenstein, E. (1961) A physical model for nucleate boiling heat transfer from a horizontal surface. Bul. Institutului Politech. Bucaresti 33(3), 79–88 (1961); Appl. Mech. Rev. 16, Rev. 6055 (1963)

Semeria, R.F.: Quelques resultats sur le mechanisme de l'ebullition, 7. J. de l'Hydraulique de la Soc. Hydrotechnique de France (1962)

Shoji, M., Zhang, L., Chatpun, S.: Nucleation site interaction in pool nucleate boiling. In: 6th World Conference on Experimental Heat Transfer, Fluid Mechanics, and Thermodynamics, Matsushima, Miyagi, Japan, April 17-21 (2005)

Siegel, R., Keshock, E.G.: Effects of reduced gravity on nucleate boiling bubble dynamics in saturated water. AIChE J. 10(4), 509–517 (1964)

Sultan, M., Judd, R.L.: Spatial distribution of active sites and bubble flux density. J. Heat Transfer, Trans. ASME 100, 56–62 (1978)

Tolubinsky, V.I., Ostrovsky, J.N.: On the mechanism of boiling heat transfer (vapor bubbles growth rise in the process of boiling in liquids, solutions, and binary mixtures). Int. J. Heat Mass Transfer 9, 1463–1470 (1966)

Ünal, H.C.: Maximum bubble diameter, maximum bubble-growth time and bubble-growth rate during the subcooled nucleate flow boiling of water up to 17.7 MN/m^2. Int. J. Heat Mass Transfer 19, 643–649 (1976)

Vachon, R.I., Tanger, G.E., Davis, D.L., Nix, G.H.: Pool boiling on polished chemically etched stainless-steel surfaces. Trans. ASME, J. Heat Transfer, 231–238 (1968)

van Krevelen, D.W., Hoftijzer, P.J.: Studies of gas-bubble formulation, calculation of interfacial area in bubble contactor. Chem. Eng. Prog. Symp. Ser. 46(1), 29–35 (1950)

van Stralen, S., Cole, R.: Boiling Phenomena. Hemisphere, Washington (1979)

van Stralen, S.J.D., Sluyter, W.M., Sohal, M.S.: Bubble growth rates in nucleate boiling of water at subatmospheric pressures. Int. J. Heat Mass transfer 18, 655–669 (1975)

Voloshko, A.A., Vurgaft, A.V.: Dynamics of vapor-bubble break off under free-convection boiling conditions. Heat Transfer-Sov. Res. 2(6), 136–141 (1970)

Wang, C.H., Dhir, V.K.: Effect of surface wettability on active nucleation site density during pool boiling of water on a vertical surface. ASME J. Heat Transfer 115, 659–669 (1993)

Wiebe, J.R.: Temperature profiles in subcooled nucleate boiling. M. Eng. thesis, Mechanical Engineering Department, McMaster University, Canada (1970)

Yamagata, K., Hirano, F., Nishikawa, K., Matsuoka, H.: Nucleate boiling of water on the horizontal heating surface. Mem. Fac. Engng. Kyushu Univ. 15, 97 (1955)

Yang, J.Y., Weisman, J.: A phenomenological model of subcooled flow boiling in the detached bubble region. Int. J. Multiphase Flow 17(1), 77–94 (1991)

Zeng, L.Z., Klausner, J.F., Mei, R.: A unified model for the prediction of bubble detachment diameters in boiling systems 1. Pool boiling. Int. J. of Heat and Mass Transfer 36(9), 2261–2270 (1993a)

Zeng, L.Z., Klausner, J.F., Bernard, D.M., Mei, R.: A unified model for the prediction of bubble detachment diameters in boiling systems 2. Flow boiling. Int. J. Heat Mass Transfer 36(9), 2271–2279 (1993b)

Zuber, N.: The dynamics of vapor bubbles in non uniform temperature fields. Int. J. Heat Mass Transfer 2, 83–98 (1961)

Zuber, N.: Nucleate boiling. The region of isolated bubbles and the similarity with natural convection. Int. J. Heat Mass Transfer 6, 53–78 (1963)

5. Bubble dynamics in single-component fluid

5.1 Introduction

If a liquid is heated for whatever reason at temperatures higher than the saturation temperature at the local pressure it generates bubbles. Vapor enclosed in a cavity with an interface colder than the saturation temperature at the bubble inside pressure condenses and causes bubble collapse. In principle, the pressure difference between the inside and outside of a bubble cause the mechanical movement resisted by the viscosity and surface tension. Simultaneously, heat and mass transfer at both sites of the interface cause pressure change and then again this pressure change has a feed back on bubble interface motion. After more than 80 years of research we finally understand the phenomena and the equations describing it. As already revived in the previous chapters the literature describing bubble radius as a function of initial parameters and time by solving these equations with different degrees of complexity is huge, mainly because of the attempts to obtain closed analytical solutions for which obtaining simplifications is inevitable. But now, computers are so powerful that direct numerical simulation of the one-dimensional problem is easily and quickly done. The subject of this chapter is to demonstrate such a solution, to relate it to the state-of-the-art and to compare it with experimental data. Finally, a discussion is provided on how to couple such small-scale physics with large-scale physics in multiphase flow computer codes.

5.2 The system of PDEs describing the problem

Consider a symmetric bubble in incompressible liquid. The equations describing the process are:

$$R_1^2 \frac{d^2 R_1}{d\tau^2} + \frac{3}{2} R_1 \left(\frac{dR_1}{d\tau} \right)^2 + 4\eta_2 \frac{dR_1}{d\tau} + \frac{1}{\varepsilon \rho_2} \left[2\sigma_2 + R_1 \left(p_2 - p_1 \right) \right] = 0, \quad (5.1)$$

$$\frac{\partial T_2}{\partial \tau} - a_2 \frac{\partial}{\partial x} \left(\frac{\partial T_2}{\partial x} \right) - 2 \frac{a_2}{r} \frac{\partial T_2}{\partial x} = 0, \quad x = r - R_1(\tau), \quad (5.2)$$

$$p_1 \approx p_{10}\left(\frac{R_{10}}{R_1}\right)^{3\kappa} \exp\left\{\frac{3\kappa\Delta\tau}{R_1}\left[\begin{array}{c} -\left(\frac{1}{\rho_1}-\frac{1}{\rho_2}\right)(\rho w)_2^{1\sigma} \\ +\dfrac{\dot{q}_1^{\prime\prime 2\sigma}-(\rho w)_2^{1\sigma} c_{p1}\left(T'(p_1)-T_1\right)}{\rho_1 c_{p1} T_1} \end{array}\right]\right\} \tag{5.3}$$

$$T_1 \approx T_{10}\left(\frac{p_1}{p_0}\right)^{(\kappa-1)/\kappa} \exp\left\{\frac{\kappa-1}{\kappa}\frac{3\Delta\tau}{R_1 p_1}\left[\dot{q}_1^{\prime\prime 2\sigma}-(\rho w)_2^{1\sigma} c_{p1}\left(T'(p_1)-T_1\right)\right]\right\}, \tag{5.4}$$

$$\frac{dR_1}{d\tau}=\left(\frac{dR_1}{d\tau}\right)_{mom}-\frac{1}{\rho_2}(\rho w)_2^{1\sigma}, \tag{5.5}$$

see *Kolev* (2010).

Equation (5.1) is the momentum equation for an incompressible liquid integrated from the radius of the bubble to infinity. This equation is known as the *Rayleigh–Plesset* equation. Appendix 5.1 gives its derivation.

Equation (5.2) is the energy-conservation equation for a symmetric incompressible liquid in spherical coordinates. The origin of the coordinate system is at the interface of the bubble. The spatial coordinate transformation used to eliminate the convective term is $x = r - R(\tau)$, see Appendix 5.2.

Equation (5.3) is the so-called pressure equation or volume-conservation equation for the bubble. Its derivation is given in Appendix 5.3.

Equation (5.4) is the energy-conservation equation in pseudoentropy form. The derivation is given in Appendix 5.4.

The change of the bubble radius with the time consists of the mechanical component $\left(dR_1/d\tau\right)_{mom} = \left(R_1^{n+1}-R_1\right)/\Delta\tau$ dictated by the force balance (Eq. (5.1)) and by a component depending on the mass transfer as given with Eq. (5.5).

The interfacial heat fluxes are defined as follows. The heat flux from the interface to the vapor is defined by Eq. (5.6). The heat flux from the interface to the liquid is defined by Eq. (5.7). The interfacial energy-jump condition is expressed by Eq. (5.7).

$$\dot{q}_1^{\prime\prime 2\sigma} = htc_1^{2\sigma}\left(T_1^{2\sigma}-T_1\right), \tag{5.6}$$

$$\dot{q}_2^{\prime\prime 1\sigma} = \lambda_2\left(\frac{\partial T_2}{\partial x}\right)_{x=0} = htc_2^{1\sigma}\left(T_2^{1\sigma}-T_2\right), \tag{5.7}$$

$$(\rho w)_2^{1\sigma} = \frac{htc_2^{1\sigma}\left(T_2^{1\sigma} - T_2\right) + htc_1^{2\sigma}\left(T_1^{2\sigma} - T_1\right)}{h_1^{2\sigma} - h_2^{1\sigma}}. \tag{5.8}$$

The gas and liquid site heat-transfer coefficients are formally $htc_1^{2\sigma}$ and $htc_2^{1\sigma}$. Some authors prefer to use the surface temperature as independent variable by enforcing a kinetic condensation and evaporation at the surface using the so-called accommodation coefficients. The reader will find such examples in *Theofanous* et al. (1969a) and *Nigmatulin* et al. (2005). The gradient of the liquid temperature field needed in Eq. (5.7) results from the solution of the liquid energy-conservation equation. The liquid-site heat transfer is essential for single-component systems. The gas-site heat transfer is important for gas bubbles and pressure oscillations not allowing mass transfer. The gas-site heat-transfer coefficient can be computed either by solving the energy-conservation equation for the gas or by means of a not very accurate empirical relation in which

$$Nu_1^{2\sigma} = \frac{htc_1^{2\sigma} 2R_1}{\lambda_1} \tag{5.9}$$

where $Nu_1^{2\sigma}$ = 1 for stagnant bubble. For moving bubble $\left|\Delta V_{12}\right| > 0$, which is not the subject of this consideration, due to the internal circulation the heat transfer improvement is a function of the relative velocity. *Celata* et al. (1991) correlated their experimental data for condensation on droplets with the following expression for the *Nusselt* number,

$$Nu_1^{2\sigma} = \max\left(1,\ 0.53 Pe_1^{*0.454}\right) \tag{5.10}$$

where the definition of the *Peclet* number is used

$$Pe_1^* = \frac{D_3\left|\Delta V_{13}\right|}{a_1}\frac{\eta_1}{\eta_1 + \eta_3}. \tag{5.11}$$

For bubble interaction one can postulate

$$Pe_1^* = \frac{D_1\left|\Delta V_{12}\right|}{a_1}\frac{\eta_2}{\eta_1 + \eta_2} \tag{5.12}$$

instead. Whether the coefficient will remain the same also for bubble internal circulation is not clear. *Nigmatulin* (1990) derived approximate formula to compute the gas-site heat transfer due to surface motion of stagnant bubble:

$$Nu_1^{2\sigma} = \sqrt{Pe} \text{ for } Pe > 100, \tag{5.13}$$

$$Nu_1^{2\sigma} = 10 \text{ for } Pe \le 100, \tag{5.14}$$

where

$$Pe = 12(\kappa - 1)\frac{T_2}{|T_2 - T_1|}\frac{R_1}{a_1}\frac{dR_1}{d\tau} = 6(\kappa - 1)\frac{T_2}{|T_2 - T_1|}\frac{\rho_1 2R_1}{\eta_1}\frac{dR_1}{d\tau}\frac{\eta_1 c_{p1}}{\lambda_1}. \tag{5.15}$$

5.3 Numerical solution method

The *Rayleigh* equation is discretized by using a second-order method

$$R_1^2 \frac{R_1^{n+1} - 2R_1 + R_1^{n-1}}{\Delta\tau^2} + \frac{3}{2}R_1\left(\frac{R_1^{n+1} - R_1^{n-1}}{2\Delta\tau}\right)^2$$

$$+4\eta_2 \frac{R_1^{n+1} - R_1^{n-1}}{2\Delta\tau} + \frac{1}{\varepsilon\rho_2}\left[2\sigma_2 + R_1(p - p_1)\right] = 0 \tag{5.16}$$

or

$$a\left(R_1^{n+1}\right)^2 + bR_1^{n+1} + c = 0, \tag{5.17}$$

where,

$$a = \frac{3}{8}R_1,$$

$$b = R_1^2 + 2\eta_2\Delta\tau - 2aR_1^{n-1},$$

$$c = \left(R_1^2 - 2\eta_2\Delta\tau\right)R_1^{n-1} - 2R_1^3 + a\left(R_1^{n-1}\right)^2 + \frac{1}{\varepsilon\rho_2}\left[2\sigma_2 + (p - p_1)R_1\right]\Delta\tau^2$$

with the solution

$$R_1^{n+1} = \left(-b + \sqrt{b^2 - 4ac}\right)\Big/(2a). \tag{5.18}$$

The liquid energy-conservation equation is solved using a second-order method of *Lax* and *Vendroff*, see in *Richtmyer* and *Morton* (1967), applied as follows. First, an explicit half-time step is made and all temperatures are computed

$$T_{2,i}^{n+1/2} - T_{2,i} - \frac{1}{2}\frac{a_2 \Delta\tau}{\Delta x^2}\left[T_{2,i+1} - 2T_{2,i} + T_{2,i-1} + \frac{\Delta x}{r_i}\left(T_{2,i+1} - T_{2,i-1}\right)\right] = 0. \quad (5.19)$$

With the new temperature values the spatial terms are computed again and a complete time step is then made

$$T_{2,i}^{n+1} - T_{2,i} - \frac{a_2 \Delta\tau}{\Delta x^2}\left[T_{2,i+1}^{n+1/2} - 2T_{2,i}^{n+1/2} + T_{2,i-1}^{n+1/2} + \frac{\Delta x}{r_i}\left(T_{2,i+1}^{n+1/2} - T_{2,i-1}^{n+1/2}\right)\right] = 0. \quad (5.20)$$

The dimensionless time $a_2 \Delta\tau / \Delta x^2$ is called the grid *Fourier* number. $\Delta x / r_i$ indicates the importance of the sphericity close to the bubble. The ordinary partial differential equations (5.3) and (5.4) are solved by the first-order *Euler* method giving the new pressure p_1 and temperatures T_1 inside the bubbles in both steps. Then, the bubble site interface temperature is the saturation temperature at vapor pressure p_1 and at the other end no heat flux is specified: $T_{2,1}^{n+1/2} = T'(p_1)$ and $T_{2,1}^{n+1} = T'(p_1)$ and $T_{2,i_{\max}}^{n+1/2} = T_{2,i_{\max}-1}^{n+1/2}$ and $T_{2,i_{\max}}^{n+1} = T_{2,i_{\max}-1}^{n+1}$. The x-region along which the liquid energy-conservation equation is integrated is associated with the density of the bubbles per cubic meter n_1 as follows

$$\sum \Delta x_i = \left(\frac{3}{4\pi n_1}\right)^{1/3} - R_{10} = R_{10}\left(\alpha_1^{-1/3} - 1\right). \quad (5.21)$$

Actually *Nigmatulin* proposed in 1978 the so-called concept of "cellular medium" considering the processes within the sphere with $R_{cell} = R_{10}\alpha_1^{-1/3}$. This allows taking into account the effective thickness of liquid layer around each bubble. The larger the bubble concentration at a given initial radius is, the smaller the thickness of the effective layer. This leads to different behavior of the bubbles depending on the bubble number density. Note the difference to *Lee* and *Merte* (1996) using $\sum \Delta x_i = 0.002$ m, which does not relate the problem to the multiphase flows.

5.4 Validation of the method

5.4.1 Bubble collapse

I will first validate the method by using the bubble collapse experiments by *Florschuetz* and *Chao* (1965) for water. Given was the initial pressure, subcooling and bubble radius, see Table 5.1.

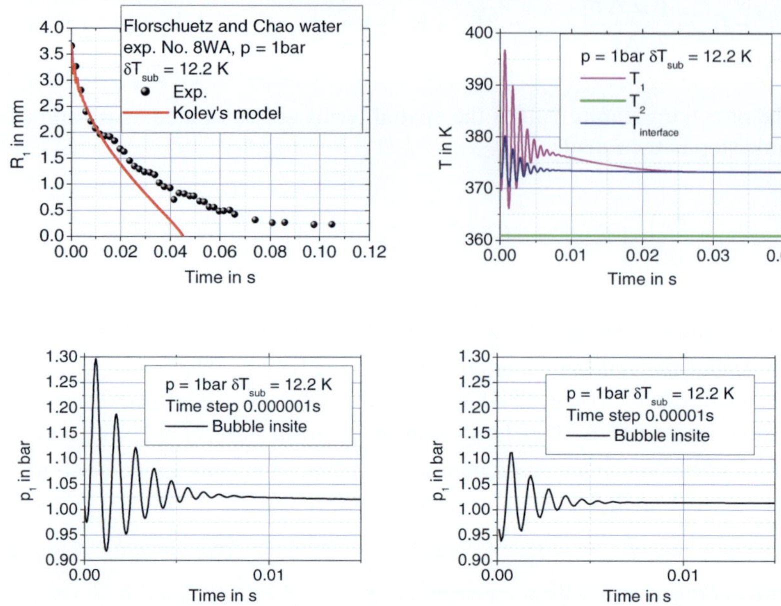

Fig. 5.1 *Florschuetz* and *Chao* (1965) experiment for water: a) Bubble radius as function of time; b) Bubble and interface temperature as a functions of time; c) Pressure inside the bubble as a function of time, time step of integration 0.001 ms; time; d) Pressure inside the bubble as a function of time, time step of integration 0.01 ms

Table 5.1 Initial and boundary conditions for the experiment by *Florschuetz* and *Chao* (1965)

Florschuetz and *Chao*'s Figs.	p in bar	R_{10} in mm	$T'-T_2$ in K
9	1	3.657580	12.2
10	1	3.185503	8.7
11	1	2.983309	8.3
12	1	4.372804	5.6
13	1	3.990081	5.0

Measured was the bubble radius as a function of time. This is also predicted by the method and compared in Figs. 5.1 to 5.5. First, we observe from Fig. 5.1 that there are residual noncondensable gases in the bubble because otherwise the stable final diameter of 0.25 mm can not be explained. Having this in mind, the very good agreement with the initial bubble collapse period and the complete condensation for a single-component system is well understood. The most remarkable feature is that the selected numerical method does not damp the natural oscillations contained in the Rayleigh equation. The inertia leads to overshooting. The surface tension acts like a spring. The viscosity introduces little damping. The accuracy of the integration influences the resolution of the pressure amplitudes as seen from Fig. 5.1. The experiment in Fig. 5.2 also indicates residual gases sufficient to stop the condensation at 0.5 mm. Similar is the situation with the experiment presented on Fig. 5.3 where the final radius is about 0.4 mm.

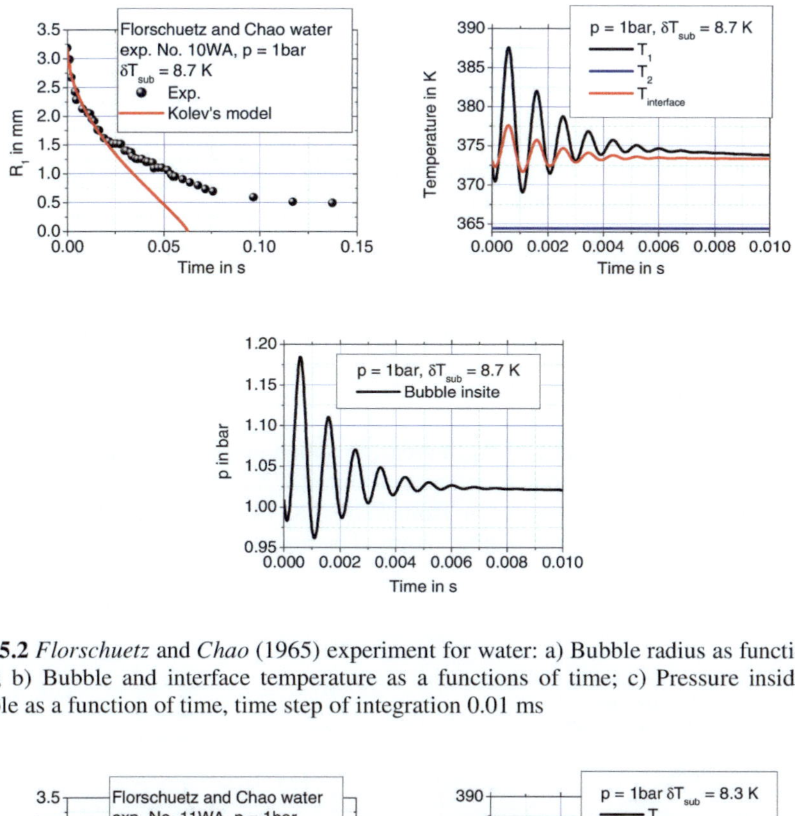

Fig. 5.2 *Florschuetz* and *Chao* (1965) experiment for water: a) Bubble radius as function of time; b) Bubble and interface temperature as a functions of time; c) Pressure inside the bubble as a function of time, time step of integration 0.01 ms

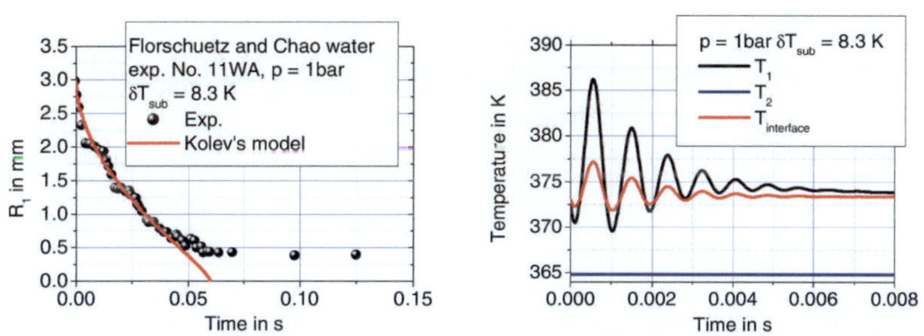

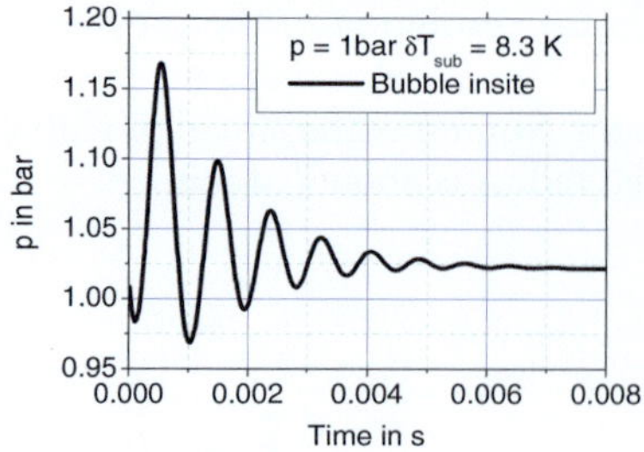

Fig. 5.3 *Florschuetz* and *Chao* (1965) experiment for water: a) Bubble radius as function of time; b) Bubble and interface temperature as a function of time; c) Pressure inside the bubble as a function of time, time step of integration 0.01 ms

Figures 5.5 and 5.6 indicate excellent agreement with the data.

Conclusion: The so-formulated numerical method is in a good position to adequately describe bubble collapse without limiting simplifying assumptions characteristic for closed analytical solutions.

Now let as see the influence of the increasing subcooling on the collapse process. Figure 5.6 presents a computed collapse of initially 5-mm bubbles in water at atmospheric pressure with subcooling from 5 to 80 K. In small subcoolings there are only slight oscillations of the bubble radius. The amplitudes increase with increasing subcooling.

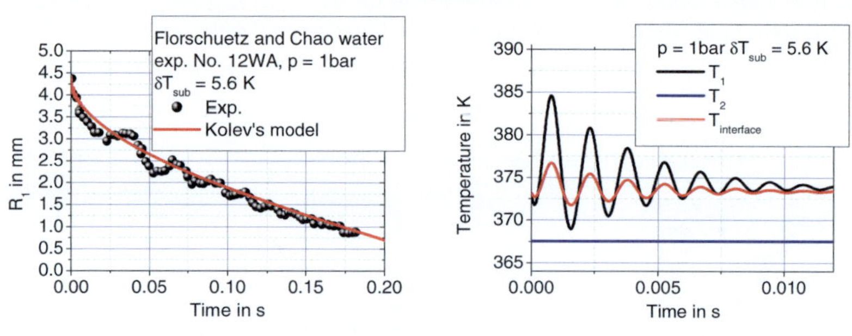

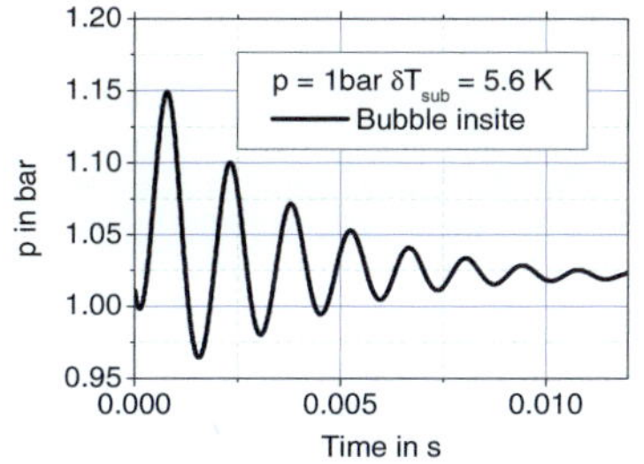

Fig. 5.4 *Florschuetz* and *Chao* (1965) experiment for water: a) Bubble radius as a function of time; b) Bubble and interface temperature as a function of time; c) Pressure inside the bubble as a function of time, time step of integration 0.01 ms

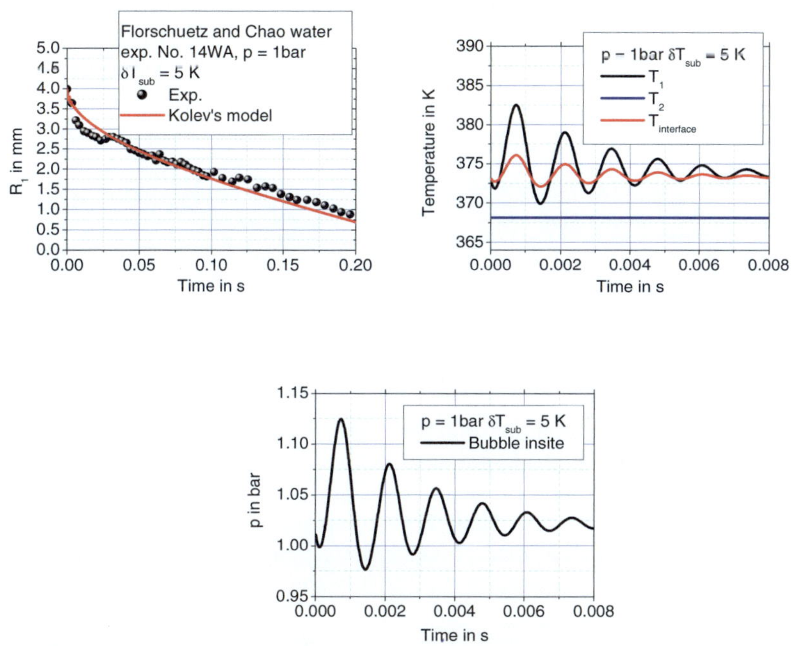

Fig. 5.5 *Florschuetz* and *Chao* (1965) experiment for water: a) Bubble radius as a function of time; b) Bubble and interface temperature as a function of time; c) Pressure inside the bubble as a function of time, time step of integration 0.01 ms

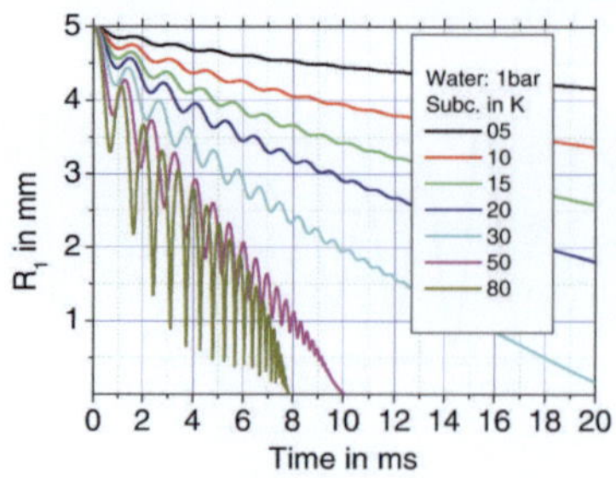

Fig. 5.6 Computed bubble radius as function of time for water at 1 bar and different subcoolings. Void 0.1

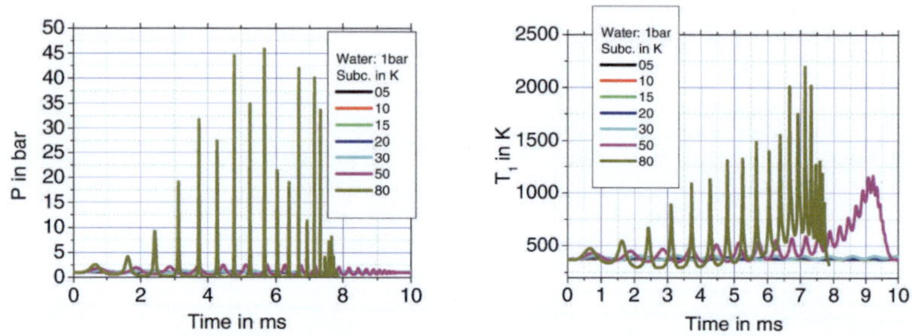

Fig. 5.7 Computed bubble pressures and temperatures as a function of time for water at 1 bar and different subcoolings. Void 0.1

Figures 5.7 show the corresponding pressures and temperatures inside the bubble. At 80 K subcooling we see strong pressure and temperature amplitudes. This explains the observed triggering of steam explosions in melt–water mixtures in which the water is strongly subcooled. The collapsing bubbles generate pressure waves that may lead to such a local pressure pulse generation that it is triggering the thermal explosion of a molten drop falling in water.

Conclusions: The presented numerical method does not damp the strong bubble oscillations in strong subcooling, which is remarkable.

5.4.2 Bubble growth

The experimental data by *Plesset* and *Zwick* (1954) presents the bubble size as a function of time for subatmospheric pressure. The bubbles are generated on heated surface at subatmospheric pressure and wall superheat given in Table 5.2.

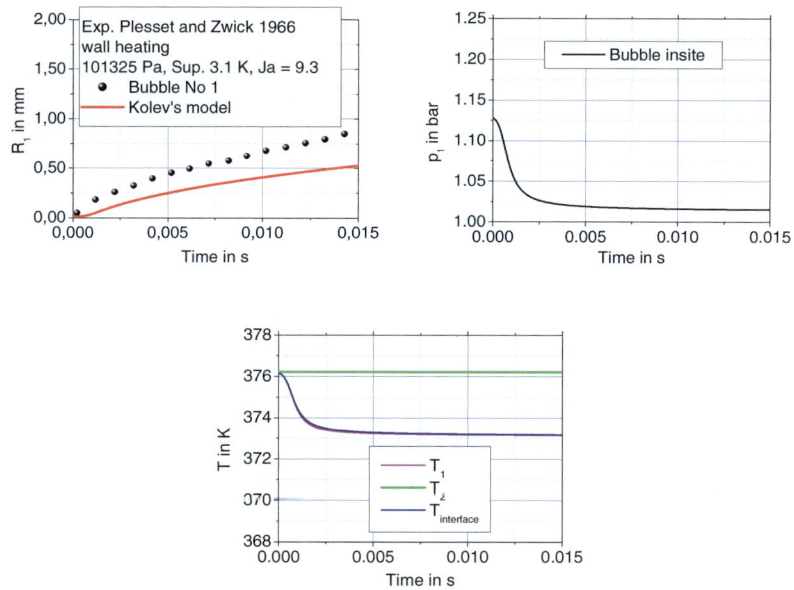

Fig. 5.8 Bubble diameter as a function of time. Bubble generated at heated wall. Experimental data for water by *Plesset* and *Zwick* (1954). Comparison with the theoretical prediction

Table 5.2 Bubble-growth experiments on heated plate performed by *Plesset* and *Zwick* 1954

p in Pa	T' in K H_2O/Na	$T_w - T'$ in K	Ja H_2O/Na
101 325	373.12/1154.6	3.1	9.3/2.78
101 325	373.12/1154.6	4.5	13.48/4.04
101 325	373.12/1154.6	5.3	15.9/4.76

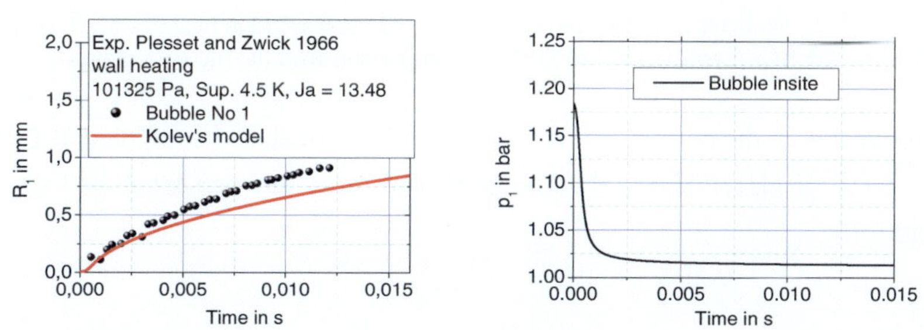

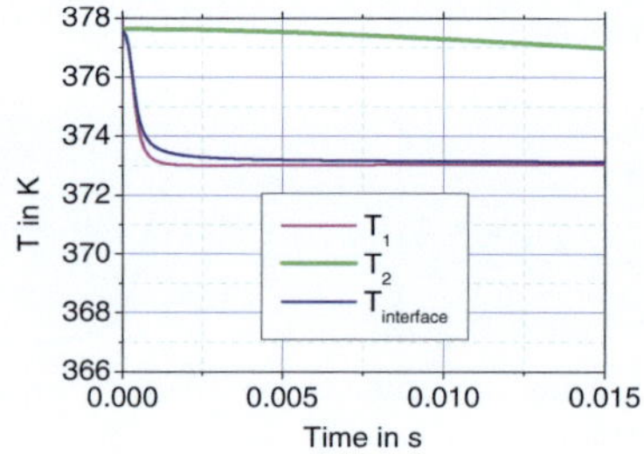

Fig. 5.9 Bubble diameter as a function of time. Bubble generated at heated wall. Experimental data for water by *Plesset* and *Zwick* (1954). Comparison with the theoretical prediction

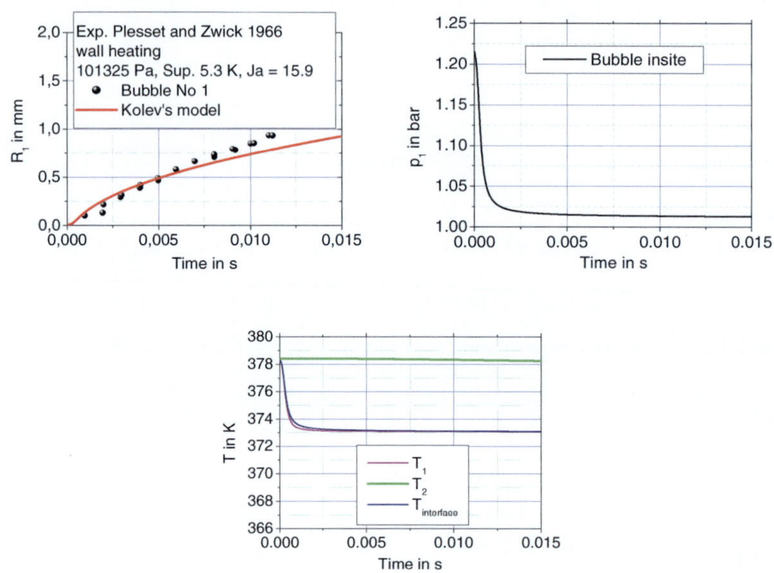

Fig. 5.10 Bubble diameter as a function of time. Bubble generated at heated wall. Experimental data for water by *Plesset* and *Zwick* (1954). Comparison with the theoretical prediction

The initial size in the computations was set to the critical size computed as follows $R_{1c} = \left[2\sigma_2 T'(p)\right] / \left(\rho'' \Delta h \Delta T_{sup}\right)$. The predicted sizes are presented in Figs. 5.8, 5.9 and 5.10.

The comparisons show that the model predicts the initial bubble growth period well and then starts to deviate at higher bubble sizes – probably because the bubbles detach and the relative motion not taken into account here improves the interfacial heat transfer. Note, that the evaporation is a much more stable process compared to the condensation.

5.5 Use in computer codes operating with large computational cells

This author does not know a system computer code simulating multiphase flows by using more than one pressure in a cell. Although 30 years ago it was found that two pressure systems recover hyperbolically, etc. this still did not find application in modern code design. Now, if we are about to resolve acoustic processes the question arises: is it possible with single pressure models? "The answer is no, it is not." See the pressure inside the bubble from Figs. 5.1–5.5. Obviously, only a separate pressure field for the vapor can have the memory along the time axis at which moment the mono-disperse bubble system in a cell is being during its dynamics. What is possible now is to trace the processes in and around the bubble with a much smaller time step, to compute the integral heat- and mass-transfer terms and to supply them to the macroscale solver. An example for a mixture with 0.1 volumetric part void is given in Figs. 5.11a and b. Figure 5.11a shows the local instantaneous mass source term and Fig 5.11b the mass source term averaged over the elapsed time since the beginning of the process. Using the second term in a system computer code using locally time-averaged equations is the correct way to reach ultimate stability. But the accuracy control remains a question of proper selection of the time step.

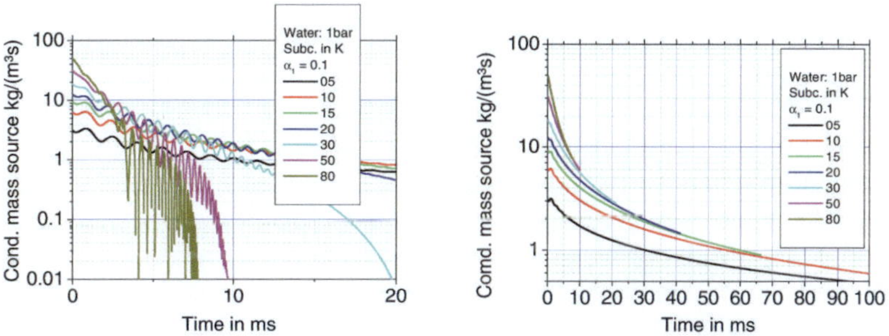

Fig. 5.11 Condensed mass per unit mixture volume and unit time as a function of the time for 0.1 volume concentration saturated void with initial bubble size 5 mm. Liquid pressure 1 bar. Parameter: water subcooling. a) Instantaneous mass source; b) Averaged over the elapsed time

In any case I should emphasize once again that without the second pressure for the vapor, an important acoustic part of the processes can not be simulated in system computer codes. Such is the case for the well-known dependence of the travel velocity of oscillations that depends on the frequencies.

5.6 Comparison with the state-of-the-art

The best review to this subject is given by *Nakoryakov*, et al. (1993) and I will not repeat it here. *Rayleigh* (1917) considered the zero-pressure cavity collapse controlled only by the equality of the compression energy and the kinetic energy of the liquid resulting in

$$\frac{3}{2}\left(\frac{dR_1}{d\tau}\right)^2+\frac{p}{\rho_2}\left(\frac{R_{10}^3}{R_1^3}-1\right)=0 \tag{5.22}$$

and after integration computed the time necessary for the complete collapse

$$\Delta\tau_{col}=R_{10}\sqrt{\frac{2\rho_2}{3p}}\int_{\beta}^{1}\frac{\beta^{3/2}}{\left(1-\beta^3\right)^{1/2}}d\beta\approx 0.9168R_{10}\sqrt{\frac{\rho_2}{p}}, \tag{5.23}$$

where $\beta=\left(R_1/R_{10}\right)^3$. Then, *Rayleigh* evaluated the pressure field around the bubble during the bubble collapse by integrating the mass and momentum equation of the liquid

$$u_2=\frac{dR_1}{d\tau}\left(\frac{R_1}{r}\right)^2, \tag{5.24}$$

$$\frac{\partial u_2}{\partial\tau}+u_2\frac{\partial u_2}{\partial r}-\frac{1}{\rho_2}\frac{\partial p_2}{\partial r}=0. \tag{5.25}$$

So, the main direction for the description of the mechanical interaction in the incompressible liquid was given by *Rayleigh* and remained up to now. Equation (5.25) is Eq. (5.1) with neglected viscosity surface tension and set bubble pressure equal to zero. Therefore, the result by *Rayleigh* is already contained in our system of PDEs.

Plesset and *Zwick* (1954) introduced the liquid energy equation for incompressible liquid. They found the following solution for the surface temperature

$$T_2^{1\sigma}(\tau) - T_2 = -\sqrt{\frac{a_2}{\pi}} \int_0^\tau \frac{R_1^2(\tau^*) \frac{dT_2^{1\sigma}}{dr}}{\left[\int_{\tau^*}^{\tau} R_1^4(\tau^{**}) d\tau^{**} \right]^{1/2}} d\tau^* , \tag{5.26}$$

which, for very small radius changes, reduces to

$$T_2^{1\sigma}(\tau) - T_2 = -\sqrt{\frac{a_2}{\pi}} \int_0^\tau \frac{\frac{dT_2^{1\sigma}}{dr}}{(\tau - \tau^*)^{1/2}} d\tau^* . \tag{5.27}$$

The combination of momentum and energy equation is at this point still not used. *Plesset* and *Zwick* (1954) assumed no difference between the surface and vapor temperature and therefore

$$(\rho w)_2^{1\sigma} = \frac{\lambda_2}{\Delta h} \left(\frac{\partial T_2}{\partial x} \right)_{x=0} . \tag{5.28}$$

In this case the vapor mass conservation

$$\frac{d}{d\tau} \left(\rho_1 \frac{4}{3} \pi R_1^3 \right) = -4\pi R_1^2 (\rho w)_2^{1\sigma} \tag{5.29}$$

gives the relation between bubble radius and temperature gradient. Neglecting the compressibility of the vapor the authors used

$$\left(\frac{\partial T_2}{\partial x} \right)_{x=0} = -\frac{\rho_1 \Delta h}{\lambda_2} \frac{dR_1}{d\tau} \tag{5.30}$$

in Eq. (5.26), resulting in

$$T_2^{1\sigma}(\tau) - T_2 = \frac{\rho_1 \Delta h}{\lambda_2} \sqrt{\frac{a_2}{\pi}} \int_0^\tau \frac{R_1^2(\tau^*) \left(\frac{dR_1}{d\tau} \right)_{\tau^*}}{\left[\int_{\tau^*}^{\tau} R_1^4(\tau^{**}) d\tau^{**} \right]^{1/2}} d\tau^* . \tag{5.31}$$

assuming that $p_1 = p'\left(T_2^{1\sigma}\right)$ in Eq. (5.1),

$$R_1^2 \frac{d^2 R_1}{d\tau^2} + \frac{3}{2} R_1 \left(\frac{dR_1}{d\tau}\right)^2 + 4\eta_2 \frac{dR_1}{d\tau} + \frac{1}{\varepsilon\rho_2}\left\{2\sigma_2 + R_1\left[p_2 - p'\left(T_2^{1\sigma}\right)\right]\right\} = 0, \quad (5.32)$$

results in the famous integrodifferential equation describing bubble dynamics. Furthermore, with several additional simplifications they come to an equation for the so-called thermal-controlled bubble growth

$$\Delta h \rho_1 \frac{dR_1}{d\tau} = \sqrt{\frac{3a_2}{\pi\tau}} \rho_2 c_{p2} \left[T_2 - T'(p)\right]. \quad (5.33)$$

Plesset and *Zwick* also estimated that the diffusion of noncondensable gases in the liquid is a much slower process and therefore has not to be considered for intensive evaporation or condensation processes.

Dergarabedian (1960) introduces the surface tension of the momentum equation and considered the noncondensable gas component inside the vapor keeping its mass constant during the process and behaving isothermally $p_{1,n} = p_{10,n}\left(R_{10}/R_1\right)^3$. He reported valuable bubble-growth experiments for water used later by many authors. He used *Plesset* and *Zwick*'s solution for the surface temperature and then used a quadratic dependence between saturation temperature and saturation pressure. Then, the pressure inside the bubble was set to the saturation pressure and the momentum equation is integrated.

Forster and *Zuber* (1954) followed the same approach as *Plesset* and *Zwick* and *Dergarabedian* but their solution of the liquid heat-conduction problem was different

$$T_2^{1\sigma}(\tau) - T_2 = T_{20}^{1\sigma} - T_2 - \frac{\rho_1}{\rho_2} \frac{\Delta h}{c_{p2}\sqrt{\pi a_2}} \int_0^\tau \frac{R_1\left(\tau^*\right)\left(\frac{dR_1}{d\tau}\right)_{\tau^*}}{R_1(\tau)\left(\tau - \tau^*\right)^{1/2}}$$

$$\times\left\{\exp\left[-\frac{\left[R_1(\tau) - R_1\left(\tau^*\right)\right]^2}{4a_2\left(\tau - \tau^*\right)}\right] - \exp\left[-\frac{\left[R_1(\tau) + R_1\left(\tau^*\right)\right]^2}{4a_2\left(\tau - \tau^*\right)}\right]\right\} d\tau^*$$

$$\approx T_{20}^{1\sigma} - T_2 - \frac{\rho_1}{\rho_2} \frac{\Delta h}{c_{p2}\sqrt{\pi a_2}} \int_0^\tau \frac{\left(dR_1/d\tau\right)_{\tau^*}}{\left(\tau - \tau^*\right)^{1/2}} d\tau^*. \quad (5.34)$$

The results compared very well with the experiments by *Dergarabedian* for water with subcooling between 1.4 and 4.5 K. Finally, the authors obtained the following approximation for bubble growth

$$\Delta h \rho_1 \frac{dR_1}{d\tau} = \sqrt{\frac{\pi a_2}{4\tau}} \rho_2 c_{p2} \left[T_2 - T'(p) \right]. \tag{5.35}$$

Scriven (1959) compared both Eqs. (5.33) and (5.35) with the data of *Dergarabedian* and concluded that *Plesset* and *Zwick*'s equation is in better agreement. The idea of how to couple analytically, not numerically, the heat-conduction problem with the momentum equation attracted the attention of *Mikic*, et al. (1970). They simplified Eq. (5.1) to

$$\frac{3}{2}\left(\frac{dR_1}{d\tau} \right)^2 + \frac{p_2 - p_1}{\varepsilon \rho_2} = 0. \tag{5.36}$$

The pressure difference can be replaced by the temperature difference using the linearized *Clausius* and *Clapeyron* equation

$$\frac{p_1 - p_2}{T'(p_1) - T'(p_2)} \approx \frac{dp'}{dT} = \frac{1}{T'(p_2)} \frac{\Delta h}{v'' - v'}, \tag{5.37}$$

resulting in $\frac{1}{A^2}\left(\frac{dR_1}{d\tau} \right)^2 = 1 - \frac{T_2 - T'(p_1)}{T_2 - T'(p_2)}$, where $A^2 = \frac{2}{3\varepsilon \rho_2}\left[T_2 - T'(p_2) \right] \frac{dp'}{dT}$. The *Plesset* and *Zwick* (1954) equation (5.33) is then used. By setting $T_2^{1\sigma} = T'(p_1)$

$$\Delta h \rho_1 \frac{dR_1}{d\tau} = \sqrt{\frac{3a_2}{\pi \tau}} \rho_2 c_{p2} \left[T_2 - T'(p_1) \right].$$

Excluding the interface temperature results in analytical expression for the time derivative of the bubble radius

$$\frac{1}{A^2}\left(\frac{dR_1}{d\tau} \right)^2 + \sqrt{\frac{\pi}{3a_2}} \frac{\Delta h \rho_1}{\left[T_2 - T'(p_2) \right] \rho_2 c_{p2}} \sqrt{\tau} \frac{dR_1}{d\tau} - 1 = 0. \tag{5.38}$$

This result reduces to *Rayleigh*'s results with strong nonequilibrium and to the thermal controlled bubble size change for small nonequilibria. In any case, the nonlinearities like nonmonotonic temperatures profiles around the surface, surface tension, viscosity and second-order terms in the momentum equation describing oscillation processes are excluded, which is the main limitation of this, in many practical cases, very useful approach.

Coming back to the 1960s: *Scriven* formulated the problem already in 1959 with Eqs. (5.1) and (5.2). In addition to the heat transfer he considered also the heat diffusion by the mass-conservation equation having the same form as the

Fourier equation. As already stated by *Plesset* and *Zwick*, mass diffusion is much slower than heat diffusion. *Scriven* uses his own substitutions and also arrives at the integral solution of the *Fourier* equation which was then estimated numerically for a variety of situations. Later, this solution serves as a benchmark for some authors for checking their approximations. In his numerical estimate the coupling with Eq. (5.1) is not made. Therefore, dynamic oscillation effects can not be derived only from the heat-transfer solution. I have to note that computationally the work for estimation of the *Scriven*'s integral is larger then for the solution of Eqs. (5.19) and (5.20).

Zuber (1961) discussed first the *Fritz* and *Ende* (1936) solution

$$\Delta h \rho_1 \frac{dR_1}{d\tau} = \sqrt{\frac{a_2}{\pi\tau}} \rho_2 c_{p2} \left[T_2 - T'(p_1) \right] \tag{5.39}$$

and then considered growth and subsequent collapse in subcooled liquid near a heated wall. With the observation that the thermal boundary layer was displaced by the growing bubble the authors come to the conclusion that as the bubble size reaches a maximum the condensation heat per unit time and unit wall surface almost equals the wall heat flux.

$$\sqrt{\pi a_2 \Delta \tau_{max}} = \lambda_2 (T_w - T') / \dot{q}''_w . \tag{5.40}$$

With this idea they obtained the maximum of the bubble size. The maximum radius is then computed by the thermal bubble growth formula. For the collapse phase the authors proposed to use Eq. (5.1) together with bubble growth equation in which the heat flux is reduced by the wall heat flux considered as a condensation heat flux. This mechanism is difficult to imagine because bubble growth and departure happens in much more complex circumstances and the collapse happens not at the place where the bubble was born.

Hewit and *Parker* (1968) performed liquid nitrogen experiments and concluded that thermally controlled bubble-growth models agree with data within a 25% error band but thermally controlled bubble collapse models do not satisfactorily represent the reality.

A complete numerical treatment was first made by *Theofanous*, et al. (1969a, b). Equation (5.1) is used coupled with the thermal solution. The authors assumed a kinetic molecular transfer across the interface and come to the conclusion that the temperature is always close to the saturation temperature at the vapor pressure that was an important finding used later by many other authors including me. The solution of the *Fourier* equation was performed by assuming a thin boundary layer and monotonic quadratic temperature profile at the boundary.

$$T - T_2^{1\sigma} = \left(T_2 - T_2^{1\sigma} \right) \left[\frac{2(r - R_1)}{\delta} - \left(\frac{r - R_1}{\delta} \right)^2 \right], \tag{5.41}$$

with

$$T\left(R_1,\tau\right)=T_2^{1\sigma},\ T\left(R_1+\delta,\tau\right)=T_2,\ \left(\frac{\partial T}{\partial r}\right)_{R_1+\delta}=0, \tag{5.42}$$

as proposed before by *Bornhorst* and *Hatsopoulos* (1967). Inserting the profile in the energy equation and integrating between R_1 and $R_1+\delta$ results in an ordinary differential equation connecting the change of the boundary layer thickness with time with the change of the bubble radius and of the bulk temperature. The so-obtained equation was then included in the model instead of the *Fourier* equation. The model very well predicted the condensation and evaporation data obtained with superheats or subcoolings of a few degrees, which were the first proof that this direction of development is very fruitful. Such a method was used also by *Casadei* (1981) who used $\left(T-T_2^{1\sigma}\right)/\left(T_2-T_2^{1\sigma}\right)=\left[\left(R_1-r\right)/\delta\right]^2$ instead of Eq. (5.41). The limitations are in the assumptions made by resolving the temperature profile. *Lee* and *Merte* (1996) used Eqs. (5.1) and (5.2) exactly in the same form. They designed a *Runge–Kutta* method for solving the momentum equation and the *Patankar* finite volume method for resolving the temperature field dropping the assumption of monotonic profile. They also designed a nonequidistant grid having better resolution close to the interface. *Mendez* and *Naude* (2007) reported that second-order discretization in time and third order in space is needed to resolve the liquid temperature profile especially in the final collapse stage. The energy-conservation equations for the bubble was not used by these authors and replaced with the saturation condition at the surface temperature at any time. This assumption does not allow having a nonequilibrium temperature in the bubble for very strong oscillations. This is the main difference from our method. Therefore, the method I use here omits the disadvantages of assuming a monotonic profile in the liquid and not using the energy conservation for the vapor. This together with the simplicity of the numerical method is the main step forward with respect to these authors.

In their detailed review of the state-of-the-art up to 1977 *Plesset* and *Prosperetti* (1977) used as a starting point the combined equation (5.1) and (5.2) as derived by *Plesset* and *Zwick* (1954) to study oscillation behavior of such systems. Assuming that the gas state change is isentropic $p_{1,n}=p_{10,n}\left(R_{10}/R_1\right)^{3\kappa}$ the authors derived the eigenfrequency of oscillation to be

$$\omega_0^2=\frac{3\kappa p}{\rho_2 R_{10}^2}-\frac{2\sigma}{\rho R_{10}^3}. \tag{5.43}$$

Jones and *Zuber* (1978) pointed out that the vapor density change in Eq. (5.29) is important especially in pressure transients and proposed an approximation to compute the saturation vapor density following the surface temperature change.

Further, they used the *Carlsaw* and *Jaeger* (1959) solution of the *Fourier* equation for heat slab with variable temperature difference at the one boundary

$$T(x,\tau) - T_2 = \frac{x}{2\sqrt{\pi a_2}} \int_0^\tau \left(T_2^{1\sigma} - T_2\right) \frac{e^{-\frac{x^2}{4a_2(\tau-\eta)}}}{(\tau-\eta)^{3/2}} d\eta \tag{5.44}$$

to compute the heat flux at the bubble surface

$$\dot{q}_2''^{1\sigma} = k_s \frac{\lambda_2}{\sqrt{\pi a_2}} \left[\frac{\Delta T(0)}{\sqrt{\tau}} + \int_0^\tau \frac{\frac{d\Delta T(\eta)}{d\tau}}{\sqrt{\tau-\eta}} d\eta \right] \tag{5.45}$$

with $T_2 = const$, $\Delta T(\tau) = T_2^{1\sigma}(\tau) - T_2$, $\Delta T(0) = T'(p_{10}) - T_2$ and k_s the sphericity correction: $\pi/2$ by *Foster* and *Zuber*, $\sqrt{3}$ by *Plesset* and *Zwick*. Assuming different functions for the pressure change and therefore for the surface temperature change the authors integrated Eq. (5.45) analytically and derived expressions for the bubble size change without connection to the momentum equation. Interestingly, the authors obtained good comparisons for bubble growth in variable-pressure fields. A very similar approach was used by *Toda* and *Kitamura* (1983). They also allow for the vapor density to change and derive an integral solution for the thin boundary layer. *Alamgir* (1984) proposed an analytical approximation for the density change combined with the *Jones* and *Zuber* solution. Again, not coupling with the momentum equation they obtained acceptable accuracy for 4 experiments performed by themselves with water at about 3 bar and superheating of about 4 to 6 K. *Wang* and *Bankoff* (1990, 1991) extended the *Jones* and *Zuber* (1978) solution to bubble growth at the wall by modifying the equivalent volume of the bubble attached at the wall. Good comparison with their data for decompression experiments is reported. *Tsung-Chang* and *Bankoff* (1986) considered the problem with linear pressure change and provided an approximate analytical solution to the *Plesset* and *Zwick* integral considering variable vapor density

$$T_2^{1\sigma}(\tau) - T_2 = \frac{\rho_1 \Delta h}{3\lambda_2} \sqrt{\frac{a_2}{\pi}} \int_0^\tau \frac{\frac{d}{d\tau^*}\left[\rho_1(\tau^*) R_1^3(\tau^*)\right]}{\left[\int_{\tau^*}^{\tau} R_1^4(\tau^{**}) d\tau^{**}\right]^{1/2}} d\tau^* . \tag{5.46}$$

Winters and *Merte* (1979) are also interested in describing the bubble behavior in variable-pressure fields. The authors concluded that for bubble growth the monotonic temperature profile is a good approximation but for bubble collapse the temperature field should be considered as nonmonotonic. In this case they write the *Carlsaw* and *Jaeger* (1959) approximate solution of the *Fourier* equation in the following form

$$\dot{q}_2''^{1\sigma} = k_s \frac{\lambda_2}{\sqrt{3a_2}} \left[\frac{T'(p_{10}) - T_2}{\sqrt{\tau}} + \sum_{m=1}^{n} \frac{T_2^{1\sigma}(\tau_m) - T_2^{1\sigma}(\tau_{m-1})}{\sqrt{\tau - \tau_m}} \right]. \tag{5.47}$$

Reynolds and *Berthoud* (1981) solved the *Fourier* equation by making the same transformation as *Plesset* and *Zwick* $h = \frac{1}{3}\left(r^3 - R_1^3\right)$, $\overline{\tau} = \int_0^{\tau} \left[R_1\left(\tau^*\right)\right]^4 d\tau^*$ and assuming a thin thermal boundary layer. Their result for the accumulated heat in J transferred from the surface into the liquid is

$$Q\left(R_1, \tau\right) = 4\lambda_2 \sqrt{\frac{\pi}{a_2}} \int_0^{\tau} \frac{R_1^4\left(\tau^*\right)\left[T_2^{1\sigma}(\tau) - T_2\right]}{\left[\int_{\tau^*}^{\tau} R_1^4\left(\tau^{**}\right) d\tau^{**}\right]^{1/2}} d\tau^*. \tag{5.48}$$

Appendix 5.1 The Raleigh–Plesset equation

Consider the liquid mass and momentum conservation equations valid in $R_1 < r < R_{2,cell}$, where $R_{2,cell} = \left[3/\left(4\pi n_1\right)\right]^{1/3}$:

$$\text{Mass: } \frac{\partial \rho_2}{\partial \tau} + \frac{1}{r^2}\frac{\partial}{\partial r}\left(r^2 \rho_2 u_2\right) = 0, \tag{5.1-1}$$

$$\text{Momentum: } \frac{\partial u_2}{\partial \tau} + u_2 \frac{\partial u_2}{\partial r} + \frac{1}{\rho_2}\frac{\partial p_2}{\partial r} + 2\eta_2 \frac{\partial^2 u_2}{\partial r^2} = 0. \tag{5.1-2}$$

The main idea by the derivation of the equation describing the bubble interface motion with respect to the center, is to integrate the momentum equation along $R_1 < r < R_{2,cell}$. It will be assumed that at $R_{2,cell}$ the pressure is the liquid pressure. First, I will present the known formalism for incompressible liquid and then I will extend the formalism for large time changes of pressure with time.

Incompressible liquid: For incompressible liquid the mass conservation simplifies to $\partial\left(r^2 u_2\right)/\partial r = 0$ or

$$\frac{\partial r}{\partial \tau} = u_2(r) = \frac{R_1^2}{r^2}\frac{\partial R_1}{\partial \tau}. \tag{5.1-3}$$

Here, the radius is considered to be only a function of time, $R_1 = R_1(\tau)$. *Rayleigh* (1917) was the first to equalize the expansion energy of a bubble from radius R_{10} to R_1

$$(p_1 - p_2)4\pi \int_{R_{10}}^{R_1} r^2 dr = (p_1 - p_2)\frac{4}{3}\pi\left(R_1^3 - R_{10}^3\right) \tag{5.1-4}$$

to the energy consumed to change the kinetic energy of the surrounding liquid

$$\frac{1}{2}\rho_2 4\pi \int_{R_1}^{R_{cell}} u_2^2 r^2 dr = \frac{1}{2}\rho_2 4\pi R_1^4\left(\frac{\partial R_1}{\partial \tau}\right)^2 \int_{R_1}^{R_{cell}} \frac{1}{r^2} dr = \frac{1}{2}\rho_2 4\pi R_1^4\left(\frac{\partial R_1}{\partial \tau}\right)^2\left(\frac{1}{R_1} - \frac{1}{R_{cell}}\right) \tag{5.1-5}$$

and obtained the equation used later in thousands of papers

$$\frac{3}{2}\left(\frac{\partial R_1}{\partial \tau}\right)^2\left(1 - \frac{R_1}{R_{cell}}\right) + \frac{p_2 - p_1}{\rho_2}\left(1 - \frac{R_{10}^3}{R_1^3}\right) = 0, \tag{5.1-6}$$

$$R_{cell} \to \infty .$$

Equation (5.1-6) is valid for no mass transfer at the surface. For mass transfer at the surface the interface velocity of the liquid is

$$u_2^{1\sigma} = u_1^{2\sigma} - \Delta u_{12}^{\sigma} = \left(1 - \frac{\rho_1^{2\sigma}}{\rho_2^{1\sigma}}\right)u_1^{2\sigma} = \left(1 - \frac{\rho_1^{2\sigma}}{\rho_2^{1\sigma}}\right)\frac{dR_1}{d\tau}. \tag{5.1-7}$$

Therefore, the velocity at any radius in the liquid is related to the bubble radius derivative as follows

$$u_2 = \left(1 - \frac{\rho_1^{2\sigma}}{\rho_2^{1\sigma}}\right)\frac{R_1^2}{r^2}\frac{dR_1}{d\tau} = \varepsilon\frac{R_1^2}{r^2}\frac{dR_1}{d\tau}. \tag{5.1-8}$$

In Eq. (5.1-6) the history of the expansion that has to satisfy the momentum equation is not reflected. This history can be considered if the momentum equation in the liquid is integrated from R_1 to R_{cell}. For this purpose we need the participating liquid velocity first and second derivatives.

Assuming noncompressible liquid they are obtained by differentiating Eq. (5.1-8) as follows:

$$\frac{\partial u_2}{\partial \tau} = \varepsilon \frac{R_1^2}{r^2}\frac{\partial^2 R_1}{\partial \tau^2} + 2\varepsilon R_1 \frac{1}{r^2}\left(\frac{\partial R_1}{\partial \tau}\right)^2 \quad (5.1\text{-}9)$$

$$\frac{\partial u_2}{\partial r} = -2\varepsilon \frac{\partial R_1}{\partial \tau} R_1^2 \frac{1}{r^3}, \quad (5.1\text{-}10)$$

$$\frac{\partial^2 u_2}{\partial r^2} = 6\varepsilon \frac{\partial R_1}{\partial \tau} R_1^2 \frac{1}{r^4}, \quad (5.1\text{-}11)$$

$$u_2 \frac{\partial u_2}{\partial r} = -2\varepsilon^2 R_1^4 \left(\frac{\partial R_1}{\partial \tau}\right)^2 \frac{1}{r^5}. \quad (5.1\text{-}12)$$

Replacing the velocity and its derivatives in the momentum equation (5.1-2) results in

$$\frac{R_1^2}{r^2}\frac{\partial^2 R_1}{\partial \tau^2} + 2\left(R_1 \frac{1}{r^2} - \varepsilon R_1^4 \frac{1}{r^5}\right)\left(\frac{\partial R_1}{\partial \tau}\right)^2 + \frac{1}{\rho_2 \varepsilon}\frac{\partial p}{\partial r} + 12\eta_2 \frac{\partial R_1}{\partial \tau} R_1^2 \frac{1}{r^4} = 0. \quad (5.1\text{-}13)$$

Integrating between R_1 to $R_{2,cell}$

$$R_1^2 \frac{\partial^2 R_1}{\partial \tau^2}\int_{R_1}^{R_{2,cell}} \frac{1}{r^2} dr + 2\left(R_1 \int_{R_1}^{R_{2,cell}} \frac{1}{r^2} dr - R_1^4 \int_{R_1}^{R_{2,cell}} \frac{1}{r^5} dr\right)\left(\frac{dR_1}{\partial \tau}\right)^2 + \frac{1}{\rho_2 \varepsilon}\int_{R_1}^{R_{2,cell}} dp_2$$

$$+12\eta_2 \frac{\partial R_1}{\partial \tau} R_1^2 \int_{R_1}^{R_{2,cell}} \frac{1}{r^4} dr = 0 \quad (5.1\text{-}14)$$

results in

$$R_1\left(1 - \frac{R_1}{R_{2,cell}}\right)\frac{\partial^2 R_1}{\partial \tau^2} + 2\left(\frac{3}{4} - \frac{R_1}{R_{2,cell}} + \frac{1}{4}\frac{R_1^4}{R_{2,cell}^4}\right)\left(\frac{\partial R_1}{\partial \tau}\right)^2 + \frac{p_2 - p_1}{\rho_2 \varepsilon}$$

$$+\frac{4\eta_2}{R_1}\frac{\partial R_1}{\partial \tau}\left(1 - \frac{R_1^3}{R_{2,cell}^3}\right) = 0. \quad (5.1\text{-}15)$$

For $R_1/R_{2,cell} = \alpha_1^{1/3}$ this equation receives the form

$$R_1\left(1-\alpha_1^{1/3}\right)\frac{\partial^2 R_1}{\partial \tau^2} + 2\left(\frac{3}{4} - \alpha_1^{1/3} + \frac{1}{4}\alpha_1^{4/3}\right)\left(\frac{\partial R_1}{\partial \tau}\right)^2 + \frac{p_2 - p_1}{\rho_2 \varepsilon}$$

$$+\left(1-\alpha_1\right)\frac{4\eta_2}{R_1}\frac{\partial R_1}{\partial \tau} = 0 \qquad (5.1\text{-}16)$$

used usually in the so-called two pressure models of bubbly flow for analyzing of acoustic processes e.g., *Akhatov* et al. (2005). For $\alpha_1 \to 0$ which means $R_{2,cell} \to \infty$ the above equation reduces to

$$R_1 \frac{\partial^2 R_1}{\partial \tau^2} + \frac{3}{2}\left(\frac{\partial R_1}{\partial \tau}\right)^2 + \frac{p_2 - p_1}{\rho_2 \varepsilon} + \frac{4\eta_2}{R_1}\frac{\partial R_1}{\partial \tau} = 0, \qquad (5.1\text{-}17)$$

which is the so-called *Rayleigh–Plesset* equation used by many authors, compare with Eq. (10) in *Scriven* (1959).

Hijikata et al. (1979) used in a one dimensional model the *Rayleigh–Plesset* equation by replacing $dR_1/d\tau = u_1\, dR_1/dx$. Following this idea

$$\frac{dR_1}{d\tau} = \frac{\partial R_1}{\partial \tau} + u_1 \nabla R_1 \qquad (5.1\text{-}18)$$

and therefore

$$R_1\left(\frac{\partial^2 R_1}{\partial \tau^2} + u_1 \frac{\partial}{\partial \tau}\nabla R_1 + \nabla R_1 \frac{\partial u_1}{\partial \tau}\right) + \frac{3}{2}\left(\frac{\partial R_1}{\partial \tau} + u_1 \nabla R_1\right)^2 + \frac{p_2 - p_1}{\rho_2 \varepsilon} + \frac{4\eta_2}{R_1}\frac{\partial R_1}{\partial \tau} = 0. \qquad (5.1\text{-}19)$$

Prosperetti and *Crum* (1988), *Kamath* and *Prosperetti* (1989) add into account some compressibility effect in the liquid by deriving the equation describing the bubble radius dynamics.

Compressible liquid: The complete system Eqs. (5.1-1) and (5.1-2) together with the energy conservation in the liquid has to be solved in this case. Replacing the density differentials in the mass-conservation equation with their equals from the differential form of the equation of state results in

$$\frac{1}{\rho_2 a_2^2}\left(\frac{\partial p_2}{\partial \tau} + u_2 \frac{\partial p_2}{\partial r}\right) + \frac{1}{\rho_2}\frac{\partial \rho_2}{\partial s_2}\left(\frac{\partial s_2}{\partial \tau} + u_2 \frac{\partial s_2}{\partial r}\right) + \frac{1}{r^2}\frac{\partial}{\partial r}\left(r^2 u_2\right) = 0.$$

Here, only the case for large pressure changes with the time will be considered resulting in

$$\frac{1}{r^2}\frac{\partial}{\partial r}\left(r^2 u_2\right) = -\frac{1}{\rho_2 a_2^2}\frac{\partial p_2}{\partial \tau}.$$

As in the case of incompressible liquid we have to derive from these equations the velocity and its derivatives, to replace in the momentum equation and to integrate it from the interface to the end of the cell. The velocity and the derivative are

$$u_2 = \varepsilon\frac{R_1^2}{r^2}\frac{dR_1}{d\tau} - \frac{1}{\rho_2 a_2^2}\frac{\partial p_2}{\partial \tau}\frac{1}{3}\left(r^3 - R_1^3\right)$$

$$\frac{\partial u_2}{\partial r} = -\frac{2}{r}u_2 - \frac{1}{\rho_2 a_2^2}\frac{\partial p_2}{\partial \tau} = -2\varepsilon\frac{R_1^2}{r^3}\frac{dR_1}{d\tau} - \frac{1}{\rho_2 a_2^2}\frac{\partial p_2}{\partial \tau}$$

$$\frac{\partial^2 u_2}{\partial r^2} = 6\varepsilon\frac{R_1^2}{r^4}\frac{dR_1}{d\tau}$$

$$\frac{\partial u_2}{\partial \tau} = \varepsilon\frac{R_1^2}{r^2}\frac{\partial^2 R_1}{\partial \tau^2} + 2\varepsilon R_1\frac{1}{r^2}\left(\frac{\partial R_1}{\partial \tau}\right)^2 - \frac{1}{\rho_2 a_2^2}\frac{\partial^2 p_2}{\partial \tau^2}\frac{1}{3}\left(r^3 - R_1^3\right)$$

and therefore

$$u_2\frac{\partial u_2}{\partial r} = -2\varepsilon^2\frac{R_1^4}{r^5}\left(\frac{dR_1}{d\tau}\right)^2 - \varepsilon\frac{R_1^2}{r^2}\frac{dR_1}{d\tau}\frac{1}{\rho_2 a_2^2}\frac{\partial p_2}{\partial \tau}$$

$$+2\varepsilon\frac{1}{3}\left(r^3 - R_1^3\right)\frac{R_1^2}{r^3}\frac{dR_1}{d\tau}\frac{1}{\rho_2 a_2^2}\frac{\partial p_2}{\partial \tau} + \left(\frac{1}{\rho_2 a_2^2}\frac{\partial p_2}{\partial \tau}\right)^2\frac{1}{3}\left(r^3 - R_1^3\right).$$

Replacing in the momentum equation I obtain

$$\frac{R_1^2}{r^2}\frac{\partial^2 R_1}{\partial \tau^2} + 2\left(R_1\frac{1}{r^2} - \varepsilon R_1^4\frac{1}{r^5}\right)\left(\frac{\partial R_1}{\partial \tau}\right)^2 + \frac{1}{\rho_2 \varepsilon}\frac{\partial p}{\partial r} + 12\eta_2\frac{\partial^2 u_2}{\partial r^2}$$

$$+\left\{\left(\frac{2}{3} - \frac{1}{r^2} - \frac{2}{3}\frac{R_1^3}{r^3}\right)R_1^2\frac{dR_1}{d\tau}\frac{\partial p_2}{\partial \tau} + \frac{1}{3\varepsilon}\left(r^3 - R_1^3\right)\left[\frac{1}{\rho_2 a_2^2}\left(\frac{\partial p_2}{\partial \tau}\right)^2 - \frac{\partial^2 p_2}{\partial \tau^2}\right]\right\}\frac{1}{\rho_2 a_2^2} = 0$$

After integration from the surface of the bubble to the end of the cell the momentum equation receives the form:

$$R_1\left(1-\frac{R_1}{R_{cell}}\right)\frac{\partial^2 R_1}{\partial\tau^2}+2\left(\frac{3}{4}-\frac{R_1}{R_{cell}}+\frac{1}{4}\frac{R_1^4}{R_{cell}^4}\right)\left(\frac{\partial R_1}{\partial\tau}\right)^2+\frac{p_2-p_1}{\rho_2\varepsilon}+\frac{4\eta_2}{R_1}\frac{\partial R_1}{\partial\tau}\left(1-\frac{R_1^3}{R_{cell}^3}\right)$$

$$\left\{\begin{array}{l}\left(\frac{2}{3}R_{cell}\left(1-\frac{R_1}{R_{cell}}\right)-\frac{1}{R_1}\left(1-\frac{R_1}{R_{cell}}\right)-\frac{1}{3}R_1\left(1-\frac{R_1^2}{R_{cell}^2}\right)\right)R_1^2\frac{dR_1}{d\tau}\frac{\partial p_2}{\partial\tau}\\ +\frac{R_{cell}^4}{3\varepsilon}\left(\frac{1}{4}-\frac{R_1^3}{R_{cell}^3}+\frac{3}{4}\frac{R_1^4}{R_{cell}^4}\right)\left[\frac{1}{\rho_2 a_2^2}\left(\frac{\partial p_2}{\partial\tau}\right)^2-\frac{\partial^2 p_2}{\partial\tau^2}\right]\end{array}\right\}\frac{1}{\rho_2 a_2^2}=0$$

Again, using $R_1/R_{cell}=\alpha_1^{1/3}$ results in

$$R_1\left(1-\alpha_1^{1/3}\right)\frac{\partial^2 R_1}{\partial\tau^2}+2\left(\frac{3}{4}-\alpha_1^{1/3}+\frac{1}{4}\alpha_1^{4/3}\right)\left(\frac{\partial R_1}{\partial\tau}\right)^2+\frac{p_2-p_1}{\rho_2\varepsilon}+\left(1-\alpha_1\right)\frac{4\eta_2}{R_1}\frac{\partial R_1}{\partial\tau}$$

$$+f\left(\frac{\partial p_2}{\partial\tau},\frac{\partial^2 p_2}{\partial\tau^2}\right)=0$$

$$f\left(\frac{\partial p_2}{\partial\tau},\frac{\partial^2 p_2}{\partial\tau^2}\right)=\frac{1}{\rho_2 a_2^2}\left\{\begin{array}{l}\left(\left(1-\alpha_1^{1/3}\right)\left(\frac{2}{3}R_{cell}-\frac{1}{R_1}\right)-\frac{1}{3}R_1\left(1-\alpha_1^{2/3}\right)\right)R_1^2\frac{dR_1}{d\tau}\frac{\partial p_2}{\partial\tau}\\ +\frac{R_{cell}^4}{3\varepsilon}\left(\frac{1}{4}-\alpha_1+\frac{3}{4}\alpha_1^{4/3}\right)\left[\frac{1}{\rho_2 a_2^2}\left(\frac{\partial p_2}{\partial\tau}\right)^2-\frac{\partial^2 p_2}{\partial\tau^2}\right]\end{array}\right\}.$$

Appendix 5.2 The liquid energy-conservation equation

The energy conservation of an incompressible liquid in a coordinate system with origin at the center of the bubble without internal heat sources is

$$\frac{\partial T_2}{\partial\tau}+\dot{r}\frac{\partial T_2}{\partial r}-\frac{a_2}{r^2}\frac{\partial}{\partial r}\left(r^2\frac{\partial T_2}{\partial r}\right)=0. \qquad (5.2\text{-}1)$$

Here, the liquid temperature is a function of time and radius $T_2 = T_2(\tau, r)$ and therefore

$$dT_2 = \frac{\partial T_2}{\partial \tau} d\tau + \frac{\partial T_2}{\partial r} dr . \tag{5.2-2}$$

Setting the coordinate system at the interface,

$$x = \frac{r - R(\tau)}{\delta}, \tag{5.2-3}$$

where $0 \le x \le 1$, $R_1(\tau) \le r \le R_1(\tau) + \delta$, $r = \delta x + R(\tau)$ the temperature is then a function of the time and the distance from the surface

$$T_2 = T_2\left[\tau, r - \delta x + R(\tau)\right] = T_2(\tau, x)$$

and

$$dT_2 = \left(\frac{\partial T_2}{\partial \tau} + \dot{r} \frac{\partial T_2}{\partial r} \right) d\tau + \delta \frac{\partial T_2}{\partial r} dx . \tag{5.2-4}$$

Comparing Eqs. (5.2-2) and (5.2-4) we obtain

$$\left(\frac{\partial T_2}{\partial \tau} \right)_x = \frac{\partial T_2}{\partial \tau} + \dot{r} \frac{\partial T_2}{\partial r}, \quad \left(\frac{\partial T_2}{\partial x} \right)_\tau = \delta \frac{\partial T_2}{\partial r} . \tag{5.2-5, 6}$$

Solving with respect to the original derivatives we have

$$\frac{\partial T_2}{\partial r} = \frac{1}{\delta} \left(\frac{\partial T_2}{\partial x} \right)_\tau, \quad \frac{\partial T_2}{\partial \tau} = \left(\frac{\partial T_2}{\partial \tau} \right)_x - \frac{\dot{r}}{\delta} \left(\frac{\partial T_2}{\partial x} \right)_\tau, \quad \frac{\partial r}{\partial \tau} = x \frac{d\delta}{d\tau} + \frac{dR_1}{d\tau} . \tag{5.2-7, 8, 9, 10}$$

Applying to the original energy-conservation equation for $\delta = const$ results in

$$\left(\frac{\partial T_2}{\partial \tau} \right)_x - \frac{a_2}{r^2} \frac{1}{\delta} \left(\frac{\partial}{\partial x} \right)_\tau \left[r^2 \frac{1}{\delta} \left(\frac{\partial T_2}{\partial x} \right)_\tau \right] = 0, \tag{5.2-11}$$

or for $\delta = 1$

$$\left(\frac{\partial T_2}{\partial \tau}\right)_x - \frac{a_2}{r^2}\left(\frac{\partial}{\partial x}\right)_\tau \left[r^2\left(\frac{\partial T_2}{\partial x}\right)_\tau\right] = 0, \tag{5.2-12}$$

see *Lee* and *Merte* (1966). As already mentioned, we see that the convection term disappears.

Appendix 5.3 Pressure equation for the bubble

Equation (5.3-1) is the so-called pressure equation or volume-conservation equation for the bubble. It is in fact Eq. (5.207) in *Kolev* (2007) written for a single gas sphere

$$\frac{1}{\rho_1 a_1^2}\frac{dp}{d\tau} = -\frac{1}{\rho_1}\frac{3}{R_1}(\rho w)_2^{1\sigma}$$
$$-\frac{1}{T_1}\frac{1}{\rho_1^2}\left(\frac{\partial \rho_1}{\partial s_1}\right)_p \frac{3}{R_1}\left\{\dot{q}_1^{\prime\prime 2\sigma} - (\rho w)_2^{1\sigma} c_{p1}\left[T'(p_1) - T_1\right]\right\} - \frac{1}{\gamma_v}\frac{d\gamma_v}{d\tau}. \tag{5.3-1}$$

Here,

$$\left(\frac{\partial \rho_1}{\partial s_1}\right)_p = \left(\frac{\partial \rho_1}{\partial T_1}\right)_p \frac{T_1}{c_{p1}}. \tag{5.3-2}$$

$$\left(\frac{\partial \rho_1}{\partial T_1}\right)_p = -\frac{\rho_1}{T_1}. \tag{5.3-3}$$

γ_v is the ratio of the bubble volume to the initial volume

$$\gamma_v = \frac{V_1}{V_{10}} = \left(\frac{R_1}{R_{10}}\right)^3, \tag{5.3-4}$$

and therefore

$$\frac{1}{\gamma_v}\frac{d\gamma_v}{d\tau} = \frac{3}{R_1}\frac{dR_1}{d\tau}. \tag{5.3-5}$$

The change of the bubble radius with the time consists of the mechanical component dictated by the force balance $\left(dR_1/d\tau\right)_{mom}$ and of a component depending on the mass transfer as follows

$$\frac{dR_1}{d\tau} = \left(\frac{dR_1}{d\tau}\right)_{mom} - \frac{1}{\rho_2}(\rho w)_2^{1\sigma} . \tag{5.3-6}$$

Replacing Eqs. (5.3-2, 3, 5, 6) into Eq. (5.3-1) results in

$$\frac{1}{\rho_1 a_1^2}\frac{dp}{d\tau} = -\frac{3}{R_1}\left[\left(\frac{1}{\rho_1} - \frac{1}{\rho_2}\right)(\rho w)_2^{1\sigma} + \left(\frac{dR_1}{d\tau}\right)_{mom}\right] + \frac{1}{\rho_1 c_{p1} T_1}\frac{3}{R_1}\left\{\dot{q}_1^{\prime\prime 2\sigma} - (\rho w)_2^{1\sigma} c_{p1}\left[T'(p_1) - T_1\right]\right\} . \tag{5.3-7}$$

Neglecting the mass transfer and assuming a perfect gas this equation reduces to the one derived by *Prosperetti* and *Crum* (1988), *Kamath* and *Prosperetti* (1989) from the mass conservation. The heat-transfer contribution on the bubble pressure change $= ... + \frac{3}{R_1}\frac{\kappa - 1}{\kappa}\frac{\dot{q}_1^{\prime\prime 2\sigma}}{p_1}$ is taken in this form into account by many authors, e.g., *Prosperetti* and *Crum* (1988), *Kamath* and *Prosperetti* (1989), *Nigmatulin* (1990), *Kim* (2002, 2004, 2005). Defining the pseudoisentropic exponent as $\kappa = \rho/(pa^2)$ and rearranging I obtain

$$d\ln\left(p_1^{\frac{1}{\kappa}} R_1^3\right) = \frac{3}{R_1}\left[-\left(\frac{1}{\rho_1} - \frac{1}{\rho_2}\right)(\rho w)_2^{1\sigma} + \frac{\dot{q}_1^{\prime\prime 2\sigma} - (\rho w)_2^{1\sigma} c_{p1}\left[T'(p_1) - T_1\right]}{\rho_1 c_{p1} T_1}\right] d\tau \tag{5.3-8}$$

or in finite-difference form

$$p_1 \approx p_{10}\left(\frac{R_{10}}{R_1}\right)^{3\kappa} \exp\left\{\frac{3\kappa\Delta\tau}{R_1}\left[\begin{array}{c} -\left(\frac{1}{\rho_1} - \frac{1}{\rho_2}\right)(\rho w)_2^{1\sigma} \\ + \frac{\dot{q}_1^{\prime\prime 2\sigma} - (\rho w)_2^{1\sigma} c_{p1}\left[T'(p_1) - T_1\right]}{\rho_1 c_{p1} T_1} \end{array}\right]\right\} . \tag{5.3-9}$$

Compared with Eq. (5.3-7) this equation has the advantage that if numerically evaluated the pressure in the bubble remains always positive.
For no heat and mass transfer across the interface the equation reduces to the well-known isentropic pressure volume relation $p_1 \approx p_{10}\left(R_{10}/R_1\right)^{3\kappa}$, *Plesset* and *Zwick*

(1954) or for isothermal state change $p_1 \approx p_{10}\left(R_{10}/R_1\right)^3$, *Dergarabedian* (1953). For isothermal state change of bubbles with constant number density $p_1 \approx p_{10}\left(\alpha_{10}/\alpha_1\right)^{\kappa}$.

Appendix 5.4 Bubble energy conservation

The pseudoentropy form of the energy conservation is

$$\rho_1 c_{p1}\frac{dT_1}{d\tau}-\frac{dp_1}{d\tau}=\frac{3}{R_1}\left[\dot{q}_1^{\prime\prime 2\sigma}-(\rho w)_2^{1\sigma}c_{p1}\left(T_1^{2\sigma}-T_1\right)\right]. \tag{5.4-1}$$

Defining the pseudoisentropic exponent $\kappa=\rho_1 a_1^2/p_1\approx c_{p1}/\left(c_{p1}-R_1\right)$ and rearranging I obtain

$$T_1\approx T_{10}\left(\frac{p_1}{p_0}\right)^{(\kappa-1)/\kappa}\exp\left\{\frac{\kappa-1}{\kappa}\frac{3\Delta\tau}{R_1 p_1}\left[\dot{q}_1^{\prime\prime 2\sigma}-(\rho w)_2^{1\sigma}c_{p1}\left(T'(p_1)-T_1\right)\right]\right\}. \tag{5.4-2}$$

Prosperetti and *Crum* (1988), *Kamath* and *Prosperetti* (1989) solved numerically the gas energy-conservation equation for the case of no mass transfer at the interface

$$\rho_1 c_{p1}\left(\frac{\partial T_1}{\partial\tau}+u_1\frac{\partial T_1}{\partial r}\right)-\frac{\partial p_1}{\partial\tau}-\frac{1}{r^2}\frac{\partial}{\partial r}\left(r^2\lambda_1\frac{\partial T_1}{\partial r}\right)=0 \tag{5.4-3}$$

by considering a uniform pressure inside the bubble in order to compute accurately the bubble site heat transfer. Considering processes without mass transfer the liquid energy-conservation equation was neglected in this case because the main heat-transfer resistance is at the gas side. This scheme was combined with a homogeneous one-dimensional model of a bubbly flow by *Kim* (2002, 2004, 2005). *Kim* has demonstrated that important nonlinear effects of the wave propagation can be considered.

Nomenclature

Latin

a temperature conductivity, m/s

c_p	specific capacity at constant pressure, J/(kg K)
D_1	bubble diameter, m
D_3	droplet diameter, m
htc	heat-transfer coefficient, W/(m² K)
h	specific enthalpy, J/kg
Nu	$= htc 2R_1/\lambda$, Nusselt number, dimensionless
Pe	Peclet number, dimensionless
p	pressure
Q	accumulated heat transferred from the surface into the liquid, J
$\dot{q}_1^{\prime\prime 2\sigma}$	heat flux from the liquid interface into the gas, W/m²
$\dot{q}_2^{\prime\prime 1\sigma}$	heat flux from the gas interface into the liquid, W/m²
R_1	bubble radius, m
r	radius, m
T	temperature, K
u	radial velocity, m/s
x	$= r - R_1(\tau)$, distance from the bubble surface, m

Greek

β	$= (R_1/R_{10})^3$, dimensionless
ε	density ratio, dimensionless
γ_v	$(R_1/R_{10})^3$, ratio of the bubble volume to the initial volume, dimensionless
$\Delta\tau$	time step, s
Δh	evaporation specific enthalpy, J/kg
Δu_{12}^{σ}	$= u_1^{2\sigma} - u_2^{1\sigma}$, difference between the gas interface velocity and the liquid interface velocity.
ΔV_{12}	bubble velocity minus liquid velocity, m/s
δ	normalizing length, m
κ	isentropic exponent, dimensionless
λ	thermal conductivity, W/(m K)
η	dynamic viscosity, kg/(m s)
ω_0	eigenfrequency of bubble oscillation, 1/s
σ	surface tension, N/m
ρ	density, kg/m³
$(\rho w)_1^{2\sigma}$	mass flow rate from the liquid interface into the gas, kg/(m² s)
$(\rho w)_2^{1\sigma}$	mass flow rate from the gas interface into the liquid, kg/(m² s)
τ	time, s

Subscripts

1	Gas, bubble
2	Liquid
0	Initial
mom	computed by use of the momentum equation

Superscripts

$'$	saturated liquid
$''$	saturated vapor
$_{2}^{1\sigma}$	inside the liquid interface to the gas
$_{1}^{2\sigma}$	inside the gas interface to the liquid

References

Akhatov, I.S., Nigmatulin, R.I., Lahey Jr., R.T.: The analysis of linear and non linear bubble cluster dynamics. Multiphase Sci. Technol. 17(3), 225–256 (2005)

Alamgir, M.: Bubble growth in exponential varying pressure fields with large density variations. In: Dhir, V.K., Schrock, V.E. (eds.) The 22th National Heat Transfer Conference and Exhibition in Basic Aspects of Two Phase Flow and Heat Transfer, Niagara Falls, New York, August 5-8, HDT, vol. 34 (1984)

Bornhorst, W.J., Hatsopoulos, G.H.: Bubble growth, calculation without neglect of the interfacial discontinuity. J. Appl. Mech. 847–853 (1967)

Carlsaw, H.S., Jaeger, J.C.: Conduction of heat in solids. Oxford University Press, Oxford (1959)

Casadei, F.: A theoretical study on the growth of large sodium vapor bubbles in liquid sodium. Report of the Research Centre Karlsruhem KfK 3171, EUR 7053e (Mai 1981)

Celata, G.P., Cumo, M., D'Annibale, F., Farello, G.E.: Direct contact condensation of steam on droplets. Int. J. Multiphase Flow 17(2), 191–211 (1991)

Dergarabedian, P.: Observations on bubble growths in various superheated liquids. J. Fluid Mech. 9(1), 39–48 (1960)

Florschuetz, L.W., Chao, B.T.: On the mechanics of the vapour bubble collapse. Trans. ASME, J. Heat Transfer, Series C 87(2), 209–220 (1965)

Forster, H.K., Zuber, N.: Growth of a vapour bubble in a superheated liquid. J. Appl. Phys. 25(4), 474–478 (1954)

Fritz, W., Ende, W.: Ueber den Verdampfungsvorgang nach kinematographischen Aufnahmen an Dampfblasen. Phys. Z. 37, 391–401 (1936)

Hewit, H.C., Parker, J.D.: Bubble growth and collapse in liquid nitrogen. Trans. ASME, J. Heat Transfer, 22–26 (1968)

Hijikata, K., Mori, Y., Nagasaki, T., Nakagawa, M.: Structure of shock waves in two-phase bubble flow. In: Bergles, A.E., Ishigai, S. (eds.) Two-Phase Flow Dynamics, Japan-I.S. Seminar. Hemisphere Publ. Corp., Washington (1979)

Jones Jr., O.C., Zuber, N.: Bubble growth in variable pressure fields. J. Heat Transfer, Trans. ASME 100, 453–459 (1978)

Kamath, V., Prosperetti, A.: Numerical integration methods in gas-bubble dynamics. J. Acoust. Soc. Am. 85(4), 1538–1548 (1989)

Kim, D.C.: A nonlinear wave dynamics model for two-phase flows and its numerical solutions. Comput. Phys. Commun. 147, 526–529 (2002)

Kim, D.C.: Hydraulic transients in two phase bubbly flows. Natural Science, KORUS, 0-7803-8383-4/04, pp. 225–229 (2004)

Kim, D.C.: Application of fourth-order compact finite difference schemes for two-phase bubbly flow problems. In: 4th International Conference on Computational Heat and Mass Transfer, Paris-Cachan, May 17–20 (2005)

Kolev, N.I.: Multiphase flow dynamics, Fundamentals, 3rd edn., vol. 1. Springer, Berlin (2007)

Kolev, N.I.: Bubble growth and collapse in single component fluid. In: International Workshop On New Horizons in Nuclear Reactor Thermal Hydraulics, Niyamak Bhavan, Atomic Energy Regulatory Board, Anushaktinagar, Mumbai-400 094, January 7–8 (2010) , Keynote Address

Lee, H.S., Merte, H.: Spherical vapour bubble growth in uniformly superheated liquids. Int. J. Heat Mass Transfer 39(12), 2427–2447 (1996)

Mendez, F., Naude, J.: Numerical analysis of an asymptotic model for the collapse of vapour bubbles. Heat Mass Transfer 43, 325–351 (2007)

Mikic, B.B., Rohsenhow, W.M., Griffith, P.: On bubble growth rates. Int. J. Heat Mass Transfer 13, 657–666 (1970)

Nakoryakov, V.E., Pokusaev, B.G., Shreiber, I.R.: Wave propagation in gas-liquid media, 2nd edn. BCRC Press, Inc., New Brighton (1993)

Nigmatulin, R.I.: The fundamentals of heterogeneous mechanics. Nauka, Moscow (1978)

Nigmatulin, R.: Heat and mass transfer in wave dynamics of gas-liquid systems. In: Hetsroni, G. (ed.) Heat Transfer 1990, Proc. Of the Ninth Int. Heat Transfer Conference, Jerusalem, Israel, vol. 1, pp. 223–235 (1990)

Nigmatulin, R., Akhatov, I.S., Topolnikov, A.S., Bolotnova, R.K., Vakhitova, N.K.: Theory of suspercompression of vapour babbles and nanoscale thermonuclear fusion. Phys. Fluids 17, 107106, 1–13 (2005)

Plesset, M.S., Zwick, S.: The growth of bubbles in superheated liquids. J. Appl. Phys. 25(4), 493–500 (1954)

Plesset, M.S., Prosperetti, A.: Bubble dynamics and cavitation. Ann. Rev. Fluid Mech. 9, 149–185 (1977)

Prosperetti, A., Crum, L.A.: Nonlinear bubble dynamics. J. Acoust. Soc. Am. 82(2), 502–514 (1988)

Rayleigh, L.: On the pressure developed in a liquid during the collapse of spherical cavity. Phil. Mag. 34, 94 (1917)

Reynolds, A.B., Berthoud, G.: Analysis of EXCOBULLE two-phase expansion tests. Nucl. Eng. Des. 67, 83–100 (1981)

Richtmyer, R.D., Morton, K.W.: Difference method for initial value problems. Wiley, New York (1967)

Scriven, L.E.: On the dynamics of phase growth. Chem. Eng. Sci. 10(1/2), 1–18 (1959)

Theofanous, T., Biasi, L., Isbin, H.S., Fauske, H.: A theoretical study on bubble growth in constant and time dependent pressure field. Chem. Eng. Sci. 24, 885–897 (1969a)

Theofanous, T., Biasi, L., Isbin, H.S., Fauske, H.: Non-equilibrium bubble collapse: A theoretical study. In: Heat Transfer – Minneapolis, Chemical Engineering Progress Symposium Series, vol. 66(102), pp. 37–47 (1969b)

Toda, S., Kitamura, M.: Bubble growth in decompression fields. In: Proc. ASME-JSME Thermal Engineering Joint Conf., vol. 3, pp. 395–402 (1983)

Tsung-Chang, G., Bankoff, S.G.: Growth of vapour bubble under flashing conditions. In: Proc. 8th Int. Heat Transfer Conference, San Francisco, Calif., New York, pp. 1901–1906 (1986)

Wang, Z., Bankoff, S.G.: Bubble growth on a solid wall in a rapidly-depressurizing liquid pool. In: Witte, L.C., Avedisian, C.T. (eds.) AIAA/ASME Thermophysics and Heat Transfer Conference in in Fundamentals of Phase Change: Boiling and Condensation, Seattle, Washington, June 18–20 (1990)

Wang, Z., Bankoff, S.G.: Bubble growth on a solid wall in a rapidly-depressurizing liquid pool. Int. J. Multiphase Flow 17(4), 425–437 (1991)

Winters Jr., W.S., Merte Jr., H.: Experimental non-equilibrium analysis of pipe blow down. Nucl. Eng. Des. 69, 411–429 (1979)

Zuber, N.: The dynamics of vapour bubbles in non-uniform temperature fields. Int. J. Heat Mass Transfer 2, 83–98 (1961)

Zwick, S.A., Plesset, M.S.: On the dynamics of small vapour bubbles in liquid. J. Math. Phys. 3, 308–330 (1955)

6. How accurately can we predict nucleate boiling?

In this chapter we present a new theory of nucleate pool boiling based on turbulence in the boundary layer induced by bubble growth and departure that incorporates the effect of the static contact angle and explains the large spread of the available experimental data. In addition, an exciting feature of the new theory is demonstrated – the capability to predict the critical heat flux, with the same correlation which was never reported before in the literature as far as the author knows. This is a slightly modified version of the work previously published in Kolev (1995).

6.1 Introduction

Nucleate boiling is probably the most investigated phenomenon in the thermal sciences over the last 60 years. Data are collected, e.g., *Jakob* and *Fritz* (1931), *Jakob* and *Linka* (1933), *Yamagata* et al. (1955), *Kurihara* and *Myers* (1960), *Gaertner* and *Westwater* (1960), *Gaertner* (1965), *Borishanskii* et al. (1961), *Rallis* and *Jawurek* (1964), *Vachon* et al. (1968), *Wiebe* (1970), *Sultan* and *Judd* (1978), *Cornwell* and *Brown* (1978), *Nishikawa* et al (1984), see also Appendix 6.2, and many theoretical models exist, e.g., *van Stralen* and *Cole* (1979), *Rohsenow* (1952), *Fritz* (1935), *Vachon* et al. (1968), *Yang* and *Weisman* (1991), *Mikic* and *Rohsenow* (1969), *Cole* and *Rohsenow* (1969), *Zuber* (1959), *Zuber* (1960), *Judd* and *Hwang* (1976), *Zuber* (1963), *von Karman* (1921), *Pohlhausen* (1921), *Brauer* (1971), *Forster* and *Zuber* (1955), *Zuber* (1958), *Chen* (1966), *Jakob* (1949) *Kocamustafaogullari* and *Zshii* (1983), *Tolubinsky* and *Ostrovsky* (1966). The practical needs are satisfied usually by empirical correlations. In these correlations the heat flux at the wall is proportional to the wall superheating with respect to the saturation temperature to a power n

$$\dot{q}''_{NB} = h_{NB}\left(p\right)\left[T_w - T'\left(p\right)\right]^n,$$

which varies between 2 and 4. The coefficient h_{NB} is found to be mainly a function of the system pressure. Some of the frequently used correlations in the literature are summarized below.

1) *Thom* (1966): $n = 2$,

$$h_{NB\_Thom} = 1942\ \exp(p/4.35\times10^6).$$

2) *Miheev* and *Miheeva* (1973): $n = 3$, valid in $1\times10^5 \le p \le 200\times10^5\,Pa$,

$$h_{NB\_Micheev} = \frac{0.07845\,p^{0.53}}{(1-4.5\times10^{-8}\,p)^3}.$$

3) *Jens* and *Lottes* (1951), $n = 4$, valid in $p \le 140\times10^5\,Pa$, $\dot{q}''_{NB} < 1.1\times10^6 W/m^2$, $G < 10^4 kg/(m^2 s)$,

$$h_{NB\_Jens-Lottes} = 2.567\exp\left(p/1.55\times10^6\right).$$

4) *Borishanskii* et al. (1964), $n = 10/3$, valid in steel/water, $9.81\times10^4 \le p \le 196.2\times10^5\,Pa$, measurement error $\mp$ 5%,

$$h_{NB\_Borishanskii} = 2.28^{10/3}\left[\left(p/10^4\right)^{0.14} + 2.53\times10^{-6}\left(p/10^4\right)^2\right]^{10/3}.$$

The data of *Borischanski* are reproduced with rms 13.7%. His data can be reproduced with rms 10%, by

$$h_{NB\_Borishanskii} = \left\{10^{\left[-24.60379\,+\,13.14641\log_{10}p - 2.30694(\log_{10}p)^2\,+\,0.13636(\log_{10}p)^3\right]}\right\}^{10/3}.$$

The same data are reproduced with rms 11.2% with a correlation that is appropriate also for other fluid: $n = 3$,

$$h_{NB\_Borishanskii} = \left\{238\frac{\left(p_c/10^4\right)^{1/3}}{T_c^{5/6}M^{1/6}}\left(\frac{p}{p_c}\right)^{0.1}\left[1+4.65\left(\frac{p}{p_c}\right)^{1.16}\right]\right\}^3.$$

Here, M is the molar weight in kg-mole and c indicates the critical state. Note that the data obtained by *Borischanski* are for a *horizontal cylinder*. *Cooper* (1984) reported that

$$\frac{\dot{q}''^{1/3}_{w,flat\ plate}}{\Delta T_w}\bigg/\frac{\dot{q}''^{1/3}_{w,hor.\ cylinder}}{\Delta T_w} = \frac{55}{95}$$

and therefore

$$\dot{q}''_{w,flat\ plate} = 0.194\dot{q}''_{w,hor.\ cylinder}.$$

This means that the heat transfer at a horizontal cylinder is about 5 times better than at a horizontal plate. I see the explanation in the induced convection around the cylinder.

6) *Arefeva* and *Aladev* (1958) , $n = 4$, valid for $0.7 < \theta < 1.08$, 1 bar

$$h_{NB\_Arefeva-Aladev} = \left(2.559\theta - 0.727\right)^4 .$$

Arefeva and *Aladev* (1958), $n = 3/2$ valid for $1.6 < \theta < 2.41$ and 1 bar,

$h_{NB\_Arefeva-Aladev} = 3287.26$. Note the differences in the predictions among the different correlations as demonstrated in Fig. 6.1 for water at 1 bar pressure.

7. *Subbotin* et al. (1968) for boiling of *liquid metals*

$$h_{NB\_Subbotin} = c\dot{q}_w''^{2/3} \left(\frac{\lambda' \Delta h \rho'}{\sigma' T'}\right)^{1/3} \left(\frac{p}{p_c}\right)^m \pm 20\%$$

where c = 8 and s = 0.45 for $0.001 \le p/p_c \le 0.02$ and c = 1 and s = 0.15 for $4\times 10^{-5} \le p/p_c \le 10^{-3}$. The critical pressures with which the data are correlated by *Subbotin* et al. are 362×10^5 Pa, 171×10^5 Pa and 104×10^5 Pa for Na, K and Cs, respectively.

8. Sodium boiling at forced convection is associated with violent pulsations. At low pressure 0.3 to 1 bar, inlet subcooling of 85 to 170 K and incipient boiling superheat 40 to 125 K, respectively, inlet velocity 1.25 to 2.5 m/s in a 200 mm tube *Schleisiek* (1970) found

$$h_{FlowB\_Schleisiek} = 5.3 + 0.018\left(\mathrm{Re}\,\mathrm{Pr}\right)^{0.85} .$$

It seems that the question of how accurately we can predict the heat flux as a function of the wall superheating for simple pool boiling is banal and boring. However, let us consider this problem in more detail.

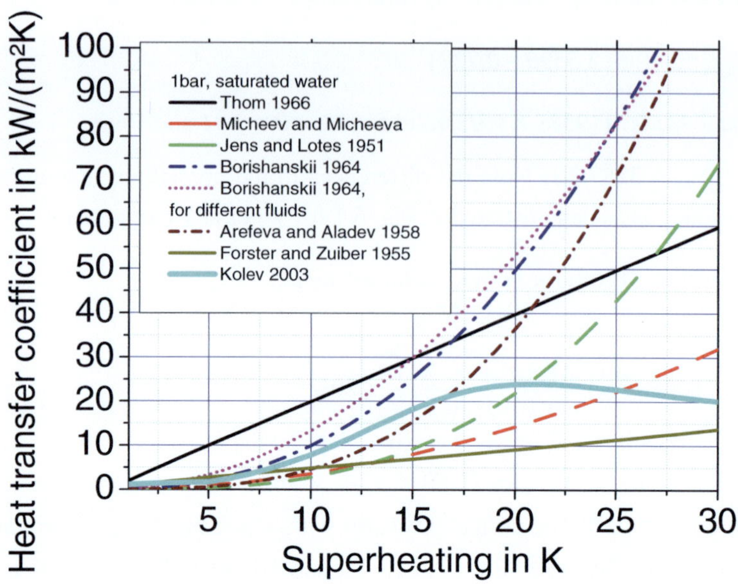

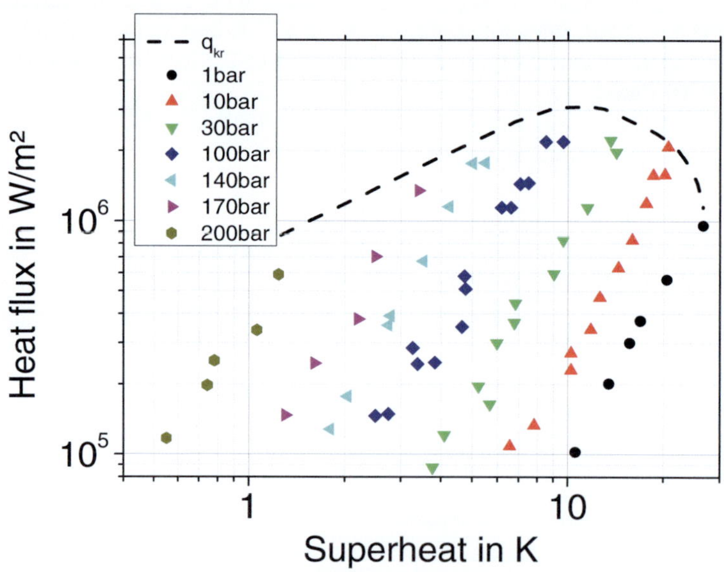

Fig. 6.1 a) Heat-transfer coefficient for pool boiling of water at 1 bar computed using different correlations; b) Heat flux for pool boiling of water at silver horizontal tube with 5 mm diameter – parameter pressure, *Golovin*, et al. (1963)

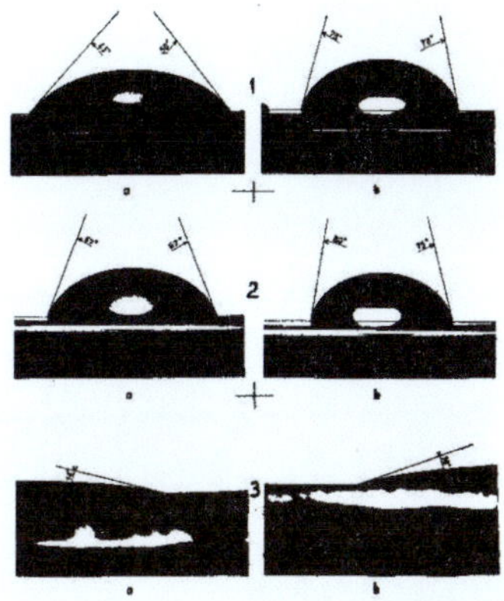

Fig. 6.2 Definition of the contact (wetting) angle: Water drops at solid surfaces: 1) stainless steel; 2) copper; 3) glass; a) surface just polished with emery paper; b) same surface after being treated with trichloroethylene, *Morin* (1964)

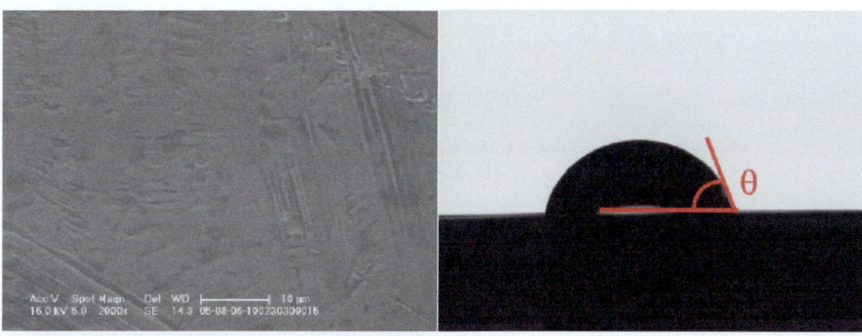

a) $\theta = 79°$

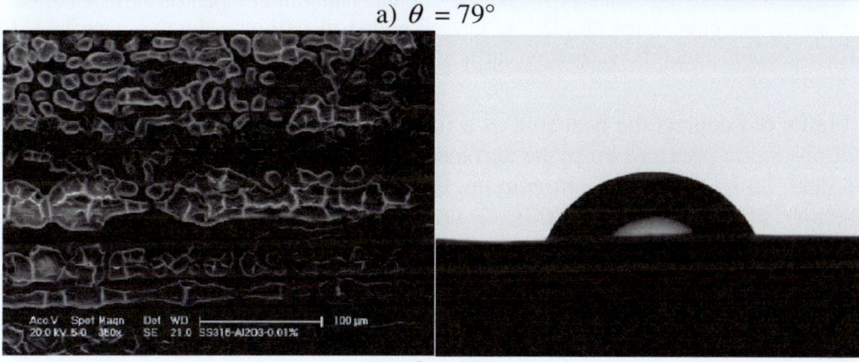

b) $\theta = 73°$

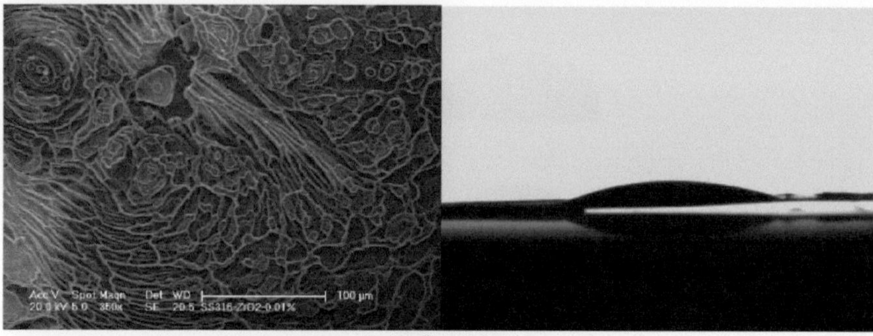

c) $\theta = 22°$

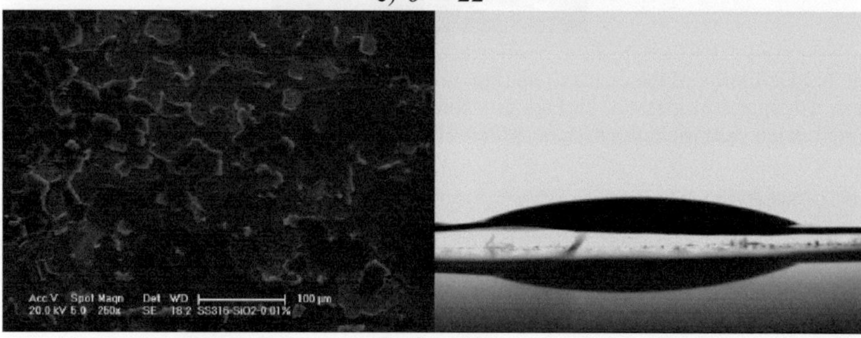

d) $\theta = 23°$

Fig. 6.3 *Kim* et al. (2007a, b): Left: SEM images of flat heater surface boiled in (a) pure water, (b) 0.01 v% alumina nanofluid, (c) 0.01 v% zirconia nanofluid, (d) 0.01 v% silica nanofluid. Right: Static contact angles of 5-. L sessile droplets on stainless steel surfaces, measured with a Krüss goniometer equipped with a camera monitor. (a) Pure water droplet on surface boiled in pure water, (b) 0.01 v% alumina nanofluid droplet on surface boiled in pure water; (c) pure water droplet on surface boiled in 0.01 v% alumina nanofluid, (d) 0.01v% alumina nanofluid droplet on surface boiled in 0.01 v% alumina nanofluid

Figure 6.4 depicts the heat flux as a function of the wall superheating for water at atmospheric pressure on plane surfaces taken from *Jakob* and *Fritz* (1931), *Jakob* and *Linka* (1933), *Yamagata* et al. (1955), *Kurihara* and *Myers* (1960), *Gaertner* and *Westwater* (1960), *Gaertner* (1965), *Borishanskii* et al. (1961), *Rallis* and *Jawurek* (1964), *Vachon* et al. (1968), *Wiebe* (1970), *Sultan* and *Judd* (1978), *Cornwell* and *Brown* (1978), *Nishikawa* et. al. (1984) for mirror-polished surfaces (data for specifically prepared rough surfaces are excluded).

By taking a close look at these data we see that while the reported error for temperature measurements is about 5 to 6% and for heat-transfer measurements about 1 to 14% in each particular experiment the spreading of different authors data is over two orders of magnitude, which cannot be explained with measurements error. This spreading corresponds to the data for the nucleation-site density (reported measurement error about 1 to 20%) as a function of wall superheat presented in Fig. 6.5 for the same data set as far as available *Avdeev* et al. (1977), *Brauer* (1971), *Chen*

(1966), *Cole* and *Rehsenow* (1969), *Fagginani* et al. (1981), *Gaertner* and *Westwater* (1960), *Gaertner* (1963). The state-of-the-art of the modeling of this phenomenon is well represented by *van Stralen* and *Cole* (1979), *Rohsenow* (1952), *Fritz* (1935), *Vachon* et al. (1968), *Yang* and *Weisman* (1991), *Mikic* and *Rohsenow* (1969), *Cole* and *Rohsenow* (1969), *Zuber* (1959), *Zuber* (1960), *Judd* and *Hwang* (1976), *Zuber* (1963), *von Karman* (1921), *Pohlhausen* (1921), *Brauer* (1971), *Forster* and *Zuber* (1955), *Zuber* (1958), *Chen* (1966), *Jakob* (1949), *Kocamustafagullari* and *Ishii* (1983), *Tolubinsky* and *Ostrovsky* (1966). Examining the main basic physical ideas on which these models rely (see Appendix 6.1) we surprisingly find that none of them predicts the main reason for the observed data spreading.

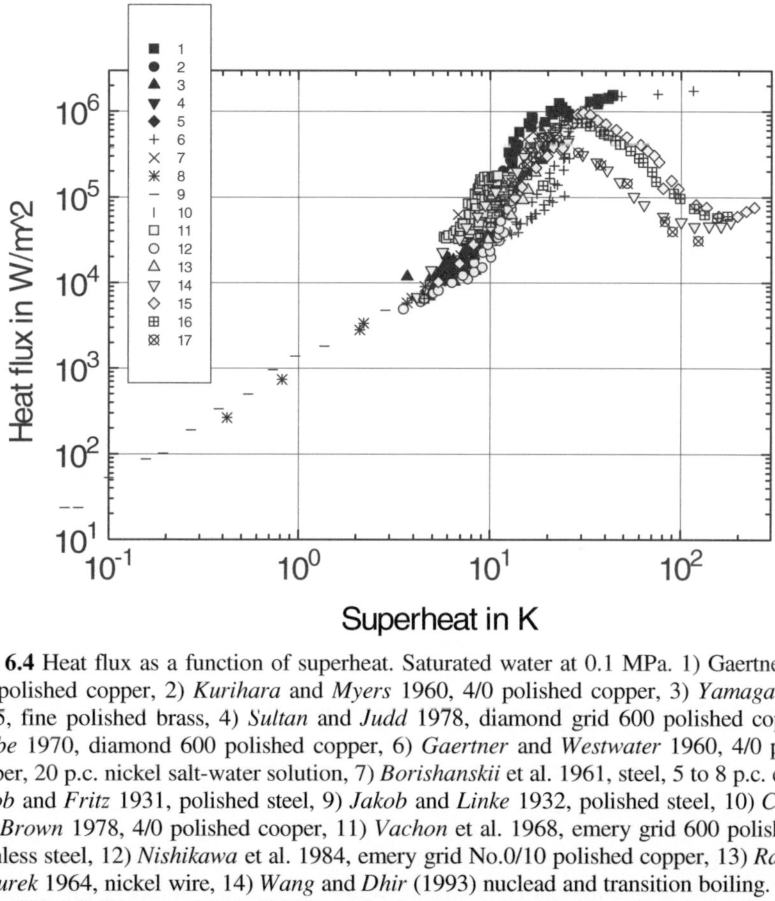

Fig. 6.4 Heat flux as a function of superheat. Saturated water at 0.1 MPa. 1) Gaertner 1965, 4/0 polished copper, 2) *Kurihara* and *Myers* 1960, 4/0 polished copper, 3) *Yamagata* et al. 1955, fine polished brass, 4) *Sultan* and *Judd* 1978, diamond grid 600 polished copper, 5) *Wiebe* 1970, diamond 600 polished copper, 6) *Gaertner* and *Westwater* 1960, 4/0 polished copper, 20 p.c. nickel salt-water solution, 7) *Borishanskii* et al. 1961, steel, 5 to 8 p.c. error, 8) *Jakob* and *Fritz* 1931, polished steel, 9) *Jakob* and *Linke* 1932, polished steel, 10) *Cornwell* and *Brown* 1978, 4/0 polished cooper, 11) *Vachon* et al. 1968, emery grid 600 polished 304 stainless steel, 12) *Nishikawa* et al. 1984, emery grid No.0/10 polished copper, 13) *Rallis* and *Jawurek* 1964, nickel wire, 14) *Wang* and *Dhir* (1993) nuclead and transition boiling. Contact angle 90°, 15) Wang and Dhir (1993) nuclead and transition boiling. Contact angle 18°, 17) Wang and Dhir (1993) nuclead and transition boiling. Contact angle 35°, 17) Wang and Dhir (1993) nucleate and transition boiling. Contact angle 35°. Liaw data

There are many qualitative hypotheses in the literature for the cause of the observed data spreading but none of them are used to quantitatively improve our predictive capability. An illustration of how a static contact angle influences the form of the solid–liquid–steam contact is demonstrated in Fig. 6.2.

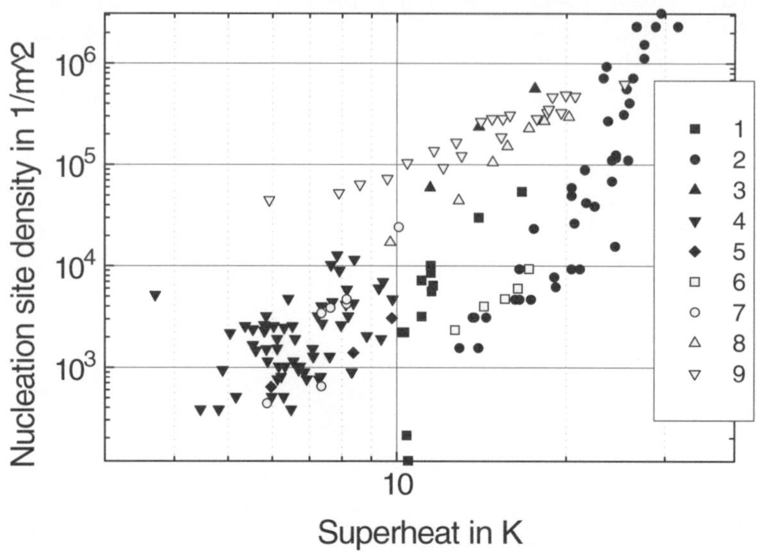

Fig. 6.5 Active nucleation-site density as a function of superheat. Saturated water at 0.1 MPa. 1) *Gaertner* 1965, 4/0 polished copper, 2) *Gaertner* and *Westwater* 1960, 4/0 polished copper, 20 p.c. nickel salt-water solution, 3) *Sultan* and *Judd* 1978, diamond grid 600 polished copper, 4) *Yamagata* et al. 1955, fine polished brass, 5) *Jakob* and *Linke* 1933, polished steel, 6) *Cornwell* and *Brown* 1978, 4/0 polished cooper, 7) *Kurihara* and *Myers* 1960, 4/0 polished copper, 8) *Rallis* and *Jawurek* 1964, nickel wire, 9) *Faggiani* et al. 1981, polished 316 steel horizontal cylinder

Wang and *Dhir* (1993) published interesting experimental work showing clearly the effect of the static contact angle on the active nucleation-site density as a function of the wall superheat. We transform the new data in the coordinates of Fig. 6.5 and plot them in Fig. 6.6. Comparing Figs. 6.5 and 6.6 it is obvious that variation of the static contact angle could account for a substantial part of the observed data spread in Fig. 6.5 and consequently in Fig. 6.4. This conclusion is supported by the following consideration. Nucleation-site density is reported to be measured with about 1 to 20% error. Static contact angle (if measured at all) is measured with ± 3 rad error. Distilled water being in contact with different polished solid surfaces not especially chemically treated possesses different static contact angles, as indicated in Tables 12.1 and 12.2. Since in Fig. 6.4 we have data sets for different static contact angles varying between very small values, as is the case with the *Gaertner* and *Westwater*'s data, and $\pi/2$, as seems to be the case for copper–water data, the data sets must differ from each other.

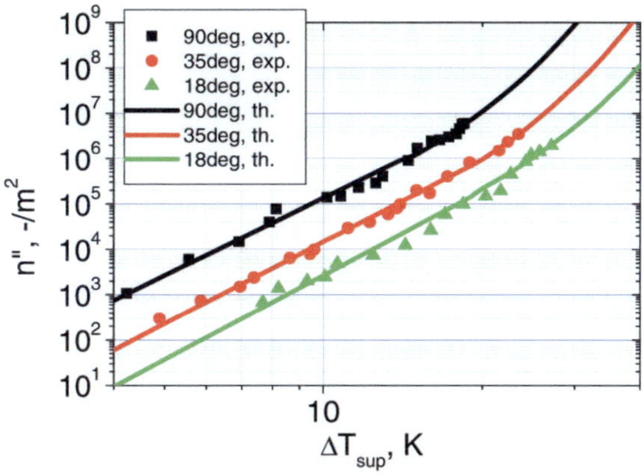

Fig. 6.6 Active nucleation-site density as a function of superheat. Saturated water at 0.1 MPa. *Wang* and *Dhir* (1993) data for three different static contact angles 1) 90°, 2) 35° and 3) 18°. Prediction of the same data with their correlation 4), 5), and 6), respectively. Larger static contact angle results of larger active nucleation-site density by the same superheating

Wang and *Dhir* correlated their data for active nucleation-site density as a function of the wall superheat as follows $n''_{1w} = 5\times 10^{-27}(1-\cos\varphi)/D_{1c}^{6}$. There is no instruction in their paper at which temperature the thermal properties for the *Laplace* equation have to be computed. Neither with the bulk temperature nor with the wall temperature, which is the logical choice, can the data be satisfactory represented by this correlation. As already mentioned in Chapter 1, because I will use these data to check my nucleate boiling theory I approximate them as accurately as possible: in the lower-temperature region linear in log-log representation and in the upper-temperature region exponential in the normal coordinate system. The result is

$$n''_{1w} = n''_{1w}(\theta, \Delta T_w), \tag{6.1}$$

where

$$n''_{1w}(18^\circ) = 0.001758368925\Delta T_2^{6.17928} \quad \text{for } \Delta T_w < 20 \text{ and}$$
$$n''_{1w}(18^\circ) = 411.05386 e^{\Delta T_w/3.18706} \quad \text{for } \Delta T_w \geq 20,$$

$$n''_{1w}(35^\circ) = 0.014362498\Delta T_2^{6.00464} \quad \text{for } \Delta T_w < 20 \text{ and}$$
$$n''_{1w}(35^\circ) = 524.77739 e^{\Delta T_w/2.66626} \quad \text{for } \Delta T_w \geq 20,$$

$$n''_{1w}(90^\circ) = 0.283302234\Delta T_2^{5.67893} \quad \text{for } \Delta T_w < 15 \text{ and}$$
$$n''_{1w}(90^\circ) = 1687.11274 e^{\Delta T_w/2.27907} \quad \text{for } \Delta T_w \geq 15.$$

Figure 6.6 shows the quality of this approximation. Note that the character of such approximation corresponds also to the measurements by *Basu* et al. (2002) presented in Fig. 1.5. Note also that the data for the lower-temperature region can be represented by the following single equation

$$n''_{1w} = 0.275181985\left[1-\cos(\theta)\right]^{1.68742} \Delta T_w^{5.64923+0.67\exp\left\{-\left[1-\cos(\theta)\right]/0.353\right\}} .$$

Of course it is possible to make the approximation in terms of

$$D_{1c} = \frac{4\sigma}{\left[p'(T_2)-p\right]\left(1-\dfrac{\rho''}{\rho_2}\right)}, \tag{6.2}$$

with $\sigma = \sigma(T_w)$ but I prefer here to approximate directly the measured dependence direct on the temperature difference to avoid any mistakes. Note that all saturation properties in this chapter are computed as a function of the wall temperature. The results of the prediction of n''_{1w} for the three different static contact angles ($\varphi =$ 90°, 35° and 18°) as a function of the wall superheating and the experimental data that they correlate are presented in Fig. 6.6. For the prediction we use the measured wall superheating $T_w - T'(p)$. For comparison see the previous correlations not taking into account the influence of the static contact angle published in *Cornwell* and *Brown* (1978), *Labuntsov* (1963), *Mickic* and *Rohsenow* (1969), *Avdeev* et al. (1977), *Hutcherson* et al. (1983), *Kocamustafaogullari* and *Ishii* (1983), *Riznic* and *Ishii* (1989), *Johov* (1969), *Jones* (1992a), *Jones* (1992b) and summarized in Appendix 6.2 – some of them are depicted together with data in Fig 6.7. The quantitative dependence on the static contact angle reported by *Wang* and *Dhir* was not previously known. It strongly supports the qualitative arguments published in 1958 by *Arefeva* and *Aladev* (1958) p.12 that are briefly summarized below:

" ... *Boiling of liquid is associated with liquid fragmentation by liquid detachment from the wall. Consequently steam must appear for lower wall superheats at places where the work for liquid detachment is lower. Quantitative measure for the influence of the wall on heat transfer is the rate of wetting i.e., the static contact angle. Therefore, (a) the nucleation site density and consequently the heat transfer coefficient increases with increasing static contact angle, and (b) increasing static contact angle decreases the critical heat flux...*"

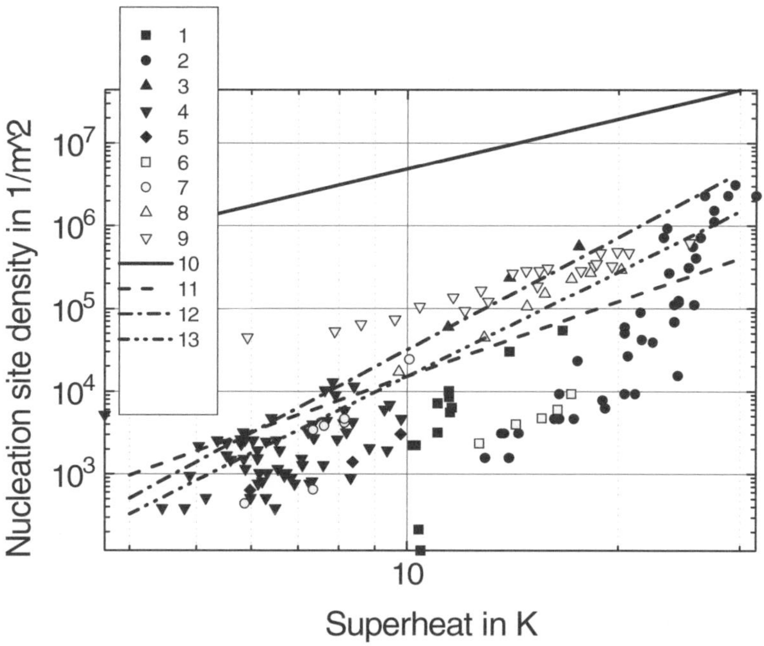

Fig. 6.7 Active nucleation-site density as a function of superheat. Saturated water at 0.1 MPa. 1) to 9) Data from Fig. 6.4. Prediction with correlations proposed by 10) *Avdeev* et al. (1977), 11) *Johov* (1969), 12) *Cornwell* and *Brown* (1978) and 13) *Kocamustafaogullari* and *Ishii* (1983)

Another important result of the measurements by *Wang* and *Dhir* is the relationship between the averaged nearest-neighbor distance and the nucleation-site density

$$\left(D_{2,\inf}^2\right)^{1/2} \approx 0.84 / n_{1w}^{\prime\prime\,1/2}, \tag{6.3}$$

which is obtained with better statistics than the earlier relation reported by *Gaertner* (1963) and *Sultan* and *Judd* (1978) $\approx 1/\left(\pi n_{1w}^{\prime\prime}\right)^{1/2}$. In what follows we shall demonstrate the importance of the experimental findings of *Wang* and *Dhir* for the prediction of heat transfer in nucleate boiling.

In this chapter, we present a new theory of nucleate pool boiling based on bubble growth and departure induced turbulence in the boundary layer that incorporates the effect of the static contact angle and explains the large spreading of the available experimental data. In addition, an exciting feature of the new theory is demonstrated – the capability to predict the critical heat flux with the same correlation. This was not previously reported in the literature as far as the author knows.

6.2 New phenomenological model for nucleate pool boiling

6.2.1 Basic assumptions

1. Wall superheating larger than a given value is necessary to initiate pool boiling. For smaller wall superheating the heat is transferred by natural convection, *Jakob* (1932)

$$\dot{q}''_{w2,nc} = 2.417\ \lambda_2 \left(\mathrm{Pr}_2\ g\beta_2 / \nu_2^2\right)^{1/4} \left(T_w - T_2\right)^{21/16}. \tag{6.4}$$

Here, indices "2" indicates bulk liquid values.

2. The heat flux from the wall into the boiling mixture during nucleate pool boiling is caused by boundary-layer turbulence due to bubble growth and departure.

3. The turbulent length scale is of the order of the *Rayleigh–Taylor* instability wavelength

$$\ell_2^t = \pi\ \lambda_{RT}\ , \tag{6.5}$$

where $\lambda_{RT} = \left\{\sigma / \left[g\left(\rho' - \rho''\right)\right]\right\}^{1/2}$ is the capillary *Laplace* constant. The imagination that boiling at a heated wall is in fact countercurrent gas liquid flow led *Zuber* (1958) and *Zuber* et al. (1958, 1961) to the conclusion that the instability of such flow is controlled by the *Rayleigh–Taylor* instability wavelength. This imagination led to a successful theory for critical heat flux. *Rohsenow* (1952) and *Tolubinsky* and *Ostrovsky* (1966) used λ_{RT} as a length scale to correlate data also for subcritical nucleate boiling heat transfer. Thus, we proceed with the idea that not only the continuous steam stream counter current to the liquid but also the train of bubbles counter current to the liquid produce macroscopic instability of order of the *Rayleigh–Taylor* instability wavelength. As will be demonstrated later by comparison with data this assumption seems quite appropriate. We found that neither the average distance between two departing bubbles, nor the bubble departure diameter is appropriate as a macroscopic length scale of turbulence for this process.

4. The nucleation-site distribution obeys the *Poisson* distribution law and the experimentally observed averaged site to site space $D_{2,inf}$ is given by Eq. (6.3).

5. The bubble growth at the wall is thermally controlled. The expression governing the bubble growth is

$$D_1 = 2B\tau^n \tag{6.6}$$

As shown in Appendix 13.1 it is generally agreed that the exponent is of the order of 1/2 for saturated liquids. For the purpose of data comparison we use in this work the *Labuntsov* approximation (1974, 1963) of the *Scriven* solution for bubble growth in a bulk liquid,

$$B = cJa'a'^{1/2}, \tag{6.7a}$$

where

$$c = \left(12/\pi\right)^{1/2}\left[1+\frac{1}{2}\left(\frac{\pi}{6Ja'}\right)^{2/3}+\left(\frac{\pi}{6Ja'}\right)^{1/2}\right]^{1/2}, \tag{6.8}$$

and n = 1/2. Here, $Ja' = \rho' c_p' \left[T_2 - T'(p)\right] / \left(\rho'' \Delta h\right)$ is the *Jacob* number. Alternatively the *Labuntsov* et al. (1964) solution for bubble growth at a heated wall

$$B = \sqrt{2\beta\, a'Ja'}, \tag{6.7b}$$

where

$$\beta \approx \frac{25.3\cos\left(\theta/2\right)}{\left(1+\cos\theta\right)^2\left(2-\cos\theta\right)},$$

validated for water boiling at silver surface at p = 1 to 100 bar, can also be used.

6. The order of magnitude of the minimum waiting time will be computed in accordance with the proposal by *Han* and *Griffith* (1965)

$$\Delta\tau_{1w} \approx \Delta\tau_{1w,\min} = \delta_{2,\min}^2 / \left(\pi a'\right), \tag{6.9}$$

where

$$\delta_{2,\min} = 12\left(T_w - T_2\right)T'(p)\sigma / \left\{\left[T_w - T'(p)\right]^2 \rho''\left(h'' - h'\right)\right\} \tag{6.10}$$

is the minimum of the thermal boundary-layer thickness.

7. The bubble departure diameter D_{1d} and consequently the bubble departure time

$$\Delta\tau_{1d} = \frac{1}{4} D_{1d}^2 / B^2, \tag{6.11}$$

are controlled by the force balance in accordance with the model

$$\left(D_{1d} / D_{1d,nc}\right)^3 + \left(D_{1d} / D_{1d,fc}\right)^2 = 1 \tag{6.12}$$

presented recently in *Kolev* (1994). The reader will find in *Kolev* (1994) the detailed derivation of Eq. (6.12) and the method for computation of the bubble departure diameter for natural convection only, $D_{1d,nc}$, and for forced convection only, $D_{1d,fc}$, which will not be repeated here. Further the model was compared in *Kolev* (1994) with the available data for saturated water for pool and flow boiling at low and elevated pressures and good agreement was found. The most exciting feature of this model is the taking into account of the influence of the mutual bubble interaction on the bubble departure diameter, something that was experimentally observed but never theoretically explained before.

The frequency of bubble departure from the wall is therefore

$$f_{1w} = 1/\left(\Delta\tau_{1d} + \Delta\tau_{1w}\right). \tag{6.13}$$

It should not be confused with the average frequency of the turbulent pulsation in the wall boundary layer.

6.2.2 Proposed model

The proposed model is strictly valid for pool boiling of saturated liquid. Nevertheless, the main ideas used here can be easily extended for nucleate flow boiling of saturated liquid and pool and flow boiling of subcooled liquid, as will be discussed in Section 6.4.

Before jumping apart from the wall, the turbulent eddies stay at the wall during the time $1/f_{1w}^t$ and receive heat from the wall by heat conduction. Therefore, the average heat flux at the wall follows the analytical solution of the *Fourier* equation averaged over the period $\Delta\tau = 1/f_{1w}^t$

$$\dot{q}''_{w2,nb}(\Delta\tau) = \frac{1}{\Delta\tau}\int_0^{\Delta\tau}\dot{q}''_w(\tau)\,d\tau = 2\left(\frac{\lambda'\rho'c'_p}{\pi\Delta\tau}\right)^{1/2}(T_w - T_2)$$

$$= \left(2/\pi^{1/2}\right)\left(f_{1w}^t a'\right)^{1/2}\rho'c'_p\left(T_w - T_2\right). \tag{6.14}$$

The idea to use the time-averaged heat flux at the wall stems from *Mikic* and *Rohsenow* (1969). In contrast to these authors we use here the turbulence renewal period rather than the bubble departure time. The time- and space-averaged pulsation velocity is

$$\overline{\overline{V}}{}'_2 = (B^2/D_{2,\inf})\frac{\Delta\tau_d}{\Delta\tau_w + \Delta\tau_d} = \frac{1}{0.84}B^2 n_{1w}''^{\,1/2}\frac{\Delta\tau_{1d}}{\Delta\tau_{1w} + \Delta\tau_{1d}}. \tag{6.15}$$

The above time- and space-averaged pulsation velocity is in fact the space-averaged microconvection velocity, first computed by *Forster* and *Zuber* (1955), see also *Zuber* (1959) p. 12, which is then time averaged. The most unexpected effect introduced by the additional time averaging is the capability of the finally obtained equation to predict the critical flux, as will be demonstrated later. The fluctuation frequency is therefore a function of the bubble departure frequency

$$f_{1w}^t = \overline{\overline{V}}{}'_2/\ell_2^t = \frac{1}{0.84}B^2 n_{1w}''^{\,1/2}\frac{\Delta\tau_{1d}}{\Delta\tau_{1w} + \Delta\tau_{1d}}\Big/\ell_2^t. \tag{6.16}$$

Substituting Eq. (6.16) into Eq. (6.14) we obtain

$$\dot{q}''_{w2,nb} = c_1\left(\frac{\lambda'\rho'c'_p}{\ell_2^t\left(1 + \Delta\tau_{1w}/\Delta\tau_{1d}\right)}\right)^{1/2} n_{1w}''^{\,1/4}B\left(T_w - T_2\right). \tag{6.17}$$

Here, $c_1 = 2/(\pi 0.84)^{1/2}$ is of order of unity. For the data comparison the value of

$$c_1 = 1.4626 \tag{6.18}$$

was used. Because $\Delta\tau_{1w}$ is a very rough estimate for the averaged waiting time and therefore the ratio $\Delta\tau_{1w}/\Delta\tau_{1d}$ is only approximately analytically known we introduce the empirical constant $c_2 \Delta\tau_{1w}/\Delta\tau_{1d}$

$$\dot{q}''_{w2,nb} = c_1 \left(\frac{\lambda' \rho' c'_p}{\ell^t_2 \left(1 + c_2 \Delta\tau_{1w}/\Delta\tau_{1d}\right)} \right)^{1/2} n''^{1/4}_{1w} B\left(T_w - T_2\right). \tag{6.19}$$

The effect of c_2 will be demonstrated by data comparison. As will be shown

$$c_2 = 0.7 \tag{6.20}$$

is appropriate. Let us now see approximately the exponent of the temperature difference dependence. If Eq. (6.1) can be written in the form $n''_{1w} \approx f(\theta)\Delta T^6$, which is valid before DNB, and using Eq. (6.7a) I obtain

$$\dot{q}''_{w2,nb} = \frac{\left[f(\theta)\right]^{1/4} c_1 c}{\sqrt{\ell^t_2 \left(1 + c_2 \Delta\tau_{1w}/\Delta\tau_{1d}\right)}} \frac{\lambda' \rho' c'_p}{\rho'' \Delta h} \left(T_w - T_2\right)^{7/2}. \tag{6.21}$$

Note the power of the temperature difference if a bubble growth equation for bulk flow is used is 7/2 = 3.5. If we use Eq. (6.7b) for bubble growing at a wall we obtain

$$\dot{q}''_{w2,nb} = \frac{c_1 \left[f(\theta)\right]^{1/4} \sqrt{2\beta}}{\sqrt{\ell^t_2 \left(1 + c_2 \Delta\tau_{1w}/\Delta\tau_{1d}\right)}} \lambda' \sqrt{\frac{\rho' c'_p}{\rho'' \Delta h}} \left(T_w - T_2\right)^3, \tag{6.22}$$

a power equal to 3. Therefore, theoretical basics dictate the dependence

$$\dot{q}''_{w2,nb} \propto \left(T_w - T'_2\right)^{3 \text{ to } 7/2}.$$

The mass of steam generated per unit time and unit flow volume is therefore

$$\mu_{21,gen} = \frac{4}{D_h} \dot{q}''_{w2,b} / (h'' - h_2). \tag{6.23}$$

The corresponding number of bubbles generated with bubble departure diameter per unit flow volume is

$$\dot{n}_{1w} = \mu_{21} \Big/ \left(\rho_1 \frac{\pi}{6} D^3_{1d} \right). \tag{6.24}$$

From that moment the bubbles start their own history inside the bulk flow and undergo condensation or further evaporation as well as collisions, coalescence or fragmentation.

6.3 Data comparison

Comparison of the prediction of Eq. (6.21) with all data available to the author for atmospheric pool boiling of water at a vertical surface is shown in Figs. 6.8 and 6.9. We see from Figs. 6.8 and 6.9 that the data in the region of pool boiling are well reproduced by Eq. (6.21) including the important effect of the static contact angle.

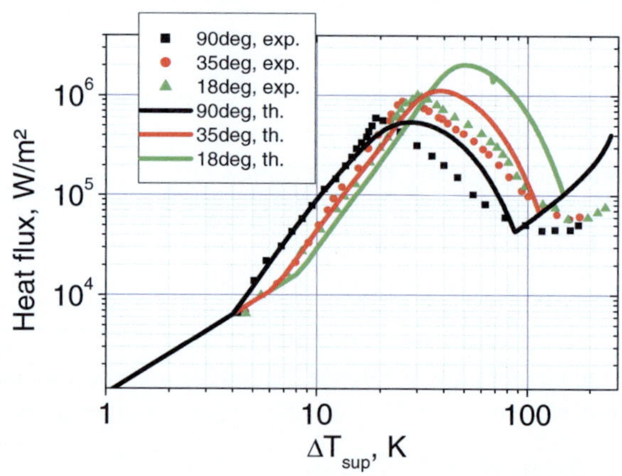

Fig. 6.8 Heat flux as a function of superheat. Saturated water at 0.1MPa. n''_{1w} best fit of corresponding data. Comparison of the prediction of the new theory with the experimental data by *Wang* and *Dhir* for three different static contact angles: 1) Exp.-90°; 2) Exp.-35°; and 3) Exp.-18°, 4) *Kolev*-90°; 5) *Kolev*-35°; and 6) *Kolev*-18°. The larger the static contact angle the smaller the critical heat flux

The film boiling part is computed using the expression for a vertical plate from Chapter 11 with averaging length equal to the Rayleigh–Taylor wavelength.

In order to demonstrate the effect of the additional time averaging we set

$$\Delta\tau_{1d} / \left(\Delta\tau_{1d} + \Delta\tau_{1w}\right) \approx 1, \tag{6.25}$$

which is equivalent to $c_2 = 0$,

$$\dot{q}''_{w2,nb} = \frac{\left[f(\theta)\right]^{1/4}}{\sqrt{\ell_2^t}} \frac{c_1 c\, \lambda' \rho' c'_p}{\rho'' \Delta h} \left(T_w - T_2\right)^{7/2}. \tag{6.26}$$

and compare only with the original data set reported by *Wang* and *Dhir*. The result is shown in Fig. 6.9.

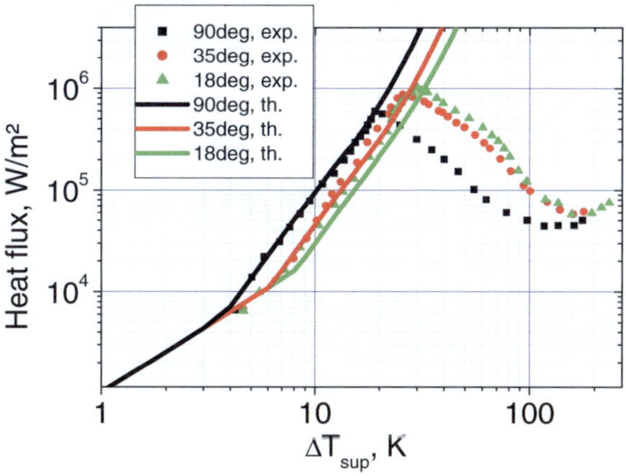

Fig. 6.9 *Wang* and *Dhir* experimental data for heat flux as a function of superheat for saturated water at 0.1 MPa and for three different static contact angles: 1) 90°; 2) 35°; and 3) 18°. 4), 5), 6) Prediction of the new theory for the corresponding static contact angles without taking into account the time averaging that is with $c_2 = 0$

From the comparison of Fig. 6.9 with Fig. 6.8 we see also that the ratio $\Delta\tau_{1w}/\Delta\tau_{1d}$ introduced by the time average is becoming important in the region of high superheating where the predicted heat flux no longer decreases with increasing superheating.

Surprisingly, the obtained equation seems to predict the critical heat flux without additional correlation.

The mechanism that works in this case is the dramatic increase of the nucleation-site density for large temperature differences, which causes a dramatic decrease of the bubble departure diameter and therefore a dramatic decrease of the bubble departure time. As a result, we reach the critical heat flux, a heat flux that does not increase any longer with increasing temperature difference. The surprising effect here is that we predict accurately the nucleate boiling heat transfer and the critical heat flux with a *single theory*. The theory also predicts the observed decreasing critical heat flux with increasing contact angle that is with decreasing wettability. This effect is observed also by *Ramilison*,

$$\dot{q}''_{CHF} / \dot{q}''_{CHF,Zuber} = 0.0336(\pi - \theta)^3 k^{0.125}, \tag{6.27}$$

where k is the RMS surface roughness.

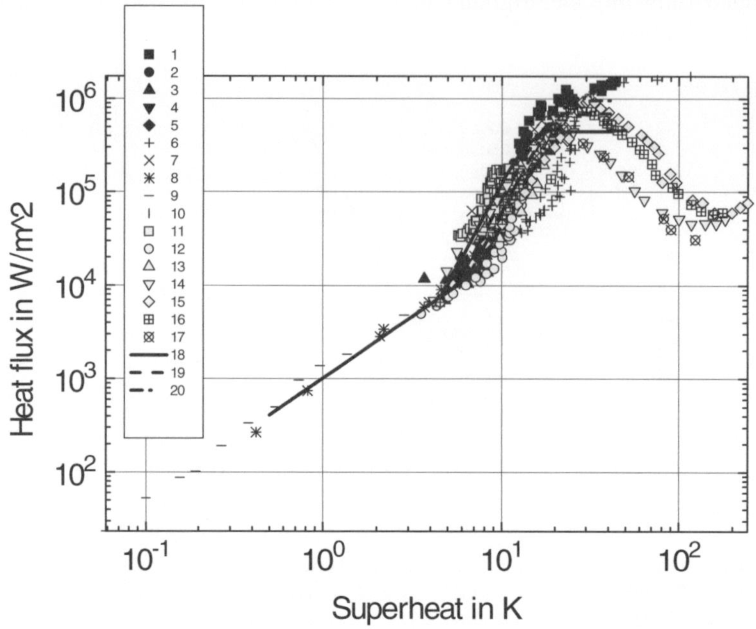

Fig. 6.10 Experimental data for heat flux as a function of superheat for saturated water at 0.1 MPa - 1) to 17) as in Fig. 6.1. Prediction of the new theory for three different static contact angles: 18) *Kolev*-90°; 19) *Kolev*-35°; and 20) *Kolev*-18°

Note the difference from the so-called "hydrodynamic theory of critical heat flux" proposed first by *Kutateladse* (1954, 1962) and later by *Zuber* (1958), *Zuber* and *Tribus* (1958) and *Zuber* et al. (1961). A brief summary of this theory is given in Section 10.4.1. The hydrodynamic theory of critical heat flux is intended to predict only the critical heat flux up to which the nucleate boiling regime is possible but it was not aimed to predict heat transfer in nucleate boiling itself. Note the two completely different approaches to the boiling crises by comparing this theory with the theory presented in Section 10.4.1. We showed here that the microscale of the nucleation processes, bubble growth and departure, and the turbulence generated by this process naturally lead to self-saturation of the boundary layer with bubbles and dramatically changes of the phenomenon. In contrast, *Kutateldse* and *Zuber* analyzed the macroscale of the instability above the layer where the microphysics happens. That is why their analysis gives an absolute upper limit that is about 5 times larger then the observed critical heat flux. This is reflected in the necessity of introducing an empirical constant of about 1/5 – see Section 10.4.1.

Recently, *Kim* et al. (2006) confirmed that microparticles in water increase the critical heat flux. For the explanation of the observed phenomena the authors considered the hydrodynamic stability theory, *Kutateladse* (1954, 1962), *Zuber* (1958, 1958) the microlayer theory, *Haramura* and *Katto* (1983) and the hot/dry spot theory *Theofanus* et al. (2002). All these theories are shown not to take into account the nucleation kinetics at the surface and its relation to active nucleation sites. Only the theory presented here was capable of explaining the change of the CHF with the change of the active nucleation-site density as presented in Fig. 6.11. In this figure Eq. (6.21) is used in the form $n''_{1w} = C(1-\cos\varphi)(T_w - T_2)^6$.

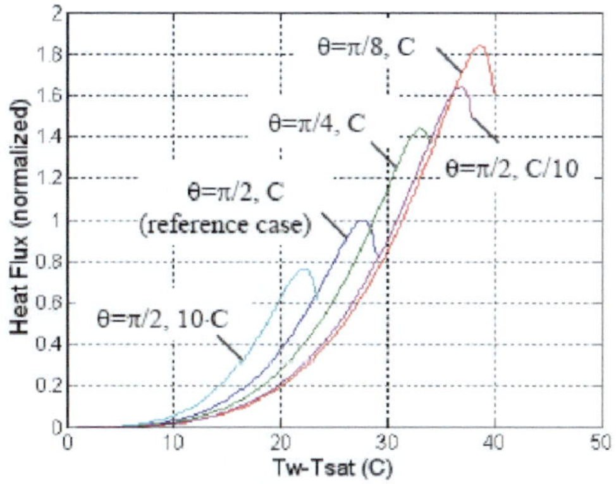

Fig. 6.11 Effect of the contact angle and the active nucleation-site density on the boiling curve, *Kim* et al. (2006)

6.3.1 Nucleation-site density at high pressures

The theory discussed here was based on the expression for the nucleation-site density derived at 1 bar experimentally. For high-pressure systematic data for water do not exist. One intermediate solution for deriving an expression for the nucleation-site density is to equalize the nucleate boiling heat flux defined by the simplified Eq. (6.17)

$$\dot{q}''_{w2,nb} = \frac{2}{\sqrt{\pi}0.84}\left(\frac{\lambda'\rho'c'_p}{\pi\,\lambda_{RT}}\right)^{1/2} n''^{\,1/4}_{1w} B\left(T_w - T_2\right) \tag{6.28}$$

to one of the empirical expressions given in the introduction

$$\dot{q}''_{NB} = h_{NB}(p)\left[T_w - T'(p)\right]^n \tag{6.29}$$

and solving with respect to the active nucleation-site density. The result is

$$n''_{1w} = 4.29\lambda^2_{RT}\left\{h_{NB}(p)\frac{\left[T_w - T'(p)\right]^{n-1}}{B\sqrt{\lambda'\rho'c'_p}}\right\}^4. \tag{6.30}$$

For $B=\sqrt{2\beta\, a'Ja'}$ we obtain

$$n''_{1w} = \frac{4.29}{(2\beta)^2}\lambda_{RT}^2 \left\{ h_{NB}(p) \frac{\left[T_w - T'(p)\right]^{n-1}}{\lambda'\sqrt{Ja'}} \right\}^4 . \tag{6.31}$$

All properties are computed at the bulk temperature. Using the correlation by *Borishanskii* in Eq. (6.31) that is appropriate for different fluids and introducing the multiplier $3.57^{-4}(1-\cos\theta)^{1.5}$ we obtain the results presented in Fig. 6.12a. Therefore we obtain a prediction method that is as accurate as the empirical correlation by *Wang* and *Dhir* at low pressure and possesses a physically based potential for prediction of the active nucleation sites up to 196.2 bar.

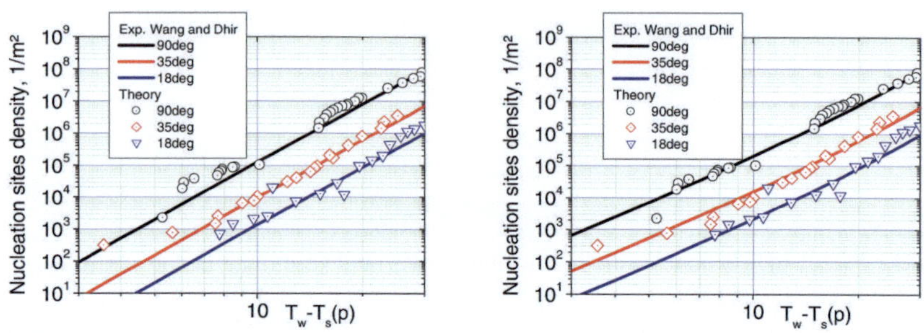

Fig. 6.12 Active nucleation-site density as a function of superheat. Saturated water at 0.1 MPa. *Wang* and *Dhir* (1993) data for three different static contact angles 1) 90°, 2) 35° and 3) 18°. a) Prediction of the same data with Eq. (6.31) and the *Borishanskii* et al. (1964) correlation for pool boiling; b) Prediction with the *Kocamustafaogullari* and *Ishii* (1983) empirical correlation multiplied by $44.54(1-\cos\theta)^{1.5}$

Note that the idea to use an expression for the heat flux in pool boiling in the form $\dot{q}''_{w2,nb} = f(p)\, n''^{\,m}_{1w}(T_w - T_2)^n$, data approximation in the form $\dot{q}''_{w2,nb} = f(T_w - T_2)$, equalizing them and solving with respect to n''_{1w} was proposed first by *Kocamustafaogullari* and *Ishii* (1983). These authors already expressed at that time the opinion that their approach is preliminary until detailed measurements for the nucleation-site density and other internal boiling characteristics are available. It is interesting to note that if their correlation is multiplied by $44.54(1-\cos\theta)^{1.5}$, it predicts an even somewhat better slope of the low-pressure data as shown in Fig. 6.12b.

6.4 Systematic inspection of all the used hypotheses

1) The used hypotheses for the development of the bubble departure theory presented in Chapter 15 are checked against data for saturated water pool boiling for

a) wall superheating up to 40 K;
b) mass flow rates up to 360 kg/(m² s); and
c) pressure up to 24 bar;

and good agreement was obtained. The trends of the dependence of the bubble departure diameter on superheating, mass flow rate and pressure are also well predicted. Thus, I could expect that the theory will also hold outside this region.

2) There are only two constants in Eq. (6.17):

a) The theoretical value of $c_1 = 1.23$ is corrected with 19% to 1.4626 after comparison with experimental data, which is not unusual in heat-transfer theory at all.
b) The c_2 constant is empirically introduced to account for uncertainties in the prediction of the ratio $\Delta\tau_{1w} / \Delta\tau_{1d}$. It was demonstrated that for superheatings giving heat flux up $\approx 70\%$ from the critical heat flux this correction is unimportant. It becomes important if one uses the new theory to improve heat-flux predictions for superheatings giving heat fluxes $\approx 70\%$ from the critical heat flux.

3) The new heat-transfer theory is strictly valid for pool boiling of saturated liquid. The bubble departure theory and therefore the theory for prediction of the fluctuation velocity in the boundary layer are correct also for flow boiling at saturated liquid. The ideas used in this chapter to reach Eq. (6.17) can be also used to extend its validity to flow boiling. The main questions that should be addressed in this case is whether the characteristic length scale of turbulence in the wall boundary layer will change very much for flow boiling with respect to $\pi\lambda_{RT}$. If no, the same theory should be used. If yes, additional investigation is *necessary* in which the flow induced fluctuations in the wall boundary layer should be superimposed onto the bubble departure generated fluctuations. This remains to be done.

4) The extension of the theory for subcooled nucleate boiling requires modification of the bubble departure model published in *Kolev* (1994). The means is: replace the bubble-growth model for saturated water with a bubble-growth model for subcooled water. This remains to be done.

6.5 Significance

1. The economical design of heat exchangers working with pool boiling depends on the accuracy of the prediction models. I believe that the proposed method will find application in this field.

2. The appropriate quantitative separation of the effects of single submechanisms into a single correlation for the prediction of pool boiling has never succeeded

before in spite of the fact that it was frequently tried. The prediction of the heat transfer in pool boiling together with the prediction of the critical heat flux for water boiling at atmospheric pressure seems to be indicative that Eq. (6.22) quantitatively predicts the appropriate separate effects.

6.6 Conclusions

1. The turbulence induced by the bubble growth and departure is the main heat-transfer mechanism during pool boiling.

2. The static contact angle is a very important parameter explaining the large spread of the experimental data for pool boiling.

3. The new theory consists of sound verified subtheories and models

- the bubble departure diameter, confirmed by data comparison (Chapter 15);
- the active nucleation-site density provided by *Wang* and *Dhir* (1993);
- the mean site to site distance provided by *Wang* and *Dhir* (1993);

and the hypothesis of bubble growth and departure introduced turbulence in the boundary layer as the main heat-transfer mechanism. These are new features that do not follow from the preceding works *van Stralen* and *Cole* (1979), *Rohsenow* (1952), *Fritz* (1935), *Vachon* et al. (1968), *Yang* and *Weisman* (1991), *Mikic* and *Rohsenow* (1969), *Cole* and *Rohsenow* (1969), *Zuber* (1959), *Zuber* (1960), *Judd* and *Hwang* (1976), *Zuber* (1963), *von Karman* (1921), *Pohlhausen* (1921), *Brauer* (1971), *Forster* and *Zuber* (1955), *Zuber* (1958), *Chen* (1966), *Jakob* (1949), *Kocamustafaogullari* and *Ishii* (1983), *Tolubinsky* and *Ostrovsky* (1966).

4. Equation (6.22) predicts the existing experimental data including the effect of the static contact angle with sufficient accuracy. It is recommended for general use. Surprisingly, Eq. (6.22) also seems to predict the critical heat flux, which was demonstrated for water at atmospheric pressure.

5. Increasing the static contact angle leads to decreasing critical heat flux.

6. From the simplified Eq. (6.26) one should easily estimate that if the measurement error for wall superheat is of order of 5 to 6% the error of the comparison with the experimental data with known static contact angle will be about 21%. Heat-transfer data with unknown static contact angle are not appropriate for validation of any nucleate boiling heat-transfer theory.

From the above conclusions we can derive some practical recommendations for the application of the new theory.

1. Equating the predictions of Eq. (6.4) and (6.22) computes the wall temperature $T_{w,NBI}$ necessary to initiate nucleate boiling

$$T_{w,NBI} = T_2 + \left\{ \frac{2.417\,\lambda_2 \left(\Pr_2 g\beta_2 / \nu_2^2 \right)^{1/4} \sqrt{\ell_2'}}{c_1 \left[f(\theta) \right]^{1/4} \sqrt{2\beta}\, \lambda' \sqrt{\dfrac{\rho' c_p'}{\rho'' \Delta h}}} \right\}^{0.595} . \tag{6.22}$$

2. If the actual wall temperature is less that $T_{w,NBI}$ we have natural circulation heat transfer. In this case Eq. (6.4) describes the heat-transfer number.

3. If the wall temperature is larger than $T_{w,NBI}$ we have two possibilities to compute the heat flux:

a) Use existing correlation for prediction of the critical heat flux; predict the heat flux by the simplified theory, Eq. (6.26), if the predicted heat flux is below the critical one the prediction of Eq. (6.26) is the desired one. Note that in this case the error is increasing in the region between 70 and 100% of the critical heat flux.
b) Use the complete theory for prediction of the heat flux, which is Eqs. (6.19) and (6.21), together with the complete algorithm described in Chapter 15 for prediction of the bubble departure diameter.

4) In accordance with the measurements by *Arefeva* and *Aladev* (1958) the influence of the contact angle on the heat flux is observed in the region between $\varphi = 0$ and $\pi/2$. Beyond this contact angle nucleate boiling heat transfer does not depend on the contact angle. That is why we recommend to use the present theory in the region $\varphi = 0$ and $\pi/2$. For larger contact angles we recommend to use $\varphi = \pi/2$ until data for the nucleation-site density as a function of superheating are available for static contact angles larger than $\pi/2$.

6.7 Extension to forced convection with nucleate boiling

The cross-section-averaged time scale of the fluctuation of *small* eddies in developed pipe flow is therefore

$$\Delta\tau_{\mu e,2,\infty} \approx b\sqrt{\nu_2/\varepsilon_{2,\infty}} ,$$

and therefore the frequency of turbulent pulsations is of order of $f_{2,\infty}' = 1/\Delta\tau_{\mu e,2,\infty}$. There is a superposition of the frequencies in turbulent flow cased by developed turbulence and by the departing bubbles. Therefore

$$f_{1w} = 1/\left(\Delta\tau_{1d} + \Delta\tau_{1w}\right) + 1/\Delta\tau_{e,2,\infty}$$

and

$$\dot{q}''_{w2,nb}(\Delta\tau) = \frac{1}{\Delta\tau}\int_0^{\Delta\tau}\dot{q}''_w(\tau)d\tau = 2\left(\frac{\lambda_2\rho_2 c_{p2}}{\pi\Delta\tau}\right)^{1/2}(T_w - T_2)$$

$$= \left(2/\pi^{1/2}\right)\left(\frac{1}{\Delta\tau_{1d}+\Delta\tau_{1w}} + \frac{1}{\Delta\tau_{e,2,\infty}}\right)^{1/2} a_2^{1/2}\rho_2 c_{p2}(T_w - T_2)\,.$$

Here, using the results of *Chandesris* et al. (2005) for computation of the developed steady-state turbulent kinetic energy and its dissipation in pipes we obtain

$$\varepsilon_{2,\infty} = \frac{\lambda_{fr}}{D_h}\frac{1}{2}\overline{w}^3\left(1 - y^+_{\lim}\sqrt{\frac{\lambda_{fr}}{8}}\right),$$

where $y^+_{\lim} = 8,\ 7,\ 16$, $c_k = 0.0306,\ 0.0367,\ 0.0368$ for channels, pipes and rod bundles, respectively. The extension of this result to multiphase flow requires an appropriate model for computation of the effective friction. The pressure loss due to friction is usually expressed in terms of the pressure loss of fictive flow consisting of the same mass flow $\rho\overline{w} = \sum_{l=1}^{l_{\max}}\alpha_l\rho_l\overline{w}_l$ but having the properties of the continuum. The so-computed pressure loss is then modified by the so-called two-phase friction multiplier Φ^2_{co} or *Martinelli–Nelson* multiplier,

$$\left(\frac{dp}{dz}\right)_{fr} = f_w = \frac{\lambda_{fr,co}\Phi^2_{co}}{D_h}\frac{1}{2}\frac{(\rho\overline{w})^2}{\rho_c} = \frac{4\tau_w}{D_h}\,.$$

Here, $\tau_w = \frac{\lambda_{fr,co}\Phi^2_{co}}{8}\frac{(\rho\overline{w})^2}{\rho_c}$ is the wall share stress. *co* stays for "continuum only". The effective friction coefficient is $\lambda_{fr,eff} = \lambda_{fr,co}\Phi^2_{co}$. The multiphase friction velocity is then $w* = \sqrt{\frac{\tau_w}{\rho_c}} = \frac{(\rho\overline{w})}{\rho_c}\sqrt{\frac{\lambda_{fr,co}\Phi^2_{c0}}{8}}$. The effective boundary layer dimensionless velocity and wall distance are $w^+ = \frac{(\rho w)}{\rho_c w*}$ and $y^+ = y\frac{w*}{\nu}$, respectively. The presence of vapor in the liquid for instance increases considerably the friction pressure loss. Recovering the profiles from averaged parameters is difficult. In any case the boundary-layer thickness in which the irreversible viscous dissipation happens should be smaller. Therefore, we introduce intuitively

$$y^+_{\lim} = \alpha_c y^+_{\lim,co}$$

in

$$\varepsilon_{2,\infty} = \frac{1}{\alpha_c} \frac{\lambda_{fr,co}\Phi^2_{c0}}{D_{hyd}} \frac{1}{2} \frac{(\rho\overline{w})^3}{\rho_c^3} \left(1 - \alpha_c y^+_{\lim,co} \sqrt{\frac{\lambda_{fr,co}\Phi^2_{c0}}{8}}\right).$$

In the limiting case of continuum only we have $y^+_{\lim} = y^+_{\lim,co}$. Here the division with the continuous volume fraction reflects the assumption that all of the dissipation is put into the continuum.

In the limiting case of no boiling we obtain after rearrangement

$$\frac{\dot{q}''_w D_h}{(T_w - T_2)\lambda_2} = \frac{2^{3/4}}{\pi^{1/2} b^{1/2}} \left[\frac{\lambda_{fr,2o}\Phi^2_{2o}}{\alpha_2}\left(1 - \alpha_2 y^+_{\lim} \sqrt{\frac{\lambda_{fr,2o}\Phi^2_{2o}}{8}}\right)\right]^{1/4} \mathrm{Re}^{3/4}\,\mathrm{Pr}_2^{1/2}.$$

This is the general form in which $\mathrm{Re} = D_h(\rho\overline{w})/\eta_2$. Comparison with *Sani*'s (1960) experimental data for single-phase flow by using the *Blasius* correlation for friction $\lambda_{fr} = 0.3164/\mathrm{Re}^{1/4}$, dictates the approximated value for b = 98.32 resulting in a mean error of 6.4% and increasing divergence for lower *Prandtl* numbers. Note the similarity to the *Dittus–Boelter* (1930) correlation

$$\frac{\dot{q}''_w D_h}{(T_w - T)\lambda_2} = 0.024\,\mathrm{Re}^{0.8}\,\mathrm{Pr}_2^{0.4},$$

which approximates *Sani*'s data with mean error of 3%. Note also the similarity to the *Avdeev*'s result

$$\frac{\dot{q}''_w D_h}{(T_w - T)\lambda_2} = 0.228 Pr_2^{1/2} Re_2^{0.7} (\Phi^2_{2ow}/\alpha_2)^{1/4}.$$

It describes his own data for recondensation in subcooled boiling at heated pipe walls within a ± 30% error band for $D_1/D_{hw} > 80/Re_2^{0.7}$. The amplification of the numerical coefficient by *Avdeev* takes into account that the bubble condensation interface is larger than the heated surface.

Appendix 6.1 State-of-the-art of nucleate pool boiling modeling

Detailed information on this topic can be found in *van Stralen* and *Cole* (1979). We confine our attention to some basic ideas on which some of the frequently used correlations in the literature rely.

Rohsenow (1952) computed the heat flux from the wall during nucleate boiling of saturated liquid as $\dot{q}''_{w2} = G_{1d}\left[h'' - h'(T_w)\right]$, where G_{1d} is the mass flow rate of departing bubbles from the surface. $G_{1d} = \dot{q}''_{w2} / \left[h'' - h'(T_w)\right]$ is used in the *Reynolds* number defined as $Re_{1d} = G_{1d}D_{1d}/\eta_2 = D_{1d}\dot{q}''_{w2} / \left\{\left[h'' - h'(T_w)\right]\eta_2\right\}$ for postulating the heat-transfer dependence $St_{1d} = Nu_{1d} / \left(Pr_2 Re_{1d}\right) \approx f\left(Re_{1d}, Pr_2\right)$, where the *Nusselt* number is defined as $Nu_{1d} = h_w D_{1d} / \lambda_2$. *Rohsenow* uses the *Fritz* equation (1935) $D_{1d} = 1.2\,\varphi\lambda_{RT}$ for the estimation of the bubble departure size. *Rohsenow* finally obtained

$$1/St_{1d} = c_{p2}\left[T_w - T'(p)\right]/\Delta h \approx const\left[\dot{q}''_w\lambda_{RT} / \left(\eta_2\Delta h\right)\right]^{1/3} Pr_2^m ,$$

where the constant and the exponent are fitted against experimental data for water: $m = 1$, $const = 0.006$ to 0.013, for other liquids: $m = 1.7$, const = 0.0025 to 0.015. Additional values for *const* and *m* are reported by *Vachon* et al. (1968) for different fluid - surface combinations and different surface preparations.

The definitions by *Rohsenow* are used later by *Kurihara* and *Myers* (1960) in the form $h_{w2}D_{1d} / \lambda_2 = const(G_{1d}D_{1d} / \eta_2)^m \, \mathrm{Pr}_2^n$, where for saturated pool boiling the authors assumed that all produced steam is removed from the wall boundary layer, i.e., $G_{id} = \left(\pi/6\right)D_{1d}\rho_1 f_{1w} n''_{1w}$. The authors assumed further $D_{1d} f_{1w} \approx const$. The comparison with experimental data for a constant pressure gives $m \approx 1/3$, which cancels the effect of the bubble departure diameter. Finally, the authors obtained

$$h_{w2} = 820\lambda_2\left(\rho_1 / \eta_2\right)^{1/3} n''^{\,1/3}_{1w} \, \mathrm{Pr}_2^{-0.89} .$$

Note that $D_{1d}f_{1w}$ is not a constant as assumed by *Kurihara* and *Myers* but a function of the pressure.

Mikic and *Rohsenow* (1969) computed the temperature as a function of time and distance y from the wall

$$T = T_2 + (T_w - T_2)erfc\left\{y / \left[2(a_2\tau)^{1/2}\right]\right\}$$

and the heat flux at the wall in the liquid

$$\dot{q}''_w(\tau) = -\lambda_2 \frac{dT}{dy}\bigg|_{y=0} = \lambda_2(T_w - T_2) / \left[(\pi a_2\tau)^{1/2}\right].$$

Thereafter, they performed time averaging of the heat flux during the time interval $\Delta\tau$

$$\dot{q}''_w(\Delta\tau) = \frac{1}{\Delta\tau}\int_0^{\Delta\tau} \dot{q}''_w(\tau)d\tau = 2\left(\frac{\lambda_2\rho_2 c_{p2}}{\pi\Delta\tau}\right)^{1/2}(T_w - T_2) .$$

The averaged heat flux over the total surface is computed assuming that the heat conduction from the wall into the liquid happens only inside a circle around the nucleation site with the diameter as the bubble departure diameter

$$\dot{q}''_w = \frac{\pi}{4} D_{1d}^2 n''_{1w} \dot{q}''_w (\Delta\tau).$$

In addition, the authors computed the bubble departure diameter in accordance with the correlation by *Cole* and *Rohsenow* (1969) $D_{1d} = c_2 \lambda_{RT} Ja^{*5/4}$, where $Ja^* = \rho_2 c_{p2} T'(p)/(\rho_1 \Delta h)$, and c_2 = 1.5×10^{-4} for water, and 4.65×10^{-4} for other liquids. The time averaging is performed during the bubble departure period only $\Delta\tau = \Delta\tau_{1d} = 1/f_{1w}$. Here, the relationship obtained by *Zuber* (1959, 1960) for free bubble rise in a pool, $f_{1d} D_{1d} \approx 0.6\left[\sigma g (\rho_2 - \rho_1)/\rho_2^2\right]^{1/4}$, was used, where f_{1d} is the bubble removal frequency from the wall boundary layer, rather than the bubble departure frequency. The bubble number density was postulated to be $n''_{1w} = c(D_1^* / D_{1c})^{3.5}$ where the critical bubble size is

$$D_{1c} = 4\sigma T'(p) \,/\left\{\rho_1 \Delta h \left[T_w - T'(p)\right]\right\}.$$

The constants $cD_1^{*3.5}$ are estimated by fitting experimental data.

Judd and *Hwang* (1976) extended this model by postulating that the area of influence of a single bubble is $C \approx 1.7$ times the projected bubble area at departure and the heat transfer outside the area of influence is due to natural convection. The final expression by *Judd* and *Hwang* is

$$\dot{q}''_{w2} = C\frac{\pi}{4} D_{1d}^2 n''_{1w}\, \dot{q}''_{w2}(\Delta\tau) + \left(1 - C\frac{\pi}{4} D_{1w}^2 n''_{1w}\right) \dot{q}''_{w2,nc},$$

or

$$\dot{q}''_{w2} - \dot{q}''_{w2,nc} = C\frac{\pi}{4} D_{1d}^2 n''_{1w} \left[\dot{q}''_{w2}(\Delta\tau) - \dot{q}''_{w2,nc}\right],$$

where the natural convection heat flux is

$$\dot{q}''_{w2,nc} = 0.18\lambda_2 \left[Pr_2 g \beta_2 / \nu_2^2\right]^{1/3} (T_w - T_2)^{4/3},$$

and $\beta_2 = -\frac{1}{\rho_2}\left(\frac{\partial \rho_2}{\partial T_2}\right)$ is the volumetric thermal expansion coefficient. The data of several investigators can be approximated using 0.272 instead 0.18 as proposed by *Zuber* (1963). Note that *Jacob* used earlier the equation

$$\dot{q}''_{w2,nc} \delta_{2,th} / \left[\lambda_2 (T_w - T_2)\right] = const\left[(T_w - T_2)\delta_{2,th}^3 g \beta_2 / \nu_2^2 Pr_2\right]^{1/4},$$

where the thermal boundary layer thickness was assumed to be

$$\delta_{2,th} \approx const / \Delta T^4 ,$$

and

$$\dot{q}''_{w2,nc} = 2.417\lambda_2 \left[Pr_2 g \beta_2 / \nu_2^2 \right]^{1/4} (T_w - T_2)^{21/16} .$$

Bubble growth and departure causes liquid fluctuation parallel to the wall, the so-called microconvection. This leads some authors to the idea to modify the well-known *Karman–Pohlhausen* equation (1921)

$$Nu = h_{w2}\ell / \lambda_2 = 0.664\,\mathrm{Re}^{1/2}\, Pr_2^{1/3}$$

for computing the averaged heat transfer at the wall for viscous flow. Note that the heat flux in the above equation is averaged along the distance length ℓ, and that ℓ is used as a scale in the *Nusselt* and *Reynolds* numbers. The velocity used in the Reynolds number is the bulk velocity of the flow parallel to the horizontal plate. The corresponding equation for turbulent flow is

$$Nu = h_{w2}\ell / \lambda_2 = 0.037 Re^{4/5} Pr_2^{1/3}$$

see *Brauer* (1971), p.182, Eq. (3.38).

Forster and *Zuber* (1955) applied the *Pohlhausen* equation using as averaging length scale the bubble departure diameter D_{1d}

$$h_{w2,mic} D_{1d} / \lambda_2 = const\ Re^m Pr_2^n .$$

The authors averaged the bubble-growth velocity along an averaged distance between two neighboring bubbles, $D_{2,inf}$, for thermally controlled bubble growth and obtained

$$\bar{V}' = 2R_1 \frac{dR_1}{d\tau} = B^2 / D_{2,\inf} ,$$

where $B = \sqrt{\pi a_2} Ja_2$. This velocity is called the microconvection velocity. The *Reynolds* number was computed as

$$Re = \bar{V}' D_{1d} / \nu_2 = \left(B^2 / \nu_2 \right) \left(D_{1d} / D_{2,\inf} \right)$$

and substituted into the *von Karman–Pohlhausen* equation with

$$Re = B^2 / \nu_2$$

setting $D_{1d}/D_{2,inf} \approx 1$. The ratio $D_{1d}/D_{2,inf} < 1$ was taken into account by the constant

$$const = 0.003$$

for the polar fluids *n*-pentane, benzene, ethanol and for water. The exponents are found to be $m = 0.62$, $n = 1/3$. For the bubble departure diameter the authors used the thermally controlled bubble-growth mechanism

$$D_{d1} \approx 2B\Delta\tau_{1d}^{1/2},$$

where the order of magnitude of the departure time

$$\Delta\tau_{1d} \approx \left\{\sigma / \left[p'(T_{2i}) - p\right]\right\}\left\{\rho_2\left[p'(T_{2i}) - p\right]\right\}^{1/2}$$

is estimated from the mechanical energy equation setting the bubble-growth work equal to the kinetic energy of the bubble environment

$$\left(dR_1 / d\tau\right)^2 \approx \left(R_{1d} / \Delta\tau_{1d}\right)^2 = (2/3)\left(1 - \rho'' / \rho_2\right)\Delta\rho / \rho_2 .$$

The authors themselves considered the use of this time constant as "not the only one that may be significant". The verification of this equation is made with data for superheat and heat flux at the point of burn out for different pressure. The exponents were questioned later by *Katz*, see the *Katz* comments to *Zuber*'s work in (1958), and new exponents are proposed in order to obtain agreement with experimental data for ethanol/chromium with *const* = 300, benzene with *const* = 1, and *n*-pentane with *const* = 0.3, where $m = 1.05$ and $n = -7$ was used for all three cases. Nevertheless, in a later work *Chen* (1966) uses the *Forster* and *Zuber* equation with $m = 1/2$, $n = 1/3$ to successfully describe the part of heat transfer due to nucleate boiling in forced convection.

In a later work *Zuber* (1960) comes back to the *von Karman–Pohlhausen* equation using again D_{1d} as a length scale but replacing the average distance between the neighboring bubbles by assuming a quadratic arrangement $1/D_{2,\inf} = n_{1w}''^{\,1/2}$. Here, n_{1w}'' is the nucleation-site density in sites per square meter. Thus, *Zuber* finally writes $h_{w,mic} D_{1d} / \lambda_2 = const\ Re^{1/2} Pr_2^{1/3}$ where $Re = B^2 D_{1d} / \left(\nu_2 D_{2,\inf}\right) = B^2 D_{1d} n''^{1/2}_{1w} / \nu_2$. This time, the $B^2 = \dfrac{4}{\pi} a_2 Ja_2^2$ function is computed in accordance with *Bosnjakovic* (1930). The intention of *Zuber* in (1959) was not to derive a correlation but to show the dependence $h_{w,mic} \approx n_{1w}''^{\,1/4}$ that was experimentally observed. Nevertheless, *Zuber* computed the heat flux at the wall during nucleate boiling as

$$\dot{q}_w'' = \dot{q}_{w,mic}''\left(1 - \frac{\pi}{4} D_{1d}^2 n_{1w}^{1/2}\right) + \dot{n}_{1w}'' f_{1w} \Delta h \frac{\pi}{6} D_{1d}^3 \rho_1 .$$

In more recent work, *Kocamustafaogullari* and *Ishii* (1983) use the idea of microconvection in the following way

$$h_{w,mic} D_{2,\inf} / \lambda_2 = 13.18 Re^m Pr_2^n \left(D_{1d} / D_{2,\inf}\right)^p ,$$

where m = 1/4, n = −0.14, p = − 1/4. The distance between the neighboring bubbles is used as a characteristic that is consistent with the length along which the fluctuation velocity was averaged. It gives a simple expression for the *Reynolds* number $Re = \overline{V}_2' D_{2,\inf} / \nu_2 = B^2 / \nu_2$. The authors again used the *Bosnjakovic* model for thermally controlled bubble growth at a wall $B^2 = \frac{4}{\pi} a_2 Ja_2^2$ and quadratic nucleation arrangement $1/D_{2,\inf} = n_{1w}''^{1/2}$. Substituting into the starting equation and rearranging the authors came to

$$h_{w,mic} = 14 Ja^{1/2} Pr_2^{-0.39} \frac{\lambda_2}{D_{1d}^{1/4}} n_{1w}''^{3/8}.$$

The constant and the powers are determined by comparison with experimental data where the bubble departure diameter was computed using a modified form of the *Fritz* equation.

Labuntsov (2000) assumed that the nucleation-site density is a function of the critical bubble radius $n_{1w}'' = c_1 / R_{1c}^2$. Using the dimensionless microconvection velocity $V_{2,\mu c}^* = R_1 \left(dR_1 / d\tau \right) / R_{1c}$, computing the thickness of the laminar boundary layer at $y^+ = 10$, $y = 10 \nu_l / V_{2,\mu c}^*$ and assuming that the heat conduction through this layer is controlling the heat transfer he obtained

$$\dot{q}_w'' = c_2 \frac{\lambda_2^2 \left(T_w - T' \right)^3}{\nu_2 \sigma_{12} T'}.$$

Later, he introduced empirically the pressure dependence and proposed the following expression for the heat-transfer coefficient at nucleate boiling

$$h_w = 0.075 \left(\frac{\lambda_2^2 \dot{q}_w''^2}{\nu_2 \sigma_{12} T'} \right)^{1/3} \left[1 + 10 \left(\frac{\rho''}{\rho' - \rho''} \right)^{2/3} \right].$$

Tolubinsky and *Ostrovsky* (1966) generalized a large number of data for water and other liquids by the correlation

$$h_w \lambda_{RT} / \lambda_2 = 75 \left[\frac{\dot{q}_{w1}'' / \left(\Delta h \rho_1 \right)}{D_{1d} f_{1w}} \right]^{0.7} \mathrm{Pr}_2^{-0.2}$$

λ_{RT} where the bubble departure velocity $D_{1d} f_{1w} = 0.36 \times 10^{-3} \left(p_c / p \right)^{1.4}$. The data are fitted within an error band of ± 25%.

Some conclusions seem to be in order after determining the above-discussed approaches to the modeling of nucleate pool boiling:

a) Obviously, heat conduction and microconvection simultaneously exist in pool boiling.
b) The appropriate scale in the *von Karman–Pohlhausen* equation for microconvection is the average distance between two neighboring bubbles. Nevertheless, the microconvection is much more similar to turbulent heat transfer than to forced convection.
c) The bubble removal frequency from the wall boundary layer is not necessarily equal to the bubble departure frequency.
d) Besides the spatial averaging of the pulsation velocity, a time averaging during the bubble generating period $1/f_{1w}$ is necessary.
e) None of the theories takes into account the influence of the static contact angle on the heat transfer.
f) All models put the unknown phenomena in empirical constants so that none of the theories is able to predict separate phenomena like bubble departure diameter, nucleation-site density, etc., appropriately.

Nomenclature

a_2	liquid thermal diffusivity, m^2/s
a'	saturation liquid thermal diffusivity, m^2/s
B^2	$= 2R_1 dR_1 / d\tau$, m^2/s
c_{p2}	specific heat of liquid, J/(kg K)
c	$= (12/\pi)^{1/2}\left[1+\frac{1}{2}\left(\frac{\pi}{6Ja'}\right)^{2/3}+\left(\frac{\pi}{6Ja'}\right)^{1/2}\right]^{1/2}$, dimensionless
c_1	= 1.4626, dimensionless
c_2	= 0.7, dimensionless
D_1	bubble diameter, m
D_{1c}	critical size, m
D_{1d}	bubble departure diameter, m
$D_{1d,fc}$	bubble departure diameter for strongly predominant forced convection, m
$D_{1d,nc}$	bubble departure diameter for natural circulation, m
$D_{2,inf}$	average center to center spacing, m
f_{1w}	bubble departure frequency, 1/s
f_{1w}^t	boundary layer turbulence fluctuation frequency, 1/s
G	mass flow rate, $kg/(m^2\ s)$
g	gravitational acceleration, m/s^2
htc	heat-transfer coefficient, $W/(m^2\ K)$
h', h''	specific saturation enthalpies for water and steam, J/(kg K)
Δh	$= h'' - h'$, J/kg
Ja	$= \rho_2 c_{p2}\left[T_2 - T'(p)\right]/\left(\rho_1 \Delta h\right)$, *Jacob* number, dimensionless
Ja'	$= \rho' c_p' \left[T_2 - T'(p)\right]/\left(\rho'' \Delta h\right)$, *Jacob* number, dimensionless

ℓ_2^t	$= \pi \ \lambda_{RT}$, turbulent length, m
n''_{1w}	Active nucleation-site density, $1/m^2$
p	pressure, Pa
p'	saturated pressure, Pa
Pr_2	$= \eta_2 /(\rho_2 a_2)$, *Prandtl* number, dimensionless
$\dot{q}''_{w2,nc}$	heat flux from the wall into the liquid during natural convection without boiling, W/m²
$\dot{q}''_{w2,nb}$	heat flux from the wall into the liquid during pool boiling, W/m²
R_1	bubble radius, m
$R_{2,inf}$	$= D_{2,inf}/2$, m
$T'(p)$	saturation temperature at system pressure p, K
T_2	liquid temperature, K
T_w	wall temperature, K
$\bar{V}_2'$	the volume-averaged fluctuation velocity, m/s
$\bar{\bar{V}}_2'$	the volume-averaged fluctuation velocity time averaged over $\Delta\tau_w + \Delta\tau_d$, m/s

Greek

β_2	$= -\dfrac{1}{\rho_2}\left(\dfrac{\partial \rho_2}{\partial T_2}\right)_p$, thermal expansion coefficient, 1/K
$\delta_{2,\min}$	minimum of the thermal boundary layer thickness, m
$\Delta\rho_{21}$	$= \rho_2 - \rho_1$, kg/m³
$\Delta\tau_d$	time needed from the initiation of a bubble with critical size to the bubble departure from the wall, s
$\Delta\tau_w$	delay time, s
η_2	dynamic viscosity of liquid, kg/(m s)
λ_2	thermal conductivity of liquid, W/(m K)
λ_{RT}	$= \left(\dfrac{\sigma}{g\Delta\rho_{21}}\right)^{1/2}$, *Rayleigh–Taylor* instability wavelength, m
ν_2	cinematic viscosity of liquid, m²/s
ρ_1	gas density, kg/m³
ρ_2	liquid density, kg/m³
ρ''	saturated steam density, kg/m³
ρ'	saturated liquid density, kg/m³
σ	surface tension, N/m
τ	time, s
θ_0	angle between the bubble axis and the wall, rad
φ	static contact angle between "liquid drop" and the wall, rad

Superscripts

$'$	saturated water
$''$	saturated steam

Subscripts

1	gas
2	liquid
boiling	boiling
DNB	departure from nucleate boiling
exp	experimental
INB	initiation of nucleate boiling
sup	superheated
sub	subcooled
exp	experimental
total	total
th	theoretical
w	wall
total	total
boiling	boiling
co	continuum only, all mass flow considered as continuum

References

Arefeva, E.I., Aladev, I.T.: O wlijanii smatchivaemosti na teploobmen pri kipenii. Injenerno - Fizitcheskij Jurnal 1(7), 11–17 (1958) (in Russian)

Avdeev, A.A., Maidanik, V.N., Selesnev, L.I., Shanin, V.K.: Calculation of the critical flow rate with saturated and subcooled water flushing through a cylindrical duct. Teploenergetika 24(4), 36–38 (1977)

Borishanskii, V., Kozyrev, A., Svetlova, L.: Heat transfer in the boiling water in a wide range of saturation pressure. High Temperature 2(1), 119–121 (1964)

Borishanskii, V., Bobrovich, G., Minchenko, F.: Heat transfer from a tube to water and to ethanol in nucleate pool boiling. In: Kutateladze, S.S. (ed.) Symposium of Heat Transfer and Hydraulics in Two-Phase Media, Gosenergoizdat, Moscow, pp. 75–93 (1961)

Brauer, H.: Stoffaustausch, Verlag Sauerländer, Aarau, Switzerland (1971)

Chandesris, M., Serre, G., Sagaut: A macroscopic turbulence model for flow in porous media suited for channel, pipe and rod bundle flows. In: 4th Int. Conf. On Computational Heat and Mass Transfer, Paris (2005)

Chen, J.C.: Correlation for boiling heat transfer to saturated fluids in convective flow. Ind. Eng. Chem. Prog. Des. Devel. 5(3), 322–329 (1966)

Cole, R., Rohsenow, W.M.: Correlation of bubble departure diameters for boiling of saturated liquids. Chem. Eng. Prog. Symp. Ser. 65(92), 211–213 (1969)

Cornwell, K., Brown, R.D.: Boiling surface topology. In: Proc. 6th Int. Heat Transfer Conf. Heat Transfer 1978, Toronto, vol. 1, pp. 157–161 (1978)

Cooper, M.G.: Saturation nucleate boiling – a simple correlation. In: Institute of Mechanical Engineers, IChemE Symposium Series, London, vol. 86, pp. 786–793 (1984)
Dittus, F.V., Boelter, L.M.K.: vol. 2(13), p. 443. Univ. of Calif. Publ., Eng. (1930)
Faggiani, S., Galbiati, P., Grassi, W.: Active site density, bubble frequency and departure on chemically etched surfaces. La Termotechnica 29(10), 511–519 (1981)
Forster, H.K., Zuber, N.: Dynamics of vapor bubbles and boiling heat transfer. AIChE J 1(4), 531–535 (1955)
Fritz, W.: Berechnung des maximalen Volumens von Dampfblasen. Phys. Z. 36(11), 379–384 (1935)
Gaertner, R.F.: Distribution of active sites in the nucleate boiling of liquids. In: Chem. Eng. Prog. Symp. Series, vol. 59(41), pp. 52–61 (1963)
Gaertner, R.F.: Photographic study of nucleate pool boiling on a horizontal surface. Transaction of the ASME J. Heat Transfer 87, 17–29 (1965)
Gaertner, R.F., Westwater, J.W.: Population of active sites in nucleate boiling heat transfer. Chem. Eng. Progr. Symp. Ser. 30(30), 39–48 (1960)
Golovin, V.S., Kalchugin, B.A., Labuntsov, D.A.: Injenerno Fizicheskiy Jurnal 6(2) (1963)
Han, C.Y., Griffith, P.: The mechanism of heat transfer in nucleate pool boiling, Part I, Bubble initiation, growth and departure. Int. J. Heat Mass Transfer 8, 887–904 (1965)
Haramura, Y., Katto, Y.: A new hydrodynamic model of CHF applicable widely to both pool and forced convection boiling on submerged bodies in saturated liquids. Int. J. Heat Mass Transfer 26, 389–399 (1983)
Hutcherson, M.N., Henry, R.E., Wollersheim, D.E.: Two-phase vessel blow-down of an initially saturated liquid - Part 2: Analytical. Trans. ASME J. Heat Transfer 105, 694–699 (1983)
Jakob, M.: Heat transfer, vol. 1, ch. 29. Wiley, New York (1949)
Jakob, M.: Kondensation und Verdampfung. Zeitschrift des Vereins Deutscher Ingenieure 76(48), 1161–1170 (1932)
Jakob, M., Fritz, W.: Versuche ueber den Verdampfungsvorgang. Forsch. Ing. - Wes. 2, 435 (1931)
Jakob, M., Linke, W.: Der Wärmeübergang von einer waagerechten Platte an siedendes Wasser. Forsch. Ing. Wes. 4, 75–81 (1933)
Jens, W.H., Lottes, P.A.: Analysis of heat transfer, burnout, pressure data and density data for high pressure water. USAEC Rep. ANL-4627 (1951)
Johov, K.A.: Nucleations number during steam production. In: Aerodynamics and Heat Transfer in the Working Elements of the Power Facilities, Proc. CKTI, Leningrad, vol. 91, pp. 131–135 (1969)
Jones, O.C.: Nonequilibrium phase change 1. Flashing inception, critical flow, and void development in ducts. In: Lahey Jr., R.T. (ed.) Boiling Heat Transfer, pp. 189–234. Elsevier Science Publishers B.V., New York (1992a)
Jones, O.C.: Nonequilibrium phase change 2. Relaxation models, general applications, and post heat transfer. In: Lahey Jr., R.T. (ed.) Boiling Heat Transfer, pp. 447–482. Elsevier Science Publishers B.V., New York (1992b)
Judd, R.L., Hwang, K.S.: A comprehensive model for nucleate pool boiling heat transfer including microlayer evaporation. Trans. ASME, J. Heat Transfer 98, 623–629 (1976)
Kim, S.J., Truong, B., Buongiorno, J., Hu, L.W., Bang, I.C.: Study of two-phase heat transfer in nanofluids for nuclear applications. In: Proceedings of ICAPP 2006, Reno, NV USA, June 4-8 (2006) (Paper 6005)
Kim, S.J., Bang, I.C., Buongiorno, J., Hu, L.W.: Study of pool boiling and critical heat flux enhancement in nanofluids. Bull. Polish Acad. Sci. Technical Sci. 55(2) (2007a)

Kim, S.J., Bang, I.C., Buongiorno, J., Hu, L.W.: Surface wettability change during pool boiling of nanofluids and its effect on critical heat flux. International, J. Heat Mass Transfer 50, 4105–4116 (2007b)

Kocamustafaogullari, G., Ishii, M.: Interfacial area and nucleation site density in boiling systems. Int. J. Heat Mass Transfer 26, 1377–1389 (1983)

Kolev, N.I.: The influence of mutual bubble interaction on the bubble departure diameter. Exp. Thermal Fluid Sci. 8, 167–174 (1994)

Kolev, N.I.: How accurate can we predict nucleate boiling. Exp. Thermal Fluid Sci. 10, 370–378 (1995)

Kurihara, H.M., Myers, J.E.: The effect of superheat and surface roughness on boiling coefficients. AIChE J. 6(1), 83–91 (1960)

Kutateladse, S.S.: A hydrodynamic theory of changes in the boiling process under free convection conditions. Izv. Akad. Nauk SSSR, Otd. Tech. Nauk 4, 529–536 (1954); AEC-tr-1991

Kutateladse, S.S.: Basics on heat transfer theory, Moscow, Mashgis, p. 456 (1962) (in Russion)

Labuntsov, D.A.: Approximate theory of heat transfer by developed nucleate boiling (Russ.). Izvestiya AN SSSR, Energetika i transport no 1 (1963)

Labuntsov, D.A.: State of the art of the nucleate boiling mechanism of liquids. Heat Transfer and Physical Hydrodynamics, 98–115 (1974) (in Russian)

Labuntsov, D.A.: Physical fundamentals of power engineering. In: Selected Works on Heat Transfer, Hydrodynamics and Thermodynamics. MPEI Publisher, Moscow (2000)

Labuntsov, D.A., Kol'chugin, V.A., Golovin, V.S., et al.: Teplofiz. Vys. Temp. 3, 446–453 (1964)

Liaw, S.P.: Experimental and analytical study of nucleate and transition boiling on a vertical surface. Ph. D. dissertation, University of California, Los Angeles, USA (1988)

Miheev, M.A., Miheeva, I.M.: Basics of heat transfer, p. 320. Energija, Moskva (1973) (in Russian)

Mikic, B.B., Rohsenow, W.M.: A new correlation of pool-boiling data including the effect of heating surface characteristics. Trans. ASME, J. Heat Transfer 91, 245–250 (1969)

Morin, R.: Wall temperature fluctuations during bubble generation in boiling. In: Borishanskii, V.M., Paleev, II. (eds.) Convective Heat Transfer in Two-phase and One-phase Flows. Energiya, Moskva (1964)

Nishikawa, K., Fujita, Y., Uchida, S., Ohta, H.: Effect of surface configuration on nucleate boiling heat transfer. Int. J. Heat Mass Transfer 27(9), 1559–1571 (1984)

Pohlhausen, K.: Zur nährungsweisen Integration der Differentialgleichung der laminaren Grenzschicht. Z. Angew. Math. Mech. 1, 252–268 (1921)

Rallis, C.J., Jawurek, H.H.: Latent heat transport in saturated nucleate boiling. Int. J. Heat Transfer 7, 1051–1068 (1964)

Ramilison, J.M., Sadasivan, P., Lienhard, J.H.: Surface factors influencing burnout on flat heaters. ASME J. Heat Transfer 114, 287–290 (1992)

Riznic, J., Ishii, M.: Bubble number density and vapor generation in flushing flow. Int. J. Heat Mass Transfer 32, 1821–1833 (1989)

Rohsenow, W.M.: A method of correlating heat transfer data for surface boiling of liquids. Trans. ASME 74, 969–975 (1952)

Sani Rle, R.: Down flow boiling and non-boiling heat transfer in a uniformly heated tube,15th edn. University of California, URL-9023, Chemistry-Gen. UC-4, TID-4500 (January 1960)

Schleisiek, K.: Heat transfer and boiling during forced convection of sodium in induction-heated tube. Nucl. Eng. Des. 14, 60–68 (1970)

Siegel, R., Keshock, E.G.: Effects of reduced gravity on nucleate boiling bubble dynamics in saturated water. AIChE J. 10(4), 509–517 (1964)

Subbotin, V.I., Sorokin, D.N., Kurjavzev, A.P.: Obobstenaja zavisimost dlja razchota teplotdachi pri razvitom kipenii stelochnych metallov. Atomnaja Energija 29(5), 45 (1968)

Sultan, M., Judd, R.L.: Spatial distribution of active sites and bubble flux density. Trans. ASME, J. Heat Transfer 100, 56–62 (1978)

Theofanus, T.G., et al.: The boiling crisis phenomenon Part II: dryout dynamics and burnout. Exp. Thermal Fluid Sci. 26, 793–810 (2002)

Thom, I.R.S., et al.: Boiling in subcooled water during up heated tubes or annuli. In: Proc. Instr. Mech. Engs., vol. 180 3C, pp. 1965–1966 (1966)

Tolubinsky, V.I., Ostrovsky, J.N.: On the mechanism of boiling heat transfer (vapor bubbles growth rate in the process of boiling in liquids, solutions, and binary mixtures). Int. J. Heat Mass Transfer 9, 1463–1470 (1966)

Vachon, R.I., Tanger, G.E., Davis, D.L., Nix, G.H.: Pool boiling on polished chemically etched stainless-steel surfaces. Trans. ASME, J. Heat Transfer 90, 231–238 (1968)

Vachon, R.I., Nix, G.H., Tanger, G.H.: Evaluation of the constants for the Rohsenow pool - boiling correlation. Trans. ASME J. Heat Transfer 90, 239–247 (1968)

van Stralen, S., Cole, R.: Boiling Phenomena. Hemisphere, Washington (1979)

von Karman, T.: Über laminare und turbulente Reibung. Z. angew. Math. Mech. 1, 233–252 (1921)

Wang, C.H., Dhir, V.K.: Effect of surface wettability on active nucleation site density during pool boiling of water on a vertical surface. ASME J. Heat Transfer 115, 659–669 (1993)

Wiebe, J.R.: Temperature profiles in subcooled nucleate boiling. M. Eng. thesis, Mechanical Engineering Department, McMaster University, Canada (1970)

Yamagata, K., Hirano, F., Nishikawa, K., Matsuoka, H.: Nucleate boiling of water on the horizontal heating Surface. Mem. Fac. Eng; Kyushu Univ. 15, 97 (1955)

Yang, J.K., Weisman, J.: A phenomenological model of subcooled flow boiling in the detached bubble region. Int. J. Multiphase Flow 17(1), 77–94 (1991)

Zuber, N.: Hydrodynamic aspect of boiling heat transfer, U.S. Atomic Energy Commission Rept., AECU - 4439, Tech. Inf. Serv. Oak Ridge, Tenn (1959)

Zuber, N.: Nucleate boiling, The region of isolated bubbles and the similarity with natural convection. Int. J. Heat Mass Transfer 6, 53–78 (1963)

Zuber, N.: On the stability of boiling heat transfer. Trans. ASME 80, 711–720 (1958)

Zuber, N.: Hydrodynamic aspect of nucleate pool boiling, Part I - The region of isolated bubbles. Research Laboratory Ramo - Wooldridge, RW-RL-164, 27 (January 1960)

Zuber, N., Tribus, M.: Further remarks on the stability of boiling heat transfer, Report 58-5, Department of Engineering, University of California, Los Angeles (1958)

Zuber, N., Tribus, M., Westwater, J.W.: The hydrodynamic crisis in pool boiling of saturated and subcooled liquids. In: International Developments in Heat Transfer, Proc. Int. Heat Transfer Conf., Boulder, Colorado, Part 2, vol. (27), pp. 230–236 (1961)

7. Heterogeneous nucleation and flashing in adiabatic pipes

A new model of heterogeneous nucleation at walls during flashing and boiling is presented. The model is based on a new method for computation of the bubble departure diameter in flows accounting for mutual bubble interactions and the limitation of the bubble production due to the limited amount of energy supplied into the boundary layer by turbulence induced by bubble growth and departure. The final model together with fragmentation and coalescence models is incorporated into the system code IVA and verified by comparison with data. This chapter is a short version of the primary published work in Kolev (1995).

7.1 Introduction

Modern theories of multiphase flows account for the dynamic generation of particles of one phase inside the other. Bubble generation at walls during boiling or flashing is a typical example discussed in Chapter 16. The number of the bubbles generated per unit time and unit surface, and the bubble departure diameter, are the most important parameters in this process, which influence the later process of bubble growth and/or collapse into the bulk liquid. There have been some publications on this field in the last decade whose main result is the so-called figure-of-merit model. This model produces agreement with data by introducing fitting parameters in the general model without verifying the submodels. Examples are *Shin and Jones* (1993), *Blinkov* et al. (1993) where the model used for bubble departure diameter is not appropriate for strong superheating *Kolev* (1994). In this chapter we extend the validity of the results obtained in Chapters 15 and 16 to adiabatic flashing of superheated liquids in pipes. The final model together with fragmentation and coalescence models is incorporated into the system code IVA4 and verified by comparison with data for flashing flows in converging-diverging nozzles and pipe blow-down of water. IVA4 is computer code describing multiphase nonequilibrium flows by means of three velocity fields in industrial networks and/or three-dimensional facilities. The system of partial differential equations, 21 PDEs for 3D and 15 PDEs for 1D network, solved by this code contains Eq. (1.62) from Volume 1, which is the local volume-averaged mass-conservation equation subsequently time averaged

$$\frac{\partial}{\partial \tau}\left(\alpha_\ell \rho_\ell \gamma_v\right)+\nabla.\left(\alpha_\ell \rho_\ell \gamma \mathbf{V}_\ell\right)=\mu_\ell \qquad (7.1)$$

and Eq. (1.102) from Volume 1, which is the local volume- and time-averaged conservation equation for the number of the particles per unit flow volume without diffusion terms

$$\frac{\partial}{\partial \tau}\left(n_\ell \gamma_v\right)+\nabla.\left(n_\ell \gamma \mathbf{V}_\ell\right)=\dot{n}_{\ell,kin}+\dot{n}_{\ell,frag}-n_{\ell,coal}. \tag{7.2}$$

Knowing the local volume- and time-averaged particle number density n_l and the volumetric fraction α_ℓ, the characteristic local volume- and time-averaged particle size D_ℓ is computed from the equation

$$n_\ell \frac{\pi}{6} D_\ell^3 = a_\ell \tag{7.3}$$

that is strictly valid for spherical bubble. Next we confine our attention only to initiation of the steam velocity field, designated by 1, in water, designated by 2.

7.2 Bubbles generated due to nucleation at the wall

Heterogeneous nucleation theory is sometimes recommended to describe nucleation at walls *Skripov* et al. (1980). This theory gives the probability of nucleation for given local conditions *Kaishev* and *Stranski* (1934) and *Volmer* (1939), but does not predict these conditions. As far the author knows there is no other attempt to describe the nucleation at walls theoretically as those presented in Chapters 4 and 5. The turbulence in the boundary layer is an important energy exchange mechanism also during flashing of superheated liquids in adiabatic pipes. This follows from the fact that no more energy for bubble generation at the wall can be consumed than is supplied by the turbulence transfer from the bulk into the boundary layer. That is why we propose to use Eqs. (5.22) to (5.24) also for flashing in adiabatic pipes with driving temperature difference $T_2 - T'$ instead of $T_w - T'$:

$$\dot{q}''_{w2,b} = \text{...Eq. (6.19)}, \tag{7.4}$$

$$\mu_{21,gen} = \frac{4}{D_h}\dot{q}''_{w2,b}/(h''-h_2), \tag{7.5}$$

and

$$\dot{n}_{1w} = \mu_{21}/\left(\rho_1 \frac{\pi}{6} D_{1d}^3\right). \tag{7.6}$$

From the moment of bubble detachment from the wall the bubble starts its own history inside the bulk flow and undergoes condensation or further evaporation as well as collisions, coalescence or fragmentation. In the next section we consider an example of such history for superheated liquids.

7.3 Bubble growth in the bulk

Consider a constant number of bubbles per unit mixture volume, n_1, during the time step $\Delta\tau$. As discussed in Chapter 2, the mass difference between the end and the beginning of the time step in a single bubble multiplied by the bubble number per unit mixture volume, gives the mass evaporated in the bulk per unit mixture volume and unit time.

$$\mu_{21,bulk} = n_1 \frac{\pi}{6}\left(\rho_1 D_1^3 - \rho_{1o} D_{1o}^3\right)/\Delta\tau = \frac{\rho_{1o}\alpha_{1o}}{\Delta\tau}\left[\frac{\rho_1}{\rho_{1o}}\left(\frac{R_1}{R_{1o}}\right)^3 - 1\right] \qquad \alpha_{1o} > 0 . \tag{7.7}$$

As already mentioned from the sixteen analytical solutions of the bubble growth problem obtained for different degrees of simplicity known to the author - see Appendix 2.1, the bubble-growth model proposed by *Mikic* et al. (1970) is used as the best one:

$$\mu_{21,bulk} = \frac{\rho''\alpha_{1o}}{\Delta\tau}\left[\left\{1+\frac{2}{3}\left[\left(\tau^+ + 1\right)^{3/2} - \left(\tau^+\right)^{3/2} - 1\right]/R_{1o}^+\right\}^3 - 1\right], \tag{7.8}$$

where

$$\tau^+ = \Delta\tau / \left(B^2 / A^2\right), \tag{7.9}$$

$$A^2 = \frac{2}{3}\frac{1}{\rho_2}\left(dp/dT\right)_{sat}\left(1 - \rho''/\rho_2\right)\left[T_2 - T'(p)\right], \tag{7.10}$$

$$R^+ = AR_1 / B^2 . \tag{7.11}$$

Thus, the steam generation is a superposition of the nucleation at the wall and of the bubble growth inside the bulk flow

$$\mu_{21} = \mu_{21,bulk} + \mu_{21,gen} . \tag{7.12}$$

7.4 Bubble fragmentation and coalescence

A review of the existing fragmentation and coalescence models for gases and liquids was given in Chapters 7 to 10 in Vol. 2. For the computation presented below we use the following gas-fragmentation model. The volume- and time-average number of generated particles per unit time and unit mixture volume due to dynamic fragmentation is

$$\dot{n}_{1,frag} = (n_{1\infty} - n_1)/\Delta\tau_{fr} \quad for \;\; n_{1\infty} > n_1 \;\; and \;\; \Delta\tau \le \Delta\tau_{fr}, \tag{7.13}$$

$$\dot{n}_{1,frag} = (n_{1\infty} - n_1)/\Delta\tau \quad for \;\; n_{1\infty} > n_1 \;\; and \;\; \Delta\tau > \Delta\tau_{fr}, \tag{7.14}$$

$$\dot{n}_{1,frag} = 0 \quad for \;\; n_{1\infty} \le n_1, \tag{7.15}$$

where $\Delta\tau$ is the time step, n_1 is the local bubble density concentration, and

$$n_{1\infty} = 6\alpha_1/(\pi D_{1\infty}^3) \tag{7.16}$$

is the stable bubble number density concentration after fragmentation. Equation (7.14) gives in fact steady-state fragmentation for large time steps. The stable bubble diameter is computed in accordance with *Achmad´s* equations (8.173) and (8.172),

$$D_{1\infty} = \lambda_{RT}\left[9\alpha_1/(1-\alpha_1)\right]^{1/3} \tag{7.17}$$

for $\alpha_1 > 0.1$, and

$$D_{1\infty} = 0.9\lambda_{RT}/\left\{1+1.34\left[(1-\alpha_1)V_2\right]^{1/3}\right\} \tag{7.18}$$

for $\alpha_1 < 0.1$, see in *Ahmad* (1970).

The fragmentation time is set equal to the natural fluctuation period of bubble flow, Eq. (8.177) in Vol. 2

$$\Delta\tau_{fr} = \left(\frac{6}{\pi\sqrt{2}}\right)^{1/3} \alpha_1^{1/3}/\Delta V_{12}^*, \tag{7.19}$$

where the relative velocity is a superposition of the difference between the mean velocity and the turbulent fluctuation velocity,

$$\Delta V_{12,}^* = \left|V_1 - V_2\right| + V_2^t \tag{7.20}$$

$$V_2^t \approx 0.3\left|V_2\right| \quad for \;\; \alpha_1 < 0.3 \tag{7.21}$$

and

$$V_2^t \approx 0.7\left|V_2\right| \quad for \;\; \alpha_1 \ge 0.3. \tag{7.22}$$

The volume- and time-averaged bubble disappearance per unit time and unit mixture volume due to collision and coalescence is computed in accordance with Eq. (7.22),

$$\dot{n}_{coal} = n_1\left[1 - 1/\exp(f_{coal}/\Delta\tau)\right]/\Delta\tau, \tag{7.23}$$

where $\Delta\tau$ is the time-averaging period. Here, the instantaneous coalescence frequency of a single bubble is

$$f_{coal} = P_{coal} f_{col} \tag{7.24}$$

where the collision frequency is computed in accordance with Eq. (7.29),

$$f_{col} = 4.5\alpha_1^{1/2} \Delta V_{12}^* / D_1 \tag{7.25}$$

and the coalescence probability P_{coal} is set to 1 for superheated liquid in order to approximate the effect of nonuniform particle distribution in the cross-section. In the case of no liquid superheating we use Eq. (7.45),

$$P_{coal} \approx 0.032\left(\Delta\tau_{col}/\Delta\tau_{coal}\right)^{1/3} \quad \text{for} \quad \Delta\tau_{col}/\Delta\tau_{coal} \geq 1, \tag{7.26}$$

where

$$\Delta\tau_{col}/\Delta\tau_{coal} = 1.56\left(\frac{D_1\sigma_2}{3\rho_1 + 2\rho_2}\right)^{1/2} \Big/ V_2'. \tag{7.27}$$

7.5 Film flashing bubble generation in adiabatic pipe flow

Boiling and flashing are governed by the same physical phenomena. Having accepted that there is departure from nucleate boiling in heated channels we have also to accept that the departure from nucleate bubble generation is also possible in adiabatic pipe flow. The steam volumetric fraction in the boundary layer with a thickness D_{1d} is

$$\alpha_{1d} = (\pi/6) n''_{1w} D_{1d}^2. \tag{7.28}$$

If $\alpha_{1d} > 0.52$ the bubbles touch each other and a film is formed. We call this bubble-generation mechanism *film* flashing because it is controlled by bubble entrainment from the vapor layer. If there is no flow velocity imposed, we have

$$D_{1d} \approx 2\pi\lambda_{RT}, \tag{7.29}$$

$$f_{1w} \approx V_{1Ku} / D_{1d}, \tag{7.30}$$

where

$$V_{1Ku} = 1.41\left[g\sigma(\rho_2 - \rho_1)/\rho_2^2\right]^{1/4}, \tag{7.31}$$

is the free rising bubble velocity (*Kutateladze* velocity). For forced convection we assume that the stability criterion

$$We_{1d} \approx 12 \tag{7.32}$$

controls the bubble entrainment size, and therefore

$$D_{1d} \approx 12\sigma/(\rho_2 \Delta V_{21}^2), \tag{7.33}$$

$$f_{1w} \approx \Delta V_{21} / D_{1d}. \tag{7.34}$$

For both cases we have

$$\dot{n}_{1w} = \frac{4}{D_h} n_{1w} f_{1w}, \tag{7.35}$$

$$\mu_{21,gen} \approx \rho_1 \dot{n}_{1w} (\pi D_{1d}^3 / 6), \tag{7.36}$$

and

$$\dot{q}'''_{\sigma 2,gen} = -\mu_{21,gen}(h'' - h_2). \tag{7.37}$$

7.6 Verification of the model

As already mentioned, the model developed in the preceding sections was incorporated into the code IVA4. Its predicting capabilities will now be verified taking the flashing flow in converging – diverging nozzles and the blow-down in pipes as examples. It is very important in such calculations to ensure that the code properly predicts the single-phase steady-state and transient flows.

The results of three BNL experiments for water flow in a converging–diverging nozzle *Abuaf* et al. (1981) are presented in Fig. 7.1, together with the results calculated with our model. We see good agreement for pressure distribution as well as for the predicted mass flow rate as a function of inlet/outlet pressure difference.

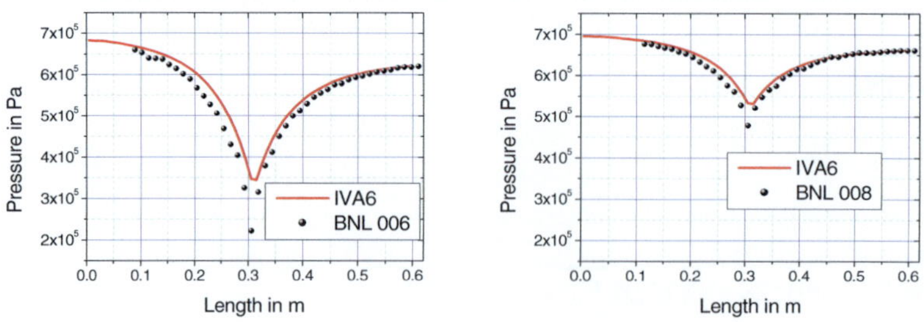

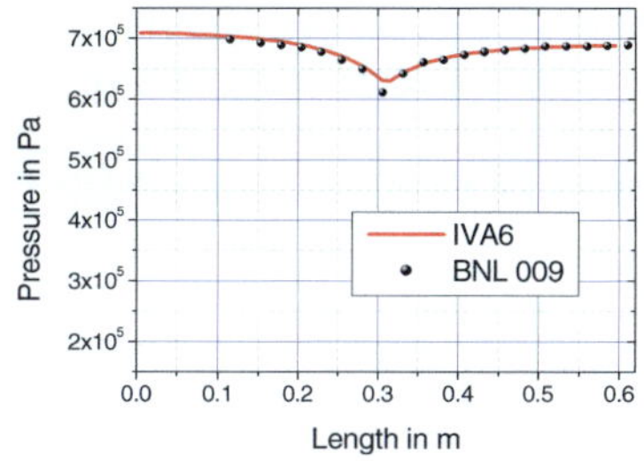

Fig. 7.1 Comparison of IVA4-NW predictions with BNL experimental data for pressure distribution. Single-phase flow. Runs 6, 8, 9. p_{in} = 682, 695, 709 kPa, p_{out} = 627, 665, 692 kPa, T_{in} = 27 °C, G_{exp} = 7010, 4710, 3130 kg/(m^2 s), G_{com} = 6319, 4362, 2991 kg/(m^2 s)

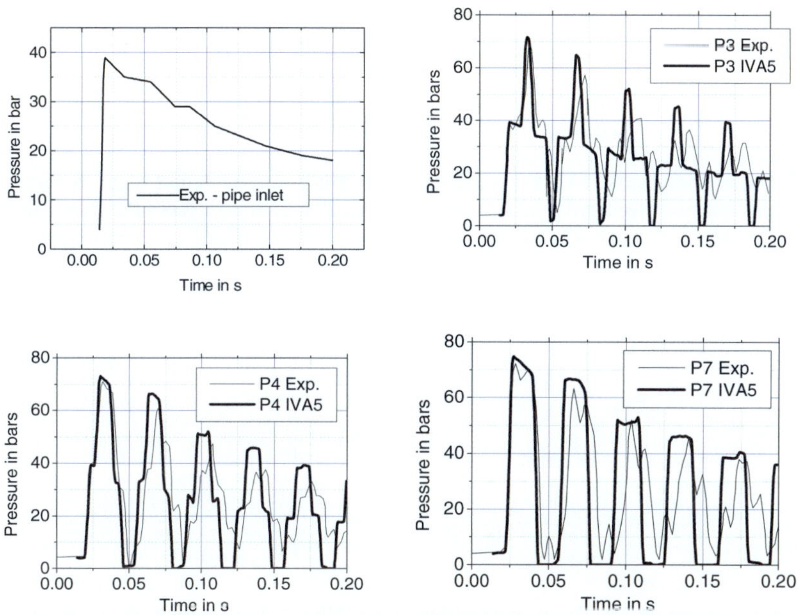

Fig. 7.2 a) Pressure at the pipe entrance as a function of time as measured by *Kellner* and *Gissler* (1984). Pipe length 13.12 m, diameter 0.1 m, 4 bends at 4.8, 10.2, 12.6, 13.64 m from the entrance. Initial conditions: water at atmospheric temperature and pressure. b) at 4.08 m from the pipe entrance. Comparison of IVA5 prediction with the experimental data by *Kellner* and *Gissler* (1984). c) at 8.18 m from the pipe entrance. d) at 12.5 m from the pipe entrance

The *Kellner* and *Gissler* experiment (1984) used for the next comparison produced pressure wave propagation in single pipes with five bends and one dead end. The imposed pressure function at the entrance of the pipe is shown in Fig. 7.2a. The comparison of the predicted and the measured responses presented in Figs. 7.2b–d shows good agreement.

Now we proceed to the BNL converging–diverging nozzle experiment *Abuaf* et al. (1981) with an entrance condition that causes flashing flow behind the nozzle throat. The results of the computation are presented in Figs. 7.3–7.6. We see that our model

a) predicts the data for pressure distribution data well;

b) slightly overpredicts the data for the volumetric void fraction distribution; resulting in

c) some underprediction of the mass flow as a function of the measured pressure difference.

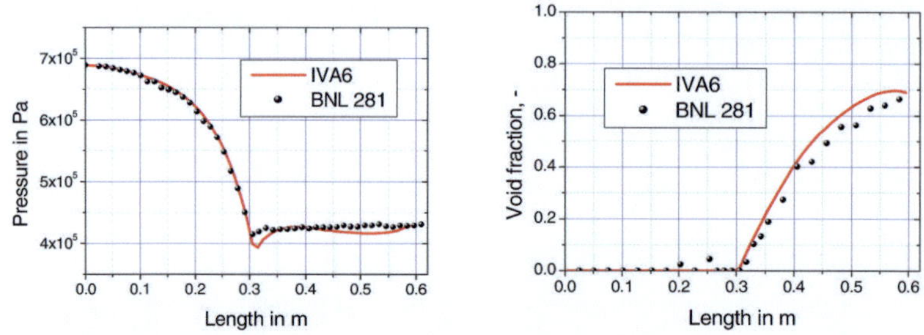

Fig. 7.3 Comparison of IVA4 prediction with BNL experimental data for pressure distributions and area averaged void fractions for Runs 281–283. p_{in} = 688 kPa, T_{in} = 148.8 °C, G_{in} = 5730 kg/(m^2 s), p_{out} = 431 kPa, p_c = 452 kPa, T_c = 148.8 °C, G_{com} = 5533 kg/(m^2 s)

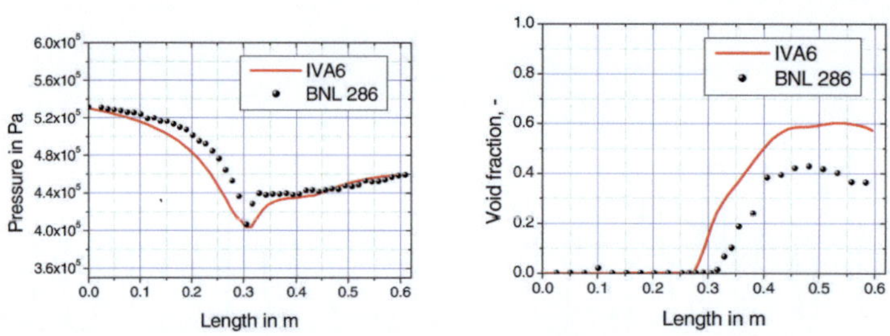

Fig. 7.4 Comparison of IVA4 predictions with BNL experimental data for pressure distributions and area average void fractions for Runs 286–288. p_{in} = 530 kPa, T_{in} = 149.2 °C, G_{in} = 3580 kg/(m^2 s), p_{out} = 459 kPa, p_c = 456 kPa, T_c = 149.2 °C, G_{com} = 2706 kg/(m^2 s)

Next, we simulate the pipe blow-down experiment documented in *Edwards* and *O'Brien* (1970), *Kolev* (1995). A horizontal pipe is initially filled with water slightly subcooled under the initial pressure. One end of the pipe is quickly opened. The pressure drops below the saturation pressure at the water temperature and the heterogeneous nucleation starts. After producing enough bubble sizes the pressure increases up to the saturation pressure corresponding to the initial temperature for a long time due to intensive flashing inside the pipe. Then it starts to decrease to the external pressure.

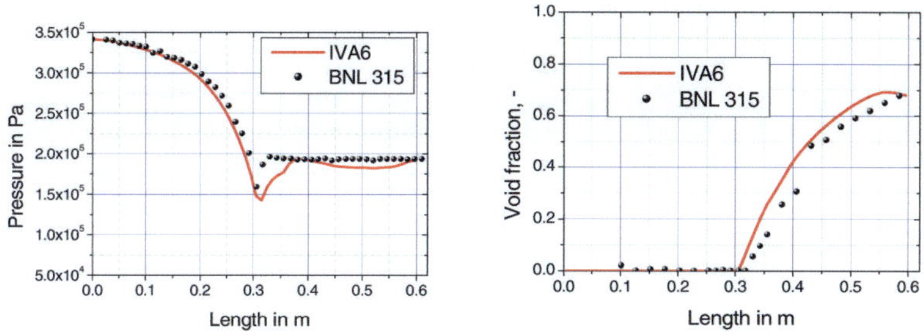

Fig. 7.5 Comparison of IVA4 predictions with BNL experimental data for pressure distributions and area average void fractions for Runs 313–315. p_{in} = 341 kPa, T_{in} = 121 °C, G_{in} = 4410 kg/(m^2 s), p_{out} = 193 kPa, p_c = 200 kPa, T_c = 119 °C, G_{com} = 4202 kg/(m^2 s)

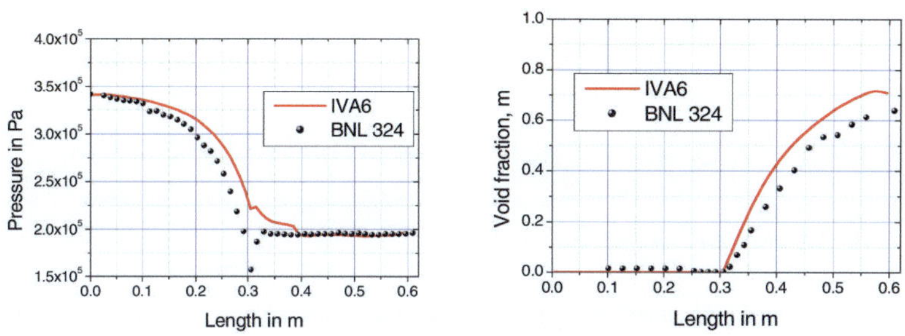

Fig. 7.6 Comparison of IVA4 predictions with BNL experimental data for pressure distributions and area average void fractions for Runs 324–325. p_{in} = 341 kPa, T_{in} = 121 °C, G_{in} = 4410 kg/(m^2 s), p_{out} = 196 kPa, p_c = 200 kPa, T_c = 121.8 °C, G_{com} =4004 kg/(m^2 s)

The results are given in Figs. 7.7–7.11. We see that the nucleation theory applied here results in an appropriate pressure prediction (Fig. 7.7). The somewhat faster emptying of the pipe in the computation in the later stage of the process (Fig. 7.8) is caused by the several interactions in the computer code like drag forces, flow pattern recognition, etc., which will not be discussed here. We see also that the computed water temperature (Fig. 7.10) decreases faster than in reality. This is evidence that more energy is released from the liquid for evaporation in the model than in reality. This is consistent with the predicted void fraction at a given position (Fig. 7.9). The large void fraction increase in the first 0.1 s is not predicted by the model. This experimental observation is in contrast with the behavior observed behind the throat of the NL nozzle.

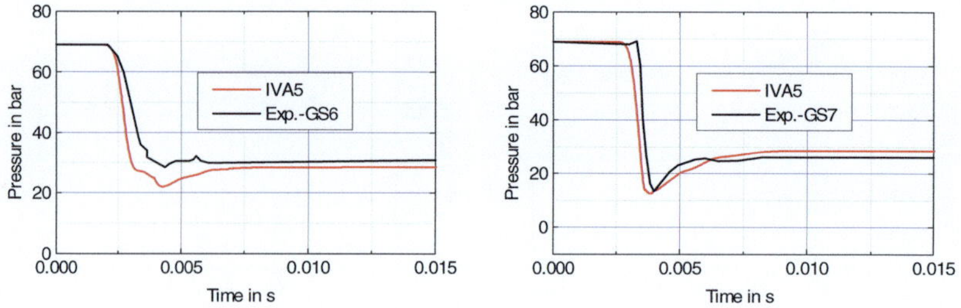

Fig. 7.7 Pressure at 0.072 and 0.914 m from the dead end of pipe as a function of time. Comparison IVA5 prediction with experimental data by *Edwards* and *O'Brien*. Pipe diameter 0.0732 m, length 4.096 m

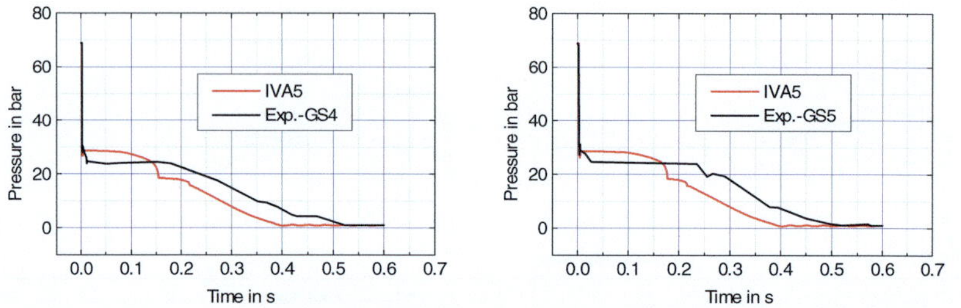

Fig. 7.8 Pressure at 2.024 and 2.469 m from the dead end of pipe as a function of time. Comparison IVA5 prediction with experimental data by *Edwards* and *O'Brien*. Pipe diameter 0.0732 m, length 4.096 m

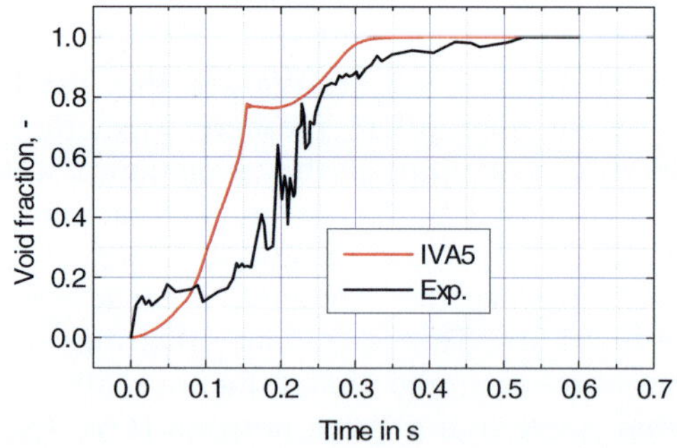

Fig. 7.9 Steam volume fraction at 1.469 m from the dead end of the pipe as a function of time. Comparison IVA5 prediction with experimental data by *Edwards* and *O'Brien*. Pipe diameter 0.0732 m, length 4.096 m

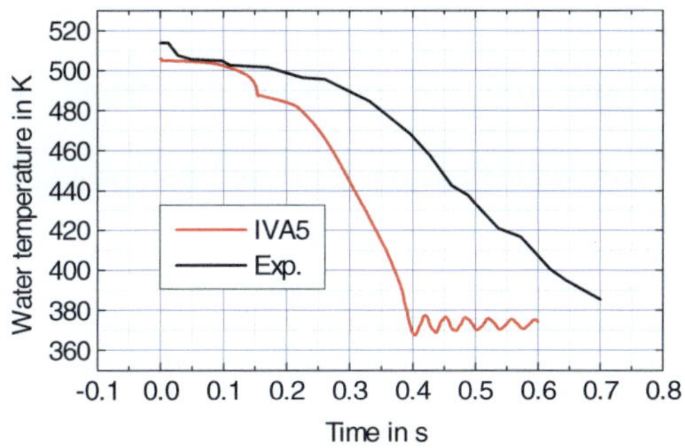

Fig. 7.10 Water temperature at 1.469 m from the dead end of the pipe as a function of time. Comparison IVA4 prediction with experimental data by *Edwards* and *O'Brien*. Pipe diameter 0.0732 m, length 4.096 m

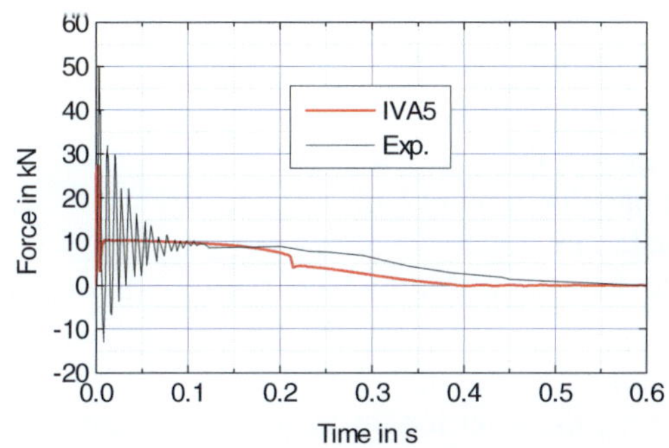

Fig. 7.11 Force as a function of time: a) IVA4 prediction without fluid–structure interactions; b) *Edwards–O'Brien* measurements reflecting fluid structure interactions.

The IVA5 force processor computes pipe forces due to hydrodynamics only as discussed in Chapter 8 of Volume 1. The result is presented in Fig. 7.11a. It corresponds to the pressure history. The oscillating character of the measurements, (Fig. 7.11b), is due to the fluid structure interactions. These interactions are not simulated by our computation.

7.7 Significance and conclusions

The proposed nucleation and flashing model consists of submodels, which where verified separately with the available data sets for each particular subphenomenon like

- the bubble departure diameter, taking into account the mutual bubble interactions at high superheatings, confirmed by data comparison in Chapter 15, *Kolev* (1993);

- the active nucleation-site density, provided by *Wang* and *Dhir* (1993) measurements;
- the mean site-to-site distance provided by *Wang* and *Dhir* (1993) measurements;
- heat transfer between wall-fluid interface and bulk liquid caused mainly by bubble-growth and departure;

introduced turbulence in the boundary layer verified in Chapter 16, *Johov* (1969) etc. As far as the author knows this is the first boiling and flashing model that gives reasonable agreement with data for complex flashing flows in addition to the good agreement in separated tests. The model has the advantage of predicting properly the separation between steam generation due to bubble production and bubble growth. I believe that it will be successfully applied in multiphase flow modeling.

Nomenclature

a_2	liquid thermal diffusivity, m²/s
B^2	$= 2R_1 dR_1 / d\tau,\ m^2/s$, m²/s
D_h	hydraulic diameter, m
D_l	particle size of velocity field l, m
D_1	bubble diameter, m
$D_{1\infty}$	stable bubble diameter under the local flow conditions, m
D_{1c}	critical size, m
D_{1d}	bubble departure diameter, m
$D_{1d,fc}$	bubble departure diameter for strongly predominant forced convection, m
$D_{1d,nc}$	bubble departure diameter for natural convection, m
$D_{2,inf}$	average center-to-center spacing, m
f_{1w}	bubble departure frequency, 1/s
f_{1w}^{t}	boundary layer turbulence fluctuation frequency, 1/s
f_{coal}	coalescence frequency, 1/s
f_{col}	collision frequency of single bubble, 1/s
P_{coal}	coalescence probability, dimensionless
G	gravitational acceleration, m/s^2
h', h''	specific saturation enthalpies for water and steam, J/kg
Δh	$= h'' - h'$, J/kg
Ja_2	$= \rho_2 c_{p2} \left[T_2 - T'(p)\right] / (\rho_1 \Delta h)$, *Jacob* number, dimensionless
n_l	particle number density of velocity field l, time averaged, 1/m^3
$n_{1\infty}$	stable particle number density concentration after fragmentation, 1/m^3
n''_{1w}	active nucleation-site density, 1/m^2
$\dot{n}_{l,kin}$	particles of velocity field l originating per unit time and unit mixture volume due to change of the state of aggregate, local volume and time averaged, 1/(s m^3)

$\dot{n}_{1w}$	bubbles originating at the wall per unit time and unit mixture volume, local volume and time averaged, 1/(s m^3)
$\dot{n}_{1,frag}$	particles of velocity field *l* originating per unit time and unit mixture volume due to fragmentation, local volume and time averaged, 1/(s m^3)
$\dot{n}_{l,coal}$	particles of velocity field *l* disappearing per unit time and unit mixture volume due to coalescence, local volume and time averaged, 1/(s m^3)
p	pressure, Pa
p'	saturated pressure, Pa
Pr_2	$=\eta_2/(\rho_2 a_2)$, Prandtl number, dimensionless
$\dot{q}'''_{2\sigma,gen}$	the energy consumed for the bubble generation until departure per unit mixture volume and unit time, W/m^3
$\dot{q}''_{2\sigma,gen}$	heat flux from the wall into the liquid during film flashing bubble generation in adiabatic pipe flow boiling, W/m^2
$\dot{q}''_{w2,b}$	heat flux from the wall into the liquid during pool boiling, W/m^2
R_1	bubble radius, m
R_{1o}	bubble radius at the beginning of the time step $\Delta\tau$, m
$R_{2,inf}$	$= D_{2,\inf}/2$, m
$T'(p)$	saturation temperature at system pressure *p*, K
T_2	liquid temperature, K
T_w	wall temperature, K
V_1	surface average and time average velocity of velocity field *l*, m/s
V_{1Ku}	free rising bubble velocity, *Kutateladze* velocity, m/s
$\overline{V}'_2$	volume-averaged fluctuation velocity, m/s
$\overline{\overline{V}}'_2$	volume-averaged fluctuation velocity time averaged over $\Delta\tau_w + \Delta\tau_d$, m/s

Greek

α_1	volume concentration of velocity field *l*, dimensionless
α_{1d}	the steam volumetric fraction in the boundary layer having thickness of D_{1d}, dimensionless
γ_v	volume porosity, dimensionless
γ	surface permeability, dimensionless
$\Delta\rho_{21}$	$=\rho_2-\rho_1$, kg/s
$\Delta\tau_d$	time elapsed from the initiation of bubble with the critical size to the bubble departure from the wall, s
$\Delta\tau_w$	delay time, s
$\Delta\tau_{1d}$	bubble departure time, s
η_2	dynamic viscosity of liquid, kg/(m s)

λ_{RT} $= \left(\dfrac{\sigma}{g \Delta \rho_{21}} \right)^{1/2}$, *Rayleigh–Taylor* instability wavelength, m

ℓ_2^t turbulent length scale, m

ν_2 kinematic viscosity of liquid, m^2/s

ρ_l density of velocity field *l*, kg/m^3

ρ_1 gas density, kg/m^3

ρ_2 liquid density, kg/m^3

ρ'' saturated steam density, kg/m^3

ρ' saturated liquid density, kg/m^3

μ_l mass introduced into the velocity field *l* per unit mixture volume and unit time, local volume and time average, $kg/(m^3\ s)$

$\mu_{21,gen}$ steam generation per unit mixture volume and unit time due to bubble generation at the wall, local volume and time average, $kg/(m^3\ s)$

$\mu_{21,bulk}$ mass evaporated in the bulk per unit mixture volume and unit time, $kg/(m^3\ s)$

σ surface tension, N/m

τ time, s

φ contact angle between "liquid drop" and the wall, rad

References

Abuaf, N., Wu, B.J.C., Zimmer, G.A., Saha, P.: A study of nonequilibrium flashing of water in a converging diverging nozzle, vol.1, Experimental, vol. 2 Modeling, NUREG/CR-1864, BNL-NUREG-51317 (June 1981)

Ahmad, S.Y.: Axial distribution of bulk temperature and void fraction in heated channel with inlet subcooling. J. Heat Transfer 92, 595 (1970)

Blinkov, V.N., Jones, O.C., Nigmatulin, B.I.: Nucleation and Flashing in Nozzles - 2. Comparison with Experiments Using a Five - Equation Model for Vapor Void Development. Int. J. Multiphase Flow 19(6), 965–986 (1993)

Edwards, A.R., O'Brien, T.P.: Studies of phenomena connected with the depressurization of water reactors. J. Br. Nucl. Energy Soc. 9(1–4), 125–135 (1970)

Johov, K.A.: Nucleations number during steam production. In: Aerodynamics and Heat Transfer in the Working Elements of the Power Facilities, Leningrad, vol. 91, pp. 131–135 (1969) (in Russian)

Kaishev, R., Stranski, I.N.: Z. Phys. Chem. 26, 317 (1934)

Kellner, H., Gissler: Program system SAPHYR: Anwendungsbeispiel II, Notiz Nr.70.02748.4. Interatom GmbH, February 7 (1984)

Kolev, N.I.: The influence of mutual bubble interaction on the bubble departure diameter. Exp. Thermal Fluid Sci. 8(2), 167–174 (1994)

Kolev, N.I.: The Code IVA4: Nucleation and flashing model. Kerntechnik 60(6), 157–164 (1995); Also in: Proc. Second Int. Conf. On Multiphase Flow, Kyoto, August 13-18 (1995); ASME & JSME Fluid Engineering Conference International Symposium on Validation of System Transient Analysis Codes, Hilton Head (SC) USA; Int. Symposium on Two-Phase Flow Modeling and Experimentation, ERGIFE Place Hotel, Rome, Italy, October 9-11 (1995)

Mikic, B.B., Rohsenow, W.M., Griffith, P.: On bubble growth rates. Int. J. Heat Mass Transfer 13, 657–666 (1970)

Shin, T.S., Jones, O.C.: Nucleation and flashing in nozzles - 1, A distributed nucleation model. Int. J. Multiphase Flow 19(6), 943–964 (1993)

Skripov, V.P., et al.: Thermophysical properties of liquids in metastable state. Atomisdat, Moskva (1980) (in Russian)

Volmer, M.: Kinetik der Phasenbildung. Dresden und Leipzig, Steinkopf (1939)

Wang, C.H., Dhir, V.K.: Effect of surface wettability on active nucleation site density during pool boiling of water on a vertical surface. ASME J. Heat Transfer 115, 659–669 (1993)

8. Boiling of subcooled liquid

8.1 Introduction

The description of the forced convection boiling of subcooled liquid started with using the equilibrium thermodynamics of two-phase flow. In this framework empirical correlations are obtained by generalizing experimental data using the *equilibrium steam mass flow rate concentration* in the flow,

$$X_{eq} = \frac{h - h'(p)}{h''(p) - h'(p)}, \tag{8.1}$$

where the specific mixture enthalpy is defined as the *mixture enthalpy of the noninert components*

$$h = \left[\sum_l \left(1 - \sum_n C_{nl}\right) \alpha_l \rho_l w_l h_{Ml}\right] \Big/ \left[\sum_l \left(1 - \sum_n C_{nl}\right) \alpha_l \rho_l w_l\right]$$

$$= \left[\sum_l \left(1 - \sum_n C_{nl}\right) X_l h_{Ml}\right] \Big/ \left[\sum_l \left(1 - \sum_n C_{nl}\right) X_l\right] \tag{8.2}$$

and the *mass flow rate* in the axial direction is defined as follows

$$G = \sum (\alpha \rho w)_l \,. \tag{8.3}$$

8.2 Initiation of visible boiling on the heated surface

Treshchev reported in (1969) measurements for the number of the vapor-formation centers on a flat surface for pressures between 5.25 and 50 bar, water temperature from 80 to 250 °C, heat fluxes between 0.4 to 5 MW and flow velocities up to 2 m/s. The heat flux at which a nucleation is observed at the surface is correlated by

$$\dot{q}''_{w0,50bar} = 15.4 \times 10^3 \left[T'(p) - T_2^*\right] w^{0.8} .$$

The exponent 0.8 resembles the dependence of the heat-transfer coefficient for single-phase flow on velocity. The corresponding equilibrium quality at which the nucleation starts is then

$$X_{1eq}^{*} = -\frac{c_{p2}\left[T'(p)-T_2^{*}\right]}{h''-h'} = -6.49\times10^{-5}\frac{c_{p2}\dot{q}''_{w0}}{\left(h''-h'\right)w^{0.8}}.$$

Here, the subscript * means onset of nucleate boiling (ONB). For larger heat fluxes the number of active nucleation seats per unit surface is

$$n''_w = \frac{1}{3}10^{-5}\left(\frac{\Delta h\rho''}{10^3}\right)^{1.7}\exp\left(-2.1\times10^{-8}\Delta h\rho''\right)\left[T'(p)-T_2^{*}\right]^{0.6}$$
$$\times\left(\dot{q}''_w - \dot{q}''_{w0,50bar}\right)\exp\left(-1.25\times10^{-3}\sqrt{\dot{q}''_w - \dot{q}''_{w0,50bar}}\right).$$

The data for the initiation of the nucleation can be correlated using dimensionless numbers within an error band of 30% as follows:

$$St =: \frac{\dot{q}''_{w0}}{Gc_{p2}\left[T'(p)-T_2^{*}\right]} = 0.01286 - 5.4553\times10^{-8}Pe_2 \quad \text{for } Pe_2 \le 166\,000,$$
$$St = 0.0038 \quad \text{for} \quad Pe_2 > 166\,000,$$

where $Pe_2 = GD_{hy}c_{p2}/\lambda_2$. The performance of the correlation is given in Fig. 8.1.

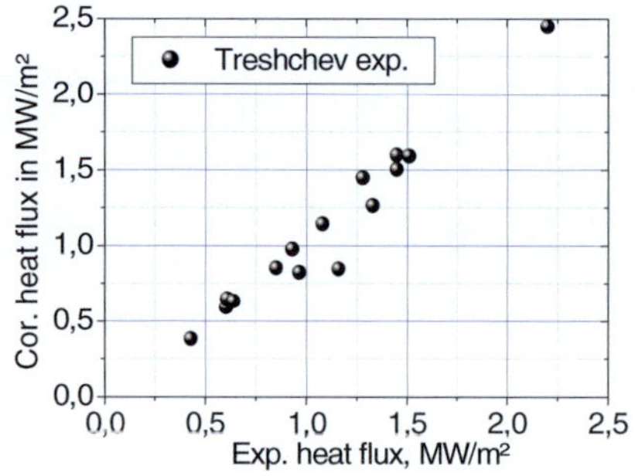

Fig. 8.1 Correlated versus measured heat fluxes at the moment of initiation of the nucleation

Note that the above information reported by *Treshchev* is based on local conditions at the surface. *Basu* et al. (2002) reported that the local onset of the nucleate boiling is dependent also on the wetting angle θ as follows

$$T_2^{*} = T'(p_1) - T'(p),$$
$$p_1 - p = 4\sigma_2/D_{1c},$$
$$D_{1c} = \left[\frac{8\sigma_2 T'(p)\lambda_2}{\rho_1\Delta h q''_{w2}}\right]^{1/2}\left[1-\exp\left(-\theta^3 - \frac{1}{2}\theta\right)\right].$$

Sekoguchi et al. (1972) correlated their experimental data for the dependence of the local equilibrium steam mass flow concentration in the flow, defining the *initiation of the visible nucleate boiling*, X_{1eq}^*, as a function of the wall heat flux and the local parameters by

$$X_{1eq}^* = -13.5\left[\frac{\dot{q}_w''}{G(h''-h')}\right]^{0.65},$$

valid in $0.3 < \dot{q}_w''/\left[G(h''-h')\right] < 1$.

Saha and *Zuber* (1974) approximated a number of experimental data by

$$X_{1eq}^* = -0.0022\frac{D_h c_{p2}}{\lambda_2}\frac{\dot{q}_w''}{h''-h'} \qquad Pe_2 \le 70\,000, \qquad (8.4)$$

or recomputed in terms of liquid temperature at which the subcooled nucleate boiling starts,

$$T_2^* = T'(p) - 0.0022\,\dot{q}_w''\,D_h/\lambda_2 \quad \text{or} \quad Nu =: \frac{\dot{q}_w''}{T'(p)-T_2^*}\frac{D_h}{\lambda_2} = 454, \qquad (8.5)$$

and

$$X_{1eq}^* = -\frac{154}{G}\frac{\dot{q}_w''}{h''-h'} \qquad Pe_2 > 70\,000 \qquad (8.6)$$

or

$$T_2^* = T'(p) - 154\frac{\dot{q}_w''}{c_{p2}G} \quad \text{or} \quad St =: \frac{\dot{q}_w''}{Gc_{p2}\left[T'(p)-T_2^*\right]} = 0.0065, \qquad (8.7)$$

where

$$Pe_2 = GD_{hy}c_{p2}/\lambda_2, \qquad (8.8)$$

and T_2^* is, as already mentioned, the liquid temperature corresponding X_{1eq}^*. The above correlation approximates experimental data in the region: $1\times10^5 < p < 138\times10^5$ Pa, $95 < G < 2760$ kg/(m^2 s), $280\,000 < \dot{q}_w'' < 1\,890\,000$ W/m^2 for channels with rectangular and circular cross-sections. For Freon the correlation is valid in $3.2\times10^5 < p < 8.5\times10^5$ Pa, $101 < G < 2073$ kg/(m^2 s), $6\,300 < \dot{q}_w'' < 5\,360\,000$ W/m^2.

Comparing the form of Eq. (8.6) with *Treshchev*'s equation we recognize the similarity. Note the differences for the application of the both correlation: *Treshchev*'s correlation is based on local parameters and the *Saha* and *Zuber* correlation is based on cross-section-averaged parameters.

Chan (1984) analyzed the net vapor generation point in horizontal flow boiling in a tube with $D_h = D_{heat} = 10.7$ mm, and length of 2 m at 4 bar pressure and came to the conclusion that the constant in Eq. (8.5) has to be equal to 380. The *Peclet* numbers of his experiments varied between 10^4 and 4×10^4.

Kawara et al. (1998) reported that at low void fraction the *Saha* and *Zuber* correlation is not accurate. They proposed the modification:

$$Nu\frac{\dot{q}''_w}{0.88\times10^7} = 454, \quad Pe_2 \le 70\,000,$$

$$St\frac{\dot{q}''_w}{0.88\times10^7} = 0.0065, \quad Pe_2 > 70\,000.$$

My own analysis, *Kolev* (2005), based on the NUPEC data (2004) demonstrated that at low void fraction the use of the *Hughes* et al. (1981) model, that will be described in the next section is superior to the use of the *Saha* and *Zuber* model in combination with the *Levy* (1967) relaxation model.

8.3 Local evaporation and condensation

8.3.1 Relaxation theory

There are different approaches proposed in the literature to describe the local evaporation and condensation processes. One of the oldest is that by *Levy*. In one of his earlier publications in this field *Levy* (1967) approximates the profile of the experimentally observed steam mass flow concentration in the flow with the following simple expressions

$$X_1 = X_{1eq} - X^*_{1eq}\exp(X_{1eq}/X^*_{1eq} - 1) \tag{8.9}$$

where X_{eq} is the *equilibrium steam mass flow concentration* in the flow defined by Eq. (8.1) for the case of no inert components

$$\sum X_l h_l = X_{1eq}h'' + (1 - X_{1eq})h' = X_{1eq}(h'' - h') + h'. \tag{8.10}$$

Now let as compute the net vapor generation rate defined by the profile described with Eq. (8.9). For one-dimensional steady-state flow in a channel with constant cross-section we write the mixture mass conservation, the vapor mass conservation and the mixture energy-conservation equations as follows

$$G = const, \tag{8.11}$$

$$\mu_{21} - \mu_{12} = G\frac{dX_1}{dz} \tag{8.12}$$

$$\frac{d}{dz}\left(\sum X_1 h_1\right) = \frac{4}{D_{heat}}\frac{\dot{q}''_w}{G}. \tag{8.13}$$

Having in mind the definition equation (8.10) for X_{1eq}

$$d\left(\sum X_1 h_1\right) = d\left[X_{1eq}(h'' - h') + h'\right] = (h'' - h')dX_{1eq} \tag{8.14}$$

Eq. (8.13) can be rewritten as follows

$$\frac{dX_{1eq}}{dz} = \frac{4}{D_{heat}} \frac{\dot{q}''_w}{G(h'' - h')}. \tag{8.15}$$

Thus, from the mass-conservation equation of the gas phase (8.12) and using Eq. (8.15) we obtain

$$\mu_{21} - \mu_{12} = G\frac{dX_1}{dz} = G\frac{dX_1}{dX_{1eq}}\frac{dX_{1eq}}{dz} = \frac{4}{D_{heat}}\frac{\dot{q}''_w}{h'' - h'}\frac{dX_1}{dX_{1eq}}, \tag{8.16}$$

or after differentiating the *Levy* approximation, Eq. (8.9),

$$\mu_{21} - \mu_{12} = \frac{4}{D_{heat}}\frac{\dot{q}''_w}{h'' - h'}\left[1 - \exp\left(\frac{X_1}{X^*_{1eq}} - 1\right)\right] \quad \text{for } X_1 > X^*_{1eq}. \tag{8.17}$$

Obviously, in accordance with this approach the heat flux at the wall causing all the net evaporation is

$$\dot{q}''_{net\_evaporation} = \dot{q}''_w\left[1 - \exp\left(\frac{X_1}{X^*_{1eq}} - 1\right)\right], \tag{8.18}$$

and the heat flux introduced into the bulk liquid due to condensation is

$$\dot{q}_2''^{\sigma 1} = \dot{q}''_w - \dot{q}''_w\left[1 - \exp\left(\frac{X_1}{X^*_{1eq}} - 1\right)\right] = \dot{q}''_w \exp\left(\frac{X_1}{X^*_{1eq}} - 1\right). \tag{8.19}$$

Therefore, the effectively condensed mass per unit time and unit mixture volume is

$$\mu_{12} = \frac{4}{D_{heat}}\frac{\dot{q}_2''^{\sigma 1}}{h'' - h'} = \frac{4}{D_{heat}}\frac{\dot{q}''_w}{h'' - h'}\exp\left(\frac{X_1}{X^*_{1eq}} - 1\right). \tag{8.20}$$

From Eqs. (8.20) and (8.17) we compute the evaporated mass per unit time and mixture volume as

$$\mu_{21} = \frac{4}{D_{heat}}\frac{\dot{q}''_w}{h'' - h'}. \tag{8.21}$$

This approach does not answer the question about the magnitude of the heat-transfer coefficient. One is forced to arbitrarily assume some wall superheat, e.g., 5

K and compute the total heat-transfer coefficient by dividing the wall heat flux by this value.

An alternative to Eq. (8.9) is the relaxation profile proposed by *Achmad* (1970)

$$X_1 = \frac{X_{1eq} - X^*_{1eq} \exp(X_{1eq} / X^*_{1eq} - 1)}{1 - X^*_{1eq} \exp(X_{1eq} / X^*_{1eq} - 1)}, \tag{8.22}$$

which results in

$$\dot{q}_2''^{\sigma 1} = \dot{q}_w'' \left\{ 1 - \frac{1 - \left(1 - X_{1eq} + X^*_{1eq}\right) \exp(X_{1eq} / X^*_{1eq} - 1)}{\left[1 - X^*_{1eq} \exp(X_{1eq} / X^*_{1eq} - 1)\right]^2} \right\}. \tag{8.23}$$

Note that there are other proposals for the approximations $X_1 = X_1(X_{1eq}, X^*_{1eq})$ – see *Kolev* (1986).

Thus, the relaxation theory provided a simple expression for the recondensation process, which does not take into account the influence of the local flow pattern and therefore of the interfacial area density. This approach is successfully used for description of steady-state boiling in a channel with constant cross-sections.

8.3.2 Boundary-layer treatment

The treatment of the boundary layers is the more sophisticated method for description of the evaporation and condensation processes. Let us illustrate this approach with the method proposed by *Hughes* et al. (1981), based on one proposal by *Lellouche* (1974) and *Lellouche* and *Zolotar* – see in *Hughes* et al. (1981). The authors consider the heat transfer at the wall as a superposition of the heat flux causing boiling for saturated liquid and a convective heat flux from the wall into the bulk liquid

$$\dot{q}_w'' = \dot{q}_{NB}'' + \dot{q}_2''^{\sigma w} = h_{NB\_T\mathrm{hom}} (T_w - T')^2 + h_{convective} \left(T_w - T_2\right). \tag{8.24}$$

Here, in accordance with *Thom* (1966) the saturated boiling heat flux $\dot{q}_{NB}''$ is a quadratic function of the wall superheat with a coefficient being a function only of the system pressure

$$h_{NB\_T\mathrm{hom}} = 1942 \, \exp(p / 4.35 \times 10^6) \tag{8.25}$$

and the convective heat-transfer coefficient is that derived by *Dittus* and *Boelter* in 1937 for single-phase liquid flow

$$h_{convective} = 0.023 Re_2^{0.8} Pr_2^{0.4} \frac{\lambda_2}{D_h}, \tag{8.26}$$

where

$$Re_2 = \rho_2 w_2 D_h / \eta_2, \tag{8.27}$$

$$Pr_2 = c_{p2}\eta_2 / \lambda_2 , \tag{8.28}$$

are the *Reynolds* and the *Prandtl* numbers respectively. The heat flux released by the recondensation into the bulk flow is in accordance with *Hancox* and *Nicoll* – see in *Hughes* et al., (1981)

$$\dot{q}_2''^{\sigma 1} = h_2^{\sigma 1}(T' - T_2) \quad \text{for} \qquad X_1 > X_{1eq}^* , \tag{8.29}$$

where

$$h_2^{\sigma 1} = 0.4 Re_2^{0.662} Pr_2 \frac{\lambda_2}{D_h} . \tag{8.30}$$

The liquid temperature defining the initiation of the visible nucleate boiling, $T_{2,NBI}$, and the corresponding wall temperature are computed from the condition that the evaporation heat flux equals the condensation one

$$h_{NB\_T\hom}(T_w - T')^2 = h_2^{\sigma 1}(T' - T_{2,NBI}) \tag{8.31}$$

and Eq. (8.24)

$$\dot{q}_w'' = h_{NB\_T\hom}(T_w - T')^2 + h_{convective}\left(T_w - T_{2,NBI}\right) . \tag{8.32}$$

Solving with respect to the liquid subcooling the authors obtained

$$T' - T_{2,NBI} = \frac{h_{NB\_T\hom}}{h_2^{\sigma 1}}(T_w - T')^2$$

$$= \left[\frac{\sqrt{h_{convective}^2 \dfrac{h_2^{\sigma 1}}{h_{NB\_T\hom}} + 4\left(h_2^{\sigma 1} + h_{convective}\right)\dot{q}''} - h_{convective}\sqrt{\dfrac{h_2^{\sigma 1}}{h_{NB\_T\hom}}}}{2\left(h_2^{\sigma 1} + h_{convective}\right)} \right]^2 . \tag{8.33}$$

The wall temperature is obtained after solving the slightly rearranged Eq. (8.32)

$$\dot{q}_w'' = h_{NB\_T\hom}(T_w - T')^2 + h_{convective}\left(T' - T_2\right) \tag{8.34}$$

with respect to the wall superheating

$$T_w - T' = \sqrt{\frac{\dot{q}_w'' - h_{convective}\left(T' - T_2\right)}{h_{NB\_T\hom}}} \quad \text{for} \quad T_2 > T_{2,NBI} . \tag{8.35}$$

The evaporated mass per unit time and unit mixture volume is then

$$\mu_{21} = \frac{4}{D_{heat}} \frac{h_{NB\_T\hom}(T_w - T')^2}{h'' - h'} . \tag{8.36}$$

The condensed mass per unit time and unit volume is

$$\mu_{12} = \frac{4}{D_{heat}} \frac{h_2^{\sigma 1}(T' - T_2)}{h'' - h'} . \tag{8.37}$$

The heat introduced into the bulk liquid due to condensation is

$$\dot{q}_2^{\prime\prime\prime\sigma 1} = \frac{4}{D_{heat}} h_2^{\sigma 1}(T' - T_2) . \tag{8.38}$$

The heat introduced from the wall into the bulk liquid due to forced convection is

$$\dot{q}_2^{\prime\prime\prime\sigma w} = \frac{4}{D_{heat}} h_{convection}(T_w - T_2) . \tag{8.39}$$

Hughes reports very good agreement with experimental data for void fraction in a vertical heated channel using the *Lellouche*'s drift flux correlation. *Kolev* (1986) compared the predictions of this approach with a number of data for void fractions in a vertical boiling channel in the region $38 < G < 2000$ kg/(m^2 s), $T' - T_2 < 132$ K, $1 < p < 140$ bar, $20.3 < \dot{q}_w'' < 1723$ kW/m^2, $0.686 < L < 1.835$, $0.0044756 < D_h < 0.0269$ m using *Chexal* and *Lellouch*'s drift flux correlation, and reported a very good agreement. The agreement was worse for very low mass flow rates.

Note that the condensation of the generated vapor happens on larger interfacial area density $6\alpha_1/D_1$ then $4/D_{hy}$ to which the condensation heat-transfer coefficient is related. Therefore, the constant in empirically obtained correlations like $h_2^{\sigma 1} = 0.4 Re_2^{0.662} Pr_2 \dfrac{\lambda_2}{D_h}$ has to be function actually of $\dfrac{3}{2}\alpha_1 \dfrac{D_{hy}}{D_1}$. Based on experiments with condensation into subcooled water after a heated section in pipe *Avdeev* (1986) obtained

$$h_2^{\sigma 1} = 0.228 Re_2^{0.7} Pr_2^{1/2} (\Phi_{2o}^2 / \alpha_2)^{1/4} \frac{\lambda_2}{D_h}, \ \pm 30\% , \ D_1 / D_{hw} > 80 / Re_2^{0.7} .$$

If the boiling heat transfer is described in general with $\dot{q}_{NB}'' = h_{NB}(T_w - T')^n$ the equation defining the temperature at which the initiation of nucleate boiling

$$\left(h_2^{\sigma 1} + h_{convective}\right)\left(T' - T_{2,NBI}\right) + h_{convective}\left[\frac{h_2^{\sigma 1}}{h_{NB}}(T' - T_{2,NBI})\right]^{1/n} - \dot{q}_w'' = 0$$

has to be solved by iteration.

Nomenclature

Latin

C_{nl} mass concentration of the inert component *n* inside the gas mixture, dimensionless

c_p specific heat at constant pressure, J/(kg K)

D_{hy} hydraulic diameter (4 times flow cross-sectional area / wet perimeter), m

D_{heat} heated diameter (4 times flow cross-sectional area / heated perimeter), m

G $=\sum(\alpha\rho w)_l$, mass flow rate in the axial direction, kg/(m² s)

h specific enthalpy, J/kg

$h_{NB\_T\hom}$ saturated boiling heat-transfer coefficient, W/(m² K)

$h_{convective}$ convective heat-transfer coefficient, W/(m² K)

$h_2^{\sigma 1}$ heat flux coefficient by the recondensation into the bulk flow based on the heated surface, W/(m² K)

Nu $\dfrac{\dot{q}''_w}{T'(p)-T_2^*}\dfrac{D_h}{\lambda_2}$, *Nusselt* number, dimensionless

h specific mixture enthalpy – mixture enthalpy of the noninert components, J/kg

L length, m

p pressure, Pa

Pe_2 $=GD_{hy}c_{p2}/\lambda_2$, *Peclet* number, dimensionless

$\Pr_2$ $=c_{p2}\eta_2/\lambda_2$, liquid *Prandtl* number, dimensionless

Re_2 $=\rho_2 w_2 D_h/\eta_2$, liquid *Reynolds* number, dimensionless

St $\dfrac{\dot{q}''_w}{c_{p2}\left[T'(p)-T_2^*\right]}$, *Stanton* number, dimensionless

T_2^* liquid temperature corresponding X_{1eq}^*, K

$\dot{q}''_w$ wall heat flux, W/m²

$\dot{q}''_{net\_evaporation}$ heat flux at the wall causing all the net evaporation, W/m²

$\dot{q}''_{NB}$ saturated boiling heat flux, W/m²

$\dot{q}_2''^{\sigma w}$ convective heat flux from the wall into the bulk liquid, W/m²

$\dot{q}_2''^{\sigma 1}$ heat flux released by the condensation into the bulk flow, W/m²

X_l mass flow rate concentration of the velocity field *l* inside the multiphase mixture, dimensionless

X_{1eq}^* local equilibrium steam mass flow concentration in the flow, defining the initiation of the visible nucleate boiling, dimensionless

X_{eq} equilibrium steam mass flow rate concentration in the flow, dimensionless
w axial velocity, m/s

Greek

α volume fraction, dimensionless
λ thermal conductivity, W/m K
ρ density, kg/m³
μ_{21} evaporation mass per unit time and unit mixture volume, kg/(m³ s)
μ_{12} condensation mass per unit time and unit mixture volume, kg/(m³ s)

Subscript

' saturated liquid
" saturated steam
1 gas
2 liquid
$M1$ noninert component in the gas mixture
$n1$ inert component in the gas mixture

References

Achmad, S.I.: Raspredelenie srednomassovoj temperatury zidkosti I istinnogo obemnogo parosoderzanie vodoi obogrevaemogo kanala s nedogrevom na vchode. Teploperedaca 4 (1970)

Avdeev, A.A.: Growth and condensation velocity of steam bubbles in turbulent flow. Teploenergetika 1, 53–55 (1986) (in Russian)

Basu, N., Warrier, G.R., Dhir, V.K.: Onset of nucleate boiling and active nucleation site density during subcooled flow boiling. Trans. ASME, J. Heat Transfer 127, 717–728 (2002)

Chan, A.M.C.: Point on net vapor generation and its movement in transient horizontal flow boiling. In: Dhir, V.K., Schrock, V.E. (eds.) Proc. of the 22th Nat. Heat Transfer Conference, Niagara Falls, New York (1984)

Dittus, F.W., Boelter, L.M.K.: Heat transfer in automobile radiators of the tubular type. University of California Publications in Engineering, vol. 2(13), p. 443 (1937)

Hughes, E.D., Paulsen, M.P., Agee, L.J.: A drift-flux model of two-phase flow for RETRAN. Nuclear Technology 54, 410–420 (1981)

Kawara, Z., Kataoka, I., Serizawa, A., Ko, Y.J., Takahashi, O.: Analysis of forced convective CHF based on two-fluid and three-fluid model. In: Proc. of the 11th IHTC, Heat Transfer 1998, Kyongju, Korea, August 23–28, vol. 2, pp. 103–108 (1998)

Kolev, N.I.: Transiente Zweiphasenstromung. Springer, Heidelberg (1986)

Kolev, N.I.: Do we have appropriate constitutive sets for sub-channel and fine-resolution 3D-analyses of two-phase flows in rod bundles? In: Mathematics and Computation, Supercomputing, Reactor Physics and Nuclear and Biological Applications Palais des Papes, Avignon, France, on CD-ROM, September 12–15. American Nuclear Society, LaGrange Park (2005)

Lellouche, G.S.: A model for predicting two-phase flow. BNL-18625 (1974)

Levy, S.: Int. J. Heat Mass Transfer 10, 351–365 (1967)

OECD/NRC Benchmark based on NUPEC BWR Full-size Fine-mesh Bundle Tests (BFBT), Assembly Specifications and Benchmark Database. Incorporated Administrative Agency, Japan Nuclear Energy Safety Organization, JNES-04N-0015, October 4 (2004)

Saha, P., Zuber, N.: Proc. Int. Heat Transfer Conf., Tokyo, Paper 134.7 (1974)

Sekoguchi, K., Nishikawa, K., Nakasatomi, M., Hirata, N., Higuchi, H.: Flow boiling in subcooled and low quality regions – heat transfer and local void fraction B4.8, 180–184 (1972)

Thom, I.R.S., et al.: Boiling in subcooled water during up heated tubes or annuli. In: Proc. Instrum. Mech. Engs., vol. 180 3C, pp. 1965–1966 (1966)

Treshchev, G.G.: The number of vapor-formation centers in surface boiling. In: Borishanskii, V.M., Pallev, I.I. (eds.) Convective Heat Transfer in Two-Phase and One-Phase Flows (1969); translated from Russian by the Israel Program of Scientific Translations

9. Natural convection film boiling

If the wall temperature at film boiling is larger then the *Leidenfrost* or the so-called minimum film boiling temperature an unstable heat-transfer regime is observed, being partially film boiling and partially nucleate boiling called transition boiling. In order to identify this regime knowledge of the so-called minimum film boiling temperature is necessary.

9.1 Minimum film boiling temperature

Groeneveld and *Stewart* (1982) derived from their measurements the following expression for the minimum film boiling temperature.

For $p \le 90 \times 10^5 Pa$

$$T_{w,MFB} = 557.85 + 0.0441 \times 10^{-3} p - 3.72 \times 10^{-12} p^2$$
$$- \min\left(0, \frac{10^4 X_{1,eq}}{2.82 + 0.00122 \times 10^{-3} p}\right), \tag{9.1}$$

and for $p > 90 \times 10^5 Pa$

$$T_{w,MFB} - T'(p) = 76.722 \frac{p_c - p}{p_{cr} - 90 \times 10^5} \tag{9.2}$$

with the limitation $T_{w,MFB} \le T_c$. Equation (9.2) guarantees that the minimum film boiling temperature at critical pressure is equal to the critical temperature. This correlation provides the best agreement with data compared to the prediction of seven other correlations re-viewed by the authors in *Groeneveld* and *Stewart* (1982). The mass flow rates in the experiments are varied between 110 and 2750 kg/(m^2 s) with no effect on the minimum film boiling temperature. This correlation is valid for pressure between 1 and 220 bar, mass flux 50 and 4000 kg/(m^2 s), equilibrium vapor mass flow fraction –19 and 64% for vertical up-flow in a circular pipe.

The minimum heat flux is defined as follows

$$q''_{w,MFB} = h_{FB}\left[T_{w,MFB} - T'(p)\right], \tag{9.3}$$

h_{FB} is the film boiling heat-transfer coefficient.

Schroeder-Richter and *Bartsch* (1990) interpreted the minimum film boiling phenomenon as a phase interface normal velocity discontinuity governed by the *Rankine* and *Hugoniot* equation *Rankine* (1870) and *Hugoniot* (1887),

$$h_2 - h_1 - \frac{1}{2}(p_2 - p_1)\left(\frac{1}{\rho_2} + \frac{1}{\rho_1}\right) = 0 \tag{9.4}$$

assuming that the liquid-side boundary layer properties are those at saturation at the minimum film boiling temperature and the vapor-side properties are those at saturation at system pressure

$$h''(p) - h'(T_{w,MFB}) - \frac{1}{2}\left[p - p'(T_{w,MFB})\right]\left[\frac{1}{\rho''(p)} + \frac{1}{\rho'(T_{w,MFB})}\right] = 0\,. \tag{9.5}$$

Henry (1974) related the minimum stable film boiling temperature to the homogeneous nucleation temperature and the wall properties

$$T_{w,MFB} = T_{hn} + (T_{hn} - T_2)\sqrt{\frac{(\lambda \rho c_p)_2}{(\lambda \rho c_p)_w}}\,. \tag{9.6}$$

Henry gives the homogeneous nucleation temperature as a curve fit to those used by *Thurgood* and *Kelley* (1979),

$$T_{hn} = \frac{5}{9}\begin{pmatrix} 459.67 + 705.44 - 4.722\times 10^{-2} p_{psi} \\ \\ +2.3907\times 10^{-5} p_{psi}^2 - 5.8193\times 10^{-9} p_{psi}^3 \end{pmatrix}, \tag{9.7}$$

where

$$p_{psi} = 3203.6 - 14.50377\times 10^{-5} p\,. \tag{9.8}$$

9.2 Film boiling in horizontal upwards-oriented plates

Film boiling in horizontal upwards-oriented plates in a large pool is possible if the temperature of the wall is larger then the so-called minimum film boiling temperature $T_{w,MFB}$,

$$T_w > T_{w,MFB}. \tag{9.9}$$

Berenson (1961) obtained the following expression for film boiling on a horizontal plane

$$h_{nc\_FB} = 0.425\left[\frac{\lambda_1^3 \rho_1 (\rho' - \rho'')(h'' - h')g}{\eta'' \Delta T \lambda_{RT}}\right]^{1/4}, \tag{9.10}$$

where

$$\lambda_{RT} = \left[\frac{\sigma}{g(\rho' - \rho'')}\right]^{1/2}, \tag{9.11}$$

and

$$\Delta T = (T_w - T'). \tag{9.12}$$

The averaged vapor film thickness can be then expressed as

$$\delta_{1F} \approx \frac{\lambda_1}{h_{nc\_FB}} = 2.35\left[\frac{\eta_1 \lambda_1 \Delta T}{(h'' - h')\rho_1 g(\rho' - \rho'')}\lambda_{RT}\right]^{1/4}, \tag{9.13}$$

where vapor properties are computed as a function of the averaged vapor film temperature $(T_w - T')/2$.

Equating the critical heat flux predicted by the *Zuber*'s correlation (1958) to the heat flux obtained for film boiling by the *Berenson* equation, an expression for the minimum film boiling temperature can be obtained

$$\Delta T_{w,MFB} = 0.127\frac{(h'' - h')\rho_{1F}}{\lambda_{1F}}\left[\frac{g(\rho_2 - \rho_1)}{\rho_2 + \rho_1}\right]^{2/3}\left[\frac{\sigma}{g(\rho_2 - \rho_1)}\right]^{1/2}\left[\frac{\eta_2}{\rho_2 - \rho_1}\right]^{1/3} \pm 10\%. \tag{9.14}$$

Leperriere (1983) modified the *Berenson* equation to obtain a correlation for forced convection film boiling in channels with very low mass flux

$$h_{Be,\mathrm{mod}} = h_{Berenson}\left(1 + 25.5\frac{T_w - T'}{T_2}\right)(1 - \alpha_{1h})^{0.5}, \tag{9.15}$$

where the homogeneous local vapor volume fraction is

$$\alpha_{1h} = \frac{X_{1,eq}/\rho''}{X_{1,eq}/\rho'' + (1 - X_{1,eq})/\rho'}. \tag{9.16}$$

9.3 Horizontal cylinder

For natural convection film boiling on a horizontal cylinder *Bromley* (1950)

$$h_{nc\_FB} = 0.62\left[\frac{\lambda_{1F}^{3}\rho_{1F}(\rho' - \rho'')g\left(h'' - h'\right)\left[1 + 0.4\frac{c_p''\left(T_w - T'\right)}{h'' - h'}\right]}{\eta'' D_h \Delta T}\right]^{1/4}, \quad (9.17)$$

obtained a very similar result to those by *Berenson* with the exception of the empirical constant. The modification of the latent heat of vaporization takes into account the superheating of the vapor. The analytical derivation of this equation is very similar to those for planes presented in Chapter 11 in detail. Later, in 1962, *Breen* and *Westwater* (1962) found better agreement with experimental data modifying the *Berenson* result by using instead of 0.62 the function

$$\left(0.59 + 0.069\,\lambda_{RT}^{*}/D_h\right)\left(D_h/\lambda_{RT}^{*}\right)^{1/4}, \quad (9.18)$$

where

$$\lambda_{RT}^{*} = 2\pi\left[\frac{3\sigma}{\left(\rho_2 - \rho''\right)g}\right]^{1/2}. \quad (9.19)$$

and instead of 0.4, using 0.34.

9.4 Sphere

For natural convection film boiling on a sphere *Bromley* (1950) computed a different constant in Eq. (9.17)

$$h_{nc\_FB} = 0.76\left\{\frac{\lambda_{1F}^{3}\rho_{1F}(\rho' - \rho'')g\left(h'' - h'\right)\left[1 + 0.4\frac{c_p''\left(T_w - T'\right)}{h'' - h'}\right]}{\eta'' D_3 \Delta T}\right\}^{1/4}. \quad (9.20)$$

For better agreement with data for liquid nitrogen *Frederking* and *Clark* recommended in 1963 the following correlation

$$h_{nc\_FB} = 0.14\left\{\frac{\lambda_{1F}^{3} g \rho_{1F}(\rho' - \rho'')\left(h'' - h'\right)}{\eta'' D_3\left(T_w - T'\right)}\left[1 + 0.5\frac{c_p''\left(T_w - T'\right)}{h'' - h'}\right]\right\}^{1/3}. \quad (9.21)$$

We devote a special Section 11.2 to this subject.

Nomenclature

Latin

c_p specific heat at constant pressure, J/(kg K)

D_{hy} hydraulic diameter (4 times flow cross-sectional area / wet perimeter), m

g gravitational acceleration, m/s²

h specific enthalpy, J/kg

h_{FB} film boiling heat-transfer coefficient, W/(m² K)

h_{nc_FB} natural convection film boiling heat-transfer coefficient, W/(m² K)

p pressure, Pa

p_{psi} $= 3203.6 - 14.50377 \times 10^{-5} p$, psi

p_{cr} critical pressure, Pa

$q''_{w,MFB}$ minimum film boiling heat flux, W/m²

$T_{w,MFB}$ minimum film boiling temperature, K

T_{hn} homogeneous nucleation temperature, K

X_{eq} equilibrium steam mass flow rate concentration in the flow, dimensionless

Greek

α volume fraction, dimensionless

α_{1h} $= \dfrac{X_{1,eq} / \rho''}{X_{1,eq} / \rho'' + \left(1 - X_{1,eq}\right) / \rho'}$, homogeneous local vapor volume fraction, dimensionless

Δ finite difference

δ_{1F} averaged vapor film thickness

λ thermal conductivity, W/(m K)

λ_{RT} $= \left[\dfrac{\sigma}{g(\rho' - \rho'')}\right]^{1/2}$, *Rayleigh–Taylor* instability wavelength, m

ρ density, kg/m³

σ surface tension, N/m

μ_{21} evaporation mass per unit time and unit mixture volume, kg/(m³ s)

μ_{12} condensation mass per unit time and unit mixture volume, kg/(m³ s)

Subscript

' saturated liquid
" saturated steam
1 gas
2 liquid
1F vapor film

References

Berenson, P.J.: Film boiling heat transfer from a horizontal surface. J. Heat Transfer, 351 (August 1961)
Breen, B.P., Westwater, J.W.: Effect of diameter of horizontal tubes on film heat transfer. Chem. Eng. Prog. 58(7), 67 (1962)
Bromley, L.A.: Heat transfer in stable film boiling. Chem. Eng. Process 46(5), 221 (1950)
Frederking, T.H.K., Clark, J.A.: Natural convection film boiling on a sphere. Adv. Cryogen. Eng. 8, 501–506 (1963)
Groeneveld, D.C., Stewart, J.C.: The minimum boiling temperature for water during film boiling collapse. In: Proc. 7th International Heat Transfer Conference, Munich, FRG, vol. 4, pp. 393–398 (1982)
Henry, R.E.: A correlation for minimum film boiling temperature. In: AIChE Symp. Series, vol. 138, pp. 81–90 (1974)
Hugoniot, P.H.: Mémoire sur la propagation du mouvement dans les corps et spécialement dans les gazes parfaits. Journal de l'École Polytechnique (1887)
Leperriere, A.: An analytical and experimental investigation of forced convective film boiling. M. A. Sc. Thesis, University of Ottawa, Ottawa (1983)
Rankine, W.J.M.: On the thermodynamic theory of waves of finite longitudinal disturbances. Philos. Trans. Roy. Soc. (1870)
Schroeder-Richter, D., Bartsch, G.: The Leidenfrost phenomenon caused by a thermo-mechanical effect of transition boiling: A revisited problem of nonequilibrium thermodynamics. In: Witte, L.C., Avedisian, C.T. (eds.) Fundamentals of Phase Change: Boiling and Condensation, HTD, vol. 136, pp. 13–20, Book No. H00589 (1990)
Thurgood, M.J., Kelly, J.M., Guidotti, T.E., Kohrt, R.J., Crowell: COBRA/TRAC - A thermal-hydraulics code for transient analysis of nuclear reactor vessels and primary coolant systems. NUREG/CR-3046, PNL-4385, U.S. Nuclear Regulatory Commission, Washington DC, vol. 1 R4 (March 1983)
Zuber, N.: On the stability of boiling heat transfer. Trans. ASME 80, 711–720 (1958)

10. Forced convection boiling

10.1 Convective boiling of saturated liquid

At some boundary conditions the liquid at the bulk flow at heated surfaces may become saturated. So, we have flow boiling of saturated liquid in the two phase mixture. The heat transfer then consists of a predominant convective part and a nucleate boiling part. The total heat flux is then consumed for evaporation.

Two successful methods can be used for computing the convective part:

(a) In the first one it is assumed that the total mixture mass flows through the channel with the properties of the liquid and the heat-transfer coefficient is computed using one of the well-known formulas, e.g., the *Dittus–Boelter* correlation

$$h_{co} = 0.024\,\mathrm{Re}_{co}^{0.8}\,\mathrm{Pr}_c^{0.4}, \tag{10.1}$$

where $Re_{co} = GD_h / \eta_c$ and $Pr_c = c_{pc}\eta_c / \lambda_c$. Then, the local convective heat-transfer coefficient is computed by correcting the so-obtained heat-transfer coefficient as follows

$$h \approx \left(\Phi_{co}^2\right)^{0.35} h_{co}. \tag{10.2}$$

This expression is very close to the theoretically obtained one in Vol. 3 of this monograph, namely

$$h \approx \left(\Phi_{co}^2\right)^{0.25} h_{co}. \tag{10.3}$$

The accuracy of this methods depends on the accuracy of the prediction of the *Martinelli–Nelson* multiplier

$$\Phi_{co}^2 = \left(dp/dz\right)_{Tph} / \left(dp/dz\right)_{co}. \tag{10.4}$$

The superposition with the boiling part can be introduced as

$$\dot{q}''_w = \left\{\left[\left(\Phi^2_{co}\right)^{0.35} h_{co}\left(T_w - T_2\right)\right]^2 + \left[h_{boiling}\left(T_w - T'\right)\right]^2\right\}^{1/2}. \qquad (10.5)$$

For imposed heat flux e.g., electrical heating, nuclear heating etc. the wall temperature is computed by using the above equation. Because the nucleate boiling heat-transfer coefficient is also a function of the wall temperature little iteration is necessary to solve the above equation with respect to the wall temperature. The total heat-transfer coefficient related to the saturation temperature is

$$h_{total} = \dot{q}''_w / \left(T_w - T'\right). \qquad (10.6)$$

In the case of saturated liquid, $T_2 = T'$, the evaporated mass per unit time and mixture volume of the flow for is

$$\mu_{21} = \frac{4}{D_{heat}} \dot{q}''_w / \left(h'' - h'\right). \qquad (10.7)$$

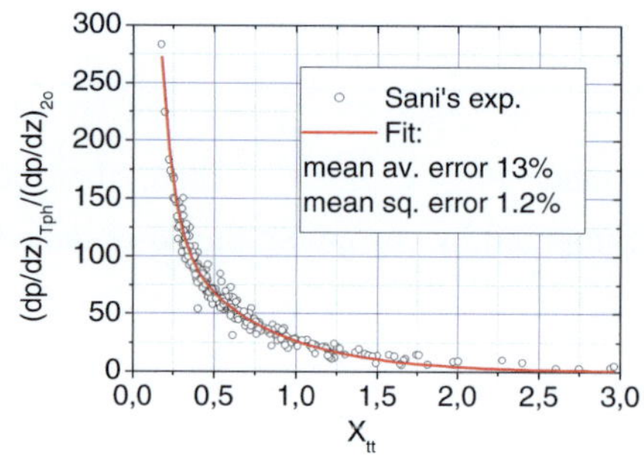

Fig. 10.1 Friction pressure drop gradient divided by the friction pressure drop gradient computed if the total mass is flowing through the channel with liquid properties as a function of the *Martinelli* parameter

This type of superposition is deduced intuitively from Section 16.7. Reducing from the *Sani*'s (1960) experiment

$$\Phi^2_{2o} = 1519.69039\exp\left(-X_{tt}/0.07719\right) + 167.21358\exp\left(-X_{tt}/0.5413\right), \qquad (10.8)$$

where

$$X_{tt} = \left(\frac{\rho''}{\rho'}\right)^{0.5} \left(\frac{\eta'}{\eta''}\right)^{0.1} \left(\frac{1 - X_{1,eq}}{X_{1,eq}}\right)^{0.9}, \qquad (10.9)$$

see Fig. 10.1, and using the *Borishanskii* et al. (1964) correlation for nucleate boiling results in 11.3% mean averaged error compared with 259 *Sani*'s (1960) data for flow boiling. Simply replacing in this model the *Borishanskii* et al. nucleate boiling model with my model, Chapter 16, results in a mean averaged error of 11%.

(b) In the second method it is assumed that only the continuum mass flows through the channel with the properties of the liquid and heat-transfer coefficient h_c is computed using one of the well-known formulas, e.g., the *Dittus–Boelter* (1937) correlation

$$h_c = 0.024\,\mathrm{Re}_c^{0.8}\,\mathrm{Pr}_c^{0.4}, \tag{10.10}$$

where $Re_{co} = X_c G D_h / \eta_c$ and $Pr_c = c_{pc}\eta_c / \lambda_c$. In this case the local convective heat-transfer coefficient is

$$h \approx \left(\Phi_c^2\right)^{0.35} h_c. \tag{10.11}$$

Reducing

$$\left(\Phi_c^2\right)^{0.35} = 1.8859 + 5.72465\exp\left(-X_{tt}/0.62374\right) \tag{10.12}$$

from the *Sani*'s data results in a very good agreement with the empirically obtained *Chen* (1963) multiplier

$$F_{Chen} = \max\left[2.35\left(X_{tt}^{-1} + 0.213\right)^{0.736}, 1\right], \tag{10.13}$$

– see Fig. 10.2.

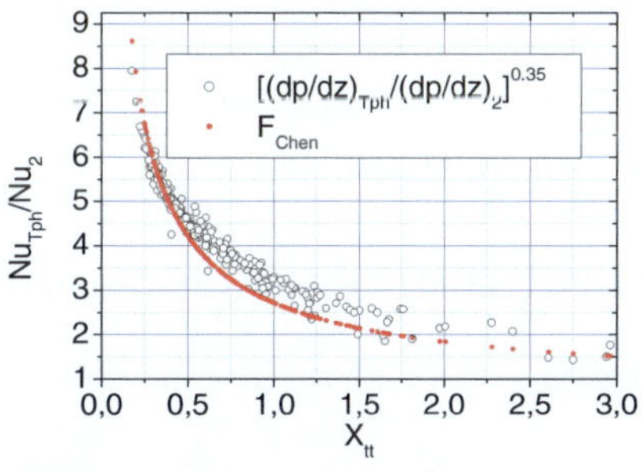

Fig. 10.2 Ratio of the convective two phase flow heat transfer to the liquid only heat transfer as a function of the *Martinelli* parameter

Chen proposed a linear superposition of the convective and boiling heat-transfer mechanisms

$$\dot{q}_w'' = F_{Chen} h_c \left(T_w - T_2\right) + Sup_{Chen} h_{boiling} \left(T_w - T'\right). \tag{10.14}$$

The wall temperature by enforced heat flux is

$$T_w = \left(\dot{q}_w'' + F_{Chen} h_c T_2 + Sup_{Chen} h_{boiling} T'\right) / \left(F_{Chen} h_c + Sup_{Chen} h_{boiling}\right). \tag{10.15}$$

Because the nucleate boiling heat-transfer coefficient is also a function of the wall temperature few iterations are necessary to solve the above equation with respect to the wall temperature. For the nucleate boiling heat-transfer coefficient *Chen* used the *Forster* and *Zuber* (1955) nucleate boiling model

$$h_{boiling} = 0.00122 \sqrt{\lambda_2 c_2 / \sigma_2} \, / \, \mathrm{Pr}_2^{0.29} \, \rho_2^{1/4} \left(\frac{\rho_2 c_{p2}}{\rho_1 \Delta h} \right)^{0.24} \left\{ p'\left[\min\left(T_w, T_c\right)\right] - p \right\}^{0.75}. \tag{10.16}$$

The larger the velocity of the mixture flow the smaller the influence of the nucleate boiling. This is taken into account by *Chen* by introducing the so-called empirical suppression factor presented by *Chen* graphically. *Bjornard* and *Grifith* (1977) approximated it by

$$Sup_{Chen} = \left(1 + 0.12 \mathrm{Re}_{Chen}^{1.14}\right)^{-1} \text{ for } \mathrm{Re}_{Chen} < 32.5, \tag{10.17}$$

$$Sup_{Chen} = \left(1 + 0.42 \mathrm{Re}_{Chen}^{0.78}\right)^{-1} \text{ for } 32.5 \le \mathrm{Re}_{Chen} < 70, \tag{10.18}$$

$$Sup_{Chen} = 0.0797 \exp\left(1 - Re_{Tp} / 70\right) \text{ for } 70 > \mathrm{Re}_{Chen}, \tag{10.19}$$

where

$$\mathrm{Re}_{Chen} = 10^{-4} F_{Chen}^{1.25} \left(1 - X_{1,eq}\right) G D_h / \eta_2 . \tag{10.20}$$

Groeneveld et al. (1989) use instead of the last expression simply

$$Sup_{Chen} = 0.1 \text{ for } 70 \le \mathrm{Re}_{Chen}. \tag{10.21}$$

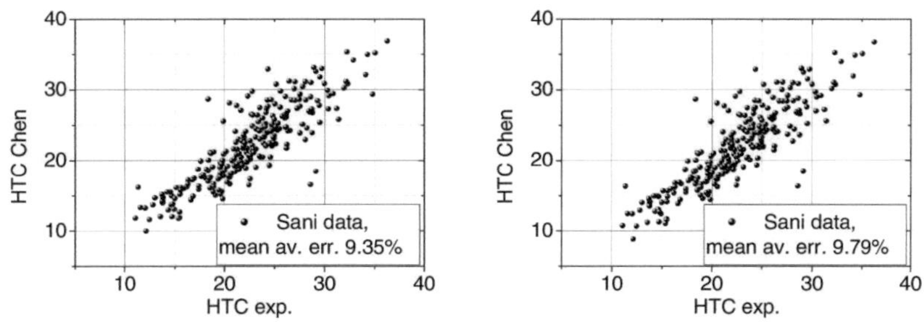

Fig. 10.3 a) Comparison between the *Chen* correlation with *Forster* and *Zuber* (1955) nucleate boiling model and the data by *Sani* (1960); b) Comparison between the *Chen* correlation with the *Kolev* (2006) nucleate boiling model with nucleation-site density function by *Wang* and *Dhir* (1993): $n''_{1w} = 5\times10^3\left[1-\cos(\theta)\right]/\left(10^5 D_{1,cr}\right)^6$ with θ the wetting angle and $D_{1,cr}$ the critical bubble size

As shown in Fig. 10.3 this correlation reproduce the *Sani*'s data with mean averaged error of 9.35%. *Chen* (1963) correlated about 600 data within an error band of 12% for convective heat transfer during forced-convection boiling in a tube. *Groeneveld* et al. (1989) validated the correlation in the region $X_1 = 0$ to 0.71, $p =$ (1.013 to 69.)$\times 10^5$ Pa, $G = 54$ to 4070 kg/(m^2 s), $\dot{q}''_w = 44$ to 2 400 kW/m².

Simply replacing in this model the *Forster* and *Zuber* (1955) nucleate boiling model with my model, Chapter 16, results in a mean averaged error of 9.79% by reproducing the *Sani*'s data.

Many engineers working in this field possess one or other library for computing the two-phase pressure drop. Depending on this, the first or the second method can be used. Remember once again that the accuracy of computing the convective part of the flow boiling depends on the accuracy of the computing of the pressure drop.

10.2 Forced-convection film boiling

In the case of vapor flow carrying droplets the heat-transfer coefficient can be computed with the method derived in vol. 3 of this book Eq. (4.46) p. 84:

$$h_1^{w\sigma} = 0.2 Re_1^{0.8} Pr_1^{0.4} (\Phi_1^2/\alpha_1)^{1/4} \lambda_1/D_h \,. \tag{10.22}$$

Here, the heat transfer is computed for a flow consisting of the total mass flow of the mixture but having the vapor properties

$$Re_1 = X_1 G D_h / \eta_1 \tag{10.23}$$

and

$$Pr_1 = c_{p1}\eta_1 / \lambda_1 , \tag{10.24}$$

and then corrected by considering the increased turbulence due to the two-phase flow effect. For the multiplier I use the approximation reported in the *Hetstrony*'s (1982) handbook

$$\Phi_{10}^2 \approx \Phi_1^2 = 1 + CX_{LM} + X_{LM}^2 ,\ C = 20, \tag{10.25}$$

$$X_{LM} = \left(\frac{\rho_1}{\rho_2}\right)^{1/2} \left(\frac{\eta_2}{\eta_1}\right)^{0.1} \left(\frac{1-X_1}{X_1}\right)^{0.9} , \tag{10.26}$$

The empirical constant 0.2 was found by comparison with the *Bennett* et al. (1967) experiments, see Fig. 10.4.

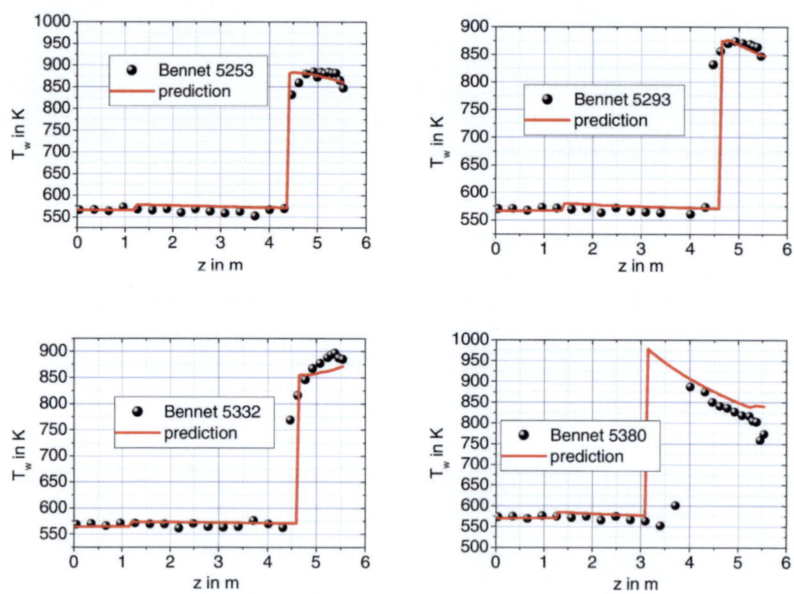

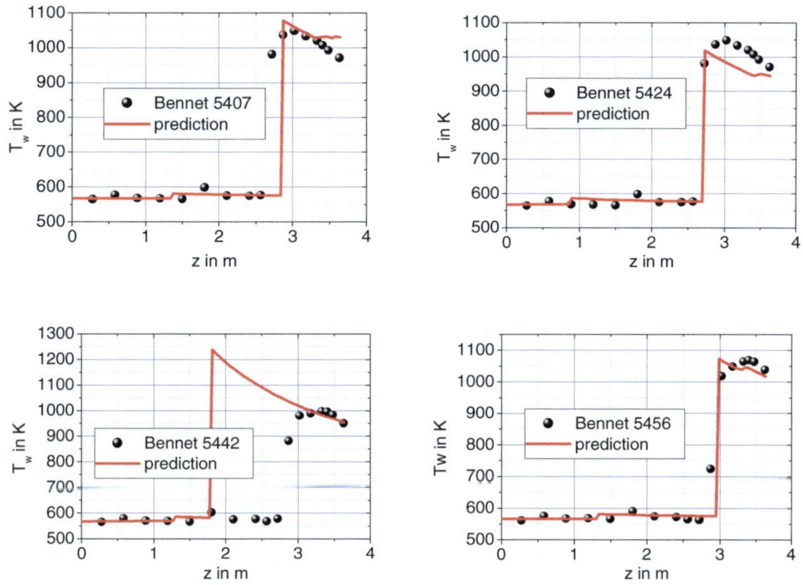

Fig. 10.4 Wall temperature as a function of length measured from the bottom. Comparison between the theoretical prediction and the data of *Bennett* et al. (1967)

In the following sections a state-of-the-art review in the literature is presented for completeness.

10.2.1 Tubes

The forced-convection film boiling description started with using the equilibrium thermodynamics of two-phase flow. In this framework the empirical correlations are obtained by generalizing experimental data using the *equilibrium steam mass flow rate concentration* in the flow,

$$X_{eq} = \frac{h - h'(p)}{h''(p) - h'(p)}, \tag{10.27}$$

where the specific mixture enthalpy is defined as the *mixture enthalpy of the noninert components*

$$h = \left[\sum_l \left(1 - \sum_n C_{nl} \right) \alpha_l \rho_l w_l h_{Ml} \right] \Big/ \left[\sum_l \left(1 - \sum_n C_{nl} \right) \alpha_l \rho_l w_l \right] \tag{10.28}$$

and the *mass flow rate* in the axial direction is defined as follows

$$G = \sum (\alpha \rho w)_l \,. \tag{10.29}$$

Dougall and *Rohsenow* proposed one of the oldest correlations in 1963 modifying the well-known single-phase heat transfer correlation of *Dittus* and *Boelter* from 1937

$$h_{fc\_FB} = 0.023 \frac{\lambda_1}{D_h} Re_1^{0.8} Pr_{1w}^{0.4} . \tag{10.30}$$

Here, the *Reynolds* number is defined by using as a characteristic velocity the center-of-mass velocity

$$w_{eq} = G / \rho_{eq} , \tag{10.31}$$

assuming that the density of the mixture is equal to the equilibrium density

$$\frac{1}{\rho_{eq}} = \frac{1 - X_{eq}}{\rho'} + \frac{X_{eq}}{\rho''} , \tag{10.32}$$

and the properties of the mixture are those of the saturated vapor

$$Re_{1\_mix} = \frac{D_h w_{eq}}{\eta'' / \rho''} = D_h G \frac{\rho''}{\eta''} \left[\frac{1 - X_{eq}}{\rho'} + \frac{X_{eq}}{\rho''} \right] . \tag{10.33}$$

The *Prandtl* number for steam

$$Pr_{1w} = c_{p1} \eta_1 / \lambda_1 , \tag{10.34}$$

is computed with the properties at the wall temperature

$$c_{p1}, \eta_1, \lambda_1 = f \left[\max\left(647,\ T_w\right),\ p \right] . \tag{10.35}$$

Note that the heat flux is then related to the saturation temperature at the local system pressure rather to the real vapor temperature

$$q''_w = h_{fc\_FB} \left[T_w - T'(p) \right] . \tag{10.36}$$

The heat is then assumed to be completely used for evaporation of the continuous liquid

$$\mu_{21} = \frac{4}{D_{heat\ w}} \frac{q''_w}{h''(p) - h'(p)} , \tag{10.37}$$

$$\mu_{31} = 0 , \tag{10.38}$$

if the continuous liquid is available, i.e., if $\alpha_2 > 0$. If there is no continuous liquid $\alpha_2 = 0$ but all the liquid is in dispersed flow $\alpha_3 > 0$ we have

$$\mu_{21} = 0 , \tag{10.39}$$

$$\mu_{31} = \frac{4}{D_{heat\ w}} \frac{q''_w}{h''(p) - h'(p)} . \tag{10.40}$$

Polomik proposed in 1967 a modification of the *Dougall* and *Rohsenow* correlation in order to take into account the liquid subcooling

$$h_{fc\_FB} = 0.023 \frac{\lambda_1}{D_h} Re_{1\_mix}^{0.8} Pr_{1w}^{0.40} \left(\frac{T_2}{T_w} \right)^{0.5} . \tag{10.41}$$

As noted in 1994 by *Andreani* and *Yadigaroglu* p. 15 this correlation provides surprisingly good agreement with data for substantial thermal nonequilibrium.

In the same year as *Dougall* and *Rohsenow*, *Miropolskij* (1963) collected several experimental data in the region of $40 \times 10^5 \le p \le 220 \times 10^5$ Pa, $400 \le G \le 2000$ kg/(m^2 s), $0 \le X_{eq} \le 1$, $0.008 < D_h < 0.029$ m and successfully reproduced them by the following correlation

$$h_{fc\_FB} = 0.023 \frac{\lambda_1}{D_h} Re_{1\_mix}^{0.8} Pr_{1w}^{0.80} Y . \tag{10.42}$$

Note the difference in the exponent of the *Prandtl* number compared to the *Dougall* and *Rohsenow* correlation and the multiplier

$$Y = 1 - 0.1 \left(\frac{\rho'}{\rho''} - 1 \right)^{0.4} \left(1 - X_{eq} \right)^{0.4} . \tag{10.43}$$

Groneveld and *Delorme* 1976 correlated their own measurements by modifying a correlation obtained by *Hadaller* and *Banerjee* (1969) for vapor only

$$h_{fc\_FB} = \max \left(3.66,\ \ 0.008348 \frac{\lambda_1}{D_{hy}} Re_{1\_mix}^{0.8774} Pr_{1w}^{0.6112} \right), \tag{10.44}$$

where the properties for the vapor *Prandtl* number are computed for an averaged temperature as follows

$$c_{p1}, \eta_1, \lambda_1 = f \left\{ \max \left[647, (T_w + T_1)/2 \right], p \right\} . \tag{10.45}$$

The authors limited the prediction of this correlation by the value of the modified *Berenson* equation

$$h_{fc\_FB} = \max(h_{nc\_FB\_Berenson}, h_{nc\_FB}) . \tag{10.46}$$

Groeneveld et al. successfully reproduced by this approach in 1989 their experimental data in the following region $D_{hy} = 0.0005$ to 0.0247 m, $p = (1.013$ to $204.99)10^5$ Pa, $G = 12.1$ to 6874 kg/(m^2 s), $X_{eq} = -0.12$ to 3.094. The predicted heat fluxes are in the region $\dot{q}''_w = 2660$ to $2\ 700\ 200$ W/m².

10.2.2 Annular channel

For prediction of the forced-convection film boiling heat-transfer coefficient in an *annular channel* in 1977 *Groeneveld* uses the *Miropolskij* multiplier in the following equation

$$h_{fc\_FB} = 0.052 \frac{\lambda_1}{D_h} Re_1^{0.688} Pr_{1w}^{1.26} / Y^{1.06} . \tag{10.47}$$

The authors reproduced successfully data in the following region: $450 \times 10^3 \le \dot{q}''_w \le 2250 \times 10^3$ W/m², $800 \le G \le 4100$ kg/(m^2 s), $34 \times 10^5 < p < 100 \times 10^5$ Pa, $0.1 < X_{eq} < 0.9$. This equation is known as *Groeneveld*'s Eq. (5-7) in the literature. *Bjornard* and *Grifith* recommended in 1977 the following modification of the correlation

$$h_{fc\_FB} = 0.052 \frac{\lambda_1}{D_h} Re_1^{0.688} Pr_{1w}^{1.26} / Y^{1.06} \quad \text{for} \quad p > 13.8 \text{ bar}, \tag{10.48}$$

$$h_{fc\_FB} = 0.052 \frac{\lambda_1}{D_h} Re_1^{0.688} Pr_{1w}^{1.26} \quad \text{for} \quad p < 13.8 \text{ bar}, \tag{10.49}$$

where

$$Re_1 = w_1 D_1 / \nu_1 , \tag{10.50}$$

which removes the singularity of the Y factor and is appropriate for direct application in the framework of the nonequilibrium models. Further the authors proposed to combine the film boiling correlation from the type given above valid for high void fractions with natural circulation film boiling in liquid pool by weighting with the void fraction.

Condie-Bengston – see in *Andreani* and *Yadigaroglu* (1994), reported the correlation

$$h_{fc\_FB} = 0.0524 \frac{\lambda_1}{D_h} \frac{D_h^{0.1905}}{\lambda_1^{0.5407}} \frac{Pr_{1w}^{2.2598}}{\left(1 + X_{eq}\right)^{2.0514}} Re_1^{\left[0.6249 + 0.2043 \ln(X_{eq}+1)\right]} \tag{10.51}$$

valid for the following region: $34 \times 10^3 \le \dot{q}''_w \le 2\,074 \times 10^3$ W/m², $16.5 \le G \le 5\,234$ kg/(m^2 s), $4.2 \times 10^5 < p < 215 \times 10^5$ Pa, $-0.12 < X_{eq} < 1.73$.

10.2.3 Tubes and annular channels

Groeneveld generalized in the same work (1977) the equation for both *tubes* and *annular channels* as follows

$$h_{fc\_FB} = 0.00327 \frac{\lambda_1}{D_h} Re^{0.901} Pr_{1w}^{1.32} / Y^{1.50} \tag{10.52}$$

This equation is known as *Groeneveld*'s Eq. (5-9) in the literature. It is reported to be valid in the following region: $120\times 10^3 \le \dot{q}''_w \le 2\ 250\times 10^3$ W/m², $700 \le G \le 5300$ kg/(m^2 s), $34\times 10^5 < p < 215\times 10^5$ Pa, $< X_{eq} < 0.9$.

10.2.4 Vertical flow around rod bundles

For vertical flow around rod bundles *Loomis* and *Schumway* (1983) reported the following correlation

$$h_{fc\_FB} = const \frac{\lambda_1}{D_h}(1-0.76\alpha_1)\left[\frac{h''-h'}{(T_w-T_1)c_{p1}}\right]^{0.25} \frac{\lambda_{RT} G}{\eta_2}\frac{\rho_2-\rho_1}{\rho_1}, \tag{10.53}$$

where

$$\lambda_{RT} = \left[\frac{1}{g}\frac{\sigma}{(\rho_2-\rho_1)}\right]^{1/2} \tag{10.54}$$

and the constant is

$$const = 18(\pm 0.4)10^{-4}. \tag{10.55}$$

The correlation is valid in the following region: $p \cong 1.27\times 105$ Pa, $31.55 \le G \le 70.9$ kg/(m^2 s), $0.1 \le \alpha_1 \le 1$. Note that the correlation is applied in the framework of nonequilibrium thermodynamics, which means that

$$q_1^{''\sigma w} = h_{fc\_FB}(T_w - T_1). \tag{10.56}$$

10.3 Transition boiling

Ramu and *Weisman* (1974) proposed the following correlation for transition boiling in tubes

$$h_{TB} = \frac{1}{2} S\ h_{CHF\_Addoms}\left\{\begin{array}{l}\exp\left[-0.01404\left(\Delta T - \Delta T_{CHF\_Addoms}\right)\right]\\ +\exp\left[-0.12564\left(\Delta T - \Delta T_{CHF\_Addoms}\right)\right]\end{array}\right\} + h_{fc\_FB}, \tag{10.57}$$

where

$$\Delta T = T_w - T'(p). \tag{10.58}$$

ΔT_{CHF_Addoms} is given in Table 10.1.

The component of the forced-convection film boiling is computed using the *Dougall* and *Rohsenow* (1963) proposal

$$h_{fc\_FB} = 0.023 \frac{\lambda_1}{D_h} Re^{0.800} Pr_{1w}^{0.40}, \tag{10.59}$$

$$Re = \frac{D_h w_{eq}}{\eta'' / \rho''} = D_h G \frac{\rho''}{\eta''} \left[\frac{1 - X_{eq}}{\rho'} + \frac{X_{eq}}{\rho''} \right]. \tag{10.60}$$

The *Prandtl* number for steam

$$Pr_{1w} = c_{p1} \eta_1 / \lambda_1, \tag{10.61}$$

is computed with the properties at the wall temperature

$$c_{p1}, \eta_1, \lambda_1 = f \left[\max \left(647, T_w \right), p \right]. \tag{10.62}$$

S is the ratio of the effective superheat to the total superheat of the wall from the *Chen* correlation.

Table 10.1 The critical heat flux and the corresponding temperature difference as a function of pressure – *Addoms* (1948) data

p	h_{CHF_Addoms}	ΔT_{CHF_Addoms}
275 800	86 877	18.9
413 700	91 874	18.9
4 137 000	227 139	16.6
6 895 000	397 478	13.9
9 653 000	438 930	12.2
16 996 175	408 835	6.1
18 478 600	384 986	5.0
19 512 850	369 087	4.4
20 478 150	348 077	3.9
21 002 170	323 661	3.3

The correlation was verified in the region $1.72 \times 10^5 < p < 2.06 \times 10^5$ Pa, $20 < G <$ *50* kg/(m^2 s), $30 < \dot{q}''_w < 260$ W/m^2, $0 < X_{eq} < 0.5$.

As noted by *Groeneveld* et al. (1989) transition-boiling correlations generally provide incorrect asymptotic trends outside their region of validity. An alternative approach uses simply a linear interpolation between the maximum and the minimum points of a log–log plot of $\dot{q}''_w$ vs. ΔT_w, i.e.,

$$\frac{\dot{q}''_{TB}}{\dot{q}''_{\min\_FB}} = \left(\frac{\dot{q}''_{CHF}}{\dot{q}''_{\min\_FB}} \right)^m, \quad \text{where} \quad m = \frac{\ln\left(\Delta T_{\min\_FB} / \Delta T_w\right)}{\ln\left(\Delta T_{\min\_FB} / \Delta T_{CHF}\right)} \tag{10.63}$$

or

$$\frac{\Delta T_w}{\Delta T_{\min\_FB}} = \left(\frac{\Delta T_{CHF}}{\Delta T_{\min\_FB}} \right)^n, \quad \text{where} \quad n = \frac{\ln\left(\dot{q}''_{TB} / \dot{q}''_{\min\_FB}\right)}{\ln\left(\dot{q}''_{CHF} / \dot{q}''_{\min\_FB}\right)}. \tag{10.64}$$

Film boiling in a large pool exists if $T_w > \Delta T_{\min\_FB}$, where $\Delta T_{\min\_FB}$ is the minimum film boiling temperature.

Note that once the value of the critical heat flux is known, the corresponding temperature is obtained from the definition equation of the heat flux for forced-convection nucleate boiling as follows

$$\Delta T_{CHF} = T_{CHF} - T'(p) = \dot{q}''_{CHF} / h_{fc\_FB}. \tag{10.65}$$

Guo and *Leung* (2005) proposed to approximate the 11 163 transition boiling data for pipes by the following relation

$$\frac{h_{TB} - h_{FB}}{h_{CHF} - h_{FB}} = \exp\left[-a \left(\frac{T_{w,TB} - T'}{T_{w,CHF} - T'} - 1 \right)^b \right], \tag{10.66}$$

valid up to 170 bar. The constants are proprietary but Fig. 5 in *Guo* and *Leung* (2005) allows the estimate of $a \approx 0.8$ and $b \approx 0.6$.

10.4 Critical heat flux

Natural- or forced-convection nucleate boiling is a very effective cooling mechanism. Increasing the local heat flux or decreasing the cooling mass flow rate may lead in technical facilities to film boiling in which the heat is transferred to the liquid through a vapor film. This is not an effective heat-transfer mechanism. This effect is manifested on the material interface by a temperature jump. At high mass flow rates the jump is not as strong as at low mass flow rates. We distinguish two different mechanisms of the occurrence of film boiling. The first one happens at low void fractions and is mainly caused by increasing the bubble population at the heated surface so that they merge together and build a film. The mechanical flow pattern is continuous liquid and dispersed void. This regime is known as departure from nucleate boiling or abbreviated DNB. The mechanism at high void fractions is different. The mechanical flow pattern is continuous liquid film at the wall and dispersed droplet cared by the void. In this case the film simply dries out and the wall experiences an intimate contact with the steam. This regime is known as the *dry-out* regime.

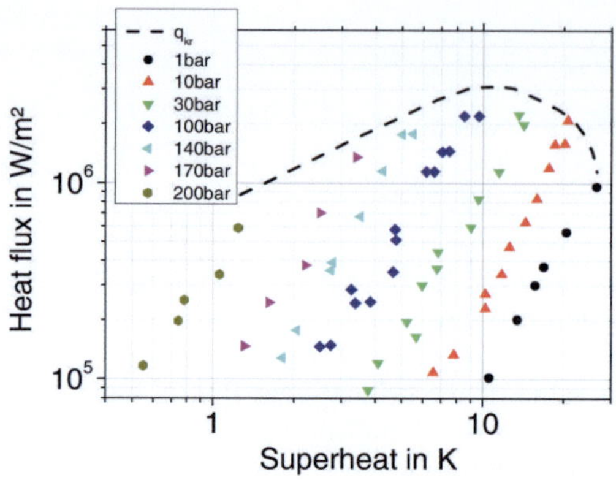

Fig. 10.5 Heat flux for pool boiling of water at silver horizontal tube with 5 mm diameter – parameter pressure, *Golovin*, et al. (1963). The upper boundary presents the critical heat flux

As illustrated on Fig. 10.5 the critical heat flux as the boiling heat transfer depends on the pressure. Figure 10.6 illustrate this dependence again. We see that the maximum is about 50–100 bar. This was the reason to select the pressure level for many boiling facilities like boiling water reactors and steam generators.

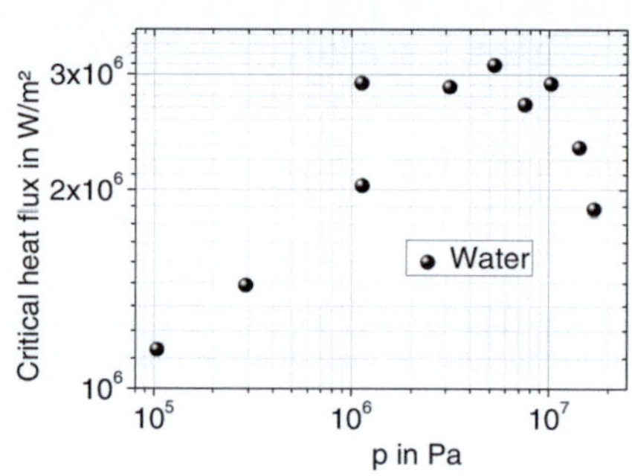

Fig. 10.6 Critical heat flux as a function of the pressure for pool boiling of water in a silver horizontal tube with 5 mm diameter, *Golovin*, et al. (1963)

10.4.1 The hydrodynamic stability theory of free-convection DNB

Kutateladze established in 1951 the hydrodynamic stability theory. The main idea on which this theory relies is as follows: The vapor film with thickness above the heated surface is governed by the capillary constant

$$\delta_{1F} \approx \left(\frac{\sigma}{g \Delta \rho_{21}} \right)^{1/2}, \tag{10.67}$$

experiences a buoyancy force per unit wall surface equal to $g \Delta \rho_{21} \delta_{1F}$. The equilibrium between the dynamic pressure forces $\rho_1 V_{1ev}^2$, and the buoyancy force is distorted if the film is not stable. The transition point is controlled by the conditions

$$\rho_1 V_{1ev}^2 / \left[g \Delta \rho_{21} \left(\frac{\sigma}{g \Delta \rho_{21}} \right)^{1/2} \right] \approx const, \tag{10.68}$$

or

$$V_{1ev} \approx const \left[\frac{g \sigma (\rho_2 - \rho_1)}{\rho_1^2} \right]^{1/4}. \tag{10.69}$$

Kutateladze set the critical mass flow rate proportional to the critical velocity

$$\dot{q}''_{CHF} / \Delta h \approx const \left[\frac{g \sigma (\rho_2 - \rho_1)}{\rho_1^2} \right]^{1/4} \tag{10.70}$$

and found from experiments $const \approx 0.16$. Later, the constant was replaced by the dimensionless group

$$const \approx 0.05 \left[10^4 \left(g \lambda_{RT} \rho_1 / p \right)^{1/2} \right]^{2/3}. \tag{10.71}$$

The obtained result is valid for saturated liquid. Subcooling increases the CHF. *Kutateladze* found in 1962 the following relationship

$$\frac{\dot{q}''_{CHF,sub}}{\dot{q}''_{CHF,sat}} = 1 + 0.065 \left(\frac{\rho_2}{\rho_1} \right)^{4/5} \frac{c_{p2}(T' - T_2)}{h'' - h'}. \tag{10.72}$$

This relationship was verified for $p = (1 \text{ to } 20) \times 10^5 Pa$, $\frac{c_{p2}(T' - T_2)}{h'' - h'} \le 0.6$ and $\rho_2 / \rho_1 = 45$ to 1650. A very similar expression was obtained in the same year by *Irvey* and *Morris* (1962),

$$\frac{\dot{q}''_{CHF,sub}}{\dot{q}''_{CHF,sat}} = 1 + 0.1 \left(\frac{\rho_2}{\rho_1} \right)^{3/4} \frac{c_{p2}(T' - T_2)}{h'' - h'}. \tag{10.73}$$

The main ideas behind the hydrodynamic theory of boiling by *Zuber* (1958) and *Zuber* and *Tribus* (1958) are slightly different from those of *Kutateladze*. *Zuber* considered the *Rayleigh–Taylor* instability as responsible for the film instability. If the vapor film possesses a wavy vapor/liquid interface the characteristic wavelength is

$$\lambda_{RT}^{*} \approx 2\pi\left(\frac{\sigma}{g\Delta\rho_{21}}\right)^{1/2}. \tag{10.74}$$

Vapor jets are then formed with a distance between the axes equal to λ_{RT}^{*}. The jets are arranged equidistantly in a rectangular grid and take a cross-section α_{w1} part of the total surface. The vertical disturbed surface of the jets is unstable is *Helmholtz* instability. The waves are traveling with velocity

$$c^2 = \frac{\sigma n_c}{\rho_2+\rho_1} - \frac{\rho_2\rho_1}{\left(\rho_2+\rho_1\right)^2}\left(V_1-V_2\right)^2, \tag{10.75}$$

where n_c is the wave number. In critical conditions $c \approx 0$ and $n_c \approx 2\pi/\lambda_{RT}^{*}$ therefore

$$V_1 - V_2 = \left(\frac{\rho_2+\rho_1}{\rho_2}\right)^{1/2}\left[\frac{g\sigma\left(\rho_2-\rho_1\right)}{\rho_1^2}\right]^{1/4} \tag{10.76}$$

The rising vapor mass flux is equal to the liquid flux (continuity) and therefore

$$\alpha_{w1}\rho_1 V_1 = -\alpha_{w2}\rho_2 V_2 \tag{10.77}$$

or

$$V_1 - V_2 = V_1\frac{\alpha_{w2}\rho_2+\alpha_{w1}\rho_1}{\alpha_{w2}\rho_{w2}}. \tag{10.78}$$

Thus, the critical velocity of the rising vapor jets is

$$V_1 = \frac{\alpha_{w2}\rho_{w2}}{\alpha_{w2}\rho_{w2}+\alpha_{w1}\rho_{w1}}\left(\frac{\rho_2+\rho_1}{\rho_2}\right)^{1/2}\left[\frac{g\sigma\left(\rho_2-\rho_1\right)}{\rho_1^2}\right]^{1/4}. \tag{10.79}$$

Thus, the critical heat flux is finally

$$\frac{\dot{q}''_{CHF}}{\rho_1(h''-h')} \approx j_1 = \alpha_{1d}V_1 = \alpha_{1d}\frac{\alpha_{w2}\rho_{w2}}{\alpha_{w2}\rho_{w2}+\alpha_{w1}\rho_{w1}}\left(\frac{\rho_2+\rho_1}{\rho_2}\right)^{1/2}\left[\frac{g\sigma\left(\rho_2-\rho_1\right)}{\rho_1^2}\right]^{1/4}$$

$$\approx \frac{\pi}{24}\left(\frac{\rho_2}{\rho_2+\rho_1}\right)^{1/2}\left[\frac{g\sigma\left(\rho_2+\rho_1\right)}{\rho_1^2}\right]^{1/4}, \tag{10.80}$$

which is in fact the *Kutateladze* equation with multiplier $\frac{\pi}{24}\left(\frac{\rho_2}{\rho_2+\rho_1}\right)^{1/2}$ instead of the constant. For low flow, $G \le 100$ kg/(m^2 s), *Zuber* et al. (1961) later modified this correlation to

$$\frac{\dot{q}''_{CHF}}{\rho_1(h''-h')} \approx 0.23164\ (0.96-\alpha_1)\left(\frac{\rho_2}{\rho_2+\rho_1}\right)^{1/2}\left[\frac{g\sigma(\rho_2-\rho_1)}{\rho_1^2}\right]^{1/4}. \qquad (10.81)$$

Note the two completely different approaches to the boiling crises by comparing this theory with the theory presented in Chapter 16. We show in Chapter 16 that the microscale of the nucleation processes, bubble growth and departure, and the turbulence generated by this process naturally lead to self-saturation of the boundary layer with bubbles and dramatically change the mechanism of the heat transfer. In contrast, *Kutateldse* and *Zuber* analyzed the macroscale above the layer where the microphysics happens. That is why their analysis gives an absolute upper limit that is about 5 times larger than the observed critical heat flux. This is reflected in the necessity of introducing an empirical constant of about 1/5.

10.4.2 Forced-convection DNB and DO correlations

For flow in heated pipes and channels the mass flow rate influences the CHF. There are more than 500 empirical correlations for description of this process, demonstrating that the final understanding of this phenomenon is not yet reached. We give here three of the successful empirical correlations. *Smolin* et al. (1977) consider separately the different flow pattern and for each particular flow pattern proposed a correlation for prediction of the CHF. For equilibrium mass flow fraction less than the one defined by

$$X_0 = 1.5\frac{\rho_1}{\rho_2+\rho_1} - 0.1 \text{ (bubble to film flow transition)} \qquad (10.82)$$

the flow is bubbly. In this case *Smolin* et al. propose to use the *Kutateladze* correlation in the form

$$\frac{\dot{q}''_{CHF,0}}{\rho_1(h''-h')} = 0.18\left(\sqrt{\frac{v''}{v'}}-1\right)\left[\frac{g\sigma(\rho_2-\rho_1)}{\rho_1^2}\right]^{1/4}. \qquad (10.83)$$

For

$$X_{eq} \ge X_0$$

we have inverted annular flow called film boiling. In this case the pipe geometry, the mass flow rate and the equilibrium mass flow concentration influence the CHF

$$\dot{q}''_{CHF,2} = \dot{q}'''_{CHF,0} / \exp\left[0.2\left(\frac{D_h}{\rho'\sigma}\right)^{1/3} G^{2/3}\left(X_{eq} - X_0\right)\right]. \tag{10.84}$$

For $X_{eq} < 0$,

$$\dot{q}''_{CHF} = \dot{q}''_{CHF,2} \,. \tag{10.85}$$

Saturated liquid boiling: For saturated liquid boiling $X_{eq} \geq 0$,

$$\dot{q}'''_{CHF} = \min\left(\dot{q}''_{CHF,2}, \dot{q}'''^{*}_{CHF,3}, \dot{q}''_{CHF,3}\right), \tag{10.86}$$

The minimum of the CHF in the following regimes is taken:

Fine dispersed flow,

$$\dot{q}''_{CHF,3} = 0.22(h'' - h')\frac{\eta''}{\eta'}\left[\frac{\eta''\rho''\rho'}{D_h(\rho' - \rho'')}\right]^{1/3} G^{2/3}(1 - X_{eq})^2 / \left[X_{eq}(\rho' - \rho'') + \rho''\right]^{1/3}; \tag{10.87}$$

Drying of the heated wall,

$$\dot{q}'''^{*}_{CHF,2} = \frac{(h'' - h')(\rho' - \rho'')}{45}\left(\frac{\sigma g}{\rho''}\right)^{1/4} \ln\left[\frac{0.9(\rho''\sigma)^{1/2}\left(\frac{v''}{v'}\right)^{2/3}\left(1 - X_{eq}\right)^{1/3}}{\eta'^{2/3} G^{2/3} D_h^{1/6} X_{eq}}\right]; \tag{10.88}$$

Continuous vapor: For bubbly flow $X_{eq} < X_0$, and $p \leq 98 \times 10^5$ we have boiling of subcooled liquid

$$\dot{q}''_{CHF} = \dot{q}''_{CHF,0}$$
$$+ 8.4 \times 10^{-3}(h'' - h')\sqrt{\rho''}\left\{[\sigma g(\rho' - \rho'')]^{1/4}\left(\frac{\eta'}{\eta''}\right)\right\}^{1.25}\left[\frac{G}{\rho'}\left(\frac{\rho' - \rho''}{\sigma g}\right)^{1/4}\right]^{2/3}(X_0 - X_{eq}). \tag{10.99}$$

For $p > 98 \times 10^5$

$$\dot{q}''_{CHF,2 \text{ at } X_{eq} = X_0} = \dot{q}''_{CHF,0} \tag{10.90}$$

film to fine dispersed flow transition (hydraulic theory)

$$\dot{q}''_{CHF,3\ at\ X_{eq}=X_0} = 0.22(h''-h')\frac{\eta''}{\eta'}\left[\frac{\eta''\rho''\rho'}{D_h(\rho'-\rho'')}\right]^{1/3} G^{2/3}\frac{(1-X_0)^2}{\left[X_0(\rho'-\rho'')+\rho''\right]^{1/3}} \quad (10.91)$$

If $\dot{q}''_{CHF,2\ at\ X_{eq}=X_0} \ge \dot{q}''_{CHF,3\ at\ X_{eq}=X_0}$, then

$$\dot{q}''_{CHF} = \text{Eq. (10.89)}, \quad (10.92)$$

boiling of subcooled or saturated liquid. Otherwise,

$$\dot{q}''_{CHF,2\ at\ X_{eq}=X_0} < \dot{q}''_{CHF,3\ at\ X_{eq}=X_0}. \quad (10.93)$$

The CHF is

$$\dot{q}''_{CHF} = \dot{q}''_{CHF,3\ at\ X_{eq}=X_0}$$
$$+0.01(h''-h')\left(\frac{\rho'}{\rho''}\right)^{1/2}\left[\sigma g(\rho'-\rho'')\right]^{1/4}\left[G\left(\frac{\rho'-\rho''}{\sigma g}\right)^{1/4}\right]^{1/2}\left(X_0-X_{eq}\right). \quad (10.94)$$

Smolin et al. (1977) verified the above set of correlations for $30\times 10^5 \le p \le 200\times 10^5$ Pa, $G \le 7500$ kg/(m^2 s), $T'-T_2 < 75$ K, $0.004 < D_h < 0.025$ m, 1 m $< L$.

Data for film boiling heat transfer in rod bundles are reported by *Morris* et al. (1985).

Biasi et al. proposed in 1967 the following correlation valid for tubes and $100 < G < 6000$ kg/(m^2 s), $0.003 < D_h < 0.0375$ m, $0.2 < L < 6$ m, $2.7\times10^5 < p < 140\times10^5$ Pa, $1/\left(1+\rho_2/\rho_1\right) < X_{eq} < 1$:

$$\dot{q}''_{CHF} = \max\left\{1000,\ \frac{1.883\times10^7}{D_h^{*n}G^{*1/6}}(FP/G^{*1/6}-X_{eq}),\ \frac{3.78\times10^7}{D_h^{*n}G^{*0.6}}HP(1-X_{eq})\right\}, \quad (10.95)$$

where

$G^* = G/10$, $D_h^* = D_h/100$, $p^* = p/10^5$, $D_h > 0.01$, $n = 0.4$, $D_h \le 0.01$, $n = 0.6$,

FP = 0.7249 + 0.099 p^*/exp(0.032 p^*),

HP = − 1.159 + 0.149 p^*/exp(0.019p^*) + 8.99 p^*/(10 + p^{*2}).

This correlation is not recommended for a rod bundle geometry, *Yoder* et al. (1985).

Bowring (1972) proposed the following correlation validated by comparison with 3800 experimental data points in the region: $2\times 10^5 < p < 190\times 10^5$ Pa, 136 < G < 18 600 kg/(m^2 s), 0.002 < D_h < 0.045 m, 0.15 < L < 3.7 m. The reported rms. error is 7%.

$$\dot{q}''_{CHF} = \left[A - B(h'' - h')X_{eq}\right]/C \tag{10.96}$$

where

$$A = 2.317\left(h'' - h'\right)B\,F_1/\left(1 + 0.0143\,F_2 D_h^{0.5} G\right),$$

$$B = 0.25 D_h G,$$

$$C = 0.077\,F_3 D_h G/\left[1 + 0.347\,F_4\,(G/1356)^{SN}\right],$$

$$SN = 2 - 0.5\bar{p},$$

$$\bar{p} = 0.14510^{-6}\,p\,.$$

$$\bar{p} \le 1$$

$$F_1 = \left\{\bar{p}^{18.942}\exp\left[20.89(1-\bar{p})\right] + 0.917\right\}/1.917\,,$$

$$F_1F_2 = \left\{\bar{p}^{1.3610}\exp\left[2.444(1-\bar{p})\right] + 0.309\right\}/1.309\,,$$

$$F_1 = F_1/\left(F_1F_2\right)$$

$$F_3 = \left\{\bar{p}^{17.023}\exp\left[16.658(1-\bar{p})\right] + 0.667\right\}/1.667\,,$$

$$F_4F_3 = \bar{p}^{1.649},$$

$$F_4 = \left(F_4F_3\right)F_3\,.$$

$$\bar{p} > 1$$

$$F_1 = \left\{\exp\left[0.648(1-\bar{p})\right]\right\}/\,\bar{p}^{0.368},$$

$$F_2 = \left\{F_1\exp\left[0.245(1-\bar{p})\right]\right\}/\,\bar{p}^{0.448},$$

$$F_3 = \bar{p}^{0.219},$$

$$F_4F_3 = \bar{p}^{1.649},$$

$$F_4 = \left(F_4F_3\right)F_3$$

10.4.3 The 1995 and 2005 look-up tables

A look-up table for critical heat flux (CHF) has been developed jointly by AECL Research (Canada) and IPPE (Obninsk, Russia) *Groeneveld* et al. (1996)

$$q''_{CHF} = q''_{CHF}\left(p, G, X_{1,eq}, D_{hy}\right). \tag{10.97}$$

It is based on an extensive data base of CHF values obtained in tubes with a vertical upward flow of steam–water mixture. While the data base covers a wide range of flow conditions, the look-up table is designed to provide CHF values for 8 mm tubes at discrete values of pressure, mass flux and dry-out quality covering the ranges 0.1 to 20.0 MPa, 0.0 to 8000 kg/(m^2 s) and -0.5 to $+1.0$, respectively. Linear interpolation is used to determine the CHF for conditions between the tabulated values, and an empirical correction factor is introduced to extend this CHF table to tubes of diameter values other than 8 mm,

$$\dot{q}''_{CHF} = \dot{q}''_{CHF\_8mm} \left(\frac{1000 D_{hy}}{8} \right)^{-1/2} . \qquad (10.98)$$

Compared against the combined AECL-IPPE world data bank (consisting of 22 946 data points after excluding duplicate data and obviously erroneous data), the 1995 look-up table predicts the data with overall average and root-mean- square errors of 0.69% and 7.82%, respectively. An assessment of various CHF tables and several empirical correlations shows that the 1995 table consistently provides the best prediction accuracy and is applicable to the widest range of conditions.

Groeneveld et al. (2005) issued the improved variant of their first table known as the 2005 look-up table containing 20% more data. The rms error was reduced to 7.1%.

Tanase et al. (2009) analyzed the accuracy of the diameter correction by comparing with 26 231 data points and found that the coefficient n = -0.4 to -0.45 instead -0.5 is very good one. Slight improvement was found with the *Wong* (1996) coefficient

$$n = -0.58\left[1 - 0.25\exp\left(-2X_{1,eq}\right)\right]\left(1 - 15D_{hy}^{-3}G\right), \qquad (10.99)$$

valid in $-0.5 < X_{1,eq} < 1$, $0.003 < D_{hy} < 0.025$, p = 100 to 20000 kPa, $300 < G <$ 30 000 kg/(m² s). For the subcooled region the authors recommended

$$n = -0{,}25 \text{ to } -0.33 \qquad (10.100)$$

and for the low-flow region G < 250 kg/(m² s) they found the *inverse* dependence defined by

$$n = 0.2 \text{ to } 0.3. \qquad (10.101)$$

Problem: Check whether it is possible to use the 2005 look-up table primarily designed for heated pipes also for heated rod bundles.

Solution: I performed this check in 2006, *Kolev* (2006). I exploit deliberately a set of simplifying assumptions usually used in the 1960s:

1. fictive subchannels corresponding to the number of rods are so defined as to be equivalent in geometry;
2. equal mass flow rate to each subchannel;
3. the critical heat flux occurs at the exit of the hot channel as a function of the exit parameters; The phenomenon is similar to pipe flow; Therefore the *Groeneveld*'s look-up table is expected to give the right dependence on mass flow rate, quality and subcooling;
4. uniform boundary conditions for all channels;
5. grid turbulization as well as the axial power distribution is not specifically treated.

The procedure I use is: Unless $\left| \frac{q''_{crit,i}\left(z_{\max}\right) - q''_i\left(z_{\max}\right)}{q''_{crit,i}\left(z_{\max}\right)} \right| < \varepsilon$ increase q with a Δq starting with q_0. Here, the power of the hot channel is $q_i = f_{lat,i}\, q/n_{rods}$, where as per the definition the bundles power is the sum of the subchannel powers $q = \sum_{i=1}^{n_{rods}} q_i$ and the lateral distribution factor obeys $\frac{1}{n_{rods}} \sum_{i=1}^{n_{rods}} f_{lat,i} = 1$. The exit equilibrium mass flow quality is $X_{1,i} = \frac{1}{\Delta h}\left(\frac{q_i}{\dot{m}_i} - \Delta h_{sub} \right) = \frac{1}{\Delta h}\left(f_{lat,i} \frac{q}{\dot{m}} - \Delta h_{sub} \right)$. Here, by virtue of assumption 2 I use $\dot{m}/\dot{m}_i = n_{rods}$. The local heat flux at the exit of the hot channel *I*, $q''_i\left(z_{\max}\right) = f_{z_{\max},i} \frac{f_{lat,i}\, q/n_{rods}}{\pi D_{rod} L_{rod}}$, is compared with the local critical heat flux. The critical heat flux at the exit of channel *i* is defined by the 2005 look-up table $q''_{crit,i}\left(z_{\max}\right) = f\left(p, G, X_{1,i}, D_{hyd}\right)$. Dividing this value by the lateral hot channel factor and recomposing the averaged flux into the bundle power we obtain the final result $q = \frac{q''_{crit,i}\left(z_{\max}\right)}{f_{lat,i}} \pi D_{rod} L_{rod} n_{rods}$. Note that the last expression is based on the intuitive assumption that this value may represent the averaged heat flux in the hot channel. It is only a *hypothesis*. It is very interesting to compare the prediction of this "primitive" approach with experimental data and to see how it works.

8×8 NUPEC BWR bundle tests: First, I simulate three data sets for experiments with the 8–8 NUPEC BWR bundle tests (Fig. 10.5) as proposed in the international benchmark by *Neykov* et al. (2005). The *first* and the *second* tests are performed with the same axial power distribution but with different lateral power peaking pattern. They serve for analysis of the influence of the lateral power distribution. The *first* and the *third* tests are performed with the same lateral power peaking pattern but with different axial power distribution. They serve for analysis of the influence of the axial power distribution.

The measured versus the predicted bundle power is presented in Figs. 10.7c, 10.8 and 10.9.

Comparing Figs. 10.7a and 10.8 we see that for flatter lateral profiles this method slightly systematically underpredicts the measurements.

Comparing Figs. 10.8 and 10.9 we see that for the two different axial power profiles this method predicts almost the same measurements.

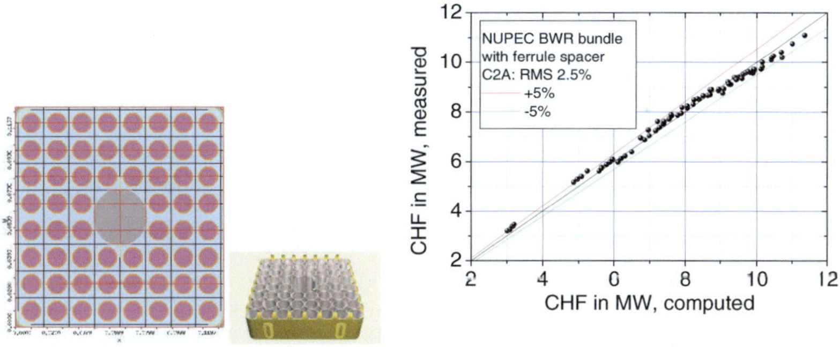

Fig. 10.7 Geometry of the 8×8 NUPEC CHF bundle tests: a) cross-section; b) ferrule spacer; c) Measured versus computed critical bundle power for *cosine* profile with $f_{z,\max} = 1.4$ and lateral peaking pattern characterized by $f_{xy} = 0.89$ to 1.3. Data by NUPEC: $p = 5.48$ to 8.67 MPa, $\Delta h_{2,sub} = 22.61$ to 137.26 kJ/kg, $\dot{m} = 9.98$ to 65.52 t/h.

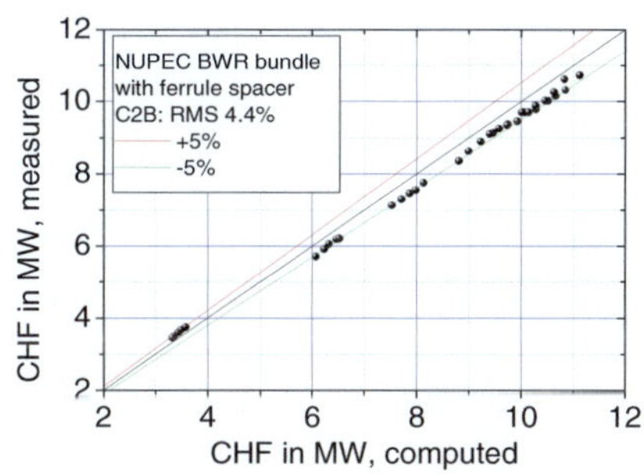

Fig. 10.8 Measured versus computed critical bundle power for *cosine* profile with $f_{z,\max} = 1.4$ and lateral peaking pattern characterized by $f_{xy} = 0.99$ to 1.18. Data by NUPEC: $p = 7.14$ to 7.20 MPa, $\Delta h_{2,sub} = 10.06$ to 128.57 kJ/kg, $\dot{m} = 9.93$ to 64.85 t/h.

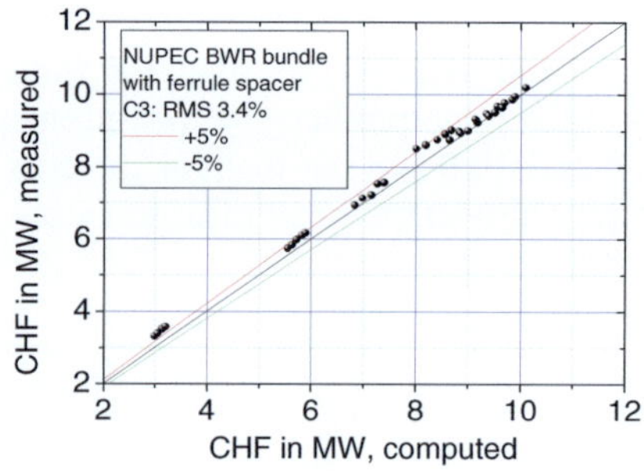

Fig. 10.9 Measured versus computed critical bundle power for *inlet peak* profile with $f_{z,\max} = 1.4$ and lateral peaking pattern characterized by $f_{xy} = 0.89$ to 1.3. Data by NUPEC: $p = 7.10$ to 7.19 MPa, $\Delta h_{2,sub} = 21.7$ to 125.74 kJ/kg, $\dot{m} = 9.93$ to 65.02 t/h.

> Having in mind the 2005 look-up table possesses a rms error of 7.1% the above predictions with rms error of 2.5, 4.4 and 3.4% can hardly be made better.

AREVA ATRIUM10 BWR tests: Figures 10.10 and 10.11 presents the measured versus the predicted bundle power for a specific tests for AREVA BWR bundles ATRIUM 10.

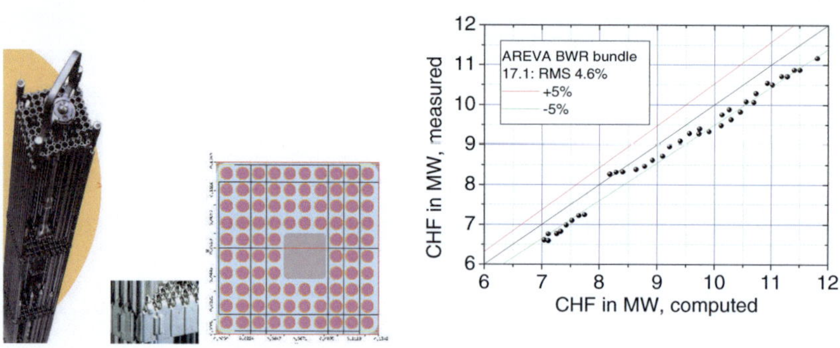

Fig. 10.10 a) BWR ATRIUM™ 10 bundle with ULTRAFLOW™ spacer; b) Measured versus computed critical bundle power for *cosine* profile with $f_{z,\max} = 1.4$ and lateral peaking pattern characterized by $f_{xy} = 0.751$ to 1.255: rms 4.6%, slight systematic underprediction. Data by *Fleiss* et al. (1992): $p = 69$ bar, $\Delta h_{2,sub} = 23$ to 184 kJ/kg, $\dot{m} = 6.3$ to 18.90 kg/s.

There are slight differences in the geometry. For both cases we see a rms error of 4.6 and 3.95%, respectively, which again lies inside the accuracy of the 2005 look-up table. Again, prediction with better accuracy can hardly be achieved.

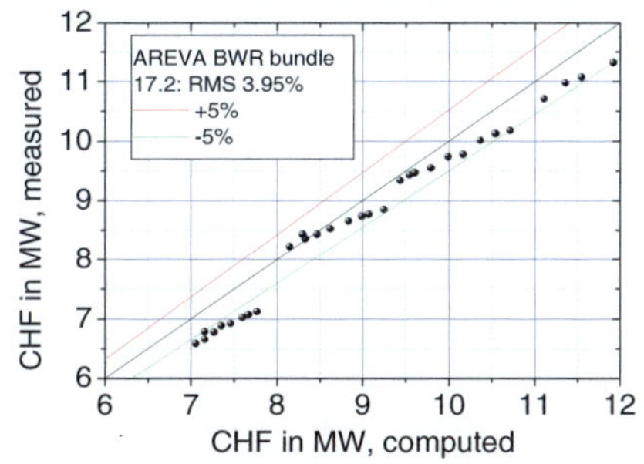

Fig. 10.11 Measured versus computed critical bundle power for *cosine* profile with $f_{z,\max} = 1.4$ and lateral peaking pattern characterized by $f_{xy}^{'} = 0.731$ to 1.251: rms 3.95%, slight systematic underprediction. Data by *Fleiss* et al. (1992): p = 68.8 to 69.5 bar, $\Delta h_{2,sub}$ = 23 to 188 kJ/kg, $\dot{m}$ = 6.33 to 15.7 kg/s.

AREVA tests for PWR bundles with FOCUS grids: Unlike the previous four data sets this one is for pressures up to 165 bar. The bundle contains 5×5 rods and 5 grids. Again, the simple method predicts the CHF with rms. error of 7.24, which is within the accuracy of the 2005 look-up table, see Fig. 10.12.

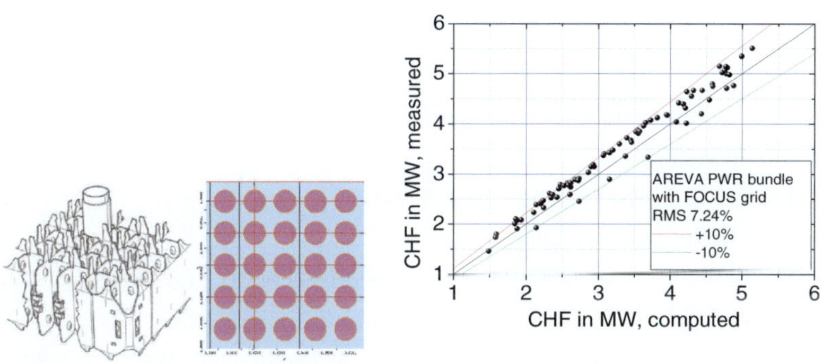

Fig. 10.12 AREVA PWR bundle with FOCUS grid: Computed versus measured bundle power: rms. 7.5%, slight systematic underprediction. Data by *Vogel* et al. (1991): p = 69.3 to 166.6 bar, $T_{2,in}$ = 179.4 to 329.5 °C, G = 1186.6 to 3629 kg/(m² s), q = 1.4643 to 5.5015 MW.

Conclusion

The surprising result of this example is that the bundle critical power for 5 data sets of three different bundles and different power distributions are predicted by a simple method described above using the 2005 look-up table within the accuracy reported by the authors of this table. Therefore, unless a better prediction method is developed it is good idea to use the above method as the best predictive instrument. The influence of the grid design can in addition be taken into account by an empirical coefficient derived from experiment.

The influence of the axial nonuniformity of the heat flux on the CHF for pipes was investigated experimentally by *Yang* et al. (2006) among others. It was found that as long as the mass fluxes are large the influence is not strong. At large mass flows and low subcooling the bottom peaking produces higher coolant velocity at the exit region and therefore better cooling compared with symmetric cases or cases with upper peaking. For large mass flows and low sub cooling the influence is within $\pm 7\%$. This confirms the conclusions of the above presented study. For the cases of large subcooling the differences are even smaller.

For small mass flows and small superheating the deviation from the data obtained with uniform flow can be up to 50%, by *Yang* et al. (2006). This is the reason for the considerable interest on developing more accurate methods based on sophisticated description of the local physical phenomena.

Nomenclature

Latin

C_{nl}	mass concentration of the inert component n inside the gas mixture, dimensionless
c_p	specific heat at constant pressure, J/(kg K)
c	wave travel velocity, m/s
D_{heat}	heated diameter (4 times flow cross-sectional area / heated perimeter), m
D_{rod}	rod diameter, m
D_{hy}	hydraulic diameter (4 times flow cross-sectional area / wet perimeter), m
$f_{lat,i}$	lateral power distribution factor, $\frac{1}{n_{rods}}\sum_{i=1}^{n_{rods}} f_{lat,i} = 1$, dimensionless
$f_{z_{\max},i}$	axial power distribution factor at the exit of the bundle for subchannel i, dimensionless
G	$=\sum(\alpha\rho w)_l$, mass flow rate in the axial direction, kg/(m² s)

g	gravitational acceleration, m/s²
h	specific enthalpy, J/kg
h	specific mixture enthalpy – mixture enthalpy of the noninert components, J/kg
Δh	evaporation enthalpy, J/kg
Δh_{sub}	subcooling enthalpy difference, J/kg
h_c	convective heat-transfer coefficient, W/(m² K)
h_{NB}	nucleate boiling heat-transfer coefficient, W/(m² K)
h_{nc_FB}	natural convection film boiling heat-transfer coefficient, W/(m² K)
$h_{nc\_FB\_Berenson}$	natural convection film boiling heat-transfer coefficient by *Berenson*, W/(m² K)
h_{fc_FB}	forced convection film boiling coefficient, W/(m² K)
h_{CHF}	critical heat-transfer coefficient, W/(m² K)
L_{rod}	rod length, m
p	pressure, Pa
q	$:= \sum_{i=1}^{n_{rods}} q_i$ bundle power, W
Δq	bundle power increment, W
q_0	initial bundle power, W
$\dot{q}''_w$	wall heat flux, W/m²
$\dot{q}''_{TB}$	transition boiling heat flux, W/m²
$\dot{q}''_{CHF}$	critical heat flux, W/m²
$\dot{q}''_{CHF,sub}$	critical heat flux for subcooled liquid, W/m²
$\dot{q}''_{CHF,sat}$	critical heat flux for saturated liquid, W/m²
$\dot{q}''_{\min\_FB}$	minimum film boiling heat flux, W/m²
$q''_{crit,i}$	critical heat flux in subchannel *i*, W/m²
q''_i	heat flux in subchannel *i*, W/m²
$\dot{m}_i$	mass flow through a subchannel *i*, kg/s
$\dot{m}$	total mass flow trough a bundle, kg/s
n_c	wave number, cm^{-1}
n_{rods}	number of rods in the bundle, dimensionless
Pr_{1w}	$= c_{p1}\eta_1 / \lambda_1$, gas *Prandtl* number at wall temperature, dimensionless
Pr_2	$= c_{p2}\eta_2 / \lambda_2$, liquid *Prandtl* number, dimensionless
Re	$= X_2 G D_{hy} / \eta_2$, liquid *Reynolds* number, dimensionless
Re_1	$= w_1 D_1 / \nu_1$, steam *Reynolds* number, dimensionless

$Re_{1\_mix} = \dfrac{D_h w_{eq}}{\eta'' / \rho''}$, mixture *Reynolds* number based on vapor viscosity, dimensionless
Re_{Tp} modified *Reynolds* number, dimensionless
S ratio of the effective superheat to the total superheat of the wall, dimensionless
T temperature, K
$T_{w,MFB}$ minimum film boiling temperature, K
ΔT_{CHF} temperature difference corresponding to the critical heat flux, K
$\Delta T_{\min\_FB}$ minimum film boiling temperature, K
X_l mass flow rate concentration in the flow of the velocity field l, dimensionless
X_{tt} *Martinelli* factor, dimensionless
X_{eq} equilibrium steam mass flow rate concentration in the flow, dimensionless
$X_{1,i}$ homogeneous equilibrium mass flow ratio of vapor, dimensionless
w axial velocity, m/s
w_{eq} $= G / \rho_{eq}$ center of mass velocity, m/s
$z_{\max}$ axial coordinate defining the exit of the bundle, m

Greek

α volume fraction, dimensionless
δ_{1F} $\approx \left(\dfrac{\sigma}{g \Delta \rho_{21}} \right)^{1/2}$, capillary constant, m
ε small number
Φ_{20}^2 the ratio of the two phase friction pressure drop to the pressure drop of the liquid field flowing alone in the channel, dimensionless
η dynamic viscosity of liquid, kg/(m s)
λ thermal conductivity, W/(m K)
λ_{RT} $= \left[\dfrac{\sigma}{g(\rho' - \rho'')} \right]^{1/2}$, *Rayleigh-Taylor* instability wavelength, m
ρ density, kg/m³
ρ_{eq} homogeneous mixture equilibrium density, kg/m³
σ surface tension, N/m
μ_{21} evaporation mass from the liquid film per unit time and unit mixture volume, kg/(m³ s)

μ_{31} evaporation mass from the droplets per unit time and unit mixture volume, kg/(m³ s)

μ_{12} condensation mass per unit time and unit mixture volume, kg/(m³ s)

Subscript

' saturated steam
" saturated liquid
1 gas
2 liquid
w wall
$M1$ noninert component in the gas mixture
$n1$ inert component in the gas mixture
$1F$ vapor film
2o liquid only, as the liquid mass flow is occupying the total cross-section

References

Addoms, J.N.: D. Sc. Thesis, M. I. T (1948)

Andreani, A., Yadigaroglu, G.: Prediction methods for dispersed flow film boiling. Int. J. Multiphase Flow 20, 1–51 (1994)

Bennett, A.W., et al.: Heat transfer to steam-water mixtures flowing in uniformly heated tubes in which the critical heat flux has been exceeded. In: Proceedings of the Institution of Mechanical Engineers 1847–1996, vol. 182(3H), A. E. R. E. R5373. Sage Publications, Thousand Oaks (1967), ISSN 0367-8849

Biasi, L., Clerici, G.C., Garribba, S., Sala, R., Tozzi, A.: Studies on burnout. Part 3: A new correlation for round ducts and uniform heating and its comparison with world data. Energ. Nucl. 14, 530–536 (1967)

Bjornard, T.A., Grifith, P.: PWR blow-down heat transfer. In: Thermal and Hydraulic Aspects of Nuclear Reactor Safety, vol. 1, pp. 17–41. American Society of Mechanical Engineers, New York (1977)

Borishanskii, V., Kozyrev, A., Svetlova, L.: Heat transfer in the boiling water in a wide range of saturation pressure. High Temperature 2(1), 119–121 (1964)

Bowring, R.W.: Simple but accurate round tube, uniform heat flux, dryout correlation over the pressure range 0.7 to 17 MN/m² (100 to 2500 psia). AEEW-R-789 (1972)

Chen, J.C.: A correlation for film boiling heat transfer to saturated fluids in convective flow. ASME Publication-63-HT-34, pp. 2–6 (1963)

Dittus, F.W., Boelter, L.M.K.: Heat transfer in automobile radiators of the tubular type, University of California Publications in Engineering, vol. 2(13), p. 443 (1937)

Dougall, R.S., Rohsenow, W.M.: Film boiling on the inside of vertical tubes with upward flow of the fluid of low qualities. MIT-TR-9079-26, Massachusetts Institute of Technology (1963)

Fleiss, Knapp, Which: STS-17.2, STS-17.1: Messung der Siedeübergangsleistungen und Druckverluste am ATRIUM-10-9Q-Testbündel, e321/92/43, e321/92/44, Karlstein, proprietary (1992)

Forster, H.K., Zuber, N.: Dynamic of the vapor bubbles and boiling heat transfer. AIChE Journal 1(4), 531–535 (1955)
Golovin, V.S., Kalchugin, B.A., Labuntsov, D.A.: Injenerno fizicheskiy jurnal 6(2) (1963)
Groeneveld, D.C.: Post-dryout heat transfer at reactor operating conditions. In: Nat. Topical Meet. Water Reactor Safety, American Nuclear Society, Conf. 730304, Rept. AECL-4513, Atomic Energy of Canada Ltd, Salt Lake City, Utah, March 26–28 (1977)
Groeneveld, D.C., Chen, S.C., Leung, L.K.H., Nguyen, C.: Computation of single and two-phase heat transfer rates suitable for water-cooled tubes and subchannels. Nucl. Eng. Des. 114, 61–77 (1989)
Groeneveld, D.C., Delorme, G.G.J.: Prediction of thermal non-equilibrium in the post-dryout regime. Nucl. Eng. Des. 36, 7–26 (1976)
Groeneveld, D.C., Leunga, L.K.H., Kirillov, P.L., Bobkov, V.P., Smogalev, I.P., Vinogradov, V.N., Huangc, X.C., Royer, E.: The 1995 look-up table for critical heat flux in tubes. Nucl. Eng. Des. 163, 1–23 (1996)
Groeneveld, D.C., Shan, J.Q., Vasi, A.Z., Leung, L.K.H., Durmayaz, A., Yang, J., Cheng, S.C., Tanase, A.: The 2005 CHF look-up table. In: The 11th Int. Top. Meeting on Nuclear Thermal-Hydraulics (NURETH11), Avignon, France, October 2–6 (2005)
Guo, Y., Leung, L.K.H.: Developing film boiling heat transfer in tubes. In: The 11th Int. Top. Meeting on Nuclear Thermal-Hydraulics (NURETH11), Avignon, France, October 2–6 (2005)
Hadaller, G., Banerjee, S.: Heat transfer to superheated steam in round tubes. AECL Unpublished Report WDI-147 (1969)
Hetstrony, G. (ed.): Handbook of multiphase systems. Hemishere Publ. Corp., McGraw-Hill Book Company, Washington, New York (1982)
Irvey, H.J., Morris, D.J.: On the relevance of the vapor-liquid exchange mechanism for subcooled boiling. Heat Transfer at High Pressure, AEEW-R127 (1962)
Kolev, N.I.: IVA Simulations to the OECD/NRC Benchmarks based on NUPEC BWR Full-size Fine-mesh Bundle Tests. In: 3th Workshop on OECD/NRC Benchmark based on NUPEC BWR full-size fine-mesh bundle tests (BFBT)-(BFBT-3), Pisa, Italy, April 26–27 (2006)
Kutateladze, S.S.: A hydrodynamic theory of changes in the boiling process under free convection conditions. Izv. Akad. Nauk SSSR, Otd. Tech. Nauk 4, 529–536 (1951); AEC-tr-1991 (1954)
Kutateladze, S.S.: Osnovy teorii teploobmena, Moskva, Mashgis, p. 456 (1962)
Loomis, G.G., Schumway, R.W.: Low-pressure transient flow film boiling in vertically oriented rod bundles. Nucl. Technology 63, 151–163 (1983)
Miropolskij, Z.L.: Heat transfer in film boiling of steam-water mixture in steam generating tubes. Teploenergetika 10(5), 49–53 (1963)
Morris, D.G., Mullins, C.B., Yoder, G.L.: An experimental study of rod bundle dispersed-flow film boiling with high pressure water. Nucl. Tech. 69, 82–93 (1985)
Neykov, B., Aydogan, F., Hochreiter, L., Ivanov, K., Utsuno, H., Fumio, K., Sartori, E., Martin, M.: NUPEC BWR full-size fine-mesh bundle test (BFBT) benchmark, vol. 1: Specifications, US NRC, OECD Nuclear Energy Agency, NEA/NSC/DOC, 5 (November 2005)
Polomik, E.E.: Transient boiling heat transfer program. Final summary report on program for February 1963–October 1967, GEAP–5563 (October 1967)
Ramu, K., Weisman, J.: A method for the correlation of transition boiling heat transfer data. In: 5th Int. Heat Transfer Conf., Heat Transfer 1974, Tokyo, vol. 4, pp. 160–164 (1974)

Sani Rle, R.: Down flow boiling and non-boiling heat transfer in a uniformly heated tube, 15th edn. University of California, URL-9023, Chemistry-Gen. UC-4, TID-4500, January 4 (1960)

Smolin, V.N., Shpanskii, S.V., Esikov, V.I., Sedova, T.K.: Method of calculating burnout in tubular fuel rods when cooled by water and a water-steam mixture. Teploenergetika 24(12), 30–35 (1977)

Tanase, A., Cheng, S.C., Groeneveld, D.C., Shang, J.Q.: Diameter effect on critical heat flux. Nucl. Eng. Des. 239, 289–294 (2009)

Vogel, Bruch, Wang: SIEMENS Test Section 52 (DTS52) Description of experiments, KWZ BT23, e 244, Erlangen, proprietary, August 19 (1991)

Wang, C.H., Dhir, V.K.: Effect of surface wettability on active nucleation site density during pool boiling of water on a vertical surface. ASME J. Heat Transfer 115, 659–669 (1993)

Wong, W.C.: Effect of the tube diameter on critical heat flux. MaSC dissertation, Ottawa, Carleton Institute for Mechanical and Aeronautical Engineering, University of Ottawa (1996)

Yang, J., Groeneveld, D.C., Leung, L.K.H., Cheng, S.C., El Nakla, M.A.: An experimental and analytical study of the effect of the axial power profile on CHF. Nucl. Eng. Des. 236, 1384–1395 (2006)

Yoder Jr., G.L., Morris, D.G., Mullins, C.B.: Rod bundle burnout data and correlation comparison. Nucl. Technol. 68, 355–369 (1985)

Zuber, N.: On the stability of boiling heat transfer. Trans. ASME 80, 711–720 (1958)

Zuber, N., et al.: The hydrodynamic crisis in pool boiling of saturated and subcooled liquids. In: Int. Heat Transfer Conf., Int. Dev. Heat Transfer, Part 2, Boulder, Colorado, vol. 27, pp. 230–236 (1961)

Zuber, N., Tribus, M.: Further remarks on the stability of boiling heat transfer. Report 58-5, Department of Engineering, University of California, Loss Angeles (1958)

11. Film boiling on vertical plates and spheres

Section 11.1 presents a closed analytical solution for mixed-convection film boiling on vertical walls. Heat-transfer coefficients predicted by the proposed model and experimental data obtained at the Royal Institute of Technology in Sweden by Okkonen et al. are compared. All data predicted are inside the $\pm 10\%$ *error band, with the mean averaged error being below 4% (using slightly modified analytical solution). The solution obtained is recommended for practical applications. The method presented here is used in Section 11.2 as a guideline for developing a model for film boiling on spheres. The new semiempirical film boiling model for spheres used in IVA computer code is compared with the experimental data base obtained by Liu and Theofanous. The data are predicted within a* $\pm 30\%$ *error band. This chapter is a slightly changed variant of the work previously published in Kolev (1997), Kolev (1998).*

11.1 Plate

11.1.1 Introduction

In nature, film boiling is the controlling heat-transfer mechanism during lava–water interaction. In the cryogenic technology in rocket engines, it is the inevitable heat-transfer mechanism in many cases. Film boiling heat transfer is of particular interest for analysis of accident processes in industry. The theory of film boiling heat transfer has become a classical part of heat- and mass-transfer text books but is often presented in the context of an unwanted heat-transfer regime that has to be avoided by design. The close scrutiny paid to accident processes in industry requires critical reevaluation of film boiling theory in order to yield a reliable prediction tool for engineering purposes. In particular, calculation of the components of the heat removed by film boiling that leads to vapor heating, liquid heating, evaporation mass flow rate, average vapor volume attached to the surface is very important for increasing the accuracy of the overall flow field prediction of melt interacting with water. As an example of the necessity for improvement of theory, it should be recalled that there is still a need for a closed analytical solution for mixed natural- and forced-convection film boiling. In this chapter a complete analytical solution for film boiling on a vertical wall will first be presented. After this, an attempt will be made to extend the model for spheres. Comparison with experimental data will indicate the success and the limitations of the theoretical results. Where necessary, empirical correction factors are introduced to give a more accurate prediction of existing experimental data.

11.1.2 State-of-the-art

The film boiling literature was reviewed by *Sakurai* in 1990, and more recently by *Liu* and *Theofanous* in 1995. There is no need to repeat this here.

In his pioneering work of 1916 *Nusselt* presented a closed model of laminar film condensation on vertical surfaces. The idea used was that the condensation is controlled by the heat conduction through the falling liquid film. This model is thus easily applied to film boiling under natural convection on vertical plates, postulating that the heat transfer is controlled by the heat conduction through the vapor film that it deserves to be called the *Nusselt theory for laminar condensation and film boiling on vertical walls due to natural convection*.

Natural-convection film boiling was first investigated by *Bromley* in 1950. He applied *Nusselt*'s idea from 1916. *Frederking* et al. demonstrated in 1965 that the treatment of vertical plates, spheres and horizontal cylinders results in the same expression multiplied by a numerical constant to characterize the appropriate one of the above geometries. Subsequent to this, many theoretical studies related to this subject have been published. The main physical ideas on which the published models rely are similar to those published by *Bromley* et al. (1950, 1953) for natural and mixed convection, and by *Witte* in 1968 for forced convection.

A rigorous formulation of the steady-state problem for *natural convection* was given by *Nishikawa* et al. (1966, 1976). Several numerical solutions of the problem defined with some simplifying assumptions are available, e.g., *Dhir* and *Purohit* (1978). An accurate numerical solution is compared with an extensive set of data for cylinders and spheres by *Sakurai* et al. (1990, 1986). *Sakurai* et al. proposed the most accurate correlation for cylinder geometry known to the author. This forms a good approximation of the accurate numerical solution and compares favorably with experimental data for saturated and subcooled water at atmospheric and elevated pressure, with radiation effects also considered. The same correlation has been recently recommended by *Liu* and *Theofanous* (1995) after comparison with an extensive experimental database.

Subsequent to *Witte*'s analytical solution (1968) for simplified *forced-convection* steady-state problems in 1968, there have only been numerical solutions provided for this problem in the literature, e.g., *Fodemski* et al. (1982), *Liu* and *Theofanous* (1995).

A third group within the literature covers the conditions of the *stagnation point*. Assuming that the stagnation-point condition is valid in the front half of the sphere, *Epstein* and *Heuser* derived in (1980) a correlation that additionally takes into account the liquid subcooling. A numerical solution to this problem is proposed by *Fodemski* (1992), *Kolev* (1995).

The effect of radiation and subcooling or superheating on the film conduction heat flux is taken into account in different ways. Some authors superposed the separate effects, recommending some reduction of the radiative component *Sakurai* (1990). The natural way to take these effects into account is to consider them within the formalism describing the local film thickness. The resulting system of differential equations is solved numerically, e.g., by *Sparrow* (1964),

Fodemski (1982), *Liu* and *Theofanous* (1995). There is no simple analytical solution to this problem available in the literature.

A numerical solution of the combined natural- and forced-convection film boiling problem for spheres in saturated, subcooled or superheated water with radiation is possible, as demonstrated in the literature, but requires a very large amount of computation effort if implemented in system computer codes. To analyze the interaction of fragmenting liquid metal in water, a simple method is needed that has the capability (a) to reproduce the available data for the heat-transfer coefficient for natural and forced convection, (b) to separate the effect of radiation, subcooling or superheating in order to compute the heat-flux component consumed to cool the hot structure, to heat or cool the liquid, and to produce steam. Dynamic system characteristics can then be predicted by using such a model that provides heat and mass source terms for the macroscopic differential equation governing the flow. Some of the existing models for this case satisfy one or two of the above conditions, but not all of them. It was for this reason that the author developed a new simplified approach that permits separation of the different effects of radiation, subcooling or superheating and computes the particular heat-flux components.

11.1.3 Problem definition

Consider the geometry presented in Fig. 11.1. Liquid of temperature T_2 flows along a vertical heated wall. The surface temperature of the wall $T_{3\sigma}$ is so high that a stable vapor film is formed. The evaporation at the liquid/vapor interface forces the interface temperature to be equal to the saturation temperature at the system pressure $T'(p)$. The thickness of the vapor film $\delta_1(z)$ varies with the height z. We introduce two coordinate systems having the same axial coordinate z. The horizontal distance from the wall to the vapor film is r_1. The horizontal distance from the vapor/liquid interface with the liquid is r_2. The vapor velocity has only a vertical component $w_1(r_1)$. The liquid in the boundary layer has a vertical component of $w_2(r_2)$. Various different heat fluxes can immediately be identified first. The radiation heat flux can be split into three different components: $\dot{q}''_{r1}$ is the radiation heat flux from the interface to the vapor. $\dot{q}''_{r2}$ is the radiation heat flux from the wall to the bulk liquid. $\dot{q}''_{r\sigma 2}$ is the radiation heat flux absorbed from the vapor/liquid interface. The partitioning depends mainly on the temperature of the emitting wall. For temperatures below 1000 K almost all radiation energy is dissipated at the liquid interface. This is not the case for the short-wavelength light corresponding to higher temperatures. At the present time, it is not clear whether the liquid/vapor interface instabilities create a boundary layer transparent to light or not. The heat flux $\dot{q}''_{FB}$ is due to conduction through the vapor film. The heat flux $\dot{q}''_{3\sigma 1}$ is from the wall into the vapor, due to the convective heating of the vapor from the saturation interface temperature to some averaged vapor film temperature. This contribution is frequently neglected in the film boiling literature. In the case of subcooled liquid there is heat flux $\dot{q}''_{\sigma 2}$ from the liquid interface into the liquid due to conduction and convection.

The problem to be solved is to compute all components of the wall heat flux.

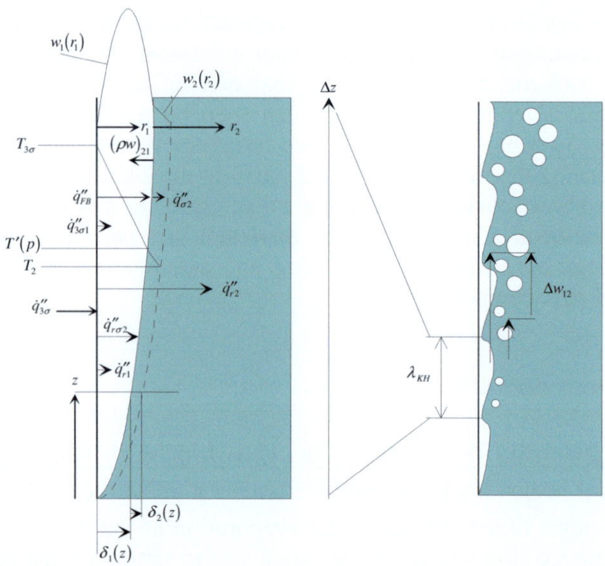

Fig. 11.1 a) Geometry definition for film boiling on vertical wall. b) Instability of film boiling

11.1.4 Simplifying assumptions

The following simplifying assumptions are introduced:

1) The vapor and the liquid are incompressible.

2) The following simplified momentum equations describe flows in the vapor and liquid boundary layers, respectively:

$$\eta_1 \frac{\partial^2 w_1}{\partial r_1^2} = \frac{\partial p}{\partial z} + \rho_1 g , \tag{11.1}$$

$$\eta_2 \frac{\partial^2 w_2'}{\partial r_2^2} = \frac{\partial p}{\partial z} + \rho_2 g \approx 0 . \tag{11.2}$$

Since the last equation simplifies to $\frac{\partial p}{\partial z} \approx -\rho_2 g$ one can write

$$\eta_1 \frac{\partial^2 w_1}{\partial r_1^2} = -(\rho_2 - \rho_1) g , \tag{11.3}$$

$$\eta_2 \frac{\partial^2 w_2'}{\partial r_2^2} = 0 . \tag{11.4}$$

The above two equations can be integrated analytically to give

$$w_1(r_1) = -\frac{1}{2\eta_1}(\rho_2 - \rho_1) g r_1^2 + C_1 r_1 + C_3 \tag{11.5}$$

$$w_2'(r_2) = C_2 r_2 + C_4 . \tag{11.6}$$

The constants are determined from the following boundary conditions

1) at $r_1 = 0$, $w_1 = 0$, (11.7)

2) at $r_1 = \delta_1$ and $r_2 = 0$, $w_1 = w_2'$, (11.8)

3) at $r_1 = \delta_1$ and $r_2 = 0$, $\eta_1 \frac{\partial w_1}{\partial r_1} = \eta_2 \frac{\partial w_2'}{\partial r_2}$, (11.9)

4) at $r_2 = \delta_2$, $w_2' = w_2$. (11.10)

The solutions to Eqs. (11.5) and (11.6) are

$$w_1 = \frac{\Delta \rho g \delta_1^2}{2\eta_1}\left[\frac{r_1}{\delta_1}\frac{f + 2\xi}{f + \xi} - \left(\frac{r_1}{\delta_1}\right)^2\right] + \frac{r_1}{\delta_1}\frac{f}{f + \xi} w_2 , \tag{11.11}$$

$$w_2' = \left(1 - \frac{r_2}{\delta_2}\right)\left[\frac{\Delta \rho g \delta_1^2}{2\eta_2}\frac{f\xi}{f + \xi} + \frac{f}{f + \xi} w_2\right] + \frac{r_2}{\delta_2} w_2 , \tag{11.12}$$

where

$$f = \eta_2 / \eta_1 , \tag{11.13}$$

$$\xi = \delta_2 / \delta_1 . \tag{11.14}$$

For the case of zero liquid velocity Eqs. (11.11) and (11.12) reduce to those reported by *Liu* and *Theofanous* (1995).

In this chapter the assumption is made that the layer thickness ratio is not a function of the axial coordinate.

The cross-section-averaged vapor velocity profile is

$$\langle w_1 \rangle_1 = \frac{1}{\delta_1}\int_0^{\delta_1} w_1 dr_1 = \int_0^1 \left\{ \frac{\Delta\rho g \delta_1^2}{2\eta_1}\left[\frac{r_1}{\delta_1}\frac{f+2\xi}{f+\xi} - \left(\frac{r_1}{\delta_1}\right)^2\right] + \frac{r_1}{\delta_1}\frac{f}{f+\xi} w_2 \right\} d\frac{r_1}{\delta_1} \tag{11.15}$$

or

$$\langle w_1 \rangle_1 = \frac{\Delta\rho g \delta_1^2}{12\eta_1}\frac{f+4\xi}{f+\xi} + \frac{1}{2}\frac{f}{f+\xi} w_2 . \tag{11.16}$$

The temperature profile in the vapor layer is assumed to be linear,

$$T_1 - T' = \Delta T_{sp}\left(1 - \frac{r_1}{\delta_1}\right). \tag{11.17}$$

This assumption results from the observation that the cyclic instabilities do not allow thick films to exist. The order of magnitude is tenths of a mm. To permit further treatment, the following average is required

$$\langle w_1 (T_1 - T') \rangle_1 = \frac{1}{\delta_1}\int_0^{\delta_1} w_1 (T_1 - T') dr_1$$

$$= \Delta T_{sp}\left[\langle w_1 \rangle_1 - \int_0^1 \left\{ \frac{\Delta\rho g \delta_1^2}{2\eta_1}\left[\left(\frac{r_1}{\delta_1}\right)^2 \frac{f+2\xi}{f+\xi} - \left(\frac{r_1}{\delta_1}\right)^3\right] + \left(\frac{r_1}{\delta_1}\right)^2 \frac{f}{f+\xi} w_2 \right\} d\frac{r_1}{\delta_1}\right]. \tag{11.18}$$

The last equation can, after integration, be written in the form:

$$\langle w_1 (T_1 - T') \rangle_1 = \frac{1}{2}\Delta T_{sp}\frac{\Delta\rho g \delta_1^2}{12\eta_1}\frac{f+3\xi}{f+\xi} + \frac{1}{6}\Delta T_{sp}\frac{f}{f+\xi} w_2 . \tag{11.19}$$

The Reynolds similarity hypothesis for the liquid boundary layer is used:

$$T_2^* - T_2 = \Delta T_{sc}\left(1 - \frac{r_2}{\delta_2}\right). \tag{11.20}$$

This means that the velocity boundary layer is similar to the temperature boundary layer. This hypothesis holds for liquids with $\mathrm{Pr} \approx 1$. For the energy conservation of the liquid boundary layer the following average is required

$$\langle w_2 (T_2' - T_2) \rangle_2 = \frac{1}{\delta_2}\int_0^{\delta_2} w_2 (T_2' - T_2) dr_2$$

$$= \Delta T_{sc} \int_0^1 \left\{ \left(1 - \frac{r_2}{\delta_2}\right) \left[\frac{\Delta \rho g \delta_1^2}{2\eta_2} \frac{f\xi}{f+\xi} + \frac{f}{f+\xi} w_2 \right] + \frac{r_2}{\delta_2} w_2 \right\} d\frac{r_2}{\delta_2}$$

$$-\Delta T_{sc} \int_0^1 \left\{ \left(\frac{r_2}{\delta_2} - \left(\frac{r_2}{\delta_2}\right)^2 \right) \left[\frac{\Delta \rho g \delta_1^2}{2\eta_2} \frac{f\xi}{f+\xi} + \frac{f}{f+\xi} w_2 \right] + \left(\frac{r_2}{\delta_2}\right)^2 w_2 \right\} d\frac{r_2}{\delta_2}$$

$$= \Delta T_{sc} \left(\frac{2\Delta \rho g \delta_1^2}{12\eta_2} \frac{f\xi}{f+\xi} + \frac{1}{6} \frac{3f+\xi}{f+\xi} w_2 \right). \tag{11.21}$$

11.1.5 Energy balance at the vapor/liquid interface, vapor film thickness, average heat-transfer coefficient

In this section the layer thickness ratio ξ is taken as known. A separate section deals with calculation of the layer thickness ratio.

We assume that no energy can be accumulated at the vapor/liquid interface. As a consequence, the energy entering the interface due to conduction from the wall through the vapor, minus the energy removed convectively for heating of the vapor plus the radiation energy dissipated into the surface, minus the energy removed from the interface into the liquid will then evaporate a certain amount of steam at the surface:

$$\left(\lambda_1 \Delta T_{sp} / \delta_1 + \dot{q}''_{r\sigma 2} - \lambda_2 \Delta T_{sc} / \delta_2\right) dA_3 - d\left[\rho_1 c_{p1} \left\langle w_1 \left[T_1(r_1) - T' \right] \right\rangle_1 A_1 \right]$$

$$= d\left(\rho_1 \Delta h \left\langle w_1 \right\rangle_1 A_1\right). \tag{11.22}$$

The film cross-section at the position defined by the coordinate z is expressed by

$$A_1 = \delta_1 L, \tag{11.23}$$

where L is the transversal dimension of the plate. The infinitesimal wall surface is expressed by

$$dA_3 = L dz. \tag{11.24}$$

By appropriate substitution within the energy-jump condition the following is then obtained

$$\lambda_1 \frac{\Delta T_{sp}}{\delta_1} \left(1 + \frac{\dot{q}''_{r\sigma 2}}{\lambda_1 \Delta T_{sp} / \delta_1} - \frac{\lambda_2 \Delta T_{sc} / \delta_2}{\lambda_1 \Delta T_{sp} / \delta_1} \right) dz$$

$$= \rho_1 \Delta h d \left\{ \frac{\left\langle w_1 \right\rangle_1 + c_{p1} \left\langle w_1 \left[T_1(r_1) - T' \right] \right\rangle_1}{\Delta h} \delta_1 \right\}. \tag{11.25}$$

Introducing the following abbreviations

$$Sp = c_{p1}\Delta T_{sp} / \Delta h \,, \tag{11.26}$$

$$r_r = \frac{\dot{q}''_{r\sigma 2}}{\lambda_1 \Delta T_{sp} / \delta_1} \approx const \,, \tag{11.27}$$

$$C_2 = \frac{\lambda_2 \Delta T_{sc}}{\lambda_1 \Delta T_{sp}} \,, \tag{11.28}$$

$$\xi = \delta_2 / \delta_1 \,, \tag{11.29}$$

$$f_{nc} = \frac{f + 4\xi}{f + \xi} \frac{1 + \frac{1}{2} Sp \frac{f + 3\xi}{f + 4\xi}}{1 + r_r - C_2 / \xi} \,, \tag{11.30}$$

$$f_{fc} = \frac{f}{f + \xi} \frac{1 + \frac{1}{3} Sp}{1 + r_r - C_2 / \xi} \,, \tag{11.31}$$

Eq. (11.25) is finally obtained in the form:

$$\frac{\lambda_1 \Delta T_{sp}}{\rho_1 \Delta h} dz = \frac{\Delta \rho g}{12 \eta_1} f_{nc} \frac{3}{4} d\delta_1^4 + \frac{1}{4} f_{fc} w_2 d\delta_1^2 \,. \tag{11.32}$$

Before integrating the last equation it is assumed that the ratio of the radiative flux and the local film conduction heat flux is constant, Eq. (11.27). At first glance this assumption would seem arbitrary but as will be discussed later, this turns out to be a good assumption. Integrating with the initial condition $z = 0$, $\delta_1 = 0$ the local vapor film thickness as a function of the axial coordinate is then obtained

$$z = \frac{\rho_1 \Delta h \Delta \rho g}{\lambda_1 \Delta T_{sp} \eta_1} f_{nc} \left(\frac{\delta_1}{2} \right)^4 + \frac{\rho_1 \Delta h w_2}{\lambda_1 \Delta T_{sp}} f_{fc} \left(\frac{\delta_1}{2} \right)^2 . \tag{11.33}$$

Comparing the above equation to Eq. (11.6) by *Bui* and *Dhir* (1985) it is evident that they are equivalent only when the conditions $f_{nc} = 1$ and $f_{fc} = 1$ hold, with the reason behind this not directly evident.

Calculation of the average heat-transfer coefficients for the limiting case of pure natural convection is given in Appendix 11.1. An estimate of the average heat-transfer coefficients for the limiting case of predominant forced convection is given in Appendix 11.2. In order to simplify Eq. (11.33) the resulting average heat-transfer coefficients are introduced

$$h_{nc} = c_{nc} \left[\frac{\lambda_1^3 \rho_1 \Delta h \Delta \rho g}{\Delta T_{sp} \eta_1 \Delta z} f_{nc} \right]^{1/4} , \tag{11.34}$$

$$h_{fc} = c_{fc} \left[\frac{\rho_1 \lambda_1 \Delta h w_2}{\Delta T_{sp} \Delta z} f_{fc} \right]^{1/2}, \tag{11.35}$$

where

$$c_{nc} = 2/3, \tag{11.36}$$

and

$$c_{fc} = 1. \tag{11.37}$$

Equations (11.34) and (11.35) can be written in terms of dimensionless numbers as follows

$$Nu_{nc} = c_{nc} \left(Gr_1 \frac{Pr_1}{Sp} f_{nc} \right)^{1/4}, \tag{11.38}$$

$$Nu_{fc} = c_{fc} \left(Re_1 \frac{Pr_1}{Sp} f_{fc} \right)^{1/2}. \tag{11.39}$$

As will be demonstrated later h_{nc} is the heat-transfer coefficient for the case of zero liquid velocity, that is natural convection only, and h_{fc} is the heat-transfer coefficient for predominant forced convection. Equation (11.33) can thus be rewritten in the following form

$$\left(\frac{\delta_1 h_{nc}}{\lambda_1 2 c_{nc}} \right)^4 + 2r^* \left(\frac{\delta_1 h_{nc}}{\lambda_1 2 c_{nc}} \right)^2 - \frac{z}{\Delta z} = 0, \tag{11.40}$$

where

$$r^* = \frac{1}{2} \left(\frac{h_{fc} c_{nc}}{h_{nc} c_{fc}} \right)^2 = \frac{1}{2} \left(\frac{\rho_1}{\Delta \rho} \frac{Pr_1}{Sp} \frac{f_{fc}^2}{f_{nc}} \right)^{1/2} Fr_2, \tag{11.41}$$

and the *Froude* number is defined as

$$Fr_2 = \frac{w_2}{\sqrt{\Delta z g}}. \tag{11.42}$$

The film thickness as a function of the axial coordinate is therefore

$$\frac{\delta_1 h_{nc}}{\lambda_1 2 c_{nc}} = \left[\left(r^{*2} + \frac{z}{\Delta z} \right)^{1/2} - r^* \right]^{1/2}. \tag{11.43}$$

At location $z = \Delta z$, the vapor thickness is at its maximum, i.e.,

$$\frac{\delta_{1,\max} h_{nc}}{\lambda_1 2c_{nc}} = \left[\left(r^{*2}+1\right)^{1/2} - r^*\right]^{1/2}. \tag{11.44}$$

The averaged film boiling heat-transfer coefficient is therefore

$$h = \frac{\dot{q}''_{FB}}{\Delta T_{sp}} = \frac{1}{\Delta z}\int_0^{\Delta z} \frac{\lambda_1}{\delta_1} dz = \frac{h_{nc}}{2c_{nc}} \int_0^1 \left[\left(r^{*2} + \frac{z}{\Delta z}\right)^{1/2} - r^*\right]^{-1/2} d\frac{z}{\Delta z}$$

$$= \frac{h_{nc}}{2c_{nc}} \frac{4}{3}\left[\left(r^{*2}+1\right)^{1/2} + 2r^*\right]\left[\left(r^{*2}+1\right)^{1/2} - r^*\right]^{1/2}$$

$$= h_{nc}\left[\left(r^{*2}+1\right)^{1/2} + 2r^*\right]\left[\left(r^{*2}+1\right)^{1/2} - r^*\right]^{1/2}. \tag{11.45}$$

When the liquid velocity is equal to zero, $r^*=0$, the expected solution $h = h_{nc}$ is obtained. Equation (11.45) is an important new result. This represents the average heat-transfer coefficient for mixed convection as a function of the average flow properties, the heat-transfer coefficient for natural convection and the *Froude* number. Note that this equation still contains one unknown variable: the layer thickness ratio ξ. The second equation is obtained in the next section by using the energy balance of the liquid boundary layer.

11.1.6 Energy balance of the liquid boundary layer, layer thickness ratio

The energy balance for the liquid boundary layer has the following form

$$\left(\lambda_2 \Delta T_{sc} / \delta_2\right) dA_3 = d\left[\rho_2 c_{p2} \left\langle w_2 \left[T_2'(r_2) - T_2\right]\right\rangle_2 A_2\right]. \tag{11.46}$$

As has already been mentioned, it is not clear at the present time whether the liquid/vapor interface instabilities create a boundary layer transparent to light or not. Observations with the naked eye indicate that the interface is rather "milky" and therefore probably not as transparent as a water/steam interface at rest. This point deserves special attention in the future.

The liquid boundary-layer cross-section at the position defined by the coordinate z is

$$A_2 = \delta_2 L, \tag{11.47}$$

Substituting Eq. (11.21) for the average used in the last equation, the following is then obtained

$$dz = \frac{\rho_2 c_{p2}}{\lambda_2}\left(\frac{2\Delta\rho g}{12\eta_2}\frac{f\xi^3}{f+\xi}\delta_1 d\delta_1^3 + \frac{1}{6}\frac{3f+\xi}{f+\xi}\xi^2 w_2 \delta_1 d\delta_1\right). \tag{11.48}$$

After integrating, this yields:

$$z = \frac{\rho_2 c_{p2}}{\lambda_2} \frac{\Delta\rho g}{\eta_2} \frac{f\xi^3}{f+\xi} 2 \left(\frac{\delta_1}{2}\right)^4 + \frac{1}{3} \frac{3f+\xi}{f+\xi} \xi^2 w_2 \frac{\rho_2 c_{p2}}{\lambda_2} \left(\frac{\delta_1}{2}\right)^2. \tag{11.49}$$

Since the form of Eq. (11.49) is the same as Eq. (11.33), the same transformation can be applied to obtain another expression for the average heat-transfer coefficient. Equating the right-hand sides of Eqs. (11.33) and (11.49), the definition equation for computation of the layer thickness ratio is then obtained. The heat-transfer coefficients as a function of the liquid-side properties will next be computed in a similar way to the previous section.

First, the following definition is made

$$h_{2,nc} = c_{2,nc} \left(\lambda_1^4 \frac{\rho_2 c_{p2}}{\lambda_2} \frac{\Delta\rho g}{\eta_2 \Delta z} 2 \frac{f\xi^3}{f+\xi} \right)^{1/4}, \tag{11.50}$$

$$h_{2,fc} = c_{2,fc} \left(\lambda_1^2 \frac{1}{3} \frac{3f+\xi}{f+\xi} \xi^2 w_2 \frac{1}{\Delta z} \frac{\rho_2 c_{p2}}{\lambda_2} \right)^{1/2}, \tag{11.51}$$

where

$$c_{2,nc} = 1, \tag{11.52}$$

and

$$c_{2,fc} = 1. \tag{11.53}$$

With this notation Eq. (11.49) divided by Δz gives

$$\left(\frac{\delta_1 h_{2,nc}}{\lambda_1 2 c_{2,nc}}\right)^4 + 2 r_2 * \left(\frac{\delta_1 h_{2,nc}}{\lambda_1 2 c_{2,nc}}\right)^2 - \frac{z}{\Delta z} = 0, \tag{11.54}$$

where

$$r_2 * = \frac{1}{2} \left(\frac{h_{2,fc} c_{2,nc}}{h_{2,nc} c_{2,fc}}\right)^2 = \frac{1}{2} \left[\frac{1}{18} \frac{\xi (3f+\xi)^2}{f(f+\xi)} \frac{\rho_2}{\Delta\rho} Pr_2 \right]^{1/2} Fr_2 . \tag{11.55}$$

Solving for the film thickness yields

$$\frac{\delta_1 h_{2,nc}}{\lambda_1 2 c_{2,nc}} = \left[\left(r_2 *^2 + \frac{z}{\Delta z} \right)^{1/2} - r_2 * \right]^{1/2} . \tag{11.56}$$

Using Eq. (11.54) the following is obtained for the average heat-transfer coefficient

$$h = h_{2,nc}\left[\left(r_2{*}^2+1\right)^{1/2}+2r_2{*}\right]\left[\left(r_2{*}^2+1\right)^{1/2}-r_2{*}\right]^{1/2}. \qquad (11.57)$$

By comparing Eqs. (11.57) and (11.45) the definition equation for the film-thickness ratio is obtained as follows:

$$h_{2,nc}\left[\left(r_2{*}^2+1\right)^{1/2}+2r_2{*}\right]\left[\left(r_2{*}^2+1\right)^{1/2}-r_2{*}\right]^{1/2}$$
$$= h_{nc}\left[\left(r{*}^2+1\right)^{1/2}+2r{*}\right]\left[\left(r{*}^2+1\right)^{1/2}-r{*}\right]^{1/2}. \qquad (11.58)$$

This is a transcendental equation with respect to the layer thickness ratio and must be solved by iteration. An analytical solution for natural circulation is available as an initial value. Setting the liquid velocity equal to zero yields $h_{2,nc} = h_{nc}$, which results in a cubic equation with respect to the unknown ξ

$$(1+r_r)C_1 f\xi^3 - fC_1C_2\xi^2 - \frac{8}{81}\left(4+\frac{3}{2}Sp\right)\xi - \frac{8}{81}\left(1+\frac{1}{2}Sp\right)f = 0. \qquad (11.59)$$

In fact, Eq. (11.59) is obtained by combining Eqs. (11.34) and (11.50). Here, the following new dimensionless parameters inevitably result

$$C_1 = \frac{\lambda_1}{\lambda_2}\frac{\nu_1}{\nu_2}\frac{c_{p2}\Delta T_{sp}}{\Delta h} = \frac{1}{f}\frac{a_1}{a_2}Sp. \qquad (11.60)$$

An analytical solution is available as follows. Substitute

$$q = -\frac{1}{3}\left\{\frac{1}{9}\left[\frac{C_2}{1+r_r}\right]^3 + \frac{4}{81}C_2\frac{4+\frac{3}{2}Sp}{(1+r_r)^2 fC_1} + \frac{4}{27}\frac{1+\frac{1}{2}Sp}{(1+r_r)C_1}\right\}, \qquad (11.61)$$

$$p = -\frac{1}{3}\left\{\frac{8}{81}\frac{4+\frac{3}{2}Sp}{(1+r_r)fC_1} + \frac{1}{3}\left[\frac{C_2}{(1+r_r)}\right]^2\right\}, \qquad (11.62)$$

$$D = q^2 + p^3. \qquad (11.63)$$

For $D > 0$ the real solution is

$$\xi = \left(-q + D^{1/2}\right)^{1/3} + \left(-q - D^{1/2}\right)^{1/3} + \frac{C_2}{3(1+r_r)}. \qquad (11.64)$$

11.1.7 Averaged heat fluxes

The film boiling heat flux is given by Eq. (11.45)

$$\dot{q}''_{FB} = h\Delta T_{sp}. \tag{11.65}$$

The average heat flux from the vapor/liquid interface into the liquid is

$$\dot{q}''_{\sigma 2} = \frac{1}{\Delta z}\int_0^{\Delta z} \lambda_2 \frac{\Delta T_{sc}}{\delta_2} dz = \frac{\lambda_2 \Delta T_{sc}}{\lambda_1 \Delta T_{sp}} \frac{1}{\xi} \frac{1}{\Delta z}\int_0^{\Delta z} \lambda_1 \frac{\Delta T_{sp}}{\delta_1} dz = \frac{C_2}{\xi} \dot{q}''_{FB}. \tag{11.66}$$

The heat flux required for heating saturated steam to a certain average temperature corresponding to the linear temperature profile is

$$\dot{q}''_{3\sigma 1}(z) dA_3 = d\left[\rho_1 c_{p1} \left\langle w_1 \left[T_1(r_1) - T'\right]\right\rangle_1 A_1\right], \tag{11.67}$$

or after rearrangement

$$\dot{q}''_{3\sigma 1}(z) = \frac{1}{2}\rho_1 c_{p1} \Delta T_{sp} \frac{\Delta \rho g}{12\eta_1} \frac{f + 3\xi}{f + \xi} \frac{d\delta_1^3}{dz} + \frac{1}{6}\rho_1 c_{p1} \Delta T_{sp} w_2 \frac{f}{f + \xi} \frac{d\delta_1}{dz}. \tag{11.68}$$

The average value can be finally expressed by

$$\begin{aligned}
\dot{q}''_{3\sigma 1} &= \frac{1}{\Delta z}\int_0^{\Delta z} \dot{q}''_{3\sigma 1}(z) dz \\
&= \frac{1}{2}\rho_1 c_{p1} \Delta T_{sp} \frac{\Delta \rho g}{12\eta_1 \Delta z} \frac{f + 3\xi}{f + \xi} \delta_{1,\max}^3 + \frac{1}{6} \frac{\rho_1 c_{p1} \Delta T_{sp} w_2}{\Delta z} \frac{f}{f + \xi} \delta_{1,\max} \\
&= \frac{1}{2} Sp \left[\frac{1}{f_{nc}} \frac{f + 3\xi}{f + \xi} \left(\frac{h_{nc}\delta_{1,\max}}{2c_{nc}\lambda_1}\right)^3 \frac{h_{nc}}{h} + \frac{4}{6} \frac{1}{f_{fc}} \frac{f}{f + \xi} \left(\frac{h_{fc}\delta_{1,\max}}{2\lambda_1 c_{fc}}\right) \frac{h_{fc}}{h}\right] \dot{q}''_{FB}.
\end{aligned} \tag{11.69}$$

The film thickness at the upper end of the integration region is determined by Eq. (11.44).

For the case of the natural convection Eq. (11.69) reduces to

$$\dot{q}''_{3\sigma 1} = \frac{1}{2} Sp \frac{1}{f_{nc}} \frac{f + 3\xi}{f + \xi} \dot{q}''_{FB}, \tag{11.70}$$

and for the case of predominant forced convection to

$$\dot{q}''_{3\sigma 1} = \frac{1}{3} Sp \frac{1}{f_{fc}} \frac{f}{f + \xi} \dot{q}''_{FB}. \tag{11.71}$$

It is evident that the energy transferred to heat up the vapor film is proportional to the energy transferred from the wall to the liquid interface by conduction.

The evaporation mass flow rate is expressed by

$$(\rho w)_{21} = \left(\dot{q}''_{FB} + \dot{q}''_{r\sigma 2} - \dot{q}''_{\sigma 2} - \dot{q}''_{3\sigma 1}\right) / \Delta h . \tag{11.72}$$

For natural convection the following is obtained

$$(\rho w)_{21} = \left(1 + r_r - \frac{C_2}{\xi} - \frac{1}{2} Sp \frac{1}{f_{nc}} \frac{f + 3\xi}{f + \xi}\right) \frac{\dot{q}''_{FB}}{\Delta h} . \tag{11.73}$$

This equation contains an interesting item of information. The condition $(\rho w)_{21} = 0$ leads to a transcendental equation with respect to superheat, as r_r and ξ are also functions of ΔT_{sp} .

The assumption that there is a linear temperature profile is not valid for higher degrees of superheat, because heating of the vapor will first bring the liquid interface to a certain average temperature but will then require more energy than could be supplied at the interface. Under such conditions only radiation heat flux has any influence on film boiling.

Thus, a complete analytical description of the heat-transfer mechanism for film boiling is now available.

11.1.8 Effect of the interfacial disturbances

The interfacial disturbances enhance the heat transfer from the interface to the liquid. This effect was recognized by *Coury* and *Dukler* (1970), *Suryanarayana* and *Merte* (1972) and discussed by *Analytis* and *Yadiragoglu* (1985) for vertical surfaces. More recent works by *Bui* and *Dhir* (1985), *Nigmatulin* et al. (1994), *Nishio* and *Ohtake* (1993) substantiate this finding. The effect is:

(a) As is evident from Fig. 11.1b, the interfacial instabilities do not allow the film to become very thick. The series of crests indicates that the characteristic length of the process is proportional to the instability wavelength λ_{KH} , rather than to the height of the heated surface so long as $\Delta z > \lambda_{KH}/2$. The classical *Kelvin–Helmholtz* instability analysis applied to this problem defines this instability wavelength by the formula

$$\lambda_{KH} = 2\pi \sqrt{\frac{2\sigma\delta_{1,z=\lambda_{KH}}}{\rho_1\left(\langle w_1\rangle_1 - w_2\right)^2}} = 2\pi \sqrt{\frac{2\sigma\delta_{1,z=\lambda_{KH}}}{\rho_1\left(\frac{\Delta\rho g \delta^2_{1,z=\lambda_{KH}}}{12\eta_1}\frac{f+4\xi}{f+\xi} - \frac{1}{2}\frac{f+2\xi}{f+\xi} w_2\right)^2}} , \tag{11.74}$$

which has to be solved together with Eq. (11.43) for

$$z = \Delta z = \lambda_{KH}/2 . \tag{11.75}$$

(b) We expect that the theory as presented up to this point will underpredict the influence of the subcooling. The instabilities at the surface produce periodical bubble entrainment that causes turbulent mixing in the liquid boundary layer and therefore additional enhancement of heat transfer into the liquid.

This effect can be included in the model by modifying the liquid thermal boundary layer thickness as follows. In place of C_2 the following is used

$$C_2^* = C_2\left(1 + \frac{2}{5}Sp\right). \tag{11.76}$$

This is valid for

$$\left(1 + r_r\right)\xi / C_2 > 1, \tag{11.77}$$

with the boundary-layer thickness left unchanged for all other conditions. For small degrees of superheating this modification is not important; it only starts to influence the computed heat transfer into the liquid in the case of higher degrees of superheat. The data comparison given in the next section provides the justification for this modification.

11.1.9 Comparison of the theory with the results of other authors

Equation (11.58) provides a new general method for layer-thickness ratio computation. Equation (11.45) provides a new general method for computation of the averaged film boiling heat-transfer coefficient for an arbitrary state of the liquid.

It is worth noting that for the case of natural convection the cubic Eq. (11.59) is obtained for the layer thickness ratio just as do *Sakurai* (1990), and *Liu* and *Theofanous* (1995), but with different coefficients.

Comparing Eq. (11.28) with Eqs. (11.68) and (11.64) the following is obtained

$$f_{nc} = 1 \Bigg/ \left[\left(1 + \frac{\dot{q}''_{r\sigma 2} - \dot{q}''_{\sigma 2}}{\dot{q}''_{FB}}\right)\frac{f + \xi}{f + 4\xi} - \frac{\dot{q}''_{3\sigma 1}}{\dot{q}''_{FB}}\right] \tag{11.78}$$

or

$$f_{nc} \approx 1 \Bigg/ \left(1 + \frac{\dot{q}''_{r\sigma 2} - \dot{q}''_{\sigma 2} - \dot{q}''_{3\sigma 1}}{\dot{q}''_{FB}}\right). \tag{11.79}$$

This is a very significant equation indicating a simple way to take into account the effects of vapor heating, liquid subcooling and radiation. As a result, Eq. (11.40) can be written in the form

$$\frac{\dot{q}''_{nc}}{\dot{q}''_{nc,0}} \approx \left(1+\frac{\dot{q}''_{r\sigma 2}-\dot{q}''_{\sigma 2}-\dot{q}''_{3\sigma 1}}{\dot{q}''_{nc}}\right)^{-1/4}, \tag{11.80}$$

where

$$\dot{q}''_{nc,0} = c_{nc}\left[\frac{\lambda_1^3 \rho_1 \Delta h \Delta \rho g}{\Delta T_{sp} \eta_1 \Delta z}\right]^{1/4} \Delta T_{sp} . \tag{11.81}$$

Similarly for the case of forced convection the following is obtained

$$\frac{\dot{q}''_{fc}}{\dot{q}''_{fc,0}} \approx \left(1+\frac{\dot{q}''_{r\sigma 2}-\dot{q}''_{\sigma 2}-\dot{q}''_{3\sigma 1}}{\dot{q}''_{fc}}\right)^{-1/2}, \tag{11.82}$$

where

$$\dot{q}''_{fc,0} = c_{fc}\left[\frac{\rho_1 \lambda_1 \Delta h w_2}{\Delta T_{sp} \Delta z}\right]^{1/2} \Delta T_{sp} , \tag{11.83}$$

is the heat flux for predominant forced convection and saturated liquid.

The notation can now be generalized as follows

$$\frac{\dot{q}''_{FB}}{\dot{q}''_{FB,0}} \approx \left(1+\frac{\dot{q}''_{r\sigma 2}-\dot{q}''_{\sigma 2}-\dot{q}''_{3\sigma 1}}{\dot{q}''_{FB,0}}\frac{\dot{q}''_{FB,0}}{\dot{q}''_{FB}}\right)^{-1/m} . \tag{11.84}$$

The value $m = 2$ applies for predominant forced convection, and $m = 4$ for natural convection. Note that for $\left(\dot{q}''_{r\sigma 2}-\dot{q}''_{\sigma 2}-\dot{q}''_{3\sigma 1}\right)/\dot{q}''_{FB} >> 1$ we have $\dot{q}''_{FB}/\dot{q}''_{FB,0} \approx \left[\left(\dot{q}''_{r\sigma 2}-\dot{q}''_{\sigma 2}-\dot{q}''_{3\sigma 1}\right)/\dot{q}''_{FB,0}\right]^{-1/(m-1)}$. This describes the way in which heat transfer from the surface to the vapor and from the interface to the liquid enhances the film boiling heat-transfer component.

It is common practice to incorporate the surface-vapor term into a modified term for latent heat of vaporization. The natural-convection film boiling heat-transfer coefficient can thus be expressed with the modified Eq. (11.34a) as follows

$$h_{nc} = c_{nc}\left[\frac{\lambda_1^3 \rho_1 \Delta h^* \Delta \rho g}{\Delta T_{sp} \eta_1 \Delta z} f_{nc}\right]^{1/4}, \tag{11.34a}$$

where

$$\Delta h^* = \Delta h\left\{(f+4\xi)/(f+\xi)+0.5\left[(f+3\xi)/(f+\xi)\right]Sp\right\} \approx \Delta h(1+0.5Sp) . \tag{11.85}$$

It is evident that this practice yields a good approximation of the general result as the ratio $\dot{q}''_{3\sigma 1} / \dot{q}''_{FB}$ is almost constant for given thermal conditions. The constant 0.5 has been also used by *Sakurai* (1990). Some authors have used other values instead of 0.5: 0.4 – *Bromley* (1952), 0.6 – *Frederking* et al. (1965) or $0.84 / \Pr_l$ – *Sparrow* (1964). If these values are compared with Eq. (11.85) it is evident that the correct value for the linear temperature profile in the vapor is very close to a constant value of 0.5. For degrees of superheat exceeding 700 K at atmospheric pressure the term $\dot{q}''_{3\sigma 1} / \dot{q}''_{FB}$ is larger than 0.4, and cannot be neglected. The modification of the latent heat of vaporization as discussed here simplifies the general equation to

$$\frac{\dot{q}''_{FB}}{\dot{q}''_{FB,0}} \approx 1 \Bigg/ \left(1 + \frac{\dot{q}''_{r\sigma 2} - \dot{q}''_{\sigma 2}}{\dot{q}''_{FB,0}} \frac{\dot{q}''_{FB,0}}{\dot{q}''_{FB}} \right)^{1/m} . \tag{11.86}$$

For predominant forced convection (m = 2 in Eq. (11.84)) the following is then obtained

$$\frac{\dot{q}''_{FB}}{\dot{q}''_{FB,0}} = \left[1 + \frac{1}{4} \left(\frac{\dot{q}''_{r\sigma 2} - \dot{q}''_{\sigma 2}}{\dot{q}''_{FB,0}} \right)^2 \right]^{1/2} - \frac{1}{2} \frac{\dot{q}''_{r\sigma 2} - \dot{q}''_{\sigma 2}}{\dot{q}''_{FB,0}} . \tag{11.87}$$

This equation can be solved numerically for natural convection. As already mentioned, the effect of subcooling or superheating is to either reduce or enhance the effect of radiation.

For the case of natural circulation, film boiling Eq. (11.86) reduces to an equation that was in fact already intuitively introduced by *Bromley*, see Eq. (9) in *Bromley* (1950), who considered it as "approximate". It was shown by *Sparrow* [(1964), p 234] that for natural-convection film boiling use of Eq. (11.85) permits the effect of the radiation on the conductive heat flux to be predicted to within a few percent of the accuracy of the exact method. This means that Eq. (11.86) is a very practical approach for the general case of film boiling with radiation in subcooled or superheated liquid as demonstrated in *Kolev* (1995).

11.1.10 Verification using the experimental data

Experimental data for averaged heat-transfer coefficients were obtained at the Royal Institute of Technology in Sweden by *Okkonen* et al. (1996). The data are for vertical heated walls with $\Delta z = 1.5$ m. Water at atmospheric pressure and subcooling of ΔT_{sc} =3 to 42 K, was used as a coolant. The wall superheat was considerable: ΔT_{sp} =487 to 1236 K. The heat-transfer coefficient was defined by *Okkonen* with respect to the wall–liquid temperature difference. This is recomputed here with respect to wall-saturation temperature difference for comparison purposes. Prediction is performed using Eqs. (11.75), (11.64), (11.30) and (11.34a). A comparison of the predicted data with the 43 data points gained by experiment is given in Fig. 11.2. It is evident that all data are predicted inside the error band of $\pm 10\%$, with a mean averaged error of less than $\pm 4\%$ when the modified theory expressed by Eq. (11.76) is used.

It is interesting to note that for this data we have $f \approx 25$ to 39.6, $\xi \approx 0.35$ to 2.3, $f_{nc} \approx 4.2$ to 205.

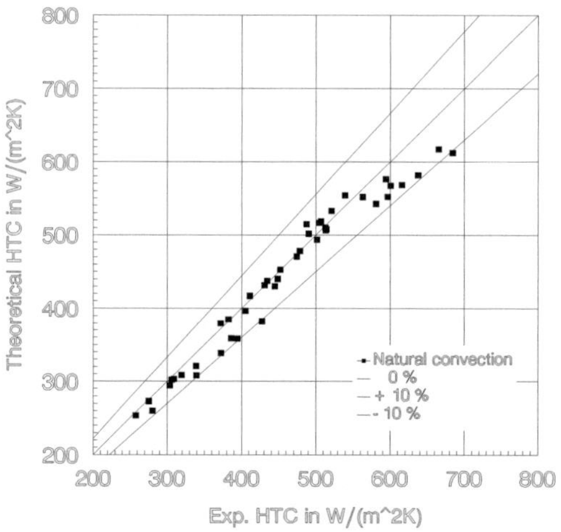

Fig. 11.2 Comparison between the film boiling model and the data base of the Royal Institute of Technology (KTH – Sweden) – *Okkonen* et al. *Vaeth's* radiation model (FzK – Karlsruhe) incorporated. $\Delta z = 1.5m$ vertical wall. Water at atmospheric pressure and subcooling $\Delta T_{sc} = 3 - 42K$. Wall superheat $\Delta T_{sp} = 487 - 1236K$

The method was analytically validated for a second time by *Vierow* (1999) and extended to hot jet heat transfer.

11.1.11 Conclusions

The approximate mathematical description proposed in this work relates to mixed-convection film boiling on a vertical heated plate, in saturated or subcooled liquid. It takes into account the effect of radiation and leads to a new analytical solution. This solution compares well with the experimental data for natural-convection film boiling. The final result is recommended for practical application.

11.1.12 Practical significance

An analytical solution of the mass-, momentum- and energy-conservation equations describing mixed-convection film boiling on vertical walls is now available. From this solution expressions are derived for the particular components of the removed heat that cause vapor heating, liquid heating, for evaporation mass flow rate, and for average vapor volume attached to the surface. This is a step forward in increasing the accuracy of the overall flow-field prediction for melt interacting with water in severe accident analysis in industry. Design calculations for vertical surfaces cooled by film boiling can now be implemented with an accuracy of about 10%.

11.2 Sphere

11.2.1 Introduction

In Section 11.1 the method was described that led to a successful analytical description of film boiling on vertical plates. Here, this method is extended to a method for description of mixed-convection film boiling at spheres.

11.2.2 Problem definition

Consider the geometry presented in Fig. 11.3. A hot sphere is flowed by liquid similarly to the case presented in Fig. 11.1. All geometry definitions are similar to those of the plate.

The problem to be solved is: compute all components of the heat flux emitted from the wall as indicated in Fig. 11.3.

11.2.3 Solution method

Encouraged by the successful model development for plane geometry we have repeated the same procedure for the spherical geometry. In addition to the assumption of a linear temperature profile in the liquid we considered also the case for a quadratic profile. The results are analytical solutions for natural and forced convection, respectively. A closed analytical solution for mixed convection could not be obtained. Therefore, interpolation between both solutions was necessary to describe the mixed convection. The final result was more complicated than the plane solution. Data comparison shows that several empirical corrections are necessary to bring the model prediction within a 30% error band compared to experiments. An ad hoc application of the plane solution to spherical geometry was much better. Therefore, we decided to modify the plane solution appropriately and to verify the solution on the data available. The motivation for this study comes from the *Meyer* and *Schumacher* experiment (1996).

11.2.4 Model

The interfacial area per unit volume of the flow

$$\frac{F_3}{Vol} = \frac{6\alpha_3}{D_3} \tag{11.88}$$

is necessary to compute power densities from heat fluxes at the surface. At the beginning of the computation we assume that the surface temperature is equal to the particle-averaged temperature $T_{\sigma 3} = T_3$. This assumption can be improved by repeating the procedure described below.

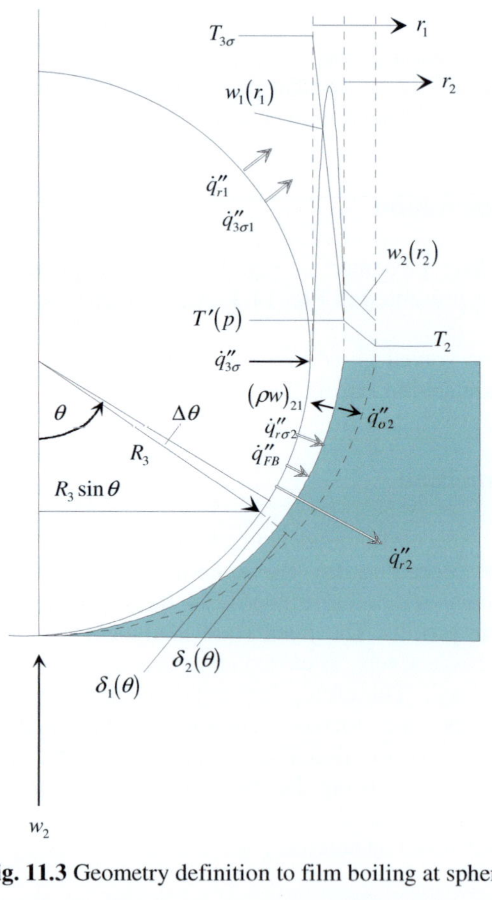

Fig. 11.3 Geometry definition to film boiling at sphere

11.2.4.1 Heat flux from the liquid interface into the liquid

In the case of subcooling there is a heat transfer from the water interface into the bulk of the water. We compute the *Nusselt* number for the natural convection only using the *Achenbach* correlation (1993) verified for $0 < Gr_2 Pr_2 < 5\times10^6$

$$Nu_{\sigma2,nc} = 3.71 + 0.402\left(Gr_2 Pr_2\right)^{1/2}, \tag{11.89}$$

where the *Grashof* number for a sphere is approximated as given in the Nomenclature. Interface turbulence due to film boiling is taken into account

$$Nu_{\sigma2,nc} = 2.1\left(1+15Sp^{*3}\right)\left[3.71 + 0.402\left(Gr_2 Pr_2\right)^{1/2}\right] \tag{11.90}$$

by introducing a multiplier depending on the superheating. The forced convection *Nusselt* number is computed by slightly modifying the method of *Gnielinski* (1978)

$$Nu_{\sigma2,fc} = \max\left(Nu_{\sigma2,fcl}, Nu_{\sigma2,fct}\right), \tag{11.91}$$

where

$$Nu_{\sigma2,fcl} = 0.664 Re_2^{1/2} Pr_2^{1/3}, \tag{11.92}$$

$$Nu_{\sigma2,fct} = 0.037\frac{Re_2^{0.8} Pr_2}{1+\dfrac{2.443}{Re_2^{0.1}}\left(Pr_2^{2/3}-1\right)}. \tag{11.93}$$

This method is verified for $Re_2 < 2\times10^4$, $0.7 < Pr_2 < 10^4$. *Achenbach* (1993) reported data that demonstrate the validity of the above approach up to $Re_2 < 7.7\times10^5$ and pressure up to 40 bar. Interface turbulence due to film boiling is taken into account by introducing a multiplier depending on the superheating

$$Nu_{\sigma2,fc} = 1.5\left(1+5Sp^{*3}\right)Nu_{\sigma2,fc}. \tag{11.94}$$

Again, as in the case of natural convection, the interface turbulence due to film boiling is taken into account by introducing an empirical multiplier $1.5\left(1+5Sp^{*3}\right)$ depending on the superheating. The coefficients are selected by comparison with experimental data, as will be discussed later.

For $Re_2 < 0.001$

$$Nu_{\sigma2} = Nu_{\sigma2,nc}, \tag{11.95}$$

otherwise

$$Nu_{\sigma2} = \max\left(Nu_{\sigma2,nc}, Nu_{\sigma2,fc}\right). \tag{11.96}$$

The thermal power transferred from the interface into the bulk per unit flow volume is therefore

$$\dot{q}'''_{\sigma 2} = \omega_{\sigma 2}\left(T' - T_2\right), \tag{11.97}$$

$$\omega_{\sigma 2} = C_{gas}\frac{F_3}{Vol}\frac{\lambda_2}{D_3}Nu_{\sigma 2}. \tag{11.98}$$

Here, the multiplier

$$C_{gas} = \left(\frac{\alpha_2}{\alpha_1 + \alpha_2}\right)^{1/3} \tag{11.99}$$

takes into account the presence of vapor in the flow outside the film. The power 1/3 rather the proposal by *Liu* and *Theofanous* (1995), ¼, is used in Eq. (11.99) for better reproduction of the experimental data for film boiling in two-phase flow with this particular model.

11.2.4.2 Radiation from the sphere

For very high temperatures radiation cannot be neglected. This is a very important topic that deserves separate and careful discussion. We will confine ourselves to referring to the formalism provided by *Lanzenberger* (1997), which gives three radiation components from the particle interface, being functions of the volumetric fractions, partial vapor pressure and the temperatures of the emitter and receivers

$$\left[\dot{q}'''_{r1}, \dot{q}'''_{r2}, \dot{q}'''_{r\sigma 2}\right]_L = \mathrm{f}\left(\alpha_1, \alpha_2, \alpha_3, p_M, T_1, T_2, T_{3\sigma}\right). \tag{11.100}$$

For convenience we introduce a splitting factor for correction of the radiation heat redistribution in water due to surface instability based on the increase of the surface absorption

$$\dot{q}'''_{r\sigma 2} = \left[\dot{q}'''_{r\sigma 2}\right]_L + c_{split}\left[\dot{q}'''_{r2}\right]_L, \tag{11.101}$$

$$\dot{q}'''_{r2} = \left(1 - c_{split}\right)\left[\dot{q}'''_{r2}\right]_L, \tag{11.102}$$

where

$$c_{split} = 0. \tag{11.103}$$

For superheated liquid

$$T_2 > T' \tag{11.104}$$

all the radiation reaching the liquid is deposited at the liquid interface, facilitating an additional spontaneous flashing condition

$$\dot{q}'''_{r\sigma 2} = \left[\dot{q}'''_{r\sigma 2}\right]_L + \left[\dot{q}'''_{r2}\right]_L \tag{11.105}$$

$$\dot{q}'''_{r2} = 0 . \tag{11.106}$$

For practical application, *Vaeth's* model (1995) in IVA4 was replaced by the new model developed by *Lanzenberger* (1997). The new model solves correctly the radiation transport equation for particular geometries in the multiphase flows without making use of some simplifying assumptions made by *Vaeth*. One of the results of the new model is, for instance, less radiation energy adsorbed in the steam. This is not significant for the discussed case, but in large-scale melt–water interaction is of importance. The comparisons with experimental data reported here are performed with the model of *Lanzenberger*.

11.2.4.3 Film boiling without radiation in saturated liquid

Here, we consider film boiling without radiation in saturated liquid. In Section 11.2.4.3.4 the prediction of the model will be extended to take into account radiation and subcooling.

11.2.4.3.1 Natural convection only

The heat-transfer coefficient for natural-convection film boiling is computed by using a somewhat modified plane solution

$$h_{nc,1} = \frac{\lambda_1}{D_3} Nu_{nc,1} \tag{11.107}$$

where

$$Nu_{nc,1} = 0.7 f\left(\bar{D}_3\right)\left(f_{nc} Gr_1 Pr_1 / Sp\right)^{1/4}, \tag{11.108}$$

$$f_{nc} = \frac{f + 4\xi}{f + \xi}\left(1 + \frac{1}{2} Sp \frac{f + 3\xi}{f + 4\xi}\right), \tag{11.109}$$

$$\xi = \left(-q + \Delta\right)^{1/3} + \left(-q - \Delta\right)^{1/3}, \quad \xi > 0, \tag{11.110}$$

$$\Delta = \sqrt{D} \quad \text{for} \quad D \ge 0, \tag{11.111}$$

$$D = q^2 + p^3, \tag{11.112}$$

$$q = -\frac{1}{3}\frac{4}{27}\frac{1 + \frac{1}{2} Sp}{C_1}, \tag{11.113}$$

$$p = -\frac{1}{3}\frac{8}{81}\frac{4 + \frac{3}{2} Sp}{fC_1}. \tag{11.114}$$

From the comparison of the data by *Liu* and *Theofanous* (1995) for saturated liquid and different particle sizes we obtain correction-function values around one.

$\bar{D}_3$	$f\left(\bar{D}_3\right)$
2.535	0.995
3.806	0.952
5.070	0.939
7.624	1.031

The natural instability length scale, the *Rayleigh–Taylor* wavelength, is used to normalize the particle size

$$\bar{D}_3 = D_3 / \lambda_{RT} . \tag{11.115}$$

For smaller values of $\bar{D}_3$ of 2.535 we use $\bar{D}_3 = 0.995$. Finally, the thermal power transferred by conduction through the vapor film per unit flow volume is

$$\dot{q}'''_{FB,nc} = \frac{F_3}{Vol} h_{nc,1} \left(T_{3\sigma} - T'\right). \tag{11.116}$$

Assuming that there is a vapor temperature profile automatically leads to the conclusion that there is some thermal power transfer into the vapor. We approximate this part

$$\dot{q}'''_{31,nc} = 0.3 Sp \frac{f + 3\xi}{f_{nc}\left(f + \xi\right)} \dot{q}'''_{FB,nc} , \tag{11.117}$$

For $q^2 + p^3 < 0$ the layers thickness ratio is theoretically not defined. If using minimum film boiling theory strictly consistent with this work no further discussion is necessary. If not, e.g., empirical correlation for minimum film boiling is used, then we set

$$\dot{q}'''_{31,nc} = 0.3\ Sp\ \dot{q}'''_{FB,nc} , \tag{11.118}$$

and $f_{nc} = 1 + \frac{1}{2} Sp$.

11.2.4.3.2 Predominant forced convection

In an earlier development *Vaeth* (1995) used for this case the *Bromley–Witte* theory (1953, 1968),

$$\dot{q}'''_{FB,fc} = \frac{F_3}{Vol} h_{fc,1} \left(T_{3\sigma} - T'\right), \tag{11.119}$$

where

$$h_{fc,1} = \frac{\lambda_1}{D_3} 2.98 \left(\frac{Re_1}{Sp} \right)^{1/2} . \quad (11.120)$$

The comparison with the forced-convection film boiling data by *Liu* and *Theofanous* (1995) resulted in replacing this model by the model of *Epstein–Heuser* (1980):

$$h_{fc,1} = \frac{\lambda_1}{D_3} 0.60 \left(\frac{\rho_2}{\rho_1} \right)^{1/4} Re_1^{1/2} / Sp^{*1/4} \quad (11.121)$$

in which the constant is changed from 0.65 to 0.6. Again, thermal power transfer into the vapor is approximated for $D < 0$ by

$$\dot{q}'''_{31,fc} = 0.05\, Sp\, \dot{q}'''_{FB,fc} \quad (11.122)$$

otherwise

$$\dot{q}'''_{31,fc} = 0.05\, Sp\, \frac{f}{f_{fc}\left(f + \xi\right)} \dot{q}'''_{FB,fc} . \quad (11.123)$$

Here,

$$f_{fc} = \frac{f}{f + \xi} \left(1 + \frac{1}{3} Sp \right) \quad (11.124)$$

is computed as Eq. (11.31) setting, $C_2 = 0$, $r_r = 0$, and ξ is computed as for natural convection.

11.2.4.3.3 Mixed convection

Mixed convection film boiling is known to exist in the region $Fr \ge 0.2$. For $Fr < 0.2$ the natural-convection-only solution

$$\dot{q}'''_{FB,0} = \dot{q}'''_{FB,nc} , \quad (11.125)$$

$$\dot{q}'''_{31} = \dot{q}'''_{31,nc} , \quad (11.126)$$

is the appropriate one. Otherwise, the approximation between natural- and forced-convection solutions is performed following *Liu* and *Theofanous* (1995)

$$\dot{q}'''_{FB,0} = \dot{q}'''_{FB,nc} \left[1 + \left(\frac{\dot{q}'''_{FB,fc}}{\dot{q}'''_{FB,nc}} \right)^5 \right]^{1/5} , \quad (11.127)$$

$$\dot{q}'''_{31} = \dot{q}'''_{31,nc} \left[1 + \left(\frac{\dot{q}'''_{31,fc}}{\dot{q}'''_{31,nc}} \right)^5 \right]^{1/5} . \quad (11.128)$$

Note that in the previous code development, *Vaeth* (1995), we used

$$\dot{q}'''_{FB,0} = \max\left(\dot{q}'''_{FB,nc}, \dot{q}'''_{FB,fc}\right).$$

11.2.4.3.4 Feedback of the radiation, subcooling and vapor heating on the vapor film conduction heat transfer

The experience with the plate solution is used here as a guideline. First, we define the ratios

$$r_r = \frac{\left|\dot{q}'''_{r\sigma 2}\right| - \dot{q}'''_{\sigma 2} - \dot{q}'''_{31}}{\dot{q}'''_{FB,0}}, \qquad (11.129)$$

$$r_r^* = \frac{1}{2} r_r, \qquad (11.130)$$

influencing the film conduction through influencing the film thickness. An analytical solution of Eq.(11.84) is available for forced convection only,

$$x_{fc} = \sqrt{1 + r_r^{*2}} - r_r^*. \qquad (11.131)$$

For natural convection Eq. (11.84) rewritten in the following form

$$x_{nc}^4 + r x_{nc}^3 + 1 = 0 \qquad (11.132)$$

has to be solving by iteration. We start with the value for forced convection and iterate up to satisfying the equation with residuals less than 0.0001. Finally, we take the maximum of both values.

$$x = \max\left(x_{fc}, x_{nc}\right), \qquad (11.133)$$

$$\dot{q}'''_{FB} = C_{gas} x \dot{q}'''_{FB,0}. \qquad (11.134)$$

The total thermal energy emitted by the interface of the particle is therefore

$$\dot{q}'''_{\sigma 3} = -\dot{q}'''_{FB} - \dot{q}'''_{31} - \dot{q}'''_{r1} - \dot{q}'''_{r2} - \dot{q}'''_{r\sigma 2}. \qquad (11.135)$$

For implementation in computer codes it is important to check whether there is enough energy that has to be exhausted from the particles. The maximum of this energy can be defined as the energy required to heat up the particle from the liquid temperature to the actual particle temperature within the computational time step,

$$\dot{q}'''_{\sigma 3,\max} = \alpha_3 \rho_3 c_{p3} \left(T_{3\sigma} - T_2\right) / \Delta\tau. \qquad (11.136)$$

Should the ratio $f_{\max} = \dot{q}'''_{\sigma 3,\max} / \dot{q}'''_{\sigma 3}$ be larger than 1 the time step has to be corrected. If the time step is small enough we correct all components as follows $\dot{q}'''_{fb} = \dot{q}'''_{fb} f_{\max}$, $\dot{q}'''_{31} = \dot{q}'''_{31} f_{\max}$, $\dot{q}'''_{r1} = \dot{q}'''_{r1} f_{\max}$, $\dot{q}'''_{r2} = \dot{q}'''_{r2} f_{\max}$, $\dot{q}'''_{r\sigma 2} = \dot{q}'''_{r\sigma 2} f_{\max}$.

Finally, the total film boiling heat-transfer coefficient is

$$h_{FB} = -\dot{q}'''_{\sigma 3} / \left[\frac{F_3}{Vol} \left(T_3 - T' \right) \right]. \tag{11.137}$$

This definition is used for comparison with the experimental data. The energy crossing the liquid interface and entering the water bulk is composed of radiation and convective components

$$\dot{q}'''_{\sigma 2,total} = \dot{q}'''_{r2} + \dot{q}'''_{\sigma 2}. \tag{11.138}$$

Similarly, the energy transferred from the particle into the gas film is composed by radiation and convective parts

$$\dot{q}'''_{3\sigma 1} = \dot{q}'''_{r1} + \dot{q}'''_{31}. \tag{11.139}$$

The energy-jump condition at the liquid interface gives for the evaporated mass per unit time and unit flow volume the following expression

$$\mu_{21} = \frac{\dot{q}'''_{r\sigma 2} + \dot{q}'''_{FB} - \dot{q}'''_{\sigma 2}}{h'' - h'}. \tag{11.140}$$

11.2.4.3.5 The interface temperature

The sphere interface temperature during film boiling may differ from the particle-averaged temperature, especially for large particles of poor heat conductors. The limitation comes from the fact that only a limited amount can be supplied from the particle bulk to the particle surface. The analytical solution of the *Fourier* equation is the tool used here to estimate the interface temperature. The temperature profile is then averaged to estimate the averaged temperature and therefore the averaged temperature change during the time step. The task is to estimate the possible energy transfer due to the conditions during the time step considered and to compare it with that computed above. The final result is

$$T_{3\sigma} = T_3 + \dot{q}'''_{\sigma 3} / \omega_3, \tag{11.141}$$

where

$$\omega_3 = \alpha_3 \rho_3 c_{p3} \frac{\Delta \overline{T}_{3\sigma}}{\Delta \tau}. \tag{11.142}$$

The averaged temperature difference is

$$\Delta \overline{T}_{3\sigma} = 1 - \frac{6}{\pi^2} \sum_{n=1}^{\infty} \frac{1}{n^2} / \exp\left(n^2 \frac{\Delta \tau}{\Delta \tau_3} \right), \tag{11.143}$$

where

$$\Delta\tau_3 = \frac{D_3^2}{4\pi^2 a_3} \tag{11.144}$$

is the time constant of the heat conduction into the spherical particle. The converging *Fourier* series is computed using n as big as necessary (but no larger then 100) to obtain an improvement in the summation of less than 0.00001. Because $\dot{q}'''_{\sigma 3}$ is a function of the interface temperature itself some iterations may be required to finally obtain the released power density.

11.2.4.3.6 Averaged film thickness in the multiphase flows

To demonstrate the importance of the averaged film thickness in the multiphase analysis, consider the question of how much vapor is associated in the film and how much is in bubble form. The heat- and mass-transfer processes at the particle interface are completely described in the case of film boiling by the method presented already. The remaining part in bubble form is subject to additional water/vapor heat and mass transfer that will not be considered here. Analysis of the QUEOS *Kolev* (1996) experiments demonstrates that at the initial phase of penetration of hot particles in water almost the entire generated vapor is in the film. This means no additional vapor/water heat and mass transfer has to be modeled. For the later phase of interaction it starts to be active.

We approximate the averaged film thickness as follow. For $\Delta V_{23} < 0.001$ we have natural-circulation film boiling leading to

$$\delta_{1F} = \delta_{1F,nc} = \left[\frac{2D_3^2 \mu_{21} \eta_1}{\alpha_3 g \rho_1 (\rho_2 - \rho_1)}\right]^{1/3}, \tag{11.145}$$

otherwise

$$\delta_{1F} = \delta_{1F,fc} = \frac{2}{9}\frac{D_3^2 \mu_{21}}{\alpha_3 \rho_1 \Delta V_{23}}, \tag{11.146}$$

with the limitation

$$\delta_{1F} = \min\left(\delta_{1F,nc}, \delta_{1F,fc}\right). \tag{11.147}$$

11.2.5 Data comparison

The sphere film boiling model presented here is implemented in the computer code IVA5. More information about the history of this development and the recent state is available in *Kolev* (1995) and *Kolev* (1996). This model is used for the data comparisons presented below. The new experimental data base provided by *Liu* and *Theofanous* in 1995 consists of single- and two-phase flow data. The single-phase data covers subcooling from 0 to 40 K, liquid velocity from 0 to 2.3 m/s, sphere superheat from 200 to 900 K and sphere diameter from 6 to 19 mm. The two-phase data are obtained for (a) upward flow having void fraction from 0.2 to 0.65, water velocity from 0.6 to 3.2 m/s, and steam velocity from 3 to 9 m/s, and (b) for downward flow having void fraction from 0.7 to 0.95, water velocity from 1.9 to 6.5 m/s, and steam velocity from 1.1 to 9 m/s.

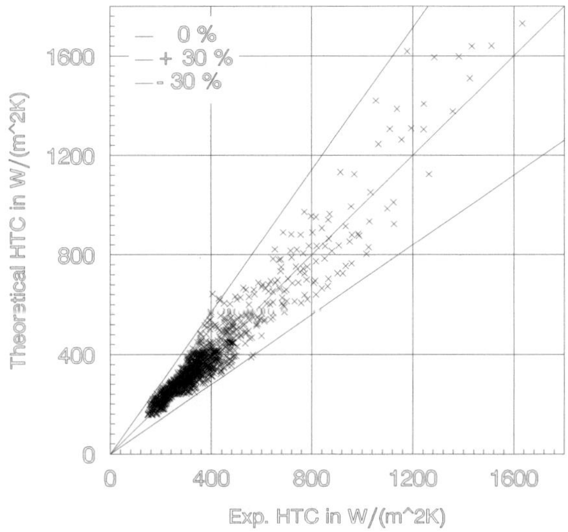

Fig. 11.4 Comparison between predicted and calculated heat-transfer coefficients for all data. Error band: ±30%

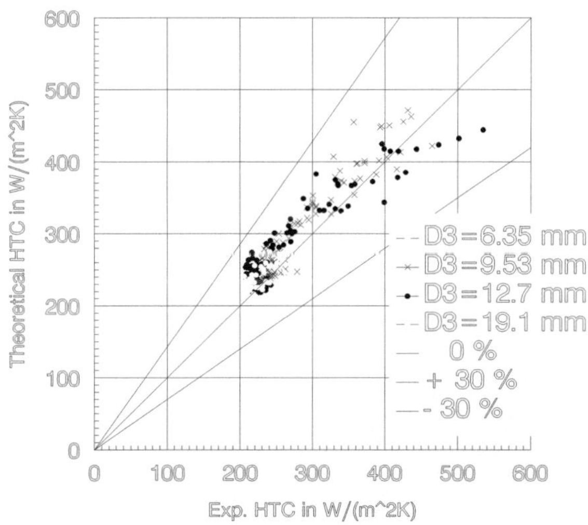

Fig. 11.5 Comparison between predicted and calculated heat-transfer coefficients for single-phase saturated water and four different sphere diameters. Error band : +20 to –15%

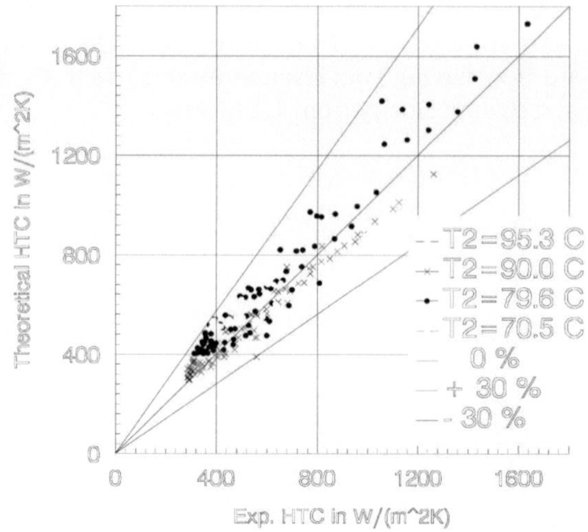

Fig. 11.6 Comparison between predicted and calculated heat-transfer coefficients for subcooled water and 12.7 mm sphere diameter. Error band: +20 to –30%.

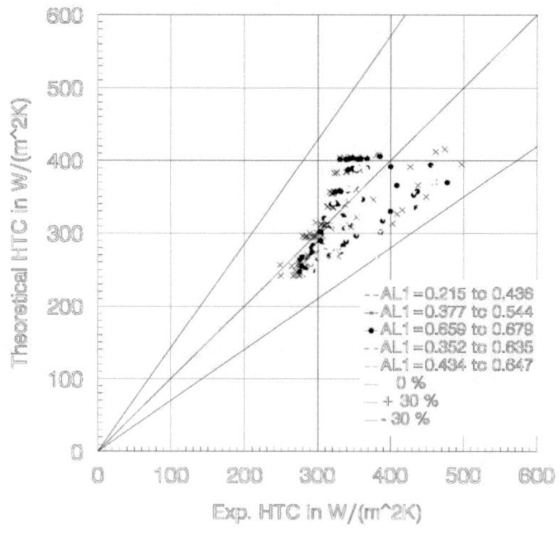

Fig. 11.7 Comparison between the predicted and calculated heat-transfer coefficients for saturated upward two phase flow, 12.7 mm sphere diameter, void fraction from 0.2 to 0.65, water velocity from 0.6 to 3.2 m/s, and steam velocity from 3 to 9 m/s. Error band: +30 to –20%

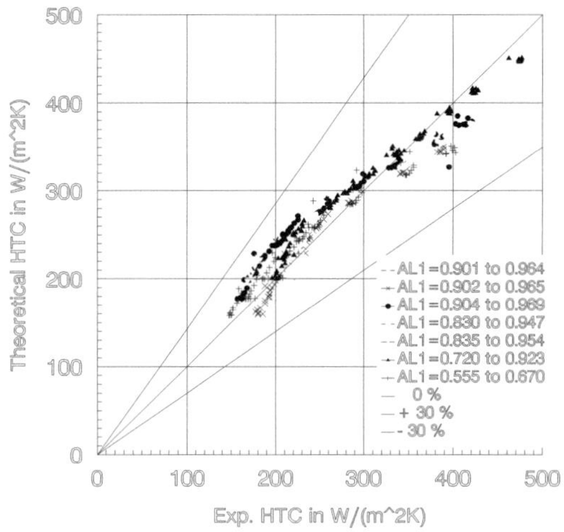

Fig. 11.8 Comparison between the predicted and calculated heat-transfer coefficients for saturated downward two phase flow, 12.7 mm sphere diameter, void fraction from 0.7 to 0.95, water velocity from 1.9 to 6.5 m/s, and steam velocity from 1.1 to 9 m/s. Error band: +20 to –20%

Figure 11.4 presents the comparison between the predicted and calculated heat-transfer coefficients for all data, *Kolev* (1997). The error band is found to be ±30% . Figure 11.5 presents the comparison between the predicted and calculated heat-transfer coefficients for single-phase saturated water and four different sphere diameters. The error band for this subset of data is found to be +20 to –15%. Figure 11.6 presents the comparison between the predicted and calculated heat-transfer coefficients for subcooled water and 12.7 mm sphere diameter. The error band for this subset of data is found to be +20 to –30%.

Figure 11.7 presents the comparison between the predicted and calculated heat-transfer coefficients for saturated upward two-phase flow, 12.7 mm sphere diameter, void fraction from 0.2 to 0.65, water velocity from 0.6 to 3.2 m/s, and steam velocity from 3 to 9 m/s. The error band for this subset of data is found to be +30 to –20%.

Figure 11.8 presents the comparison between the predicted and calculated heat-transfer coefficients for saturated downward two-phase flow, 12.7 mm sphere diameter, void fraction from 0.7 to 0.95, water velocity from 1.9 to 6.5 m/s, and steam velocity from 1.1 to 9 m/s. The error band for this subset of data is found to be +20 to –20%.

11.2.6 Conclusions

The semiempirical film boiling model used in IVA5 computer code predicts the data base provided by *Liu* and *Theofanous* within $\pm 30\%$.

Appendix 11.1 Natural convection at vertical plate

From Eq. (11.31) the local vapor film thickness as a function of the axial coordinate for natural-convection film boiling at a vertical plate is then obtained

$$\frac{1}{\delta_1} = \frac{1}{2}\left[\frac{\rho_1 \Delta h \Delta \rho g}{\lambda_1 \Delta T_{sp} \eta_1 z} f_{nc}\right]^{1/4}. \qquad \text{(A11.1-1)}$$

The film boiling heat flux averaged over the height Δz is therefore

$$\dot{q}''_{FB} = \frac{1}{\Delta z}\int_0^{\Delta z} \lambda_1 \frac{\Delta T_{sp}}{\delta_1} dz = \Delta T_{sp} \frac{1}{2}\left[\frac{\lambda_1^3 \rho_1 \Delta h \Delta \rho g}{\Delta T_{sp} \eta_1} f_{nc}\right]^{1/4} \frac{1}{\Delta z}\int_0^{\Delta z} dz / z^{1/4}$$

$$= \Delta T_{sp} \frac{2}{3}\left[\frac{\lambda_1^3 \rho_1 \Delta h \Delta \rho g}{\Delta T_{sp} \eta_1 \Delta z} f_{nc}\right]^{1/4}. \qquad \text{(A11.1-2)}$$

The heat-transfer coefficient is

$$h_{nc} = c_{nc}\left[\frac{\lambda_1^3 \rho_1 \Delta h \Delta \rho g}{\Delta T_{sp} \eta_1 \Delta z} f_{nc}\right]^{1/4}, \qquad \text{(A11.1-3)}$$

where

$$c_{nc} = 2/3. \qquad \text{(A11.1-4)}$$

Equation (A11.1-3) for $f_{nc} = 1$ reduces to the *Nusselt* equation for laminar film condensation at a cold wall *Nusselt* (1916) with correspondingly inverted subscripts.

Appendix 11.2 Predominant forced convection only at vertical plate

Consider the predominant forced-convection film boiling at a vertical plate. Neglecting the natural convection terms as small against the predominant forced convection terms in Eq. (11.31), the following is then obtained

$$\frac{1}{\delta_1} = \frac{1}{2}\left[\frac{\rho_1 \Delta h w_2}{\lambda_1 \Delta T_{sp} z} f_{fc}\right]^{1/2}. \qquad \text{(A11.2-1)}$$

The film boiling heat flux averaged over the height Δz is therefore

$$\dot{q}''_{FB} = \frac{1}{\Delta z}\int_0^{\Delta z} \lambda_1 \frac{\Delta T_{sp}}{\delta_1} dz = \Delta T_{sp} \frac{1}{2}\left[\frac{\rho_1 \lambda_1 \Delta h w_2}{\Delta T_{sp}} f_{fc}\right]^{1/2} \frac{1}{\Delta z}\int_0^{\Delta z} dz / z^{1/2}$$

$$= \Delta T_{sp}\left[\frac{\rho_1 \lambda_1 \Delta h w_2}{\Delta T_{sp} \Delta z} f_{fc}\right]^{1/2}. \qquad \text{(A11.2-2)}$$

The heat-transfer coefficient is

$$h_{fc} = c_{fc}\left[\frac{\rho_1 \lambda_1 \Delta h w_2}{\Delta T_{sp} \Delta z} f_{fc}\right]^{1/2}, \qquad \text{(A11.2-3)}$$

where

$$c_{fc} = 1. \qquad \text{(A11.2-4)}$$

Nomenclature

a_1 $= \dfrac{\lambda_1}{\rho_1 c_{p1}}$, vapor temperature diffusivity, m²/s

a_2 $= \dfrac{\lambda_2}{\rho_2 c_{p2}}$, liquid temperature diffusivity m²/s

a_3 $= \dfrac{\lambda_3}{\rho_3 c_{p3}}$, particle temperature diffusivity, m²/s

C_1 $= \dfrac{\lambda_1}{\lambda_2} \dfrac{\nu_1}{\nu_2} \dfrac{c_{p2} \Delta T_{sp}}{\Delta h}$, superheat property number, dimensionless

C_2 $= \dfrac{\lambda_2 \Delta T_{sc}}{\lambda_1 \Delta T_{sp}}$, superheat to subcooling ratio, dimensionless

C_3 $= \dfrac{\rho_2 \lambda_1}{\rho_1 \lambda_2} \dfrac{c_{p2} \Delta T_{sp}}{\Delta h} = \dfrac{a_1}{a_2} Sp$, superheat property number, dimensionless

c_{p1} vapor specific heat at constant pressure, J/(kg K)

c_{p2} liquid specific heat at constant pressure, J/(kg K)

c_{nc} numerical constant for heat-transfer coefficient for pure natural convection, dimensionless

c_{fc} numerical constant for heat-transfer coefficient for pure forced convection, dimensionless

Fr $= \dfrac{|\Delta V_{23}|}{\sqrt{gD_3}}$, *Froude* number for sphere, dimensionless

Fr_2 $= \dfrac{w_2}{\sqrt{2R_3 g}}$, *Froude* number for sphere and cylinder, dimensionless

Fr_2 $= \dfrac{w_2}{\sqrt{\Delta z g}}$, *Froude* number for vertical plane, dimensionless

f $= \eta_2 / \eta_1$, dynamic viscosity ratio, dimensionless

Gr_1 $= g(2R_3)^3 \rho_1(\rho_2 - \rho_1)/\eta_1^2$, *Grashof* number for sphere and cylinder, dimensionless

Gr_1 $= g\Delta z^3 \rho_1(\rho_2 - \rho_1)/\eta_1^2$, *Grashof* number for vertical plane, dimensionless

Gr_2 $= gD_3^3 \left|\dfrac{T' - T_2}{T_2}\right| \left(\dfrac{\rho_2}{\eta_2}\right)^2$, *Grashof* number for sphere, dimensionless

g acceleration due to gravity, m/s²

h heat-transfer coefficient, W/(m² K)

h_{nc} heat-transfer coefficient for pure natural convection, W/(m² K)

h_{fc} heat-transfer coefficient for pure forced convection, W/(m² K)

$\dot{q}''_{FB}$ heat flux due to conduction through the vapor film, W/m²

$\dot{q}''_{3\sigma 1}$ heat flux from the wall into the vapor due to convective heating of the vapor from the saturation interface temperature to an averaged vapor film temperature, W/m²

$\dot{q}''_{\sigma 2}$ heat flux from the liquid interface into the liquid due to convection, W/m²

$\dot{q}''_{r2}$ radiation heat flux from the wall into the bulk liquid, W/m²

$\dot{q}''_{3\sigma}$ total heat flux from the wall, W/m²

$\dot{q}''_{r\sigma 2}$ radiation heat flux absorbed from the vapor/liquid interface, W/m²

$\dot{q}''_{r1}$ radiation heat flux from the interface into the vapor, W/m²

Pr_1 $= \dfrac{\eta_1 c_{p1}}{\lambda_1}$, vapor *Prandtl* number, dimensionless

Pr_2 $= \dfrac{c_{p2}\eta_2}{\lambda_2}$, liquid *Prandtl* number, dimensionless

R_3 radius of sphere or cylinder, m

Re_1 $= \rho_1 \Delta w_{23} 2R_3 / \eta_1$, *Reynolds* number, dimensionless

Re_2 $= |\Delta V_{23}| D_3 \rho_2 / \eta_2$, liquid *Reynolds* numbers, dimensionless

r^* $= \left(\frac{h_{fc} c_{nc}}{h_{nc} c_{fc}} \right)^2 / 2 = \frac{2}{9} r^2$, dimensionless

r $= h_{fc} / h_{nc}$, dimensionless

r_r $= \frac{\dot{q}''_{r\sigma 2}}{\lambda_1 \Delta T_{sp} / \delta_1}$, ratio of the radiation heat flux to the conduction heat flux for plates, dimensionless

r_r $= \frac{\left|\dot{q}'''_{r\sigma 2}\right| - \dot{q}'''_{\sigma 2} - \dot{q}'''_{31}}{\dot{q}'''_{FB,0}}$, for spheres, dimensionless

r_r^* $= \frac{1}{2} r_r$, for spheres, dimensionless

r_1 distance from the wall, m

r_2 distance from the interface, m

Nu $= h 2 R_3 / \lambda_1$, *Nusselt* number for sphere and cylinder, dimensionless

Nu $= h \Delta z / \lambda_1$, *Nusselt* number for vertical plane, dimensionless

p pressure, Pa

p_M partial vapor pressure in a mixture of vapor and inert components, Pa

Sp $= c_{p1} \Delta T_{sp} / \Delta h$, superheat number, dimensionless

Sp^* $= \frac{c_{p1} \left(T_{\sigma 3} - T' \right)}{\Delta h^* \Pr_1}$, modified superheat number, dimensionless

T_2 liquid temperature, K

$T_{3\sigma}$ temperature at the wall surface, K

$T'(p)$ vapor/liquid interface temperature equal to the saturation temperature of the liquid at the system pressure, K

$w_1(r_1)$ vapor film velocity as a function of the distance from the wall, m/s

$w_2(r_2)$ liquid boundary-layer velocity as function of the distance from the interface, m/s

z axial coordinate, m

$\langle \ \rangle_1$ $= \frac{1}{\delta_1} \int_0^{\delta_1} dr_1$, cross-section average, dimensionless

α_1 gas volume fraction in three-phase flow, dimensionless

α_2 water volume fraction in three-phase flow, dimensionless

α_3 volume fraction in three-phase flow, dimensionless

ΔT_{sp} $T_{3\sigma} - T'$, wall superheat, K

ΔT_{sc} $T' - T_2$, liquid subcooling, K

Δh latent heat of evaporation, J/kg

Δh^* $= \left(h'' - h'\right)\left(1 + \frac{1}{2} Sp\right)$, modified latent heat of evaporation, J/kg
Δz total height of the wall, m
Δw_{12} vapor film–liquid velocity difference, m
$\Delta \theta$ infinitesimal change of the angular coordinate, rad
$\delta_1(z)$ vapor layer thickness as a function of the axial coordinate, m
$\delta_2(z)$ liquid boundary layer thickness as a function of the axial coordinate, m
$\delta_1(\theta)$ vapor layer thickness as a function of the angular coordinate, m
$\delta_2(\theta)$ liquid boundary layer thickness as a function of the angular coordinate, m
η_1 dynamic viscosity of vapor, Pa s
η_2 dynamic viscosity of liquid, Pa s
ρ_1 density of vapor, kg/m³
ρ_2 density of liquid, kg/m³
$(\rho w)_{21}$ evaporation mass flow rate, kg/m²
σ surface tension, N/m
ω_2 heat-transport coefficient liquid bulk/liquid surface, W/(K m³)
ω_3 heat-transport coefficient particle bulk/particle surface, W/(K m³)
θ angular coordinate, rad
λ_{KH} *Kelvin–Helmholtz* instability wavelength, m
λ_{RT} $= \sqrt{\frac{\sigma_2}{g\left(\rho_2 - \rho_1\right)}}$, *Rayleigh–Taylor* wavelength, m
ξ $= \delta_2 / \delta_1$, film thickness ratio, dimensionless

Subscripts

nc natural convection
fc forced convection
l laminar
t turbulent
1 gas
2 water
3 particles
mn from field *m* to field *n*
FB film boiling
σ surface

References

Achenbach, E.: Heat and flow characteristics of packed beds. In: Kelleher, M.D., et al. (eds.) Experimental Heat Transfer, Fluid Mechanics and Thermodynamics, pp. 287–293. Elsevier, New York (1993)

Analytis, G.T., Yadiragoglu, G.: Analytical modeling of IAFB. In: Chiu, C., Brown, G. (eds.) Proc. of the Third Int. Topical Meeting on Reactor Thermal Hydraulics, Newport, Rhode Island, USA, October 15-18, vol. 1, paper no 2.B (1985)

Bromley, L.A.: Heat transfer in stable film boiling. Chem. Eng. Prog. 46(5), 221–227 (1950)

Bromley, L.A.: Effect of heat capacity of condensate. Indust. Eng. Chem. 44(12), 2966–2969 (1952)

Bromley, L.A., LeRoy, N.R., Rollers, J.A.: Heat transfer in forced convection film boiling. Indust. Eng. Chem. 45, 2639–2646 (1953)

Bui, T.D., Dhir, V.K.: Film boiling heat transfer on isothermal vertical surface. ASME J. Heat Transfer 107, 764–771 (1985)

Coury, G.E., Dukler, A.E.: Turbulent film boiling on vertical surfaces – A study including the interface waves. In: 4th International Heat Transfer Conference, Paper no B.3.6 (1970)

Dhir, V.K., Purohit, G.P.: Subcooled film-boiling heat transfer from spheres. Nucl. Eng. Des. 47, 49–66 (1978)

Epstein, M., Heuser, G.M.: Subcooled forced-convection film boiling in the forward stagnation region of a sphere or cylinder. Int. J. Heat Mass Transfer 23, 179–189 (1980)

Fodemski, T.R.: Forced convection film boiling in the stagnation region of molten drop and its application to vapor explosions. Int. J. Heat Mass Transfer 35, 2005–2016 (1992)

Fodemski, T.R., Hall, W.B.: Forced convection film boiling on a sphere immersed in (a) sub-cooled or (b) superheated liquid. In: Proc. 7th Int. Heat Transfer Conference, Muenchen, vol. 4, pp. 375–739 (1982)

Frederking, T.H.K., Chapman, R.C., Wang, S.: Heat transport and fluid motion during cooldown of single bodies to low temperatures. In: Timmerhaus, K.D. (ed.) International Advances in Cryogenic Engineering, pp. 353–360. Plenum Press, New York (1965)

Gnielinski, V.: Gleichungen zur Berechnung des Waerme und Stoffaustausches in durchstroemten ruhenden Kugelschuettungen bei mittleren und grossen Pecletzahlen. Verfahrenstechnik 12, 363–366 (1978)

Kolev, N.I.: Film boiling on vertical plates and spheres. Exp. Thermal Fluid Sci. 18, 97–115 (1998)

Kolev, N.I.: Verification of the IVA4 film boiling model with the data base of Liu and Theofanous. In: Proceedings of OECD/CSNI Specialists Meeting on Fuel-Coolant Interactions (FCI), JAERI-Tokai Research Establishment, Japan, May 19-21 (1997)

Kolev, N.I.: IVA4 Computer code: An universal analyzer for multi-phase flows and its applicability to melt water interaction. In: Proc. of the Technical Committee Meeting on Advances in and Experiences with Accidents Consequences Analysis. IAEA-TC-961, Vienna, Austria, September 27-29 (1995)

Kolev, N.I.: IVA4 Computer code: The model for film boiling on a sphere in subcooled, saturated and superheated water. In: Proc. Second Int. Conference On Multiphase Flow, Kyoto, Japan, April 3-7 (1995); Presented also in Workshop zur Kühlmittel/Schmelze – Wechselwirkung, Köln, Germany, November 14-15 (1994)

Kolev, N.I.: Film boiling: vertical plates. In: Proceedings of 4th World Conference on Experimental Heat Transfer, Fluid Mechanics and Thermodynamics EXHFT 4, Brussels, Belgium, June 2-6 (1997)

Kolev, N.I.: Three fluid modeling with dynamic fragmentation and coalescence fiction or daily practice? 7th FARO Experts Group Meeting Ispra, October 15-16 (1996); Proceedings of OECD/CSNI Workshop on Transient Thermal-hydraulic and Neutronic Codes Requirements, Annapolis, MD, USA, November 5-8 (1996); 4th World Conference on Experimental Heat Transfer, Fluid Mechanics and Thermodynamics, ExHFT 4, Brussels, June 2-6 (1997); ASME Fluids Engineering Conference & Exhibition, The Hyatt Regency Vancouver, Vancouver, British Columbia, CANADA, June 22-26 (1997) (invited paper)

Lanzenberger, K.: Radiation models for multi-phase flows. Part 1: Film boiling. Siemens Report, KWU NA-M/97/E024 (1997)

Liu, C., Theofanous, T.G.: Film boiling on spheres in single- and two-phase flows. In: Part 1: Experimental Studies ANS Proceedings, Part 2: A Theoretical Study, National Heat Transfer Conference, Portland (August 1995)

Meyer, L., Schumacher, G.: QUEOS a simulation-experiment of the premixing phase of steam explosion with hot spheres in water base case experiments, Forschungs-zentrum Karlsruhe Technik und Umwelt, Karlsruhe, Wissenschaftliche Berichte FZKA 5612 (April 1996)

Nigmatulin, N.I., et al.: Interface oscillations and heat transfer mechanism at film boiling. In: Proc. of the Tenth International Heat Transfer Conference, Brighton, UK (1994)

Nishikawa, K., Ito, T.: Two-phase boundary-layer treatment of free convection film boiling. Int. J. Heat Mass Transfer 9, 103–115 (1966)

Nishikawa, K., Ito, T., Matsumoto, K.: Investigation of variable thermophysical Property problem concerning pool film boiling from vertical plate with prescribed uniform temperature. Int. J. Heat Mass Transfer 19, 1173–1182 (1976)

Nishio, S., Ohtake, H.: Vapor film unit model and heat transfer correlation for natural convection film boiling with wave motion under subcooled conditions. Int. J. Heat Mass Transfer 36(10), 2541–2552 (1993)

Nusselt, W.Z.: Z. Ver. deut. Ing. 60, 541–569 (1916)

Okkonen, T., et al.: 6, 1996) Film boiling on a long vertical surface under high heat flux and water subcooling conditions. In: Proc. of the 31st Nat. Heat Transfer Conference, Houston, Texas, August 3-6 (1996)

Sakurai, A.: Film boiling heat transfer. In: Hetstroni, G. (ed.) Proc. of the Ninth Int. Heat Transfer Conference, Jerusalem, Israel, vol. 1, pp. 157–187 (1990)

Sakurai, A., Shiotsu, M., Hata, K.: Effect of subcooling on film boiling heat transfer from horizontal cylinder in a pool water. In: Proc. 8th Int. Heat Transfer Conf., vol. 4, pp. 2043–2048 (1986)

Sparrow, E.M.: The effect of radiation on film boiling heat transfer. Int. J. Heat Mass Transfer 7, 229–238 (1964)

Suryanarayana, N.V., Merte, H.: Film boiling on vertical surfaces. J. Heat Transfer 94, 337–383 (1972)

Vaeth, L.: Radiative heat transfer for the transient three phase three component flow model IVA3, Kernforschungszentrum Karlsruhe GmbH, INR, Interner Bericht 32.21.02/10A, INR-1914, PSF-3212 (March 1995)

Vierow, K.: Integrated steam explosion analysis with VESUVIUS code. PhD thesis, University of Tokyo (July 1999)

Witte, L.C.: Film boiling from a sphere. I & EC Fundamentals 7(3), 517–518 (1968)

12. Liquid droplets

12.1 Spontaneous condensation of pure subcooled steam – nucleation

Consider steam at local temperature T_1 and local static pressure p. The saturation temperature at the system pressure is

$$T' = T'(p). \tag{12.1}$$

The saturation pressure as a function of the steam temperature is

$$p' = p'(T_1). \tag{12.2}$$

The subcooling is defined as a temperature difference

$$\Delta T_{1,sub} = T'(p) - T_1. \tag{12.3}$$

For steam temperature larger than the saturation temperature there is no condensation. One expects intuitively condensation if the local steam temperature is lower than the saturation temperature at the system pressure, that is if

$$\Delta T_{1,sub} > 0. \tag{12.4}$$

The originating nucleus releases energy at saturation temperature $T'(p)$. The released energy is then transferred into the subcooled vapor acting toward reducing subcooling and toward stopping condensation. Therefore, in nature subcooling is a necessary condition to observe condensation. The condensation itself starts around a preferred number of points in the unit volume n_3. Around these points the predominantly low-energy molecules form microscopic drops called activated clusters – nucleus with size R_3. Very small clusters are not stable and degrade spontaneously. Clusters with larger sizes may become stable. Intuitively we expect increasing origination frequency of clusters if the subcooling of the vapor is larger. We also intuitively cannot imagine that arbitrary large metastable subcooling is sustainable in nature.

In order to describe nucleation it is necessary to answer the following questions:

a) what is the nucleation size separating the stable from the unstable clusters;
b) how does nucleation site density depend on the subcooling; and
c) what is the maximum allowable subcooling in the steam.

The purpose of this section is to give answers to these questions for pure vapors.

12.1.1 Critical nucleation size

The considerations leading to estimate of the critical nucleation size, probably first formulated by *Gibbs* in 1878, will be next summarized. Consider a subcooled non-stable spherical steam volume, which is just about forming a nucleus, having an initial radius R_{30}, and volume $\frac{4}{3}\pi R_{30}^3$, pressure p, temperature T_1, and density $\rho_1 = \rho_1\left(p, T_1\right)$. Note that this density is computed for unstable steam. For practical application attention should be paid to this point because usually the state equation is obtained for stable fluids and an extrapolation into the unstable region is not straightforward for all approximations of the equation of state. During the initiation of the nucleus the sphere collapse from the initial radius to R_3, that is to the volume $\frac{4}{3}\pi R_3^3$. The final state is characterized by the same temperature T_1 but the state is stable, that is, the local pressure inside the nucleus is $p' = p'(T_1)$. Because of the small nucleus size the pressure inside is considered to be uniform. The density is therefore $\rho' = \rho'(T_1)$. During the initiation of the nucleus the initial and the end mass are the same. Therefore

$$\left(\frac{R_{30}}{R_3}\right)^3 = \frac{\rho'(T_1)}{\rho_1}. \tag{12.5}$$

Thus, the initial volume of the sphere is changed by

$$\frac{4}{3}\pi\left(R_3^3 - R_{30}^3\right) = \frac{4}{3}\pi R_3^3\left[1 - \frac{\rho'(T_1)}{\rho_1}\right]. \tag{12.6}$$

Volume reduction requires mechanical work. The amount of this work

$$4\pi\int_{R_{30}}^{R_3}\left[p'(T_1) - p\right]r^2 dr \approx \frac{4}{3}\pi\left(R_3^3 - R_{30}^3\right)\left[p'(T_1) - p\right]$$

$$= \frac{4}{3}\pi R_3^3\left[1 - \frac{\rho'(T_1)}{\rho_1}\right]\left[p'(T_1) - p\right] \tag{12.7}$$

is reducing the internal energy of sphere. The initial sphere interface was not a real one. The final sphere interface is a real one. The interface is a prominent characteristic of the origination of new liquid phase inside the steam. This well-defined surface requires additional mechanical work

$$4\pi \int_0^{R_3} \frac{4\sigma}{r} r^2 dr = 4\pi R_3^2 \sigma \tag{12.8}$$

to be created. The surface tension σ is a function of the local temperature that may strongly vary during transient processes. It is usually measured at macroscopic droplets. Whether the so-obtained information is still valid for a very small nucleus is not clear. We will assume that the so-measured surface tension holds also for very small droplets. One example of the dependence of the surface tension on the temperature is given here for water surrounded by its steam,

$$\sigma = 0.2358\left(1 - T_3/T_c\right)^{1.256}\left[1 - 0.625\left(1 - T_3/T_c\right)\right], \tag{12.9}$$

Lienhard. We see that if the droplet surface approaches the critical temperature for water, $T_{c,H_2O} = 647.2K$, the surface tension diminishes. Within the region 366 to 566 K the above equation can be approximated by

$$\sigma = 0.14783\left(1 - T_3/T_c\right)^{1.053} \tag{12.10}$$

with an error of ±1 %.

Thus, the work required to create a single nucleus

$$\Delta E_3 = 4\pi R_3^2 \sigma - \frac{4}{3}\pi R_3^3 \left(1 - \frac{\rho'(T_1)}{\rho_1}\right)\left[p'(T_1) - p\right] = 4\pi\sigma\left(R_3^2 - \frac{2}{3}\frac{R_3^3}{R_{3c}}\right), \tag{12.11}$$

where

$$R_{3c} = 2\sigma \Big/ \left\{\left[1 - \frac{\rho'(T_1)}{\rho_1}\right]\left[p'(T_1) - p\right]\right\}, \tag{12.12}$$

possesses a maximum at

$$\Delta E_{3c} = \frac{4}{3}\pi\sigma R_{3c}^2, \tag{12.13}$$

"*...which does not involve any geometrical magnitudes*", see *Gibbs* 1878. Equation (12.12) is known in the literature as the *Laplace* and *Kelvin* equation. Clusters with sizes smaller than the critical one are unstable and disappear quickly after creation, absorbing the excess energy from the surrounding vapor. Droplets with sizes larger than the critical one remain stable and release heat into the surrounding vapor. Obviously to create a nucleus a deficiency of internal energy of the steam is necessary, which has to be greater than the critical energy. The energy required to create a stable nucleus with critical size divided by the vapor temperature defines the entropy change of the system between the initial and final state.

$$\Delta S_{13c} = \Delta E_{3c}/T_1 . \tag{12.14}$$

The entropy change is made dimensionless by dividing by the *Boltzmann* constant $k = 13.805\times10^{-24}$. The resulting dimensionless number

$$Gb_1 = \frac{\Delta E_{3c}}{kT_1}, \tag{12.15}$$

is called the *Gibbs* number and is frequently used in the literature as a measure of the subcooling. We see that the larger the subcooling the smaller the *Gibbs* number. Two consequences of Eqs. (12.12)–(12.15) are of practical importance.

a) If the steam temperature approaches the critical temperature the surface tension diminishes and therefore the energy required for creation of a nucleus is zero, which means that all molecules are nuclei for themselves and the initiation of liquid inside the steam is impossible. The *Gibbs* number in this case is also equal to zero.

b) Rapid decompression can lead to large subcooling. The maximum possible subcoolings experimentally observed are in the range of *Gibbs* numbers from 6.8 to 9.6, *Skripov* et al. (1980), p. 136. It follows from the Eqs. (12.12)–(12.14)

$$p'\left(T_{1,spin}\right) - p = (1.57\_to\_1.32)\frac{\sigma^{3/2}}{\left(kT_{1,spin}\right)^{1/2}\left(1-\dfrac{\rho'\left(T_{1,spin}\right)}{\rho_1\left(p,T_{1,spin}\right)}\right)}. \tag{12.16}$$

This equation is known as the *Furth* equation (1941). It describe a dependence in the *p*,*T*-plane called a *spinoidal line*.

12.1.2 Nucleation kinetics, homogeneous nucleation

Initiation of nucleation clusters within the steam is called homogeneous nucleation. It has to be distinguished from the so-called heterogeneous nucleation. In the heterogeneous nucleation the origination of the liquid drops happens at the surfaces of other materials, e.g., walls, solid particles, etc. Statistical thermodynamics provides the mathematical tools for the derivation of an equation defining the generated nucleus per unit volume and unit time, $\dot{n}_{3,kin}$, as a function of the local parameters. If we observe over the time $\Delta\tau$ a point in the space occupied by subcooled vapor, we will see that the stronger the subcooling the more frequently the system departs from the averaged state. With other words over an integral time $\Delta\tau_{13}$ the system is outside of the averaged state. *Volmer* (1939) p.81 following *Boltzman* and *Einstein* defined the probability of departure from the averaged state as $\Delta\tau_{13}/\Delta\tau$ and assumed that this probability is proportional to $\exp\left(-Gb_1\right)$, that is

$$\Delta\tau_{13}/\Delta\tau = \exp(-Gb_1). \tag{12.17}$$

This mathematical expression says that the smaller the entropy change needed for a departure from the averaged state, the higher the probability of its occurrence. The probability of departure from the averaged state is usually set equal to the probability of origination of a single nucleus, that is

$$\frac{nucleation\_frequency}{molecular\_colisions\_frequency} = \exp(-Gb_1). \tag{12.18}$$

Zeldovich (1962) obtained the final expression for the generated nuclei per unit volume and unit time in the form

$$\dot{n}_{3,kin} = c_T Ze_1 \left(\frac{1}{4}N_1 a_1 F_{3c}\right) N_1 \exp(-Gb_1), \tag{12.19}$$

see also in *Fieder* et al. (1966). The assumptions used to derive this expression are at the same time the limitation of the model as discussed by *Ehrler* in VDI (1984):

a) The clusters are distributed in the vapor and are assumed to have a degree of freedom characteristic for big molecules in the vapor.

b) The nuclei are solid nonoscillating spheres with respect to the center of the mass of the system.

c) Macroscopic properties like density and surface tension are applicable to clusters having sizes of the order of nanometers.

Here, the nucleation frequency per unit volume is considered as proportional to the number of the clusters with critical size originating per unit time and unit volume, $N_1 \exp(-Gb_1)$, and the number of the molecules striking the cluster, $\frac{1}{4}N_1 a_1 F_{3c}$, and remaining there $c_c \frac{1}{4}N_1 a_1 F_{3c}$. The proportionality factor was found to be $c_T Ze_1$, where

$$Ze_1 = \sqrt{\frac{Gb_1}{3\pi n_{3c}^2}} \tag{12.20}$$

is the so-called *Zeldovich* number (1942) and c_T is the so-called temperature correction factor taking into account the heat transfer from the nucleus to the surrounding vapor. For pure water steam $c_T \approx 0.001$. The components of these two expressions are computed as follows. The molecular mass, m_μ, and the *Avogadro* number, $N_A = 6.02\times10^{23}$, uniquely define the mass of a single molecule $m_1 = m_\mu / N_A$. The number of the vapor molecules per unit volume is therefore $N_1 = \rho_1 / m_1$. The thermal velocity of the vapor molecules is defined by

$$a_1 = \sqrt{\frac{8}{\pi} R_1 T_1} . \tag{12.21}$$

After its initiation the surface F_{3c} of the cluster is struck by the following number of molecules per unit time $\frac{1}{4} N_1 a_1 F_{3c}$. Only a fraction of them, $c_c \frac{1}{4} N_1 a_1 F_{3c}$ remain there, where $c_c \approx 0.1$ for water-steam plus noncondensable gases, and $c_c \approx 0.5$ to 1, for pure water-steam. The number of the molecules in a single nucleus with critical size is

$$n_{3c} = \rho'(T_1) \frac{4}{3} \pi R_{3c}^3 / m_1 . \tag{12.22}$$

For successful prediction of several experiments with steam flows in *Laval* nozzles *Deitsch* and *Philiphoff* (1981) p.153 provided the following modification of one expression obtained by *Frenkel* (1973)

$$\dot{n}_{3,kin} = \left(\frac{p}{kT_1} \right)^2 \frac{c_T}{\rho'(T_1)} \sqrt{\frac{2\sigma m_1}{\pi}} \exp(-\beta G b_1) . \tag{12.23}$$

The modification contains the following empirical factor

$$\beta = 0.25 + 6.7 \left\{ \left[p'(T_1) + \frac{1}{2} \rho_1 w^2 \right] \Big/ p_c \right\}^{1/3} \tag{12.24}$$

for $c_T = 1$. The validity of Eq. (12.23) is proven for pressures between 0.25 and 4 MPa.

The mass condensed per unit volume and time due to nucleation is

$$\mu_{13} = \rho'(T_1) \frac{d}{d\tau} (n_3 V_3) = \rho'(T_1) V_{3c} \dot{n}_{3,kin} + \rho'(T_1) n_3 \frac{dV_3}{d\tau} . \tag{12.25}$$

We see that it consists of two parts, one due to nucleation, and the other due to growth of the droplets. The energy released during the condensation into the surrounding vapor is therefore

$$\dot{q}_1^{'''\sigma 3} = \dot{n}_{3,kin} \Delta E_{3c} + \rho'(T_1) n_3 \frac{dV_3}{d\tau} \left[h_1 - h'(T_1) \right] . \tag{12.26}$$

12.1.3 Droplet growth

Small droplets: For very small droplets with sizes comparable with the molecular free path length the kinetic gas theory provides the basis for computation of the heat and mass transfer. The net mass flow rate of condensation is then

$$\left(\rho w\right)_{13}=\frac{p_{M1}}{\sqrt{2\pi R_{M1}T_1}}-\frac{p'\left(T_3^{\sigma 1}\right)}{\sqrt{2\pi R_{M1}T_3^{\sigma 1}}}. \tag{12.27}$$

or having in mind that

$$\dot{q}_3^{\prime\prime\prime\sigma 1}=a_{13}h_c\left(T_1-T_3^{\sigma 1}\right)=a_{13}\left(\rho w\right)_{13}\left[h_1-h'\left(T_1\right)\right],$$

we obtain for the effective heat-transfer coefficient the following expression

$$h_c=\left(\rho w\right)_{13}\frac{h_1-h'\left(T_1\right)}{T_1-T_3^{\sigma 1}}=\frac{h_1-h'\left(T_1\right)}{T_1-T_{3\sigma}}\left(\frac{p_{M1}}{\sqrt{2\pi R_{M1}T_1}}-\frac{p'\left(T_3^{\sigma 1}\right)}{\sqrt{2\pi R_{M1}T_3^{\sigma 1}}}\right). \tag{12.28}$$

Mason (1951, 1957), *Ludvig* (1975) proposed the following equation describing the change of the droplet radius under such conditions

$$\rho_3\frac{dR_3}{d\tau}=\frac{1-\dfrac{R_{3c}}{R_3}}{R_3+1.59\ell_1}\frac{T'\left(p_{M1}\right)-T_1}{\dfrac{h''-h'}{\lambda_1}+\dfrac{p-p_{M1}}{p}\dfrac{R_{M1}T'\left(p_{M1}\right)p_{M1}^2}{\rho_{M1}\left(h''-h'\right)D_{M\to\Sigma n}}}, \tag{12.29}$$

for $R_3>R_{3c}$, which is equivalent to the equation

$$\rho_3\frac{dR_3}{d\tau}=\frac{\mu_{13}}{a_{13}}=\frac{\dot{q}_1^{\prime\prime\prime\sigma 3}}{h_{M1}-h'}, \tag{12.30}$$

where

$$h_c=\left(h_{M1}-h'\right)\frac{1-\dfrac{R_{3c}}{R_3}}{R_3+1.59\ell_1}\frac{T'\left(p_{M1}\right)-T_1}{\dfrac{h''-h'}{\lambda_1}+\dfrac{p-p_{M1}}{p}\dfrac{R_{M1}T'\left(p_{M1}\right)p_{M1}^2}{\rho_{M1}\left(h''-h'\right)D_{M\to\Sigma n}}}, \tag{12.31}$$

is the heat-transfer coefficient. Here,

$$\ell_1=\frac{\eta_1}{p}\sqrt{\frac{\pi R_{M1}T_1}{2m_{\mu,M1}}} \tag{12.32}$$

is the *free path of a steam molecule* and $m_{\mu,M1}$ is the *mole weight*. For water we have $m_{\mu,M1}=m_{\mu,H_2O}=18$. Note that this equation is valid for steam–air mixtures.

Large droplets: For *larger macroscopic droplets* heat conduction and convection govern the heat and mass transfer. In this case, the droplet grows due to condensation at the surface, which depends on the properties of the environment to absorb the released heat. For a droplet in stagnant steam the results obtained by the modeling of *thermally controlled bubble growth* can be applied. Note that the continuous phase is the gas phase in this case. This means that in the relationship

$$R_3 = f\sqrt{a_1\tau^*}\,, \tag{12.33}$$

where $f = f(..., Ja)$, $a_1 = \lambda_1/(\rho_1 c_{p1})$, and

$$Ja = \frac{\rho_1 c_{p1}(T' - T_1)}{\rho'(h'' - h')} \tag{12.34}$$

is the *Jakob* number. Remember that the denominator of the *Jakob* number is obtained from the boundary condition on the droplet surface, namely the equality of the heat released by the condensation, and of the heat taken away by heat conduction from the environment. The mass corresponding to this heat changes the radius of the droplet. The numerator in the *Jakob* number is obtained from the energy-conservation equation for the continuum.

Analogously for convective heat transfer we can compute a heat-transfer coefficient for a moving droplet and the corresponding mass and energy source terms.

12.1.4 Self-condensation stop

As already mentioned, the released heat during the condensation increases the gas temperature, which for closed systems may decrease the condensation to zero. For a closed control volume or negligible convection and the diffusion from the neighboring elementary cells, the approximate form of the energy-conservation equation is

$$\alpha_1 \rho_1 c_{p1} \frac{dT_1}{d\tau} = \dot{q}_1'''^{\sigma 3} = \frac{3\alpha_3}{R_3} h_c \left(T_3^{\sigma 1} - T_1\right) \tag{12.35}$$

or

$$\frac{dT_1}{d\tau} = \frac{T_3^{\sigma 1} - T_1}{\Delta\tau_1^*}, \tag{12.36}$$

where

$$\Delta\tau_1^* = \frac{\alpha_1 \rho_1 c_{p1} R_3}{3\alpha_3 h_c} \tag{12.37}$$

approximates the *time constant of the phenomena*. The product of the density and the specific heat at constant pressure for a gas is less than the analogous product for a liquid. That is why the time constant for the analyzed case is smaller compared to the time constant for processes controlled by the heat accumulation in a liquid. This fact is very important and has to be taken into account for the construction of stable solution algorithms especially for condensing flows.

Thus, under these circumstances the difference between the saturation temperature and the gas temperature decreases exponentially

$$T_3^{\sigma 1} - T_1 = \left(T_{30}^{\sigma 1} - T_{10}\right) e^{-\Delta\tau/\Delta\tau_1^*} . \tag{12.38}$$

Therefore, the *average* heat released during the condensation within the time step $\Delta\tau$ per unit mixture volume and unit time is

$$\dot{q}_1^{\prime\prime\prime\sigma 3} = \frac{3\alpha_3}{R_3} h_c \left(T_3^{\sigma 1} - T_1\right) f , \tag{12.39}$$

where

$$f = \frac{\Delta\tau_1^*}{\Delta\tau}\left(1 - e^{-\Delta\tau/\Delta\tau_1^*}\right). \tag{12.40}$$

The corresponding *averaged* mass source term is

$$\mu_{13} = \frac{\dot{q}_1^{\prime\prime\prime\sigma 3}}{h_{M1} - h'} . \tag{12.41}$$

12.2 Heat transfer across droplet interface without mass transfer

Consider a family of mono-disperse droplet designated with *d*, moving in a continuum, designated with *c*, with relative velocity ΔV_{cd}. We are interested in how much thermal energy is transferred between the surface of a droplet and the surrounding continuum. We will consider first the steady-state case, thereafter the transient heat conduction problem within a droplet and finally we will give some approximate solutions for the effective interface temperature and the average energy transferred.

The energy transported between the drop velocity field and the continuum per unit time and unit mixture volume for a steady-state case is frequently approximated by

$$\dot{q}_{cd}^{\prime\prime\prime} = a_{cd} h_{cd}\left(T_c - T_d\right) = 6\alpha_d \lambda_c Nu_d \left(T_c - T_d\right)/D_d^2 . \tag{12.42}$$

Here $a_{cd} = 6\alpha_d / D_d$ is the interfacial area density, $h_{cd} = \lambda_c Nu_d / D_d$ is the heat-transfer coefficient and Nu_d is the *Nusselt* number defined as given in Table 12.1. Table 12.1 contains analytically and experimentally obtained solutions for the *Nusselt* number as a function of the relative velocity and the continuum properties.

Table 12.1 Heat-transfer coefficient on the surface of moving solid sphere and water droplets

$$Nu_d = \frac{D_d h_c}{\lambda_c},\ Pe_c = \frac{D_d \left|\Delta V_{cd}\right|}{a_c} = Re_d Pr_c,\ \ Re_d = \frac{D_d \rho_c \left|\Delta V_{cd}\right|}{\eta_c},\ \ Pr_c = \frac{\eta_c}{\rho_c a_c} = \frac{\eta_c c_{pc}}{\lambda_c},$$

$$a_c = \frac{\lambda_c}{\rho_c c_{pc}}$$

$Re_d << 1$		Potential flow
$Pe_c << 1$		
$Nu_d = 2 + \frac{1}{2} Pe_c + \frac{1}{6} Pe_c^2 + ...$		*Soo* (1969)
$Pe_c >> 1$		
$Nu_d = 0.98 Pe_c^{1/3}$	$\eta_d / \eta_c = \infty$	*Nigmatulin* (1978)
$Nu_d = \frac{2}{\pi^{1/2}} Pe_c^{1/3}$		*Boussinesq* (1905), isothermal sphere
$Nu_d = \left(\frac{3}{4\pi} \frac{1}{1+\eta_d/\eta_c} Pe_c \right)^{1/2}$		*Levich* (1962)
$Nu_d = 0.922 + 0.991 Pe_c^{1/3}$		*Acrivos–Goddard* (1965)
$Nu_d = 4.73 + 1.156 Pe_c^{1/2}$		*Watt* (1972) isothermal sphere
$Re_d < 1$,		
$Pe_c < 10^3$		
$Nu_d = 2 + \dfrac{\frac{1}{3} Pe_c^{0.84}}{1 + \frac{1}{3} Pe_c^{0.51}}$	$\eta_d / \eta_c = \infty$	*Nigmatulin* (1978)
$Pe_c < 1$		
$Nu_d = 2 + \frac{1}{2} Pe_c^{1/3}$		*Acrivos* and *Taylor* (1962) only the resistance in the gas is taken into account

$1 < \mathrm{Re}_d < 7\times10^4$, $0.6 < Pr_c < 400$

$Nu_d = 2 + (0.55 \text{ to } 0.7) Re_d^{1/2} Pr_c^{1/3}$ *Soo* (1969)

Droplets and bubbles *Michaelides* (2003)

$0 < \mathrm{Re}_d < 1$, $Pe_c > 10$

$$Nu_d = \left(\frac{0.651}{1+0.95\eta_d/\eta_c} Pe_c^{1/2} + \frac{0.991\eta_d/\eta_c}{1+\eta_d/\eta_c} Pe_c^{1/3} \right)(1.032 + A)$$

$$+ \frac{1.651(0.968 - A)}{1+0.95\eta_d/\eta_c} + \frac{\eta_d/\eta_c}{1+\eta_d/\eta_c}, \quad A = \frac{0.61Re_d}{21+Re_d}.$$

$10 \le Pe_c \le 1000$, $\mathrm{Re}_d > 1$

$$Nu_d = \frac{2-\eta_d/\eta_c}{2} Nu_{d,0} + \frac{4\eta_d/\eta_c}{6+\eta_d/\eta_c} Nu_{d,2} \quad \text{for } 0 \le \eta_d/\eta_c \le 2$$

$$Nu_d = \frac{4}{\eta_d/\eta_c+2} Nu_{d,2} + \frac{\eta_d/\eta_c - 2}{\eta_d/\eta_c+2} Nu_{d,\infty} \quad \text{for } 2 < \eta_d/\eta_c < \infty,$$

where

$$Nu_{d,0} = 0.651 Pe_c^{1/2}(1.032 + A) + 1.6 - A \quad \text{for } \eta_d/\eta_c = 0,$$

$$Nu_{d,\infty} = 0.852 Pe_c^{1/3}\left(1 + 0.233\mathrm{Re}_d^{0.287}\right) + 1.3 - 0.182\mathrm{Re}_d^{0.355} \quad \text{for } \eta_d/\eta_c = \infty,$$

$$Nu_{d,2} = 0.64 Pe_c^{0.43}\left(1 + 0.233\mathrm{Re}_d^{0.287}\right) + 1.41 - 0.15\mathrm{Re}_d^{0.287} \quad \text{for } \eta_d/\eta_c = 2.$$

Solid sphere undergoing a step temperature change, *Feng* and *Michaelides* (1986)

$$Nu_d = 2\left\{1 + 2Pe_c^2 \ln(2Pe_c) + Pe_c \left[\frac{1}{2} erf\left(2Pe_c\sqrt{\tau^*}\right) + \frac{\exp\left(-4Pe_c^2\tau^*\right)}{4Pe_c\sqrt{\pi\tau^*}}\right]\right\}.$$

Water droplets

$Nu_d = 2 + 0.6 Re_d^{1/2} Pr_c^{1/3}$ *Ranz* and *Marschal* (1952)

$64 < \mathrm{Re}_d < 250$, $0.00023 < D_d < 0.00113m$, $1.03\times10^5 < p < 2.03\times10^5 Pa$,
$2.8 < T_c - T_d < 36K$, $2.7 < \Delta V_{cd} < 11.7m/s$

$Nu_d = 2 + 0.738 Re_d^{1/2} Pr_c^{1/3}$. *Lee* and *Ryley* (1968)

In nature, the temperature of both media changes with the time. That is why we consider next the transient heat-conduction problem inside the droplet associated with the heat transfer at the surface.

Consider the case for which the continuum temperature is T_c and the droplet temperature at the beginning of the time $\Delta\tau$ is T_d. The *Fourier* equation

$$\frac{\partial T}{\partial \tau} = \frac{\lambda_d}{\rho_d c_{cd}} \left(\frac{\partial^2 T}{\partial r^2} + \frac{2}{r} \frac{\partial T}{\partial r} \right) \tag{12.43}$$

describes the temperature profile in the droplet. Usually the equation is written in terms of the dimensionless temperature

$$\bar{T} = \frac{T(r,\tau) - T_d}{T_d^{\sigma c} - T_d} \tag{12.44}$$

for $r = 0$, R_d, $\tau = \tau, \tau + \Delta\tau$ in the following form

$$\frac{\partial \bar{T}}{\partial \tau} = a_d \left(\frac{\partial^2 \bar{T}}{\partial r^2} + \frac{2}{r} \frac{\partial \bar{T}}{\partial r} \right), \tag{12.45}$$

where $a_c = \lambda_c / (\rho_c c_{pc})$ is the droplet thermal diffusivity. A text book analytical solution of the *Fourier* equation is available for the following boundary conditions:

(a) Drop initially at uniform temperature: $\bar{T}(r,0) = 0$;

(b) Drop surface at $r = R_d$ immediately reaches the temperatures $T_d^{\sigma c}$, $\bar{T}(R_d, \tau) = 1$;

(c) Symmetry of the temperature profile: $\left. \frac{\partial \bar{T}}{\partial r} \right|_{r=0} = 0$.

The solution is represented by the converging *Fourier* series

$$\bar{T}(r,\tau) = 1 - \frac{2}{\pi} \frac{R_d}{r} \sum_{n=1}^{\infty} \frac{(-1)^n}{n} \sin\left(n\pi \frac{r}{R_d} \right) \exp\left(-n^2 \tau / \Delta\tau_d \right). \tag{12.46}$$

Carslaw and *Jaeger* (1996), where $\Delta\tau_d = D_d^2 / (4\pi^2 a_d)$ is the characteristic time constant of the heat-conduction process. The volume-averaged nondimensional temperature is

$$\bar{T}_{dm} = 1 - \frac{6}{\pi^2} \sum_{n=1}^{\infty} \frac{1}{n^2} \exp\left(-n^2 \tau / \Delta\tau_d \right). \tag{12.47}$$

Convection inside the droplet caused by the interfacial shear due to relative velocity can improve the thermal conductivity of the droplet a_d by a factor $f \geq 1$. Therefore

$$\overline{T}_{dm} = 1 - \frac{6}{\pi^2} \sum_{n=1}^{\infty} \frac{1}{n^2} \exp\left(-fn^2 \tau / \Delta\tau_d\right). \tag{12.48}$$

Celata et al. (1991) correlated their experimental data for condensation with the following expression for *f*,

$$f = 0.53 Pe_c^{*0.454}, \tag{12.49}$$

where a special definition of the *Peclet* number is used $Pe_c^* = \frac{D_d \left|\Delta V_{cd}\right|}{a_c} \frac{\eta_c}{\eta_c + \eta_d}$.

For a water droplet with $D_d = 0.00002, 0.0002, 0.001$ m and $T_c - T_d = 20K$ the time necessary to reach $\overline{T}_{dm} = 0.99$ is 0.00006, 0.00613, 0.612 s, respectively. For values of $\overline{T}_{dm}$ greater than 0.95 (long contact times) the first term only of Eq. (12.48) is significant

$$\overline{T}_{dm} \approx 1 - \frac{6}{\pi^2} \exp\left(-fn^2 \tau / \Delta\tau_d\right). \tag{12.50}$$

One alternative solution of Eq. (12.48) can be obtained by the method of *Laplace* transformations that takes the form of a diverging infinite series. For values of $\overline{T}_{dm}$ less than 0.4 (short contact times) the first term only of this series

$$\overline{T}_{dm} = \frac{3\sqrt{\pi}}{D_d \sqrt{a_d \tau}} \tag{12.51}$$

is significant. This equation applies for short contact times when the temperature gradient of the surface surroundings has not penetrated to the center of the sphere that consequently behaves as a semi-infinite body.

From the energy balance of the droplet velocity field we obtain the time-averaged energy transfer from the surface into the droplet per unit time and unit mixture volume

$$\dot{q}_d^{\prime\prime\prime\sigma c} = \alpha_d \rho_d c_{pd} \frac{1}{\Delta\tau} \int_0^{\Delta\tau} \frac{d\overline{T}_{dm}}{d\tau} d\tau = \alpha_d \rho_d c_{pd} \frac{1}{\Delta\tau} \left(T_d^{\sigma c} - T_d\right) \left[\overline{T}_{dm}(\Delta\tau) - \overline{T}_{dm}(0)\right]$$

leading to

$$\dot{q}_d^{\prime\prime\prime\sigma c} = \alpha_d \rho_d c_{pd} \frac{\overline{T}_{dm}(\Delta\tau)}{\Delta\tau} \left(T_d^{\sigma c} - T_d\right) \equiv \omega_d \left(T_d^{\sigma c} - T_d\right) \tag{12.52}$$

where

$$\omega_d = \alpha_d \rho_d c_{pd} \frac{\overline{T}_{dm}(\Delta\tau)}{\Delta\tau}.$$

If the surface temperature differs from the continuum temperature there is a convective component of the energy transported from the continuum to the surface for $T_c > T_d^{\sigma c}$ or from the surface to the continuum for $T_c < T_d^{\sigma c}$, namely

$$\dot{q}_c^{\prime\prime\prime\sigma d} = 6\alpha_d \lambda_c Nu_d \left(T_d^{\sigma c} - T_c\right) / D_d^2 = \omega_c \left(T_d^{\sigma c} - T_c\right), \qquad (12.53)$$

where Nu_d is the *Nusselt* number for convective heat transfer between a sphere and the continuum, and $\omega_c = 6\alpha_d \lambda_c Nu_d / D_d^2$.

For $T_c \approx const$ the effective interface temperature is obtained from the energy-jump condition $\dot{q}_d^{\prime\prime\prime\sigma c} = -\dot{q}_c^{\prime\prime\prime\sigma d}$ after solving for $T_d^{\sigma c}$

$$T_d^{\sigma c} = \frac{\omega_d T_d + \omega_c T_c}{\omega_d + \omega_c}. \qquad (12.54)$$

For the case when $T_c \neq const$ the heat transfer causes changes in the continuum temperature too. If we neglect the convection and the diffusion from the neighboring elementary cells, which is valid for a heat-transfer time constant considerably smaller than the flow time constant, the continuum energy conservation can be approximated by

$$\alpha_c \rho_c c_{pc} \frac{dT_c}{d\tau} = 6\alpha_d \lambda_c Nu_d \left(T_d^{\sigma c} - T_c\right) / D_d^2 \qquad (12.55)$$

or

$$\frac{dT_c}{d\tau} = \frac{T_d^{\sigma c} - T_c}{\Delta\tau_c^*}, \qquad (12.56)$$

where

$$\Delta\tau_c^* = \frac{\alpha_c \rho_c c_{pc} D_d^2}{6\alpha_d \lambda_c Nu_d} \qquad (12.57)$$

approximates the characterizing time constant of the continuum heating or cooling. Note that the product of the density and the specific heat at constant pressure for gas is less than the analogous product for liquid. That is why the time constant for gases as a continuum should be smaller compared to the time constant for liquids as a continuum. This makes the gas temperature very sensitive to the heat- and mass-transfer process.

For constant surface temperature the equation (12.56) has a solution $T_c = T_d^{\sigma c} - \left(T_d^{\sigma c} - T_{c0}\right)\exp\left(-\tau/\Delta\tau_c^*\right)$. The time-averaged change of the gas temperature would be

$$\frac{1}{\Delta\tau}\int_0^{\Delta\tau}\frac{dT_c}{d\tau}d\tau = \frac{T_c - T_{c0}}{\Delta\tau} = \frac{T_d^{\sigma c} - T_{c0}}{\Delta\tau}\left[1-\exp\left(-\tau/\Delta\tau_c^*\right)\right]$$

and therefore the heat transferred into the gas per unit time and unit mixture volume over the time step $\Delta\tau$ is

$$\dot{q}_c^{\prime\prime\prime\sigma d} = \alpha_c\rho_c c_{pc}\frac{1}{\Delta\tau}\int_0^{\Delta\tau}\frac{dT_c}{d\tau}d\tau = \frac{\alpha_c\rho_c c_{pc}}{\Delta\tau}\left[1-\exp\left(-\tau/\Delta\tau_c^*\right)\right]\left(T_d^{\sigma c} - T_{c0}\right)$$

$$= \omega_c\left(T_d^{\sigma c} - T_{c0}\right), \quad (12.59)$$

with

$$\omega_c = \frac{\alpha_c\rho_c c_{pc}}{\Delta\tau}\left[1-\exp\left(-\tau/\Delta\tau_c^*\right)\right] = \frac{6\alpha_d\lambda_c Nu_d}{D_d^2}f_c,$$

$$f_c = \left[1-\exp\left(-\frac{\Delta\tau}{\Delta\tau_c^*}\right)\right]\frac{\Delta\tau_c^*}{\Delta\tau}. \quad (12.60)$$

Note that for $\Delta\tau \to 0$, $f_c \to 1$.

Thus, from the energy-jump conditions $\dot{q}_d^{\prime\prime\prime\sigma c} = -\dot{q}_c^{\prime\prime\prime\sigma d}$ we obtain the effective interface temperature by Eq. (12.54). Having this temperature we can compute the average heat power density $\dot{q}_c^{\prime\prime\prime\sigma d}$ by Eq. (12.59).

12.3 Direct contact condensation of pure steam on subcooled droplet

Consider the case where the steam temperature is equal to the saturation temperature by the system pressure

$$T_1 = T'(p), \quad (12.61)$$

and the droplet temperature at the beginning of the time step considered is less than the saturation temperature

$$T_3 < T'(p). \tag{12.62}$$

The heat conduction from the surface, which has the saturation temperature, to the bulk of the droplet is the condensation-controlling mechanism in this case. Therefore, the *Fourier* equation describes the temperature profile inside the droplet. *Ford* and *Lekic* (1973) considered additionally the following effects

(a) convection inside the droplet;
(b) limitation of the steam supply toward the droplet surface;

and introduced the corrector *f* in Eq. (12.48),

$$\overline{T}_m \approx 1 - \exp(-f\tau/\Delta\tau). \tag{12.63}$$

From the energy balance of the droplet velocity field we obtain the time-averaged energy transfer from the surface into the droplet per unit time

$$\dot{q}_3^{\prime\prime\prime\sigma 1} = \alpha_3 \rho_3 c_{p3} \left[T'(p) - T_3\right] \overline{T}_{3m}(\Delta\tau)/\Delta\tau. \tag{12.64}$$

If the steam temperature differs from the saturation temperature there is a convective component of the energy transported from the steam to the surface for $T_1 > T'(p)$ or from the surface to the steam for $T_1 < T'(p)$, namely

$$\dot{q}_1^{\prime\prime\prime\sigma 3} = a_{13} \frac{\lambda_1'}{D_3} Nu_{13} (T' - T_1), \tag{12.65}$$

where Nu_{13} is the *Nusselt* number for convective heat transfer between sphere and steam. For the case when $T_1 < T'(p)$ part of the released heat during the condensation increases the gas temperature. If we neglect the convection and the diffusion from the neighboring elementary cells, the energy conservation can be approximated by

$$\alpha_1 \rho_1 c_{p1} \frac{dT_1}{d\tau} = 6\alpha_3 \lambda_1 Nu_3 \left(T_3^{\sigma 1} - T_1\right)/D_3^2, \tag{12.66}$$

or

$$\frac{dT_1}{d\tau} = \frac{T_3^{\sigma 1} - T_1}{\Delta\tau_1^*}, \tag{12.67}$$

where

$$\Delta\tau_1^* = \frac{\alpha_1 \rho_1 c_{p1} D_3^2}{6\alpha_3 \lambda_1 Nu_3} \tag{12.68}$$

approximates the time constant of the steam heating. For nearly constant pressure, and therefore constant T', the difference between the saturation temperature and the gas temperature decreases exponentially during the time interval $\Delta\tau$,

$$\overline{T}_{cm}\left(\Delta\tau\right)=\frac{T_3^{\sigma 1}-T_1}{T_3^{\sigma 1}-T_{10}}=\exp\left(-\frac{\Delta\tau}{\Delta\tau_1^*}\right). \tag{12.69}$$

Therefore, the time-averaged heat transferred from the surface into the gas per unit mixture volume and unit time is

$$\dot{q}_1^{\prime\prime\prime\sigma 3}=6\alpha_3\frac{\lambda_1'}{D_3^2}Nu_{13}\left(T'-T_1\right) \tag{12.70}$$

where

$$f=\left[1-\exp\left(-\frac{\Delta\tau}{\Delta\tau_1^*}\right)\right]\frac{\Delta\tau_1^*}{\Delta\tau}. \tag{12.71}$$

Thus, from the assumption that there is no energy and mass accumulation the interface (energy- and mass-jump conditions) we obtain for the averaged condensing mass per unit mixture volume and unit time the following expression

$$\mu_{13}=\frac{\dot{q}_3^{\prime\prime\prime\sigma 1}-\dot{q}_1^{\prime\prime\prime\sigma 3}}{h''-h'}. \tag{12.72}$$

12.4 Spontaneous flashing of superheated droplet

Next, we consider a single droplet in a gas environment having pressure p. The averaged droplet temperature is T_3. The surface temperature of the droplet, $T_3^{\sigma 1}$, is *higher* than the saturated temperature $T'\left(p\right)$. In the first instant of the process considered here, the spontaneous emission of mass will be *explosive*. In the next moment the surface temperature would drop and reach the saturation temperature, if no heat is transferred from the bulk region of the droplet to the surface. The heat transferred from the bulk region to the surface maintains the surface temperature *higher* than the saturation temperature and causes further evaporation. The *thermal resistance inside the drop* has a delaying effect on the evaporation, because the heat conduction is the slower process compared to the mass emission from the surface.

The net mass flow rate of evaporation from the droplet surface can be calculated by solving numerically the *Boltzmann* kinetic equation for such conditions. This is quite a complicated task. For the practical application the theory first proposed by *Hertz* is used. *Hertz* (1882) computed the evaporation mass flow rate from a surface with temperature $T_3^{\sigma 1}$ and surface pressure $p'\left(T_3^{\sigma 1}\right)$ multiplying the frequency of the molecules leaving unit surface $N_1 a_1 / 4$ by the mass of a single molecule m_1. Here, N_1 is the number of vapor molecules in unit volume. The averaged thermal velocity of the molecules at temperature $T_3^{\sigma 1}$ is

$$a_1 = \sqrt{\frac{8}{\pi} R_{M1} T_3^{\sigma 1}} . \tag{12.73}$$

The result is

$$(\rho w)_{31}^* = m_1 \frac{1}{4} N_1 a_1 . \tag{12.74}$$

Having in mind that

$$m_1 N_1 = \rho_1 \left[p'\left(T_3^{\sigma 1}\right), T_3^{\sigma 1} \right], \tag{12.75}$$

$$p'\left(T_3^{\sigma 1}\right) / \rho_1 \left[p'\left(T_3^{\sigma 1}\right), T_3^{\sigma 1} \right] \approx R_{M1} T_3^{\sigma 1}, \tag{12.76}$$

Hertz obtain finally

$$(\rho w)_{31}^* = \frac{p'\left(T_3^{\sigma 1}\right)}{\sqrt{2\pi R_{M1} T_3^{\sigma 1}}} . \tag{12.77}$$

Similarly, *Hertz* computed the condensing mass flow rate from the steam with temperature T_1 equal to the gas temperature and partial pressure p_{M1} multiplying the frequency of the molecules striking unit surface by the mass of a single molecule. The averaged thermal velocity of the molecules at temperature T_1 across the plane surface in local thermodynamical equilibrium is

$$a_1 = \sqrt{\frac{8}{\pi} R_{M1} T_1} . \tag{12.78}$$

The result is

$$(\rho w)_{13}^* = \frac{p_{M1}}{\sqrt{2\pi R_{M1} T_1}} . \tag{12.79}$$

Assuming (i) reversible process, (ii) $T_1 \approx T_3^{\sigma 1}$, and (iii) condensation and evaporation flows do not affect each other, *Hertz* obtained finally

$$(\rho w)_{31} = (\rho w)_{31}^* - (\rho w)_{13}^* = \frac{1}{\sqrt{2\pi R_{M1} T_1}} \left[\sqrt{\frac{T_1}{T_3^{\sigma 1}}} p'\left(T_3^{\sigma 1}\right) - p_{M1} \right]$$

$$\approx c \frac{1}{\sqrt{2\pi R_{M1} T_1}} \left[p'\left(T_3^{\sigma 1}\right) - p_{M1} \right], \tag{12.80}$$

where c is the so-called accommodation coefficient, being of the order of

$$c = 0.1 \text{ to } 1, \tag{12.81}$$

as shown by *Volmer*'s review in *Volmer* (1939). For condensation of vapors of liquid metals where the thermal resistance of the liquid is almost zero the accommodation coefficient was found to be a function of pressure

$$c \approx \beta/(1-0.48\beta), \quad \beta \approx \left[400/p'(T_3^{\sigma 1})\right]^{1/2},$$

Aladyev et al. (1966), *Brauer* (1961). For $p < 1000$ Pa, $\beta \approx 1$ and for increasing pressures β decreases, *Subbotin* et al. (1966). In the presence of noncondensable gases or impurities at the surface β dramatically decreases, *Subbotin* et al. (1966).

Knudsen (1915) and *Langmuir* (1913), *Langmuer* et al. (1927) arrived at the following formula for the *net mass flow rate of the spontaneous evaporation*

$$(\rho w)_{31} = (\rho w)_{31}^* - (\rho w)_{13}^* = c_e \frac{p'(T_3^{\sigma 1})}{\sqrt{2\pi R_{M1} T_3^{\sigma 1}}} - c_c \frac{p_{M1}}{\sqrt{2\pi R_{M1} T_1}}. \tag{12.82}$$

Here, c_e is the *probability of escape* for a liquid molecule at the interface, and c_c is the *probability of capture* for a vapor molecule at the interface. This relationship is known in the literature as the equation of *Hertz*, *Knudsen* and *Langmuir* (see in *Ranz* and *Marschal* (1952). For no sound physical reason, Eq. (12.81) is generally approximated in the literature by

$$(\rho w)_{31} = c \left(\frac{p'(T_3^{\sigma 1})}{\sqrt{2\pi R_{M1} T_3^{\sigma 1}}} - \frac{p_{M1}}{\sqrt{2\pi R_{M1} T_1}} \right). \tag{12.83}$$

For pressures below 0.1 bar measurements reported by *Fedorovich* and *Rohsenow* (1968) for liquid metals tightly clustered about unity. For larger pressure Eq. (12.81) is a good approximation – see the discussion by *Mills* and *Mills* and *Seban* (1967). For water the discussion and the measurements in *Mills* and *Seban* (1967), *Nabavian* and *Bromley* (1963), *Berman* (1961) give

$$c = 0.35 \text{ to } 1\,. \tag{12.84}$$

Paul (1962) reported for acetone values of 0.2 to 1. *Knowles* (1985) derived an analytical expression taking into account the influence of the condensation due to the relative motion of the interface with respect to the steam. The result is

$$(\rho w)_{31} = c \left(\frac{p'(T_3^{\sigma 1})}{\sqrt{2\pi R_{M1} T_3^{\sigma 1}}} - f\left(\Delta|\mathbf{V}_{21}|\right) \frac{p_{M1}}{\sqrt{2\pi R_{M1} T_1}} \right), \tag{12.85}$$

where the so-called *transport enhancement factor* approximating the analytical solution is

$$f\left(\Delta\mathbf{V}_{21}\right)=1+1.7\frac{\Delta\left|\mathbf{V}_{21}\right|}{\sqrt{2R_{M1}T_1}}+0.7\left(\frac{\Delta\left|\mathbf{V}_{21}\right|}{\sqrt{2R_{M1}T_1}}\right)^2 \tag{12.86}$$

for $\frac{\Delta\left|\mathbf{V}_{21}\right|}{\sqrt{2R_{M1}T_1}}\geq -1$ and

$$f\left(\Delta\left|\mathbf{V}_{21}\right|\right)=0 \tag{12.87}$$

otherwise. For completeness we give some other modifications given in the literature. *Schrage*, see in *Bankoff* (1980), used the following form:

$$\left(\rho w\right)_{d1}=\frac{2c}{2-c}\frac{1}{\sqrt{2\pi R_{M1}T_3^{\sigma 1}}}\left[p'\left(T_3^{\sigma 1}\right)-p_{M1}-\frac{p'\left(T_3^{\sigma 1}\right)}{2T_3^{\sigma 1}}\left(T_3^{\sigma 1}-T_1\right)\right], \tag{12.88}$$

Here, a modified *accommodation* coefficient is used having values for water c_{H_2O} = *0.01* to *1*. *Samson* and *Springer* (1969) used the following form for very small droplets, for $\ell_1/D_d >> 1$,

$$\left(\rho w\right)_{d1}=\frac{1}{\sqrt{2\pi R_{M1}T_1}}\left[p'\left(T_d^{\sigma c}\right)-p_{M1}-\frac{p_{M1}}{2T_1}\left(T_d^{\sigma c}-T_1\right)\right], \tag{12.89}$$

and for large droplets, $\ell_1/D_d << 1$

$$\left(\rho w\right)_{d1}=\frac{1}{\sqrt{2\pi R_{M1}T_1}}\left[p'\left(T_d^{\sigma c}\right)-p_{M1}\right]. \tag{12.90}$$

Here

$$\ell_1=\frac{\eta_1}{p}\sqrt{\frac{\pi R_{M1}T_1}{2M_{M1}}}, \tag{12.91}$$

is the free path of a steam molecule and $M_{M1}=M_{H_2O}=18$ is the mole weight. *Labunzov* and *Krjukov* (1977) used the following form:

$$\left(\rho w\right)_{d1}=2\left[\rho''\left(T_{di}\right)-\rho_{M1}\left(T_1,p_{M1}\right)\right]\sqrt{\frac{R_{M1}T_{di}\rho_{M1}\left(T_1,p_{M1}\right)}{2\pi\rho''\left(T_{di}\right)}}. \tag{12.92}$$

The spontaneously evaporating mass from the superheated velocity field per unit mixture volume is

$$\mu_{d1}=\left(6\alpha_d/D_d\right)\left(\rho w\right)_{d1}. \tag{12.93}$$

In the case of strong emission of steam from the surface into the surrounding gas the heat transfer from the surface to the surrounding continuum can be neglected and therefore

$$\dot{q}_1'''^{\sigma 3} = 0 . \tag{12.94}$$

The energy-jump condition for this situation reads

$$\mu_{d1}\left[h''\left(T_d^{\sigma c}\right) - h'\left(T_d^{\sigma c}\right)\right] = -\dot{q}_d'''^{\sigma c} . \tag{12.95}$$

If we use the approximate solution of the *Fourier* equation we obtain the one equation with respect the unknown surface temperature, namely

$$\frac{6}{D_d}(\rho w)_{d1}\left[h''\left(T_d^{\sigma c}\right) - h'\left(T_d^{\sigma c}\right)\right] = \rho_d c_{pd}\left(T_d - T_d^{\sigma c}\right)\frac{1}{\Delta\tau}\left[1 - \frac{6}{\pi^2}\exp\left(-\frac{\Delta\tau}{\Delta\tau_d}\right)\right]. \tag{12.96}$$

This equation can be solved by iteration starting with

$$T_d^{\sigma c} \approx T_d . \tag{12.97}$$

Having the surface temperature it is very easy to compute the mass source term from Eqs. (12.85) and (12.93), and the corresponding heat power density from Eq. (12.95).

The spontaneous evaporation will cease if the droplet surface temperature reaches the saturation temperature corresponding to the system pressure. If the saturation temperature corresponding to the partial steam pressure $T'\left(p_{M1}\right)$ is lower than the surface temperature the slow evaporation continues due to molecular diffusion. This process will be considered in Section 12.5.

12.5 Evaporation of saturated droplets in superheated gas

Let us consider the evaporation of saturated droplets in superheated gas. *Lee* and *Ryley* (1968) found that the heat needed for the evaporation is convectively transported from the gas to the surface, and the emitted mass from the surface does not influence considerably this heat-transfer process. In this case we have

$$\mu_{1d} = 0 , \tag{12.98}$$

$$\dot{q}_3'''^{\sigma 1} = 0 . \tag{12.99}$$

During the droplet evaporation the gas is *cooled* with

$$\dot{q}_1'''^{\sigma d} = -a_{d1}h_c\left[T_1 - T'(p)\right] = -\left(6\alpha_d / D_d\right)h_c\left[T_1 - T'(p)\right]. \tag{12.100}$$

This is the instantaneous power density. From the energy-jump condition (no energy accumulation on the droplet interface) we obtain for the instantaneous evaporating mass per unit mixture volume and unit time

$$\mu_{31} = (6\alpha_d / D_d) h_c \left[T_1 - T'(p)\right] / \left[h''(p) - h_{Md}\right]. \tag{12.101}$$

Actually, we are interested in the *averaged* values of the mass source term during the time step $\Delta\tau$. For their approximate estimation we use the following method. Because the mass change of the droplet is caused by the evaporated mass per unit time, we have the following form of the mass-conservation equation for a single droplet

$$\rho' \frac{dR_d}{d\tau} = -h_c \left[T_1 - T'(p)\right] / \left[h''(p) - h_{Md}\right] \tag{12.102}$$

or

$$\frac{dR_d}{d\tau} = -\frac{h_c}{\rho'} \left[T_1 - T'(p)\right] / \left[h''(p) - h_{Md}\right]$$
$$= -\frac{Nu_{d1}\lambda_1}{\rho' 2R_d} \left[T_1 - T'(p)\right] / \left[h''(p) - h_{Md}\right] = -c / R_3. \tag{12.103}$$

This is the relationship between the change of the radius, the *Nusselt* number, and the temperature difference driving the evaporation. Integrating the above equation we obtain

$$\frac{R_d}{R_{d0}} = \sqrt{1 - 2c\Delta\tau / R_{d0}^2} = \sqrt{1 - \Delta\tau / \Delta\tau^*}. \tag{12.104}$$

If sufficient internal energy of the gas is available the *evaporation* of the entire droplet $R_d = 0$ happens after the time interval

$$\Delta\tau^* = R_{d0}^2 / (2c). \tag{12.105}$$

After this time interval Eq. (12.103) is no longer valid.

The third power of the radii ratio

$$\left(\frac{R_d}{R_{d0}}\right)^3 = \left(1 - \frac{\Delta\tau}{\Delta\tau^*}\right)^{3/2}. \tag{12.106}$$

is necessary to compute the averaged value of the evaporated mass per unit mixture volume and unit time within the time interval $\Delta\tau$

$$\mu_{d1} = \rho' n_d \frac{V_{d0} - V_d}{\Delta\tau} = \frac{\rho' n_d V_{d0}}{\Delta\tau}\left(1 - \frac{V_d}{V_{d0}}\right) = \frac{\rho' \alpha_d}{\Delta\tau}\left[1 - \left(\frac{R_d}{R_{d0}}\right)^3\right] \tag{12.107}$$

for $\Delta\tau < \Delta\tau^*$ and

$$\mu_{d1} = \frac{\rho'\alpha_d}{\Delta\tau} \tag{12.108}$$

for $\Delta\tau \geq \Delta\tau^*$. The last equation is practically the equation of the entire evaporation of the available droplets inside the time interval $\Delta\tau$.

Having in mind that during the evaporation

$$\mu_{d1} = 0 \tag{12.109}$$

$$\dot{q}_d^{m\sigma c} = 0, \tag{12.110}$$

we obtain from the energy-jump condition the following relationship

$$\dot{q}_1^{m\sigma d} = -\mu_{d1}\left[h''(p) - h'(p)\right]. \tag{12.111}$$

The above analysis was performed under the assumption that the *Nusselt* number does not depend on the radius during the considered time interval. In fact, the radius decreases in accordance with the relationship of *Lee* and *Ryley* (1968),

$$Nu_{d1} = 2 + 0.738 Re_{d1}^{1/2} Pr_1^{1/3} = 2 + 0.738\left(2\left|\Delta V_{1d}\right|/\nu_1\right)^{1/2} Pr_1^{1/3} D_d^{1/2} = 2 + c_3 R_d^{1/2}, \tag{12.112}$$

and therefore, the *Nusselt* number decreases too. This makes the process slower than that previously described (assuming the *Nusselt* number not depending on the radius). Taking into account this relationship

$$\frac{R_d}{2 + c_3 R_d^{1/2}} \frac{dR_d}{d\tau} = c_2, \tag{12.113}$$

Lee and *Ryley* (1968) obtained an analytical integral of the above equation, namely

$$\Delta\tau = \Delta\tau^* - \frac{1}{c_2}\left[\frac{2}{3}\frac{\sqrt{R_d}}{c_3} - \frac{D_d}{c_3^2} + \frac{8\sqrt{D_d}}{c_3^3} - \frac{16}{c_3^4}\ln\left(\frac{2}{2 + c_3 D_d^{1/2}}\right)\right]. \tag{12.114}$$

For a known time interval $\Delta\tau$ we can compute the corresponding droplet diameter after solving the above transcendental equation iteratively. The time required for the full droplet evaporation (if sufficient excess internal gas energy is available), i.e., $R_d = 0$ is easily obtained

$$\Delta\tau^* = \frac{1}{c_2}\left[\frac{2}{3}\frac{\sqrt{R_{d0}}}{c_3} - \frac{D_{d0}}{c_3^2} + \frac{8\sqrt{D_{d0}}}{c_3^3} - \frac{16}{c_3^4}\ln\left(\frac{2}{2 + c_3 D_{d0}^{1/2}}\right)\right]. \tag{12.115}$$

Finally for completeness let us mention the empirical correlation proposed by *Saha* (1980), describing the droplet evaporation in superheated steam in dispersed nonadiabatic flow, in dry-out condition on the heated wall

$$\alpha_{1d} h_c = 6300 \alpha_d \frac{\lambda_1}{D_h^2} \left(1 - \frac{p}{p_{kr}}\right)^2 \left(\frac{\rho_1 V_1^2 D_h}{\sigma}\right)^{1/2} . \tag{12.116}$$

Condie et al. (1984) show that this correlation overestimates the evaporation rate compared to the experimental data and recommend *Webb*'s correlation (see *Condie* et al. (1984))

$$\alpha_{1d} h_c = 1.32 \left(1 - \alpha_1^*\right)^{2/3} \frac{\lambda_1}{D_h^2} \left(1 - \frac{p}{p_{kr}}\right)^2 \left(\frac{\rho_1 V_1^2 D_h}{\sigma}\right)^{1/2} \tag{12.117}$$

approximating better a number of experimental data (α_1^* is the homogeneous void fraction).

12.6 Droplet evaporation in gas mixture

Consider a gas mixture with inert mass concentration $\sum C_{1n}$ and vapor concentration C_{1M}, having total pressure p and gas temperature T_1 see Fig. 12.1. A family of mono-dispersed droplets is flowing through the gas with relative velocity ΔV_{13}. At the beginning of the time interval considered, $\Delta \tau$, the droplets possess temperature T_3 and diameter D_3. Consider the case when the droplet temperature T_3 is greater than the saturation temperature T' corresponding to the partial vapor pressure p_{M1}

$$T_3 > T'(p_{M1}). \tag{12.118}$$

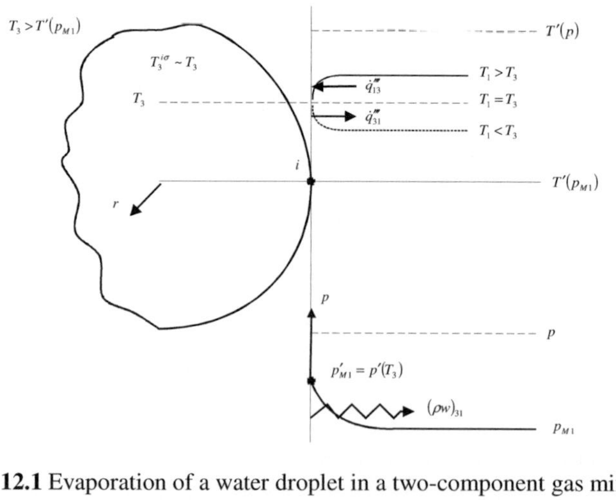

Fig. 12.1 Evaporation of a water droplet in a two-component gas mixture

In this case the interface temperature $T_3^{1\sigma}$ is somewhere between $T'(p_{M1})$ and T_1

$$T'(p_{M1}) < T_3^{1\sigma} < T_1 . \tag{12.119}$$

The surface is emitting vapor molecules that are struggling through the gas mixture boundary layer. The molecules of the inert gas present considerable resistance to their movement. The transport through the boundary layer is diffusion controlled. The evaporation mass flow per unit surface and unit time is $(\rho w)_{31}$. The heat needed for the evaporation is absorbed partially from the bulk of the droplet by thermal diffusion. Heat is transferred from or to the surface to or from the surrounding gas due to convection. The purpose of this section is to describe a method for computation of the evaporation mass flow rate.

The vapor pressure of the immediate neighborhood of the interface is equal to the saturation pressure corresponding to the surface temperature

$$p_{1M}^{3\sigma} = p'(T_1^{3\sigma}). \tag{12.120}$$

Therefore, there is a pressure difference $p_{1M}^{3\sigma} - p_{1M}$ across the diffusion boundary layer driving the mass flow rate $(\rho w)_{31}$. The simplest way to compute $(\rho w)_{31}$ is to use the assumption for the similarity between temperature and concentration profile around the droplets and therefore the analogy between the heat and mass transfer, namely

$$(\rho w)_{31} = \rho_1 \beta (C_{1M}^{3\sigma} - C_{1M}), \tag{12.121}$$

where

$$\beta = -\frac{D_{M\rightarrow\Sigma n}}{C_{1M}^{3\sigma} - C_{1M}} (\nabla C_{M1})_1^{3\sigma} \qquad \text{(Definition)} \tag{12.122}$$

$$C_{1M}^{3\sigma} = \frac{\rho_{1M}^{3\sigma}}{\rho_{1M}^{3\sigma} + \sum_{n=1}^{n_{max}} \rho_{1n}^{3\sigma}} = \frac{1}{1 + \frac{1}{\rho_{1M}^{3\sigma}} \sum_{n=1}^{n_{max}} \rho_{1n}^{3\sigma}}, \tag{12.123}$$

$$\rho_{1M}^{3\sigma} = \rho''(T_{1M}^{3\sigma}), \tag{12.124}$$

$$\sum_{n=1}^{n_{max}} \rho_{1n}^{3\sigma} = \frac{p - p'(T_1^{3\sigma})}{R_{\Sigma n1} T_1^{3\sigma}}, \tag{12.125}$$

where

$$R_{\Sigma n1} = 287.04 \tag{12.126}$$

is the gas constant for air.

In these cases the mass-transfer coefficient is computed from an appropriate heat-transfer correlation, e.g., the correlation of *Ranz* and *Marschall* (1952) replacing the *Nu* and *Pr* numbers with *Sh* and *Sc* numbers, respectively

$$Sh = \frac{\beta D_3}{D_{M \to \Sigma n}} = 2 + 0.6 Re_{13}^{1/2} Sc_1^{1/3}, \tag{12.127}$$

where

$$Re_{13} = \rho_1 D_1 \Delta V_{13} / \eta_1 , \tag{12.128}$$

$$Sc_1 = \eta_1 / \left(\rho_1 D_{M \to \Sigma n} \right). \tag{12.129}$$

An additional simplifying assumption

$$T_{1M}^{3\sigma} \approx T_3 \tag{12.130}$$

completes this model.

Having in minded that

$$\mu_{13} = 0, \tag{12.131}$$

$$\dot{q}_1^{m3\sigma} = 0, \tag{12.132}$$

we obtain from the energy-jump condition the heat extracted from the droplet during the slow evaporation per unit mixture volume and unit time

$$\dot{q}_3^{m1\sigma} = -\mu_{31} \left[h''(p) - h_3^{1\sigma} \right]. \tag{12.133}$$

In order to obtain the averaged value of the mass source term, we use the sum of the mass-conservation equations for the inert components in the gas mixture, neglecting convection and diffusion from the neighbor control volumes,

$$\alpha_1 \rho_1 \frac{d}{d\tau} \sum C_{1n} = -\mu_{31} \sum C_{1n} = a_{13} \rho_1 \beta \sum C_{1n} \left(\sum C_{1n}^{3\sigma} - \sum C_{1n} \right) \tag{12.134}$$

or after the substitution

$$\overline{\tau} = \frac{\tau a_{13} \rho_1 \beta \sum C_{1n}^{3\sigma}}{\alpha_1}, \tag{12.135}$$

$$C = \frac{\sum C_{n1}}{\sum C_{1n}^{3\sigma}}, \tag{12.136}$$

$$\frac{dC}{d\overline{\tau}} = C(1 - C). \tag{12.137}$$

The solution of this equation for $\tau = 0$, $C = C_0$ is

$$\frac{1-C}{1-C_0} = \frac{1}{1+C_0\left(e^{\bar{\tau}}-1\right)}. \tag{12.138}$$

Using the so-obtained solution, we compute the *averaged* value of the evaporated mass per unit mixture volume and unit time within the time interval $\Delta\tau$,

$$\mu_{31} = -a_{13}\rho_1\beta\left(\sum C_{1n}^{3\sigma} - \sum C_{n1}\right)f, \tag{12.139}$$

where

$$f = \frac{1}{1-C_0}\left\{1 - \frac{1}{\Delta\bar{\tau}}\ln\left[1 - C_0\left(1-e^{\bar{\tau}}\right)\right]\right\} \le 1. \tag{12.140}$$

This expression takes into account the *reduction of the driving forces* of the evaporation under the above-discussed conditions.

There are other proposals in the literature on how to compute $(\rho w)_{31}$. The simplest of them is derived by *Spalding* (1953) as an introduction to this subject from the conservation noninert mass and energy

$$C_{1M}\left(\rho w\right)_{1,conv} + \left(\rho w\right)_{1M}^{3\sigma} - C_{1M}^{3\sigma}\left[\left(\rho w\right)_{1,conv} + \left(\rho w\right)_{1M}^{3\sigma}\right] = 0,$$

$$\left(\rho w\right)_{1,conv} h_1 + \left(\rho w\right)_{1M}^{3\sigma} h_1^{3\sigma} - \left[\left(\rho w\right)_{1,conv} + \left(\rho w\right)_{1M}^{3\sigma}\right] h_1^{3\sigma} = -\dot{q}_1^{\prime\prime 3\sigma},$$

written for a control volume containing the gas-site boundary layer. The total mass flow rate from the gas bulk is $\left(\rho w\right)_{1,conv}$ enters the boundary layer with bulk properties C_{1M} and h_1. The diffusion-controlled evaporation mass flow rate enters the control volume with $h_1^{3\sigma}$. Both mass flow rates then leave the control volume with $h_1^{3\sigma}$. Solving with respect to the both mass flow rates *Spalding* obtained simply

$$\left(\rho w\right)_{1M}^{3\sigma} = \frac{C_{1M} - C_{1M}^{3\sigma}}{C_{1M}^{3\sigma} - 1}\left(\rho w\right)_{1,conv},$$

where the so-called *Reynolds* mass flow rate is $\left(\rho w\right)_{1,conv} = h_1^{3\sigma}/c_{p1}$ and therefore

$$\left(\rho w\right)_{1M}^{3\sigma} = \frac{C_{1M} - C_{1M}^{3\sigma}}{C_{1M}^{3\sigma} - 1}\frac{h_1^{3\sigma}}{c_{p1}}.$$

Spalding (1953) also computed the diffusion mass flow rate at macroscopic surface using *Stefan*'s equation in terms of mass concentrations

$$\left(\rho w\right)_{1M}^{3\sigma} = \rho_1 \beta \ln\left(1+Sp\right) = \rho_1 \beta \ln C \,, \tag{12.141}$$

where

$$Sp = \frac{C_{1M}^{3\sigma} - C_{1M}}{1 - C_{1M}^{3\sigma}} \,, \tag{12.142}$$

is the *Spalding* mass transfer number and $C_{1M}^{3\sigma}$ is the vapor mass concentration at the boundary layer. Finally, this equation solved with the energy-jump condition across the interface with respect to the interface temperature and evaporation mass flow rate is the requested solution of the problem.

Clift et al. (1978) use for the region $0.25 < Sc_1 < 100$ and $\mathrm{Re}_{13} < 400$ the following correlation to compute

$$Sh = 1 + \left(1 + Re_{13} Sc_1\right)^{1/3} f\left(Re_{13}\right), \tag{12.143}$$

$$f\left(Re_{13}\right) = 1 \qquad \text{for} \qquad Re_{13} < 1 \,, \tag{12.144}$$

$$f\left(Re_{13}\right) = Re_{13}^{0.077} \qquad \text{for } 1 < Re_{13} < 400 \,. \tag{12.145}$$

Approximating numerical results *Chiang* and *Sirignano* (1990) use the following correlation

$$\left(\rho w\right)_{31} = \rho_1 \beta \frac{Sp}{\left(1+Sp\right)^{0.7}} \,, \tag{12.146}$$

where β is computed by

$$Sh = 2 + 0.46 Re_{13}^{0.6} Sc_1^{1/3} \,. \tag{12.147}$$

The validity of this approach is recommended by the authors within the region $0.3 < Sp < 4.5$, $30 < Re_{13} < 250$.

Tanaka (1980) uses the following equation

$$\left(\rho w\right)_{31} = \rho_1 \beta \left(C_{1M}^{3\sigma} - C_{1M}\right) 1.39 \frac{\left(\dfrac{\sum C_{1n}^{3\sigma}}{\sum C_{1n}}\right)^{0.52}}{\left(1 + \dfrac{\sum C_{1n}^{3\sigma}}{\sum C_{1n}}\right)^{0.48} \sum C_{1n}} \tag{12.148}$$

where the mass-transfer coefficient β was computed by means of the *Ranz* and *Marchal* (1952) correlation.

Abramson and *Sirignano* (1989) used the *Spalding* correlation by computing the mass-transfer coefficient taking into account the heat transfer from the droplet bulk to the surface by modifying the *Ranz* and *Marchal* equation for the *Sh* number as follows

$$Sh = 2 + B\frac{Sh_{RM} - 2}{(1+Sp)^{0.7}\ln(1+Sp)}, \tag{12.149}$$

where

$$B = \frac{c_{p3}(T_1 - T_3)}{h'' - h'}. \tag{12.150}$$

All properties for Re_{13}, Sc_1, except ρ_1 are computed at reference parameters, C_{ref}, defined as follows

$$C_{ref} = \frac{1}{n}\left[(n-1)C_{1M}^{3\sigma} + C_{1M}\right], \; n = 3. \tag{12.151}$$

Next, we remove the simplifying assumption $T_3^{\sigma 1} \approx T_3$.

The time-averaged heat transfer from the bulk of the droplet to the surface is

$$\dot{q}_3'''^{1\sigma} = \alpha_3\rho_3 c_{p3}\frac{\overline{T}_{3m}(\Delta\tau)}{\Delta\tau}\left(T_3^{1\sigma} - T_3\right) \equiv \chi_3^{1\sigma}\left(T_3^{1\sigma} - T_3\right). \tag{12.152}$$

The time-averaged convective heat transfer from the surface to the gas is

$$\dot{q}_1'''^{3\sigma} = 6\alpha_3\lambda_1 Nu_{13}\left(T_3^{1\sigma} - T_1\right) f_1 / D_3^2 = \chi_1^{3\sigma}\left(T_3^{1\sigma} - T_1\right). \tag{12.153}$$

In this case the energy-jump condition also has to be satisfied

$$(\rho w)_{1M}^{3\sigma}\frac{6\alpha_3}{D_3} = -\frac{\chi_3^{1\sigma}\left(T_3^{1\sigma} - T_3\right) + \chi_1^{3\sigma}\left(T_1^{3\sigma} - T_1\right)}{h_{M1}^{3\sigma} - h_{M3}^{1\sigma}}, \tag{12.154}$$

where $T_3^{1\sigma} = T_1^{3\sigma}$ and $T_1^{3\sigma} = T'\left(p_1^{3\sigma}\right)$. Using *Stefan*'s equation and solving with respect to the surface vapor mass concentration results in

$$C_{1M}^{2\sigma} = 1 - (1 - C_{1M})\exp\left[\frac{D_3^2}{6\alpha_3\rho_1 Sh D_{M\to\Sigma n}}\frac{\chi_3^{1\sigma}\left(T_3^{1\sigma} - T_3\right) + \chi_1^{3\sigma}\left(T_1^{3\sigma} - T_1\right)}{h_{M1}^{3\sigma} - h_{M3}^{1\sigma}}\right]. \tag{12.155}$$

Because the interface temperature is a function of the interface concentration the above equation has to be solved by iteration. A method that works is: compute the partial surface pressure of the vapor by

$$p_{1M}^{3\sigma} = p\frac{C_{1M}^{3\sigma}}{M_{1M}}\Bigg/\left[\frac{C_{1M}^{3\sigma}}{M_{1M}} + \frac{1 - C_{1M}^{3\sigma}}{M_{1n}}\right], \tag{12.156}$$

and then compute the surface temperature $T_1^{3\sigma} = T'\left(p_{1M}^{3\sigma}\right)$. Only view cycles are necessary. Note also the evaporation enthalpy

$$h_{M1}^{3\sigma} - h_{M3}^{1\sigma} = h''\left(T_1^{3\sigma}\right) - h'\left(T_1^{3\sigma}\right) = f\left(T_1^{3\sigma}\right) \tag{12.157}$$

at the interface is also a function of the interface temperature $T_1^{3\sigma}$. Having $T_3^{\sigma 1}$ it is a simple matter to compute $\left(\rho w\right)_{31}$, $\dot{q}_3^{'''\sigma 1}$ and $\dot{q}_1^{'''\sigma 3}$.

The method to describe vapor *condensation* from a gas mixture on a liquid droplet is similar and will not be described separately.

Nomenclature

Latin

a_1	Temperature conductivity of gas, m/s
a_3	Temperature conductivity of droplet, m/s
a_1	$= \sqrt{\dfrac{8}{\pi} R_1 T_1}$, thermal velocity of the vapor molecules, m/s
C_{M1}	mass concentration of the noninert component in the gas, dimensionless
C_{M3}	mass concentration of the noninert component in the droplet, dimensionless
c_p	specific capacity at constant pressure, J/(kg K)
c_c	≈ 0.1 for water steam plus noncondensable gases, ≈ 0.5 to 1, for pure water-steam, part of the molecules striking the nucleus and remaining there, dimensionless
c_T	temperature correction factor taking into account the heat transfer from the nucleus to the surrounding vapor, dimensionless
D_3	droplet diameter, m
$D_{M\to\Sigma n}$	diffusions coefficient, m/s
F_{3c}	surface of the cluster, m²
f	factor improving the conduction due to internal recirculation, dimensionless
Gb_1	$= \dfrac{\Delta E_{3c}}{kT_1}$, Gibbs number, dimensionless
k	$= 13.805 \times 10^{-24}$ is the Boltzmann constant, J/K

m_μ molar mean molecular mass, as many kilograms of a given substance as its molecular mass is, e.g., $m_{\mu,H_2O}=18$, $m_{\mu,Air}=28.9$, kg/kmol

m_1 $=\dfrac{m_\mu}{N_A}$ mass of a single molecule, kg

n =1, 2, 3, 4…

N_A $=6.02\times10^{23}$, *Avogadro* number, 1/mol

N_1 $=\rho_1/m_1$, number of the vapor molecules per unit volume, 1/m³

$\dot{n}_{3,kin}$ generated nuclei per unit volume and unit time, 1/(m³ s)

n_3 number of droplets per unit flow, 1/m³

n_{3c} $=\rho'(T_1)\dfrac{4}{3}\pi R_{3c}^3/m_1$, number of the molecules in a single nucleus with critical size, 1/m³

Nu Nusselt number, dimensionless

p local static pressure, Pa

p' $=p'(T_1)$, saturation pressure as a function of the steam temperature, Pa

p_c critical pressure, for water 221.2×10^5, Pa

Pe_1^* $=\dfrac{D_3\left|\Delta V_{13}\right|}{a_1}\dfrac{\eta_1}{\eta_1+\eta_3}$, *Peclet* number is used, dimensionless

p local static pressure, Pa

p' $=p'(T_1)$, saturation pressure as a function of the steam temperature, Pa

$\dot{q}_3^{\prime\prime\prime 1\sigma}$ power per unit volume of the gas–droplet mixture introduced into the droplet field from the interface with the gas, W/m³

$\dot{q}_1^{\prime\prime\prime 3\sigma}$ power per unit volume of the gas–droplet mixture introduced into the droplet field from the interface with the gas, W/m³

r radius, m

R_3 droplet radius, m

R_{30} radius of a subcooled nonstable spherical steam volume, m

R_{3c} critical nucleus radius, m

Re_{13} $=\rho_1 D_1\Delta V_{13}/\eta_1$, gas Reynolds number, dimensionless

Sc_1 $=\eta_1/\left(\rho_1 D_{M\to\Sigma n}\right)$, gas Schmidt number, dimensionless

Sh $=\dfrac{\beta D_3}{D_{M\to\Sigma n}}$, Sherwood number, dimensionless

T_1 steam at local temperature, K

T' $=T'(p)$, saturation temperature at the system pressure, K

T_c critical temperature, K

T_{c,H_2O} $=647.2$, critical temperature of water, K

$T_{1,spin}$ temperature at the spinoidal line, K

T_1 steam at local temperature, K

T' $=T'(p)$, saturation temperature at the system pressure, K

$\bar{T}(r,\tau)$ $=\dfrac{T(r,\tau)-T_3}{T_3^{1\sigma}-T_3}$, droplet temperature as a function of radius and time, dimensionless

w local velocity, m/s

Ze_1 $=\sqrt{\dfrac{Gb_1}{3\pi n_{3c}^2}}$, *Zeldovich* number, dimensionless

Greek

α volume fraction, dimensionless

β mass-transfer coefficient, m/s

β empirical coefficient, dimensionless

ΔE_3 work required to create a single nucleus, J

ΔE_{3c} work required to create a stable single nucleus, J

ΔS_{13c} $=\Delta E_{3c}/T_1$, entropy increase required to create a stable single nucleus, J/K

$\Delta T_{1,sub}$ $=T'(p)-T_1$, subcooling, K

$\Delta\tau_3$ $=D_3^2/\left(4\pi^2 a_3\right)$ Characteristic heat conduction droplet time constant, s

$\Delta\tau_1^*$ $=\dfrac{\alpha_1\rho_1 c_{p1} D_3^2}{6\alpha_3\lambda_1 Nu_3}$, characterizing time constant of the continuum heating or cooling, s

ΔV_{13} difference between the velocity of gas and velocity of droplet, m/s

λ thermal conductivity, W/(m² K)

μ_{13} condensation mass per unit time and unit mixture volume, nonnegative, kg/(m³ s)

μ_{31} evaporation mass per unit time and unit mixture volume, nonnegative, kg/(m³ s)

η_1 dynamic viscosity of gas, kg/(m s)

η_3 dynamic viscosity of droplet, kg/(m s)

ρ density, kg/m³

$(\rho w)_{1M}^{3\sigma}$ $=\rho_1\beta\ln\dfrac{1-C_{1M}}{1-C_{1M}^{2\sigma}}$, diffusion-controlled evaporation mass flow rate, kg/(m² s)

ρ_1 $=\rho_1(p,T_1)$, steam density, kg/m³

σ	surface tension, N m
τ	time, s
$\chi_3^{1\sigma}$	$=\alpha_3\rho_3 c_{p3}\dfrac{\overline{T}_{3m}(\Delta\tau)}{\Delta\tau}$, volumetric heat-transfer coefficient surface of the droplet into the droplet, W/(m³ K)
$\chi_1^{3\sigma}$	volumetric heat-transfer coefficient surface of the gas into the gas, W/(m³ K)
ψ_{13}	digital switch, dimensionless

Subscripts

1	gas
3	droplets
3m	averaged over the droplet volume

Superscripts

$^{1\sigma}$	at the surface to gas
$^{3\sigma}$	at the surface to droplet

References

Abramson, B., Sirignano, W.A.: Droplet vaporization model for spray combustion calculations. Int. Heat Mass Transfer 32, 1605–1618 (1989)

Acrivos, A., Goddard, J.: Asymptotic expansion for laminar forced-convection heat and mass transfer. Part 1, J. Fluid Mech. 23, 273 (1965)

Acrivos, A., Taylor, T.D.: Heat and mass transfer from single spheres in Stokes flow. Phys. Fluids 5(4), 387–394 (1962)

Aladyev, I.T., Kondratiev, N.S., Mukhin, V.A., Kipshidze, M.E., Parfentyva, Y., Kisselev V.V.: Paper Third Int. Heat Transfer Conf., USA, Paper 74 (1966)

Bankoff, S.S.: Some Condensation Studies Pertinent to LWR Safety. Int. J. Multiphase Flow 6, 51–67 (1980)

Berman, L.D.: Soprotivlenie na granize razdela fas pri plenochnoi kondensazii para nizkogo davleniya, Tr. Vses. N-i, i Konstrukt In-t Khim. Mashinost 36, 66 (1961)

Boussinesq, M.: Calcul du pourvoir retroidissant des courant fluids. J. Math. Pure Appl. 1, 285 (1905)

Brauer, H.: Atomkernenergie 5(4) (1961)

Carslaw, H.S., Jaeger, J.C.: Conduction of heat in solids, 2nd edn. Oxford Science Publications, Clarendon Press, Oxford (1996)

Celata, G.P., Cumo, M., D'Annibale, F., Farello, G.E.: Direct contact condensation of steam on droplets. Int. J. Multiphase Flow 17(2), 191–211 (1991)

Chiang, C.H., Sirignano, W.A.: Numerical analysis of convecting and interacting vaporizing fuel droplets with variable properties. Presented at the 28th AIAA Aerospace Sciences, Mtg, Reno (November 1990)

Clift, R., Ggarce, J.R., Weber, M.E.: Bubbles, drops, and particles. Academic Press, New York (1978)
Condie, K.G., et al.: Comparison of heat and mass transfer correlation with forced convective non equilibrium post-CHF experimental data. In: Dhir, V.K., Schrock, V.E. (eds.) Proc. of 22nd Nat. Heat Transfer Conf. & Exhibition, Basic Aspects of Two-Phase Flow and Heat Transfer, Niagara Falls, New York, August 5-8, pp. 57–65 (1984)
Deitsch, M.E., Philiphoff, G.A.: Two phase flow gas dynamics. Energoisdat, Moscu (1981) (in Russian)
Fedorovich, E.D., Rohsenow, W.M.: The effect of vapor subcooling on film condensation of metals. Int. J. Heat Mass Transfer 12, 1525–1529 (1968)
Feng, Z.-G., Michaelides, E.E.: Unsteady heat transfer from a spherical particle at finite Peclet numbers. ASME J. Fluid Eng. 118, 96–102 (1986)
Fieder, J., Russel, K.C., Lothe, J., Pound, G.M.: Homogeneous nucleation and growth of droplets in vapours. Adv. Phys. 15, 111–178 (1966)
Ford, J.D., Lekic, A.: Rate of growth of drops during condensation. Int. J. Heat Mass Transfer 16, 61–64 (1973)
Frenkel, F.I.: Selected works in gasdynamics. Nauka, Moscu (1973) (in Russian)
Furth, R.: Proc. Cambr. Philos. Soc. 37, 252 (1941)
Gibbs, J.W.: Thermodynamische Studien, Leipzig 1982. Am. J. Sci. Arts XVI, 454–455 (1878)
Hertz, H.: Wied. Ann 17, 193 (1882)
Knowles, J.B.: A mathematical model of vapor film destabilization. Report AEEW-R-1933 (1985)
Knudsen, M.: Ann. Physik 47, 697 (1915)
Labunzov, D.A., Krjukov, A.P.: Processes of intensive flushing. Thermal Engineering 24(4), 8–11 (1977) (in Russian)
Langmuir, I.: Physik. Z. 14, 1273 (1913)
Langmuir, I., Jones, H.A., Mackay, G.M.J.: Phys. Rev. 30, 201 (1927)
Lee, K., Ryley, D.J.: The evaporation of water droplets in superheated steam. J. Heat Transfer 90 (November 1968)
Levich, V.G.: Physicochemical Hydrodynamics. Prentice-Hall, Englewood Cliffs (1962)
Lienhard, J.H.: A heat transfer textbook. Prentice-Hall, Inc., Englewood Cliffs, New Jersey 07632
Ludvig, A.: Untersuchungen zur spontaneous Kondensation von Wasserdampf bei stationaerer Ueberschalllstroemung unter Beruecksichtigung des Realgasverhaltens. Dissertation, Universitaet Karlsruhe, TH (1975)
Mason, B.J.: Spontaneous condensation of water vapor in expansion chamber experiments. Proc. Phys. Soc. London, Serie B 64, 773–779 (1951)
Mason, B.J.: The Physics of Clouds. Clarendon Press, Oxford (1957)
Michaelides, E.E.: Hydrodynamic force and heat/mass transfer from particles, bubbles and drops – The Freeman Scholar Lecture. ASME J. Fluid Eng. 125, 209–238 (2003)
Mills, A.F.: The condensation of steam at low pressure. Techn. Report Series No. 6(39). Space Sciences Laboratory, University of California, Berkeley
Mills, A.F., Seban, R.A.: The condensation coefficient of water. J. Heat Transfer 10, 1815–1827 (1967)
Nabavian, K., Bromley, L.A.: Condensation coefficient of water. Chem. Eng. Sci. 18, 651–660 (1963)
Nigmatulin, R.I.: Basics of the mechanics of the heterogeneous fluids. Nauka, Moskva (1978) (in Russian)
Paul, B.: Compilation of evaporation coefficients. J. Am. Rocket Soc. 32, 1321 (1962)

Ranz, W., Marschall Jr., W.: Evaporation from drops. Chem. Eng. Prog. 48, 141–146 (1952)

Saha, P.: Int. J. Heat Mass Transfer 23, 481 (1980)

Samson, R.E., Springer, G.S.: Condensation on and evaporation from droplets by a moment method. J. Fluid Mech. 36, 577–584 (1969)

Skripov, W.P., Sinizyn, E.N., Pavlov, P.A., Ermakov, G.W., Muratov, G.N., Bulanov, N.B., Bajdakov, W.G.: Thermophysical properties of liquids in metastable state. Atomisdat, Moscu (1980) (in Russia)

Soo, S.L.: Fluid dynamics of multiphase systems. Woltham, Massachusetts (1969)

Spalding, D.B.: The combustion of liquid fuels. In: Proc. 4th Symp (Int.) on Combustion, pp. 847–864. Williams & Wilkins, Baltimore (1953)

Subbotin, V.I., Ivanovskiy, M.H., Sorokin, V.P., Chulkov, B.: Obstie vaprosy teplo i massobmena, Nauka i technika, Minsk (1966)

Tanaka, M.: Heat transfer of a spray droplet in a nuclear reactor containment. Nucl. Technol. 47, 268 (1980)

VDI-Waermeatlas, 4. Auflage VDI-Verlag, Dusseldorf (1984)

Volmer, M.: Kinetik der Phasenbildung, Dresden und Leipzig, Verlag von Theodor Steinkopff (1939)

Watts: Vlijanie krivisnji na teplo- I massootdachu ot izotemicheskoj sfere. Teploperedacha, Mir 1, 1 (1972)

Zeldovich, J.B.: To the theory of origination of the new phase, cavitation. J. Exp. Theor. Phys. 12(11/12), 525–538 (1942)

13. Heat and mass transfer at the film/gas interface

13.1 Geometrical film–gas characteristics

There are different flow patterns leading to a film/gas interface in multiphase flows. Stratified and annular flows in channels with different geometry, walls of large pools with different orientations and part of the stratified liquid gas configuration in large pools are the flow patterns of practical interest. The problem of the mathematical modeling of interface heat and mass transfer consists of (a) macroscopic predictions of the geometrical sizes and local averaged temperatures, velocities, concentrations and pressure and (b) microscopic modeling of the interface heat and mass transfer. The purpose of this chapter is to review the state-of-the-art for modeling of the interface heat and mass transfer for known geometry.

Next, we summarize some important geometrical characteristics needed further.

The cross-section occupied by gas and liquid in a channel flow is

$$F_1 = \alpha_1 F \tag{13.1}$$

and

$$F_2 = \alpha_2 F \, , \tag{13.2}$$

respectively, where F is the channel cross-section. Depending on the gas velocity the film structure can be

(a) symmetric; or
(b) asymmetric.

The symmetric film structure is characterized by uniformly distributed film on the wet perimeter in the plane perpendicular to the main flow direction. For this case the film thickness is

$$\delta_{2F} = D_h \left(1 - \sqrt{1 - \alpha_2}\right)/2 \, , \tag{13.3}$$

where D_h is the hydraulic diameter of the channel in a plane perpendicular to the main flow direction.

The liquid/gas interface per unit flow volume that is called the interfacial area density is

$$a_{12} = \frac{4}{D_h}\sqrt{1-\alpha_2} \,. \tag{13.4}$$

The hydraulic diameter of the gas is

$$D_{h1} = D_h \sqrt{1-\alpha_2} \,. \tag{13.5}$$

Asymmetric film structure results in the case of dominance of the gravitation force. The flow can be characterized by three perimeters, one of the gas–wall contact, Per_{1w}, one of the gas–liquid contact, Per_{12}, and one of the liquid-wall contact, Per_{2w}. To these three perimeters correspond three shear stresses, one between the gas and wall, τ_{1w}, one between gas and liquid, τ_{12}, and one between the liquid and wall, τ_{2w}. The hydraulic diameters for computation of the friction pressure loss of the both fluids can be defined as

$$D_{h1} = 4\alpha_1 F / \left(Per_{1w} + Per_{12}\right), \tag{13.6}$$

$$D_{h1} = 4\alpha_2 F / \left(Per_{2w} + Per_{12}\right). \tag{13.7}$$

Two characteristic *Reynolds* numbers can be defined by using these length scales

$$Re_1 = \frac{\rho_1 V_1 D_{h1}}{\eta_1} = \frac{\alpha_1 \rho_1 V_1}{\eta_1} \frac{4F}{Per_{1w} + Per_{12}}, \tag{13.8}$$

$$Re_2 = \frac{\rho_2 V_2 D_{h2}}{\eta_2} = \frac{\alpha_2 \rho_2 V_2}{\eta_2} \frac{4F}{Per_{2w} + Per_{12}}. \tag{13.9}$$

In the three-dimensional space it is possible that the liquid in a computational cell is identified to occupy the lower part of the cell. In this case the gas/liquid interfacial area density is

$$a_{12} = 1/\Delta z \,. \tag{13.10}$$

The film thickness in this case is

$$\delta_{2F} = \alpha_2 \Delta z \,, \tag{13.11}$$

where the α_2 is the local liquid volume fraction in the computational cell. In the case of film attached at the vertical wall of radius r of a control volume in cylindrical coordinates

$$\delta_{2F} = r\left\{1-\sqrt{1-\alpha_2\left[1-\left(r_{i-1}/r\right)^2\right]}\right\}. \tag{13.12}$$

For the limiting case of $r_{i-1} = 0$ Eq. (13.12) reduces to Eq. (13.3). The interfacial area density in this case is

$$a_{12} = 2\frac{1-\delta_{2F}/r}{r\left[1-\left(r_{i-1}/r\right)^2\right]} = 2\frac{\sqrt{1-\alpha_2\left[1-\left(r_{i-1}/r\right)^2\right]}}{r\left[1-\left(r_{i-1}/r\right)^2\right]}. \tag{13.13}$$

For the case of $r_{i-1} = 0$ Eq. (13.13) reduces to Eq. (13.4). For such flow patterns the following film *Reynolds* number is used

$$Re_{2F} = \frac{\rho_2 V_2 \delta_{2F}}{\eta_2}.$$

13.2 Convective heat transfer

Next, we consider a flow without mass transfer. Detailed consideration of the convective heat transfer is useful not only for flows without mass transfer but also for flow with mass transfer at the film/gas interface.

The gas and liquid temperatures are T_1 and T_2, respectively. The interface temperature, which plays an important role in the mathematical description is $T_2^{1\sigma}$. The total heat-transfer mechanism can be considered as

(a) heat transfer from the interface into the gas, $\dot{q}_1'''^{2\sigma}$; and

(b) heat transfer from the interface into the bulk liquid $\dot{q}_2'''^{1\sigma}$.

13.2.1 Gas-side heat transfer

To model the heat transfer between the interface and gas we can use known correlations for heat transfer from gas to solid surface

$$\dot{q}_1'''^{2\sigma} = a_{12} h_1^{2\sigma}\left(T_1^{2\sigma} - T_1\right). \tag{13.14}$$

Here, $h_1^{2\sigma}$ is the heat-transfer coefficient.

For laminar gas flow in channels, $Re_1 < 1450$ in accordance with *Hausen*, see in (1958), the maximum *Nusselt* (1915) number is 3.66, corrected for the effect of the temperature gradients,

$$h_1^{2\sigma} = 3.66 \frac{\lambda_1}{D_{h1}} \left(\frac{T_1}{T_1^{2\sigma}} \right)^{1/4} . \tag{13.15}$$

For turbulent gas flow in channels, $Re_1 < 1450$ in accordance with *McEligot*, see (1958), the *Nusselt* number is computed by modified similarity to momentum transfer

$$h_1^{2\sigma} = 0.021 \frac{\lambda_1}{D_{h1}} Re_1^{0.8} Pr_1^{0.4} \left(\frac{T_1}{T_1^{2\sigma}} \right)^{1/2} . \tag{13.16}$$

Sometimes this correlation is used in the form

$$\beta_1 = \frac{h_{1\sigma}}{\rho_1 c_{p1}} = \frac{Nu_1}{Re_1 Pr_1} V_1 = Sr_1 V_1 \tag{13.17}$$

or

$$\beta_1 = 0.021 Pr_1^{-0.6} Re_1^{-1/5} \left(\frac{T_1}{T_1^{2\sigma}} \right)^{1/2} V_1 . \tag{13.18}$$

Similar is the equation introduced by *Bunker* and *Carey* (1986) for gas-side heat transfer, namely

$$\beta_1 = 0.037 Pr_1^{-1/3} Re_1^{-1/5} V_1 . \tag{13.19}$$

The order of magnitude of the constants in the above equations, 0.021 to 0.037, was derived by comparison with experiments.

Vertical wall: In the case of natural circulation of gas at a vertical wall we can use the proposal made by *Miheev* and *Miheeva* (1960)

$$h_{1\sigma} \Delta z / \lambda_1 = 0.76 \left(Gr_1 Pr_1 \right)^{1/4} \left(\mathrm{Pr}_1 / \mathrm{Pr}_{1,w} \right)^{0.25}$$

for laminar flow, $10^3 < Gr_1 Pr_1 < 10^9$ and

$$h_{1\sigma} \Delta z / \lambda_1 = 0.15 \left(Gr_1 Pr_1 \right)^{1/3} \left(\mathrm{Pr}_1 / \mathrm{Pr}_{1,w} \right)^{0.25}$$

for turbulent flow, $10^9 < Gr_1 Pr_1$, where the *Grashof* number

$$Gr_1 = g \left(-\frac{1}{\rho_1} \frac{\partial \rho_1}{\partial T_1} \right) \left| T_1 - T_1^{2\sigma} \right| \Delta z^3 / \nu_1^2 \approx g \left(\left| T_1 - T_1^{2\sigma} \right| / T_1 \right) \Delta z^3 \left(\rho_1 / \eta_1 \right)^2 \tag{13.20}$$

is defined using the vertical size Δz of the considered wall and the ratio of the fluid Prandtl number to the fluid Prandtl number at wall temperature is $\Pr_1/\Pr_{1,w}$. The validation data base is given by *Miheev* and *Miheeva* (1960) on their Fig. 3-29, p.90.

One year later *Rohsenow* and *Choi* (1961) reported

$$h_{1\sigma}\Delta z/\lambda_1 = 0.56\left(Gr_1 Pr_1\right)^{1/4} \quad \left(10^4 < Gr_1 Pr_1 < 10^8 \text{ laminar}\right), \tag{13.21}$$

$$h_{1\sigma}\Delta z/\lambda_1 = 0.13\left(Gr_1 Pr_1\right)^{1/3} \quad \left(10^8 < Gr_1 Pr_1 < 10^{10} \text{ transition}\right), \tag{13.22}$$

$$h_{1\sigma}\Delta z/\lambda_1 = 0.021\left(Gr_1 Pr_1\right)^{2/5} \quad \left(10^{10} < Gr_1 Pr_1 \text{ turbulent}\right). \tag{13.23}$$

For a horizontal plane the *Jakob* correlation (1933) for natural circulation can be used

$$h_1^{2\sigma} = 2.417\lambda_1\left[g\left(\left|T_1 - T_1^{2\sigma}\right|^{20/16}/T_1\right)\left(\rho_1/\eta_1\right)^2 \Pr_1\right]^{1/4}. \tag{13.24}$$

The heat-transfer coefficient for forced convection at a plane surface can be computed using the classical *Reynolds* analogy, assuming similarity of the velocity and temperature profiles and no distortion of the boundary layer due to mass transfer

$$\beta_1 = \frac{\dot{q}_1''^{2\sigma}/\left(T_1 - T_1^{2\sigma}\right)}{\rho_1 c_{p1}} = \frac{1}{2}\lambda_{w1}V_1 \tag{13.25}$$

or

$$\beta_1 = \frac{h_{1\sigma}}{\rho_1 c_{p1}} = V_1\lambda_{w1}/2 \tag{13.26a}$$

or

$$Nu_{1z} = \mathrm{Re}_{1z}\Pr_1\lambda_{w1}/2, \tag{13.26b}$$

where

$$Nu_{1z} = h_{1\sigma}\Delta z/\lambda_1, \tag{13.27}$$

$$Re_{1z} = V_1\Delta z/\nu_1. \tag{13.28}$$

For plane walls the average shear stress along Δz is

$$\tau_{w1} = \lambda_{w1}\tfrac{1}{2}\rho_1 V_1^2, \tag{13.29}$$

where for laminar flow

$$Re_{1z} < (1 \text{ to } 3)\ 10^5, \tag{13.30}$$

in accordance with *Prandtl* and *Blasius*

$$\lambda_{w1} = 1.372 / Re_{1z}^{1/2}, \tag{13.31}$$

and for turbulent flow - see in *Albring* (1970)

$$\lambda_{w1} = 0.072 / Re_{1z}^{1/5}. \tag{13.32}$$

For the laminar flow Eq. (13.31) substituted in Eq. (13.26) gives

$$Nu_{1z} = 0.686 Re_{1z}^{1/2} Pr_1, \tag{13.33}$$

which is very similar to the widely used *Pohlhausen* equation (1921)

$$Nu_{1z} = 0.664 Re_{1z}^{1/2} Pr_1^{1/3} \tag{13.34}$$

recommended by *Gnielinski* (1976) after comparison with data for the laminar flow. For turbulent flow Eq. (13.26) was improved by *Petukhov* and *Popov* (1963)

$$Nu_{1z} = \left(Re_{1z} Pr_1 \lambda_{w1} / 2\right) / \left[1 + 12.7 \left(\lambda_{w1} / 2\right)^{1/2} \left(Pr_1^{2/3} - 1\right)\right]. \tag{13.35}$$

The influence of the change of the gas properties at the wall can be taken into account in accordance with *Churchill* and *Brier* (1955) by multiplying the results by $\left(T_1 / T_{1\sigma}\right)^{0.12}$. After comparing with a large number of data for turbulent flows *Gnielinski* recommended Eqs. (13.35) and (13.32). For comparison we give here the recently successfully used modification of the *Petukhov* and *Popov* result, by *Kim* and *Corradini* (1990)

$$Nu_{1z} = \left(Re_{1z} Pr_1 \lambda_{w1} / 2\right) / Pr_1^t, \tag{13.36}$$

where the turbulent *Prandtl* number is

$$Pr_1^t = c_1 + c_2 / Pr_1 \tag{13.37}$$

$$c_1 = 0.85, \tag{13.38}$$

$$c_2 = 0.012 \text{ to } 0.05 \text{ for } Re_{1z} = 2\times 10^4, \tag{13.39}$$

$$c_2 = 0.005 \text{ to } 0.015 \text{ for } Re_{1z} = 10^5. \tag{13.40}$$

In pure vapor condensation the condensate film provides the only heat-transfer resistance.

13.2.2 Liquid-side heat transfer due to conduction

Next, we consider the heat transfer from the interface into the bulk liquid. The liquid can be laminar or turbulent. For laminar liquid the heat is transferred from the interface into the liquid due to heat conduction described by the *Fourier* equation, (1822),

$$\frac{\partial T}{\partial \tau} = -a_2 \frac{\partial^2 T}{\partial y^2}, \tag{13.39}$$

where the positive y direction is defined from the interface into the bulk. The text book solution for the following boundary conditions

$$\tau \geq 0,\ y = 0,\ T = T \text{ (interface)}, \tag{13.40}$$

$$\tau = 0,\ y > 0,\ T = T_2 \text{ (bulk liquid)}, \tag{13.41}$$

$$\tau > 0,\ y \rightarrow 1,\ T = T_2 \text{ (bulk liquid)}, \tag{13.42}$$

is

$$T - T_2 = \left(T_2^{1\sigma} - T_2\right) erfc \frac{y}{2\left(a_2 \tau\right)^{1/2}}. \tag{13.43}$$

Using the temperature gradient

$$\frac{\partial T}{\partial \tau} = -\frac{T_2^{1\sigma} - T_2}{\left(\pi a_2 \tau\right)^{1/2}} \exp\left[-y^2 / \left(4\alpha_2 \tau\right)\right] \tag{13.44}$$

at $y = 0$ we compute the heat flux at the interface

$$\dot{q}_2^{\prime\prime 1\sigma} = -\lambda_2 \frac{\partial T}{\partial y}\bigg|_{y=0} = \sqrt{\frac{2a_2}{\pi \tau}} \rho_2 c_{p2} \left(T_2^{1\sigma} - T_2\right) = \frac{B}{\sqrt{\tau}} \rho_2 c_{p2} \left(T_{2\sigma} - T_2\right)$$
$$= \beta_2 \rho_2 c_{p2} \left(T_2^{1\sigma} - T_2\right). \tag{13.45}$$

The averaged heat flux over the time period $\Delta \tau$ is

$$\dot{q}_2^{\prime\prime 1\sigma} = B \rho_2 c_{p2} \left(T_2^{1\sigma} - T_2\right) 2 / \sqrt{\Delta \tau}. \tag{13.46}$$

Assuming $T_1 \approx$ const, $T_2 \approx$ const and using the energy-jump condition $\dot{q}_1^{\prime\prime 2\sigma} + \dot{q}_2^{\prime\prime 1\sigma} = 0$ in or

$$h_1^{2\sigma}\left(T_1^{2\sigma} - T_1\right) + \frac{2B}{\sqrt{\Delta \tau}} \rho_2 c_{p2} \left(T_2^{1\sigma} - T_2\right) = 0, \tag{13.47}$$

and assuming $T_2^{1\sigma} = T_1^{2\sigma}$, which is valid for the case of no mass transfer across the interface we obtain

$$T_2^{1\sigma} = \left(h_1^{2\sigma} T_1 + \frac{2B}{\sqrt{\Delta \tau}} \rho_2 c_{p2} T_2 \right) \Big/ \left(h_1^{2\sigma} + \frac{2B}{\sqrt{\Delta \tau}} \rho_2 c_{p2} \right). \tag{13.48}$$

Thus, having $T_2^{1\sigma}$ we compute for Eq. (13.46) the heat flux from the interface into the bulk liquid

$$\dot{q}_2''^{1\sigma} = h_{12}^* (T_1 - T_2), \tag{13.49}$$

where

$$\frac{1}{h_{12}^*} = \frac{1}{h_1^{2\sigma}} + \frac{1}{\dfrac{2B}{\sqrt{\Delta \tau}} \rho_2 c_{p2}}. \tag{13.50}$$

The heat transferred per unit control volume and unit time is

$$\dot{q}_2'''^{1\sigma} = -\dot{q}_1'''^{2\sigma} = a_{12} \dot{q}_2''^{1\sigma} . \tag{13.51}$$

13.2.3 Liquid-side heat conduction due to turbulence

Let $\Delta \tau_2$ be the average time in which a turbulent eddy stays in the neighborhood of the free surface before jumping apart, sometimes called the renewal period. During this time the average heat flux from the interface to the eddies by heat conduction is

$$\dot{q}_2''^{1\sigma} = \left(2B \big/ \sqrt{\Delta \tau_2} \right) \rho_2 c_{p2} \left(T_2^{1\sigma} - T_2 \right) = \beta_2 \rho_2 c_{p2} \left(T_2^{1\sigma} - T_2 \right). \tag{13.52}$$

If the volume-averaged pulsation velocity is V_2' and the turbulent length scale in the film is ℓ_{e2} the time in which the eddy stays at the interface is

$$\Delta \tau_2 = f \left(\ell_{e2}, V_2' \right). \tag{13.53}$$

Thus, the task to model turbulent heat conduction from the interface into the liquid is reduced to the task to model

(i) pulsation velocity V_2'; and
(ii) turbulent length scale ℓ_{e2}.

Usually, β_2 is written as a function of the dimensionless turbulent *Reynolds* number

$$Re_2^t = \rho_2 V_2' \ell_{e2} / \eta_2, \tag{13.54}$$

and *Prandtl* number Pr_2,

$$\beta_2 = \sqrt{\frac{2a_2}{\pi \Delta\tau_2}} = \sqrt{\frac{2}{\pi}} Pr_2^{-1/2} \sqrt{\frac{\eta_2}{\rho_2 \Delta\tau_2}} = \sqrt{\frac{2}{\pi}} Pr_2^{-1/2} Re_2^{t-1/2} \sqrt{\frac{V_2' \ell_{e2}}{\Delta\tau_2}}, \qquad (13.55)$$

and is a subject of modeling work and verification with experiments. Once we know β_2 we can compute some effective conductivity by the equation

$$\sqrt{\frac{2\lambda_{2eff}}{\rho_2 c_{p2} \pi \Delta\tau_2}} \rho_2 c_{p2} \left(T_{2\sigma} - T_2\right) = \beta_2 \rho_2 c_{p2} \left(T_{2\sigma} - T_2\right) \qquad (13.56)$$

or

$$\lambda_{2eff} = \frac{\pi}{2} \rho_2 c_{p2} \Delta\tau_2 \beta_2^2, \qquad (13.57)$$

and use Eq. (13.46) for both laminar and turbulent heat transfer.

13.2.3.1 High Reynolds number

One of the possible ways for computation of the renewal period $\Delta\tau_2$ for high turbulent *Reynolds* numbers

$$Re_2^t > 500 \qquad (13.58)$$

is the use the hypothesis by *Kolmogoroff* for isotropic turbulence

$$\Delta\tau_2 \approx c_1 \left(\nu_2 \ell_{e2} / V_2'^3\right)^{1/2}, \qquad (13.59)$$

where

$$\ell_{e2} \approx c_1 \delta_{2F}, \qquad (13.60)$$

where c_1 is a constant. In this region the turbulence energy is concentrated in microscopic eddies (mechanical energy dissipating motion). Substituting $\Delta\tau_2$ from Eq. (13.61) into Eq. (13.57) we obtain

$$\beta_2 = \sqrt{\frac{2}{\pi}} Pr_2^{-1/2} Re_2^{t-1/4} V_2'. \qquad (13.61)$$

The qualitative relationship $\beta_2 Pr_2^{1/2} \approx Re_2^{t-1/4}$ was originally proposed by *Banerjee* et al. in (1968) and experimentally confirmed by *Banerjee* et al. (1990). *Lamont* and *Scot*, see in *Lamont* and *Yuen* (1982), describe successfully heat transfer from the film surface into the flowing turbulent film for high Reynolds numbers using Eq. (13.61) with a constant 0.25 instead of $\sqrt{2/\pi}$. *Magnaudet* and *Calmet* (2006) confirmed the constant 0.25 by direct numerical simulation. *Hobbhahn* (1989) obtained experimental data for condensation on a free surface, which are successfully described by Eq. (13.61) modified as follows

$$\beta_2 = 0.07 Pr_2^{-0.6} Re_2^{t-1/5} V_2', \tag{13.62}$$

where the discrepancy between the assumption made by the authors

$$\ell_{e2} \approx \delta_{2F} \tag{13.63}$$

and

$$V_2' \approx |V_2| + \sqrt{\rho_1 / \rho_2}\, |V_1 - V_2| \tag{13.64}$$

and the reality are compensated by the constant 0.07.

13.2.3.2 Low Reynolds number

For low turbulence *Reynolds* number

$$Re_2^t < 500 \tag{13.65}$$

the *Kolmogoroff* time scale is no longer applicable. In this region the turbulence energy is concentrated in macroscopic eddies (mechanical energy containing motion). The choice of the liquid film thickness, as a length scale of turbulence

$$\ell_{e2} \approx \delta_{2F} \tag{13.66}$$

is reasonable. Therefore, the renewal period is

$$\Delta\tau_2 \approx \delta_{2F} / V_2' \tag{13.67}$$

and Eq. (13.55) reduces to

$$\beta_2 = \sqrt{\frac{2}{\pi}} Pr_2^{-1/2} Re_2^{t\ 1/2} V_2'. \tag{13.68}$$

For low *Reynolds* numbers *Fortescue* and *Pearson* (1967) recommended instead of $\sqrt{2/\pi}$ in Eq. (13.68) to use

$$\text{const} = 0.7(1+0.44 / \Delta\tau_2^*), \tag{13.69}$$

where

$$\Delta\tau_2^* = \Delta\tau_2 V_2' / \ell_{e2} \approx 1, \tag{13.70}$$

which is > 0.85, as recommended by *Brumfield* et al. (1975).

Theofanous et al. (1976) show that most of their experimental data for condensation on channels with free surface can be described by using Eqs. (13.61) and (13.68) with the above-discussed corrections introduced by *Lamont* and *Scot*, and *Fortescue* and *Person*, respectively.

13.2.3.3 Pulsation velocity

a) Time scale of turbulence pulsation velocity based on average liquid velocity.

A very rough estimate of the pulsation velocity is

$$V' = c_1 V_2 \approx (0.1 \ to \ 0.3) \ V_2 . \tag{13.71}$$

With this approach and

$$\ell_{e2} \approx c_2 \delta_{2F} \tag{13.72}$$

the two equations for β_2 read

$$Re_{2F} < c_1 500 , \tag{13.73}$$

$$\beta_2 = c_2^{-1/4} c_1^{3/4} \sqrt{\frac{2}{\pi}} Pr_2^{-1/2} Re_{2F}^{-1/4} V_2 = const Pr_2^{-1/2} Re_{2F}^{-1/4} V_2 . \tag{13.74}$$

Comparing this equation with the *McEligot* equation we see that

$$const \approx 0.021 \text{ to } 0.037 . \tag{13.75}$$

For

$$Re_2 < c_1 500, \tag{13.76}$$

$$\beta_2 = c_1^{1/2} \sqrt{\frac{2}{\pi}} Pr_2^{-1/2} Re_{2F}^{-1/2} V_2 . \tag{13.77}$$

b) The scale of turbulence pulsation velocity based on friction velocity.

Improvement of the above theory requires a close look at the reasons for the existence of turbulence in the liquid film. This is either the wall shear stress, τ_{2w}, or the shear stress acting at the gas/liquid interface, τ_{12}, or both simultaneously. The shear stress at the wall for channel flow is

$$\tau_{w2} = \frac{1}{4} \lambda_{w2} \frac{1}{2} \rho_2 V_2^2 , \tag{13.78}$$

where

$$\lambda_{w2} = \lambda_{friction} \left(\frac{\rho_2 V_2 D_{h2w}}{\eta_2}, k_w / D_{h2w} \right) . \tag{13.79}$$

Here,

$$D_{h2w} = \frac{4 \alpha_2 F}{Per_{2w}} . \tag{13.80}$$

is the hydraulic diameter of the channel for the liquid, and k_w / D_{h2w} is the relative roughness. The shear stress of the gas liquid interface in channel flow is

$$\tau_{12} = \frac{1}{4}\lambda_{12}\frac{1}{2}\rho_1 \Delta V_{12}^2, \tag{13.81}$$

where

$$\lambda_{w2} = \lambda_{friction}\left(\frac{\rho_1 V_{12} D_{h1}}{\eta_2}, \frac{\delta_{2F}/4}{D_{h1}}\right). \tag{13.82}$$

Here

$$D_{h1} = \frac{4\alpha_1 F}{Per_{1w} + Per_{12}}, \tag{13.83}$$

is the hydraulic diameter of the "gas channel" and $(\delta_{2F}/4)/D_{h1}$ is the relative roughness for the "gas channel" taken to be a function of the waviness of the film.

Vertical wall: For vertical plane walls the average shear stress at the wall along Δz is

$$\tau_{w2} = \lambda_{w2}\frac{1}{2}\rho_2 V_2^2 \tag{13.84}$$

where for laminar flow

$$Re_{2z} < (1 \text{ to } 3)\times 10^5 \tag{13.85}$$

in accordance with *Prandtl* and *Blasius*

$$\lambda_{w2} = 1.372 / Re_{2z}^{1/2} \tag{13.86}$$

and

$$Re_{2z} = V_2 \Delta z / \nu_2 \tag{13.87}$$

and for turbulent flow – see in *Albring* (1970)

$$\lambda_{w2} = 0.072 / Re_{2z}^{1/5}. \tag{13.88}$$

Similarly, the gas-side averaged shear stress is computed using

$$Re_{1z} = \Delta V_{12} \Delta z / \nu_2. \tag{13.89}$$

Thus, the effective shear stress in the film is

$$\tau_{2eff} = \frac{Per_{2w}\tau_{2w} + Per_{12}\tau_{12}}{Per_{2w} + Per_{12}}. \tag{13.90}$$

Now, we can estimate the time scale of the turbulence in the shear flow

$$\Delta\tau_2 \approx \left(\nu_2\delta_{2F}/V_2'^3\right)^{1/2}, \tag{13.91}$$

using the dynamic friction velocity

$$V_{2eff}^* = \sqrt{\tau_{2eff}/\rho_2}\,. \tag{13.92}$$

Assuming that the pulsation velocity is of the order of magnitude of the friction velocity

$$V_2' \approx const\ V_{2eff}^*, \tag{13.93}$$

where the

$$\text{const} \approx 2.9, \tag{13.94}$$

we obtain for

$$Re_{2F}^* > \text{const } 500, \tag{13.95}$$

$$\beta_2 = const^{3/4}\sqrt{\frac{2}{\pi}}Pr_2^{-1/2}Re_{2F}^{*-1/4}V_{2eff}^*, \tag{13.96}$$

and for

$$Re_2 < \text{const } 500, \tag{13.97}$$

$$\beta_2 = const^{1/2}\sqrt{\frac{2}{\pi}}Pr_2^{-1/2}Re_{2F}^{*-1/2}V_{2eff}^*, \tag{13.98}$$

where

$$Re_{2F}^* = \frac{\rho_2 V_{2eff}^*\delta_{2F}}{\eta_2}. \tag{13.99}$$

More careful modeling of the turbulent length scale for the derivation of Eq. (13.98) was done by *Kim* and *Bankoff* (1983). The authors used the assumption

$$V_2' \approx V_{2eff}^*, \tag{13.100}$$

and modified Eq. (13.98) as follows

$$\beta_2 = 0.061 Pr_2^{-1/2} Re_2^{t0.12} V_{2eff}^*, \tag{13.101}$$

where

$$\ell_{e2} = \left[\frac{\sigma}{(\rho_2-\rho_1)g}\right]^{1/2} 3.03\times10^{-8} Re_1^{*1.85} Re_2^{*0.006} Pr_2^{-0.23}, \tag{13.102}$$

for 3000 < $Re_1^* < 18\,000$ and 800 < $Re_2^* < 5000$, $\lambda_{12} = 0.0524 + 0.92\times 10^{-5}\, Re_2^*$ valid for $Re_2^* > 340$, and $\tau_{12} >> \tau_{2w}$. Note the special definition of the *Reynolds* numbers as mass flow per unit width of the film

$$Re_1^* = \frac{\alpha_1 \rho_1 V_1}{\eta_1} \frac{F}{Per_{12}}, \tag{13.103}$$

$$Re_2^* = \frac{\alpha_2 \rho_2 V_2}{\eta_2} \frac{F}{Per_{12}}. \tag{13.104}$$

Assuming that the pulsation velocity is of order of the magnitude of the friction velocity as before but the time scale of the turbulence is

$$\Delta\tau_2 \approx \left(\nu_2 / \varepsilon_2\right)^{1/2} = \nu_2 / V_{2eff}^{*2}, \tag{13.105}$$

results for

$$Re_{2F}^* > \text{const } 500, \tag{13.106}$$

in

$$\beta_2 = \sqrt{\frac{2}{\pi}} Pr_2^{-1/2} V_{2eff}^* . \tag{13.107}$$

This equation is recommended by *Jensen* and *Yuen* (1982) with a constant 0.14 instead $\sqrt{2/\pi}$ for $\tau_{12} >> \tau_{2w}$. *Hughes* and *Duffey* reproduced an excellent agreement with experimental data for steam condensation in horizontal liquid films by using Eq. (13.96) and the assumption

$$\tau_{2eff} = \left(\tau_{12} + \tau_{2w}\right)/2 . \tag{13.108}$$

Nevertheless, one should bear in mind that Eq. (13.90) is more general.

An example of detailed modeling of the turbulence structure in the film during film condensation from stagnant steam on vertical cooled surfaces is given by *Mitrovich* et al. (1961). For this case

$$\dot q_2^{\prime\prime\prime 1\sigma} = h_2^{1\sigma} \left(T_2^{1\sigma} - T_w\right), \tag{13.109}$$

where

$$h_2^{1\sigma} \delta_{2F} / \lambda_2 = 1.05 Re_{2F}^{-0.33} \left[1 + C^{1.9} Re_{2F}^{1.267} Pr_2^{1.11}\right]^{0.526}, \tag{13.110}$$

$$C = 8.8\ 10^{-3}/(1 + 2.29\times 10^{-5} Ka_2^{0.269}), \tag{13.111}$$

$$Ka_2 = \rho_2 \sigma^3 / \left(g \eta_2^4\right). \tag{13.112}$$

Summarizing the results discussed above we can say the following:

(a) Gas in the two-phase film flow behaves as a gas in a channel. Therefore, the gas-side heat transfer can be considered as a heat transfer between gas and the "channel wall" taking into account that the "wall roughness" is influenced by the waves at the liquid surface.

(b) The liquid-side heat transfer at the interface is due to molecular and turbulent heat conduction. The modeling of the turbulent heat conduction can be performed by modeling the time and length scale of the turbulence taking into account that turbulence is produced mainly

(i) at the wall/liquid interface; and
(ii) at the gas/liquid interface.

(c) Gas-side heat transfer in a pool flow can be considered as a heat transfer at a plane interface.

(d) The liquid-side heat transfer from the interface into the bulk liquid is governed by the solution of the transient *Fourier* equation where in the case of turbulence the use of effective heat conductivity instead of the molecular conductivity is recommended.

13.3 Spontaneous flashing of superheated film

The considerations for the estimation of the spontaneous evaporation of the superheated film are identical to those presented in Section 12.5 for the spontaneous evaporation of a superheated droplet. In order to calculate the source terms we simply replace the subscripts 3 with 2 and use the relationships from Chapter 22 Section 5.

Having in mind that

$$\mu_{12} = 0, \tag{13.113}$$

$$\dot{q}_1^{\prime\prime\prime 2\sigma} = 0, \tag{13.114}$$

we obtain for the energy-jump condition

$$(\rho w)_{21}\left[h''\left(T_2^{1\sigma}\right) - h'\left(T_2^{1\sigma}\right)\right] = -\beta_2 \rho_2 c_{p2}\left(T_2^{1\sigma} - T_2\right), \tag{13.115}$$

or

$$T_2^{1\sigma} = T_2 - (\rho w)_{21}\left[h''\left(T_2^{1\sigma}\right) - h'\left(T_2^{1\sigma}\right)\right] / \left(\beta_2 \rho_2 c_{p2}\right). \tag{13.116}$$

From this equation the interface temperature can be computed by iterations starting with $T_2^{1\sigma} \approx T'(p)$. The heat transfer from the interface to the bulk liquid is

$$\dot{q}_2^{\prime\prime 1\sigma} = \beta_2 \rho_2 c_{p2} \left(T_2^{1\sigma} - T_2 \right), \tag{13.117}$$

and the evaporated mass per unit time and unit mixture volume is

$$\mu_{21} = a_{12} \left(\rho w \right)_{21}. \tag{13.118}$$

13.4 Evaporation of saturated film in superheated gas

If the flow consists of superheated gas, $T_1 > T'(p)$, and saturated liquid, $T_2 = T'(p)$ the heat transferred from the gas to the interface by convection is

$$\dot{q}_2^{\prime\prime\prime 1\sigma} = -a_{12} h_c \left[T_1 - T'(p) \right]. \tag{13.119}$$

The energy-jump condition at the surface is

$$\mu_{21} = -\dot{q}_2^{\prime\prime\prime 1\sigma} / \left[h''(p) - h'(p) \right] = a_{12} h_c \left[T_1 - T'(p) \right] / \left[h''(p) - h'(p) \right]. \tag{13.120}$$

The energy balance for the gas at $p \approx const$ and negligible gas diffusion and convection from the neighboring cells is

$$\alpha_1 \rho_1 c_{p1} \frac{dT_1}{d\tau} = -\dot{q}_2^{\prime\prime\prime 1\sigma} \frac{h_1 - h''(p)}{h''(p) - h'(p)} \approx -\dot{q}_2^{\prime\prime\prime 1\sigma} \frac{c_{p1} [T_1 - T'(p)]}{h''(p) - h'(p)}, \tag{13.121}$$

or

$$\frac{dT_1}{d\tau} = [T_1 - T'(p)]^2 / \Delta\tau_1, \tag{13.122}$$

where

$$\Delta\tau_1 = [h''(p) - h'(p)] \alpha_1 \rho_1 / a_{12}.$$

We see from the above equation that the colder steam entering the bulk steam has a cooling effect on the bulk steam. For $\tau = 0$, $T_1 = T_{1a}$ the analytical solution gives

$$\frac{T_1 - T'(p)_1}{T_{1a} - T'(p)} = \frac{1}{1 - [T_1 - T'(p)] \dfrac{\Delta\tau}{\Delta\tau_1}}. \tag{13.123}$$

The averaged temperature difference $T_1 - T'(p)$ along the time $\Delta\tau$ is

$$T_1 - T'(p) = -\frac{\Delta\tau_1}{\Delta\tau} \ln \left\{ 1 - [T_{1a} - T'(p)] \frac{\Delta\tau}{\Delta\tau_1} \right\}. \tag{13.124}$$

Thus, for evaporation of a saturated film in superheated gas the instantaneous heat and mass transfer is defined by Eqs. (13.119) and (13.120) and the average heat and mass transfer along the time $\Delta\tau$ is computed as follows:

(i) Compute the averaged heat temperature difference using Eq. (13.124).
(ii) Compute the average heat transfer using Eq. (13.119).
(iii) Compute the average mass transfer using the energy-jump condition, Eq. (13.120).

13.5 Condensation of pure steam on subcooled film

Consider steam having temperature T_1 interacting with a liquid film having temperature T_2 where

$$T_2 < T'(p). \tag{13.125}$$

The steam-side interface temperature is $T_1^{2\sigma}$. The heat transferred from the interface into the gas per unit time and unit mixture volume is

$$\dot{q}_1'''^{2\sigma} = a_{12} h_c \left(T_1^{2\sigma} - T_1\right) \tag{13.126}$$

where h_c is computed as discussed in Section 13.1. The heat transferred from the interface into the bulk liquid is

$$\dot{q}_2'''^{1\sigma} = a_{12} \beta_2 \rho_2 c_{p2} \left(T_2^{1\sigma} - T_2\right). \tag{13.127}$$

A simplified way to compute the condensing mass per unit time and unit mixture volume is to assume

$$T_2^{1\sigma} = T_1^{2\sigma} \approx T'(p). \tag{13.128}$$

Using the energy-jump condition we obtain

$$\mu_{12} = a_{12} \left\{ h_c \left[T'(p) - T_1\right] + \beta_2 \rho_2 c_{p2} \left[T'(p) - T_2\right] \right\} / \left[h''(p) - h'(p)\right]. \tag{13.129}$$

An improvement of the description can be achieved by dropping the assumption (13.128). In this case one must compute the condensing mass flow rate $(\rho w)_{12}$ as a function of the interface temperature using the kinetic theory as discussed in Section 12.1.2. Consequently, we have

$$(\rho w)_{12} = f\left(T_2^{1\sigma}, T_1, p\right) \tag{13.130}$$

The energy-jump condition

$$(\rho w)_{12}\left[h''(T_{2i})-h'(T_{2i})\right]=h_c\left(T_1^{2\sigma}-T_1\right)+\beta_2\rho_2 c_{p2}\left(T_2^{1\sigma}-T_2\right), \qquad (13.131)$$

governs the interface temperature

$$T_2^{1\sigma}=\left\{(\rho w)_{12}\left[h''(T_2^{1\sigma})-h'(T_2^{1\sigma})\right]+h_c T_1+\beta_2\rho_2 c_{p2}T_2\right\}/\left(h_c+\beta_2\rho_2 c_{p2}\right). \qquad (13.132)$$

Obviously $T_2^{1\sigma}$ has to be obtained by iterations starting with $T_2^{1\sigma} \approx T'(p)$.

13.6 Evaporation or condensation in presence of noncondensable gases

In pure vapor condensation the condensate film provides the only heat-transfer resistance, whereas if a small amount of noncondensable gas is present, then the main resistance to heat transfer is the gas/vapor boundary layer. Air mass fractions as low as 0.5% can reduce the condensation by 50% or more, see *Pohlhausen* (1921), *Potukho* and *Popov* (1963), *Gnielinski* (1975), *Al-Diwani* and *Rose* (1973), *Siddique* and *Golay* (1994), *Uchida* et al. (1964).

Next we consider gas–film flow. The gas has temperature T_1 and concentration of steam C_{M1}. The system pressure is p and the steam partial pressure in the bulk of the gas is p_{M1}. The liquid has temperature T_2, that is less than the saturation temperature $T'(p)$ for the system pressure. This means that the liquid is subcooled and spontaneous flashing is not possible. In this case of diffusion-controlled evaporation or condensation at the interface is possible depending on the interface temperature satisfying the energy-jump condition. If the partial pressure in the gas boundary layer $p'(T_2^{1\sigma}) > p_{M1}$ steam is transported from the interface into the gas bulk. The opposite is the case if $p'(T_2^{1\sigma}) < p_{M1}$.

13.6.1 Useful definitions for describing the diffusion mass transfer

In the following read the superscript and subscript $_1^{\sigma 2}$ as follows: from surface 2 into the bulk of 1. The heat flux in W/m² from the interface of field 2 into the bulk of field 1 is defined by the *Fourier*'s low,

$$\dot{q}_1''^{\sigma 2}=-\lambda_1\frac{\partial T_1}{\partial y}=h_{conv}\left(T_{1M}^{2\sigma}-T_1\right). \qquad (13.133)$$

Here, h_{conv} is the heat-transfer coefficient due to convection. If the thickness of the thermal boundary layer of the undisturbed flow δ_0 is known the heat-transfer coefficient for flow without diffusion mass transfer is

$$h_{conv,0} = \lambda_1/\delta_0 . \tag{13.134}$$

The mass flux of non inert specie M in kg/m² from the interface of field 2 into the bulk of the field 1 is defined by *Fick*'s law (1855),

$$(\rho w)_{1M}^{2\sigma} = -\frac{D_{1,M\to\Sigma n}}{R_1 T_1}\frac{\partial p_{1M}}{\partial y} = -\frac{D_{1,M\to\Sigma n}}{\delta_m}\frac{\Delta p_{1M}}{R_1 T_1} = -\beta_{12}\frac{\Delta p_{1M}}{R_1 T_1} = \beta_{12}\rho_1\left(C_{1M}^{2\sigma} - C_{1M}\right). \tag{13.135}$$

Here the definition of the mass-transfer coefficient for low driving forces is analogous to the heat-transfer coefficient

$$\beta_{12,0} = \frac{D_{1,M\to\Sigma n}}{\delta_0} . \tag{13.136}$$

The last equation part of *Fick*'s law is valid for constant temperature and constant total pressure along y. In this case

$$p_{1M} = \rho_{1M} R_1 T_1 \tag{13.137}$$

and

$$(\rho w)_{1M}^{2\sigma} = -D_{1,M\to\Sigma n}\rho_1 \frac{1}{\rho_1}\frac{d\rho_{1M}}{dy} = \beta_{12}\rho_1\left(C_{1M}^{2\sigma} - C_{1M}\right). \tag{13.138}$$

There are large numbers of empirical correlation for the heat-transfer coefficient in the form

$$h_{conv} = \frac{\lambda_1}{\delta_m} = cRe_1^m Pr_1^n \lambda_1/D_{h1} . \tag{13.139}$$

h_{conv} is the convective heat-transfer coefficient not depending on the direction of the heat flux in W/(m² K) defined by the *Nusselt* (1915) equation. Because the form of the heat and the mass transfer equations is similar it is believed that the solutions are also similar. For this reason the mass-transfer coefficient for convection in a pipe not depending on the direction of the mass flux in m/s is also defined by the *Nusselt* equation in a form

$$\beta_{12} = \frac{D_{1,M \to \Sigma n}}{\delta_m} = cRe_1^m Sc_1^n \, D_{1,M \to \Sigma n} / D_{h1} . \quad (13.140)$$

Because the constants and the exponents are the same for heat and mass transfer the following relation is valid

$$\frac{\beta_{12}}{h_{conv}} = \left(\frac{Sc_1}{Pr_1} \right)^n \frac{D_{1,M \to \Sigma n}}{\lambda_1} = \left(\frac{\lambda_1}{\rho_1 c_{p1} D_{1,M \to \Sigma n}} \right)^n \frac{D_{1,M \to \Sigma n}}{\lambda_1} = \frac{D_{1,M \to \Sigma n}}{\lambda_1} \Big/ Le_{1M}^n . \quad (13.141)$$

This is *Ackerman*'s equation (34), *Akermann* (1937). Here,

$$Le_{1,M} = Sc_{1,M} / Pr_1 = \lambda_1 / \left(\rho_1 c_{p1} D_{1,M \to \Sigma n} \right) \quad (13.142)$$

is called the *Lewis* number. *Ackermann* demonstrated that the *Lewis* number can be assumed unity only as a very rough approximation. In reality it is different from 1. At very high *Reynolds* numbers $n = 1$ and $\beta_{12} / h_{conv} = 1 / \left(\rho_1 c_{p1} \right)$. Equation (13.141) can be used alternatively instead of Eq. (13.140) in order to compute the mass-transfer coefficient by knowing the heat-transfer coefficient, but there is no more information in it. Some authors use this form to correlate experimental results instead of using Eq. (13.140).

Note that if the form of the empirical approximations for the heat transfer has other forms, e.g., the *Gnielinski* (1976) improvement of the *Petukhov* and *Kirillov* (1958) correlation, validated in the range $2300 < \mathrm{Re} < 10^6$ and $0.6 < \mathrm{Pr} < 10^5$,

$$\frac{Nu}{\mathrm{Re}\,\mathrm{Pr}} \frac{8}{\lambda_{fr}} = \left[1.07 + 12.7 \sqrt{\frac{\lambda_{fr}}{8}} \left(\mathrm{Pr}^{2/3} - 1 \right) \right]^{-1} \left(1 - 1000 / \mathrm{Re} \right),$$

Eq. (13.141) will receive another form.

13.6.2 Modification of the velocity profile in the boundary layer due to diffusion mass transfer, *Ackerman*'s multiplier

In fact, the mass transfer perpendicular to the flow direction influences the solution of the conservation equation. Therefore, effective condensation or evaporation changes the mass-transfer coefficient. This is easily demonstrated by adding to the boundary-layer momentum equation an additional term $-\left(\rho w\right)_{1M}^{2\sigma}\dfrac{dw}{dy}=0$ as follows:

$$\eta\frac{d^2w}{dy^2}-\left(\rho w\right)_{1M}^{2\sigma}\frac{dw}{dy}=0. \qquad (13.143)$$

This evaporation term changes the thickness of the boundary layer by making it larger and therefore reducing the normal velocity gradient at the wall and consequently reducing the drag coefficient. If the crossflow is directed from the bulk into the wall (condensation) the boundary-layer thickness decreases, the normal velocity gradient increases and therefore the drag coefficient increases. The solution of the above equation for y = 0, w = 0 and $y = \delta_m$, $w = \overline{w}$ is Eq. (44) of *Ackermann* (1937)

$$\frac{w}{\overline{w}}=\frac{e^{cy}-1}{e^{c\delta_m}-1}, \qquad (13.144)$$

$$\frac{dw}{dy}=\overline{w}c\frac{e^{cy}}{e^{c\delta_m}-1}, \qquad (13.145)$$

where

$$c=\left(\rho w\right)_{1M}^{2\sigma}/\eta. \qquad (13.146)$$

In order to obtain the relation between two boundary layers thicknesses for flows *with* and *without* crossflow *Ackermann* required the share stress at the end of the boundary layer with crossflow to be the same as without crossflow.

This is a hypothesis. It means that

$$\tau_m=\tau_0 \text{ at } y=\delta_m \qquad (13.147)$$

where

$$\tau_0=\eta\frac{dw}{dy}=\eta\frac{\overline{w}}{\delta_0} \qquad (13.148)$$

or

$$\overline{w} c\, e^{c\delta_m} / \left(e^{c\delta_m} - 1\right) = \overline{w} / \delta_0 \tag{13.149}$$

resulting in

$$c\delta_m = \ln \frac{1}{1 - c\delta_0} . \tag{13.150}$$

The product

$$\mathrm{Re}_0 = c\delta_0 \tag{13.151}$$

is in fact the *crossflow Reynolds* number using as a characteristic length the boundary layer thickness of the undisturbed flow. For pipe flow for instance the share stress at the wall without crossflow is

$$\tau_0 = \eta\, \overline{w} / \delta_0 = \frac{\lambda_{fr}}{8} \rho_1 \overline{w}^2 \tag{13.152}$$

and therefore the thickness of the boundary layer

$$\boxed{\delta_0 = \frac{8\eta}{\lambda_{fr} \rho_1 \overline{w}},} \tag{13.153}$$

and finally

$$\mathrm{Re}_0 = c\delta_0 = c \frac{8\eta}{\lambda_{fr} \rho_1 \overline{w}} = \frac{8 (\rho w)_{1M}^{2\sigma}}{\lambda_{fr} \rho_1 \overline{w}} . \tag{13.154}$$

The boundary-layer thickness ratio is then

$$\boxed{\frac{\delta_m}{\delta_0} = \frac{1}{\mathrm{Re}_0} \ln \frac{1}{1 - \mathrm{Re}_0} .} \tag{13.155}$$

For the share stress of the flow with crossflow *Ackerman* obtained finally

$$\tau_m = \eta\, \overline{w} / \delta_m = \eta \frac{\overline{w}}{\delta_0} \frac{\delta_0}{\delta_m} = \eta \frac{\overline{w}}{\delta_0} Ac = \tau_0 Ac . \tag{13.156}$$

I will call the multiplier

$$Ac := \frac{\tau_m}{\tau_0} = \frac{\delta_0}{\delta_m} = -\frac{\mathrm{Re}_0}{\ln\left(1-\mathrm{Re}_0\right)} \approx \frac{1}{1+\frac{1}{2}\mathrm{Re}_0} \approx 1 - \frac{1}{2}\mathrm{Re}_0 , \tag{13.157}$$

the *Ackermann multiplier* in honor of its author. The last two variants of the above equation are valid for $\mathrm{Re}_0 << 1$. They are useful for explicitly resolving the corrected diffusion mass flow rates.

Similarly the mass-transfer coefficients

$$\beta_{12,0} = D_{1,M\to\Sigma n} / \delta_0 , \tag{13.158}$$

$$\beta_{12,m} = D_{1,M\to\Sigma n} / \delta_m , \tag{13.159}$$

are related to each other by

$$\beta_{12,m} / \beta_{12,0} = \delta_0 / \delta_m = Ac . \tag{13.160}$$

One has to be careful for the direction of the effective crossflow because diffusion at one components cause the transport of the other component in the opposite direction. So, for instance, in an air steam mixture if the steam wall pressure is higher than the bulk pressure the steam diffusion mass flow rate is

$$\left(\rho w\right)_{1M}^{2\sigma} = -\beta_{12,m} \frac{p_{1M,w} - p_{1M}}{RT_1} M_{1M} . \tag{13.161}$$

This equation is implicit with respect to the evaporation mass flow rate. Replacing the mass-transport coefficient and solving with respect to the evaporation mass flow rate results in

$$\left(\rho w\right)_{1M}^{2\sigma} = \lambda_{fr} \rho_1 \overline{w} \left[1 - \exp\left(\frac{D_{1,M\to\Sigma n}}{\eta} \frac{p_{1M,w} - p_{1M}}{RT_1} M_{1M}\right)\right] \Big/ 8 .$$

The air diffusion mass flow rate is

$$\left(\rho w\right)_{1n}^{2\sigma} = -\beta_{12,m} \frac{p_{1n,w} - p_{1n}}{RT_1} M_{1n} = \beta_{12,m} \frac{p_{1M,w} - p_{1M}}{RT_1} M_{1n} \tag{13.163}$$

and the sum of both mass flow rates is

$$\left(\rho w\right)_{1M}^{2\sigma} + \left(\rho w\right)_{1n}^{2\sigma} = -\beta_{12,m} \frac{p_{1M,w} - p_{1M}}{RT_1} \left(M_{1M} - M_{1n}\right) . \tag{13.164}$$

Here, the mole masses are M_{1M} , M_{1n} , and the universal gas constant is R.

13.6.3 *Stefan*'s theory of diffusion mass transport

We owe the following considerations to *Josef Stefan*, a Slovenian physicists who lived between 1835–1893. Consider evaporation from a film of liquid 2 into the bulk of the flowing mixture consisting of the vapor of this liquid, designated with subscript M and of inert component, designated with subscript n. For small pressure gradients the velocity boundary layer will not be modified and the mass transfer is defined by

$$(\rho w)_{1M}^{2\sigma} = -\frac{D_{1,M\to\Sigma n}}{R_{1,M}T_1}\frac{dp_{1M}}{dy}. \tag{13.165}$$

Because the diffusion is a process with zero net molecule flux the inert-gas diffusion mass flow rate is

$$(\rho w)_{1n}^{2\sigma} = -\frac{D_{1,M\to\Sigma n}}{R_{1,n}T_1}\frac{dp_{1n}}{dy} = \frac{D_{1,M\to\Sigma n}}{R_{1,n}T_1}\frac{dp_{1M}}{dy}. \tag{13.166}$$

At constant sum of the partial pressures we have

$$\frac{dp_{1n}}{dy} = -\frac{dp_{1M}}{dy}, \tag{13.167}$$

leading to the last form of the above equation. Interestingly, the inert mass flow towards the film can not penetrate the film and is reflected back by the same amount. The reflected velocity is then

$$u_{1n,Stefan} = -\frac{(\rho w)_{1n}^{2\sigma}}{\rho_{1n}} = -\frac{D_{1,M\to\Sigma n}}{\rho_{1n}R_{1,n}T_1}\frac{dp_{1M}}{dy} = -\frac{D_{1,M\to\Sigma n}}{p_{1n}}\frac{dp_{1M}}{dy} = -\frac{D_{1,M\to\Sigma n}}{p-p_{1M}}\frac{dp_{1M}}{dy}. \tag{13.168}$$

At this point *Stefan* (1874) assumed that this velocity will carry additionally also vapor with the mass flow rate

$$(\rho w)_{1M,Stefan}^{2\sigma} = \rho_{1M}u_{1n,Stefan} = -\frac{D_{1,M\to\Sigma n}\rho_{1M}}{p-p_{1M}}\frac{dp_{1M}}{dy} = -\frac{D_{1,M\to\Sigma n}}{R_{1,M}T_1}\frac{p_{1M}}{p-p_{1M}}\frac{dp_{1M}}{dy}, \tag{13.169}$$

thus introducing additional evaporation. Further, *Stefan* assumed that the net mass flux of vapor will be at any place in the boundary layer the same and is the sum of both fluxes

$$\left(\rho w\right)_{1M,0}^{2\sigma} = \left(\rho w\right)_{1M}^{2\sigma} + \left(\rho w\right)_{1M,Stefan}^{2\sigma} = -\frac{D_{1,M\to\Sigma n}}{R_{1,M}T_1}\frac{p}{p-p_{1M}}\frac{dp_{1M}}{dy}$$

$$= \frac{D_{1,M\to\Sigma n}p}{R_{1,M}T_1}\frac{d\ln\left(p-p_{1M}\right)}{dy}. \tag{13.170}$$

The finite-difference approximation of the pressure differential across the boundary layer δ_0 is

$$\frac{d\ln\left(p-p_{1M}\right)}{dy} \approx \frac{1}{\delta_0}\ln\frac{p-p_{1M,bulk}}{p-p_{1M}^{\sigma 2}}. \tag{13.171}$$

With this, *Stefan* obtained his well-known result in 1874:

$$\left(\rho w\right)_{1M,0}^{2\sigma} = \frac{D_{1,M\to\Sigma n}}{\delta_0}\frac{p}{R_{1,M}T_1}\ln\frac{p-p_{1M,bulk}}{p-p_{1M}^{\sigma 2}}. \tag{13.172}$$

Stefan's equation can be found in the literature in a variety of forms like

$$\left(\rho w\right)_{1M,eff}^{2\sigma} = \frac{D_{1,M\to\Sigma n}}{\delta_0}\rho_1\frac{R_1}{R_{1,M}}\ln\frac{p-p_{1M,bulk}}{p-p_{1M}^{\sigma 2}} = \frac{D_{1,M\to\Sigma n}}{\delta_0}\rho_1\frac{M_{1,M}}{M_1}\ln\frac{p-p_{1M,bulk}}{p-p_{1M}^{\sigma 2}}$$

$$= \beta_{12,0}\rho_1\frac{M_{1,M}}{M_1}\ln\frac{p-p_{1M,bulk}}{p-p_{1M}^{\sigma 2}} = \left(h_{conv,0}\rho_1\frac{D_{1,M\to\Sigma n}}{\lambda_1}\frac{M_{1,M}}{M_1}\ln\frac{p-p_{1M,bulk}}{p-p_{1M}^{\sigma 2}}\right)\Big/ Le_{1M}^{n}$$

$$= \left(\frac{D_{1,M\to\Sigma n}}{\delta_0}\frac{p}{p_{1M}^{\sigma 2}-p_{1M,bulk}}\ln\frac{p-p_{1M,bulk}}{p-p_{1M}^{\sigma 2}}\right)\frac{p_{1M}^{\sigma 2}-p_{1M,bulk}}{R_{1,M}T_1}$$

$$= \frac{D_{1,M\to\Sigma n}}{\delta_0}St_p\frac{p_{1M}^{\sigma 2}-p_{1M,bulk}}{R_{1,M}T_1} = \rho_1\frac{D_{1,M\to\Sigma n}}{\delta_0}\frac{1}{\overline{Y}_{1n}}\frac{M_{1,M}}{M_1}\frac{p_{1M}^{\sigma 2}-p_{1M,bulk}}{p_1}. \tag{13.173}$$

Peterson, et al. (1993) reported that the temperature form of the surface mass flow rate

$$\left(\rho w\right)_{1M,eff}^{2\sigma} = \frac{\rho_1 D_{1,M\to\Sigma n}M_{1,M}\Delta h}{\delta_0\overline{Y}_{1n}M_1\left(v''-v'\right)T_{1,av}p_1}\left(T_{1M}^{2\sigma}-T_{1M,bulk}\right) \tag{13.174}$$

obtained by using the finite difference form of the *Clapeyron*'s equation

$$\frac{p_{1M}^{\sigma 2}-p_{1M,bilk}}{T_{1M}^{\sigma 2}-T_{1M,bulk}}=\frac{1}{T_{1,av}}\frac{\Delta h}{v''-v'} \tag{13.175}$$

deviate only 3% from the exact pressure form. I call the new multiplier

$$St_p=\frac{p}{p_{1M}^{\sigma 2}-p_{1M,bulk}}\ln\frac{p-p_{1M,bulk}}{p-p_{1M}^{\sigma 2}}=\frac{\ln\left(Y_{1n,bulk}/Y_{1n,w}\right)}{Y_{1n,bulk}-Y_{1n,w}}=\frac{1}{\overline{Y}_{1n}}>0, \tag{13.176}$$

the *Stefan*'s number in honor of its inventor. Remarkably, *Stefan*'s number is always positive because

$$Y_{1n,bulk}<\overline{Y}_{1n}<Y_{1n,w}. \tag{13.177}$$

This means that taking into account *Stefan*'s flow always results in more diffusion mass transfer independent of whether it is a condensation or evaporation.

In the recent days *Stefan*'s statement is reached simply by writing the vapor flux

$$(\rho w)_{1M,0}^{2\sigma}=\left[(\rho w)_{1M,0}^{2\sigma}+(\rho w)_{1n,0}^{2\sigma}\right]C_{1M}-\rho_1 D_{1,M\to\Sigma n}\frac{dC_{1M}}{dy} \tag{13.178}$$

at the liquid interface where $(\rho w)_{1n,0}^{2\sigma}=0$,

$$(\rho w)_{1M,0}^{2\sigma}=-\frac{\rho_1 D_{1,M\to\Sigma n}}{1-C_{1M}}\frac{dC_{1M}}{dy}=\rho_1 D_{1,M\to\Sigma n}\frac{d\ln\left(1-C_{1M}\right)}{dy} \tag{13.179}$$

in finite-difference form across the boundary layer

$$(\rho w)_{1M,0}^{2\sigma}=\rho_1\frac{D_{1,M\to\Sigma n}}{\delta_0}\ln\frac{1-C_{1M,bulk}}{1-C_{1M}^{2\sigma}}$$

$$=\rho_1\frac{D_{1,M\to\Sigma n}}{\delta_0}\frac{\ln\dfrac{1-C_{1M,bulk}}{1-C_{1M}^{2\sigma}}}{C_{1M}^{2\sigma}-C_{1M,bulk}}\left(C_{1M}^{2\sigma}-C_{1M,bulk}\right)$$

$$=\rho_1\frac{D_{1,M\to\Sigma n}}{\delta_0}\frac{\ln\left(C_{1n,bulk}/C_{1n}^{2\sigma}\right)}{C_{1n,bulk}-C_{1n}^{2\sigma}}\left(C_{1M}^{2\sigma}-C_{1M,bulk}\right), \tag{13.180}$$

e.g., *Siddique*, et al. (1994), *Liao* and *Vierow* (2006) among many others. Knowing the convective cross gas-mixture velocity,

$$w_{10} = \frac{(\rho w)^{2\sigma}_{1M,0}}{\rho_1} = -\frac{D_{1,M\to\Sigma n}}{1-C_{1M}} \frac{dC_{1M}}{dy}, \tag{13.181}$$

the concentration profile in the boundary layer can be derived by solving the equation for conservation of vapor mass

$$w_{10} \frac{dC_{1M}}{dy} = D_{1,M\to\Sigma n} \frac{d^2 C_{1M}}{dy^2}. \tag{13.182}$$

Eliminating the gas convection velocity from the both equations results in

$$(1 \quad C_{1M}) \frac{d^2 C_{1M}}{dy^2} + \left(\frac{dC_{1M}}{dy} \right)^2 = 0. \tag{13.183}$$

Surprisingly the equation does not contain the diffusion constant and therefore the solution will not depend on it. The nonlinear partial differential equation is solved by introducing the substitution

$$\eta = C_{1M} - 1, \tag{13.184}$$

resulting in

$$\left(\frac{d\eta}{dy} \right)^2 - \eta \frac{d^2 \eta}{dy^2} = 0. \tag{13.185}$$

Fitting the solution constants in $\eta = c_1 e^{c_2 y}$ to the boundary conditions $y = 0$, $\eta = C^{2\sigma}_{1M} - 1$, $c_1 = C^{2\sigma}_{1M} - 1$ and $y = \delta$, $\eta = C_{1M,bulk} - 1$, $c_2 = \frac{1}{\delta} \ln \frac{C_{1M,bulk} - 1}{C^{2\sigma}_{1M} - 1}$ results in the profile

$$\frac{C_{1M}(y)-1}{C_{1M}^{2\sigma}-1}=\exp\left(\frac{y}{\delta_0}\ln\frac{C_{1M,bulk}-1}{C_{1M}^{2\sigma}-1}\right). \qquad (13.186)$$

At this point, *Ackerman* continued in 1937 by observing that for large gradients the crossflow will modify the velocity boundary layer from δ_0 to δ_m, so that

$$(\rho w)_{1M}^{\sigma 2}=\frac{\delta_0}{\delta_m}(\rho w)_{1M,0}^{\sigma 2}. \qquad (13.187)$$

Using the approximate form of the *Ackermann*'s equation

$$Ac:=\frac{\delta_0}{\delta_m}=-\frac{\mathrm{Re}_0}{\ln(1-\mathrm{Re}_0)}\approx\frac{1}{1+\frac{1}{2}\mathrm{Re}_0}\approx 1-\frac{1}{2}\mathrm{Re}_0, \qquad (13.188)$$

where

$$\mathrm{Re}_0=\delta_0(\rho w)_{1M}/\eta, \qquad (13.189)$$

allows explicitly to compute the effective evaporation mass flow rate from

$$(\rho w)_{1M}^{\sigma 2}=\left[1-\frac{\delta_0}{2\eta}(\rho w)_{1M}\right](\rho w)_{1M,0}^{\sigma 2}. \qquad (13.190)$$

The final *Ackermann*'s result is

$$(\rho w)_{1M}^{\sigma 2}=\frac{(\rho w)_{1M,0}^{\sigma 2}}{1+\frac{\delta_0}{2\eta}(\rho w)_{1M,0}^{\sigma 2}}. \qquad (13.191)$$

13.6.4 Energy transport into or from the bulk flow due to the diffusion mass flow rate

For positive $(\rho w)_{1M}$ there is effectively an energy transfer from the wall into the bulk flow

$$\dot{q}_{1,diff}^{\prime\prime 2\sigma}=(\rho w)_{1M}^{2\sigma}(h_{1w}-h_1)=(\rho w)_{1M}^{2\sigma}c_p(T_{1w}-T_1). \qquad (13.192)$$

In the case of condensation $\dot{q}''^{2\sigma}_{1,diff} = 0$ because the mass flow leaves the bulk with its bulk temperature. Therefore,

$$\dot{q}''^{2\sigma}_{1} = -\lambda \frac{dT}{dy} + (\rho w)^{2\sigma}_{1M} c_p (T_{1w} - T_1) \text{ if } (\rho w)^{2\sigma}_{1M} > 0 \tag{13.193}$$

$$\dot{q}''^{2\sigma}_{1} = -\lambda \frac{dT}{dy} \text{ if } (\rho w)^{2\sigma}_{1M} < 0 . \tag{13.194}$$

13.6.5 Modification of the temperature boundary layer due to the diffusion mass flow rate

The diffusion mass flow rate influences the boundary-layer temperature field. This is reflected in the boundary-layer energy-conservation equation

$$\lambda \frac{d^2T}{dy^2} - (\rho w)^{2\sigma}_{1M} c_p \frac{dT}{dy} = 0 \tag{13.195}$$

or

$$\frac{d^2T}{dy^2} - c_T \frac{dT}{dy} = 0 , \tag{13.196}$$

where

$$c_T = (\rho w)^{2\sigma}_{1M} c_p / \lambda . \tag{13.197}$$

The solution of this equation

$$T(y) = c_1 \frac{e^{c_T y}}{c_T} + c_2 \tag{13.198}$$

or fitted to the boundary conditions $y = 0$, $T = T_{1w}$ and $y = \delta_0$, $T = T_1$ is found by *Ackermann*:

$$T(y) = -\frac{T_{1w} - T_1}{e^{c_T \delta_0} - 1} e^{c_T y} + \frac{T_{1w} e^{c_T \delta_0} - T_1}{e^{c_T \delta_0} - 1} , \tag{13.199}$$

$$\frac{dT}{dy} = -\frac{T_{1w} - T_1}{e^{c_T \delta_0} - 1} c_T e^{c_T y} . \tag{13.200}$$

The heat flux at $y = 0$ is then

$$\dot{q}_1''^{2\sigma} = \frac{e^{c_T \delta_0}}{e^{c_T \delta_0} - 1} (\rho w)_{1M}^{2\sigma} c_p (T_{1w} - T_1) \text{ if } (\rho w)_{1M}^{2\sigma} > 0 , \tag{13.201}$$

$$\dot{q}_1''^{2\sigma} = \frac{(\rho w)_{1M}^{2\sigma} c_p}{e^{c_T \delta_0} - 1} (T_{1w} - T_1) \text{ if } (\rho w)_{1M}^{2\sigma} < 0 . \tag{13.202}$$

Introducing the undisturbed heat-transfer coefficient

$$h_{conv,0} = \lambda / \delta_0 \tag{13.203}$$

results in

$$\dot{q}_1''^{2\sigma} = h_{conv,m} (T_{1w} - T_1) , \tag{13.204}$$

where

$$\frac{h_{conv,m}}{h_{conv,0}} = \frac{Pe_0 e^{Pe_0}}{e^{Pe_0} - 1} \text{ if } (\rho w)_{1M}^{2\sigma} > 0 , \tag{13.205}$$

$$\frac{h_{conv,m}}{h_{conv,0}} = \frac{Pe_0}{e^{Pe_0} - 1} \text{ if } (\rho w)_{1M}^{2\sigma} < 0 . \tag{13.206}$$

The dimensionless *crossflow Peclet* number

$$Pe_0 = \delta_0 c_T = \frac{\delta_0 (\rho w)_{1M}^{2\sigma} c_p}{\lambda} \tag{13.207}$$

is built with the thickness of the undisturbed boundary layer and the diffusion mass flow rate.

13.6.6 Some practical recipes

Problem 1: Compute the local diffusion mass flow rate from an air–steam mixture flowing inside a pipe with diameter D_h. The liquid film thickness is negligible. The wall temperature is T_{1w}. The gas temperature is T_1. The steam mass concentration in the gas is $C_{1,M}$. The mass flow rate of the air–steam mixture is *G*.

Solution to problem 1:

1. First the friction coefficient is computed from known empirical correlations as a function of the local *Reynolds* number and roughness of the pipe for undisturbed boundary layer $\lambda_{R,0} = \lambda_R \left(\mathrm{Re} = GD_h / \eta_1,\ k / D_h\right)$, e.g.,

$$\lambda_{R,0} = \max\left(64/\mathrm{Re},\ 0.316/\mathrm{Re}^{1/4}\right). \qquad (13.208)$$

2. Then, the boundary-layer thickness of the undisturbed flow is estimated

$$\delta_0 = \frac{8\eta_1}{\lambda_{fr} G}. \qquad (13.209)$$

3. Then, the undisturbed diffusion mass coefficient is computed $\beta_{12,0} = cRe_1^m Sc_1^n D_{1,M\to\Sigma n} / D_h$, from known empirical correlations as a function of the local *Reynolds*- and *Schmidt* number, e.g.,

$$\beta_{12,0} = \left(D_{1,M\to\Sigma n} / D_h\right) \max\left(3.66,\ 0.22\,\mathrm{Re}_1^{0.8}\, Sc_1^{0.5}\right). \qquad (13.210)$$

4. Using *Stefan*'s result compute the uncorrected diffusion mass flow rate

$$\left(\rho w\right)_{1M,0}^{\sigma 2} = \beta_{12,0} St \frac{p_{1M,w} - p_{1M,bulk}}{R_{1,M} T_1}, \qquad (13.211)$$

$$St = \frac{p}{p_{1M,w} - p_{1M,bulk}} \ln \frac{p - p_{1M,bulk}}{p - p_{1M,w}} > 0. \qquad (13.212)$$

Here, the partial steam pressure at the wall is assumed to be the saturation pressure at the wall temperature

$$p_{1M,w} = p'\left(T_{1w}\right). \qquad (13.213)$$

The bulk partial pressure is a function of the concentrations. First, the mole concentration is computed from the mass concentrations

$$Y_{1,M} \approx \frac{C_{1,M}/M_{1,M}}{C_{1,M}/M_{1,M} + (1 - C_{1,M})/M_{1,n}}. \tag{13.214}$$

and then the corresponding partial pressure

$$p_{1,M} = Y_{1,M}\, p. \tag{13.215}$$

Here, $M_{1,n}$ and $M_{1,M}$ are the mole masses of the inert components and of the steam, respectively. $M_{1,n}$ = 28.96 kg/kmol for air and $M_{1,M}$ = 18.96 kg/kmol for water steam.

$$M_1 = \frac{p_{1,n}}{p} M_{1,n} + \frac{p_{1,M}}{p} M_{1,M} \tag{13.216}$$

is the mixture mole mass.

5. Compute the *crossflow Reynolds* number

$$\mathrm{Re}_0 = \delta_0 (\rho w)^{\sigma 2}_{1M,0} / \eta. \tag{13.217}$$

6. Compute *Ackermann*'s multiplier

$$Ac = -\frac{\mathrm{Re}_0}{\ln(1 - \mathrm{Re}_0)} \approx \frac{1}{1 + \frac{1}{2}\mathrm{Re}_0} \approx 1 - \frac{1}{2}\mathrm{Re}_0. \tag{13.218}$$

7. Use *Ackermann*'s result to compute finally the friction coefficient

$$\lambda_R = Ac \lambda_{R,0}. \tag{13.219}$$

8. Use *Ackermann*'s result to compute finally the diffusion mass flow rate

$$(\rho w)^{\sigma 2}_{1M} = Ac (\rho w)^{\sigma 2}_{1M,0}. \tag{13.220}$$

9. Compute the undisturbed heat-transfer coefficient $h_{conv,0} = \lambda_1/\delta_m = c Re_1^m Pr_1^n \lambda_1 / D_{h1}$, from known empirical correlations as a function of the local *Reynolds* and *Prandtl* numbers, e.g.,

$$h_{conv,0} = \left(\lambda_1 / D_h\right) \max\left(3.66,\ 0.22\,\mathrm{Re}_1^{0.8}\,\mathrm{Pr}_1^{0.5}\right). \tag{13.221}$$

10. Compute the *crossflow Peclet* number

$$Pe_0 = \frac{\delta_0 \left(\rho w\right)_{1M}^{\sigma 2} c_p}{\lambda} \tag{13.222}$$

using the thickness of the undisturbed boundary layer and the diffusion mass flow rate.

11. Correct the local heat-transfer coefficients due to crossflow by using *Ackermann*'s solution

$$h_{conv} = h_{conv,0} \frac{Pe_0 e^{Pe_0}}{e^{Pe_0} - 1} \text{ if } \left(\rho w\right)_{1M} > 0, \tag{13.223}$$

$$h_{conv} = h_{conv,0} \frac{Pe_0}{e^{Pe_0} - 1} \text{ if } \left(\rho w\right)_{1M} < 0. \tag{13.224}$$

70 years after his invention this approach is still in use. *Zhou* and *Revankar* (2007) for instance compared data and prediction for condensation in a vertical pipe at 1.8 to 3.4 bar with acceptable accuracy.

Problem 2: Consider the same problem as in the task 2 but take into account that the liquid film thickness is *not* negligible.

Solution to problem 2: The heat entering into or removed from the film influences the interface temperature and therefore influences the evaporation or condensation mass flow rate. The most important part of the solution of this task is the finding of the interface temperature from the instantaneous energy balance on a control volume containing both sides of the interface and being very thin. The way I propose here is:

1. Assume as a first estimation that the interface temperature is equal to the bulk temperature of the film $T_2^{1\sigma} \approx T_2$.

2. Compute the steam partial pressure at the interface $p'\left(T_2^{1\sigma}\right)$.

3. Compute the diffusion-controlled evaporation mass flow rate $\left(\rho w\right)_{21}$.

4. Compute the gas and the liquid heat flux from the interface

$$\dot{q}_1''^{2\sigma} = h_{conv,m}\left(T_1^{2\sigma} - T_1\right), \tag{13.225}$$

$$\dot{q}_2''^{1\sigma} = \beta_2 \rho_2 c_{p2}\left(T_2^{1\sigma} - T_2\right). \tag{13.226}$$

5. Compute by iterations the interface temperature $T_2^{1\sigma}$ from the energy-jump condition

$$Ac\left(\rho w\right)_{1M,0}^{\sigma 2} = -\frac{\beta_2 \rho_2 c_{p2}\left(T_1^{2\sigma} - T_2\right) + h_{conv,m}\left(T_1^{2\sigma} - T_1\right)}{\Delta h} \tag{13.227}$$

starting with $T_2^{1\sigma} = T_2$. An algorithm that works is: Using *Stefan*'s equation and solving with respect to the surface vapor mass concentration results in

$$C_{1M}^{2\sigma} = 1 - \left(1 - C_{1M}\right)\exp\left[\frac{\delta_0}{D_{1,M\to\Sigma n}} \frac{\beta_2 \rho_2 c_{p2}\left(T_1^{2\sigma} - T_2\right) + h_{conv,m}\left(T_1^{2\sigma} - T_1\right)}{Ac\rho_1 \Delta h}\right]. \tag{13.228}$$

Because the interface temperature is a function of the interface concentration the above equation has to be solved by iteration. A method that works is: compute the partial surface pressure of the vapor by

$$p_{1M}^{2\sigma} = p\frac{C_{1M}^{2\sigma}}{M_{1M}} \Bigg/ \left[\frac{C_{1M}^{2\sigma}}{M_{1M}} + \frac{1 - C_{1M}^{2\sigma}}{M_{1n}}\right], \tag{13.229}$$

and then compute the surface temperature $T_1^{2\sigma} = T'\left(p_{1M}^{2\sigma}\right)$. Only view cycles are necessary. Note also that the evaporation enthalpy

$$h_{M1}^{2\sigma} - h_{M2}^{1\sigma} = h''\left(T_1^{2\sigma}\right) - h'\left(T_1^{2\sigma}\right) = f\left(T_1^{2\sigma}\right) \tag{13.230}$$

at the interface is also a function of the interface temperature $T_1^{2\sigma}$. Having $T_2^{1\sigma}$ it is a simple matter to compute $\left(\rho w\right)_{1M}^{\sigma 2}$, $\dot{q}_2'''^{1\sigma}$ and $\dot{q}_1'''^{2\sigma}$.

Nomenclature

Latin

a	$=\lambda/(\rho c_p)$, temperature conductivity, m²/s
a_{12}	$=1/\Delta z$ gas liquid interfacial area density in Cartesian coordinates, 1/m
D_{h1}	hydraulic diameter for the gas, m
D_{h2}	hydraulic diameter for the liquid, m
D_{h12}	hydraulic diameter for computation of the gas friction pressure loss component in a gas-liquid stratified flow, m
F	channel cross-section, *m*
Gr_1	$=g\left(-\frac{1}{\rho_1}\frac{\partial\rho_1}{\partial T_1}\right)\left\|T_1-T_1^{2\sigma}\right\|\Delta z^3/\nu_1^2$, gas-site *Grashof* number, dimensionless
$h_1^{2\sigma}$	gas site interface heat-transfer coefficient, W/(m² K)
h_{conv}	$=\lambda_1/\delta_0=cRe_1^m Pr_1^n\,\lambda_1/D_{h1}$, convective heat-transfer coefficient not depending on the direction of the heat flux, W/(m² K)
k_w	wall roughness, *m*
Le	$=\lambda_1/\left(\rho_1 c_{p1} D_{M\to\sum n}\right)$ *Lewis* number, dimensionless
$Le_{1,M}$	$=Sc_{1,M}/Pr_1=\lambda_1/\left(\rho_1 c_{p1} D_{M\to\sum n}\right)$, *Lewis* number, dimensionless
$D_{M\to\sum n}$	diffusion coefficient for steam in the gas mixture, m²/s
ℓ_{e2}	turbulent length scale in the film, m
M_1	mixture mole mass, kg/1_mol
M_{n1}	mole masses as the inert components, kg/mol
M_{M1}	mole masses of the steam, kg/mol
Nu_{1z}	$=h_{1\sigma}\Delta z/\lambda_1$, gas *Nusselt* number, dimensionless
Nu_{12}	$=h_{conv}D_{h1}/\lambda_1$, *Nusselt* number, dimensionless
Pe_0	$=\frac{\delta_0(\rho w)_{1M,real}c_p}{\lambda}$, crossflow *Peclet* number, dimensionless
Per_w	perimeter of the rectangular channel, m
Per_{1w}	wetted perimeters for the gas, m
Per_{2w}	wetted perimeters for the liquid, m
Pr_1^t	turbulent gas *Prandtl* number, dimensionless
Pr_2	liquid *Prandtl* number, dimensionless
Pr_1	$=\eta_1 c_{p1}/\lambda_1$, gas *Prandtl* number, dimensionless

p pressure, Pa

$\dot{q}_1^{'''2\sigma}$ $= -\lambda_1 \dfrac{\partial T_1}{\partial y} = h_{conv}\left(T_{1M}^{\sigma 2} - T_1\right)$, heat flux density from the film interface into the gas defined by *Fourier*'s law, dimensionless, W/m³

$\dot{q}_2^{'''1\sigma}$ heat flux density from the interface into the bulk liquid, W/m³

Re_1 $= \rho_1 w_1 D_{h1} / \eta_1$, gas *Reynolds* number, dimensionless

Re_2 $= \rho_2 w_2 D_{h2} / \eta_2$, liquid *Reynolds* number, dimensionless

Re_1^* $= \dfrac{\alpha_1 \rho_1 V_1}{\eta_1} \dfrac{F}{Per_{12}}$, modified gas *Reynolds* number, dimensionless

Re_2^* $= \dfrac{\alpha_2 \rho_2 V_2}{\eta_2} \dfrac{F}{Per_{12}}$, modified liquid *Reynolds* number, dimensionless

Re_{2F} $= \dfrac{\rho_2 V_2 \delta_{2F}}{\eta_2}$ film *Reynolds* number, dimensionless

Re_{1z} $= V_1 \Delta z / \nu_1$, gas *Reynolds* number, dimensionless

Re_2^t $= \rho_2 V_2' \ell_{e2} / \eta_2$, liquid turbulent *Reynolds* number, dimensionless

Re_{2F}^* $= \dfrac{\rho_2 V_{2eff}^* \delta_{2F}}{\eta_2}$, liquid turbulent *Reynolds* number based on the friction velocity, dimensionless

Re_1 $= \rho_1 w_1 D_{h1} / \eta_1$, gas *Reynolds* number, dimensionless

Re_0 $= \delta_0 (\rho w)_{wb} / \eta$ crossflow Reynolds number, dimensionless

r radius, m

$Sc_{1,M}$ $= \eta_1 / \left(\rho_1 D_{1,M}^*\right)$, *Schmidt* number, dimensionless

$Sh_{12,M}$ $= \beta_{12,i} D_{h1} / D_{1,M}^*$, *Sherwood* number, dimensionless

Sr_1 $= \dfrac{Nu_1}{\mathrm{Re}_1 \mathrm{Pr}_1}$, *Struhal* number, dimensionless

T temperature, K

$T_2^{1\sigma}$ liquid-side interface temperature, K

$\mathbf{V}_1$ gas velocity, m/s

$\mathbf{V}_2$ liquid velocity, m/s

V_2' volume-averaged liquid pulsation velocity, m/s

V_{2eff}^* $= \sqrt{\tau_{2eff} / \rho_2}$, dynamic friction velocity, m/s

w_1 axial gas velocity, m/s

w_2 axial liquid velocity, m/s

y distance, m

z axial coordinate, m

Greek

α_2	local liquid volume fraction in the computational cell, dimensionless
β_1	$=\dfrac{h_{1\sigma}}{\rho_1 c_{p1}}=\dfrac{Nu_1}{Re_1 Pr_1}V_1 = Sr_1 V_1$, weighted heat-transfer coefficient, m/s
β_{12}	$= cRe_1^m Sc_1^n D_{1,M}^* / D_{h1}$, convective mass-transfer coefficient not depending on the direction of the mass flux, m/s
Δ	finite difference, dimensionless
$\Delta\tau_2$	time for which the eddy stays at the interface, renewal period, s
δ_{2F}	film thickness, m
δ_0	thickness of the thermal boundary layer of the undisturbed flow, *m*
η	dynamic viscosity of liquid, kg/(m s)
λ	thermal conductivity, W/(m K)
λ_{2eff}	effective conductivity, W/(m K)
λ_{w1}	friction coefficient, dimensionless
μ_{21}	steam mass generation per unit mixture volume and unit time due to evaporation, kg/(s m³)
μ_{12}	liquid mass generation per unit mixture volume and unit time due to condensation, kg/(s m³)
ρ	density, kg/m³
$(\rho w)_{21}$	steam mass generation per unit surface and unit time due to evaporation, kg/(m² s)
$(\rho w)_{1M}^{\sigma 2}$	$=-\dfrac{D_{1,M}^*}{R_1 T_1}\dfrac{\partial p_M}{\partial y}=\beta_{12}\rho_1\left(C_{1M}^{\sigma 2}-C_{1M}\right)$, mass flux of specie *M* from the interface of field 2 into the bulk of field 1 defined by *Fick*'s law, dimensionless
$(\rho w)_{12}$	liquid mass generation per unit surface and unit time due to condensation, kg/(m² s)
τ_{w1}	$=\lambda_{w1}\frac{1}{2}\rho_1 V_1^2$, shear stress, N/m²
τ_{2w}	liquid–wall shear stress, N/m²
τ_{12}	gas/liquid interface shear stress, N/m²
τ	time, s

Subscripts

1	gas
2	liquid film
2F	liquid film

$M1$, $1M$ vapor inside the gas mixture
$n1$, $1n$ inert components inside the gas
i component i

References

Ackermann, G.: Wärmeübergang und molekulare Stoffübertragung im gleichen Feld bei großen Temperatur- und Parzialdruckdifferenzen, Forschungsheft 382, Ausgabe B Band 8 (January, February 1937)

Albring, W.: Angewandte Stroemungslehre, Verlag Theodor Steinkopf, Dresden, 4. Auflage (1970)

Al-Diwani, H.K., Rose, J.W.: Free convection film condensation of steam in presence of non condensing gases. Int. J. Heat Mass Transfer 16, 1959 (1973)

Banerjee, S.: Turbulence structure and transport mechanisms at interfaces. In: Proc. Ninth Int. Heat Transfer Conference, Jerusalem, Israel, vol. 1, pp. 395–418 (1990)

Brumfield, L.K., Houze, K.N., Theofanous, T.G.: Turbulent mass transfer at free, Gas-Liquid Interfaces, with Applications to Film Flows. Int. J. Heat Mass Transfer 18, 1077–1081 (1975)

Bunker, R.S., Carey, V.P.: Modeling of turbulent condensation heat transfer in the boiling water reactor primary containment. Nucl. Eng. Des. 91, 297–304 (1986)

Churchill, S.W., Brier, J.C.: Convective heat transfer from a gas stream at high temperature to a circular cylinder normal to the flow. In: Chem. Eng. Prog. Symp. Ser. 51, vol. 17, pp. 57–65 (1955)

Fick, A.: Ann. Phys. 94, 59–86 (1855)

Fortescue, G.E., Pearson, J.R.A.: On gas absorption into a turbulent liquid. Chem. Eng. Sci. 22, 1163–1176 (1967)

Fourier, J.: Théorie analytique de la chaleur (1822); see in Fourier, J.: (1878); The analytical theory of heat. Cambridge University Press, Cambridge (reissued by Cambridge University Press) (2009), ISBN 978-1-108-00178-6)

Gnielinski, V.: Berechnung mittlerer Waerme- und Stoffuebertragungskoeffizienten an laminar und turbulent ueberstroemenden Einzelkoerpern mit Hilfe einer einheitlichen Gleichung. Forsch. Ing. Wes 41(5), 145–153 (1975)

Gnielinski, V.: New equations for heat and mass transfer in turbulent pipe and channel flow. Int. Chem. Eng. 16(2), 359–368 (1976)

Hausen, H.: Darstellung des Waermeueberganges in Roehren durch verallgemeinerte Potenzgleichungen. Verfahrenstechnik 9(4/5), 75–79 (1958)

Hobbhahn, W.K.: Modeling of condensation in light water reactor safety. In: Mueller, U., Rehme, K., Rust, K., Braun, G. (eds.) Proc. of the Fourth International Topical Meeting on Nuclear Reactor Thermal-Hydraulics, Karlsruhe, October 10-13, vol. 2, pp. 1047–1053 (1989)

Jakob, M., Linke, W.: Der Waermeübergang von einer waagerechten Platte an sidendes Wasser. Forsch. Ing. Wes. 4, 75–81 (1933)

Jensen, R.J., Yuen, M.C.: Interphase transport in horizontal stratified concurrent flow, U.S. Nuclear Regulatory Commission Report NUREG/CR-2334 (1982)

Kim, H.J., Bankoff, S.G.: Local heat transfer coefficients for condensation in stratified countercurrent steam - water flows. Trans. ASME 105, 706–712 (1983)

Kim, M.H., Corradini, M.L.: Modeling of condensation heat transfer in a reactor containment. Nucl. Eng. Des. 118, 193–212 (1990)

Lamont, J.C., Yuen, M.C.: Interface transport in horizontal stratified concurrent flow, U. S. Nuclear Regulatory Commission Report NUREG/CR-2334 (1982)

Liao, Y., Vierow, K.: Application of a generalized diffusion layer theory to predict experimental data on condensation of vapour-air mixture. In: Proc. of ICAPP 2006, Reno, NV USA, June 4-8 (2006)

Magnaudet, J., Calmet, I.: Turbulent mass transfer through a flat shear-free surface. J. Fluid Mech. 553, 155–185 (2006)

McEligot, D.M.: The effect of large temperature gradients on turbulent flow of gases in the downstream region of tubes. Ph.D. thesis, Stanford Univ., USA (1963)

Miheev, Miheeva: Kratkiy kurs teploperedachi. Gidroenergoisdat, Moskva-Leningrad (1960)

Nusselt, W.: Das Grundgesetz des Wärmeüberganges, Gesundh. Gesundh. Ing. Bd. 38, S.477 (1915)

Peterson, P.F., Schrock, V.E., Kageyama, T.: Diffusion layer theory for turbulent vapor condensation with noncondensable gases. J. Heat Transfer 115, 998–1003 (1993)

Petukhov, B.S., Kirillov, P.L.: About heat transfer at turbulent fluid flow in tubes. Thermal Eng. 4, 63–68 (1958) (in Russian)

Petukhov, B.S., Popov, V.N.: Theoretical calculation of heat exchange and friction resistance in turbulent flow in tubes of an incompressible fluid with variable physical properties. High Temperature 1, 69–83 (1963)

Pohlhausen, E.: Der Waermeaustausch zwischen festen Koerpern und Fluessigkeiten mit kkleiner Reibung und kleiner Waermeleitung. Z. angew. Math. Mech. 1(2), 115–121 (1921)

Rohsenow, W.M., Choi, H.: Heat, mass and momentum transfer. Prentice-Hall Publishers, New Jersey (1961)

Siddique, M., Golay, M.W., Kazimi, M.S.: Theoretical modelling of forced convection condensation of steam in a vertical tube in the presence of non condensable gas. Nucl. Technol. 106, 202–215 (1994)

Stefan, J.: Versuche über die Verdampfung, Wiener Berichte, Bd. 68, S. 385 (1874)

Theofanous, T.G., Houze, R.N., Brumfield, L.K.: Turbulent mass transfer at free, gas-liquid interfaces, with application to open- channel, bubble and jet flows. Int. J. Heat Mass Transfer 19, 613–624 (1976)

Uchida, U., Oyama, A., Togo, Y.: Evolution of post - incident cooling system of light water reactors. In: Proc. 3th Int. Conf. Peaceful Uses of Atomic Energy, International Atomic Energy Agency, Vienna, Austria, vol. 13, p. 93 (1964)

VDI-Waermeatlas, 4. Auflage, VDI-Verlag, Dusseldorf (1984)

Zhou, W., Revankar, S.: Single and multi-tube PCCS condensation heat transfer. In: 12th Int. Top. Meeting on Nuclear Reactor Thermal Hydraulic, Pittsburg, Pa, USA, September 30-October 4 (2007)

14. Condensation at cooled walls

14.1 Pure steam condensation

14.1.1 Onset of the condensation

Consider gas flow in a pipe. The heat-transfer coefficient at the wall for turbulent flow without correction of the properties in the wall boundary layer is

$$h_{convective} = 0.023 Re_1^{0.8} Pr_1^{0.4} \frac{\lambda_1}{D_{hy}}, \tag{14.1}$$

where the *Reynolds* and the *Prandtl* numbers are defined as functions of the gas-mixture properties

$$Re_1 = D_{hy} \rho_1 \left| w_1 \right| / \eta_1, \tag{14.2}$$

$$Pr_1 = \eta_1 c_{p1} / \lambda_1. \tag{14.3}$$

For prescribed heat flux the wall temperature is

$$T_w = T_1 + \dot{q}''_w / h_{convective}. \tag{14.4}$$

If the wall temperature is less then the saturation temperature at the partial pressure of the steam

$$T_w \le T_{cond}, \tag{14.5}$$

where

$$T_{cond} = T'(p_{M1}), \tag{14.6}$$

condensation at the wall occurs.

14.1.2 Gravitation films on plates

Gravitational films along inclined surfaces are common in nature and technology. That is why they attracted the attention of scientists like *Nusselt* already in 1916. The source of the film is usually defined as the flow rate per unit of film width $\dot{V}'$ in m^2/s. This amount is either prescribed or subject to analyses, for instance, in the case of mass transfer due to condensation, droplet deposition on the wall, etc. If the source of the liquid is known then the independent values characterizing the film flow are (ν_2, σ_2, ρ_2, g, φ). Isothermal films without heat and mass transfer are easy to describe. For this purpose, the conservation of momentum is solved for steady developed flow neglecting entrance effects. The solutions are presented in the form of the dependence between the share stress, film thickness and flow rate per unit of the film width,

$$f\left(\tau_{2w}, \delta_2, \dot{V}_2'\right) = 0, \tag{14.7}$$

called in the literature *triangular* relationships. In our case the wall share stress is easily obtained by setting the weight of the liquid attached to a surface divided by this surface,

$$\tau_{2w} = weight/\left(\Delta x \Delta z\right) = \rho_2 g \delta_2, \tag{14.8}$$

so that the relation defining the film thickness depends only on the volumetric flow rate per unit of film width

$$\delta_2 = f\left(\dot{V}', ...\right). \tag{14.9}$$

We see that for given $\dot{V}'$ the velocity profile of the film being either laminar or turbulent determines the film thickness. In order to decide which type of velocity profile we have to expect the classification of the different film-flow regimes is helpful.

14.1.2.1 Film-flow regimes

The waviness of the surface depends (a) on the properties of the liquid collected in a dimensionless group called the *Kapitza* number

$$Ka_2 = \sigma_2^3/\left(\rho_2^3 g \nu_2^4\right), \tag{14.10}$$

and (b) on the mass flow rate represented by the film *Reynolds* number

$$\mathrm{Re}_{2\delta} = \dot{V}'/\nu. \tag{14.11}$$

Usually, engineers consider laminar flow for Reynolds number less then 400 and turbulent for larger then 400. More precisely, a variety of wave types lead to different classifications of the flow regimes. *Ishigai* et al. (1972), based on the analysis of surface perturbations, discriminate five regimes in the flow of a vertical liquid film, as given in Table 14.1. The table contains also the regimes for water at atmospheric conditions.

Table 14.1 *Ishigai* et al. (1972) classification of the film regimes

Type	Region	Water at atmospheric conditions
Laminar	$Re_{2\delta} \le 0.47 Ka_2^{1/10}$	$Re_{2\delta} \le 5.7$
First transient regime	$0.47 Ka_2^{1/10} < Re_{2\delta} \le 2.2 Ka_2^{1/10}$	$5.7 < Re_{2\delta} \le 27$
Stable wavy	$2.2 Ka_2^{1/10} < Re_{2\delta} \le 75$	$27 < Re_{2\delta} \le 75$
Second transient regime	$75 < Re_{2\delta} \le 400$	$75 < Re_{2\delta} \le 400$
Fully turbulent	$400 < Re_{2\delta}$	$400 < Re_{2\delta}$

Note that the initiation of the first waves was already observed by *Kapitza* in 1948 for $Re_{2\delta} > 0.61 Ka_2^{1/11}$. *Brauer* (1971) reported for this limit $Re_{2\delta} > 0.261 Ka_2^{1/8}$.

14.1.2.2 Film thickness without mass transfer

We concentrate our attention on liquid films without evaporation, condensation, gas absorption or gas release. In this case the mass flow rate is delivered to the wall from an external source.

Steady-state laminar films (*Nusselt* (1916)): For constant properties the solution of the steady state *Navier–Stokes* equation $\nu_2 \dfrac{d^2 u_2}{dy^2} + g\cos\varphi = 0$ with boundary conditions $u_2 = 0$ at $y = 0$ (adherence at the wall), $du_2/dy = 0$ at $y = \delta_2$ (absence of share stresses on the free surface) is

$$u_2 = \frac{g\cos\varphi}{\nu_2}\left(y\delta_2 - \frac{1}{2} y^2 \right). \tag{14.12}$$

The volumetric flow rate per unit of film width is then

$$\dot{V}_2' = \int_0^{\delta_2} u_2 dy = \frac{g\delta_2^3}{3\nu_2}\cos\varphi . \tag{14.13}$$

The film thickness then depends on the volumetric flow rate per unit of film width as follows

$$\delta_2 = \left(\frac{\dot{V}_2' 3\nu_2}{g\cos\varphi} \right)^{1/3} = \left(\frac{3\nu_2^2}{g\cos\varphi} \right)^{1/3} Re_{2\delta}^{1/3} . \tag{14.14}$$

This expression was confirmed by several authors to agree well with experimental data – see *Fulford* (1964), *Ganchev* and *Kozlov* (1973), *Alekseenko* et al. (1973),

Tananaiko and *Vorontsov* (1975), *Gimbutis* (1988), *Boyadjiev* and *Beschkov* (1984).

Steady-state turbulent film flow on a smooth plate: *Alekseenko* et al. (1996) assuming smooth surface postulated the validity of the *Blasius* law for wall shear stress

$$\tau_{2w} = \frac{0.045}{\dfrac{u_2 \delta_2}{\nu_2}} \frac{1}{2} \rho u_2^2 . \tag{14.15}$$

Assuming that the 1/7th profile

$$\frac{u'}{u} = \left(\frac{y}{\delta} \right)^{1/7} , \tag{14.16}$$

represents the real turbulent velocity distribution in the film the authors obtained $u_2 = \dfrac{8}{7} \dfrac{\dot{V}_2'}{\delta_2}$. Taking into account that the wall share stress is known, $\tau_{2w} = \rho_2 g \delta_2$, and replacing into the *Blasius* equation the authors come to the result

$$\delta_2 = 0.305 \frac{\dot{V}_2'^{7/12} \nu_2^{1/12}}{\left(g \cos \varphi \right)^{1/3}} = 0.305 \left(\frac{\nu_2^2}{g \cos \varphi} \right)^{1/3} \mathrm{Re}_{2\delta}^{7/12} . \tag{14.17}$$

Correspondingly, the shear stress is

$$\tau_{2w} = \rho_2 g \delta_2 = 0.305 \rho_2 \left(g \cos \varphi \right)^{2/3} \nu_2^{1/12} \dot{V}_2'^{7/12} = 0.305 \rho_2 \left(\nu_2 g \cos \varphi \right)^{2/3} \mathrm{Re}_{2\delta}^{7/12} . \tag{14.18}$$

Note that *Brauer* (1971) recommend for the turbulent film the following empirical relation

$$\delta_2 = 0.302 \left(\frac{3 \nu_2^2}{g \cos \varphi} \right)^{1/3} \mathrm{Re}_{2\delta}^{8/15} , \tag{14.19}$$

which is obviously very close to the $1/7^{th}$ profile solution. *Kutateladze* (1979) and *Dukler* and *Bergelin* (1952) have applied similar approaches using instead of $1/7^{th}$ profiles logarithmic profiles for two-layered schemes. As can be seen from Fig. 14.1 for the data for shear stress as well as from the analysis by *Gimbutis* (1988) for the data for thickness, various approaches differ insignificantly and generalize fairly well the experimental results for $Re > 500$.

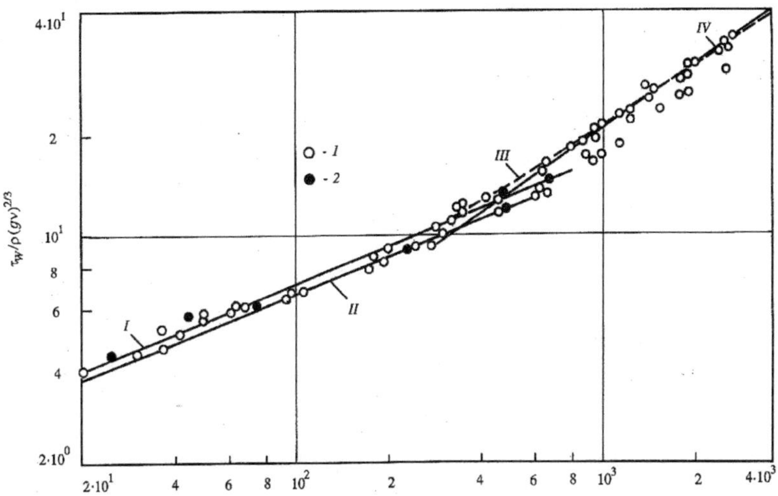

Fig. 14.1 Wall shear stresses as a function of the film Reynolds number for water falling film on a vertical wall taken from Berman and Tumanov (1962). Prediction: I – *Nusselt* (1916); II – *Kapitza* (1948); III – *Kutateladze* (1979); IV – *Alekseenko* et al. (1996); Experiment by *Alekseenko* et al. (1996) – points 1, 2

Conclusion: For engineering applications it is recommended to compute the thickness of a falling film on inclined smooth walls as follows:

$$\delta_2 = \left(\frac{\dot{V}_2' 3\nu_2}{g\cos\varphi}\right)^{1/3} = \left(\frac{3\nu_2^2}{g\cos\varphi}\right)^{1/3} \mathrm{Re}_{2\delta}^{1/3} \text{ for } \mathrm{Re}_{2\delta} \le 400\,, \tag{14.20}$$

$$\delta_2 = 0.305\left(\frac{\nu_2^2}{g\cos\varphi}\right)^{1/3} \mathrm{Re}_{2\delta}^{7/12} \text{ for } 400 > \mathrm{Re}_{2\delta}\,. \tag{14.21}$$

14.1.2.3 Film thickness during condensation

Now we concentrate our attention on liquid films for which the source of mass is a *condensation process* from a steam being in saturation. The surface temperature of the film during the condensation is the saturation temperature at the environmental pressure T' and the wall temperature is T_w. Assuming a linear temperature profile in the film the heat flux release by the condensation and removed into the wall is

$$\dot{q}_w''^{2\sigma} = \lambda_2 \frac{T' - T_w}{\delta_2} \tag{14.22}$$

and the heat-transfer coefficient

$$h_{stagnant} = \lambda_2 / \delta_2\,. \tag{14.23}$$

Note that for this case the Reynolds number can be expressed by the averaged heat flux over the height H as follows

$$Re_{2\delta} = \frac{4\rho_2 w_2 \delta_2}{\eta_2} = \frac{4\dot{q}_w''^{2\sigma} H}{\eta_2 \Delta h}. \qquad (14.24)$$

For a prescribed wall temperature Eqs. (14.20) or (14.21) together with Eqs. (14.22)–(14.24) can be solved iteratively with respect to the heat flux and the film thickness. While for laminar flow this is accurate, for turbulent flow this is an approximation because the thermal eddy diffusivity inside the film is not a constant. More sophisticated approaches are described below. The improvements take into account the variation of the thermal properties in the film, the waviness of the film that promote mass transfer, the turbulent structure of the film and some adjustments to experimental data.

The *Nusselt* water skin theory for laminar condensation from 1916: Assuming that the thermal resistance controlling the condensation is the heat conduction through the film and the flow in the film is laminar *Nusselt* obtained the following analytical solution for the averaged heat-transfer coefficient for laminar condensation on a vertical wall with high H from stagnant steam

$$h_{stagnant} := \frac{\dot{q}_w''^{2\sigma}}{T_{cond} - T_w} = \frac{2}{3}\sqrt{2}\left[\frac{\rho'(\rho' - \rho'')\lambda'^3 g \cos\varphi \Delta h}{H\eta'(T_{cond} - T_w)}\right]^{1/4} f_{prop} f_{wave}, \qquad (14.25)$$

or

$$T_w = T_{cond} - \left(-\dot{q}_w''^{2\sigma}/C\right)^{4/3}, \qquad (14.26)$$

where

$$C = \frac{2}{3}\sqrt{2}\left[\frac{\rho'(\rho' - \rho'')\lambda'^3 g \cos\varphi \Delta h}{H\eta'}\right]^{1/4} f_{prop} f_{wave}, \qquad (14.27)$$

$\Delta h = h'' - h'$ is the latent heat of vaporization and $f_{prop} = 1$, $f_{wave} = 1$. The geometrical constant for plate $\frac{2}{3}\sqrt{2} = 0.94$ is for sphere 0.826, *Popiel* and *Boguslawski* (1975), and for an external horizontal tube 0.729 *Dhir* and *Lienhard* (1971). For condensation on the external surface of the N horizontal pipes being above each other *Incopera* and *DeWitt* (2002) p. 624 proposed the modification $h_{stagnant} = \left(h_{stagnant}\right)_{\text{single row}} N^{-1/4}$.

In order to take into account the variation of the thermal properties inside the film and the wave structure for $\mathrm{Re} \le 400$ *Labuntzov* (1957) proposed to correct the *Nusselt* solution by

$$f_{prop} = \left\{ \left[\frac{\lambda_2(T_w)}{\lambda'} \right]^3 \frac{\eta'}{\eta(T_w)} \right\}^{1/8}, \tag{14.28}$$

and

$$f_{wave} = \mathrm{Re}_{2\delta}^{0.04}, \tag{14.29}$$

respectively. Correction for the influence of the film subcooling is introduced by *Rohsenow* et al. (1985) p. 115 as follows:

$$\Delta h = h'' - h' + \frac{3}{8} c_p' \left(T' - T_w \right). \tag{14.30}$$

Condensation from stagnant steam at turbulent liquid film: For large vertical surfaces the heat-transfer coefficient may exceed the computed one by 70% due to ripples that attain the greatest amplitude even for $Re_{\delta 2} < 400$. Some authors write thc *Nusselt* result in the form

$$\frac{h_{satgnant}}{\lambda'} \left[\frac{\eta'^2}{\rho'(\rho' - \rho'') g \cos\varphi} \right]^{1/3} = 0.95 Re_{2\delta}^{-1/3}, \tag{14.31}$$

and seek an appropriate modification to make it valid for high film *Reynolds* numbers. For the turbulent region, $Re_{\delta 2} > 400$, *Labuntzov* used in 1957 variable distribution of the eddy viscosity to describe the turbulent velocity profile. He finally approximated the analytical integral solution within $1 \le \mathrm{Pr}_2 \le 25$ and $1.5 \times 10^3 \le \mathrm{Re} \le 6.9 \times 10^4$ by

$$\frac{h_{satgnant}}{\lambda'} \left[\frac{\eta'^2}{\rho'(\rho' - \rho'') g \cos\varphi} \right]^{1/3} = 0.0325 Re_{2\delta}^{0.25} \mathrm{Pr}_2^{0.5}, \tag{14.32}$$

reproducing experimental data of several authors within 12% . *Isachenko* extended in 1977 this result by taking into account also the upper laminar part to

$$\frac{h_{satgnant}}{\lambda'} \left[\frac{\eta'^2}{\rho'(\rho' - \rho'') g \cos\varphi} \right]^{1/3} = \frac{Re_{2\delta}}{2300 + 41 \mathrm{Pr}_2^{0.5} \left(Re_{2\delta}^{3/4} - 89 \right)}. \tag{14.33}$$

For comparison we give here also the *Kutateladze* correlation from 1963

$$\frac{h_{satgnant}}{\lambda'} \left[\frac{\eta'^2}{\rho'(\rho' - \rho'') g \cos\varphi} \right]^{1/3} = \frac{Re_{2\delta}}{389 + 18.5 \mathrm{Pr}_2^{-0.4} \left(Re_{2\delta}^{5/6} - 47 \right)}. \tag{14.34}$$

For $Re_{\delta 2} > 2000$ the right-hand side simplifies to $0.054 Re_{2\delta} \mathrm{Pr}_2^{0.4}$. Note the differences in the *Prandtl* number dependence.

Butterworth proposed in 1981 for a vertical plane the following set of correlations for the three regions:

1. Laminar film: *Nusselt* theory.
2. Wavy film: $8.25 < Re_{\delta 2} < 400$,

$$\frac{h_{satgnant}}{\lambda'}\left[\frac{\eta'^2}{\rho'(\rho'-\rho'')}\right]^{1/3} = \frac{Re_{2\delta}}{1.465 Re_{2\delta}^{1.22} - 1.3}. \tag{14.35}$$

3. Turbulent film: $Re_{\delta 2} > 400$,

$$\frac{h_{satgnant}}{\lambda'}\left[\frac{\eta'^2}{\rho'(\rho'-\rho'')}\right]^{1/3} = \frac{Re_{2\delta}}{2187 + 41 Pr'^{-1/2}\left(Re_{2\delta}^{3/4} - 89.4\right)}. \tag{14.36}$$

Incopera and *DeWitt* (2002) p. 620 proposed the following set of correlations that differ slightly from the

$$\frac{h_{satgnant}}{\lambda'}\left[\frac{\eta'^2}{\rho'(\rho'-\rho'')g\cos\varphi}\right]^{1/3} = 1.47 Re_{2\delta}^{-1/3}, \quad Re_{\delta 2} < 30, \tag{14.37}$$

$$\frac{h_{satgnant}}{\lambda'}\left[\frac{\eta'^2}{\rho'(\rho'-\rho'')}\right]^{1/3} = \frac{Re_{2\delta}}{1.08 Re_{2\delta}^{1.22} - 5.2}, \quad 30 \le Re_{\delta 2} \le 1800, \tag{14.38}$$

$$\frac{h_{satgnant}}{\lambda'}\left[\frac{\eta'^2}{\rho'(\rho'-\rho'')}\right]^{1/3} = \frac{Re_{2\delta}}{8750 + 58 Pr'^{-1/2}\left(Re_{2\delta}^{3/4} - 253\right)}, \quad Re_{\delta 2} > 1800, \tag{14.39}$$

verified for water under $1 \le Re_{\delta 2} \le 7200$ by *Gregorig*, et al. (1974).

Problem 1: Consider a room filled with 100 °C steam at 1 bar. The vertical 10 m walls of the room consist of 10 cm steel followed by 95 cm concrete with an initial temperature of 50 °C. The external wall surface is considered as adiabatic. Compute the temperature profile in the wall, the film Reynolds number and the film thickness.

Solution to problem 1: We create a simple algorithm solving the Fourier equation for the direction normal to the wall. At the internal site we first postulate that the condensing film is laminar. Therefore, we use Eq. (14.25) to compute the heat-transfer coefficient, Eq. (14.24) to compute the Reynolds number and Eq. (14.20) to compute the film thickness. The results are presented in Figs. 14.2 and 14.3. We see that (a) after a half an hour the steel is heated almost to the saturation temperature but the concrete, due to its poor heat conductivity, follows slowly; (b) Within the first 100 s the film Reynolds number exceeds 400, which indicates that in this region the film is turbulent and the *Nusselt* theory is not appropriate; (c) In almost all the time the film is laminar with a weave surface.

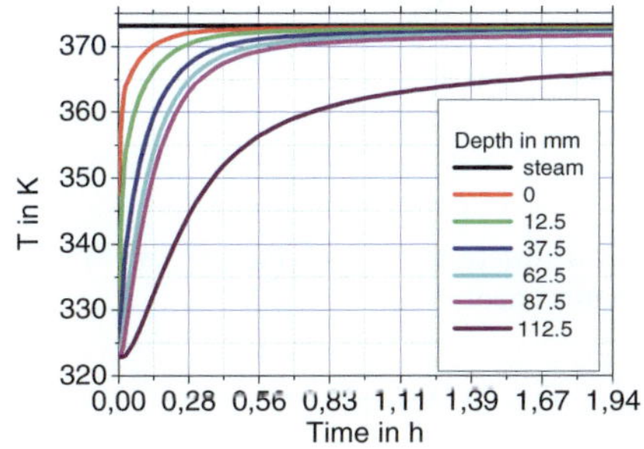

Fig. 14.2 Temperatures as a function of time

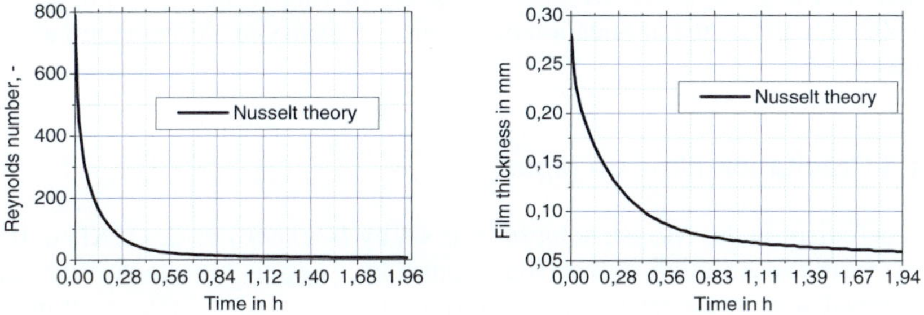

Fig. 14.3 a) Film Reynolds number as a function of time; b) Film thickness as a function of time

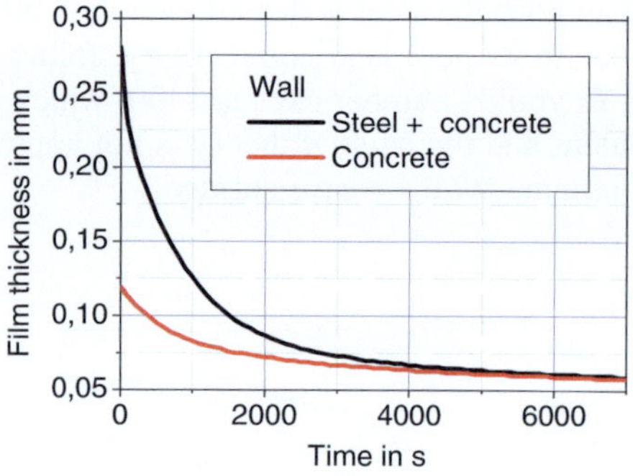

Fig. 14.4 Film thickness as a function of time. Parameter: the thermal conductivity of the wall removing the condensation heat

Problem 2: Repeat the analysis for problem 1 by considering only a concrete wall.

Solution to problem 2: The solution is presented in Fig. 14.4. As expected, the worst conduction of the concrete controls the rate of condensation resulting in a smaller film thickness. The *Reynolds* number, varying from 58 to 5.8, was less than 400. Therefore, the application of the *Nusselt* theory in all the region of interest was correct.

14.1.3 Gravitation films on pipes

The coefficient in the *Nusselt* solution was set by *Isachenko* et al. (1981) to 0.728 for external condensation on horizontal pipes by using as a characteristic length the external wetted diameter of the pipe $H = D_h$. *Bird* et al. (1960) recommended the constant 0.725 for *external condensation on a horizontal single pipe* reporting $\pm$ 10% error compared with experiments. The correlation was also recommended for *bundles of horizontal pipes Bird* et al. (1960) p 417. The *Nusselt* equation was modified for condensation *inside inclined pipe* by using a constant 0.296 taking into account the partial occupation of the pipe cross-section by the flowing condensate and $H = D_h$ by *Collier* in 1972. *Chato* (1962) arrived at 0.68 instead to 3/8, for $H = D_h$, and recommend for small inclinations $\Delta z/\Delta x < 0.002$ the constant in the *Nusselt* equation to be

$$0.468\, f\left[\mathrm{Pr}_2, \frac{c_p'\left(T_k' - T_w\right)}{h'' - h'}\right]$$

a function of the liquid *Prandtl* number given graphically in his work. For larger inclinations $0.002 < \Delta z/\Delta x < 0.6$, 0.3 is used instead of 0.468.

14.2 Condensation from forced-convection two-phase flow at liquid film

14.2.1 Down-flow of vapor across horizontal tubes

As already mentioned, the *Nusselt* theory is also valid for condensation of stagnant vapor on a horizontal cylinder with the modification of the numerical coefficient to 0.728 by *Isachenko*. Several authors correlated their data by using as a measure the *Nusselt* heat-transfer coefficient as follows:

Fijii et al. (1979): $$\frac{h}{h_{Nu}} = 1.24\left(\frac{w_1 D_{hy}}{\nu_2}\right)\left(\frac{h_{Nu} D_{hy}}{\lambda'}\right)^{-0.2}. \quad (14.40)$$

Berman and Tumanov (1962): $$\frac{h}{h_{Nu}} = 1 + 0.0095\left(\frac{w_1 D_{hy}}{\nu_1}\right)^{11.8/\sqrt{\frac{h_{Nu} D_{hy}}{\lambda'}}}. \quad (14.41)$$

Shekriladze and *Geomelauri* (1966): $$\frac{h}{h_{Nu}} = \left\{\frac{1}{2}\left(\frac{h_{Sh}}{h_{Nu}}\right)^2 + \frac{1}{4}\left[\left(\frac{h_{Sh}}{h_{Nu}}\right)^4 + 1\right]^{1/2}\right\}^{1/2}, \quad (14.42)$$

where

$$h_{Sh} = 0.9\frac{\lambda_2}{D_{pipe}}\left(\frac{w_1 D_{hy}}{\nu_2}\right)^{1/2}. \quad (14.43)$$

Kutateladze (1962): $$\frac{h}{h_{Nu}} = 1 + 0.004\left(\frac{w_1^2 \rho_1 h_{Nu}}{g \rho_2 \lambda_2}\right)^{0.8}. \quad (14.44)$$

14.2.2 *Collier* correlation

For forced convection in an inclined pipe in the case of a *laminar film* inside *Collier* (1972) proposed

$$h_{lam} = 0.825\left[\frac{\rho'(\rho' - \rho'')\lambda'^3 g \cos\varphi}{G D_h}\right]^{1/3}, \quad (14.45)$$

and in the case of *turbulent film*

$$h_{turb} = 0.065\frac{\lambda' \rho'^{1/2}}{\eta'}\frac{0.023\rho' w_1^2}{Re_1^{1/4}} Pr'^{1/2}, \quad (14.46)$$

where the liquid properties are those of saturated liquid at the local pressure,

$$Pr' = \eta' c_p' / \lambda' . \tag{14.47}$$

In practical computations the maximum of all three expressions is taken,

$$h = \max\left(h_{stagnant}, h_{lam}, h_{turb}\right), \tag{14.48}$$

which is in fact using the *Nusselt* theory in cases where the forced convection correlations gives lower values than the values for stagnant steam condensation.

14.2.3 *Boyko* and *Krujilin* approach

Boyko and *Krujilin* recommended in 1966 for forced-convection condensation the following correlation

$$h = h_{20}\left[1 + X_1\left(\frac{\rho'}{\rho''} - 1\right)\right]^{1/2}, \tag{14.49}$$

where

$$h_{20} = 0.024 Re_{20}^{0.8} Pr'^{0.4} \frac{\lambda_2}{D_{hy}}, \tag{14.50}$$

$$Re_{20} = D_{hy} G / \eta_2 . \tag{14.51}$$

The correlation was verified with experimental data for $12.3 < p < 88$ bar, $1 < X_1 < 0.26$, $162\ 000 < \dot{q}_w''^{\sigma 2} < 1\ 570\ 000$ W/m². The constant 0.024 was recommended for steel pipes and 0.032 for cupper pipes.

Miropolskij (1960) obtained experimentally for vertical steel tubes

$$h = 0.023 Re_2^{*0.8} Pr'^{0.4} \frac{\lambda'}{D_{hy}}\left[1 + 0.63\left(\frac{\rho'}{\rho''}\right)^{0.67} X_1\right], \tag{14.52}$$

where

$$Re_2^* = D_{hy} \rho' w_1 / \eta' . \tag{14.53}$$

This correlation has a mean error of 15% for pressures 7 to 220 atm, mass flow rates 400 to 2000 kg/(m² s) and heat fluxes 0.15 to 6×10^6 kcal/(m² h). For cupper tubes the coefficient was found to be 0.03.

14.2.4 The *Shah* modification of the *Boyko* and *Krujilin* approach

Shah (1979) reported a modification of this correlation

$$h = h_{20}\left[\left(1 - X_1\right)^{0.8} + \frac{3.8X_1^{0.76}\left(1 - X_1\right)^{0.04}}{\left(p/p_{cr}\right)^{0.38}}\right]. \tag{14.54}$$

The correlation is reported to reproduce 474 data points within $0.002 < p/p_{cr} < 0.44$, $21 < T'(p) < 310°C$, $3 < w_1 < 300m/s$, $0 < X_1 < 1$, 17.5 < *G* < 210.6*kg/s*, and 158 < $\dot{q}_w^{\prime\prime\sigma 2}$ < 1 893 000*W/m²* with mean deviation 15.4%.

The condensing mass per unit flow volume is then

$$\mu_{12} = \frac{4}{D_{hy}} \frac{\dot{q}_w''}{\Delta h}. \tag{14.55}$$

14.2.5 Flow regime associated models

In recent years more attention has been paid to using the available information regarding the local flow pattern to design appropriate condensation models. As an example let us consider the proposal made by *Jaster* and *Kosky* (1976). The authors introduced the flow pattern transition criterion

$$F = \frac{\textit{Axial shear force}}{\textit{Gravitational body force}} = \frac{\tau_{2w}}{\rho_2 \mathbf{g} \delta_2}$$

$$= 7.85\times 10^{-3}\left(\frac{\eta_1}{D_h}\right)^{3/8} \frac{\left(\alpha_1 w_1 \rho_1\right)^{21/24} \rho_2 \left(1 + X_{LM}^{2/m}\right)^{3m/2}}{\left(\rho_1 \rho_2\right)^{3/2} \mathbf{g} \delta_2^{+} \eta_2}, \tag{14.56}$$

where

$$X_{LM} = X_{LM}^2 = \frac{\rho_1}{\rho_2}\left(\frac{\eta_2}{\eta_1}\right)^{0.2}\left(\frac{1 - X_1}{X_1}\right)^{1.75} \tag{14.57}$$

is the *Lockhardt* and *Martinelli* factor, *m* = 5.13, and the dimensionless film thickness is computed by a semiempirical two-layer model

$$\delta_2^+ = 0.7071 \mathrm{Re}_2^{0.5} \text{ for } \mathrm{Re}_2 \le 1250, \tag{14.58}$$

$$\delta_2^+ = 0.0504 \mathrm{Re}_2^{0.875} \text{ for } 1250 < \mathrm{Re}_2, \tag{14.59}$$

with

$$\delta_2^+ = \delta_2 \sqrt{\tau_{2w}/\rho_2} / \nu_2 , \tag{14.60}$$

$$\mathrm{Re}_2 = (\rho w)_2 D_h / \eta_2 , \tag{14.61}$$

with D_h the pipe diameter. The authors found experimentally that for $F > 29$ the flow is annular, for $5 < F \le 29$ the flow is a transition flow and for $F \le 5$ the flow is stratified. The physical message of this criterion is easily understandable: If the wall share stress is larger than 29 times the gravity force trying to move the film downwards the flow pattern is film flow almost equally distributed at the wall. If the wall share force is less than 5 times the gravitational force the wall film collects at the bottom and form stratified flow. Consequently $F \le 5$ part of the heat transfer happens at the upper part following the main mechanism of the *Nusselt* water skin theory, Nu_{up}, and the lower part is occupied by much thicker layer and enforce Nu_{da}. The combination of the two regimes results then in weighting the heat-transfer contributions using the portion of the surface experiencing the corresponding regime

$$Nu = \frac{1}{\pi}\left[\theta Nu_{up} + (\pi - \theta) Nu_{da}\right]. \tag{14.62}$$

The angle θ with the origin the pipe axis is defined between the upwards-oriented vertical and the liquid–gas–wall triple point and is a function of the liquid volume fraction defined by

$$1 - \alpha_2 = \frac{\theta - \sin\theta\cos\theta}{\pi}. \tag{14.63}$$

A good explicit approximation for $\theta = f(\alpha_2)$ is given in Chapter 2 of vol. 2. For the transition between the regimes *Jaster* and *Kosky* (1976) proposed

$$Nu = Nu_{up} + \frac{F - 29}{24}\left(Nu_{up} - Nu_{da}\right), \tag{14.64}$$

Thome, et al. (2003) proposed to interpolate between two angles using as the dependent variable the mass flow at both limits

$$\theta = \theta_{strat}\left(\frac{G_{wavy} - G}{G_{wavy} - G_{strat}}\right)^{1/2}. \tag{14.65}$$

These authors found that

$$Nu_{da} = 0.003\,\mathrm{Re}_{2\delta}^{0.74}\,\mathrm{Pr}_2^{0.5}\,f_{waves}, \tag{14.66}$$

where $\mathrm{Re}_{2\delta} = 4\delta_2 \rho_2 w_2 / \eta_2$ and (14.67)

$$f_{waves} = 1 + \left(\frac{w_1}{w_2}\right)^{1/2} \left[\frac{(\rho_2 - \rho_1) g \delta^2}{\sigma_2}\right]^{1/4} \frac{G}{G_{stratified}}. \tag{14.68}$$

Comparing with a large number of data the agreement was within ± 20% error band.

14.3 Steam condensation from mixture containing noncondensing gases

Consider forced convection in a pipe. Given is the gas temperature, the wall temperature and the local parameters. We ask for the mass transferred per unit time and mixture volume and for the power density removed from the wall.

Before moving to the solution we will need methods for estimation of the molecular diffusion coefficient. Two methods will be given here. *Slattery* and *Bird* (1958), reported the following method for computation of the diffusion constant for solute A into solvent B

$$D_{1A\to B} = 3.64\times10^{-4}\left(\frac{T_1}{\sqrt{T_{1,A,c}T_{1,B,c}}}\right)^{2.334}\left(T_{1,A,c}T_{1,B,c}\right)^{5/12}\frac{\left(p_{1,A,c}\,p_{1,B,c}\right)^{1/2}}{p}\left(\frac{1}{M_A}+\frac{1}{M_B}\right)^{1/2}, \tag{14.69}$$

see *Bird*, et al. (1960) p. 505, where the mole masses are in g-mol units and the temperature in K. The critical pressures and temperatures for mixtures themselves are computed using the mole concentrations and the specific values for each particular component

$$p_c = \sum_i Y_i\, p_{c,i}, \tag{14.70}$$

$$T_c = \sum_i Y_i\, T_{c,i}. \tag{14.71}$$

Problem 3: Derive the appropriate expression for computation of the diffusivity coefficient of air into steam.

Solution to problem 3: Having in mind that for air

$Y_{1,air,N_2} = 0.7812$, $p_{c,N_2} = 33.999$ bar, $T_{c,N_2} = 126.15$ K,

$Y_{1,air,O_2} = 0.2188$, $p_{c,O_2} = 50.430$ bar, $T_{c,O_2} = 154.55$ K, we obtain $p_{c,air}$ = 37.594 bar, $T_{c,air}$ = 132.36 K. For mole masses M_{air} = 28.96 g/mol, and M_s = 18.96 g/mole the final expression is

$$D_{1,steam\to air} = 1.95\times10^{-5} T_1^{2.334}/p. \tag{14.72}$$

Perry and *Green* (1985) p. 2-370 proposed the following model valid for binary air–hydrocarbons or air–nonhydrocarbons mixtures at low pressure. Written for air and steam it takes the form

$$D_{1,steam\to air} = 0.10131\frac{T_1^{1.75}}{p\left(v_{air}^{1/3}+v_s^{1/3}\right)^2}\left(\frac{1}{M_{air}}+\frac{1}{M_s}\right)^{1/2} = 1.284\times10^{-3} T_1^{1.75}/p, \tag{14.73}$$

where v_{air} = 20.1, v_{H_2O} = 2×1.98 + 1×5.481 = 9.441 taken from Table 2-400 in *Perry* and *Green* (1985) p. 2-370 and the mole masses are in gmol. Note that computation of the diffusion coefficients is still not very accurate.

The overall heat-transfer coefficient can also be estimated in this case by using the *Boyko* and *Krujilin* (1966) heat-transfer correlation. What changes in this case is the driving temperature difference across the water film. There is convective heat transfer between the gas and the interface

$$h_{convective}\left(T_1 - T_1^{\sigma2}\right), \tag{14.74}$$

where

$$h_{convective} = 0.023 Re_1^{0.8} Pr_1^{0.4}\frac{\lambda_1}{D_h}. \tag{14.75}$$

A diffusion-controlled condensation mass flux is defined as follows

$$(\rho w)_{12} = \beta_{12}\rho_1\left(C_{1M} - C_{1M}^{\sigma2}\right), \tag{14.76}$$

where the condensation coefficient β_{12} is defined depending on the geometry and local conditions. The procedure for its computation will be discussed later. The energy-jump condition at the gas/liquid interface requires

$$h\left(T_1^{\sigma2} - T_w\right) = (\rho w)_{12}\Delta h + h_{convective}\left(T_1 - T_1^{\sigma2}\right). \tag{14.77}$$

Equation (14.59) has to be solved by iteration. To construct the *Newton* iteration we rewrite it in the form

$$f\left(T_1^{\sigma2}\right) = h\left(T_1^{\sigma2} - T_w\right) - (\rho w)_{12}\Delta h - h_{convective}\left(T_1 - T_1^{\sigma2}\right) = 0, \tag{14.78}$$

and search for the zeros. For this purpose we need the derivative

$$\frac{df\left(T_1^{\sigma 2}\right)}{dT_1^{\sigma 2}} = h + \frac{d\left(\rho w\right)_{12} \Delta h}{dT} + h_{convective}, \tag{14.79}$$

which is used in

$$T_1^{\sigma 2} = T_{1,old}^{\sigma 2} - f \Big/ \frac{df\left(T_1^{\sigma 2}\right)}{dT_1^{\sigma 2}}. \tag{14.80}$$

Note that the iterative solution is very sensitive to the chosen initial values, and it has to be carefully controlled.

14.3.1 Computation of the mass-transfer coefficient

14.3.1.1 Gas flow in a pipe

Bobe and *Solouhin* (1972), see also in *Isachenko* (1977) p. 136, correlated experimental data for steam condensation from gas–steam mixtures inside pipes as follows:

$$\beta_{12} = \beta_{12,0} \left(\frac{p_{1\sum n}}{p_{1M} - p_{1M}^{\sigma 2}} \right)^m, \tag{14.81}$$

where

$$\beta_{12,0} = 0.0191 \left(\frac{w_1 D_h}{\nu_1} \right)^{0.8} \left(\frac{\nu_1}{D_{\sum n \to M}} \right)^{0.43} \frac{D_{1\sum n \to M}}{D_h} \left(\frac{p}{p_{1\sum n}} \right)^n \tag{14.82}$$

or

$$\left(\rho w\right)_{12} = \rho_1 \beta_{12,0} \left[\frac{p_{1M} - p'\left(T_1^{\sigma 2}\right)}{p} \right]^{1-m}, \tag{14.83}$$

where for $0.1 < \dfrac{p_{1\sum n}}{p_{1M} - p'\left(T_1^{\sigma 2}\right)} < 1$, n = 0.6, m = 0.4, and for $\dfrac{p_{1\sum n}}{p_{1M} - p_{1M}^{\sigma 2}} \ge 1$, n = 0.9, $m = 0.1$. The derivative required for the iterative solution is then

$$\frac{d\left(\rho w\right)_{12}}{dT_1^{\sigma 2}} = -\beta_{12,0}\left(1-m\right)\frac{\rho_1}{p}\left[\frac{p}{p_{1M} - p'\left(T_1^{\sigma 2}\right)} \right]^m \frac{dp'\left(T_1^{\sigma 2}\right)}{dT_1^{\sigma 2}}. \tag{14.84}$$

14.3.1.2 Gas flow across a bundle of pipes

Berman and *Fuks* (1958) correlated experimental data for steam condensation from gas–steam mixtures flowing perpendicular to horizontal internally cooled tubes:

$$\beta_{12} = const \left(\frac{w_1 D_h}{\nu_1} \right)^{0.8} \frac{D_{1\sum n \to M}}{D_h} \left(\frac{p}{p_{1\sum n}} \right)^{0.6} \left(\frac{p}{p_{1M} - p_{1M}^{\sigma 2}} \right)^{1/3}, \tag{14.85}$$

or

$$(\rho w)_{12} = const\ \rho_1 \left(\frac{w_1 D_h}{\nu_1} \right)^{0.8} \left(\frac{p}{p_{1\sum n}} \right)^{0.6} \left(\frac{p_{1M} - p'\left(T_1^{\sigma 2}\right)}{p} \right)^{2/3} \frac{D_{1\sum n \to M}}{D_h}, \tag{14.86}$$

where *const* = 0.94 for single pipe, 1.06 for the first row of pipes, 1.64 for the third and the subsequent row of pipes.

14.3.1.3 The Vierow and Shrock explicit correlation

Figure 14.5 shows the ratio of the real heat-transfer coefficient for an air–steam mixture to the heat-transfer coefficient computed under the same conditions without air in accordance with *Gudymchuk* and *Konstantinov* (1936). Obviously for $\rho_{1n} C_{1n} / \left(\rho_{1M} C_{1M} \right) = 0.01$ the condensation is reduced to about 40%.

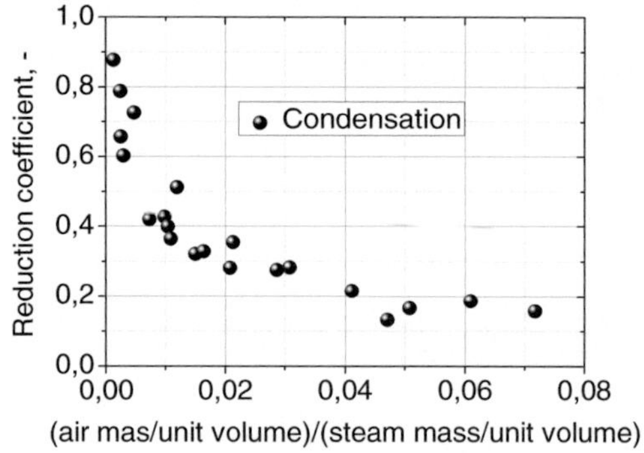

Fig. 14.5. The ratio of the real heat-transfer coefficient for an air–steam mixture to the heat-transfer coefficient computed under the same conditions without air, *Gudymchuk* and *Konstantinov* (1936), $f_2 = 0.15546 + 0.81496 \exp\left[-\frac{\rho_{1n} C_{1n}}{\rho_{1m}\left(1 - C_{1n}\right)} \Big/ 0.01074 \right]$

The simplest method for taking into account the resistance of the noncondensable gases was proposed by *Vierow* and *Shrock* (1991). It consists of three steps: 1) Compute the condensation from stagnant pure steam; 2) Correct the expression for the influence of the gas mixture forced convection; 3) Modify the resulting heat-transfer coefficient by a multiplier, being a function of the inert mass concentration. In particular, the *Nusselt* theory is used in the first step to compute $h_{stagnant}$. In the second step the multiplier

$$f_1 = \min\left(2,\ 1 + 2.88^{-5} Re_1^{1.18}\right) \tag{14.87}$$

is used, where $Re_1 = D_h \rho_1 |w_1| / \eta_1$. In the third step an explicit correlation of experimental data is generated in the simple form by the second multiplier. Recently, *Choi* et al. (2002) modified the originally proposed multiplier to

$$f_2 = 1 - 1.00962 C_{1n}^{0.13354} \quad \text{for} \quad 0.1 \le C_{1n} \le 0.65 \tag{14.88}$$

in order to obtain a better approximation of their own data base for $h_{cond} D_{hy,film} / \lambda_{film} < 30$ for stratified flow in inclined tubes (2.1° and 5°) – averaged percentage error 20 to 35%.

For comparison, we give here the recent correlation proposed by *Goldbrunner* (2003) for condensation on films in pipes

$$\frac{h_2^{1\sigma} \delta_2}{\lambda_2} = 0.0388 \left(\frac{u_2^t \delta_2}{\nu_2} \right)^{1.53} \mathrm{Pr}_2^{1/2}\, 0.5024 \left(1 - 1.5701 C_{1n}^{0.5987}\right), \tag{14.89}$$

where $u_2^t = \left(\varepsilon_2 \nu_2\right)^{1/4}$, $\varepsilon_2 = u_2 dp/dz$, valid for horizontal water film flows in an 80×64 mm rectangular channel as follows. The standard deviation from the measurements reported was 22%.

parameters	cocurrent	countercurrent
p, bar	4 to 10	4 to 10
N_2-mass conc., -	0 to 0.1	0 to 0.15
steam velocity, m/s	1.3 to 5.8	0.6 to 1.6
water velocity, m/s	0.25 to 0.30	0.25 to 0.30

Nomenclature

Latin

a $= \lambda / \left(\rho c_p\right)$, temperature conductivity, m²/s

a_{12} $= 1/\Delta z$ gas/liquid interfacial area density in Cartesian coordinates, 1/m

C_{1M} vapor mass concentration inside the gas mixture, dimensionless

C_{1n} mass concentration of the noncondensing gas n inside the gas mixture, dimensionless

$C_{1M}^{\sigma 2}$ vapor mass concentration inside the gas mixture at the liquid interface, dimensionless

$D_{1\sum n\rightarrow M}$ diffusion gas coefficient for the vapor inside the mixture consisting of condensable and noncondensable gases, m²/s

D_{hy} pipe hydraulic diameter, m

g acceleration due to gravity, m/s²

R_1 mixture gas constant, J/(kg K)

$Re_{2\delta}$ $=\dfrac{\rho_2 w_2 4\delta_2}{\eta_2}$, *Reynolds* number based on the hydraulic diameter of the film, dimensionless

H height of the plate, m

h specific enthalpy, J/kg

$h_{convective}$ convective heat-transfer coefficient, W/(m² K)

$h_{stagnant}$ heat-transfer coefficient for stagnant steam, W/(m² K)

$h_{stagnat\_steam/turbulent\_film}$ heat-transfer coefficient for stagnant steam and turbulent condensate film, W/(m² K)

h_{Nu} heat-transfer coefficient for stagnant steam in accordance with the *Nusselt* water skin theory, W/(m² K)

h_{lam} heat-transfer coefficient for forced convection inclined pipe – laminar film, W/(m² K)

h_{turb} heat-transfer coefficient for forced convection inclined pipe – turbulent film, W/(m² K)

h_{20} heat-transfer coefficient for forced convection computed for flow having the mixture mass flow but consisting of liquid only, W/(m² K)

$\dot{q}''_w$ wall heat flux from the interface into the gas, W/m³

p pressure, Pa

p_{cr} critical pressure, Pa

p_{1M} vapor partial pressure inside the gas mixture, Pa

$p_{1M}^{\sigma 2}$ vapor partial pressure inside the gas mixture at the liquid interface, Pa

$p_{1\sum n}$ partial pressure of the noncondensing gases inside the gas mixture, Pa

$p'\left(T_1^{\sigma 2}\right)$ vapor partial pressure inside the gas mixture at the liquid surface, Pa

Re_1 $= D_{hy}\rho_1 \left|w_1\right| / \eta_1$, gas *Reynolds* number, dimensionless

Re_{20} $= D_{hy} G / \eta_2$ total flow *Reynolds* number based on the liquid viscosity, dimensionless

Pr $= \eta c_p / \lambda$, *Prandtl* number, dimensionless

T temperature, K

$T_2^{1\sigma}$ liquid-side interface temperature, K

$T_1^{\sigma 2}$	gas-side interface temperature, K
T_{cond}	condensation initiation temperature, K
$\mathbf{V}_1$	gas velocity, m/s
$\mathbf{V}_2$	liquid velocity, m/s
$\dot{V}'$	volumetric flow rate per unit of film width, m^2/s
w	axial velocity, m/s
X_l	mass flow concentration of the field l inside the mixture, dimensionless

Greek

β_{12}	condensation coefficient, m/s
δ	film thickness, m
φ	angle between the positive flow direction and the upwards directed vertical, rad
Δh	$= h'' - h'$, latent heat of vaporization, J/kg
η	dynamic viscosity of liquid, kg/(m s)
λ	thermal conductivity, W/(m K)
μ_{21}	steam mass generation per unit mixture volume and unit time due to evaporation, kg/(s m³)
μ_{12}	liquid mass generation per unit mixture volume and unit time due to condensation, kg/(sm³)
ρ	density, kg/m³
$(\rho w)_{21}$	steam mass generation per unit surface and unit time due to evaporation, kg/(m² s)
$(\rho w)_{12}$	liquid mass generation per unit surface and unit time due to condensation, kg/(m² s)

Subscripts

1	gas
2	liquid film
2F	liquid film
M1	vapor inside the gas mixture
n1	inert components inside the gas

Superscripts

´	time fluctuation
'	saturated steam
"	saturated liquid

References

Alekseenko, S.V., Nakoryakov, V.E., Pokusaev, B.G.: Wave flow of liquid films. Begell House, New York (1996) (English ed. Fukano T)

Alekseenko, S.V., Nakoryakov, V.E., Pokusaev, B.G., Khristoforov, V.V.: Shear stress at film falling down a vertical wall (in Russian). Inzh. -Fiz. Zhurn. 24(5), 824–830 (1973)

Berman, L.D., Fuks, S.N.: Teploenergetika 8, 66–74 (1958): Teploenergetika 7, 74–83 (1959)

Berman, L.D., Tumanov, V.A.: Heat transfer during condensation of flowing steam in a tube bundle. Teploenergetika 9(10), 77–83 (1962)

Bird, R.B., Steward, W.E., Lightfoot, E.N.: Transport phenomena. John Wiley & Sons, New York (1960)

Bobe, L.S., Solouhin, V.A.: Heat and mass transfer during steam condensation from steam-gas mixture by turbulent pipe flow. Teploenergetika 9, 27–30 (1972)

Boyadjiev, C., Beschkov, V.: Mass Transfer in Liquid Film Flows. Publishing House of the Bulgarian Academy of Sciences, Sofia (1984)

Boyko, L.D., Krujilin, G.N.: Izvestia akademii nauk SSSR Energetika I transport, vol. (2), pp. 113–128 (1966): See also Boyko, L.D.: Teploobmen w elementach energeticheskih ustanovak, pp. 197–212. Nauka, Moskwa (1966)

Brauer, H.: Stofaustausch, Verlag Sauerländer, Aarau und Franfurt am Main (1971)

Butterworth, D.: Condensers: Heat transfer and fluid flow. In: Kakac, S., Bergeles, A.E., Mayinger, F. (eds.) Heat Exchanger, pp. 89–314. Hemisphere, Washington (1981)

Chato, J.C.: Laminar condensation inside horizontal and inclined tubes. ASHREE J. 4, 52–60 (1962)

Choi, K.Y., Chung, H.J., No, H.C.: Nucl. Eng. Des. 211, 139–151 (2002)

Collier, J.G.: Convection boiling and condensation. McGraw-Hill Book Company Inc., London (1972)

Dhir, V.K., Lienhard, J.H.: Laminar film condensation on plane and axisymmetric bodies in non uniform gravity. J. Heat Transfer 93, 97 (1971)

Dukler, A.E., Berglin, O.P.: Characteristics of flow in falling liquid films. Chem. Eng. Prog. 48, 558 (1952)

Fijii, T., Honda, H., Oda, K.: Condensation of steam on horizontal tubes – influence of the oncoming velocity, Condensation heat transfer. In: 18th Nat. Heat Transfer Conf., San Diego, August 1979, pp. 35–43 (1979)

Fulford, G.D.: The flow of liquids in thin films. Adv. Chem. Eng. 5, 151–236 (1964)

Ganchev, B.G., Kozlov, V.M.: Investigation into gravitational flow of liquid film along the walls of vertical channel with large length (in Russian). Zhurn. Prikl. Mekh. Tekh. Fiz. 1, 128–135 (1973)

Gimbutis, G.: Heat Transfer in Gravitational Flow of Liquid Film, 232 p. Vilnyus, Mokslas (1988) (in Russian)

Goldbrunner, M.: Kondensation bei stratifizierter Strömung in Gegenwart nichtkondensierbarer Gase, Final Report, Reaktor Safety Research Project No: 1501207, TU München (June 2003)

Gregorig, R., Kern, J., Turek, K.: Improved correlation of film condensation based on more rigorous application of similarity parameters. Wärme Stoffübertragung 1, 1 (1974)

Gudymchuk, V.A., Konstantinov, V.A.: Jurnal Teplofisiki (JTF) 6(9) (1936)

Incopera, F.P., DeWitt, D.P.: Fundamentals of heat and mass transfer, 5th edn. John Wiley & Sons, New York (2002)

Isachenko, V.P., Osipova, V.A., Sukomel, A.S.: Teploperedacha, 4th edn. Energoisdat, Moskva (1981)

Isachenko, V.P.: Teploobmen pri kondensacii. Energia, Moskva (1977)
Ishigai, S., Nakanisi, S., Koizumi, T., Oyabi, Z.: Hydrodynamics and heat transfer of vertical falling liquid films. Bull. JSME 15(83), 594–602 (1972)
Jaster, H., Kosky, P.G.: Condensation heat transfer in a mixed flow regime. Int. J. Heat Mass Transfer 19, 95–99 (1976)
Kapitza, P.L.: Wave flow of thin layer of viscous fluid (in Russian). Zhurn. Eksper. Teor. Fiz. 18(1), 3–28 (1948)
Kapitza, P.L.: Zhur. Exper. Teoret. Fiz. 18(1) (1948)
Kitateladze, S.S.: Heat transfer in condensation and boiling, AEC-tr-3778, ch. 5-7 (1962)
Kutateladze, S.S.: Fundamentals of heat transfer. Academic, New York (1963)
Kutateladze, S.S.: Fundamentals of heat transfer theory (in Russian). Atomizdat. Moskva (1979)
Labuntzov, D.A.: Heat transfer in film condensation of pure steam. Teploenergetica 4(7), 72–80 (1957)
Miropolskij, S.L.: Teplootdaza pri kondensazii para vysokogo davlenija vnutri trub. Teploenergetika (3), 79–83 (1960)
Nusselt, W.: Die Oberflächenkondensation des Wasserdampfes. Zeitschrift VDI 60, 541–546 (1916)
Perry, R.H., Green, D.: Perry's chemical engineer's handbook, 6th edn., pp. 3–285. McGraw-Hill, New York (1985)
Popiel, C.O., Boguslawski, L.: Heat transfer by laminar film condensation on spherical surface. Int. J. Heat Mass Transfer 18, 1485 (1975)
Rohsenow, W.M., Hartnett, J.P., Ganic, E.N.: Handbook of heat transfer fundamentals, 2nd edn. McGraw-Hill Book Company, New York (1985)
Shah, M.M.: A general correlation for heat transfer during film condensation inside pipes. Int. J. Heat Mass Transfer 22, 547–556 (1979)
Shekriladze, I.G., Gomelauri, V.I.: Laminar condensation on flowing vapor. Int. J. Heat Mass Transfer 9, 581–591 (1966)
Slattery, J.C., Bird, R.B.: AIChE J. 4, 137–142 (1958)
Tananaiko, Y.M., Vorontsov, E.G.: Methods of Calculation and Investigation into Film Processes (in Russian). Tekhnika, Kiev, 312 (1975)
Thome, J.R., El Hajal, J., Cavallini, A.: Condensation in horizontal tubes, part 2: new heat transfer model based on flow regimes. Int. J. Heat Mass Transfer 46(18), 3365–3387 (2003)
Vierow, K.M., Shrock, K.E.: Condensation in a natural circulation loop with non condensable gases, Part 1 – Heat transfer. In: International Conference on Multi-Phase Flows, Tsukuba, Japan (1991)

15. Discrete ordinate method for radiation transport in multiphase computer codes

15.1 Introduction

Several applications in technology require radiation transport analysis (RTA). Examples are processes with combustion and interaction of high-temperature molten material with water in the safety analysis in the nuclear industry and in metallurgy. The RTA is always a part of the general flow simulation and of the analysis of the flow interaction with the structure. Radiation transport in high-temperature media may be the governing process of energy transfer. This chapter is not intended to replace the famous book of *Siegel* and *Howell* (1996) on this subject. The interested reader intending to reach a high level of skill in this field has to go through this book. The chapter rather gives some of the main ideas required to understand how the radiation-transport theory can be implemented in multiphase flow analysis. The chapter is based on the experience gained by creating a radiation transport model for the computer code IVA *Kolev* (1999). This work started in 1998. In that year *Lanzenberger* (1997) created, as far as I know, the first computer code for modeling of 3D radiation transport in three-phase flow as a part of the IVA4 code development. The code uses the first-order *step* scheme setting the positive wall coefficient equal to zero. Strictly speaking the code was developed for simulation of Cartesian coordinates without any obstacles inside the computational region. The further development of this technique in IVA6 went through the following direction: (a) rewriting the code from Fortran 77 to Fortran 90, improving macroscopic data and coupling with the IVA6 computer code; (b) changing the algorithm so that it allows one to handle internal obstacles.

15.1.1 Dimensions of the problem

For technical application on the ground we consider each beam as being linear. It can travel in an arbitrary direction through a single point. Therefore, the first three dimensions are coming from the spatial description. The volume of interest is also three-dimensional. It introduces another three dimensions in the problem. The electromagnetic radiation energy flow transported per unit time across unit normal surface is dependent on the wavelength. The continuous spectrum is usually split in n groups with prescribed boundaries. This gives the other n-dimensions of the problem. In the discrete ordinate radiation-transport method the governing energy-conservation equation is replaced by a set of equations for a finite number of ordinate directions, for a finite number of control volumes, and for a finite number of spectral regions. It is the large number of dimensions that makes the RTA complicated.

15.1.2 Micro- versus macrointeractions

Any interactions of electromagnetic radiation with matter are described in physics by theoretical and experimental observation on a microscale. The engineer integrates the interactions over a control volume and prepares the information necessary for microscopic radiation transport on a macroscale. If the characteristic interaction length is much smaller than the scale of the control volume, RTA on a macroscale is not necessary because all the interaction happens inside the cell and no energy leaves the volume. An example is a relatively dense suspension of hot particles, which are shielding effectively the radiation inside large cells. If the characteristic interaction length is larger than the scale of the control volume then RTA is necessary because energy is leaving and entering the control volume. Examples are small concentrations of very hot spheres in water. The radiation energy may be transported up to 10 m in water up to full absorption. If the volume of interest is few cubic meters and the control volume of the discretization is of the order of liters the RTA is necessary.

15.1.3 The radiation-transport equation (RTE)

The basic equation for the description of the radiation transport is the steady-state energy-conservation equation called radiation transport equation (RTE)

$$\frac{di'_\lambda(\mathbf{r},\mathbf{s})}{ds} = -a_\lambda i'_\lambda(\mathbf{r},\mathbf{s}) - \sigma_{s\lambda} i'_\lambda(\mathbf{r},\mathbf{s}) + a_\lambda i'_{\lambda b}(\mathbf{r},\mathbf{s})$$

$$+\frac{\sigma_{s\lambda}}{4\pi}\int_{4\pi} i_\lambda(\mathbf{r},\omega_i)\Phi(\lambda,\omega,\omega_i)d\omega_i \, . \tag{15.1}$$

Here, $i'_\lambda(\mathbf{r},\mathbf{s})$ is the spectral radiation intensity at the spatial location $\mathbf{r}$ along the direction $\mathbf{s}$ that is within an infinitesimal solid angle $d\omega$. Spectral – because it is associated with each particular wavelength λ. The dimensions of i'_λ are those of power flux through a square meter. It is in fact a heat flux. $a_\lambda, \sigma_{s\lambda}$ are the spectral absorption and scattering coefficients, $\mathbf{\Phi}$ is the spectral scattering phase function and ω_i is the scattering angle. The scattering phase function describes that amount of incident radiation is scattered in all other directions, Fig. 15.1.

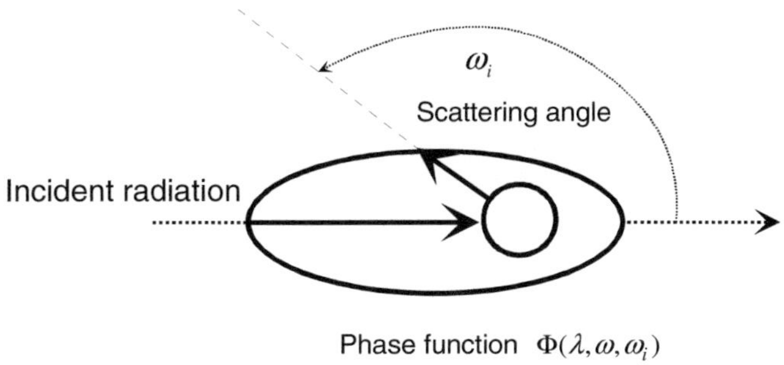

Fig. 15.1 Scattering of incident radiation

The intensity of a ray passing through an infinitesimal volume is attenuated by absorption. This is reflected by the first term on the right-hand side of Eq. (15.1). The intensity of a ray passing through an infinitesimal volume is attenuated also by scattering. It is taken into account in the second term on the RHS. The intensity is increased by the emitted intensity of the medium reflected by the third term on the RHS, and by the scattering from any other direction to the considered direction of the vector **s**. For numerical integration of Eq. (15.1) it is convenient to introduce the so-called spectral extinction coefficient

$$\beta_\lambda = a_\lambda + \sigma_{s\lambda}, \tag{15.2}$$

and the source term

$$S' = a_\lambda i'_{\lambda b}(\mathbf{r}, \mathbf{s}) + \frac{\sigma_{s\lambda}}{4\pi} \int_{4\pi} i_\lambda(\mathbf{r}, \omega_i) \Phi(\lambda, \omega, \omega_i) d\omega_i, \tag{15.3}$$

which simplify the notation

$$\frac{di'_\lambda(\mathbf{r}, \mathbf{s})}{ds} = -\beta_\lambda i'_\lambda(\mathbf{r}, \mathbf{s}) + S'. \tag{15.4}$$

In the discrete ordinate radiation-transport method Eq. (15.4) will be replaced by a set of equations for a finite number of ordinate directions, for a finite number of control volumes, and for a finite number of spectral regions.

15.2 Discrete ordinate method

The spatial location in the space is defined by the position vector **r**. In a polar coordinate system as presented in Fig. 15.2 with radius $r = 1$, azimuthal angle φ, and polar angle θ the components of the unit vectors of **s** in a Cartesian coordinate system with coordinates x, y and z are defined as the direction cosines

$$\mathbf{s} = (\sin\theta\cos\varphi, \quad \sin\theta\sin\varphi, \quad \cos\theta) \equiv (s_x, \quad s_y, \quad s_z). \tag{15.5}$$

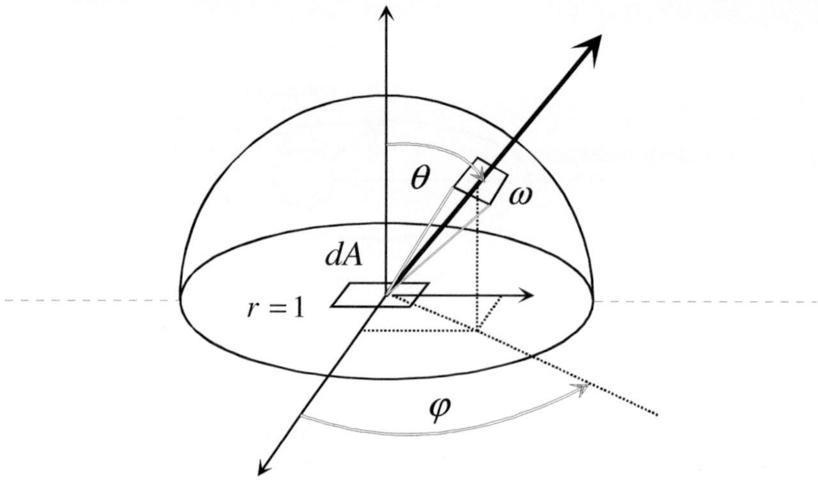

Fig. 15.2 Unit hemisphere, solid angle ω, angle φ and θ

The infinitesimal solid angle $d\omega$ is equal to the infinitesimal area on the surface of the unit hemisphere

$$d\omega = \sin\theta d\theta d\varphi . \qquad (15.6)$$

Integrating over all possible solid angle gives

$$\int_{\varphi=0}^{2\pi}\int_{\theta=-\pi/2}^{\pi/2} \sin\theta d\theta d\varphi = 4\pi . \qquad (15.7)$$

The total surface area of the unit hemisphere is 2π, and is denoted as the *total solid angle*. Let us divide the space into a discrete number of solid angles $\Delta\omega_{a,b}$, where

$$a = 1, a_{\max}, \qquad (15.8)$$

$$b = 1, b_{\max}, \qquad (15.9)$$

with $a_{\max}$ and $b_{\max}$ being even numbers. For equidistant discretization we have

$$\Delta\varphi = 2\pi / a_{\max}, \ \varphi_a = \left(a - \frac{1}{2}\right)\Delta\varphi, \ a = 1, a_{\max} \qquad (15.10)$$

$$\Delta\theta = 2\pi / b_{\max}, \ \theta_b = \left(b - \frac{1}{2}\right)\Delta\theta, \ b = 1, b_{\max}, \qquad (15.11)$$

and

$$\Delta\omega_{a,b} = \int_{\varphi_a-\Delta\varphi/2}^{\varphi_a+\Delta\varphi_a/2} \int_{\theta_b-\Delta\theta_b/2}^{\theta_b+\Delta\theta_b/2} \sin\theta \, d\theta \, d\varphi = 2\Delta\varphi_a \sin\theta_b \sin\frac{\Delta\theta_b}{2}$$

$$= \Delta\varphi_a \left[\cos\left(\theta_b - \frac{\Delta\theta_b}{2}\right) - \cos\left(\theta_b + \frac{\Delta\theta_b}{2}\right) \right]. \tag{15.12}$$

Here, φ_a and θ_b define the specific ordinate direction denoted with subscripts *a,b*. Other notation is

$$\mathbf{s}_{a,b} = \left(s_{x,a,b},\; s_{\mathrm{y,a,b}},\; s_{\mathrm{z,a,b}}\right). \tag{15.13}$$

There is an alternative notation used frequently in the literature

$$\mathbf{s}_{a,b} = \left(\mu_{a,b},\; \eta_{\mathrm{a,b}},\; \xi_{\mathrm{a,b}}\right). \tag{15.15}$$

15.2.1 Discretization of the computational domain for the description of the flow

The discretization of the computational domain for the description of the flow is performed by using a curvilinear coordinate system. In many practical cases either a Cartesian or a cylindrical coordinate system is used. We give here examples with the Cartesian systems for simplicity, but this is not a limitation of the approach. The discretization is structured and nonequidistant in each direction. Each control volume has 6 faces with surface A_m. The surfaces are numbered 1 for high *i*, 2 for low *i*, 3 for high *j*, 4 for low *j*, 5 for high *k* and 6 for low *k*. The orientation of each surface is defined by the outwards directed unit vector

$$\mathbf{e} = \left(e_x,\; e_{\mathrm{y}},\; e_{\mathrm{z}}\right). \tag{15.16}$$

For a Cartesian coordinate system the 6 surfaces have the surface vectors

$$\mathbf{e}_1 = (1,\; 0,\; 0), \qquad \mathbf{e}_2 = (-1,\; 0,\; 0),$$
$$\mathbf{e}_3 = (0,\; 1,\; 0), \qquad \mathbf{e}_4 = (0,\; -1,\; 0),$$
$$\mathbf{e}_5 = (0,\; 0,\; 1), \qquad \mathbf{e}_6 = (0,\; 0,\; -1). \tag{15.17-22}$$

For discretization of the cylindrical computational domain the 6 surfaces have the surface vectors given below

$$\mathbf{e}_m = (\cos\theta_m,\; \sin\theta_m,\; 0),\; m = 1,4,\; \mathbf{e}_5 = (0,\; 0,\; 1),\; \mathbf{e}_6 = (0,\; 0,\; -1). \tag{15.23-28}$$

The angles θ_m are defined as follows $\theta_1 = \theta$, $\theta_1 = \pi + \theta$, $\theta_1 = \frac{3}{4}\pi + \theta + \frac{1}{2}\Delta\theta$, $\theta_1 = \frac{3}{4}\pi + \theta - \frac{1}{2}\Delta\theta$. Here, θ is the azimuthal angle of the center of the control volume in the fluid domain. $\Delta\theta$ is the size of the control volume in the azimuthal direction. Note the difference between these two angles and the angles used in the polar coordinate system for definition of the discrete ordinates.

15.2.2 Finite-volume representation of the radiation-transport equation

Integrating the RTE over the control volume $dVol$ and over the angle $\Delta\omega_{a,b}$ we obtain

$$\iint\limits_{\Delta\omega_{\mathrm{a,b}}} \int\limits_{\Delta Vol} \frac{di'_\lambda(\mathbf{r},\mathbf{s})}{ds} dVol\, d\omega = -\iint\limits_{\Delta\omega_{\mathrm{a,b}}} \int\limits_{\Delta Vol} \beta_\lambda i'_\lambda(\mathbf{r},\mathbf{s}) dVol\, d\omega + \iint\limits_{\Delta\omega_{\mathrm{a,b}}} \int\limits_{\Delta Vol} S' dVol\, d\omega \tag{15.29}$$

or

$$\sum_{m=1}^{6} i'_{\lambda,a,b,m} A_m \iint\limits_{\Delta\omega_{\mathrm{a,b}}} (\mathbf{s}\cdot\mathbf{e})_m\, d\omega + \beta_\lambda i'_{\lambda,a,b} \Delta Vol\, \Delta\omega_{\mathrm{a,b}} = S'_{a,b} \Delta Vol\, \Delta\omega_{\mathrm{a,b}}. \tag{15.30}$$

Scalar product of the discrete directions vector and given surface vector: Now, we are interested in the estimation of the integral

$$D_{a,b} = \iint\limits_{\Delta\omega_{\mathrm{a,b}}} \mathbf{s}\cdot\mathbf{e}\, d\omega = \int\limits_{\varphi_a - \Delta\varphi/2}^{\varphi_a + \Delta\varphi_a/2} \int\limits_{\theta_b - \Delta\theta_b/2}^{\theta_b + \Delta\theta_b/2} (\mathbf{s}\cdot\mathbf{e}) \sin\theta\, d\theta\, d\varphi =$$

$$\int\limits_{\varphi_a - \Delta\varphi/2}^{\varphi_a + \Delta\varphi_a/2} \int\limits_{\theta_b - \Delta\theta_b/2}^{\theta_b + \Delta\theta_b/2} \left(e_x \sin\theta\cos\varphi + e_y \sin\theta\sin\varphi + e_z \cos\theta \right) \sin\theta\, d\theta\, d\varphi$$

$$= \left(e_x \int\limits_{\varphi_a - \Delta\varphi/2}^{\varphi_a + \Delta\varphi_a/2} \cos\varphi\, d\varphi + e_y \int\limits_{\varphi_a - \Delta\varphi/2}^{\varphi_a + \Delta\varphi_a/2} \sin\varphi\, d\varphi \right) \int\limits_{\theta_b - \Delta\theta_b/2}^{\theta_b + \Delta\theta_b/2} \sin^2\theta\, d\theta$$

$$+ e_z \Delta\varphi_a \int\limits_{\theta_b - \Delta\theta_b/2}^{\theta_b + \Delta\theta_b/2} \cos\theta \sin\theta\, d\theta\ . \tag{15.31}$$

Integrating we obtain

$$D_{a,b} = \left(e_x \cos\varphi_a + e_y \sin\varphi_a\right) 2\sin\frac{\Delta\varphi_a}{2}\left[\left(1-2\cos^2\theta_b\right)\sin\frac{\Delta\theta_b}{2}\cos\frac{\Delta\theta_b}{2} + \frac{\Delta\theta_b}{2}\right]$$

$$+ e_z \Delta\varphi_a 2\cos\theta_b \sin\theta_b \cos\frac{\Delta\theta_b}{2}\sin\frac{\Delta\theta_b}{2}. \qquad (15.32)$$

Working form of the finite differential form of the RTE: Using the above results the finite differential form of the RTE becomes

$$\sum_{m=1}^{6} i'_{\lambda,a,b,m} A_m D_{a,b,m} + \beta_\lambda i'_{\lambda,a,b} \Delta Vol\ \Delta\omega_{\text{a,b}} = S'_{a,b} \Delta Vol\ \Delta\omega_{\text{a,b}}. \qquad (15.33)$$

It is convenient to divide both sides by $\Delta Vol\ \Delta\omega_{\text{a,b}}$. The result is

$$\sum_{m=1}^{6} a_{a,b,m} i'_{\lambda,a,b,m} + \beta_\lambda i'_{\lambda,a,b} = S'_{a,b}, \qquad (15.34)$$

where

$$a_{a,b,m} = \frac{A_m}{\Delta Vol}\frac{D_{a,b,m}}{\Delta\omega_{\text{a,b}}}, \qquad (15.35)$$

is geometry dependent only and is computed once at the beginning of the computation. Note that the ratio $D_{a,b,m}/\Delta\omega_{\text{a,b}}$ is dimensionless and the coefficients $a_{a,b,m}$ have dimension 1/m, which is the same as that of the extinction coefficient. For Cartesian coordinates using Eqs. (15.17)–(15.22) the above equation simplifies as follows

$$\left(i'_{\lambda,a,b,1} - i'_{\lambda,a,b,2}\right)\frac{A_1}{\Delta Vol}\frac{s_{x,a,b}}{\Delta\omega_{\text{a,b}}} + \left(i'_{\lambda,a,b,3} - i'_{\lambda,a,b,4}\right)\frac{A_3}{\Delta Vol}\frac{s_{y,a,b}}{\Delta\omega_{\text{a,b}}}$$

$$+ \left(i'_{\lambda,a,b,5} - i'_{\lambda,a,b,6}\right)\frac{A_5}{\Delta Vol}\frac{s_{z,a,b}}{\Delta\omega_{\text{a,b}}} + \beta_\lambda\ i'_{\lambda,a,b} = S'_{a,b}. \qquad (15.36)$$

The general form is preferred here.

Note that the spectral intensities $i'_{\lambda,a,b,m}$ are defined at the cell boundaries and the spectral intensity $i'_{\lambda,a,b}$ is defined in the cell center. There are numerous methods giving different approximations for the relationship between $i'_{\lambda,a,b,m}$ and $i'_{\lambda,a,b}$. Three of them are discussed in the next section.

Weighting factors: Step (donor cell), diamond and positive weighting: The general form of the weighting factors for the first octant $s_x > 0$, $s_y > 0$, $s_z > 0$

$$i'_{\lambda,a,b} = f_x i'_{\lambda,a,b,1} + (1-f_x) i'_{\lambda,a,b,2} = f_y i'_{\lambda,a,b,3} + (1-f_y) i'_{\lambda,a,b,4}$$
$$= f_z i'_{\lambda,a,b,5} + (1-f_z) i'_{\lambda,a,b,6}, \tag{15.37}$$

or

$$i'_{\lambda,a,b,1} = \frac{1}{f_x} i'_{\lambda,a,b} + \left(1 - \frac{1}{f_x}\right) i'_{\lambda,a,b,2}, \tag{15.38}$$

$$i'_{\lambda,a,b,3} = \frac{1}{f_y} i'_{\lambda,a,b} + \left(1 - \frac{1}{f_y}\right) i'_{\lambda,a,b,4}, \tag{15.39}$$

$$i'_{\lambda,a,b,5} = \frac{1}{f_z} i'_{\lambda,a,b} + \left(1 - \frac{1}{f_z}\right) i'_{\lambda,a,b,6}. \tag{15.40}$$

Substituting in the discretized RTE and solving with respect to the intensity in the center of the cell $i'_{\lambda,a,b}$ we obtain finally

$$i'_{\lambda,a,b} = \left\{ \begin{array}{l} S'_{a,b} - \left[a_{a,b,2} + a_{a,b,1}\left(1 - \frac{1}{f_x}\right)\right] i'_{\lambda,a,b,2} \\ - \left[a_{a,b,4} + a_{a,b,3}\left(1 - \frac{1}{f_y}\right)\right] i'_{\lambda,a,b,4} - \left[a_{a,b,6} + a_{a,b,5}\left(1 - \frac{1}{f_z}\right)\right] i'_{\lambda,a,b,6} \end{array} \right\}$$
$$\times \left(a_{a,b,1}\frac{1}{f_x} + a_{a,b,3}\frac{1}{f_y} + a_{a,b,5}\frac{1}{f_z} + \beta_\lambda \right)^{-1}. \tag{15.41}$$

Step difference scheme: Setting

$$f_x = f_y = f_z = 1 \tag{15.42}$$

results in a first-order *step difference scheme*

$$i'_{\lambda,a,b,1} = i'_{\lambda,a,b}, \quad i'_{\lambda,a,b,3} = i'_{\lambda,a,b}, \quad i'_{\lambda,a,b,5} = i'_{\lambda,a,b}, \tag{15.43}$$

which resembles the *donor cell* scheme in the CFD. Consequently, we have

$$i'_{\lambda,a,b} = \frac{S'_{a,b} - a_{a,b,2} i'_{\lambda,a,b,2} - a_{a,b,4} i'_{\lambda,a,b,4} - a_{a,b,6} i'_{\lambda,a,b,6}}{\beta_\lambda + a_{a,b,1} + a_{a,b,3} + a_{a,b,5}}$$
$$\equiv \frac{S'_{a,b} - a_{a,b,2} i'_{\lambda,a,b,i-1} - a_{a,b,4} i'_{\lambda,a,b,j-1} - a_{a,b,6} i'_{\lambda,a,b,k-1}}{\beta_\lambda + a_{a,b,1} + a_{a,b,3} + a_{a,b,5}}. \tag{15.44}$$

A similar strategy is used for the other octants.

The optimum marching strategy: If the point *Gauss–Seidel* method is used to solve the above equation there is an optimum strategy of visiting the cells, and the ordinate directions – called the optimum marching strategy. The optimum marching strategy is direction dependent. First the octants are defined in which the discrete ordinate is, as given in Table 15.1, and then the incident and the outlet walls of the cell. Then the marching strategy is defined in Tables 15.2 and 15.3.

Table 15.1 Definition of the octants. Incident and outlet walls in the cell

Angles	Octant	Incident wall cells i1,i2,i3	Outlet wall cells o1,o2,o3
$0 \le \theta_b < \pi/2$			
$0 \le \varphi_a < \pi/2$	1	2,4,6	1,3,5
$\pi/2 \le \varphi_a < \pi$	2	1,4,6	2,3,5
$\pi \le \varphi_a < 3\pi/2$	3	1,3,6	2,4,5
$3\pi/2 \le \varphi_a < 2\pi$	4	2,3,6	1,4,5
$\pi/2 \le \theta_b < \pi$			
$0 \le \varphi_a < \pi/2$	5	2,4,5	1,3,6
$\pi/2 \le \varphi_a < \pi$	6	1,4,5	2,3,6
$\pi \le \varphi_a < 3\pi/2$	7	1,3,5	2,4,6
$3\pi/2 \le \varphi_a < 2\pi$	8	2,3,5	1,4,6

Table 15.2 Marching strategy depending on the discrete ordinate direction

Octant	Marching strategy *x* direction	*y* direction	*z* direction
1	$i = 2, i_{\max} - 1$	$j = 2, j_{\max} - 1$	$k = 2, k_{\max} - 1$
2	$i = i_{\max} - 1, 2, -1$	$j = 2, j_{\max} - 1$	$k = 2, k_{\max} - 1$
3	$i = i_{\max} - 1, 2, -1$	$j = j_{\max} - 1, 2, -1$	$k = 2, k_{\max} - 1$
4	$i = 2, i_{\max} - 1$	$j = j_{\max} - 1, 2, -1$	$k = 2, k_{\max} - 1$
5	$i = 2, i_{\max} - 1$	$j = 2, j_{\max} - 1$	$k = k_{\max} - 1, 2, -1$
6	$i = i_{\max} - 1, 2, -1$	$j = 2, j_{\max} - 1$	$k = k_{\max} - 1, 2, -1$
7	$i = i_{\max} - 1, 2, -1$	$j = j_{\max} - 1, 2, -1$	$k = k_{\max} - 1, 2, -1$
8	$i = 2, i_{\max} - 1$	$j = j_{\max} - 1, 2, -1$	$k = k_{\max} - 1, 2, -1$

Table 15.3 Marching strategy for the discrete ordinate

Octant	Marching strategy *a* direction	*b* direction
1	$a = 1, a_{max}/4$	$b = 1, b_{max}/2$
2	$a = a_{max}/4+1, a_{max}/2$	$b = 1, b_{max}/2$
3	$a = a_{max}/2+1, 3a_{max}/4$	$b = 1, b_{max}/2$
4	$a = 3a_{max}/4+1, a_{max}$	$b = 1, b_{max}/2$
5	$a = 1, a_{max}/4$	$b = b_{max}/2+1, b_{max}$
6	$a = a_{max}/4+1, a_{max}/2$	$b = b_{max}/2+1, b_{max}$
7	$a = a_{max}/2+1, 3a_{max}/4$	$b = b_{max}/2+1, b_{max}$
8	$a = 3a_{max}/4+1, a_{max}$	$b = b_{max}/2+1, b_{max}$

Therefore, for each octant we have

$$i'_{\lambda,a,b} = \frac{S'_{a,b} - a_{a,b,i1} i'_{\lambda,a,b,i1} - a_{a,b,i2} i'_{\lambda,a,b,i2} - a_{a,b,i3} i'_{\lambda,a,b,i3}}{\beta_\lambda + a_{a,b,o1} + a_{a,b,o2} + a_{a,b,o3}}, \tag{15.45}$$

where $i'_{\lambda,a,b,o1} = i'_{\lambda,a,b}$, $i'_{\lambda,a,b,o2} = i'_{\lambda,a,b}$, $i'_{\lambda,a,b,o3} = i'_{\lambda,a,b}$.

This method has been in use for a long time. It is very simple in design and stable in performance regardless of the ordinate set chosen or the local radiative properties, but it is reported to be inaccurate *Mathews* (1999).

Lathrop's diamond difference scheme: Setting

$$f_x = f_y = f_z = \frac{1}{2} \tag{15.47}$$

results in the second-order *Lathrop* diamond difference scheme. The diamond schemes are known to be unstable without additional improvement of the weighting coefficients *Mathews* (1999).

Fiveland scheme: Fiveland (1988) proposed to use

$$f_x = f_y = f_z = f, \tag{15.48}$$

satisfying the following conditions

$$\Delta x < \frac{|s_{x,a,b}|}{\beta(1-f)}\phi, \quad \Delta y < \frac{|s_{y,a,b}|}{\beta(1-f)}\phi, \quad \Delta z < \frac{|s_{z,a,b}|}{\beta(1-f)}\phi, \tag{15.49-51}$$

where

$$\phi = \frac{f^3 + (1-f)^2 (2-5f)}{f}. \quad (15.53)$$

It is thinkable to use a set of nonequal weighting factors like

$$f_x = \max\left[\left(1 - \frac{|s_{x,a,b}|}{\beta \Delta x}\right), \frac{1}{2}\right], \quad (15.54)$$

$$f_y = \max\left[\left(1 - \frac{|s_{y,a,b}|}{\beta \Delta y}\right), \frac{1}{2}\right], \quad (15.55)$$

$$f_z = \max\left[\left(1 - \frac{|s_{x,a,b}|}{\beta \Delta z}\right), \frac{1}{2}\right], \quad (15.56)$$

to assure the positive scheme.

15.2.3 Boundary conditions

The radiation heat flux across the surface m is that computed by integrating over the discrete ordinates

$$\dot{q}''_{\lambda,m} = \Delta Vol \sum_{b=1}^{b_{\max}} \sum_{a=1}^{a_{\max}} a_{a,b,m} \Delta\omega_{\mathrm{a,b}} i'_{\lambda,a,b,m} . \quad (15.57)$$

The incident radiation heat flux from the inside of the computational cell is

$$\dot{q}''_{\lambda,m-incident} = \Delta Vol \sum_{b=1}^{b_{\max}} \sum_{a=1}^{a_{\max}} a_{a_i,b_i,m} \Delta\omega_{\mathrm{a_i,b_i}} i'_{\lambda,a_i,b_i,m} . \quad (15.58)$$

For Cartesian coordinate system the incident ordinate directions a_i, b_i are summarized in Table 15.4. At *diffusely emitting and reflecting* walls with temperature $T_{w,m}$ and emissivity coefficient ε_w the emitting intensity is

$$i'_{\lambda,a_o,b_o,m} = \varepsilon_w \frac{\sigma T_{w,m}}{\pi} + (1-\varepsilon_w)\frac{\dot{q}''_{\lambda,m-incident}}{\pi} . \quad (15.59)$$

The outlet ordinate directions are specified for each wall in Table 15.5 for a Cartesian coordinate system.

Setting

$$i'_{\lambda,a_o,b_o,m} = 0 \quad (15.60)$$

means a *transparent* wall. For *mirroring* walls we have

$$i'_{\lambda,a_{\max}-a+1,b_{\max}-b+1,m} = \varepsilon_w \frac{\sigma T_{w,m}}{\pi} + i'_{\lambda,a,b}, \quad (15.61)$$

where $a,b \in m - incident$.

Table 15.4 Marching strategy for computation of the incident radiation at the cell walls required for definition of boundary conditions

Wall Nr.	Marching strategy, incident *a* direction	*b* direction
1	$a = 1, a_{\max}/4$	$b = 1, b_{\max}$
	$a = 3a_{\max}/4+1, a_{\max}$	
2	$a = a_{\max}/4+1, 3a_{\max}/4$	$b = 1, b_{\max}$
3	$a = 1, a_{\max}/2$	$b = 1, b_{\max}$
4	$a = a_{\max}/2+1, a_{\max}$	$b = 1, b_{\max}$
5	$a = 1, a_{\max}$	$b = 1, b_{\max}/2$
6	$a = 1, a_{\max}$	$b = b_{\max}/2+1, b_{\max}$

Table 15.5 Marching strategy for computation of the outlet radiation at the cell walls required for definition of boundary conditions

Wall Nr.	Marching strategy, outlet *a* direction	*b* direction
1 (e)	$a = a_{\max}/4+1, 3a_{\max}/4$	$b = 1, b_{\max}$
2 (w)	$a = 1, a_{\max}/4$	$b = 1, b_{\max}$
	$a = 3a_{\max}/4+1, a_{\max}$	
3 (n)	$a = a_{\max}/2+1, a_{\max}$	$b = 1, b_{\max}$
4 (s)	$a = 1, a_{\max}/2$	$b = 1, b_{\max}$
5 (u)	$a = 1, a_{\max}$	$b = b_{\max}/2+1, b_{\max}$
6 (d)	$a = 1, a_{\max}$	$b = 1, b_{\max}/2$

For mirroring the incident directions required are $a = a_{\max} - a + 1$, $b = b_{\max} - b + 1$

15.3 Material properties

15.3.1 Source terms – emission from hot surfaces with known temperature

The black body intensity in any direction is computed by integrating *Planck*'s spectral distribution emissive power within the prescribed boundaries as follows

$$i'_{1b} = \frac{1}{\pi}\int_{\lambda_1}^{\lambda_2} \frac{2\pi C_1}{\lambda^5\left(e^{\frac{C_2}{\lambda T}} - 1\right)} d\lambda , \qquad (15.62)$$

where

$$C_1 = 0.59552197 \times 10^{-16} m^2 / sr , \qquad (15.63)$$

$$C_2 = 0.014338769 \times 10^{-16} mK . \qquad (15.64)$$

Here, index b stands for black body. We use three bands for the source terms. The boundaries are selected as follows

$$\lambda_1 = 7 \times 10^{-7} m , \ \lambda_2 = 1.2 \times 10^{-6} m , \qquad (15.65,66)$$

for the first band,

$$\lambda_1 = 1.2 \times 10^{-6} m , \ \lambda_2 = 1.8 \times 10^{-6} m , \qquad (15.67,68)$$

for the second band, and

$$\lambda_1 = 1.8 \times 10^{-6} m , \ \lambda_2 = 20 \times 10^{-6} m , \qquad (15.69,70)$$

for the third band. Other choices are of course possible. Thus, the source term due to radiating melt particles assuming

$$\Phi(\lambda, \omega, \omega_i) = 1 \qquad (15.71)$$

is

$$S' = f_{r,3} a_{3\lambda} i'_b(T_3) + \frac{\sigma_{3s\lambda}}{4\pi} \sum_{a=1}^{a_{\max}} \sum_{b=1}^{b_{\max}} i'_{\lambda,a,b} = f_{r,3} a_{3\lambda} i'_b(T_3) + \frac{\sigma_{3s\lambda}}{4\pi} i'_{\lambda,total} . \qquad (15.72)$$

15.3.2 Spectral absorption coefficient of water

The experimentally measured index of refraction, *n*, and the index of absorption, *k*, as a function of the incident wavelength are taken from *Hale* and *Marvin* (1973) and *Zolotarev* and *Dyomin* (1977). The data of *Hale* and *Marvin* are used for wavelengths in the region of 0.7 to 7 μm and the data of *Zolotarev* and *Dyomin* for the region of 7 to 17 μm. The spectral absorption coefficient is then computed as follows

$$a_\lambda(\lambda) = \frac{4\pi k}{\lambda}. \tag{15.73}$$

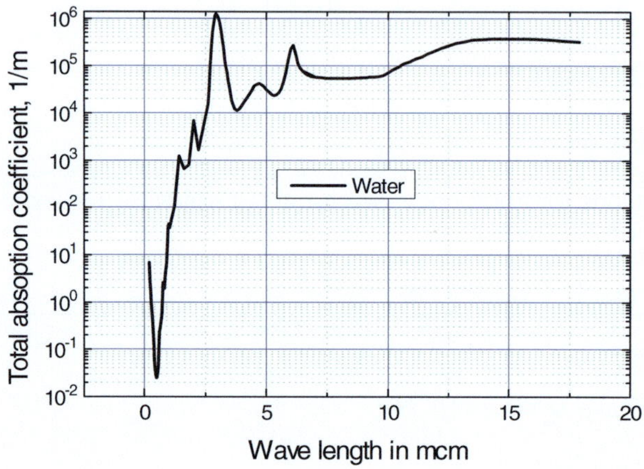

Fig. 15.3 Spectral absoption coefficient for water as a function of the wavelength

The so-obtained absorption coefficient is presented in Fig. 15.3. The argumentation that the spectral properties of water are not strong functions of water temperature and pressure summarized by *Fletcher* (1999) is acceptable. The total absorption coefficient is then computed as follows

$$a'(T_3) = \frac{\int\limits_{\lambda_{\min}}^{\lambda_{\max}} i'_{1b}(T_3,\lambda)\left[1-e^{-a_\lambda(\lambda)s}\right]d\lambda}{\int\limits_{\lambda_{\min}}^{\lambda_{\max}} i'_{1b}(T_3,\lambda)\,d\lambda}. \tag{15.74}$$

We use for the limits the following values $\lambda_{\min} = 0.7$ μm and $\lambda_{\max} = 33$ μm and perform the integration numerically using 1000 steps. This is a very time-consuming procedure. To avoid computation cost *Fletcher* (1985) tabulated the results obtained in a similar manner for three different radiation surface temperatures, namely 1000, 2500 and 3500 K as a function of the water thickness *s*. *Vaeth* (1995) approximated the absorption coefficients obtained by *Fletcher* (1985) for three different radiation surface temperature, namely 1000, 2500 and 3500 K by an analytical function. We compared *Vaeth*'s approximation with the analytically obtained coefficients and found that the coefficients for 3500 K are with about 30% lower and for 2500 K about 10% lower then those computed analytically. The modified *Vaeth* approximation is then recommended here

$$a(T_3,s) = a(T_3 = 1000K, s)\frac{1}{15}\left[32 + 4\frac{T_3}{1000}\left(-6 + \frac{T_3}{1000}\right)\right]$$
$$+a(T_3 = 2500K, s)\frac{1}{3}\left[-7 + \frac{T_3}{1000}\left(9 - 2\frac{T_3}{1000}\right)\right]$$
$$+a(T_3 = 3500K, s)\frac{1}{5}\left[5 + \frac{T_3}{1000}\left(-7 + 2\frac{T_3}{1000}\right)\right], \tag{15.75}$$

where

$$a(T_3 = 1000K, s) = 1.395086\ln\left(s^{0.01374877} + 1\right), \tag{15.76}$$

$$a(T_3 = 2500K, s) = 1.1\times 0.8641743\ln\left(s^{0.1427472} + 1\right), \tag{15.77}$$

$$a(T_3 = 3500K, s) = 1.3\times 0.4948444\ln\left(s^{0.2694667} + 1\right). \tag{15.78}$$

Note that this correlation cannot be extrapolated for water thickness less than 1 mm because of the unaceptable error increase - see Fig. 15.5. Values for this region are given in Figs. 15.4 and 15.6. We see that for water thickness between 1 and 30 μm the error of the modified *Vaeth* approximation is less than 8%.

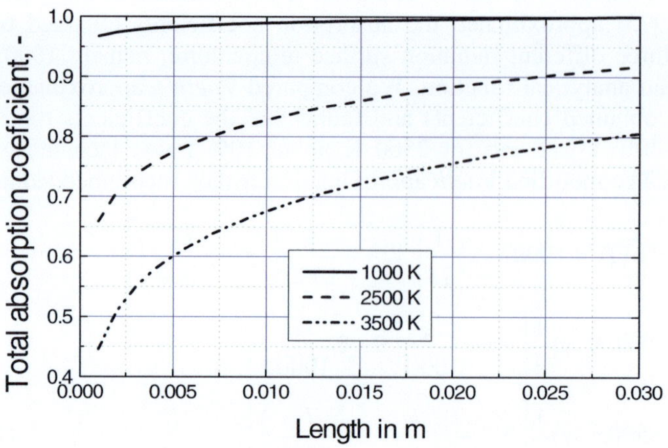

Fig. 15.4 Analytically computed total absorption coefficient as a function of the water-layer thickness. Parameter – temperature of the radiation sources

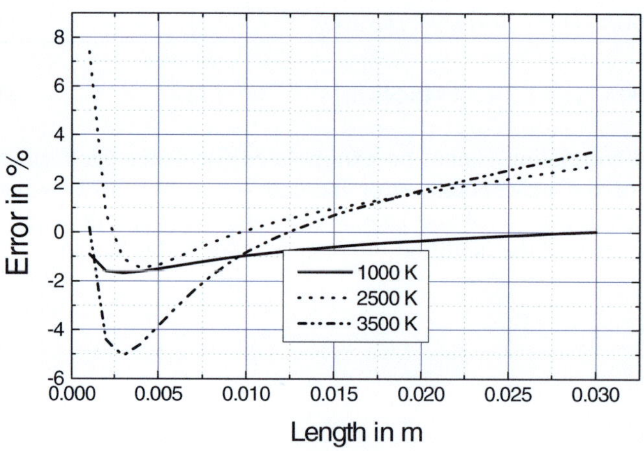

Fig. 15.5 Error of the modified Vaeth approximation compared to the analytically computed total absorption coefficient as a function of the water-layer thickness. Parameter – temperature of the radiation sources

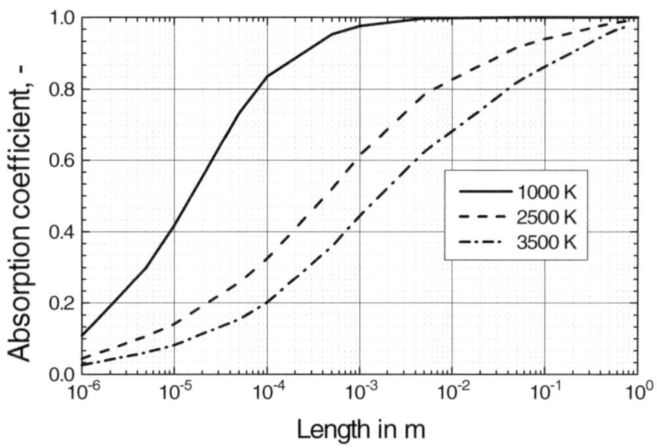

Fig. 15.6 Analytically computed total absorption coefficient as a function of the water-layer thickness. Parameter – temperature of the radiation sources

Partitioning of the absorbed energy between the bulk and the surface: Now let us estimate the band absorption coefficients for three different bands as presented in Figs. 15.7a–c. The figures demonstrated the strong heterogeneity of the absorption depending on the wavelength. Thermal radiation in the first band can cover much longer distance without being absorbed than the radiation in the second and in the third band. In band no.2 the temperature dependence is not as strong as in the other two bands.

The problem of how much radiation energy is deposited at the surface and immediately transferred into evaporation mass is not solved up to now and has to be attacked in the future. Figure 15.7d demonstrates that thermal radiation with wavelengths larger 3 μm are completely absorbed within a 0.5 mm water layer. *Vaeth* (1995) proposed to consider all this energy deposited onto the surface that is obviously true only for lengths larger then 0.5 mm. The *Vaeth* approximation of the part of the energy deposited onto the surface is

$$f_{surface} = \min\left(1, \frac{\left(\dfrac{T_3}{1000}\right)}{\left(\dfrac{T_3}{1000}\right)^2 - 0.4611115\left(\dfrac{T_3}{1000}\right) + 0.8366274}\right). \tag{15.79}$$

If we arbitrary take all the deposit energy in 0.01 mm as surface energy the absorbed fractions of the incident energy analytically computed are 0.417, 0.141 and 0.0815 for the temperatures 1000, 2500 and 3500 K, respectively.

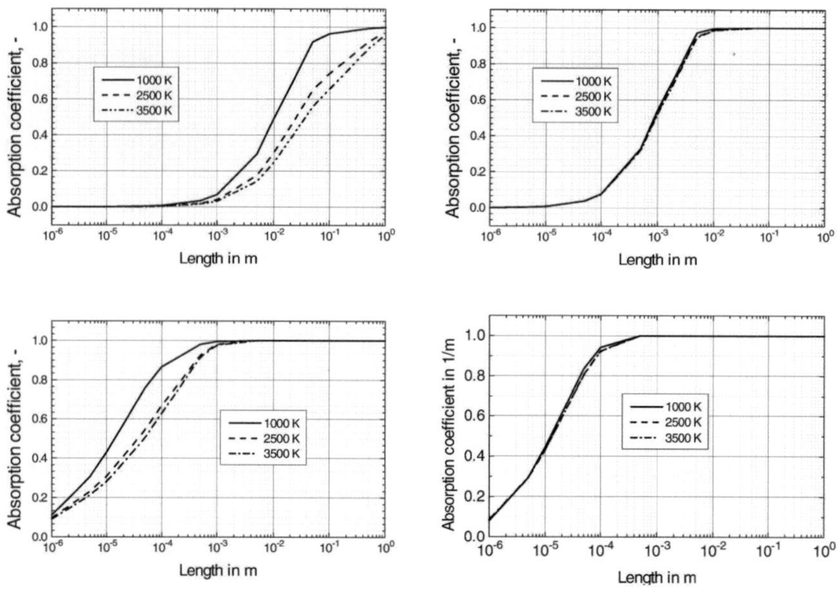

Fig. 15.7 Analytically computed partial absorption coefficient as a function of the water-layer thickness. Parameter – temperature of the radiation sources. a) Band no.1 7×10^{-7} to 1.2×10^{-6} m; b) Band no.2 1.2×10^{-6} to 1.8×10^{-6} m; c) Band no.3 1.8×10^{-6} to 20×10^{-6} m; d) Infrared 3×10^{-6} to 200×10^{-6} m.

15.3.3 Spectral absorption coefficient of water vapor and other gases

Information about the total emissivity and absorptivity for carbon dioxide, water vapor and their mixtures is provided by *Steward* and *Kocaefe* (1986). Information about the gas emissivity for wide range of process conditions is provided by *Docherty* (1983).

15.4 Averaged properties for some particular cases occurring in melt–water interaction

During interaction of molten material with water some specific geometrical arrangements called flow pattern may happen. For some of them we provide the averaged properties across a control volume with finite size. We will consider three different cases as examples, starting with the simplest and ending with a more complicated one:

1) Gas sphere inside a molten material.
2) Concentric spheres of water droplets, surrounded by vapor, surrounded by molten material.
3) Spherical particles of hot radiating material surrounded by a layer of vapor surrounded by water.

The first two cases are typical examples of local interaction. Whatever happens with the radiation transport it happens within the cavity. The third one is a typical example of interactions across the faces of the control volumes. It is the idealized film boiling case.

15.4.1 Spherical cavity of gas inside a molten material

Consider the case presented in Fig. 15.8.

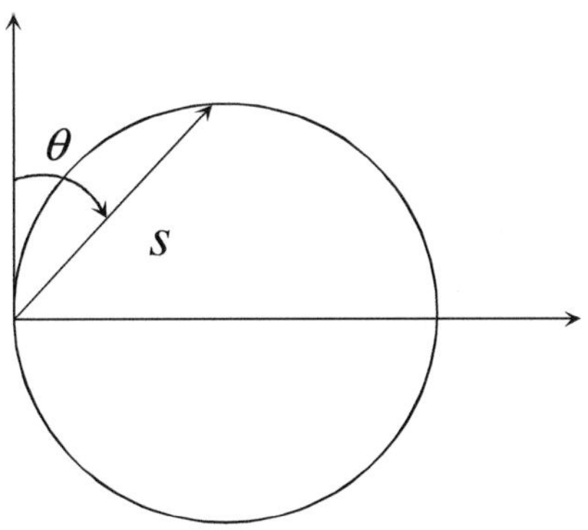

Fig. 15. 8 Spherical cavity of gas inside a molten material

The gas cavity is surrounded completely by melt. Given is the diameter of the cavity D_1, the temperatures of the melt T_3 and of the gas T_1 and their radiation properties. The task is to estimate how much energy per unit time Q_A is transferred from the melt into the gas.

We select an arbitrary point at the interface and draw one axis going through the point an the center of the sphere and one vertical axis perpendicular to it. With respect to the vertical axis we define the angle θ. All possible beam directions in the horizontal cross-section are defined within the angle interval: $\theta_1 = 0$, $\theta_2 = \frac{\pi}{2}$.

The beam length s is then depending only on the angle $s = D_1 \sin\theta$. The averaged beam length is then defined as follows

$$\overline{s} = \frac{\int_0^{2\pi}\int_{\theta_1}^{\theta_2} s_1 \cos\theta \sin\theta d\theta d\varphi}{\frac{1}{2}\left(\sin^2\theta_2 - \sin^2\theta_1\right)} = 2\int_0^{\frac{\pi}{2}} s\cos\theta\sin\theta d\theta = \frac{2}{3}D_1 . \tag{15.80}$$

The integration over the angle φ is in the boundaries 0 and 2π because we have axis symmetry. Note in this case that $\sin^2\theta_2 - \sin^2\theta_1 = 1$. With this averaged beam length we compute the $\alpha_1'(\overline{s})$ spectral-averaged absorption coefficient and the spectral-averaged emissivity coefficient of the vapor $\varepsilon_1'(\overline{s})$. The transferred power through the vapor sphere is therefore

$$Q = \pi D_1^2 \sigma\left\{[1-\alpha_1'(\overline{s})]\varepsilon_3 T_3^4 + \varepsilon_1'(\overline{s})T_1^4\right\}. \tag{15.81}$$

The absorbed power inside the sphere is

$$Q_A = \pi D_1^2 \sigma\left[\alpha_1'(\overline{s})\varepsilon_3 T_3^4 - \varepsilon_1'(\overline{s})T_1^4\right]. \tag{15.82}$$

15.4.2 Concentric spheres of water droplets, surrounded by vapor, surrounded by molten material

Consider a continuous melt with water and vapor enclosed inside the melt as presented in Fig. 15.9.

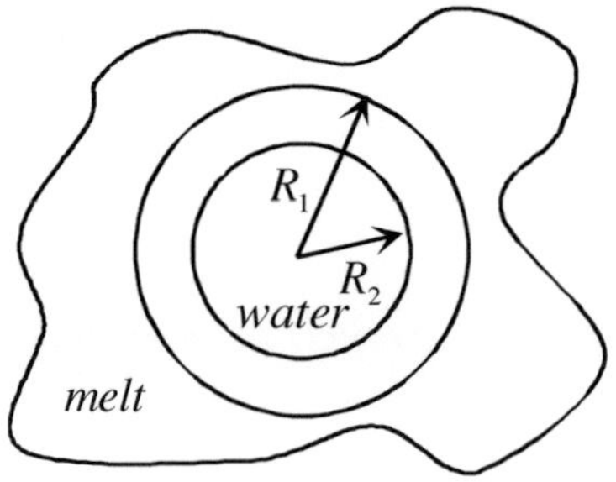

Fig. 15.9 Sphere of water droplet, surrounded by vapor, surrounded by molten material

The volume fraction of the vapor α_1, and of the water α_2 are known. The water diameter is known. We assume that the vapor is surrounding the water concentrically. The vapor diameter is then

$$D_1 = D_2\sqrt[3]{1+\frac{\alpha_1}{\alpha_2}}, \ \alpha_2 > 0 . \tag{15.83}$$

The melt radiates through the vapor and through the droplet. It is to be computed how much power radiated from the melt is absorbed by the vapor and how much is absorbed by the water droplet. Again, we select an arbitrary point at the melt interface and draw an axis going through the point to the center of the sphere and one vertical axis perpendicular to it. With respect to the vertical axis we define the angle θ. All possible beam directions in the horizontal cross-section are defined within the angle interval: $\theta_1 = 0$, $\theta_2 = \frac{\pi}{2}$ but in contrast with the previous case there are two characteristic zones as shown in Fig. 15.10. In the first the beam crosses gas only and in the second gas, water, gas successively. We perform the integration within these two zones separately.

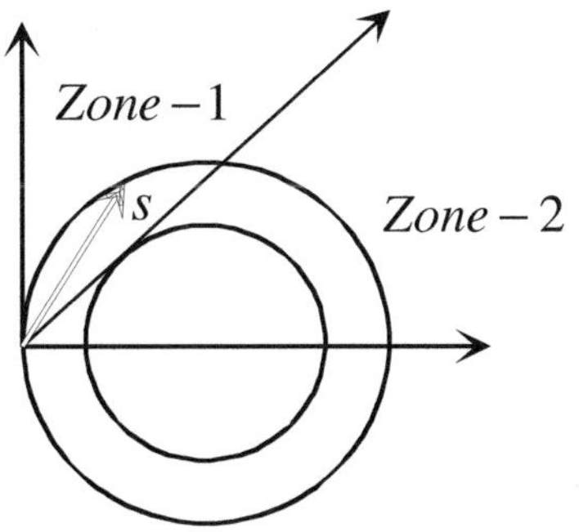

Fig. 15.10 Zone 1 – beam crosses vapor only, zone 2 – beam crosses vapor–liquid–vapor

First zone – vapor only: This zone is defined within the angles

$$\theta_1 = 0,\ \theta_2 = \frac{\pi}{2} - \arcsin\left(\frac{D_2}{D_1}\right),\ \sin^2\theta_2 - \sin^2\theta_1 = 1 - \left(\frac{D_2}{D_1}\right)^2 .$$

The beam length inside the zone is

$$s_{v,1.zone} = D_1 \sin\theta . \tag{15.84}$$

The averaged beam length is

$$\overline{s}_{v,1.zone} = \frac{\int_0^{2\pi}\int_{\theta_1}^{\theta_2} s_1 \cos\theta \sin\theta d\theta d\varphi}{\frac{1}{2}\left(\sin^2\theta_2 - \sin^2\theta_1\right)}$$

$$= 2\frac{\int_0^{\frac{\pi}{2}-\arcsin\left(\frac{D_2}{D_1}\right)} s_{2,1.zone}\cos\theta\sin\theta d\theta}{1-\left(\frac{D_2}{D_1}\right)^2} = \frac{2}{3}D_1\left[1-\left(\frac{D_2}{D_1}\right)^2\right]^{1/2} . \tag{15.85}$$

The transferred power through the vapor is

$$Q = \pi D_1^2 \sigma \left\{ [1 - \alpha_v'(\overline{s}_{v,1.zone})]\varepsilon_3 T_3^4 + \varepsilon_v'(\overline{s}_{v,1.zone}) T_1^4 \right\} \left(1 - \frac{D_2^2}{D_1^2} \right). \quad (15.86)$$

The absorbed power inside the vapor is

$$Q_A = \pi D_1^2 \sigma \left\{ \alpha_v'(\overline{s}_{v,1.zone})\varepsilon_3 T_3^4 - \varepsilon_v'(\overline{s}_{v,1.zone}) T_1^4 \right\} \left(1 - \frac{D_2^2}{D_1^2} \right). \quad (15.87)$$

Second zone – vapor/droplet: The second zone is defined within the interval

$$\theta_1 = \frac{\pi}{2} - \arcsin\left(\frac{D_2}{D_1}\right) \text{ and } \theta_2 = \frac{\pi}{2}, \ \sin^2\theta_2 - \sin^2\theta_1 = \left(\frac{D_2}{D_1}\right)^2.$$

Vapor: The beam length inside the vapor is

$$s_{v,2.zone} = \frac{1}{2} D_1 \left[\sin\theta - \left(\left(\frac{D_2}{D_1}\right)^2 - \cos^2\theta \right)^{1/2} \right]. \quad (15.88)$$

The averaged beam length results then in

$$\overline{s}_{v,2.zone} = \frac{\int_{\theta_1}^{\theta_2} s_{v,2.zone} \cos\theta \sin\theta d\theta}{\frac{1}{2}\left(\sin^2\theta_2 - \sin^2\theta_1\right)} = \frac{2\int_{\theta_1}^{\theta_2} s_{v,2.zone} \cos\theta \sin\theta d\theta}{\left(\frac{D_2}{D_1}\right)^2}$$

$$= \frac{1}{3} D_1 \frac{1 - \left(\frac{D_2}{D_1}\right)^2 - \left(1 - \left(\frac{D_2}{D_1}\right)^2\right)^{3/2}}{\left(\frac{D_2}{D_1}\right)^2}. \quad (15.89)$$

The transferred power through the vapor is

$$Q = \pi D_1^2 \sigma \left\{ [1 - \alpha_v'(\overline{s}_{v,2.zone})]\varepsilon_3 T_3^4 + \varepsilon_v'(\overline{s}_{v,2.zone}) T_1^4 \right\} \left(\frac{D_2}{D_1} \right)^2. \quad (15.90)$$

The absorbed power inside the vapor is

$$Q_A = \pi D_1^2 \sigma \left\{ \alpha_v'(\overline{s}_{v,2.zone})\varepsilon_3 T_3^4 - \varepsilon_v'(\overline{s}_{v,2.zone}) T_1^4 \right\} \left(\frac{D_2}{D_1} \right)^2. \quad (15.91)$$

Water: The beam length inside the water is

$$\Delta s_{w,2.zone} = D_1 \left(\left(\frac{D_2}{D_1} \right)^2 - \cos^2 \theta \right)^{1/2} . \tag{15.92}$$

The averaged beam length is then

$$\Delta \overline{s}_{w,2.zone} = 2 \frac{\int_{\theta_1}^{\theta_2} s_{w,2.zone} \cos\theta \sin\theta d\theta}{\left(\frac{D_2}{D_1} \right)^2} = \frac{2}{3} D_2 . \tag{15.93}$$

The transferred power through the water droplet is

$$Q = \pi D_1^2 \sigma \left\{ \begin{array}{l} [1-\alpha_w'(\overline{s}_{w,2.zone})][1-\alpha_v'(\overline{s}_{v,2.zone})]\varepsilon_3 T_3^4 \\ +[1-\alpha_w'(\overline{s}_{w,2.zone})]\varepsilon_v'(\overline{s}_{v,2.zone})T_1^4 \\ +\varepsilon_w'(\overline{s}_{w,2.zone})T_2^4 \end{array} \right\} \left(\frac{D_2}{D_1} \right)^2 . \tag{15.94}$$

The absorbed power by the water droplet is

$$Q_{Aw} = \pi D_1^2 \sigma \left\{ \begin{array}{l} \alpha_w'(\overline{s}_{w,2.zone})[1-\alpha_v'(\overline{s}_{v,2.zone})]\varepsilon_3 T_3^4 \\ +\alpha_w'(\overline{s}_{w,2.zone})\varepsilon_v'(\overline{s}_{v,2.zone})T_1^4 - \varepsilon_w'(\overline{s}_{w,2.zone})T_2^4 \end{array} \right\} \left(\frac{D_2}{D_1} \right)^2 . \tag{15.95}$$

The absorbed total power by the vapor is the sum of the absorbed power in the both vapor regions

$$Q_{Av} = \pi D_1^2 \sigma \left[\begin{array}{l} \left\{ \alpha_v'(\overline{s}_{v,1.zone})\varepsilon_3 T_3^4 - \varepsilon_v'(\overline{s}_{v,1.zone})T_1^4 \right\} \left(1 - \frac{D_2^2}{D_1^2} \right) \\ +\left\{ \alpha_v'(\overline{s}_{v,2.zone})\varepsilon_3 T_3^4 - \varepsilon_v'(\overline{s}_{v,2.zone})T_1^4 \right\} \left(\frac{D_2}{D_1} \right)^2 \end{array} \right] . \tag{15.96}$$

15.4.3 Clouds of spherical particles of radiating material surrounded by a layer of vapor surrounded by water – *Lanzenberger's* solution

Consider a cloud of equidistantly arranged hot spherical particles. The cloud is flying in a water and gas mixture. Compute the power radiated by a hot spherical particle and its environment. Compute what fraction of this energy is absorbed by water and what fraction is absorbed in the gas as a function of the characteristic scales inside the three-phase mixture. Compute the length required for almost complete length absorption.

The task can formally be described in the following way: Consider an artificially defined sphere with diameter D_2 containing concentrically two other spheres as presented in Fig. 15.11. This resembles film boiling of hot particles in water. We use this structure as an abstract model for computing the radiation properties of the mixture consisting of particles, vapor and water. This elementary cell is called sometimes a *Wigner* cell. It contains as much vapor and water as the total vapor and water in the computational cell divided by the number of the particles inside the cell, respectively. We will first compute what happens inside the *Wigner* cell and how much radiation energy is leaving the cell. Then, we will ask the question what happens with the radiation energy in consequently interacting *Wigner* cells.

The inner most sphere, having diameter D_3, is radiating. We call this sphere a "particle". The space between the first and the second spherical surfaces is filled with vapor. The second sphere has diameter D_1. The space between the second and the external sphere (*Wigner* cell) is filled with water. Now, consider a space filled equidistantly with particles having volume fraction $\alpha_3 > 0$ surrounded by vapor with volume fraction $\alpha_1 > 0$ and water having volume fraction $\alpha_2 > 0$.

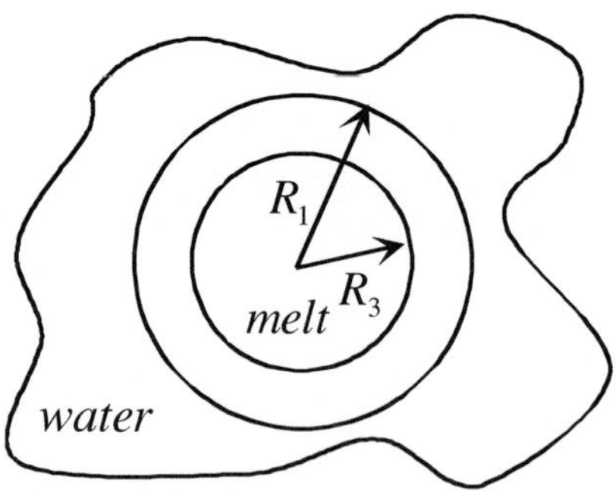

Fig. 15.11 Radiating sphere concentrically surrounded by vapor and water

For homogeneously distributed particles we have

$$D_1 = D_2 \sqrt[3]{\alpha_1 + \alpha_3} = D_3 \sqrt[3]{1 + \frac{\alpha_1}{\alpha_3}}, \quad \alpha_3 > 0, \tag{15.97}$$

$$D_2 = \frac{D_3}{\sqrt[3]{\alpha_3}}, \; \alpha_3 > 0. \tag{15.98}$$

Frequently used ratios in the following text are

$$\frac{D_1}{D_2} = (\alpha_1 + \alpha_3)^{1/3}, \tag{15.99}$$

$$\frac{D_3}{D_2} = \alpha_3^{1/3}. \tag{15.100}$$

In this section we will present *Lanzenberger's* results reported in (1997) simplifying in some places the resulting integrals for convenient use.

15.4.3.1 Space-averaged beam lengths for the concentric-spheres radiation problem

The averaging procedure for computing the averaged beam length results from the radiation theory is

$$\overline{s} = \frac{\int_0^{2\pi}\int_{\theta_1}^{\theta_2} s\,(\varphi,\theta)\cos\theta\sin\theta d\theta d\varphi}{\int_0^{2\pi}\int_{\theta_1}^{\theta_2}\cos\theta\sin\theta d\theta d\varphi} = \frac{\int_{\theta_1}^{\theta_2} s\,(\theta)\cos\theta\sin\theta d\theta}{\frac{1}{2}\left(\sin^2\theta_2 - \sin^2\theta_1\right)}. \tag{15.101}$$

It gives the averaged beam length over the angles $0 < \varphi < 2\pi$ and $\theta_1 < \theta < \theta_2$. The formal procedure contains (a) specification of θ_1 and θ_2, (b) derivation of the relationship $s_1(\varphi,\theta)$ and (c) finally integration in order to obtain $\overline{s}$. Two specific geometries presented in Fig. 15.12 and Fig. 15.13 represent all cases. The results of the computations are presented below:

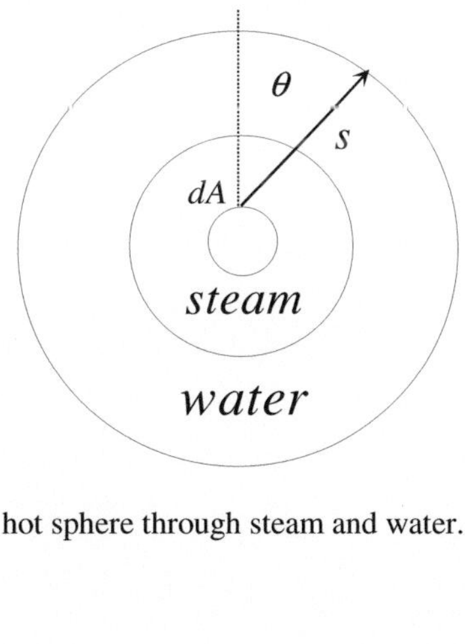

Fig. 15.12 Radiating hot sphere through steam and water. The first Wigner cell.

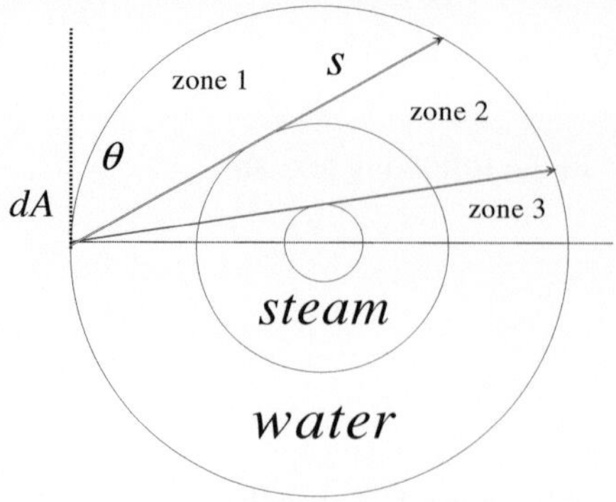

Fig. 15. 13 Radiation entering the second Wigner cell from outside sources

15.4.3.1.1 Beam emitted from the particle surface

Angle interval: $\theta_1 = 0$ and $\theta_2 = \pi/2$, $\sin^2\theta_2 - \sin^2\theta_1 = 1$.

Vapor: $0 < s < s_1$

Beam length: $s_1 = \frac{1}{2} D_3 \left[\left(\left(\frac{D_1}{D_3} \right)^2 - \sin^2\theta \right)^{1/2} - \cos\theta \right]$. (15.102)

Averaged beam length:

$$\overline{s_1} = 2\int_0^{\pi/2} s_1 \cos\theta \sin\theta d\theta = \frac{1}{3} D_3 \left[\left(\frac{D_1}{D_3} \right)^3 - \left(\left(\frac{D_1}{D_3} \right)^2 - 1 \right)^{3/2} - 1 \right]. \quad (15.103)$$

Water: $s_1 < s < s_2$

Beam length:

$$s_2 - s_1 = \frac{1}{2} D_3 \left[\left(\left(\frac{D_2}{D_3} \right)^2 - \sin^2\theta \right)^{1/2} - \left(\left(\frac{D_1}{D_3} \right)^2 - \sin^2\theta \right)^{1/2} \right]. \quad (15.104)$$

Averaged beam length:

$$\Delta \overline{s}_2 = 2\int_0^{\pi/2} \left(s_2 - s_1\right)\cos\theta \sin\theta d\theta$$

$$= \frac{1}{3}D_3 \left\{ \left(\frac{D_2}{D_3}\right)^3 - \left(\frac{D_1}{D_3}\right)^3 + \left(\left(\frac{D_1}{D_3}\right)^2 - 1\right)^{3/2} - \left(\left(\frac{D_2}{D_3}\right)^2 - 1\right)^{3/2} \right\}. \quad (15.105)$$

15.4.3.1.2 Beam entering the external sphere (*Wigner* cell) surface

First zone – water only:

Angle interval: $\theta_1 = 0$, $\theta_2 = \frac{\pi}{2} - \arcsin\left(\frac{D_1}{D_2}\right)$, $\sin^2\theta_2 - \sin^2\theta_1 = 1 - \left(\frac{D_1}{D_2}\right)^2$.

Beam length: $s_{2,1.zone} = D_2 \sin\theta$. (15.106)

Averaged beam length:

$$\overline{s}_{2,1.zone} = \frac{\int_0^{2\pi}\int_{\theta_1}^{\theta_2} s_1 \cos\theta \sin\theta d\theta d\varphi}{\frac{1}{2}\left(\sin^2\theta_2 - \sin^2\theta_1\right)} =$$

$$= 2\frac{\int_0^{\frac{\pi}{2}-\arcsin\left(\frac{D_1}{D_2}\right)} s_{2,1.zone} \cos\theta \sin\theta d\theta}{1 - \left(\frac{D_1}{D_2}\right)^2} = \frac{2}{3}D_2\left[1 - \left(\frac{D_1}{D_2}\right)^2\right]^{1/2}. \quad (15.107)$$

Second zone:

Angle interval: $\theta_1 = \frac{\pi}{2} - \arcsin\left(\frac{D_1}{D_2}\right)$ and $\theta_2 = \frac{\pi}{2} - \arcsin\left(\frac{D_3}{D_2}\right)$,

$$\sin^2\theta_2 - \sin^2\theta_1 = \left(\frac{D_1}{D_2}\right)^2 - \left(\frac{D_3}{D_2}\right)^2.$$

Region 1 – water:

Beam length: $s_{21,2.zone} = \frac{1}{2}D_2\left[\sin\theta - \left(\left(\frac{D_1}{D_2}\right)^2 - \cos^2\theta\right)^{1/2}\right]$. (15.108)

Averaged beam length:

$$\bar{s}_{21,2.zone} = \frac{\int_{\theta_1}^{\theta_2} s_{21,2.zone} \cos\theta \sin\theta d\theta}{\frac{1}{2}\left(\sin^2\theta_2 - \sin^2\theta_1\right)} = \frac{2\int_{\theta_1}^{\theta_2} s_{21,2.zone} \cos\theta \sin\theta d\theta}{\left(\frac{D_1}{D_2}\right)^2 - \left(\frac{D_3}{D_2}\right)^2}$$

$$= \frac{1}{3} D_2 \frac{\left(1-\left(\frac{D_3}{D_2}\right)^2\right)^{3/2} - \left[\left(\frac{D_1}{D_2}\right)^2 - \left(\frac{D_3}{D_2}\right)^2\right]^{3/2} - \left(1-\left(\frac{D_1}{D_2}\right)^2\right)^{3/2}}{\left(\frac{D_1}{D_2}\right)^2 - \left(\frac{D_3}{D_2}\right)^2}. \quad (15.109)$$

Region 2 – vapor:

Beam length: $$\Delta s_{1,2.zone} = D_2 \left(\left(\frac{D_1}{D_2}\right)^2 - \cos^2\theta\right)^{1/2}. \quad (15.110)$$

Averaged beam length:

$$\Delta\bar{s}_{1,2.zone} = 2\frac{\int_{\theta_1}^{\theta_2} s_{21,2.zone} \cos\theta \sin\theta d\theta}{\left(\frac{D_1}{D_2}\right)^2 - \left(\frac{D_3}{D_2}\right)^2} = \frac{2}{3} D_2 \left[\left(\frac{D_1}{D_2}\right)^2 - \left(\frac{D_3}{D_2}\right)^2\right]^{1/2}.$$

(15.111)

Region 3 – water:

Beam length: The same as in region 1.

Averaged beam length: $$\Delta\bar{s}_{22,2.zone} = \bar{s}_{21,2.zone}. \quad (15.112)$$

Third zone:

Angle interval: $$\theta_1 = \frac{\pi}{2} - \arcsin\left(\frac{D_3}{D_2}\right) \text{ and } \theta_2 = \frac{\pi}{2},\ \sin^2\theta_2 - \sin^2\theta_1 = \left(\frac{D_3}{D_2}\right)^2.$$

Region 1 –water only:

Beam length: $$s_{2,3.zone} = \frac{1}{2} D_2 \left[\sin\theta - \left(\left(\frac{D_1}{D_2}\right)^2 - \cos^2\theta\right)^{1/2}\right]. \quad (15.113)$$

Averaged beam length:

$$\bar{s}_{2,3.zone} = \frac{\int_{\theta_1}^{\theta_2} s_{2,3.zone}\cos\theta\sin\theta d\theta d\varphi}{\frac{1}{2}\left(\sin^2\theta_2 - \sin^2\theta_1\right)} = \frac{2\int_{\theta_1}^{\theta_2} s_{2,3.zone}\cos\theta\sin\theta d\theta d\varphi}{\left(\frac{D_3}{D_2}\right)^2}$$

$$= \frac{1}{3}D_2 \frac{1-\left(\frac{D_1}{D_2}\right)^3 + \left[\left(\frac{D_1}{D_2}\right)^2 - \left(\frac{D_3}{D_2}\right)^2\right]^{3/2} - \left(1-\left(\frac{D_3}{D_2}\right)^2\right)^{3/2}}{\left(\frac{D_3}{D_2}\right)^2}. \tag{15.114}$$

Region 2 –vapor only:

Beam length: $$s_{1,3.zone} = \frac{1}{2}D_2\left\{\left(\left(\frac{D_1}{D_2}\right)^2 - \cos^2\theta\right)^{1/2} - \left(\left(\frac{D_3}{D_2}\right)^2 - \cos^2\theta\right)^{1/2}\right\}. \tag{15.115}$$

Averaged beam length:

$$\bar{s}_{1,3.zone} = \frac{\int_{\theta_1}^{\theta_2} s_{1,3.zone}\cos\theta\sin\theta d\theta}{\frac{1}{2}\left(\sin^2\theta_2 - \sin^2\theta_1\right)} = \frac{2\int_{\theta_1}^{\theta_2} s_{1,3.zone}\cos\theta\sin\theta d\theta d\varphi}{\left(\frac{D_3}{D_2}\right)^2}$$

$$= \frac{1}{3}D_2 \frac{\left(\frac{D_1}{D_2}\right)^3 - \left(\frac{D_3}{D_2}\right)^3 - \left(\left(\frac{D_1}{D_2}\right)^2 - \left(\frac{D_3}{D_2}\right)^2\right)^{3/2}}{\left(\frac{D_3}{D_2}\right)^2}. \tag{15.116}$$

15.4.3.2 Solution of the RTE

15.4.3.2.1 Directional spectral intensity inside the vapor

Along the beam $0 < s < s_1$ the analytical solution of the RTE equation without scattering gives the heat flux per unit solid angle, per unit wavelength and per unit area as a function of the beam length

$$i'_\lambda\left(\lambda, s, T_1, T_3\right) = i'_{3,\lambda}\left(\lambda, T_3\right)e^{-a_{1,\lambda}s} + i'_{1,\lambda}\left(\lambda, s, T_1\right)\left(1-e^{-a_{1,\lambda}s}\right). \tag{15.117}$$

Across the infinitesimal cross-section $dA\cos\theta$, for the infinitesimal solid angle $d\omega$, and within the wavelength region $d\lambda$ we have

$$d^3Q = i'_\lambda\left(\lambda, s, T_1, T_3\right) dA\cos\theta d\omega d\lambda$$
$$= \left[i'_\lambda\left(\lambda, 0, T_3\right) e^{-a_{1,\lambda}s} + i'_{\lambda b}\left(\lambda, T_1\right)\left(1 - e^{-a_{1,\lambda}s}\right)\right] dA\cos\theta d\omega d\lambda . \tag{15.118}$$

Integrated over all wavelengths of the emitting media we have

$$d^2Q = \left[\int_0^\infty i'_\lambda\left(\lambda, 0, T_3\right) e^{-a_{1,\lambda}s} d\lambda + \int_0^\infty i'_{\lambda b}\left(\lambda, T_1\right)\left(1 - e^{-a_{1,\lambda}s}\right) d\lambda \right] dA\cos\theta d\omega . \tag{15.119}$$

Estimation of the first integral: Having in mind the definition of the emissivity coefficient of the particle surface

$$\varepsilon_3\left(T_3\right) = \frac{\int_0^\infty i'_\lambda\left(\lambda, T_3\right) d\lambda}{\int_0^\infty i'_{\lambda b}\left(\lambda, T_3\right) d\lambda}, \tag{15.120}$$

and the definition of the vapor absorption coefficient of black-body radiation with temperature T_3

$$a'_1\left(T_3, s\right) = \frac{\int\limits_0^\infty i'_{\lambda b}\left(T_3, \lambda\right)\left[1 - e^{-a_{1,\lambda}s}\right] d\lambda}{\int\limits_0^\infty i'_{\lambda b}\left(T_3, \lambda\right) d\lambda}$$

$$= 1 - \frac{\int\limits_0^\infty i'_{\lambda b}\left(T_3, \lambda\right) e^{-a_{1,\lambda}s} d\lambda}{\int\limits_0^\infty i'_{\lambda b}\left(T_3, \lambda\right) d\lambda} = 1 - \frac{\int\limits_0^\infty i'_{\lambda b}\left(T_3, \lambda\right) e^{-a_{1,\lambda}s} d\lambda}{\dfrac{\sigma T_3}{\pi}} \tag{15.122}$$

we obtain

$$\int_0^\infty i'_\lambda\left(\lambda, 0, T_3\right) e^{-a_{1,\lambda}s} d\lambda = \varepsilon_3\left(T_3\right) \int_0^\infty i'_{\lambda b}\left(\lambda, T_3\right) e^{-a_{1,\lambda}s} d\lambda$$
$$= \varepsilon_3\left(T_3\right)\left[1 - a'_1\left(T_3, s\right)\right] \frac{\sigma T_3^4}{\pi} . \tag{15.123}$$

Estimation of the second integral: For the second integral we again use the definition of the total emittance for vapor radiating with temperature T_1

$$\int_0^\infty i'_{\lambda b}\left(\lambda,T_1\right)\left(1-e^{-a_{1,\lambda}s}\right)d\lambda=\varepsilon_1\left(T_1,s\right)\frac{\sigma T_1^4}{\pi}. \qquad (15.124)$$

Thus, after the integration over the total wavelength region we obtain

$$d^2Q=\left[\varepsilon_3\left(T_3\right)\left[1-a'_1\left(T_3,s\right)\right]\frac{\sigma T_3^4}{\pi}+\varepsilon_1\left(T_1,s\right)\frac{\sigma T_1^4}{\pi}\right]dA\cos\theta d\omega. \qquad (15.125)$$

The first term gives the intensity at the point s coming from the particle with temperature T_3. The second term is the net contribution of the gas radiating with temperature T_1 at point s.

15.4.3.2.2 Directional spectral intensity inside the water

Along the beam $s_1 < s < s_2$ the analytical solution of the RTE equation without scattering gives the heat flux per unit solid angle, per unit wavelength and per unit area as a function of the beam length

$$i'_\lambda\left(\lambda,s,T_1,T_2,T_3\right)=i'_\lambda\left(\lambda,s_1\right)e^{-a_{2,\lambda}s}+i'_{2,\lambda}\left(\lambda,s,T_2\right)\left(1-e^{-a_{2,\lambda}s}\right). \qquad (15.126)$$

Across the infinitesimal cross-section $dA\cos\theta$, for the infinitesimal solid angle $d\omega$, and within the wavelength region $d\lambda$ results in

$$d^3Q=i'_\lambda\left(\lambda,s,T_1,T_2,T_3\right)dA\cos\theta d\omega d\lambda$$
$$=\left[i'_\lambda\left(\lambda,s_1\right)e^{-a_{2,\lambda}s}+i'_{2,\lambda}\left(\lambda,s,T_2\right)\left(1-e^{-a_{2,\lambda}s}\right)\right]dA\cos\theta d\omega d\lambda. \qquad (15.127)$$

Integrated over all wavelengths of the emitting media results in

$$d^2Q=\left[\int_0^\infty i'_\lambda\left(\lambda,s_1\right)e^{-a_{2,\lambda}s}d\lambda+\int_0^\infty i'_{\lambda b}\left(\lambda,T_2\right)\left(1-e^{-a_{2,\lambda}s}\right)d\lambda\right]dA\cos\theta d\omega.$$

Estimation of the first integral: The directional spectral intensity entering the water at point s_1 is

$$i'_\lambda\left(\lambda,s_1,T_1,T_3\right)=i'_{3,\lambda}\left(\lambda,T_3\right)e^{-a_{1,\lambda}s_1}+i'_{b\lambda}\left(\lambda,s_1,T_1\right)\left(1-e^{-a_{1,\lambda}s_1}\right). \qquad (15.128)$$

The total intensity at the same point is

$$i'\left(s_1,T_1,T_3\right)=\int_0^\infty i'_\lambda\left(\lambda,s_1\right)d\lambda=\varepsilon_3\left(T_3\right)\left[1-a'_1\left(T_3,s_1\right)\right]\frac{\sigma T_3^4}{\pi}+\varepsilon_1\left(T_1,s_1\right)\frac{\sigma T_1^4}{\pi}.$$
$$(15.129)$$

Now note the special definition of the absorption coefficient in water required here

$$\overline{a}_2'(s,T_1,T_3)=\frac{\int_0^\infty i_\lambda'(\lambda,s_1)\left(1-e^{-a_{2,\lambda}s}\right)d\lambda}{\int_0^\infty i_\lambda'(\lambda,s_1)d\lambda}\approx a_2'(s,T_3), \tag{15.130}$$

which can be replaced by the usual definition of the water absorption coefficient only as an *approximation*. Thus, we obtain finally

$$\int_0^\infty i_\lambda'(\lambda,s_1)e^{-a_{2,\lambda}s}d\lambda$$
$$=\left[1-a_2'(T_3,s)\right]\left\{\varepsilon_3(T_3)\left[1-a_1'(T_3,s_1)\right]\frac{\sigma T_3}{\pi}+\varepsilon_1(T_1,s_1)\frac{\sigma T_1}{\pi}\right\}. \tag{15.131}$$

Estimation of the second integral: For the second integral we again use the definition of the total emittance of the water radiating with temperature T_2

$$\int_0^\infty i_{\lambda b}'(\lambda,T_2)\left(1-e^{-a_{2,\lambda}s}\right)d\lambda=\varepsilon_2(T_2,s)\frac{\sigma T_2^4}{\pi}. \tag{15.132}$$

Thus, after the integration over the total wavelength region we obtain

$$d^2Q=\left[\left[1-a_2'(T_3,s)\right]\left\{\begin{array}{l}\varepsilon_3(T_3)\left[1-a_1'(T_3,s_1)\right]\dfrac{\sigma T_3^4}{\pi}\\ \\ +\varepsilon_1(T_1,s_1)\dfrac{\sigma T_1^4}{\pi}\end{array}\right\}+\varepsilon_2(T_2,s)\frac{\sigma T_2^4}{\pi}\right]dA\cos\theta d\omega. \tag{15.133}$$

The first term gives the intensity at the point s coming from the particle with temperature T_3 and crossing the vapor with temperature T_1. The second term is the net contribution of the water radiating with temperature T_2 at point s.

15.4.3.2.3 Total absorbed intensity

Vapor: The total absorbed directional intensity in the vapor is therefore that emitted by the particle minus that leaving the vapor sphere at s_1

$$d^2Q_{1a}=\left[\varepsilon_3(T_3)a_1'(T_3,s_1)\frac{\sigma T_3^4}{\pi}-\varepsilon_1(T_1,s_1)\frac{\sigma T_1^4}{\pi}\right]dA\cos\theta d\omega. \tag{15.134}$$

Water: The total absorbed directional intensity in the water is therefore that entering from the internal sphere minus that leaving at s_2.

$$d^2Q_{2a} = \left\{ \begin{array}{l} \varepsilon_3(T_3)\left[1-a_1'(T_3,s_1)\right]a_2'(T_3,s_2)\dfrac{\sigma T_3^4}{\pi} \\ \\ +a_2'(T_3,s_2)\varepsilon_1(T_1,s_1)\dfrac{\sigma T_1^4}{\pi} - \varepsilon_2(T_2,s_2)\dfrac{\sigma T_2^4}{\pi} \end{array} \right\} dA\cos\theta d\omega . \quad (15.135)$$

Integration over the solid angle

The integration over the solid angle gives an expression in which optical properties have to be averaged over the solid angle. For the case of the absorption coefficient it will result in

$$\int_0^{2\pi}\int_{\theta_1}^{\theta_2} a'(T,s)\cos\theta\sin\theta d\theta d\varphi \approx \overline{a}'(T)\int_0^{2\pi}\int_{\theta_1}^{\theta_2}\cos\theta\sin\theta d\theta d\varphi , \quad (15.136)$$

where

$$\overline{a}'(T) = \frac{\int_0^{2\pi}\int_{\theta_1}^{\theta_2} a'(T,s)\cos\theta\sin\theta d\theta d\varphi}{\int_0^{2\pi}\int_{\theta_1}^{\theta_2}\cos\theta\sin\theta d\theta d\varphi} . \quad (15.137)$$

The computation of the averaged absorption coefficient is very expensive. That is why the following approximation will be used. The averaged optical property is set as the optical property, being a function of the averaged beam distance. We illustrate this in the following example.

$$\overline{a}'(T) \approx a'(T,\overline{s}), \quad (15.138)$$

where

$$\overline{s} = \frac{\int_0^{2\pi}\int_{\theta_1}^{\theta_2} s(\theta,\varphi)\cos\theta\sin\theta d\theta d\varphi}{\int_0^{2\pi}\int_{\theta_1}^{\theta_2}\cos\theta\sin\theta d\theta d\varphi} . \quad (15.139)$$

The appropriate averaged beam lengths were already given in Section 15.4.3.1.

Vapor:

Performing the averaging of the optical properties over the solid angle results in

$$\int_0^{2\pi}\int_{\theta_1}^{\theta_2} a_1'(T_3,s_1)\cos\theta\sin\theta d\theta d\varphi \approx a_1'(T_3,\overline{s}_1)\int_0^{2\pi}\int_{\theta_1}^{\theta_2}\cos\theta\sin\theta d\theta d\varphi$$

$$= a_1'(T_3,\overline{s}_1)\pi\left(\sin^2\theta_2 - \sin^2\theta_1\right) \quad (15.140)$$

$$\int_0^{2\pi}\int_{\theta_1}^{\theta_2} \varepsilon_1\left(T_1, s_1\right)\cos\theta\sin\theta d\theta d\varphi \approx \varepsilon_1\left(T_1, \overline{s}_1\right)\pi\left(\sin^2\theta_2 - \sin^2\theta_1\right). \quad (15.141)$$

In particular, integrating for $\theta_1 = 0$ and $\theta_1 = \pi/2$ we obtain

$$\int_0^{2\pi}\int_{\theta_1}^{\theta_2} a_1'\left(T_3, s_1\right)\cos\theta\sin\theta d\theta d\varphi \approx a_1'\left(T_3, \overline{s}_1\right)\pi, \quad (15.142)$$

$$\int_0^{2\pi}\int_{\theta_1}^{\theta_2} \varepsilon_1\left(T_3, s_1\right)\cos\theta\sin\theta d\theta d\varphi \approx \varepsilon_1\left(T_3, \overline{s}_1\right)\pi, \quad (15.143)$$

and consequently

$$d^2 Q_{1a} = \left[\varepsilon_3\left(T_3\right) a_1'\left(T_3, \overline{s}_1\right)\sigma T_3^4 - \varepsilon_1\left(T_1, \overline{s}_1\right)\sigma T_1^4\right] dA. \quad (15.144)$$

Integrating over the surface of the particle results in

$$Q_{1a} = \pi D_3^2\left[\varepsilon_3\left(T_3\right) a_1'\left(T_3, \overline{s}_1\right)\sigma T_3^4 - \varepsilon_1\left(T_1, \overline{s}_1\right)\sigma T_1^4\right]. \quad (15.145)$$

The emitted total power from the particle is

$$Q_{3\to\infty} = \pi D_3^2 \varepsilon_3\left(T_3\right)\sigma T_3^4. \quad (15.146)$$

The emitted power from the vapor interface into the water is

$$Q_{3\to\infty} - Q_{1a} = \pi D_3^2 \sigma\left\{\left[1 - a_1'\left(T_3, \overline{s}_1\right)\right]\varepsilon_3\left(T_3\right) T_3^4 + \varepsilon_1\left(T_1, \overline{s}_1\right) T_1^4\right\}. \quad (15.147)$$

Water:

Similarly integrating for $\theta_1 = 0$ and $\theta_1 = \pi/2$ we obtain for the averages over the solid angle of the water optical properties

$$\int_0^{2\pi}\int_{\theta_1}^{\theta_2} a_2'\left(T_3, s_2\right)\cos\theta\sin\theta d\theta d\varphi \approx a_2'\left(T_3, \Delta\overline{s}_2\right)\pi, \quad (15.148)$$

$$\int_0^{2\pi}\int_{\theta_1}^{\theta_2} \varepsilon_2\left(T_3, s_2\right)\cos\theta\sin\theta d\theta d\varphi \approx \varepsilon_2\left(T_3, \Delta\overline{s}_2\right)\pi, \quad (15.149)$$

and consequently

$$d^2 Q_{2a} = \sigma\left\{\begin{array}{l}\varepsilon_3\left(T_3\right)\left[1 - a_1'\left(T_3, \overline{s}_1\right)\right] a_2'\left(T_3, \Delta\overline{s}_2\right) T_3^4 \\ \\ + a_2'\left(T_3, \Delta\overline{s}_2\right)\varepsilon_1\left(T_1, \overline{s}_1\right) T_1^4 - \varepsilon_2\left(T_2, \Delta\overline{s}_2\right) T_2^4\end{array}\right\} dA. \quad (15.150)$$

Integrating over the surface of the particle results in

$$Q_{2a} = \pi D_3^2 \sigma \left\{ \begin{array}{l} \varepsilon_3 (T_3) \left[1 - a_1' (T_3, \overline{s}_1) \right] a_2' (T_3, \Delta \overline{s}_2) T_3^4 \\ + a_2' (T_3, \Delta \overline{s}_2) \varepsilon_1 (T_1, \overline{s}_1) T_1^4 - \varepsilon_2 (T_2, \Delta \overline{s}_2) T_2^4 \end{array} \right\}. \quad (15.151)$$

The emitted power from the vapor sphere is

$$Q_{3\to\infty} - Q_{1a} = \pi D_3^2 \sigma \left\{ \left[1 - a_1' (T_3, \overline{s}_1) \right] \varepsilon_3 (T_3) T_3^4 + \varepsilon_1 (T_1, \overline{s}_1) T_1^4 \right\}. \quad (15.152)$$

The difference between this power and that absorbed by the water is the power radiated from the water surface into the environment of the first cell

$$Q_{3\to\infty} - Q_{1a} - Q_{2a}$$

$$= \pi D_3^2 \sigma \left\{ \begin{array}{l} \left[1 - a_2' (T_3, \Delta \overline{s}_2) \right] \left[1 - a_1' (T_3, \overline{s}_1) \right] \varepsilon_3 (T_3) T_3^4 \\ + \left[1 - a_2' (T_3, \Delta \overline{s}_2) \right] \varepsilon_1 (T_1, \overline{s}_1) T_1^4 + \varepsilon_2 (T_2, \Delta \overline{s}_2) T_2^4 \end{array} \right\}. \quad (15.153)$$

Second cell

The power emitted by the first *Wigner* cell is entering the second one – Fig. 15.13. There are three characteristic space angle regions for computing the absorbed and the transmitted power in the second *Wigner* cell. We consider the three regions separately. The total directional intensity of the entering radiation is

$$\int_0^\infty i_\lambda' (\lambda, T_1) d\lambda = \frac{Q_{3\to\infty} - Q_{1a} - Q_{2a}}{\pi D_2^2 \pi}. \quad (15.154)$$

First zone: Water only

$$i_\lambda' (\lambda, s) = i_\lambda' (\lambda) e^{-a_\lambda s} + i_\lambda' (\lambda, s) \left(1 - e^{-a_\lambda s} \right) \quad (15.155)$$

$$Q = \pi D_2^2 ([1 - \alpha_w' (\overline{s}_{11})] \frac{Q_1}{\pi D_2^2} + \varepsilon_w' (\overline{s}_{11}) \sigma T_2^4) \left(1 - \frac{D_1^2}{D_2^2} \right), \quad (15.156)$$

$$Q_A = \pi D_2^2 (\alpha_w' (\overline{s}_{11}) \frac{Q_1}{\pi D_2^2} - \varepsilon_w' (\overline{s}_{11}) \sigma T_2^4) \left(1 - \frac{D_1^2}{D_2^2} \right). \quad (15.157)$$

Second zone:

Region 1: water

$$Q = \pi D_2^2 ([1 - \alpha_w' (\overline{s}_{21})] \frac{Q_1}{\pi D_2^2} + \varepsilon_w' (\overline{s}_{21}) \sigma T_2^4) \left(\frac{D_1^2}{D_2^2} - \frac{D_3^2}{D_2^2} \right) \quad (15.158)$$

$$Q_A = \pi D_2^2 \left(\alpha_w'(\overline{s}_{21}) \frac{Q_1}{\pi D_2^2} - \varepsilon_w'(\overline{s}_{21}) \sigma T_2^4 \right) \left(\frac{D_1^2}{D_2^2} - \frac{D_3^2}{D_2^2} \right). \tag{15.159}$$

Region 2: vapor

$$Q = \pi D_2^2 \Big([1-\alpha_v'(\overline{s}_{22})]\,[1-\alpha_w'(\overline{s}_{21})] \frac{Q_1}{\pi D_2^2}$$
$$+ [1-\alpha_v'(\overline{s}_{22})]\, \varepsilon_w'(\overline{s}_{21})\, \sigma T_2^4 + \varepsilon_v'(\overline{s}_{22})\, \sigma T_1^4 \Big) \left(\frac{D_1^2}{D_2^2} - \frac{D_3^2}{D_2^2} \right) \tag{15.160}$$

$$Q_A = \pi D_2^2 \Big(\alpha_v'(\overline{s}_{22})[1-\alpha_w'(\overline{s}_{21})] \frac{Q_1}{\pi D_2^2}$$
$$+ \alpha_v'(\overline{s}_{22})\, \varepsilon_w'(\overline{s}_{21})\, \sigma T_2^4 - \varepsilon_v'(\overline{s}_{22})\, \sigma T_1^4 \Big) \left(\frac{D_1^2}{D_2^2} - \frac{D_3^2}{D_2^2} \right) \tag{15.161}$$

Region 3: water

$$Q = \pi D_2^2 \left([1-\alpha_w'(\overline{s}_{23})] Q^* + \varepsilon_w'(\overline{s}_{23}) \sigma T_2^4 \right) \left(\frac{D_1^2}{D_2^2} - \frac{D_3^2}{D_2^2} \right) \tag{15.162}$$

$$Q_A = \pi D_2^2 \left(\alpha_w'(\overline{s}_{23}) Q^* - \varepsilon_w'(\overline{s}_{23}) \sigma T_2^4 \right) \left(\frac{D_1^2}{D_2^2} - \frac{D_3^2}{D_2^2} \right) \tag{15.163}$$

with

$$Q^* = [1-\alpha_v'(\overline{s}_{22})][1-\alpha_w'(\overline{s}_{21})] \frac{Q_1}{\pi D_2^2} + [1-\alpha_v'(\overline{s}_{22})] \varepsilon_w'(\overline{s}_{21}) \sigma T_2^4 + \varepsilon_v'(\overline{s}_{22}) \sigma T_1^4 . \tag{15.163}$$

Third zone:

Region 1: water

$$Q = \pi D_2^2 \left([1-\alpha_w'(\overline{s}_{31})] \frac{Q_1}{\pi D_2^2} + \varepsilon_w'(\overline{s}_{31}) \sigma T_2^4 \right) \frac{D_3^2}{D_2^2}, \tag{15.164}$$

$$Q_A = \pi D_2^2 \left(\alpha_w'(\overline{s}_{31}) \frac{Q_1}{\pi D_2^2} - \varepsilon_w'(\overline{s}_{31}) \sigma T_2^4 \right) \frac{D_3^2}{D_2^2}. \tag{15.165}$$

Region 2: vapor

$$Q = \pi D_2^2 ([1-\alpha_v'(\overline{s}_{32})][1-\alpha_w'(\overline{s}_{31})]\frac{Q_1}{\pi D_2^2}$$

$$+[1-\alpha_v'(\overline{s}_{32})]\varepsilon_w'(\overline{s}_{31})\sigma T_2^4 + \varepsilon_v'(\overline{s}_{32})\sigma T_1^4)\frac{D_3^2}{D_2^2}, \tag{15.166}$$

$$Q_A = \pi D_2^2 (\alpha_v'(\overline{s}_{32})[1-\alpha_w'(\overline{s}_{31})]\frac{Q_1}{\pi D_2^2} + \alpha_v'(\overline{s}_{32})\varepsilon_w'(\overline{s}_{31})\sigma T_2^4 - \varepsilon_v'(\overline{s}_{32})\sigma T_1^4)\frac{D_3^2}{D_2^2}. \tag{15.167}$$

15.4.4 Chain of infinite number of *Wigner* cells

Knowing what happens inside a single cell we can now compute the energy that is transported $Q_{T,1}$ and that is absorbed by the vapor $Q_{av,1}$ and by the water $Q_{aw,1}$ if such cells are virtually interacting in a subsequent chain. The *Lanzenberger* solution is given below

$$Q_{T,i} = a_T Q_{T,i-1} + b_T,\ i = 2, n, \tag{15.168}$$

$$Q_{av,i} = a_v Q_{T,i-1} + b_v,\ i = 2, n, \tag{15.169}$$

$$Q_{aw,i} = a_w Q_{T,i-1} + b_w,\ i = 2, n, \tag{15.170}$$

$$Q_{T,2} = a_T Q_{T,1} + b_T, \tag{15.171}$$

$$Q_{T,3} = a_T a_T Q_{T,1} + a_T b_T + b_T, \tag{15.172}$$

$$Q_{T,4} = a_T a_T a_T Q_{T,1} + a_T a_T b_T + a_T b_T + b_T, \tag{15.173}$$

$$Q_{T,i} = a_T^{i-1} Q_{T,1} + b_T \sum_{m=2}^{i} a_T^{m-2}, \tag{15.174}$$

$$Q_{av,i} = a_v Q_{T,i-1} + b_v,\quad i \ge 2, \tag{15.175}$$

$$Q_{av,2} = a_v Q_{T,1} + b_v, \tag{15.176}$$

$$Q_{av,i} = a_v a_T^{i-1} Q_{T,1} + a_v b_T \sum_{m=2}^{i} a_T^{i-2} + b_v,\quad i \ge 3, \tag{15.177}$$

$$Q_{av} = Q_{av,1} + Q_{av,2} + \sum_{i=3}^{n} Q_{av,i} ,$$

$$= Q_{av,1} + a_v Q_{T,1} + b_v + \sum_{i=3}^{n} a_v a_T^{i-1} Q_{T,1} + \sum_{i=3}^{n} a_v b_T \sum_{m=2}^{i} a_T^{i-2} + \sum_{i=3}^{n} b_v$$

$$= Q_{av,1} + (n-1) b_v + a_v \left[Q_{T,1} \left(1 + \sum_{i=3}^{n} a_T^{i-1} \right) + b_T \sum_{i=3}^{n} \sum_{m=2}^{i} a_T^{i-2} \right] . \qquad (15.178)$$

15.4.5 Application of *Lanzenbergers*'s solution

Consider a cloud of equidistantly arranged hot spherical particles. The cloud is flying in a water and gas mixture. Compute the radiated power by a hot spherical particle and its environment. Compute what fraction of this energy is absorbed by water and what in the gas, as a function of the characteristic scales inside the three-phase mixture. Compute the length required for almost complete light absorption. Figure 15.15 presents the total absobtion depth in three-phase particle–vapor–water flow as a function of the particle volume fraction. The gas volume fraction is 1%. The hot particle size is 0.005 m. We see from Figs. 15.14 and 15.15 an important behavior: increasing the particle concentration leads to reduction of the length neccesary for complete absorbtion of the radiation. That is why for cases with large particle concentrations the discrete ordinate method is not necessary and a local in-cell description of the radiated energy is sufficient. This is not the case for small particle concentrations where the absorbtion length can be considerably larger than the cell size. In this case doing so as every thing happens in the cell leads to an overestimation of the local energy transport and therefore to larger local vapor production with all the consequences to the final macroscopic results.

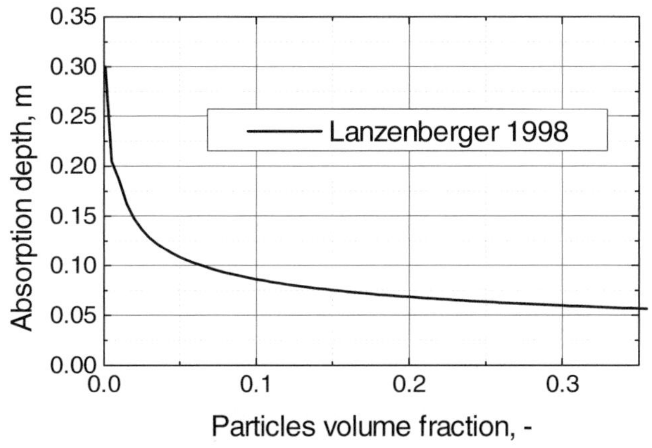

Fig. 15.14 Total absobtion depth in three-phase particle–vapor–water flow as a function of the particle volume fraction. Gas volume fraction 1%. Particle size 0.005 m

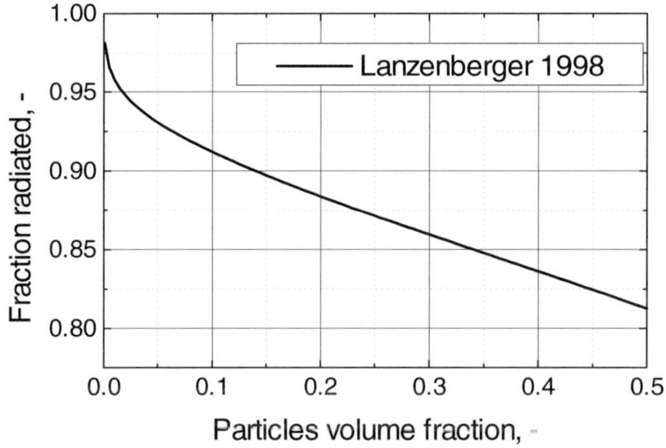

Fig. 15.15 Radiated specific power of the cloud divided by the radiated specific power of a single particle as a function of the partcles volume fraction

Nomenclature

Latin

A_m surface area at side m of the cell, m^2

a_{max} number of the discrete angle in the azimuthal direction

$a_\lambda(\lambda)$ $= \frac{4\pi k}{\lambda}$, spectral absorption coefficient, 1/m

$a_{a,b,m}$ $= \frac{A_m}{\Delta Vol} \frac{D_{a,b,m}}{\Delta \omega_{a,b}}$, m-surface geometry coefficient, 1/m

b_{max} number of the discrete angles in the polar direction

c_0 $= 2.998 \cdot 10^8$, speed of light in vacuum, m/s

c speed of light, m/s

$D_{a,b,m}$ $\iint\limits_{\Delta\omega_{a,b}} (\mathbf{s} \cdot \mathbf{e})_m \, d\omega$, scalar product of the (a,b)-directed unit vector and the outward directed m-surface unit vector integrated over the angle $\Delta\omega_{a,b}$, sr.

D_3 diameter, m

$\mathbf{e}$ $= (e_x, e_y, e_z)$, outward directed unit normal to a surface

$i'_\lambda(\mathbf{r}, \mathbf{s})$ spectral radiation intensity at the spatial location $\mathbf{r}$ along the direction $\mathbf{s}$ that is within an infinitesimal solid angle $d\omega$, in other words heat flux per unit solid angle, per unit wavelength, and per unit area, W/(m^3 sr)

k index of absorption

n index of refraction – ratio of propagation velocity c_0 in vacuum related to the propagation velocity c inside the medium

$\dot{q}''_{\lambda,m-incident}$ incident radiation heat flux from the inside of the computational cell, W/m²

$\mathbf{r}$ position vector, m

r radius, m

$\mathbf{s}$ direction vector

$T_{w,m}$ wall temperature at the m-face of the control volume, K

x x-coordinate, m

y y-coordinate, m

z z-coordinate, m

Greek

Δ	finite difference
ΔVol	cell volume, m³
β_λ	$= a_\lambda + \sigma_{s\lambda}$, spectral extinction coefficient, 1/m
ε_w	wall emissivity coefficient
Φ	spectral scattering phase function
φ	azimuthal angle, rad
$\sigma_{s\lambda}$	spectral scattering coefficients, 1/m
σ	$= \dfrac{2C_1\pi^5}{15C_2^4} = 5.67051\times10^{-8}$, *Stefan–Boltzmann* constant, W/(m² K⁴)
θ	polar angle, rad
ω	solid angle, rad
$d\omega$	infinitesimal solid angle, rad

Indices

a	integer index for the discrete angle in the azimuthal direction
b	integer index for the discrete angle in the polar direction
b	black body
w	wall
x	in *x*-coordinate
y	in *y*-coordinate
z	in *z*-coordinate
1	vapor, gas
2	water
3	melt, solid particle

References

Docherty, P.: Prediction of gas emissivity for wide range of process conditions. In: Proc. of the 7th Heat Transfer Conference, Munich, vol. 2, pp. 481–485. Hemisphere Publ. Corp., Washington (1982)

Fiveland, W.A.: Three-dimensional radiative heat transfer solutions by discrete-ordinates method. J. Thermophys. Heat Transfer 2(4), 309–316 (1988)

Fletcher, D.F.: Assessment and development of the Bankoff and Han coarse mixing model, Culham Laboratory Report, CLM-R252 (1985)

Fletcher, D.F.: Radiation absorption during premixing. Nucl. Eng. Des. 189, 435–440 (1999)

Hale, G.M., Marvin, R.Q.: Optical constants in water in the 200-nm to 200-μm wavelength region. Appl. Opt. 12(3), 555–563 (1973)

Kolev, N.I.: Verification of IVA5 computer code for melt–water interaction analysis, Part 1: Single phase flow, Part 2: Two-phase flow, three-phase flow with cold and hot solid spheres, Part 3: Three-phase flow with dynamic fragmentation and coalescence, Part 4: Three-phase flow with dynamic fragmentation and coalescence – alumna experiments. In: Proc. of the Ninth International Topical Meeting on Nuclear Reactor Thermal Hydraulics (NURETH-9), San Francisco, California, Log. Nr. 315, October 3-8 (1999)

Lanzenberger, K.: Thermal radiation in multiphase flow - Application to the severe accident scenario of molten fuel coolant interaction (MFCI), Siemens AG, Power Generation (KWU), Karl Wirtz Prize for young scientists (1997)

Mathews, K.A.: On the propagation of rays in discrete ordinates. Nucl. Sci. Eng. 132, 155–180 (1999)

Siegel, R., Howell, J.R.: Thermal radiation heat transfer, 3rd edn. Taylor & Francis, Washington (1996)

Steward, F.R., Kocaefe, Y.S.: Total emissivity and absorptivity for carbon dioxide, water vapor and their mixtures. In: Proc. 8th Heat Transfer Conf., San Francisco, vol. 2, pp. 735–740. Springer, Berlin (1986)

Vaeth, L.: Radiative heat transfer for the transient three-phase, three-component flow model IVA-KA, Forschungszentrum Karlsruhe, Internal report, INR-1914, PSF-3212 (1995)

Zolotarev, V.M., Dyomin, A.V.: Optical constants of water on wide wavelength range 0.1 A to 1 m. Opt. Spectrosc. 43(2) (1977)

16. Validation of multiphase flow models

Engineering science is associated with solving the right equations with known accuracy. But how to judge whether the equations are right and how to estimate the accuracy of the solution?

16.1 Introduction

Creating computer codes for numerical integration of systems of partial differential equations governing multiphase multicomponent flows is an important part of scientific activities in the last 30 years. Whatever the result of such activity is, it has to be validated for a class of practical applications. The validation is a complex process consisting of many elements. The first question asked in this case is: is the adequate equation set solved properly? To answer the question several iteration steps are necessary. The formal check whether the numerical algorithm provides the expected results is usually performed on simplified mathematical benchmarks that possess analytical solutions. Order-of-magnitude estimates and global balances that have to be fulfilled are very important in more complex cases. The verification of computer codes for more complex processes is mainly associated with an experimental data base on two different levels. The closure correlations are usually based on separate effect experiments. The system behavior in which a variety of the elements of the computer code are interacting is checked on system experiments.

There is no absolute proof for the validity of a computer code solving complex problems in the multiphase dynamics. There is only increasing trust in the results obtained by an increasing number of successful predicted experiments.

Note that the falsification for a given class of problems is much easier than the verification. That is why this critical step has to be taken first. Then verification in general is not possible because of the variety of complex combinations between processes. The so-called fitness for purpose strategy is the common one used in science and technology. This strategy is usually performed after the basic formal tests and the separate effect tests are completed. It contains selection of experimental data for the class of problems of interest and comparisons of the code predictions with them. In the case of success, or after removing misconceptions and errors and then reaching successful prediction, the code is considered to be verified for this particular class of applications. There is an enormous amount of experimental activity worldwide for collecting experimental data for different applications. The review of this process is beyond of the scope of this chapter. This chapter provides a small number of selected analytical benchmarks and experimental tests and gives an example of how predictions of the computer code IVA compare with these data. The code is based on the methods described in the five volumes of this monograph.

16.2 Quick look at IVA computer code

Transient 3D-multiphase flows consisting of many chemical constituents in nature and technology (Figs. 16.2.1 and 16.2.2) are the common case of flows. In many technical applications we have to deal with particular realization of the multiphase flows like steady-state flows, or single-component flows or single-phase flows, etc.

Fig. 16.2.1 The fascinating picture of the discovery start, the peace of the universe, the tornado, the volcano, the flows in the human heart or even the "pure" water or the sky in the picture of Van Gogh are in fact different forms of multiphase flows

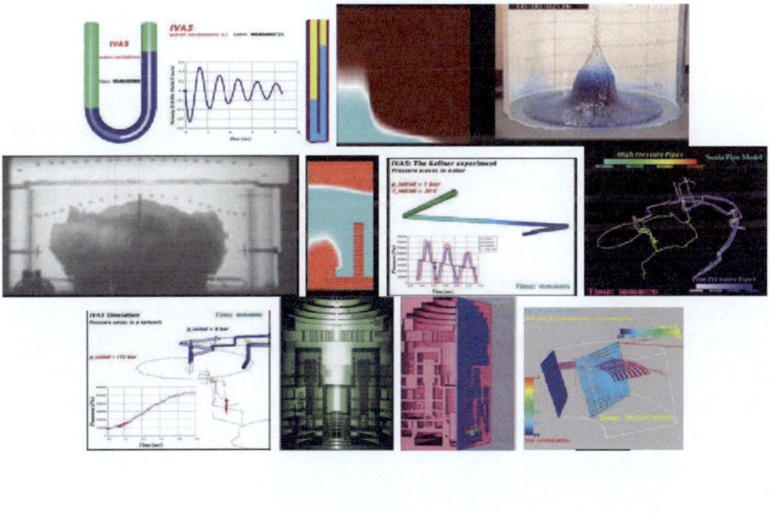

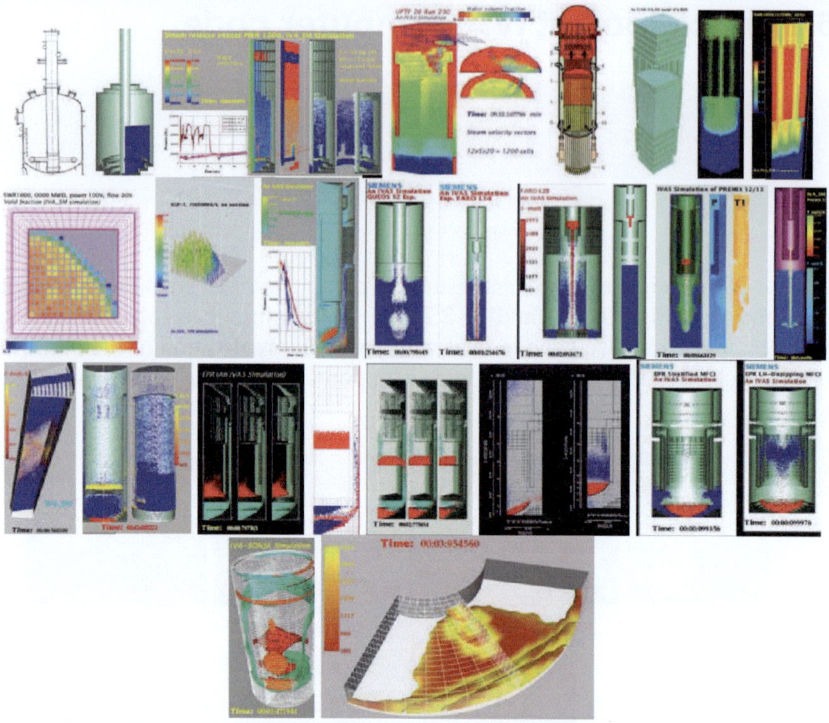

Fig. 16.2.2 Examples of multiphase flows in the nuclear technology, see http://www.herzovision.de/kolev-nikolay/

IVA is a computer code for computational simulation of transient multiphase flows based on three multicomponent velocity fields in porous structure that may change its geometry in time. The foundations of the computer code IVA allows applications for mathematical simulation of variety of processes. From two-phase gas-plasma multicomponent hydrogen detonation in a pipe network with dissociation of the gases, Fig. 16.2.3, through condensation water–steam shock waves in complex pipe networks, Fig. 16.2.4, gas solution and dissolution in liquids, dissolved gas release from water in pipe network and gas-slug formation and transport, Fig. 16.2.5, diesel injection problems, Fig. 16.2.6, particles sedimentation in water, Fig. 16.2.7, turbulent mixing and transport in a NPP sump, Fig. 16.2.8, thermite injection by high-pressure steam–hydrogen mixture into air environment, melt–water interaction in postulated SWR 1000 severe accidents, alumina melt jet dropped into a subcooled water, urania melt jet dropped in water, Fig. 16.2.9, void formation in existing-, Fig. 16.2.10, or future boiling water reactors, Fig. 16.2.11, all this demonstrates the capability of a single model architecture to handle different material systems, different intensities of interactions, and a large variety of the spatial and temporal scales of the simulated processes.

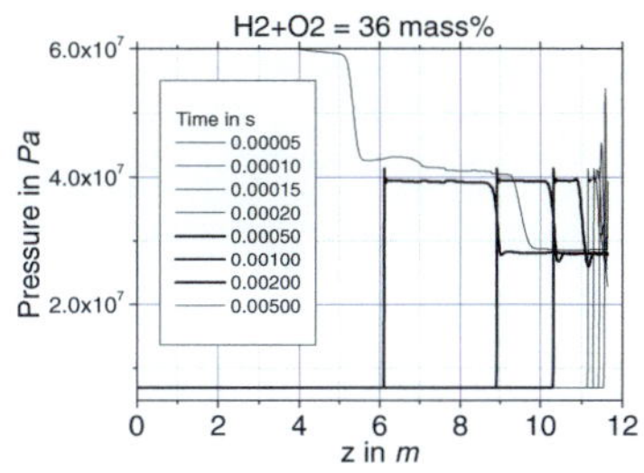

Fig. 16.2.3 Pressure as a function of the distance for different times after the ignition [1.11] p. 432

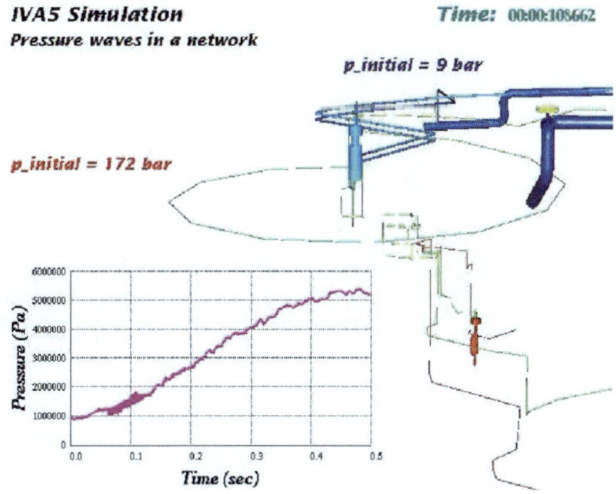

Fig. 16.2.4 Tube ruptures into a main recirculation pump seal water heat exchanger

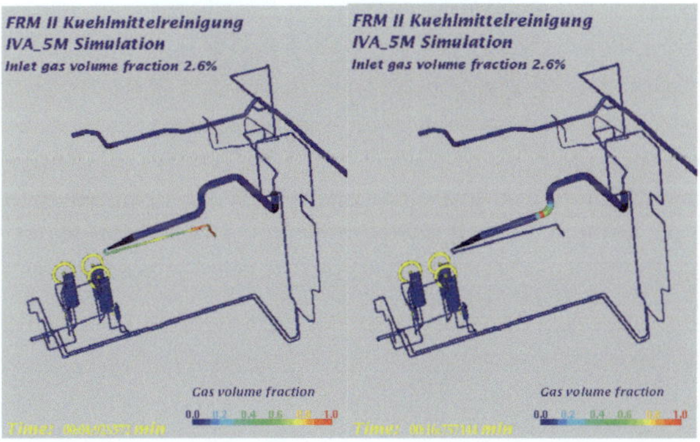

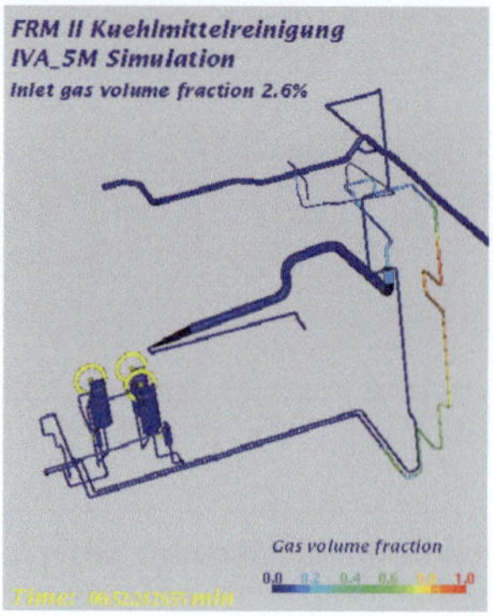

Fig. 16.2.5 Gas release from water in pipe network – gas-slug formation and transport

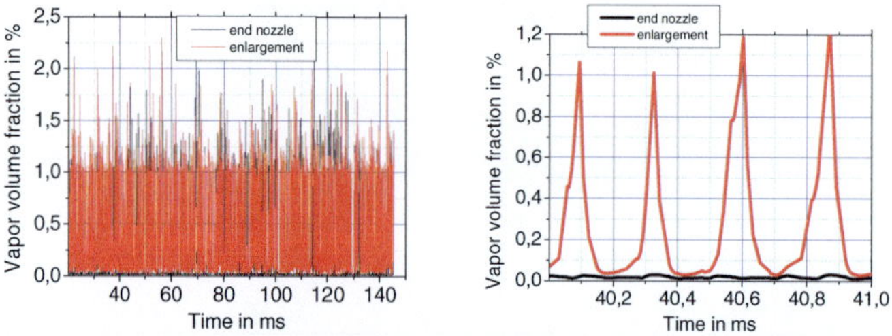

Fig. 16.2.6 Diesel vapor volume fraction at the end of a 300-μm diesel injection nozzle and at the enlargement immediately after the nozzle. The collapses damage the surface

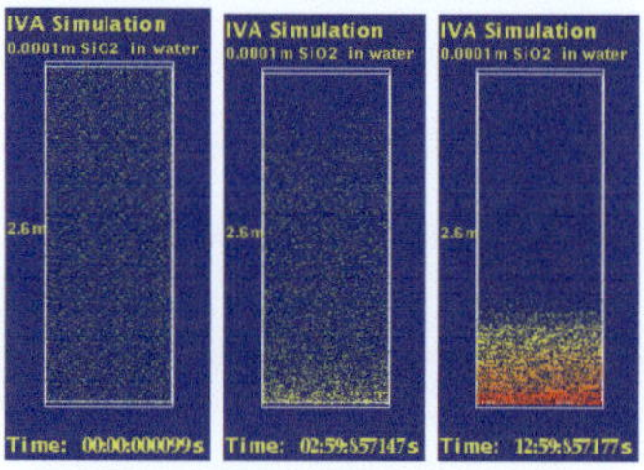

Fig. 16.2.7 Sedimentation of 100-μm sand particles (SiO_2). a) Initial state of 10% volume fraction of initially homogeneously distributed in a 2.6-m deep water pool; b) after 3 min; c) 13 min

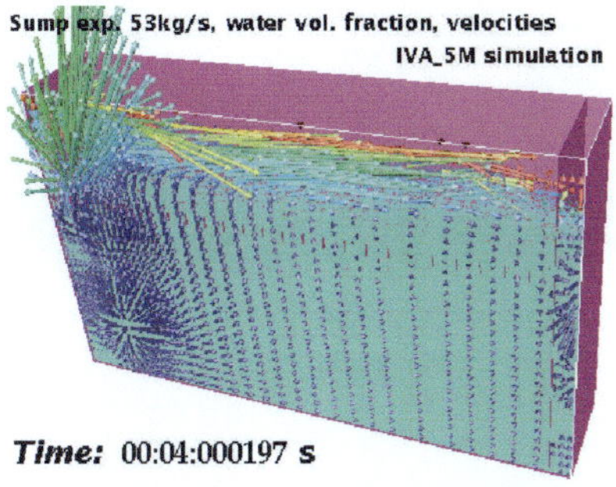

Fig. 16.2.8 Free fall of two phase jet into the left end of a water reservoir causing sloshing and mixing. Vortex burst and reflection at the wall

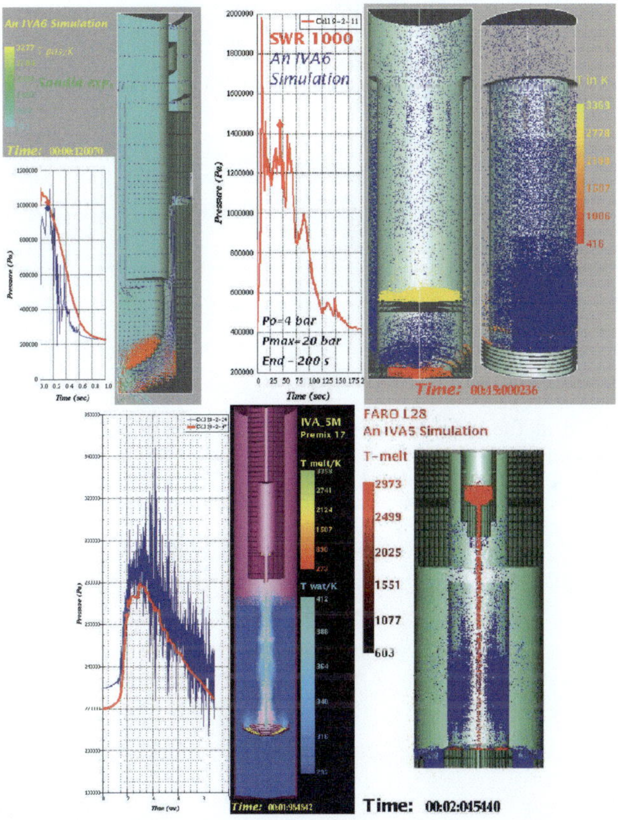

Fig. 16.2.9 a) Thermite injection by high-pressure steam–hydrogen mixture into air environment; b) Melt–water interaction in postulated SWR 1000 severe accidents; c) Alumina melt jet dropped into a subcooled water; d) Urania melt jet dropped in water

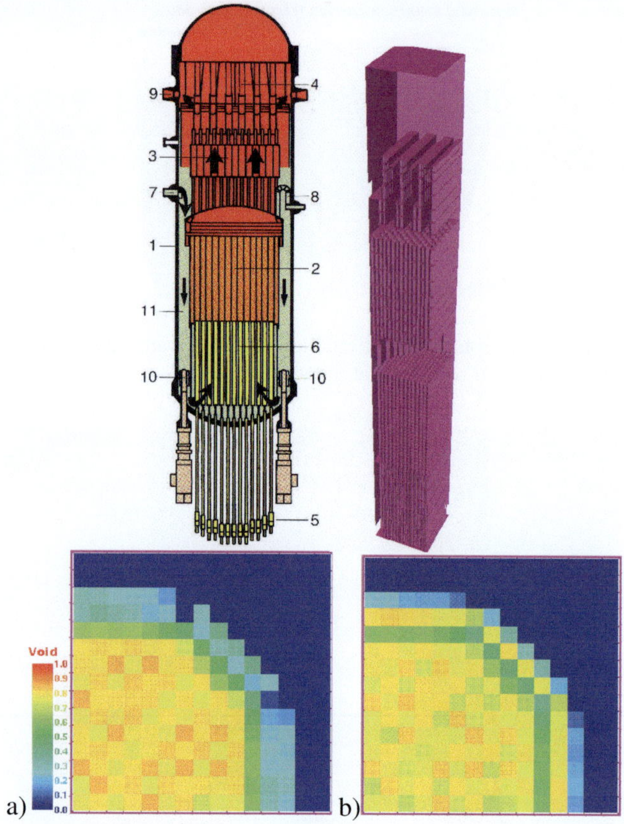

Fig. 16.2.10 Void at the exit of a core of BWR for the case (a) without additional nozzles and (b) with nozzles

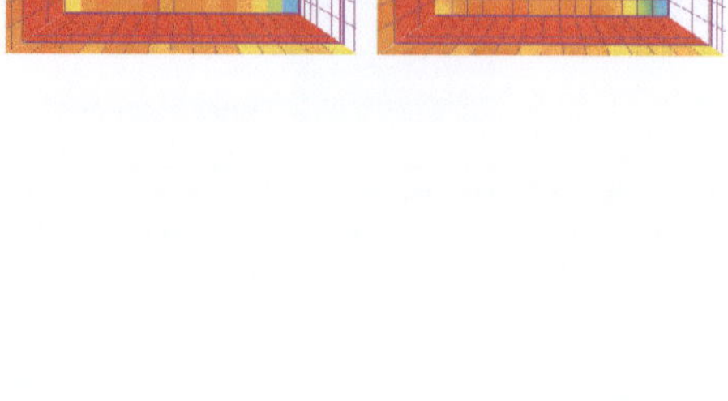

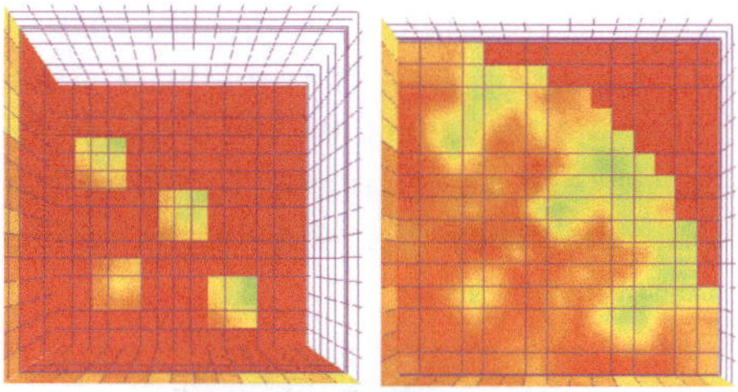

Fig. 16.2.11 Void fraction at different elevation inside a SWR 1000, case studies for different geometry

As already mentioned the flow in IVA model is described by means of three velocity fields. Each of the fields consists of several chemical components.

The first field presents gas mixtures of up to 13 real gases in addition to water vapor. The water-vapor properties are valid up to 2000 bar pressure and 6000 K temperature. The thermal dissociation of the vapor is taken into account. The gases may be air, nitrogen, helium, oxygen, carbon dioxide, methane, carbon oxide, hydrogen, ethane, propane, n-butane, sulfur dioxide and argon, in arbitrary combination. The number and the kind of the appearing components have to be specified as a part of the input. The components are allowed to move by turbulent or molecular diffusion inside the gas field in addition to the macro-scale movement in the space of the gas field. Hydrogen combustion single-step kinetics is available.

The gas components O_2, N_2, H_2 are allowed to be solved or dissolved by the liquid water component in the fields 2 and 3.

The second and the third velocity fields consist of liquid water and inert components of different species. In the limiting case of absence of water in one of these fields or in both of them the inert components are allowed to be a material either in a molten state, or in a solid–liquid equilibrium state, or in liquid state. One of the following inert materials may be chosen: uranium dioxide, nuclear reactor corium (mixture consisting of 76% UO_2, 24% ZrO_2), zirconium, zirconium dioxide, stainless steel, aluminum dioxide, silicon dioxide, iron oxide, molybdenum, aluminum and boron oxide. If water is present in the fields 2 or 3 the inert component inside this field is solid, being in thermal equilibrium with the water (having the temperature of the water) and is allowed to move by turbulent or molecular diffusion inside the corresponding fields in addition to the macro-scale movement in the space. An example of such situation is boron oxide and water mixture used frequently in the nuclear rector technology.

Each of the three velocity field possesses its own velocity and temperature. Thus, the flow is in a complete thermodynamical and mechanical nonequilibrium with all consequences for interfacial heat, mass and momentum transfer.

The fields are allowed to be continuous or disperse. The transition of the fields between continuum and dispersion as well as the local size of the dispersions are modeled by means of dynamic fragmentation and coalescence techniques.

$k-\varepsilon$ models for each field are available for modeling of multiphase flow turbulence. For the time being they are used for the continuous field with two-ways coupling to the dispersed phase.

Some special capabilities like computing of critical heat flux for bundles using a simplified method and successive initialization of the 3D model for refining the analysis is available. Another special feature is the modeling of cells as a cyclone and labyrinth-type separators with predefined characteristics.

The geometry of the space within which the flow is simulated is described either in Cartesian or cylindrical coordinates in a single block.

Another capability of the code is the description of multiphase flows in multiple interconnected blocks with boundary-fitted orthogonal grids.

In addition, pipe networks are simulated consisting of an arbitrary number of pipes and components like, pumps, valves, etc. The pipe network is connected with the three-dimensional space.

Local surface permeabilities in the three main directions and local volumetric porosities are defined as a function of time as for the 3D space. The same is valid also for the pipe network.

Inside the blocks different kinds of structure types can be simulated like heat-conducting 3D structures and/or heat transmitting 1D structures. The heat conducting 3D structures my have internal heat sources prescribed by the user. The structures are connected to the 3D flow in the 3D space, or to the pipe flow by complete heat-transfer mechanism including all known heat-transfer regimes to single-, two- and multiphase flows. Special kinds of structure like a nuclear reactor core can also be defined and simulated.

The numerical method used for integration of the system of the governing partial differential equations is: first-order donor-cell discretization for the convective terms, second-order central differencing for the diffusion terms, first-order time discretization, implicit. The method is characterized by a strong coupling between the velocity fields obtained by analytical reduction of the algebraic problem to a pressure equation and successive substitutions.

The code possesses powerful visualization systems, SONJA, for input processing, post-processing, on-line visualization and movie production.

IVA is written in FORTRAN 95 language and can run on all modern computers and platforms having this compiler. SONJA is written in C and works on platforms having this compiler and OpenGL. These may be either UNIX workstations or PCs under LINUX.

References documenting the IVA computer code development are given below.

16.3 References documenting the IVA-code development

1.1. Kolev NI (May 17–19 1984) Model of transient equilibrium three phase three component flow, 6-th National Conf. "Heat and Nuclear Energy Problems of People's Republic of Bulgaria" held in Varna, Bulgaria, p 37–42, in Bulgarian

1.2. Kolev NI (June 23–27, 1985) Transient three-phase three-component non equilibrium non homogeneous flow described by 1D-three-velocity fields slip model, Proc. Specialists Meeting on Small Break LOCA Analysis in LWRs, Pisa, vol 3, p 363–378

1.3. Kolev NI (1985), Transient one-dimensional equilibrium three-phase, three-component flow, Atomkernenergie/Kerntechnik, vol 47 no 3, p 198–203

1.4. Kolev NI (1986) Transient three phase three component non equilibrium non homogeneous flow, Nucl. Eng. Des., vol 91 p 373–390

1.5. Kolev NI (1986) Transient three-dimensional three-phase three-component nonequilibrium flow in porous bodies described by three-velocity fields, Kernenergie, vol 29 no 10, p 383–392. Also in (March 1985) Transiente Drephasen Dreikomponenten Stroemung", Teil 1: Formulierung des Differentialgleichungssystems, KfK 3910, in German

1.6. Kolev NI (1987) Transient three-phase three-component non equilibrium non homogeneous flow described by 1D-three-velocity fields slip model, Atomkernenergie/Kerntechnik, vol 49 no 4, p 217–221. Also in (August 1985) Transiente Dreiphasen Dreikomponenten Stroemung", Teil 2: Eindimensionales Schlupfmodell Vergleich Theorie-Experiment, KfK 3926, in German

1.7. Kolev NI (August 1987) A three field-diffusion model of three-phase, three-component flow for the transient 3D-computer code IVA2/01, Nuclear Technology, vol 78, p 95–131. Also in (1986) Transiente Dreiphasen Dreikomponenten Stroemung, Teil 3: 3D-Dreifluid-Diffusionsmodell, KfK 4080, in German, Also in (April 27–30, 1987) Proc. of the Int. Top. Meeting on "Advances In Reactor Physics, Mathcmatics And Computation", Paris, France, vol 3, p 1899–1911

1.8. Kolev NI (1987) IVA-2 A computer code for modeling of three dimensional, three-phase, three-component flow by means of three velocity fields in cylindrical geometry with arbitrary internal structure including nuclear reactor core, Atom Energy, vol 63, pp 381–386. Also in (October 1986) Proc. of the int. seminar "Thermal physics 86" held in Rostock, German Democratic Republic, in Russian. Also in (September 26–30, 1988) Proc. of the Int. Seminar "Thermal Physics 88" held in Warsaw, Poland, in Russian. Also in (1986) KfK 4088, in German

1.9. Kolev NI (October 1988) IVA2 Verification: High pressure gas injection in a liquid pool, Nuclear Technology, vol 83 no 1 p 65–80. Also in (4–9 September 1988), Proc. of thc First World Conf. Experimental Heat Transfer, Fluid Mechanics And Thermodynamics, Dubrovnik, Yugoslavia. Shah RK, Ganic EN, Yang KT eds, p 827–881. Also in (1987) KfK 4310, September in German

1.10. Kolev NI (March 23–24, 1989) IVA2 Application: Three dimensional effects during APWR LOCA, 7th Eurotherm Seminar, Thermal Non-Equilibrium in Two-Phase Flow, Rome, p 235–252. Also in (September 18–21, 1989) Multiphase Flow and Heat Transfer, Second International Symposium, Sian, China. Xue-Jun Chen, Veziroglu ZN, Tien CL eds., vol 2 p 949–958. Also in (September 1987) KfK 3419

1.11. Kolev NI (1989) A practical description of three-phase multi-component nonequilibrium flow by means of one dimensional multi-velocity field slip model, Kerntechnik, vol 54 no 3 p 186–190

1.12. Kolev NI (1990) Derivatives for the state equations of multi-component mixtures for universal multi-component flow models, Nucl. Sci. Eng., vol 108 p 74–87

1.13. Kolev NI (Sept. 1991) A three-field model of transient 3D multi-phase, three-component flow for the computer code IVA3, Part 1: Theoretical Basics: Conservation and state equations, numerics, KfK 4948, Kernforschungszentrum Karlsruhe

1.14. Kolev NI (Sept. 1991) A three-field model of transient 3D multi-phase, three-component flow for the computer code IVA3, Part 2: Models for the interfacial transport phenomena, code validation, KfK 4949, Kernforschungszentrum Karlsruhe

1.15. Kolev NI (Sept. 1991) IVA3: Computer code for modeling of three dimensional three phase flow in complicated geometry, Program Documentation: Input Description, KfK 4950, Kernforschungszentrum Karlsruhe

1.16. Kolev NI, Tomiyama A, Sakaguchi T (1991) Modeling of the mechanical interaction between the velocity fields in three phase flow, Exp. Thermal Fluid Sci., vol 4 no 5 p 525–545

1.17. Kolev NI (June 23–28, 1991) IVA3 Verification: Drag forces for three phase flow - comparison with experimental data, Proc. of the First World Conf. On Experimental Heat Transfer, Fluid Mechanics and Thermodynamics, Dubrovnik, Yugoslavia

1.18. Kolev NI (July 21–25, 1991) IVA3, A transient 3D three-phase, three-component flow analyzer, Proc. of the Int. Top. Meeting on Safety of Thermal Reactors, Portland, Oregon, p. 171–180. The same presented at (August 27 to 29, 1991) 7th Meeting of the IAHR Working Group on Advanced Nuclear Reactor Thermal-Hydraulics, Kernforschungszentrum Karlsruhe

1.19. Kolev NI (August 27 to 29, 1991) IVA3 Modeling of corium water interaction in deep pool, Presented at 7th Meeting of the IAHR Working Group on Advanced Nuclear Reactor Thermal-Hydraulics, Kernforschungszentrum Karlsruhe

1.20. Kolev NI (1993) Fragmentation and coalescence dynamics in multi-phase flows, Exp. Thermal Fluid Sci., vol 6 p 211–251

1.21. Kolev NI (1993), The code IVA3 for modelling of transient three-phase flows in complicated 3D geometry, Kerntechnik, vol 58 no. 3 p 147–156, Also in (October 5–8, 1993) Proc. of the Sixth Int. Top. Meeting on Nuclear Reactor Thermal Hydraulics, Grenoble, France

1.22. Kolev NI (1994) The influence of the mutual bubble interaction on the bubble departure diameter, Exp. Thermal Fluid Sci., vol 8 p 167–174

1.23. Kolev NI (1994) The code IVA5: Modeling of mass conservation in multi-phase multi component flows in heterogeneous porous media, Kerntechnik, vol 59 no 4–5 p 226–237

1.24. Kolev NI (1994) The code IVA5: Modeling of momentum conservation in multi phase flows in heterogeneous porous media, Kerntechnik, vol 59 no 6 p 249–258

1.25. Kolev NI (1995) The code IVA5: Second law of thermodynamics for multi-phase multi-component flows in heterogeneous media, Kerntechnik, vol 60 no 1 p 28–39

1.26. Kolev NI (1995) The code IVA5: Nucleation and flashing model, Kerntechnik, vol 60 no 2 p 157–164. Also in (April 3–7, 1995) Proc. Second Int. Conf. On Multiphase Flow, Kyoto; (August13–18, 1995) ASME & JSME Fluid Engineering Conference International Symposium on Validation of System Transient Analysis Codes, Hilton Head (SC) USA; (October 9–11, 1995) Int. Symposium on Two-Phase Flow Modeling and Experimentation, ERGIFE Place Hotel, Rome, Italy
1.27. Kolev NI (1995) How accurately can we predict nucleate boiling, Exp. Thermal Fluid Sci., vol 10 p 370–378
1.28. Kolev NI (April 3–7 1995) IVA5 Computer code: The model for film boiling on a sphere in subcooled, saturated and superheated water, Proc. of Second Int. Conference On Multiphase Flow, Kyoto, Japan. Also presented in (November 14–15, 1994) Workshop zur Kühlmittel/Schmelze - Wechselwirkung, Cologne, Germany
1.29. Kolev NI (April 23–27, 1995) IVA5 computer code: Dynamic fragmentation model for liquid and its application to melt–water interaction, Proc. ICONE-3, The Third International Conf. on Nucl. Engineering, "Nuclear Power and Energy Future", Kyoto, Japan. Also presented in (November 14–15, 1994) Workshop zur Kühlmittel/Schmelze - Wechselwirkung, Cologne, Germany
1.30. Stewart HB (November 2, 1994) Assessment of the IVA3 code for multifield flow simulation, Brookhaven Nation Laboratory, Report BNL 52473 FZ Karlsruhe Report FZKA 5591 (July 1995)
1.31. Vaeth L (March 1995) Radiative heat transfer for transient three-phase, three-component flow model IVA-KA, Forschungszentrum Karlsruhe, Internal Report 32.21.02/10A
1.32. Jacobs H (January 9–13 1993) Analysis of large-scale melt-water mixing events, Proc. CSNI Specialists Meeting of Fuel-Coolant Interactions, Santa Barbara, CA. U.S. Nuclear Regulatory Commission Report NUREG/CP-0127, NEA/CSNI/R(93) (March 1994), p 14–26
1.33. Jacobs H, Lummer M, Meyer L, Stehle B, Thurnay K, Väth L (March 9–13, 1995) Multifield simulations of premixing experiments, NSFIJSPS Seminar, Santa Barbara
1.34. Meyer L, Schumacher G, QUEOS, a simulation experiment of the premixing phase of steam explosion with hot spheres in water base case experiments, Forschungszentrum Karlsruhe, FZKA 5612
1.35. Kolev NI (1995) IVA5 computer code: Modeling of melt-water droplet interaction for very high melt temperatures, unpublished. Later some of this material is used in volume two of this monograph
1.36. Kolev NI and E Hofer (August 1996) Uncertainty and sensitivity analysis of post experimental IVA5 simulations of melt water interaction, Exp. Thermal Fluid Sci., vol 13 p 98–116
1.37. Kolev NI (October 15–16, 1996) Three fluid modeling with dynamic fragmentation and coalescence - fiction or daily practice? 7th FARO Experts Group Meeting Ispra; (5th–8th November 1996) Proceedings of OECD/CSNI Workshop on Transient thermal-hydraulic and neutronic codes requirements, Annapolis, Md, U.S.A.; (June 2–6, 1997) 4th World Conference on Experimental Heat Transfer, Fluid Mechanics and Thermodynamics, ExHFT 4, Brussels; (June 22–26, 1997) ASME Fluids Engineering Conference & Exhibition, The Hyatt Regency Vancouver, Vancouver, British Columbia, CANADA Invited Paper; (May 22–24, 1997) Proc. of the International Seminar on Vapor Explosions and Explosive Eruptions (AMIGO-IMI), Aoba Kinen Kaikan of Tohoku University, Sendai-City, Japan

1.38. Kolev NI (19th–21st May 1997) Verification of the IVA4 film boiling model with the data base of Liu and Theofanous, Proceedings of OECD/CSNI Specialists Meeting on Fuel-Coolant Interactions (FCI), JAERI-Tokai Research Establishment, Japan
1.39. Kolev NI (Japan 19th–21st May 1997) Numerical modeling of in-vessel melt water interaction in large scale PWRs, Proceedings of OECD/CSNI Specialists Meeting on Fuel-Coolant Interactions (FCI), JAERI-Tokai Research Establishment
1.40. Kolev NI (May 22–24, 1997) IVA4 Analysis of the FARO L14 Experiment (ISP 39), Proceedings of the International Seminar on Vapor Explosions and Explosive Eruptions (AMIGO-IMI), Aoba Kinen Kaikan of Tohoku University, Sendai-City, Japan
1.41. Kolev NI (May 26–30, 1997) Is large melt water mixing possible under low pressure, Proceedings of ICON 5: 5th International Conference on Nuclear Engineering, Nice, France
1.42. Kolev NI (June 2–6, 1997) Film boiling: vertical plates, Proceedings of 4th World Conference on Experimental Heat Transfer, Fluid Mechanics and Thermodynamics EXHFT 4, Brussels, Belgium
1.43. Kolev NI (1997) Comments on the entropy concept, Kerntechnik, vol 62 no 1 pp 67–70
1.44. Kolev NI (1998) On the variety of notation of the energy conservation principle for single phase flow, Kerntechnik, vol 63 no 3 pp 145–156
1.45. Kolev NI (1998) Film boiling on vertical plates and spheres, Exp. Thermal Fluid Sci., vol 18 pp 97–115
1.46. Kolev NI (August 1999) In-vessel melt–water interaction caused by core support plate failure under molten pool, Part1: Choice of the solution method. Kerntechnick vol 64 no 4 pp 209–215
1.47. Kolev NI (1999) In-vessel melt–water interaction caused by core support plate failure under molten pool, Part 2: Analysis, Kerntechnick vol 64 no 5 pp 278–283
1.48. Kolev NI (2000) Detonation waves in melt-coolant interaction, Part.1: Theory, Kerntechnik vol 65 no 5–6 pp 254–260. Also in EU Nr. INV-MFCI(99)-D038
1.49. Kolev NI and Hulin H (2001) Detonation waves in melt-coolant interaction, Part. 2: Applied analysis, Kerntechnik vol 66 no 1–2 pp 21–25. Also in MFCI Project, 6th progress meeting, CEA, Grenoble, 23–24 June 1999, EU Nr. INV-MFCI(99)-D038
1.50. Kolev NI, Chen T, Kollmann T, Schlicht G (June 8–12, 1998) Visual multi-phase flow analysis, Third International Conference On Multiphase Flow, Lyon, France
1.51. Kolev NI (17 December 1998) Verification of IVA5 computer code for melt–water interaction analysis, 5th MFCI Project Meeting, Forschungszentrum Karlsruhe, Germany 1998
1.52. Magallon D, Berthoud G, Schuetz W, Alemberti A, Seghal R, Buerger M, Colombo G, Turland B, Zero S (Nov. 1997) Characterization of processes which govern quenching of molten corium in water including steam explosions, Proc. FISA-EU research on severe accidents, Luxemburg, EUR 18258 EN (1998)
1.53. Kolev NI (April 19–23, 1999) Verification of IVA5 computer code for melt–water interaction analysis, Part 1: Single-, two-, and three-phase flows without melt fragmentation, Proc. of the 7th International Conference on Nuclear Engineering, Tokyo, Japan, ICONE-7359
1.54. Kolev NI (April 19–23, 1999) Verification of IVA5 computer code for melt–water interaction analysis, Part 2: Three-phase flows with melt fragmentation, Proc. of the 7th International Conference on Nuclear Engineering, Tokyo, Japan, ICONE-7507

1.55. Kolev NI (April 19–23, 1999) In-vessel melt–water interaction caused by core support plate failure under molten pool, Proc. of the 7th International Conference on Nuclear Engineering, Tokyo, Japan, ICONE-7360
1.56. Kolev NI (17 December 1998) Ex-vessel melt–water interaction caused by symmetric lower head unzipping within the reactor pit, 5th MFCI Project Meeting, Forschungszentrum Karlsruhe, Germany. Slightly modified in (April 19–23, 1999) Proc. of the 7th International Conference on Nuclear Engineering, Tokyo, Japan, ICONE-7361. Extended version in (October 3–8,1999) Proc. of the Ninth International Topical Meeting on Nuclear Reactor Thermal Hydraulics (NURETH-9), San Francisco, California, Log. Nr. 317
1.57. Kolev NI (August 15–17, 1999) Verification of IVA5 computer code for melt–water interaction analysis: Part 1 and 2, Experimental Study of Multiphase Flow, Proc. of the 33rd National Heat Transfer Conference, Albuquerque, New Mexico
1.58. Kolev NI (October 3–8,1999) Verification of IVA5 computer code for melt–water interaction analysis, Part 1: Single phase flow, Proc of the Ninth International Topical Meeting on Nuclear Reactor Thermal Hydraulics (NURETH-9), San Francisco, California, Log. Nr. 315_1
1.59. Kolev NI (October 3–8,1999) Verification of IVA5 computer code for melt–water interaction analysis, Part 2: Two-phase flow, three-phase flow with cold and hot solid spheres, Proc. of the Ninth International Topical Meeting on Nuclear Reactor Thermal Hydraulics (NURETH-9), San Francisco, California, Log. Nr. 315_2
1.60. Kolev NI (October 3–8,1999) Verification of IVA5 computer code for melt–water interaction analysis, Part 3: Three-phase flow with dynamic fragmentation and coalescence, Proc. of the Ninth International Topical Meeting on Nuclear Reactor Thermal Hydraulics (NURETH-9), San Francisco, California, Log. Nr. 315_3
1.61. Kolev NI (October 3–8,1999) Verification of IVA5 computer code for melt–water interaction analysis, Part 4: Three-phase flow with dynamic fragmentation and coalescence – alumna experiments, Proc. of the Ninth International Topical Meeting on Nuclear Reactor Thermal Hydraulics (NURETH-9), San Francisco, California, Log. Nr. 315_4
1.62. Kolev NI (1988) In-vessel melt–water interaction caused by core support plate failure under molten pool, Part 1: Choice of the solution method, 5th MFCI Project Meeting, 17 December 1998, Forschungszentrum Karlsruhe, Germany. Also in (October 3–8, 1999) Proc. of the Ninth International Topical Meeting on Nuclear Reactor Thermal Hydraulics (NURETH-9), San Francisco, California, Log. Nr. 316_1. Also in (August 1999) Kerntechnick vol 64 no 4 pp 209–215
1.63. Kolev NI (1998) In-vessel melt–water interaction caused by core support plate failure under molten pool, Part 2: Analysis, 5th MFCI Project Meeting, 17 December 1998, Forschungszentrum Karlsruhe, Germany. Also in (October 3–8,1999) Proc. of the Ninth International Topical Meeting on Nuclear Reactor Thermal Hydraulics (NURETH-9), San Francisco, California, Log. Nr. 316_2. Also in (1999) Kerntechnick vol 64 no 5 pp 278–283
1.64. Kolev NI (5–10 September, 1999) Applied multi-phase flow analysis and its relation to constitutive physics, 8th International Symposium on Computational Fluid Dynamics, ISCFD '99 Bremen, Germany. Also in (April 2000) Japan Journal for Computational Fluid Dynamics, vol 9 no 1 pp 549–561, Invited Lecture
1.65. Magalon D, Huhtiniemi H, Dietrich D, Bertoud G, Valette M, Schuetz W, Jacobs H, Kolev N, Graziosi G, Sehgal R, Buerger M, Buch M, Colombo G, Turland B, Dobson G, Monhardt D (29 Nov.–1 Dec. 1999) Molten fuel coolant interaction (MFCI), FISA-99 Symposium, EC, Jean Monnet Building, Luxemburg

1.66. Kolev NI and Roloff-Bock I (1998) IVA5 Computer code: Relief and back pressing valve models, KWU NA-T/1998/E000, Project R&D
1.67. Kolev NI (1998) Can melt–water interaction jeopardize the containment integrity of the EPR? Part 3: Fragmentation and coalescence dynamics in multi-phase flows, KWU NA-T/1998/E083a, Project EPR
1.68. Lanzenberger K (1996) Comments to the entropy equation and its discretisation in the numerical code IVA4, KWU-NA-M/96/E046b
1.69. Lanzenberger K (1996) IVA4 Computer Code: Material properties of 13 pure gases and their mixtures, KWU-NA-M/96/E017
1.70. Lanzenberger K (1996) IVA4 Computer Code: Mixture properties of core melt, KWU-NA-M/96/E045
1.71. Lanzenberger K (1997) Gas propagation in a reactor containment computed with the system code IVA4, Jahrestagung Kerntechnik '97 in Aachen, Analyse und Beherrschung schwerer Störfälle
1.72. Dombrovsky L (1997) Radiation heat transfer between nuclear fuel rods and in dispersed corium, KWU NA-T/1997/E047
1.73. Lanzenberger K (1998) Thermal radiation in multiphase flow, KWU NA-T/98/E011. Karl Wirtz Prize für Junge Wissenschaftler – 1997: Lanzenberger K, Thermal radiation in multiphase flow - Application to the severe accident scenario of molten fuel coolant interaction (MFCI), Siemens AG, Power Generation (KWU)
1.74. Fjodorow A (1998) Entwicklung eines verbrennungsmodels für den computercode IVA-5, KWU NA-T/1998/101
1.75. Wittmaack R (10.12.1997) SURTRAC, a subprogram to determine surfaces of discontinuity inside of control volumes. KWU NA-T/1997/E062, Erlangen
1.76. Roloff-Bock I (1998) IVA5: Interface description for graphical analysis, KWU NA-T/1998/E093r
1.77. Roloff-Bock I (1998) IVA5 Computer Code: Pipe network input generation, KWU NA-T/1998/E017
1.78. Kolev NI and Roloff-Bock I (1998) IVA5 Computer Code: Relief and back pressing valve model, KWU NA-T/1998/E058
1.79. Roloff-Bock I (1998) IVA4-NW: Interface description for graphical analysis, KWU NA-T/1998/E059
1.80. Roloff-Bock I (1999) IVA5 related programs, KWU NA-T/1999/E040.
1.81. Kollmann T (1999) MovieView – Short Description, Siemens AG, Erlangen
1.82. Kollmann T (1999) MovieTrans – Short Description, Siemens AG, Erlangen
1.83. Kollmann T (1999) DataGraph – Short Description, Siemens AG, Erlangen
1.84. Kolev NI (April 2–6, 2000) Computational analysis of transient 3D-melt–water interactions, Proc. of the 8th International Conference on Nuclear Engineering, Baltimore, Maryland USA, ICONE-8809. Also in (May 22–25, 2000) CFD 2000 in Trondheim Norway. Also in abbreviated form in (November 5–11, 2000) Symposium on "Dispersed Flows in Combustion, Incineration and Propulsion Systems", ASME International Mechanical Engineering Congress& Exposition, Orlando, FL
1.85. Kolev NI (June 11–15, 2000) Needs of industrial fluid dynamics applications, Invited lecture, 2000 ASME Fluids Engineering Division Summer Meeting (FEDSM), Industry Exchange Program, Sheraton Boston Hotel, Boston, Massachusetts,

1.86. Kolev NI, Seitz H and Roloff I (May 27–June 1, 2001) Hot Leg Injection: IVA 3D Versus 1D Three Velocity Fields Modeling and Comparison with UPTF 26 Run 230 Experiment, ICMF-2001, CD-Proceedings of the 4th International Conference on Multiphase Flow, New Orleans, Louisiana, U.S.A. Also in (September 24–28, 2001) Proc. ExHFT-5, 5th World Conference on Experimental Heat Transfer, Fluid Mechanics and Thermodynamics, Thessaloniki, Greece

1.87. Kolev NI (May 29–June 1, 2001) Conservation equations for multi-phase multi-component multi-velocity fields in general curvilinear coordinate systems, 2001 ASME Fluids Engineering Division Summer Meeting (FEDSM), Keynote Lecture, New Orleans, Louisiana

1.88. Kolev NI (April 2–12, 2001) SWR 1000 Severe accident control through in-vessel melt retention by external RPV cooling, 9th International Conference on Nuclear Engineering, Nice, France

1.89. Kolev NI (May 29–June 1, 2001) International cooperation to develop computational fluid dynamic technology: opportunities and obstacles, 2001 ASME Fluids Engineering Division Summer Meeting (FEDSM), New Orleans, Louisiana, Invited Lecture

1.90. Kolev NI (5–10 September 1999) Applied multi-phase flow analysis and its relation to constitutive physics, 8th International Symposium on Computational Fluid Dynamics, ISCFD '99, Bremen, Germany. Also (April 2000) Japan Journal for Computational Fluid Dynamics, vol 9 no 1 pp 549–561, Invited Lecture

1.91. Kolev NI (June 10–13, 2002) IVA_5M numerical method for analysis of three-fluid multi-component flows in boundary-fitted multi-blocks, The 40th European Two-Phase Flow Group Meeting, Stockholm, Sweden

1.92. Kolev NI (April 14–18, 2002) Simple analytical U-tube benchmarks appropriate for testing of pipe-network computer codes, Proceedings of ICONE10: Tenth International Conference on Nuclear Engineering - Arlington, Virginia (Washington, D.C.), USA ICONE10-22711

1.93. Kolev NI (2002, 2004) Multiphase Flow Dynamics, Vol. 1 Fundamentals, 2nd ed., with 114 Figures and CD-ROM Springer, Berlin, New York, Tokyo, ISBN 3-540-2206-0, see the content in http://www.springeronline.com/east/3-540-22106-9

1.94. Kolev NI (2002, 2004) Multiphase Flow Dynamics, Vol. 2 Thermal and mechanical interactions, 2nd ed. with 81 Figures, Springer, Berlin, New York, Tokyo, ISBN 3-540-22107-7, see the content in http://www.springeronline.com/east/3-540-22107-7

1.95. Kolev NI (April 20–23, 2003) Controlling the moisture content in the steam by using nozzles in BWR'S, Eleventh International Conference on Nuclear Engineering, ICONE 11, Keio Plaza Inter-Continental, Shinjuku, Tokyo, Japan, ICONE11-36306; ETPFG, European Two-Phase Flow Group Meeting 2003 Norway; The 10th International Topical Meeting on Nuclear Reactor Thermal Hydraulics (NURETH-10) Seoul, Korea, October 5–9, 2003.

1.96. Kolev NI (27–29 November 2002) 3D-Multiphase-Fluid-Dynamik: Reduzierung der Dampffeuchte am RDB-Austritt der Siedewasserreaktoren, Betriebsleitertagung, Geneve; in English: Controlling the moisture content in the steam by using nozzles in BWRs, Eleventh International Conference on Nuclear Engineering, ICONE 11, Keio Plaza Inter- Continental, Shinjuku, Tokyo, Japan, April 20–23, 2003, ICONE11-36306, The 10th International Topical Meeting on Nuclear Reactor Thermal Hydraulics, Seoul, Korea, October 5–9, 2003, E00301

1.97. Kolev NI (17–20 June 2003) IVA_5M numerical method for analysis of three-fluid multi-component flows in boundary-fitted multi-blocks, Second M.I.T. Conference on Computational Fluid and Solid Mechanics, Boston

1.98. Kolev NI (5–9 October 2003) Analysis pf the PREMIX 17 and 18 experiments with IVA_5M computer code, The 10th International Topical Meeting on Nuclear Reactor Thermal Hydraulics, Seoul, Korea, E00106

1.99. Kolev NI (June 13–17, 2004) External cooling - the SWR 1000 severe accident management strategy, Proceedings of ICONE-12 '04 Arlington VA, USA, April 25–29, 2004, Paper ICONE12-49055, Presented first as SWR 1000 In-Vessel Melt Retention, STUK Meeting held at 13.8.2003 in Helsinki, Finland; (Nov.17–18, 2005) European BWR Forum, 11th Seminar on SWR1000 Design Features, Framatome ANP, Offenbach, Germany

1.100. Kolev NI (April 25–29, 2004) 25-Years three-fluid modeling-experience: successes and limits, Proceedings of ICONE-12 '04 Arlington VA, USA, Paper ICONE12-49054. Also in 4th ECCOMAS (European Congress on Computational Methods in Applied Sciences and Engineering, Finland July 24–28, 2004. Also in 6th WCCM (World Congress of Computational Mechanics) Beijin Sept. 5–10 2004. In extended form in 6th International Topical Meeting on Nuclear Reactor Thermal Hydraulics, Operations and Safety, October 4–8, 2004, Nara-Ken New Public Hall, Nara, Japan, NUTHOS6-N6P380

1.101 Kolev NI (October 4–8, 2004) External cooling - the SWR 1000 severe accident management strategy, Part 2: Analysis: vessel-, penetrations-, and containment integrity during the external cooling, 6th International Topical Meeting on Nuclear Reactor Thermal Hydraulics, Operations and Safety, Nara-Ken New Public Hall, Nara, Japan, NUTHOS6-N6P382

1.102 Kolev NI (October 4–8, 2004) External cooling - the SWR 1000 severe accident management strategy Part 1: Motivation, strategy, analysis: melt phase, vessel integrity during melt–water interaction, Proceedings of 6th International Topical Meeting on Nuclear Reactor Thermal Hydraulics, Operations and Safety, Nara-Ken New Public Hall, Nara, Japan, NUTHOS6-N6P381

1.103 Kolev NI (April 17–21, 2005) Common features of heterogeneous nucleate boiling & flashing for water, 6th World Conference on Experimental Heat Transfer, Fluid Mechanics, and Thermodynamics Matsushima, Miyagi, Japan

1.104 Kolev NI (May 15–19, 2005) Uniqueness of the elementary physics driving heterogeneous nucleate boiling and flashing, Proceedings of ICAPP '05, Seoul, KOREA, Paper 5678. In abbreviated form ICCHMT' 05 Fourth International Conference on Computational Heat and Mass Transfer, Paris (France) 17 to 20 may 2005, Paper reference No: ICCHMT'05 – 491

1.105 Kolev NI (30 May – 3rd June 2005) Is it possible to design universal multi-phase flow analyzer? 6th World Congress on Structural and Multidisciplinary Optimization WCSMO6 Rio de Janeiro; International Conference "Nuclear Energy for New Europe 2005", September 5–8, 2005, Bled (Slovenia)

1.106 Kolev NI, Roloff-Bock I and Schlicht G (June 14–17, 2005) Gas release driven dynamics in research reactors piping, Third M.I.T Conference on Computational Fluid and Solid Mechanics, Cambridge, Massachusetts. ICAPP '06 Reno, NV USA, June 4–8, 2006, ICAP06 Paper 6053, submitted for publication

1.107 Kolev NI (October 2–6, 2005) Flow boiling in bundles: Comparison of IVA predictions for 153 experiments for variety of bundles, flow regimes including DO, steady state and transients, The 11th International Topical Meeting on Nuclear Reactor Thermal-Hydraulics (NURETH-11) Log Number: 547, Popes Palace Conference Center, Avignon, France

1.108 Kolev NI (October 2–6, 2005) Strictly conservative limiter for fourth order CIP methods for multi-fluid analyses, The 11th International Topical Meeting on Nuclear Reactor Thermal-Hydraulics (NURETH-11) Log Number: 006, Popes Palace Conference Center, Avignon, France

1.109 Kolev NI (September 12–15, 2005) Do we have appropriate constitutive sets for sub-channel and fine-resolution 3D-analyses of two-phase flows in rod bundles? Mathematics and Computation, Supercomputing, Reactor Physics and Nuclear and Biological Applications Palais des Papes, Avignon, France, on CD-ROM, American Nuclear Society, LaGrange Park, IL (2005)

1.110 Kolev NI (27–29 June 2005) IVA Simulations of exercises 1 of phase 1 and 2 of the OECD/NRC benchmark based on NUPEC BWR full-size fine-mesh bundle tests (preliminary), Second Workshop on OECD/NRC Benchmark based on NUPEC BWR Full-size Fine-mesh Bundle Tests (BFBT), State College, PA, USA

1.111 Kolev NI (2006) Uniqueness of the elementary physics driving heterogeneous nucleate boiling and flashing, Nuclear Engineering and Technology, vol 38 no 1, pp 33–42, Seoul, Korea

1.112 Kolev NI (2003, 2005) IVA_5M numerical method for analysis of three-fluid multi-component flows in boundary-fitted multi-blocks, Computers & Structures, vol 83 (2005) pp 499–523, USA. Presented in Second M.I.T. Conference on Computational Fluid and Solid Mechanics, (17–20 June 2003) Boston

1.113 Kolev NI (July 17–20, 2006) Verification of IVA computer code for flow boiling stability analysis, Proceedings of ICONE-14 '06, Miami, Florida, USA, Paper ICONE14-89034, submitted for publication

1.114 Kolev NI (2006) Check of the 2005 look-up table for prediction of CHF in bundles, Nucl. Eng. Des., doi:10.1016/j.nucengdes.2006.10.021

1.115 Kolev NI (2009) To the nucleate boiling theory, Nucl. Eng. Des., vol 239 pp 187–192

1.116 Kolev NI (May 16–19, 2011) Analysis of boiling, Proceedings of ICONE19, 19th International Conference on Nuclear Engineering, Chiba, Japan, ICONE19-43770

1.117 Kolev NI (May 16–19, 2011) Bubble dynamics in single component fluid, Proceedings of ICONE19, 19th International Conference on Nuclear Engineering, Chiba, Japan, ICONE19-43771

16.4 IVA-validation database

Example of the strategy of the validation. Consider that a given computer code has to be applied for the analysis of melt–water interaction. Some important elements of the models of melt–water interaction can be illustrated by a simple example. Let us start with the simple example of a closed volume filled with water and gas in which melt initially separated from the water is dropped into the water. The overall pressure increase is governed by the simplified form of the volume-conservation equation $dp/d\tau \approx \rho a^2 \mu_1 / \rho_1$ derived in Chapter 5, see also Ref. [2.43], p. 37, Eq. (103). Here, μ_1 is the evaporating mass per unit time and unit volume, ρ_1 is the density of the steam, ρ is the mixture density and a is the velocity of sound for the mixture. For gas volume fractions α_1 of 10% or more we can write as a very good approximation

$$dp/d\tau \approx a_1^2 \mu_1 / \alpha_1 . \quad (16.4.1)$$

This simple equation is remarkable. Bearing in mind that during the short time in which melt–water interaction takes place, e.g., 0.001 to 3 s, the volume of steam in the closed system and the steam temperature do not change very much, we see that 1) the pressure change is directly proportional to the evaporation rate and 2) the smaller the void fraction the stronger this dependence.

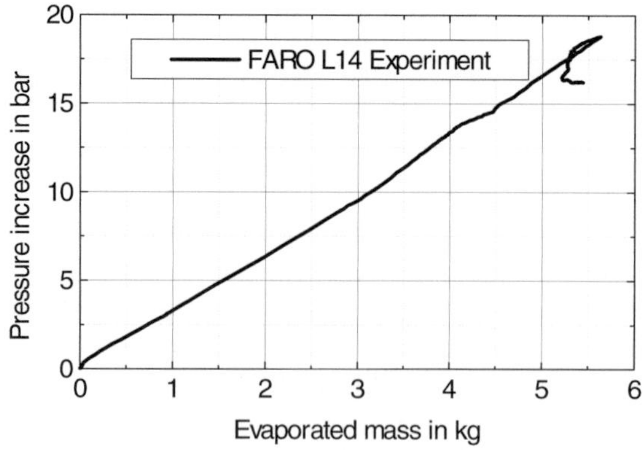

Fig. 16.4.1 Validity of Eq. (16.4.1) for the FARO L14 experiment - IVA simulation

Figure 16.4.1 shows the pressure increase versus the evaporated mass for the FARO L14 experiment simulated with the IVA computer code. The linear dependence as prescribed by Eq. (16.4.1) is reflected surprisingly well even after solving 21 complicated partial differential equations in time and space.

The evaporation rate caused by the energy transfer from the melt into the water is $\mu_1 \approx \left(F_3^{2\sigma}/Vol\right)\chi_3^{2\sigma}\left(T_3 - T_2\right)/\Delta h_{12}$. Here, $\left(F_3^{2\sigma}/Vol\right)$ is the surface of the melt that is surrounded by water divided by the system volume. For dispersed melt we have $F_3^{2\sigma}/Vol = f_{wet} 6\alpha_3 / D_3$. Here, f_{wet} is the portion of the melt able to produce steam, which means the portion of the melt not surrounded by gas but by water. $\chi_3^{2\sigma}$ is some averaged heat-transfer coefficient, $T_3 - T_2$ is the driving temperature difference and Δh_{12} is the latent heat of evaporation. With this, the expression for the pressure generation is

$$\frac{dp}{d\tau} \approx f_{wet}\frac{\alpha_3}{\alpha_1}\frac{6a_1^2}{D_3}\chi_3^{2\sigma}\frac{T_3 - T_2}{\Delta h_{12}} \approx f_{wet}\frac{\alpha_3}{\alpha_1}\frac{6}{D_3}\chi_3^{2\sigma}\left(\kappa_1 - 1\right)c_{p1}T_1\frac{T_3 - T_2}{\Delta h_{12}}. \quad (16.4.2)$$

We learn from this simple expression that 1) the geometrical pattern of the material distribution in the space during the interaction reflected by $f_{wet}\dfrac{\alpha_3}{\alpha_1}$ called material relocation is very important, 2) the melt particle size D_3 is very important, 3) the local heat-transfer conditions reflected by $\chi_3^{2\sigma}$ are very important, and 4) the history of the phase temperatures

$$a_1^2\left(T_3 - T_2\right) \approx \left(\kappa_1 - 1\right)c_{p1}T_1\left(T_3 - T_2\right), \quad (16.4.3)$$

is very important. Here κ_1 and c_{p1} are the vapor isentropic exponent and the specific vapor heat capacity at constant pressure, respectively.

We readily accept that mathematical models created to reflect the melt–water interactions have to be validated on experimental information for each of those particular groups of physical phenomena.

Another important point is the multidimensional nature of the interaction. The material relocation and the local flow conditions are obviously strong functions of time and space. Thus, the lumped-parameter approach, as introduced here for the sake of simplicity, holds true only after averaging over the total volume.

In particular, the phenomena identified to be important for the melt–water interaction scenarios are summarized in Tables 16.4.1 to 16.4.11. Many phenomena for the so-called classical applications of the multiphase flow analysis are subdomains of the melt–water interaction. Therefore, codes undergoing this very challenging verification program are of course applicable in the subdomains, e.g., of missing melt, or missing components, or missing dimensions, etc.

Table 16.4.1 IVA: Tested physical phenomena against experimental data and analytical benchmarks: **Gravitational material relocation**

No	Phenomena successfully reproduced by IVA computer code	Experiment (e) or analytical benchmark (a)
1	2D waves	(e) *Maschek* [2.72]
26	Single-phase natural convection in uniformly heated vertical part of a U-tube	(a) *Kolev* [2.67, 1.93, 1.94]
27	Single-phase natural convection in uniformly heated inclined part of a U-tube	(a) *Kolev* [2.67, 1.93, 1.94]
28	Single-phase natural convection in U-tube with inclined part heated by steam condensation	(a) *Kolev* [2.67, 1.93, 1.94]
45	Critical gas flow in nozzles and pipe	(e) *Frössel* [2.113]
46	Discharge from gas vessel through pipe	(a) *Kolev*, vol. 1

Table 16.4.2 IVA: Tested physical phenomena against experimental data and analytical benchmarks: **Single-phase flow**

No	Phenomena successfully reproduced by IVA	Experiment (e) or analytical benchmark (a)
20	Steady-state single-phase nozzle flow	(e) *BNL* [2.1]
	1D pressure waves	
2	In gas	(a) *Oertel* [2.85]
3	In water	(a) *Kellner* et al. [2.26]
4	2D pressure waves in gas	(e) *SGI* [2.73, 2.92]

Table 16.4.3 IVA: Tested physical phenomena against experimental data and analytical benchmarks: **Adiabatic two phase flow**

No	Phenomena successfully reproduced by IVA	Experiment (e) or analytical benchmark (a)
4	2D pressure waves	(e) *SGI* [2.73, 2.92]
	1D pressure waves in a single pipe	
5	- without intensive mass exchange	(e) *Kellner* et al. [2.26]
6	- with intensive evaporation – flashing	(e) *Edwards* et al. [2.17]
25	Pressure waves in complex networks	(a) *Kolev* [1.93, 1.94]
33	Droplet fragmentation for Weber numbers of 1 to 1000	(e) see in *Kolev* [1.94]
24	Heterogeneous nucleation, bubble generation frequency	(e) see in *Kolev* [1.94]
21	Film entrainment in pipe flow	(e) *UPTF* [2.65]
23	Water flashing in nozzle flow	(e) *BNL* [2.1]
29	3D small break loss of coolant	(e) *Kevchishvili* et al. [2.27]
30	3D hot leg emergency water injection against steam flow	(e) *UPTF* [2.65]

Table 16.4.4 IVA: Tested physical phenomena against experimental data and analytical benchmarks: **Diabatic two-phase flow**

No	Phenomena successfully reproduced by IVA	Experiment (e) or analytical benchmark (a)
31	Bubble departure diameter, nucleation frequency, bubble generation frequency, nucleate boiling	(e) see in *Kolev* [1.94]
22	Boiling in pipes	(e) *Sabotinov* [2.88], *Pierre* [2.87], *Egen* et al. [2.18], *Bartolomei* et al. (1980) [2.112]
22	Boiling in rod bundles	(e) *Nylund* [2.84],
7	Film boiling in pipes	(e) *Bennet* [2.12]
7	Film boiling at plates	(e) *Okkonen* et al. [2.86]
7	Film boiling at spheres	(e) *Liu* et al. [2.70]

Table 16.4.5 IVA: Tested physical phenomena against experimental data and analytical benchmarks: **Three-phase solid sphere flow**

No	Phenomena successfully reproduced by IVA	Experiment (e) or analytical benchmark (a)
32	Drag forces in solid-spheres–water–air bubble flows	(e) *Sakaguchi* [2.32]
8	2D dropping cold spheres into water	(e) *Kolev* et al. [2.32], *QUEOS* [2.75–79]
9,10	2D dropping extremely hot spheres into water	(e) *QUEOS* [2.75–79]

Table 16.4.6 IVA: Tested physical phenomena against experimental data and analytical benchmarks: **3D-effects in single-phase flow**

No	Phenomena successfully reproduced by IVA	Experiment (e) or analytical benchmark (a)
17	Rigid-body steady-rotation problem	(a) *Harison* et al. [2.21]
18	Pure radial symmetric flow	(a) *Harison* et al. [2.21]
19	Radial-azimuthal symmetric flow	(a) *Harison* et al. [2.21]

Table 16.4.7 IVA: *Tested physical phenomena against experimental data and analytical benchmarks:* **Melt–water interaction with dynamic fragmentation and coalescence**

No	Phenomena successfully reproduced by IVA	Experiment (e) or analytical benchmark (a)
11,12,13	Fragmentation of interpenetrating water, melt and vapor using corium as melt	(e) FARO L14, L20, L24 [2.71, 2.13, 2.22, 2.2, 2.3, 2.4]
14	Uncertainty in the prediction of nonexplosive melt–water interactions	(a, e) FARO L14 and [2.48]

15	Effect of the geometry	(e) FARO L28, L31 [2.91, 2.90]
16	Fragmentation of interpenetrating saturated water, melt and vapor using alumina as melt	(e) PREMIX 13, 15, [2.24, 2.25]
34	Fragmentation of interpenetrating subcooled water, melt and vapor using alumina as melt	(e) PREMIX 17, 18 [2.94, 2.95, 2.96, 2.97]

Table 16.4.8 IVA: *Tested physical phenomena against experimental data and analytical benchmarks:* **Flow boiling instability in natural circulation loop containing rod bundle**

No	Phenomena successfully reproduced by IVA	Experiment (e) or analytical benchmark (a)
35	Mass flow rate as a function of power, frequencies, decay factors	(e) *Schäfer* and *Beisiegel* [2.94]

Table 16.4.9 IVA: *Tested physical phenomena against experimental data and analytical benchmarks:* **3D steady-state flow boiling in rod bundle**

No	Phenomena successfully reproduced by IVA	Experiment (e) or analytical benchmark (a)
36	8×8 BWR rods, void fractions, NUPEC experiments	*Morooka* et al. [2.95], *Yagi* et al. [2.96], *Inoue* et al. [2.97–99]
37	8×8 BWR rods, void fractions, 15 NUPEC experiments in a framework of an ECD/USNRC, different power profiles	*Kolev* [2.100]
38	36 rods, void fractions, FRIGG experiments, different power profiles	*Nylund* et al. [2.101, 2.102]
39	8×8 PWR rods, void profiles, dry out position	*Anklama* and *Miller* [2.103]
41	Regional void fraction in ATRIUM 10XP at various peaking patterns	*Spierling* [2.105]

Table 16.4.10 IVA: *Tested physical phenomena against experimental data and analytical benchmarks:* **3D transient flow boiling in rod bundle subchannel**

No	Phenomena successfully reproduced by IVA	Experiment (e) or analytical benchmark (a)
43	Void for mass flow or pressure or inlet temperature variations	(e) *Hori* et al. [2.106–109]
44	5–5 PWR bundle, void for power or mass flow variations	(e) *Hori* et al. [2.106, 109], *Akiyama* et al. [2.110, 2.111]

Table 16.4.11 IVA: *Tested physical phenomena against experimental data and analytical benchmarks:* **3D dry out in rod bundle**

No	Phenomena successfully reproduced by IVA	Experiment (e) or analytical benchmark (a)
40	4×4 BWR rods	(e) *Lucchini* and *Marinelli* [2.104]
42	ATRIUM 10-9Q 10×10 bundle	(e) AREVA proprietary

IVA was extensively used during the licensing process of the research reactor FRM II in Munich for melt–water-interaction analysis. The German state authority accepted the analysis as appropriate and correct.

The code was used during the melt–water interaction study for the EPR design.

The code was used for design of the concept for handling severe accidents in the BWR1000 design. The Finish authority STUK judged positive this concept.

The code is being used for solving daily problems in engineering practice at Siemens.

The references used during the verification procedure and the documents containing the results of the verification are summarized in the list given below.

16.5 References of data sources and references documenting the validation of IVA

2.1 Abuaf N, Wu BJC, Zimmer GA, and Saha P (June 1981) A study of non equilibrium flashing of water in a converging diverging nozzle, vol 1 Experimental, vol 2 Modeling, NUREG/CR-1864, BNL-NUREG-51317

2.2 Annunziato A, Addabbo C, Leva G (1996) OECD-CSNI International standard problem 39 on FARO test L-14, Reference specification, JRC Technical Note I.96.xx, Draft

2.3 Annununziato A, Addabbo C, Magallon D (Oct. 1996) FARO test L-20 quick look report, Technical Note No. I.96.163, JRC Ispra, Italy

2.4 Annununziato A, Addabbo C, Magallon D, FARO Test L-24 Quick Look Report, Technical Note No. I.97.185, JRC Ispra, Italy

2.5 Batchelor GK (ed) (1958) Collected works of G. I. Taylor, Cambridge University Press, Cambridge, MA

2.6 Basu S, Ginsberg T (Aug. 1996) A reassessment of the potential for an alpha-mode containment failure and a review of the current understanding of broader fuel-coolant interaction issues, NUREG-1524

2.7 Bracco FV (Feb. 25–March 1, 1985) Modeling of engine sprays, Proc. International Congress & Exposition Detroit, Michigan, pp 113–136

2.8 Bürger M, Buck M, Saied-Ahmad A, Schatz A (December 1996) Experimental and theoretical investigation on the fragmentation of melt drops in relative flow, Report INV-MFC (98)-D016, Institut für Kernenrgetik und Energiesysteme, Universität Stutgart, IKE 2-135

2.9 Bürger M (July 1998) Comparison and theoretical interpretation of thc experiments on hydrodynamic drop fragmentation, Report INV-MFC (98)-D014, Institut für Kernenergetik und Energiesysteme, Universität Stutgart, IKE 2-FB-16

2.10 Bürger M, Cho SH, von Berg E, A Schatz (Nov. 1998) Modeling of drop fragmentation in thermal detonation waves and experimental verification, Report INV-MFC (98)-D016, Institut für Kernenrgetik und Energiesysteme, Universität Stutgart, IKE 2-FB-15

2.11 Coste P, Valette M and Berthoud G (23–24 June, 1999) Pb-Bi droplets fragmentation in liquid water (Li et al. experiment, ICMF 98): comparison of the existing code correlations and calculations with MC3D, 6th MFCI Project Meeting, CEA Grenoble, France

2.12 Bennett AW et al. (1967) Heat transfer to steam-water mixtures flowing in uniformly heated tubes in which the critical heat flux has been exceeded, AERE-R5373

2.13 Benuzzi A, Magalon D (December 1994) FARO-LWR programme L-14 test quick look report, JRC Technical Note I.94.171

2.14 Chen JC (1963) A correlation for film boiling heat transfer to saturated fluids in convective flow, ASME Publication-63-HT-34, p 2–6

2.15 Corradini ML, Rohsenow WM and Todreas NE (1980) The effect of sodium entrainment and heat transfer with two-phase UO2 during a hypothetical core disruptive accident, NSE, vol 73 pp 242–258
2.16 Crowe CT, Sharma MP, and Stock DE,
(June 1977) The particle-in-cell (PSI-Cell) model for gas droplet flows, ASME, Transactions, Series I – J. Fluids Eng., vol 99 pp 325–332
2.17 Edwards AR and O'Brien TP (1970) Studies of phenomena connected with the depressurization of water reactors, J. Brit. Nucl. Energy Soc., vol 9 nos 1–4 pp 125–135
2.18 Egen RA, Dingee DA and Chastain JW
(1957) Vapor formation and behavior in boiling heat transfer, AEC Report BMI – 1167
2.19 Groeneveld DC et al. (1996) The 1995 look-up table for critical heat flux in tubes, Nucl. Eng. Des., vol 163 pp 1–23
2.20 Haraldsson HO, Li HX, Dinh TN, Green JA and BR Sehgal (July 6, 1999) Effect of solidification and coolant voiding on breakup of molten metal jet in water, Royal Institut of Technology (KTH), NPS, EU report INV-MFCI(00)-D048, Report EU-KTH/NPS/MFCI1-9907-SE, Stockholm
2.21 Harison RG and Johnson EC (eds) (July 1992) Developmental assessment of the multidimensional component in RELAP5 for Savannah River Site thermal hydraulic analysis, EGG-EAST-9803, DE93 001942, EG&G Idaho, Inc., Idaho Falls, Idaho 83415
2.22 Hohmann H, Magallon D, Huhtiniemi I, Annunziato A, Yerkess A (October 24–26, 1994) Advance in the FARO/KROTOS Melt Quenching Test Series, 22nd Water Reactor Safety Meeting, Bethesda (Maryland)
2.23 Hughes ED, MP Paulsen and LJ Agee (1981) A drift-flux model of two phase flow for RETRAN, Nuclear Technology, vol 54 p 410
2.24 Kaiser A, Schütz W, Will H (1998) PREMIX experiment, EC 4th Framework Program on Reactor Safety Shared Cost Actions, Contract no FI4s-CT96-0037, MFCI PROJECT, Yearly Progress Report no 1
2.25 Kaiser A, Schütz W, Will H (17 December 1998) PREMIX: Investigation of premixing behavior of a hot melt being poured into water, Test PM15, 16.9.1998, 5th MFCI Project Meeting, Forschungszentrum Karlsruhe, Germany
2.26 Kellner A and Gissler D (7.2.1984) Programsystem SAPHYR: Anwendungsbeispiel II, Notiz Nr.70.02748.4, Interatom GmbH
2.27 Kevchishvili NA and Dementev BS (1985) Investigation of the influence of the decay heat on the blow down characteristics of steam-water mixtures, Teploenergetika, vol 7 p 67
2.28 Kolev NI (1986) Transiente Zweiphasenströmung (Transient Two-Phase Flow), Springer Verlag
2.29 Kolev NI (1986) Transiente Dreiphasen Dreikomponenten Strömung, Teil 3: 3D-Dreifluid-Diffusionsmodell, KfK 4080
2.30 Kolev NI (1986) Ein Computerprogramm zur Modellierung transienter 3D - Dreiphasen Dreikomponenten Strömungen mittels drei Geschwindigkeitsfeldern in zylindrischer Geometrie mit beliebigen Einbauten einschließlich der Spaltzone eines PWR/BWR. KfK 4088
2.31 Kolev NI (September 1987) IVA2 verification expansion phase experiment in SNR geometry, KfK 4310
2.32 Kolev NI, Tomiyama A, Sakaguchi T (Sept. 1991) Modeling of the mechanical interaction between the velocity fields in three phase flow, Exp. Thermal Fluid Sci., vol 4 no 5 pp 525–545

2.33 Kolev NI (Sept. 1991) A three-field model of transient 3D multi-phase, three-component flow for the computer code IVA3, Part 1: Theoretical basics: Conservation and state equations, numerics. KfK 4948, Kernforschungszentrum Karlsruhe
2.34 Kolev NI (Sept. 1991) A three-field model of transient 3D multi-phase, three-component flow for the computer code IVA3, Part 2: Models for the interfacial transport phenomena. Code validation. KfK 4949, Kernforschungszentrum Karlsruhe
2.35 Kolev NI (1993) Fragmentation and coalescence dynamics in multi-phase flows, Exp. Thermal Fluid Sci., vol 6 pp 211 – 251
2.36 Kolev NI (1993) The code IVA3 for modeling of transient three-phase flows in complicated 3D geometry, Kerntechnik, vol 58 no 3 pp 147–156
2.37 Kolev NI (1994) The code IVA4: Modeling of mass conservation in multi-phase multi component flows in heterogeneous porous media, Kerntechnik, vol 59 no 4–5 pp 226–237
2.38 Kolev NI (1994) The code IVA4: Modeling of momentum conservation in multi-phase multi component flows in heterogeneous porous media, Kerntechnik, vol 59 no 6, pp 249–258
2.39 Kolev NI (1994) The influence of the mutual bubble interaction on the bubble departure diameter, Exp. Thermal Fluid Sci., vol 8 pp 167–174
2.40 Kolev NI (14–15 Nov. 1994) IVA4 computer code: Dynamic fragmentation model for liquid and its application to melt water interaction, Presented at the Workshop zur Kühlmittel/Schmelze - Wechselwirkung, Köln, Germany. (Apr. 23–27, 1995) Proc. ICONE-3, The Third International Conf. on Nucl. Engineering, "Nuclear Power and Energy Future", Kyoto, Japan
2.41 Kolev NI (27–29 September 1995) IVA4 computer code: An universal flow analyzer for multiphase flows and its applicability to melt water interaction, Technical Committee Meeting on Advances in and Experience with Accident Consequences Analysis IAEA Headquarters, Vienna, Austria
2.42 Kolev NI (1995) The code IVA4: Nucleation and flashing model, Kerntechnik, vol 60 no 6 pp157–164. Also in: (Apr. 3–7, 1995) Proc. Second Int. Conf. On Multiphase Flow, Kyoto; (Aug.13–18, 1995) ASME & JSME Fluid Engineering Conference International Symposium on Validation of System Transient Analysis Codes - Hilton Head (SC) USA; (October 9–11, 1995) Int. Symposium on Two-Phase Flow Modeling and Experimentation, ERGIFE Place Hotel, Rome, Italy
2.43 Kolev N I (1995) The code IVA4: Second law of thermodynamics for multi phase flows in heterogeneous porous media, Kerntechnik, vol 60 no 1 pp 1–39
2.44 Kolev NI (1995) The code IVA4: Nucleation and flashing model, Kerntechnik vol 60 no 6 pp 157–164. Also in: (Apr. 3–7, 1995) Proc. Second Int. Conf. On Multiphase Flow, Kyoto; (Aug.13–18, 1995) ASME & JSME Fluid Engineering Conference International Symposium on Validation of System Transient Analysis Codes - Hilton Head (SC) USA; (October 9–11, 1995) Int. Symposium on Two-Phase Flow Modeling and Experimentation, ERGIFE Place Hotel, Rome, Italy
2.45 Kolev NI (1995) How accurate can we predict nucleate boiling, Exp. Thermal Fluid Sci., Exp. Thermal Fluid Sci., vol 10 pp 370–378
2.46 Kolev NI (Apr.3–7, 1995) IVA4 computer code: The model for film boiling on a sphere in subcooled, saturated and superheated water, Proc. Second Int. Conference On Multiphase Flow, Kyoto, Japan. Presented also in (Nov. 14–15, 1994) "Workshop zur Kühlmittel/Schmelze - Wechselwirkung", Köln, Germany

2.47 Kolev NI (1995) IVA4 Computer code: Dynamic fragmentation model for liquid and its application to melt water interaction, Proc. ICONE-3, The Third International Conf. on Nucl. Engineering, "Nuclear Power and Energy Future", Apr.23–27 1995, Kyoto, Japan. Presented at the "Workshop zur Kühlmittel/ Schmelze - Wechselwirkung", 14–15 Nov. 1994, Köln, Germany

2.48 Kolev NI and Hofer E (1996) Uncertainty and sensitivity analysis of post experimental IVA4 simulations of melt Water Interaction, Exp. Thermal Fluid Sci., vol 13 pp 98–116

2.49 Kolev NI (October 15–16, 1996) Three fluid modeling with dynamic fragmentation and coalescence, fiction or daily practice? 7th FARO Experts Group Meeting Ispra; (5th–8th November 1996) Proceedings of OECD/CSNI Workshop on Transient thermal-hydraulic and neutronic codes requirements, Annapolis, MD, USA; (June 2–6, 1997) 4th World Conference on Experimental Heat Transfer, Fluid Mechanics and Thermodynamics, ExHFT 4, Brussels; (June 22–26, 1997) ASME Fluids Engineering Conference & Exhibition, The Hyatt Regency Vancouver, Vancouver, British Columbia, CANADA Invited Paper; (May 22–24, 1997) Proceedings of 1997 International Seminar on Vapor Explosions and Explosive Eruptions (AMIGO-IMI), Aoba Kinen Kaikan of Tohoku University, Sendai-City, Japan

2.50 Kolev NI (May 19–21, 1997) Numerical modeling of in-vessel melt water interaction in large scale PWR's, Proceedings of OECD/CSNI Specialists Meeting on Fuel-Coolant Interactions (FCI), JAERI-Tokai Research Establishment, Japan

2.51 Kolev NI (May 26–30, 1997) Is large melt water mixing possible under low pressure, Proceedings of ICON 5: 5th International Conference on Nuclear Engineering, Nice, France

2.52 Kolev NI (1997) Comments on the entropy concept, Kerntechnik, vol 62 no 1 pp 67–70

2.53 Kolev NI (19th–21st May 1997) Verification of the IVA4 film boiling model with the data base of Liu and Theofanous, Proceedings of OECD/CSNI Specialists Meeting on Fuel-Coolant Interactions (FCI), JAERI-Tokai Research Establishment, Japan

2.54 Kolev NI (May 22–24, 1997) IVA4 analysis of the FARO L14 experiment (ISP 39), International Seminar on Vapor Explosions and Explosive Eruptions (AMIGO-IMI), Aoba Kinen Kaikan of Tohoku University, Sendai-City, Japan

2.55 Kolev NI (June 2–6, 1997) Film boiling: vertical plates, Proceedings of 4th World Conference on Experimental Heat Transfer, Fluid Mechanics and Thermodynamics EXHFT 4, Brussels, Belgium

2.56 Kolev NI (1998) Film boiling on vertical plates and spheres, Exp. Thermal Fluid Sci., vol 18 pp 97–115

2.57 Kolev NI (1998) On the variety of notation of the energy conservation principle for single phase flow, Kerntechnik, vol 63 no3 pp 145–156

2.58 Kolev NI (17 December 1998) In-vessel melt–water interaction caused by core support plate failure under molten pool, Part 1: Choice of the solution method, 5th MFCI Project Meeting, Forschungszentrum Karlsruhe, Germany. (October 3–8,1999) Proc. of the Ninth International Topical Meeting on Nuclear Reactor Thermal Hydraulics (NURETH-9), San Francisco, California. (August 1999) Kerntechnick, vol 64 no 4

2.59 Kolev NI (17 December 1998) In-vessel melt–water interaction caused by core support plate failure under molten pool, Part 2: Analysis, 5th MFCI Project Meeting, Forschungszentrum Karlsruhe, Germany. (October 3–8, 1999) Proc. of the Ninth International Topical Meeting on Nuclear Reactor Thermal Hydraulics (NURETH-9), San Francisco, California. (1999) Kerntechnick, vol 64 no 5

2.60 Kolev NI and Hulin H (23–24 June 1999) Detonation waves in melt-coolant interaction, Part 2: Applied analysis, MFCI Project, 6th progress meeting, CEA, Grenoble. EU Nr. INV-MFCI(99)-D038. (2001) Kerntechnik, vol 66 no 1–2 pp 21–25
2.61 Kolev NI (1999) The IVA5 dynamic droplet fragmentation and coalescence models and their verification data base: Part 1 (INV-MFCI(99)-D027) and Part 2 (INV-MFCI(99)-D028)
2.62 Kolev NI (5–10, September 1999) Applied multi-phase flow analysis and its relation to constitutive physics, 8th International Symposium on Computational Fluid Dynamics, ISCFD '99 Bremen, Germany. (April 2000) Jpn. J. Comput. Fluid Dyn., vol 9 no 1
2.63 Kolev NI (October 3–8, 1999) Verification of IVA5 computer code for melt–water interaction analysis, Part 1: Single phase flow, Part 2: Two-phase flow, three-phase flow with cold and hot solid spheres, Part 3: Three-phase flow with dynamic fragmentation and coalescence, Part 4: Three-phase flow with dynamic fragmentation and coalescence – alumina experiments, Proc of the Ninth International Topical Meeting on Nuclear Reactor Thermal Hydraulics (NURETH-9), San Francisco, California
2.64 Kolev NI (2000) Detonation waves in melt-coolant interaction, Part 1: Kernenergie, vol 65 pp 254–260
2.65 Kolev NI, Seitz H and Roloff-Bock I (2000) Hot-leg injection: 3D versus 1D three velocity fields modeling and comparison with UPTF experiment, Siemens report KWU NA-T/00/E027, R&D Project 903060, Verification IVA6
2.66 Kolev NI (April 2–6, 2000) Computational analysis of transient 3D-melt–water interactions, 8th International Conference on Nuclear Engineering, Baltimore, Maryland, USA, ICONE-8809. See also (November 5–11, 2000) ASME International Mechanical Engineering Congress & Exposition Dispersed Flow in Combustion, Incineration and Propulsion Systems
2.67 Kolev NI (April 14–18, 2002) Simple analytical U-tube benchmarks appropriate for testing of pipe-network computer codes, Proceedings of ICONE10: Tenth International Conference on Nuclear Engineering, Arlington, Virginia (Washington, D.C.), USA ICONE10-22711
2.68 Lanzenberger K (1998) Thermal radiation in multiphase flow, KWU NA-T/98/E011. 1998 Karl Wirtz Award for Young Scientists: K. Lanzenberger, Thermal radiation in multiphase flow - Application to the severe accident scenario of molten fuel coolant interaction (MFCI), Siemens AG, Power Generation (KWU)
2.69 Levy S (1967) Int. J. Heat Transfer, vol 10 pp 351–365
2.70 Liu C and Theofanous TG (August 1995) Film boiling on spheres in single- and two-phase flows part 1: Experimental studies ANS Proceedings, Part 2: A Theoretical Study, National Heat Transfer Conference, Portland
2.71 Magallon D, Leva G (February 1996) FARO LWR programme test L-14 data report, JRC Technical Note I.96.25
2.72 Maschek W, Roth A, Kirstahler M, Meyer L (Dec. 1992) Simulation experiments for centralized liquid sloshing motions, Kernforschungszentrum Karlsruhe, KfK 5090
2.73 Meyer L, Kirstahler M (September 4–9, 1988) Experimental investigation of the fluid dynamic of gas jet expansion in a liquid pool, Proc. 1st World Conf. Experimental Heat Transfer, Fluid Mechanics, and Thermodynamics, Dubrovnik, Yugoslavia. (1988) Shah K, Ganic EN, Yang KT (eds), Elsevier, New York, Amsterdam, London
2.74 Meyer L and Wilhelm D (March 1994) Investigation of the fluid dynamics of gas jet expansion in a liquid pool, Kernforschungszentrum Karlsruhe, KfK 5307

2.75 Meyer L and Schumacher G (April 1996) QUEOS a simulation-experiment of the premixing phase of steam explosion with hot spheres in water base case experiments, Wissenschaftliche Berichte FZKA 5612, Forschungszentrum Karlsruhe Technik und Umwelt, Karlsruhe

2.76 Meyer L (Aug.3–6, 1996) The interaction of falling mass of hot spheres with water, ANS Proceedings 1996 National Heat Transfer Conference, Houston, Tx., vol 9, pp 105–114, American Nuclear Society, La Grange Park, Illinois

2.77 Meyer L (July 1997) QUEOS a simulation-experiment of the premixing phase of steam explosion with hot spheres in water, Results of the Second Test Series, Internal Report 32.21.02/INR1962/PSF3267, Forschungszentrum Karlsruhe Technik und Umwelt, Karlsruhe

2.78 Meyer L (1998) QUEOS, an experimental investigation of premixing phase with hot spheres, Nucl. Eng. Des., vol 18

2.79 Meyer L and Kuhn D (May 23–25, 1999) The interaction of very hot particles falling into water, 2nd Int. Symp. on Two-Phase Flow Modeling and Experimentation Pisa, Italy

2.80 Miropolskij ZL (1963) Heat transfer in film boiling of steam-water mixture in steam generating tubes. Teploenergetika, vol 10 no 5 pp 49–53

2.81 Mitchel et al. DE (1981) Intermediate scales steam explosion phenomena: Experimental analysis SAND8-0124

2.82 Nelson LS and Duda PM (September 1981) Steam explosion experiments with single droplet of CO_2 laser melted with iron oxide. Report NUREG/CR-2295 (SAND 81-1346)

2.83 Nelson LS and Duda PM (September 1985) Steam explosion experiments with single droplet of CO_2 laser melted with iron oxide. Part 2. Parametric Studies. Report NUREG/CR-2718 (SAND 81-1105)

2.84 Nylund D et al. (1968) Hydrodynamic and heat transfer measurements on a full-scale simulated 36-rod Marviken fuel element with uniform heat flux distribution, FRIG-2, Danish Atomic Energy Commission

2.85 Oertel H (1966) Stossrohre, Springer Verlag, Wien-New York

2.86 Okkonen T, et al. (August 3–6, 1996) Film boiling on a long vertical surface under high heat flux and water subcooling conditions, Proc. of the 31st Nat. Heat Transfer Conference, Houston, Texas

2.87 Pierre CCS (1965) ANL-7041

2.88 Sabotinov LS (1974) Experimental investigation of void fraction in subcooled boiling for different power distribution laws along the channel. Moskva PhD Thesis in Russian

2.89 Saha P and Zuber N (1974) Proc. Int. Heat Transfer Conference Tokyo Paper B4.7

2.90 Silverii R, Magallon D (June 1999) FARO LWR Programme Test L 31 Data Report, Technical Note No. I.99.100, INV-MFCI(99)-D035, JRC, ISIS, Ispra (VA), Italy

2.91 Silverii R and Annunziato A, FARO LWR program, Test L-28 data report, Technical Note No.I.99.76, INV-MFCI(99)-D033, JRC Ispra, Italy

2.92 Wilchelm D, Kirstahler M (1987) Versuchsanlage zur schnellen Gasinjektion in einen wassergefüllten Behälter, KfK-4321, Kernforschungszentrum Karlsruhe

2.93 Yamano N, Sugimoto J, Moryama Y and Soda K (1992) Studies of fuel coolant interactions during core melt accident of nuclear power plants, NURETH-6th Int. Meeting On Thermal-Hydraulics, pp 271–281

2.94 Schäfer H and Beisiegel A (17.03.1992) Feasibility of stability tests under natural circulation conditions in the Karlstein test loop, Siemens technical report E32/92/e14a, proprietary

2.95 Morooka S, Inoue A, Oishi M, Aoki T, Nagaoka K and Yoshida H (1991) In-bundle void measurement of BWR fuel assembly by X-ray CT Scanner, The 1st JSME/ASME Joint International Conference on Nuclear Engineering, pp 237–243

2.96 Yagi M, Mitsutake T, Morooka S and Inoue A (1992) Void fraction distribution in BWR fuel assembly and the evaluation of subchannel code. In: Ninokata H and Aritomi M eds., Subchannel Analysis in Nuclear Reactors, pp 141–167, Tokyo, Japan

2.97 Inoue A, Futakuchi M, Yagi M, Kurosu T, Mitsutake T and Morooka S (1995) Void fraction distribution in a boiling water reactor fuel assembly and the evaluation of subchannel analysis codes, Nuclear Technology, vol 112 no 3 pp 388–400

2.98 Inoue A, Kurosu T, Oaki T, Futaguchi M, Yagi M, Morooka S, Hoshide A, Ishizuka T, Yoshimura K, Yoshida H, Mitsutake T, Abe N, Kozima A, Nakajima T, Nakamura S and Sawa T (1995) Void measurement of BWR fuel assembly, J. At. Energy Soc. Japan, vol 37 no 8 pp 710–720 (in Japanese)

2.99 Inoue A, Kurosu T, Oaki T, Yagi M, Mitsutake T and Morooka S (1995) Void fraction distribution in boiling water reactor fuel assembly and evaluation of subchannel code", J. Nucl. Sci. Technol., 32 no 7, 629–640

2.100 Kolev NI (27–29 June 2005) IVA Simulations of exercises 1 of phase 1 and 2 of the OECD/NRC benchmark based on NUPEC BWR full-size fine-mesh bundle tests (preliminary), Second Workshop on OECD/NRC Benchmark based on NUPEC BWR Full-size Fine-mesh Bundle Tests (BFBT), State College, PA, USA

2.101 Nylund O et al. (1968) Hydrodynamic and heat transfer measurements on a full scale simulated 36-rod Marviken fuel element with uniform heat flux distribution, FRIGG-2, AB Atomenergi, Stockholm, Sweden, ASEA, VÄSTERAS, R-4-447/RTL-1007

2.102 Nylund O et al. (1970) Hydrodynamic and heat transfer measurements on a full scale simulated 36-rod BHWR fuel element with non-uniform axial and radial heat flux distribution, FRIGG-4, AB Atomenergi, Stockholm, Sweden, ASEA, VÄSTERAS,R4-502/RL-1253

2.103 Anklama TM and Miller RF (1982) Void fraction under high pressure, low flow conditions in rod bundle geometry, Nucl. Eng. Des., vol 75 pp 99–108

2.104 Lucchini F and Marinelli V (1974) Experimental data on burn-out in simulated BWR fuel bundle, Nucl. Eng. Des. vol 31 pp 371–378

2.105 Spierling H (Sep. 24, 2002) Regional void fraction in ATRIUM 10XP at various peaking patterns, Framatome-ANP report, FGT, A1C-1311669-1, proprietary

2.106 Hori K, Miyazaki K, Kurosu T, Sugiyama S, Matsumoto J and Akiyama Y (1993) In bundle void fraction measurement of PWR fuel assembly, Proceedings of ICONE 2, vol 1 pp 69–76

2.107 Hori K, Akiyama Y, Miyazaki K, Kurosu T and Sugiyama S (1994) Transient void fraction measurement in a single Channel simulating one channel for a PWR Fuel Assembly, 10th Proceedings of Nuclear Thermal-Hydraulics, ANS Winter Meeting, pp 56–68

2.108 Hori K, Akiyama Y, Miyazaki K, Kurosu T and Sugiyama S (1995) Void fraction in a single channel simulating one subchannel of a PWR fuel assembly, Proceedings of the First International Symposium an Two-Phase Flow Modelling and Experimentation

2.109 Hori K, Akiyama Y, Miyazaki K, Nishioka H and Takeda N (1996) Total evaluation of in bundle void fraction measurement test of PWR fuel assembly, Proceedings of ICONE-4, vol 1, Part B, pp 801–811

2.110 Akiyama Y, Hori K, Miyazaki K, Nishioka H and Sugiyama S (23–27 April 1995) Evaluation of the bundle void fraction measurement data of PWR fuel assembly, The 3rd JSME/ASME Joint International Conference on Nuclear Engineering, Kyoto International Conference Hall, Kyoto, Japan, vol 1, pp 211–216

2.111 Akiyama Y, Hori K, Miyazaki K, Nishioka H and Sugiyama S (Dec. 2005) Pressurized water reactor fuel assembly subchannel void measurement, Nucl. Technol., vol 112, pp 412–421

2.112 Bartolomei GG, Batashova GN, Brantov VG et al. (1980) Heat and Mass Transfer IV, Izd. ITMO AN BSSR, Minsk, vol 5, p 38, in Russian

2.213 Frössel W (1936) Strömung in glatten, geraden Rohren mit Über- und unterschalgeschwindigkeit, Forsch. vol 7 no 2 pp 75–84

16.6 Material relocation: gravitational waves (1D, 2D), density waves in boiling and condensing systems

16.6.1 Flow through straight microchannel with constant cross-section

One of the simplest benchmarks is the computation of the mass flow through a straight channel as a function of the pressure difference. Before starting any thing more complex we consider the following simple task.

Consider *fully developed flow* in rectangular channel (300 μm × 303 μm) over a length of 1 mm. Compute the mass flow through this channel as a function of pressure difference for the diesel fuel once analytically,

$$\rho w F = F\sqrt{2\rho\Delta p\, D_h/(\lambda L)}\,,$$

and once using the computer code, and then compare the so-obtained results. The inlet pressure is 100 bar and the fuel temperature is 30°C. The density at these conditions is 829.088 kg/m³ and the cinematic viscosity 4×10^{-6} m²/s. For the analytical computation the properties are taken as constant. The computation with the computer code takes into account the variable properties. The results are given in Table 16.6.1 and presented in Fig. 16.6.1. The friction coefficient $\lambda = \lambda\left(\rho w D_h/\eta, k/D_h\right)$, is a function of the Reynolds number and the relative roughness. The resulting transcendent system of equations in the analytical model is solved by a few iterations. It is computed in both the computer code and the analytical solution. The friction coefficient is computed in accordance with the algorithms presented in *Idelchik* (1993) p.113. The larger the pressure difference, the larger the density change. This is not considered in the analytical solution, but is taken into account in the code. If we "hard wire" the density as being constant in the code the predictions are very close to the analytical computation – see Fig. 16.6.1. This clear demonstrates that a) the code predicts properly mass flows for *developed flow* as function of pressure difference for microchannels with constant cross-section, and that b) the variation of the properties is getting important as the pressure variations become larger than 80–100 bar.

Table 16.6.1 Mass flow as a function of the pressure difference for a straight channel with constant cross-section

Δp bar	Mass flow, kg/s analytic	Mass flow, kg/s IVA_5M	Re, -	Friction coefficient, -
25	0.01282	0.012799	12 867	0.0618
50	0.01844	0.018127	18 515	0.0597
75	0.02259	0.021882	22 676	0.0597
99	0.02595	0.024800	26 053	0.0597

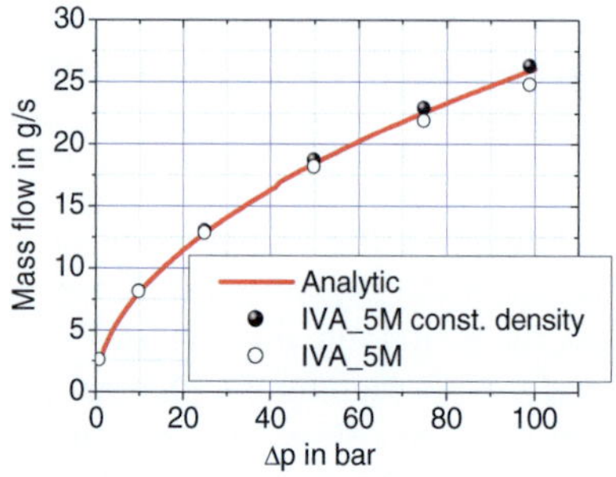

Fig. 16.6.1 Mass flow as function of pressure difference for rectangular channel (300 μm × 303 μm) width and 1 mm length

16.6.2 Critical gas flow through nozzles and pipes with constant cross-section

Another classical example is the gas flow through short nozzle and pipes with a given length. Consider air flow through a 2-cm diameter pipe with inlet condition 1bar and 20 °C. Compute the critical mass flow rate for pipes with $L/D_h = 0$, 100 and 350 and compare the results with the measurements reported by *Frössel* (1936).

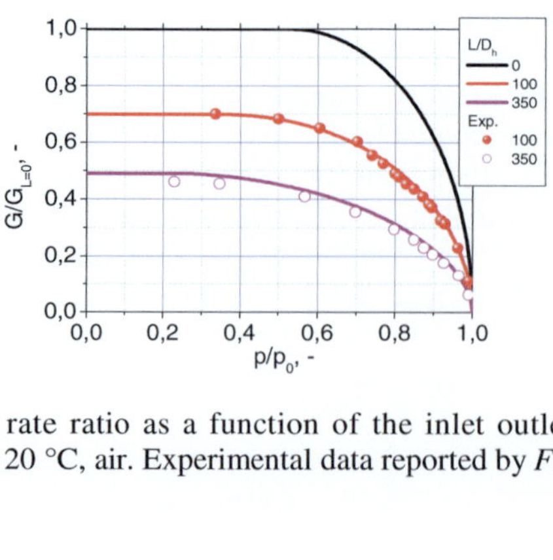

Fig. 16.6.2 Mass flow rate ratio as a function of the inlet outlet pressure ratio. D_h = 0.02 m, p_0 = 1 bar, T_0 = 20 °C, air. Experimental data reported by *Frössel* (1936).

Using the perfect gas theory we perform this computation and present the results in Fig. 16.6.2. We see that the prefect gas theory works well in this particular case.

Now, use the system computer code IVA and perform the same analysis for $L/D_h = 350$. Compare the results with the prediction of the perfect gas theory. The comparison is presented in Fig. 16.6.3. The slight deviations are explained with the use of variable properties in the computer code along the tube.

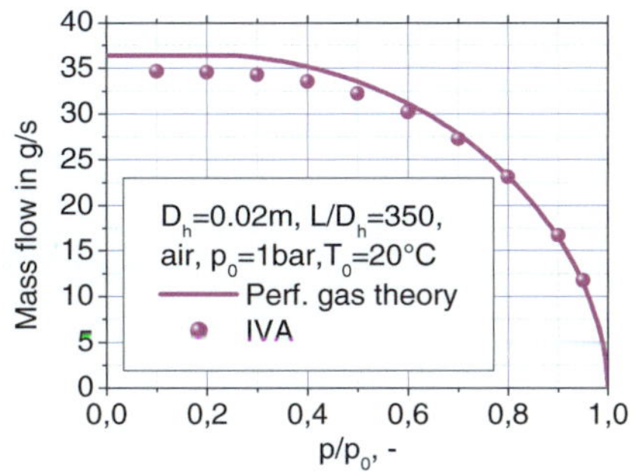

Fig. 16.6.3 Comparison between the prediction of the perfect gas theory and IVA computer code

Now consider a 1-m² vessel filled initially with 10 bar and 100 °C air connected with the environment by a 1-cm diameter pipe with length L/D_h = 100 and 350. Compute with the system computer code the discharge process with low spatial resolution, 20 computational cells only, and compare it with the solution resulting from the perfect gas theory. The solution is presented in Fig. 16.6.4. We see that even with very low resolution the computer code prediction is reasonable.

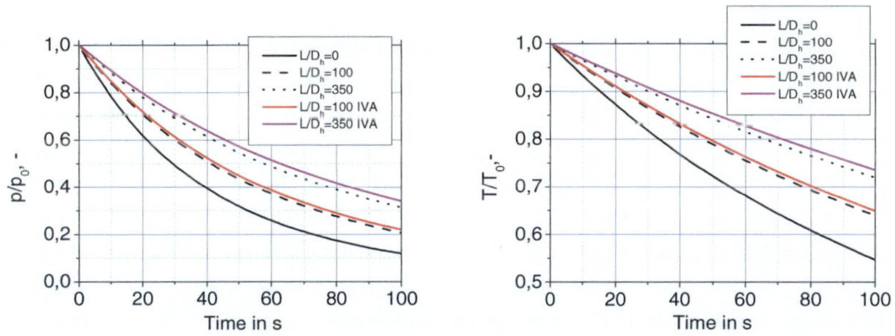

Fig. 16.6.4 Air release from a 1-m³ vessel through 1-cm diameter pipes. Pressure and temperature as functions of time.

16.6.3 U-tube benchmarks

16.6.3.1 Adiabatic oscillations

Consider the U-tube presented in Fig. 16.6.5a filled with water. The initial level deviates from the equilibrium one. The resulting oscillation period can be analytically computed for the frictionless case $\Delta\tau_{analytical} = 2\pi\sqrt{L/(2g)} = 1.608$ s. This makes this benchmark very suitable for verifying the dynamic performance of the code for gravitationally driven flow. The computed velocity is given in Fig. 16.6.5b.

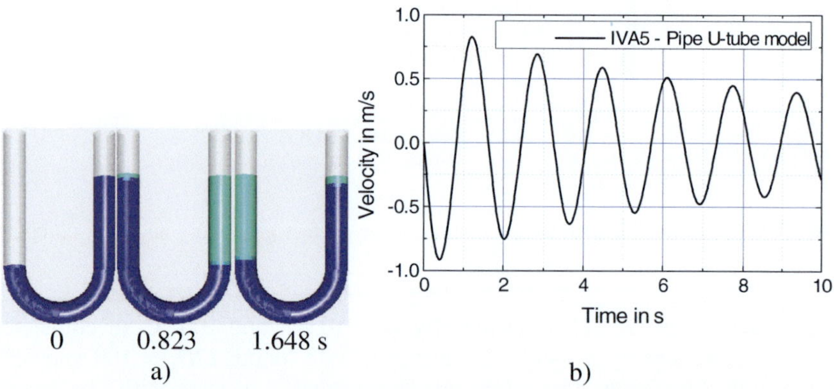

Fig. 16.6.5 IVA5 simulation of the U-tube benchmark. Length of the water axis L = 1.285 m. Analytical oscillation period 1.608 s neglecting the friction. Computed oscillation period 1.63 s (with friction + numerical diffusion). a) The first water oscillation. b) Velocity as a function of time. The process visualization was performed with SONJA

A similar case is modeled as a two-dimensional flow having the same cross-section of the vertical channels – see Fig. 16.6.6a. The resulting velocity as a time function is presented in Fig. 16.6.6b. We see that IVA5 predicts pretty well the period for the first case. Note that the period is only slightly influenced by the friction. In the second case the predicted period is 6.5% larger than the theoretical one. In the second case the cross-section normal to the flow velocity is not constant in the lower part. This is one of the reasons for this difference. In general, the damping of the amplitude is a result of the friction and of the numerical diffusion. The numerical diffusion is predominant compared to the friction for both cases.

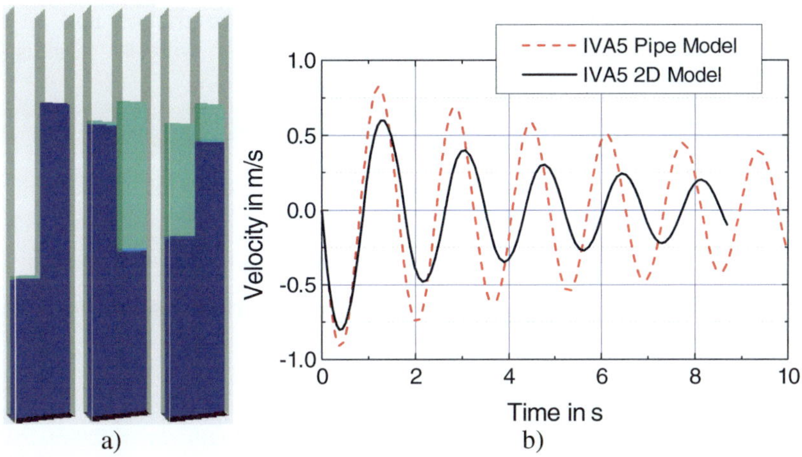

Fig. 16.6.6 IVA5 simulation of the 2D U-tube benchmark. Length of the water axis L = 1.282 m. Analytical oscillation period for 1D tube 1.606 s neglecting the friction. Computed averaged oscillation period 1.71 s. a) The first water oscillation. b) Velocity as a function of time. The process visualization was performed with SONJA

16.6.3.2 Single-phase natural convection in uniformly heated vertical part of a U-tube

We consider a U-tube with the following parameters: heated vertical part $\Delta z_{heat} = 1m$, total length $L = 2.1m$, pipe diameter $D_{pipe} = 0.032m$, constant heat flux density $\dot{q}''$ = 78946 W/m^2. The fluid is water at atmospheric pressure with thermal properties: density $\rho_{fr} \approx 993 kg/m^3$, thermal expansion $d\rho/dT = -0.408 kg/(m^3 K)$, specific capacity at constant pressure $c_p = 4169 J/(kgK)$. The friction factor is $\lambda_R = 0.0357$ for a turbulent flow with $\mathrm{Re} = 6912$. In this case the analytical solution *Kolev* (2002) for the mass flow rate is

$$G = \left(-4\Delta z_{heat}^2 \frac{g\rho_{in}}{\lambda_R L} \frac{d\rho}{dT} \frac{\dot{q}''}{c_p} \right)^{1/3}$$

$$= \left(4 \times 1^2 \frac{9.81 \times 993}{0.0357 \times 2.1} 0.408 \frac{78946}{4169} \right)^{1/3} = 158.95 kg/(m^2 s).$$

With this mass flow rate the *Reynolds* number takes the value already mentioned. The relative density change in this case is

$$B = \frac{1}{\rho_{in}} \frac{d\rho}{dT} \frac{\dot{q}''}{Gc_p} \frac{4\Delta z_{heat}}{D_h} = -\frac{0.408 \times 78946 \times 4 \times 1}{993 \times 158.95 \times 4169 \times 0.032} = -6.1625 \times 10^{-2},$$

which makes the influence of the density change on the friction and on the acceleration inertia negligible. We simulate the process with 42 equidistant computational cells. The prediction of IVA6 is

$$G_{IVA6} = 157.82 kg/\left(m^2 s\right).$$

The relative difference is less than 0.7%. We consider this agreement as excellent.

16.6.3.3 Single-phase natural convection in uniformly heated inclined part of a U-tube

We consider a U-tube consisting of three parts: the first is vertical and starts at $\Delta z_0 = 2.94m$ below a given horizontal plane. It has a length $\Delta z_1 = 2.94m$. The second part of length $\Delta z_{heated} = 4.2m$ is inclined at an angle 1.47063 rad with respect to the upwards directed vertical. The third part has a length $\Delta z_3 = 5.46m$, is vertical and ends at the above-mentioned horizontal plane. The total length of the pipe is $L = 12.6m$. The pipe has a diameter $D_{pipe} = 0.032m$. The heated part receives the heat flux density $\dot{q}'' = 78946\ W/m^2$. The fluid is water with thermal properties $\rho_{in} = 996.4 kg/m^3$, $\frac{d\rho}{dT} = -0.465\ kg/\left(m^3 K\right)$, $c_p = 4175 J/(kgK)$. The friction factors for the three parts are $\lambda_{R1} = 0.0316$, $\lambda_{R2} = 0.02951$, $\lambda_{R3} = 0.02809$ for a turbulent flow with $\mathrm{Re} = 16396$. The length-weighted averaging of the friction coefficient gives $\bar{\lambda}_R = 0.02938$. In this case the analytical solution *Kolev* (2002) gives

$$G = \left[-g \frac{8}{\bar{\lambda}_R} \rho_{in} \frac{d\rho}{dT} \frac{\dot{q}''}{c_p} \frac{\Delta z_{heated}}{L} \left(\Delta z_3 + \Delta z_{heated} \frac{1}{2} \cos\varphi \right) \right]^{1/3}$$

$$= \left(9.81 \frac{8}{0.02938} 996.4 \times 0.465 \frac{78946}{4175} \frac{4.2}{12.6} \left[5.46 + \frac{4.2}{2} \cos\left(1.47063\right) \right] \right)^{1/3}$$

$$= 353.65 kg/\left(m^2 s\right).$$

Neglecting the contribution of the inclined heated part results in $349.23 kg/\left(m^2 s\right)$, which is only 1.25% less, which means that the main contribution comes from the vertical nonheated part filled with hot fluid. With this mass flow rate the *Reynolds* number takes the value already mentioned. The relative density change in this case is

$$B = \frac{1}{\rho_{in}} \frac{d\rho}{dT} \frac{\dot{q}'' 4 \Delta z_{heated}}{G c_p D_1} - \frac{0.465 \times 78946 \times 4 \times 4.2}{996 \times 354.65 \times 4175 \times 0.032} = -1.307 \times 10^{-2},$$

which justifies the use of length-weighted averaging of the friction coefficient and the neglect of the inertia effect due to the density change. We simulate the process using IVA6 with 126 equidistant computational cells and obtain

$$G_{IVA6}\left(126cells\right) = 336.30kg/\left(m^2s\right),$$

which is 4.9% less than the analytical prediction. We learn from this example that appropriate discretization is crucial especially in the case of a sensible balance between small driving forces, friction and momentum loss due to numerical diffusion.

16.6.3.4 Single-phase natural convection in a U-tube with inclined part heated by steam condensation

As in the previous case we consider a U-tube consisting of three parts: the first is a vertical one and starts at $\Delta z_0 - 2.94m$ below a given horizontal plane. It has a length $\Delta z_1 = 2.94m$. The second part of length $\Delta z_{heated} = 4.2m$ is inclined at an angle 1.47063 rad with respect to the upwards directed vertical. The third part has a length $\Delta z_3 = 5.46m$, is vertical and ends at the above mentioned horizontal plane. The total length of the pipe is $L = 12.6m$. The pipe has a diameter $D_{pipe} = 0.032m$. The thickness of the pipe is 3 mm. The wall heat conduction coefficient used is that for steel, $\lambda_w = 14.93W/\left(mK\right)$. The heated part receives the heat flux density $\dot{q}'' = 78946\ W/m^2$. The fluid is water with thermal properties $\rho_{in} = 995.78kg/m^3$, $\frac{d\rho}{dT} = -0.465\ kg/\left(m^3K\right)$, $c_p = 4175J/(kgK)$. The friction factors for the three parts are $\lambda_{R1} = 0.0312$, $\lambda_{R2} = 0.0293$, $\lambda_{R3} = 0.0283$, respectively, for a turbulent flow with $\mathrm{Re} = 11886$. The length-weighted averaging of the friction coefficient gives $\bar{\lambda}_R = 0.02926$. For both heat-transfer coefficients we use the values $\alpha_{conv} = 1727W/\left(m^2K\right)$ and $\alpha_{cond} = 10396W/\left(m^2K\right)$. In this case, the analytical solution *Kolev* (2002) gives

$$G_1 = \sqrt{-2g\rho_{in}\frac{d\rho}{dT}\left(T_{cond} - T_{in}\right)\left(1 - e^{-a}\right)\frac{D_1}{\bar{\lambda}_R L}\left[\Delta z_3 + \Delta z_{heated}\left(\frac{1}{1 - e^{-a}} - \frac{1}{a}\right)\cos\varphi\right]},$$

or $G = 296.27kg/\left(m^2s\right)$, with

$$\bar{\lambda}_R = \lambda_{R1}\frac{\Delta z_1}{L} + \lambda_{R2}\frac{\Delta z_{heated}}{L}\frac{\ln\left\{\varepsilon_0\left[\left(1/\varepsilon_0 + 1\right)e^a - 1\right]\right\}}{a\left(1 + \varepsilon_0\right)} + \lambda_{R3}\frac{\Delta z_3}{L}\frac{1}{1 + B}.$$

The thermal power computed for 4 times 120 pipes is $P(4 \times 120 pipes) = 13.151 MW$. Note that the expression

$$\frac{\ln\left\{\varepsilon_0\left[\left(1/\varepsilon_0+1\right)e^a-1\right]\right\}}{a\left(1+\varepsilon_0\right)}=1.00498,$$

is very close to one. With this mass flow rate the *Reynolds* number takes the value already mentioned. The relative density change in this case is $B = -9.14 \times 10^{-3}$, which justifies the use of length-weighted averaging of the friction coefficient and the neglect of the inertia effect due to the density change. The computed outlet water temperature is 330.68 K. The coefficient "*a*" is about ½, which defines the local distribution of the water temperature.

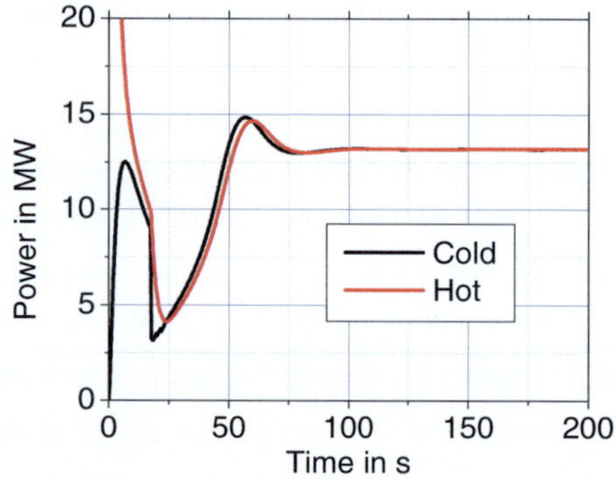

Fig. 16.6.7 Power of a condenser with 4 times 120 pipes as a function of time at the two sides of the heat-conducting pipe

We simulate the process with 126 equidistant computational cells on the primary side. The secondary size is simulated by a large-diameter pipe (0.1 m) which is concentric with the heated part and has a bend at the lower end going downwards. The secondary pipe is then equidistantly discretized with 52 cells. The boundary conditions at the secondary pipe are atmospheric pressure and saturated steam. The transient behavior is given in Fig. 16.6.7. The steady state solution is

$$G_{IVA6}\left(126 cells\right) = 312.17 kg/\left(m^2 s\right),$$

$$P_{IVA6}\left(4 \times 120 pipes\right) = 13.1932 MW,$$

which is in very good agreement with the analytical prediction. The computed outlet water temperature is 329.36 K, which compared to the analytical solution of 330.68 K is also in a very good agreement. Note that to allow for correct comparison with the analytical solution, we took the averaged values of the friction coefficients, and of the heat-transfer coefficients, as well as the thermal properties as computed by IVA6 and used them in the analytical solution.

16.6.4 Flow boiling stability

Natural circulation loops are known to behave like the already-discussed U-tube manometers. They have their own eigenfrequency of oscillations, depending on the square of the length of the liquid axis. In addition, heat input into the fluid without or with boiling can excite damped or not damped oscillations depending on the input power. In boiling loops density waves interact with their own "U-tube" oscillations. Therefore, each natural circulation loop has its own signature. For many practical reasons it is always of interest for a given geometry and inlet flow properties to find the mass-flow–power relation and the stability region. Analytical benchmarks for loops heated by external condensation and the successful validation of the IVA performance on them are provided in the previous section (see also in *Kolev* (2002)). The specific physics in those cases was that condensation deposits the latent heat into the wall and that the feedback of the power to the mass flow is very sensitive.

Our intention is to verify the computer code IVA for stability analysis of boiling bundles in loops. Thermal hydraulic instabilities in boiling systems are reviewed by *Yadigaroglu* (1978, 2005), *Nakanishi* (1981) and *Kakac* (1985). In the last 50 years the stability of boiling systems has usually been analyzed by using simplified linearized models applying perturbation theory. It is outside the scope of this section to review this subject but let us recall some of the representative works:

1) Linear stability analysis using homogeneous equilibrium models are reported in *Wallis* and *Heasley* (1961), *Stenning* and *Veziroglu* (1965, 1967), *Lahey* and *Yadigaroglu* (1973), *Yadigaroglu* and *Lahey* (1975), *Furutera* (1986), *Lahey* and *Moody* (1993), *Nayak* et al. (1998), *Delmastro* et al. (1991), *Lee* and *Lee* (1991), *Wang* et al. (1994) among others. *Ambrosini* and *Ferreri* (2006) demonstrated that numerical accuracy is crucial to obtain consistent results. Bad spatial resolution for instance either degrades or shifts the boundaries in the stability maps.
2) Linear stability analysis using slip or drift flux equilibrium models are reported in *Fukuda* et al. (1984), *Ishii* (1971), *Saha* and *Zuber* (1978), *Park* et al. (1986), *Rizwan-Uddin* and *Doming* (1986), *van Bragt* et al. (2000) among others.
3) *Saha* (1974) and *Saha* et al. (1976) extended the earlier work by adding a subcooled boiling model. Using such types of models *Chan* (1979) and *Chan* and *Yadigaroglu* (1981) analyzed the stability of steam generators, and *Wolf* et al. (1982) analyzed stability of solar-thermal steam generators. Allowing for flushing in such models by enforcing equilibrium outside the subcooled boiling region *Zhou* and *Yadigaroglu* (1988), *Inada* et al. (2000) analyzed the stability of small heating reactors with flashing driven by pressure changes. Coupling this type of coolant modeling with nodal neutron kinetics

was done by *Park* et al. (1983a, b, c), *Peng* et al. (1985, 1986), *Taleyarkhan* (1982), *Taleyarkhan* et al. (1985, 1994). Finally, one should mention the recent development of the MATSTAB program *Hänggi* (2001) *Hänggi* et al. (1999) that is a linearized version of the RAMONA computer program – a time-domain code which is often used for BWR stability analysis.

The advantage of this method – its simplicity – is also its drawback. Nonlinearities that come from the complex physics are not taken into account and therefore essential instability modes may not be revealed. The computational power of the resent computers allows applying complete nonlinear models to instability analyses and overcoming these difficulties. Moreover, the instability analysis is also a fundamental tests for thermal-hydraulics computer codes, allowing to judge whether the codes possess the appropriate mathematical correctness to reflect real oscillations of the process parameters.

In this section we select one experimental data set obtained by *Schäfer* and *Beisiegel* in 1992 on the AREVA/FANP KATHY loop and compare the results of the simulation with IVA computer code. Finally, conclusions are drawn.

In the case considered here a test series with no. 11 for ATRIUM-10B fuel bundle, see Fig. 16.6.8, performed on the AREVA/FANP KATHY loop, *Schäfer* and *Beisiegel* (1992), is used. An electrically heated 1:1 simulator for this bundle is used. The schematic loop representation of the geometry is given in Fig. 16.6.9a.

Fig. 16.6.8 71-MWd/kgU burnup ATRIM 10-rod bundle

This geometry was modeled using the network capability of IVA. Some inevitable approximations of the geometry are done here and there. So for instance several irreversible friction contributions before the entrance of the bundle section are lumped in an irreversible friction coefficient 25.62 related to the flow cross-section in the bundle. The inlet temperature was assumed to be 560.89 K although some small variations from experiment to experiment are recorded.

We prescribe for the computation the time evolution of the electrical power of the bundle as shown in Fig. 16.6.9b. After each jump the power remains constant to see the steady-state solution as shown in Fig. 16.6.10. Figure 16.6.10 shows the mass flow as a function of time.

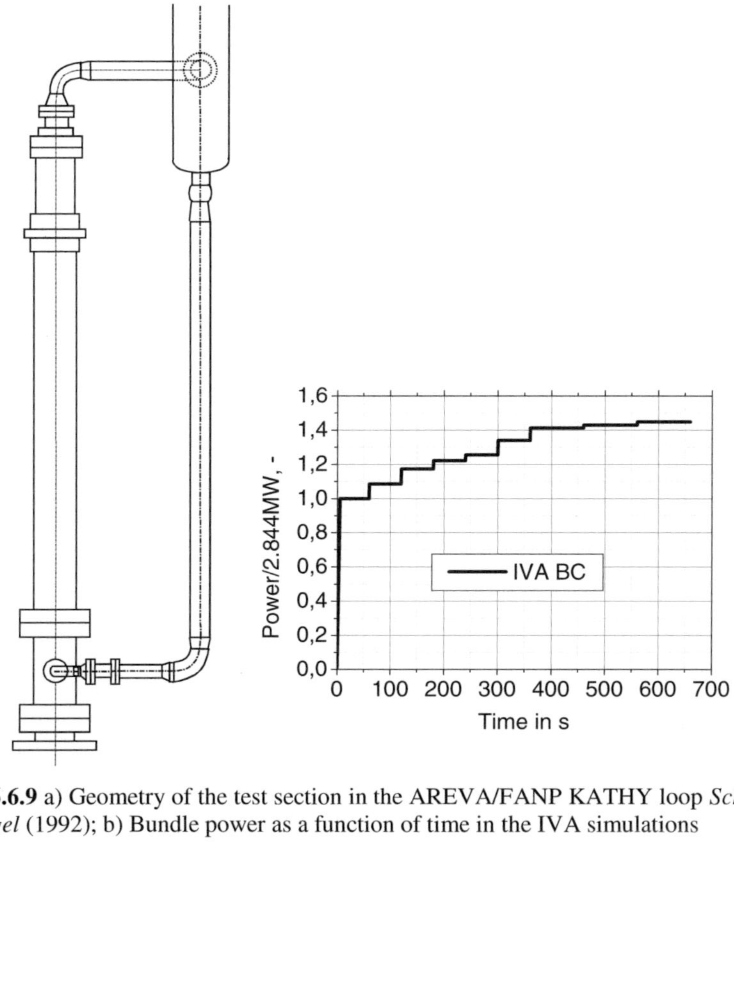

Fig. 16.6.9 a) Geometry of the test section in the AREVA/FANP KATHY loop *Schäfer* and *Beisiegel* (1992); b) Bundle power as a function of time in the IVA simulations

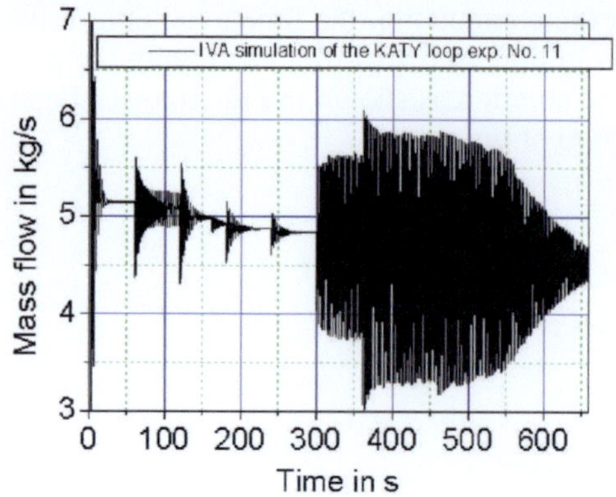

Fig. 16.6.10 Mass flow as a function of time for changing bundle power as given in Fig. 16.6.9b

Analyzing the solution we obtain the results presented in Figs. 16.6.11a and 16.6.12a.

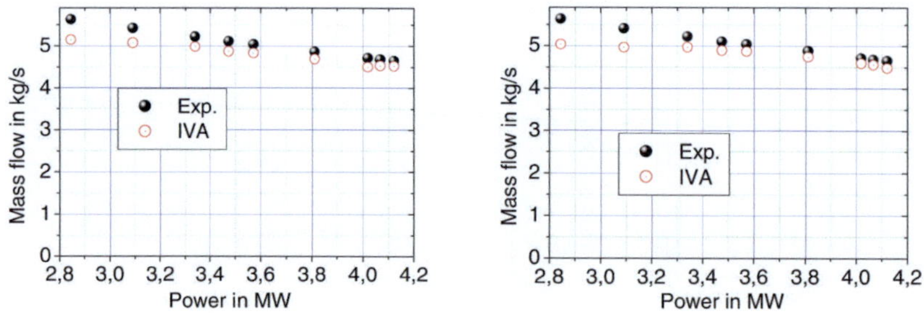

Fig. 16.6.11 Mass flow as a function of the bundle power, *Kolev* (2006): a) Power change as given in Fig. 16.6.9b; b) Power jump from zero.

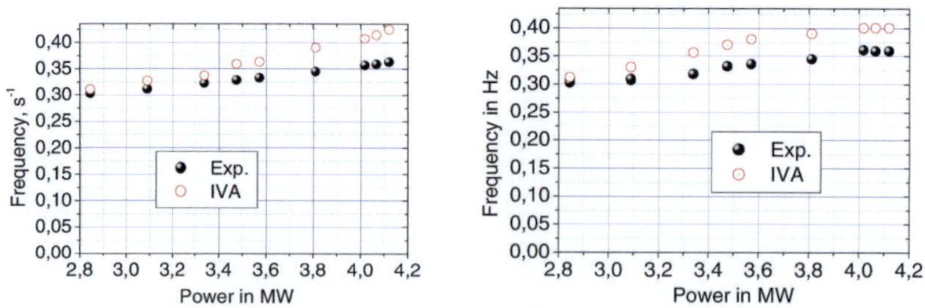

Fig. 16.6.12 Frequency of the oscillation of the total mass flow, *Kolev* (2006): a) Power change as given in Fig. 16.6.9b; b) Power jump from zero.

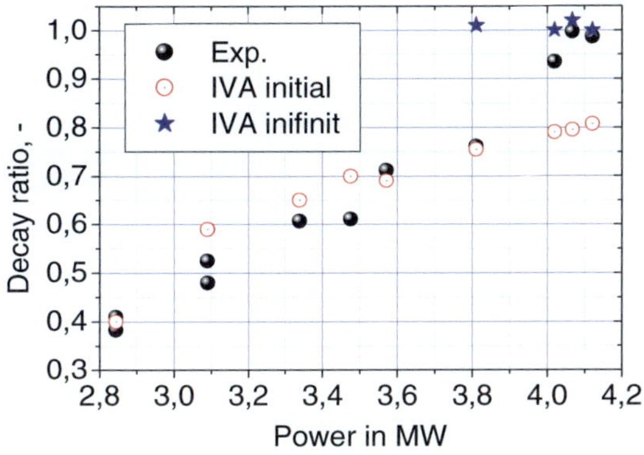

Fig. 16.6.13 Decay ratio of the oscillation as a function of power, *Kolev* (2006)

In order to check whether the obtained solutions are dependent on the way the power disturbance is imposed we repeat the computation with imposing a jump of the power from zero to each measured value.

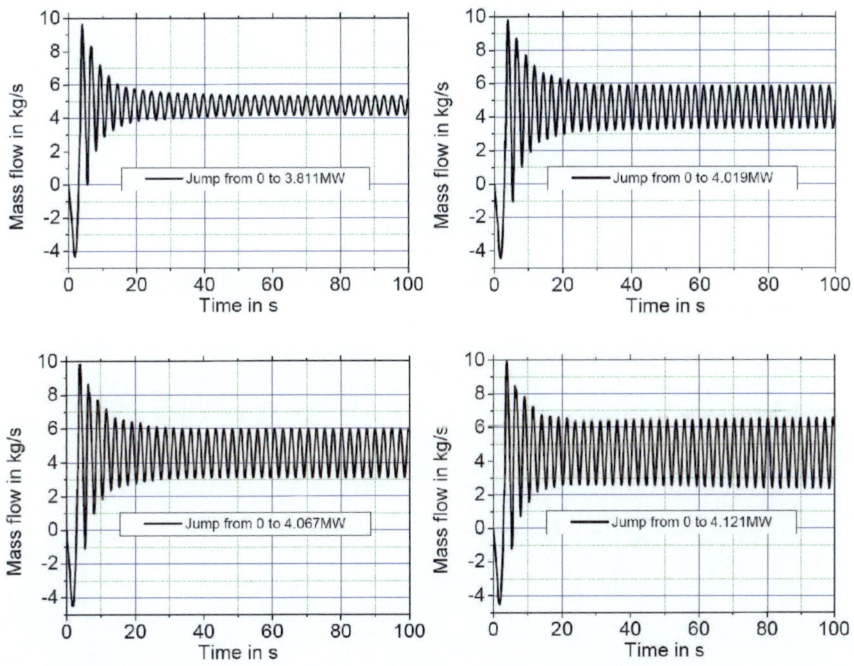

Fig. 16.6.14 Mass flow as a function of time for the cases approaching the stability limit

The results are presented in Figs. 16.6.11b and 16.6.12b. Figure 16.6.11 shows the mass flow as a function of the power. Increasing the power increases the void over the heated section and after it and causes a larger pressure loss that itself reduces the mass flow. The measured trend is properly predicted. The selected irreversible friction coefficient is obviously somewhat larger than the real one. The observed frequencies of the oscillations slightly increase with the power increase, as shown in Fig. 16.6.12. This trend is also properly predicted by IVA. The difference between the computed and observed oscillation frequencies slightly increases with the power. The decay ratio for each case is presented in Fig. 16.6.13. We see that up to 3.6 MW the predicted and the measured decay ratios agree well with the experiment. Up 3.811 MW, as shown in Fig. 16.6.14, the initial perturbation is damped but the final steady state is a harmonic oscillation. Increasing of the power results in exponential increasing of the amplitude as shown in Fig. 16.6.15. The structure of the density waves for the case of 4.121 MW is illustrated on Fig. 16.6.16, where mass flow rates and void fractions at different elevations are presented as a time functions.

$$Mass\ flow\ in\ kg/s = 0.48556 + 3.70165\times10^{-15}\exp\left[\left(Power\ in\ MW\right)/0.12228\right]$$

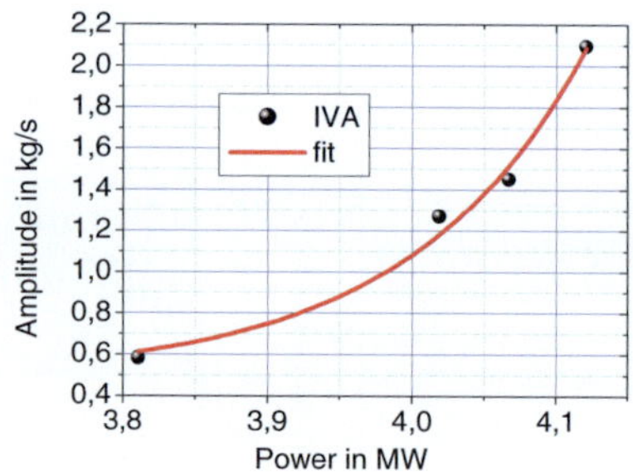

Fig. 16.6.15 Amplitude of the oscillations as a function of the bundle power close to the stability limit.

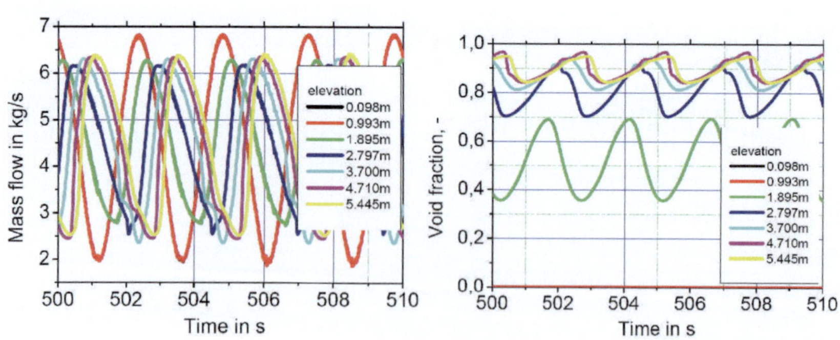

Fig. 16.6.16 Structure of the density waves for the 4.121-MW case: a) Mass flow as a function of time at different elevations; b) Void fractions as a function of time at different elevations.

At powers above 4.15 MW nondamped oscillations are expected. We see a correct prediction of the measured phenomena by IVA computer code.

Conclusions

There are two important conclusions drawn from this analysis: one regarding the mathematical modeling of the complex boiling dynamics in bundles in oscillating systems and the other regarding the practical usefulness of the computer tool for such analyses.

1. The complete simulation of the dumped and not dumped instability observed in the eleven experiments with all complexities of the interactions of different physical mechanisms, mathematical methods, geometry descriptions, etc. is a severe challenge for any computer code simulating flow boiling. The IVA computer code successfully reproduces the complexity of the observed processes.
2. Besides the condensation-induced instability benchmarks reported in the previous section this benchmark demonstrates the capability of IVA to be successfully used for stability analyses in boiling loops and in particular for boiling loops of BWR type.

Acknowledgment: This study was motivated by the fruitful discussion with *Douglas Pruitt* and *Yousef Farawila* from Framatome-ANP in the US. The data processing done by *Eva Ziegerer* from the University of Erlangen is highly appreciated.

16.6.5 Flow condensation stability

Condensation processes are similar to the boiling processes in the sense of occurrence of density-wave oscillations. I will consider here experimentally observed in-pipe condensation and will try to reproduce the experiments performed in the Research Center Jülich by *Schaffrath* (1996) with the IVA computer code. Many interesting features will be discussed in the course of this comparison. 8 condenser steel pipes with averaged length 9.8m, 0.0378 m internal diameter and 0.0029 m thickness as shown in Fig. 16.6.17 are used. The pipes are placed as shown in Fig. 16.6.18 in a 1 bar secondary water reservoir containing saturated water. The condensate pipes are connected with a high-pressure vertical cylinder as shown in Fig. 16.6.19. The nozzle at the entrance reduces the cross-section to 0.462 of the inlet pipe cross-section and at the exit to 0.432. This is taken into account.

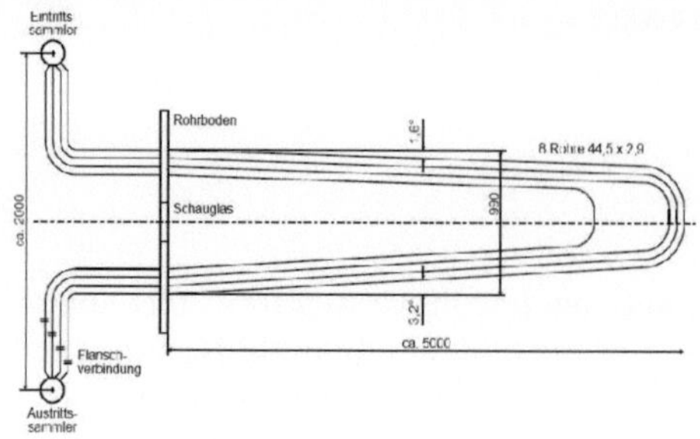

Fig. 16.6.17 Condenser pipes

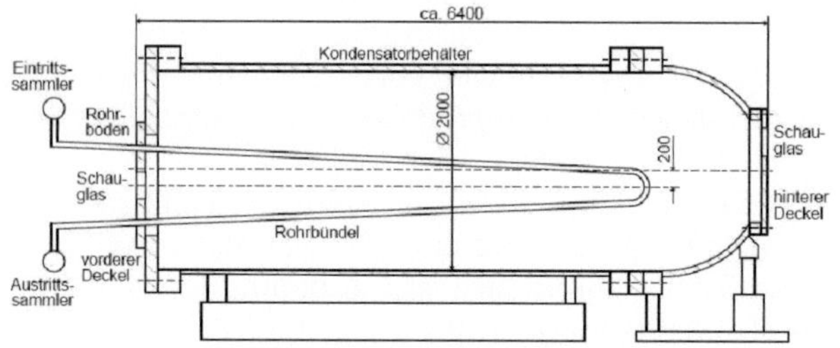

Fig. 16.6.18 Condenser pipes in secondary vessel

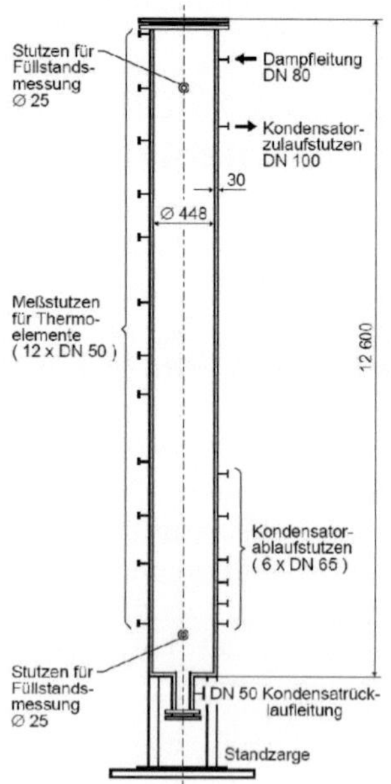

Fig. 16.6.19 Pressure vessel

The experiments designated with A are performed with steam at 70, 50 30, 10 and 3 bar and different levels of water above the condensate outflow nozzle. After adjusting the water level the power was measured in different ways. The secondary temperature was the saturation temperature at 1 bar pressure. Vertical plates are used to direct the secondary flow vertically. The main result of the experiment is the thermal power as a function of the water level above the condensate exit pipe.

The simulation performed with the pipe network features of IVA has a 10-cm spatial resolution. The secondary site is very approximately modeled by enforcing water flow through a countercurrent environment allowing almost constant external wall temperature corresponding to boiling. The initial conditions are: condenser system filled with water. At time zero the condensate exit pressure is set to the pressure corresponding to the water level. Computationally, the simulated empting process which takes about 10 s initiates the condensation process. Then, a quasisteady state is computationally established and the power reached is recorded. The first observation is that the steady state is a stable oscillatory state. The time-averaged thermal power is used for comparison with the measurements.

Next, I will discuss in some detail the dynamic behavior of such a system based on the A3 experiment series. Figures 16.6.20a–e show the primary and the secondary power as a function of time. The obtained dynamical steady state is clearly visible.

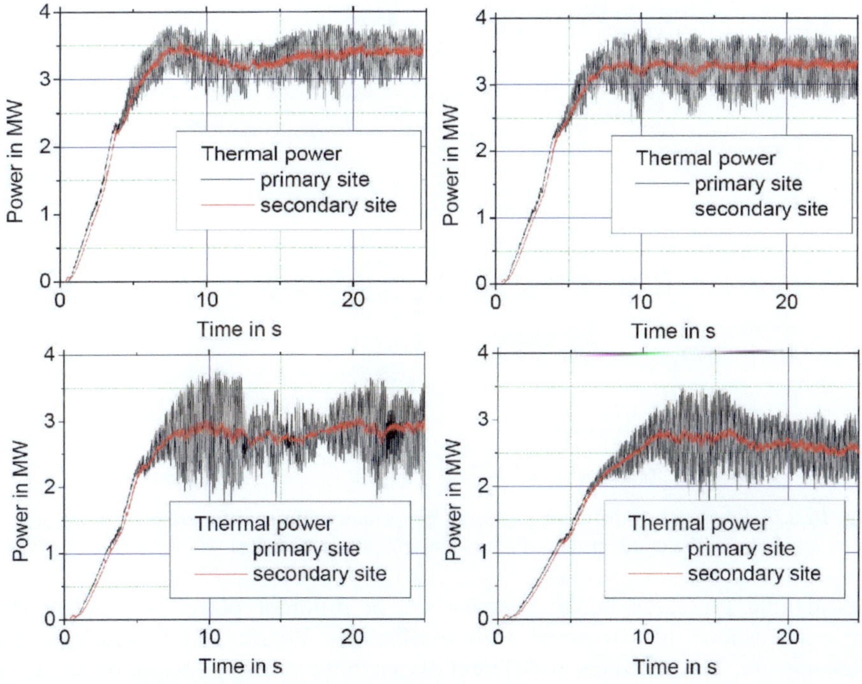

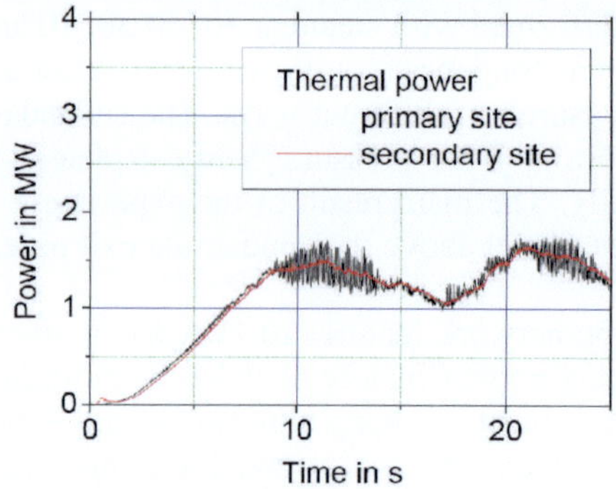

Fig. 16.6.20 Jülich emergency condenser experiment: Thermal power as a function of time. Vessel pressure 70 bar. Water level over the drainage nozzle: a) 0.88 m; b) 1.8 m; c) 3.4 m; d) 4.9 m; e) 6.3 m

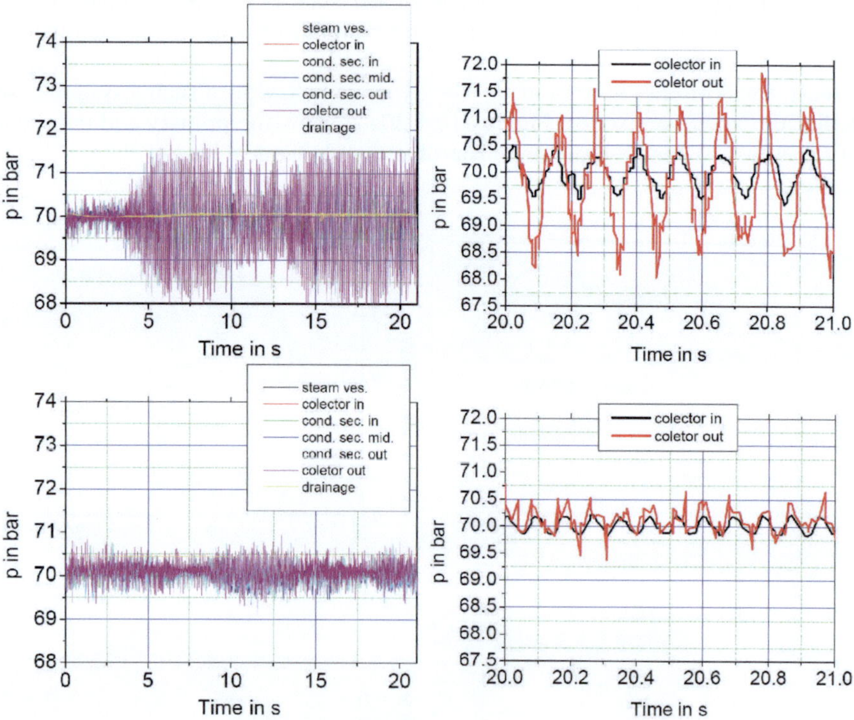

Fig. 16.6.21 Pressures at different places in the primary site as a function of time for 0 m water level above the condensate outlet nozzle: a) 0.88 m; b) 6.3 m

Plotting the pressures in the primary site at different places we realize that the condensation is associated with oscillations. Figure 16.621 illustrate this phenomenon. The pressures at different places in the primary site are presented as a function of time for 0.88 m and 6.5m water level above the condensate outlet nozzle. Oscillations of 0.2–2 bar amplitudes with frequency about 8 Hz are expected. Increasing the level up to 6.3 m increases the frequency up to about 12 Hz by reducing the amplitude.

Figure 16.6.22a shows the computed thermal power as a function of the water level above the exit nozzle. It is as expected, a decreasing function of the submergence of the exit pipe. The experimental data points are also entered in this figure.

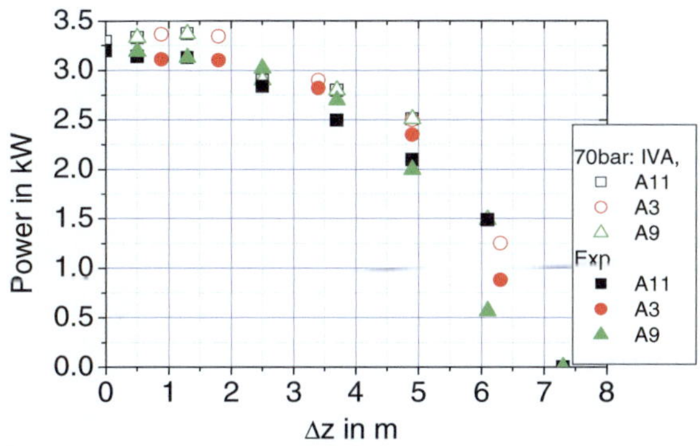

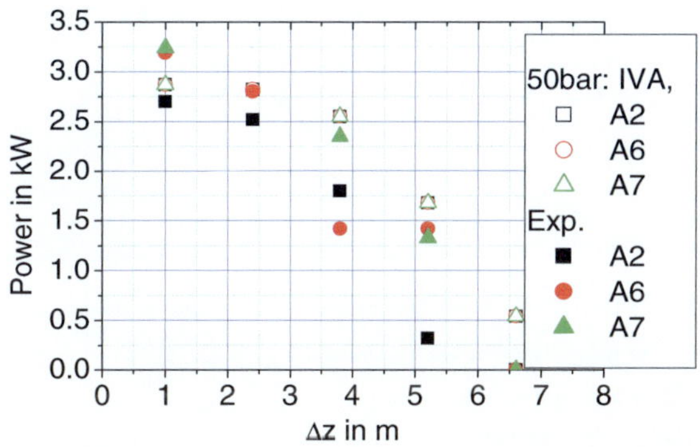

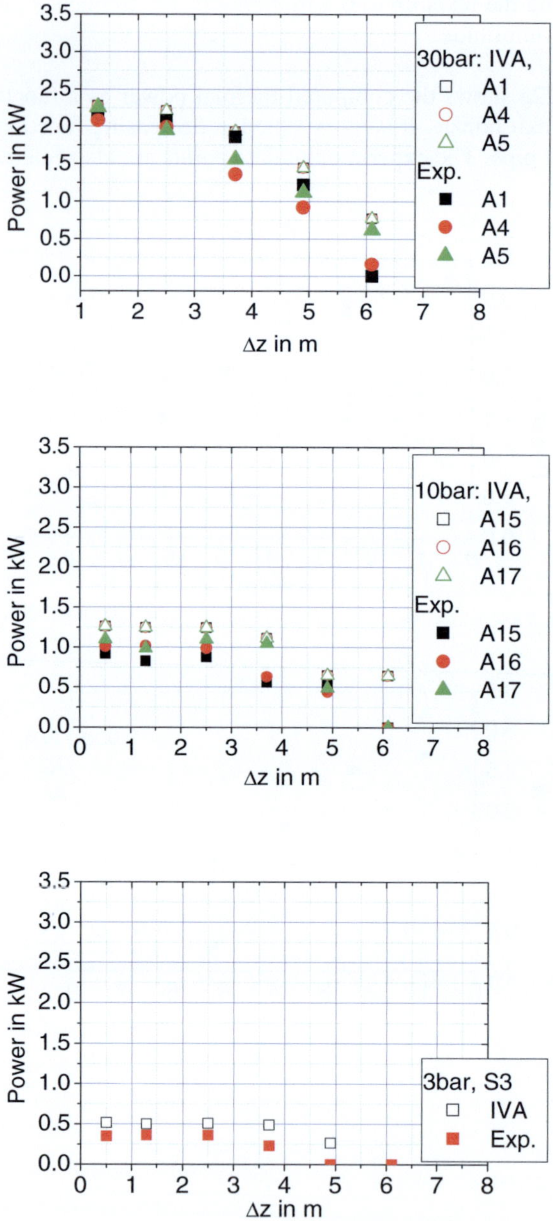

Fig. 16.6.22 Power as a function of the water level above the condensate exit nozzle

I perform similar analysis for all 73 available states for pressures of 70, 50, 30, 10 and 3 bar. The groups of data are plotted on Figs. 16.6.22a–e. Before judging the accuracy of the computation let us see Fig. 16.6.23 that is taken from p. 68 of *Schaffrath*'s work. It shows that the error is below 10% for powers larger then 1.5 MW. For smaller power the error exponentially increases.

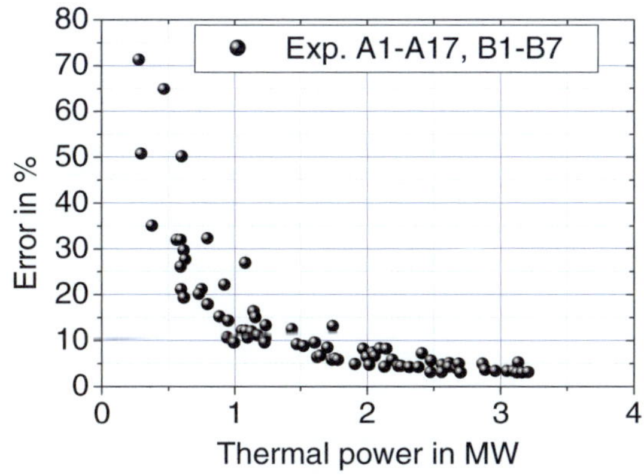

Fig. 16.6.23 *Schaffrath* (1996): Measurement error in % estimated by three different methods

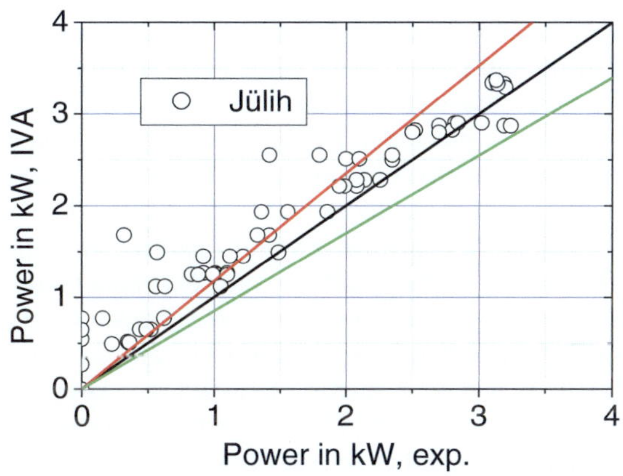

Fig. 16.6.24 Power estimated using IVA computer code compared to the real measured experimental power

Figure 16.6.24 shows the power estimated using IVA computer code compared to the real measured experimental power for all 73 points. The 15% error band is also given.

One source of error of the modeling is the not very accurate simulation of the secondary site. But I believe that this has a minor contribution in this particular case because the secondary heat transfer is nucleate boiling. Due to the very high secondary heat-transfer coefficient the main thermal resistance is on the condensation site.

> Comparing Figs. 16.6.23 and 16.6.24 we realize that as long as the measurements are accurate, the agreement with the analysis is very good.

With decreasing thermal power the measurement error increases exponentially making data comparison not informative.

The important output of such analysis is obtaining the oscillating characteristics of the gravitationally removed condensate and the interaction with the condensing steam. At small power, the amplitudes are low but with increasing power the amplitudes increases. Therefore, the structural design of such facilities has to take this into account.

16.6.6 Gravitational 2D waves

An experiment performed by *Maschek* et al. (1992) was used to study the capability of IVA5 to model gravitational relocation of water, the so-called dam-break problem. The predicted water volume fraction as a function of radius and height for different times is presented in Fig. 26.6.25. Good agreement between IVA5 prediction and experiment is obtained from the start of the experiment to the moment when the water reaches the external boundary. The next part of the experiment is associated with growth of the surface instabilities and strong turbulization of the water. Note that in the experimentally observed data circumferential instability occurs, with 16 to 18 eruptions taking place with peaks located approximately equidistantly in one circle. 2D modeling of the three-dimensional process does not predict this 3D effect as expected. Here, the code predicts relatively compact movement of the water with weaker "turbulence" and fragmentation of the surface than that actually observed in the experiment.

Fig. 16.6.25 Comparison between the IVA5 prediction of the water volume fraction as a function of radius and height and the experimental observation by Maschek et al. with the time used as a parameter: 0, 0.08, 0.16, 0.20, 0.40, 0.56, 0.64, 0.72 s (from left to right)

Note that *Fannelop* and *Waldman* (1972) reported an analytical solution for the position of the triple point at the bottom as a function of time:

$$r(\tau) = 2\left[\frac{g\ Vol\ Fr}{\pi(4 - Fr)}\right]^{0.25} \sqrt{\tau}\ .$$

Here *Vol* is the initial liquid volume of the cylinder and the *Froude* number $Fr = V/\sqrt{ge}$ is defined as a function of the bottom spreading velocity V and the thickness of the depth of the fluid at the leading edge e.

Conclusion. From this comparison it can be concluded that the hydraulic model of IVA5 works properly for cases in which gravitational acceleration is driving the liquid flow, and that it needs improvement for cases in which strong surface turbulence and surface droplet entrainment occur. Note that turbulence modeling for general multiphase flows is at the very beginning of its development. High-order modeling of the free surface, as already done by some authors, will substantially improve the results.

References

Ambrosini, W., Ferreri, J.C.: Analysis of basic phenomena in boiling channel instabilities with different flow models and numerical schemes. In: Proc. of ICONE14, Int. Conf. on Nuclear Engineering, Miami, FA, USA, July 17-20 (2006)

Chan, K.C.: Thermal-hydraulic stability analysis of steam generators. Ph.D. Thesis, Department of Nuclear Engineering, University of California Berkeley (1979)

Chan, K.C., Yadigaroglu, G.: Two-phase flow stability of steam generators. In: Bergles, A.E., Ishigai, S. (eds.) Two-Phase Flow Dynamics, pp. 365–377. Hemisphere Publishing Corp., Washington (1981)

Delmastro, D.F., Clausse, A., Converti, J.: The influence of gravity on the stability of boiling flows. Nucl. Eng. Des. 127, 29–139 (1991)

Fannelop, T.K., Waldman, G.D.: Dynamics of oil slicks. AIAA J 10(4), 506–510 (1972)
Frössel, W.: Strömung in glatten, geraden Rohren mit Über-und unterschalgeschwindigkeit. Forsch. 7(2), 75–84 (1936)
Fukuda, K., Kato, A., Hasegawa, S.: Two-phase flow instability at low flow rate conditions. J. Nucl. Sci. Technol. 21, 491–500 (1984)
Furutera, M.: Validity of homogeneous flow model for instability analysis. Nucl. Eng. Des. 95, 65–77 (1986)
Hänggi, P., Smed, T., Lansäker, P.: A fast frequency domain based code to predict boiling water reactor stability using detailed three dimensional model. In: NURETH-9, San Francisco (1999)
Hänggi, P.: Investigating BWR stability with a new linear frequency-domain method and detailed 3D neutronics. Doctoral dissertation. Swiss Federal Institute of Technology, Zurich BTHZ (2001)
Idelchik, I.E.: Handbook of hydraulic resistance, 3rd edn. Begell House, New York (1993)
Inada, F., Furuya, M., Yusuo, A.: Thermo-hydraulic instability of boiling natural circulation loop induced by flashing (analytical consideration). Nucl. Eng. Des., 187–199 (2000)
Ishii, M., Zuber, N.: Thermally induced flow instabilities in two-phase mixtures. In: 4th International Heat Transfer Conference, Paris, vol. 5 (1970)
Ishii, M.: Thermally induced flow instabilities in two-phase mixtures in thermal equilibrium. PhD Thesis. School of Mechanical Engineering, Georgia Institute of Technology (1971)
Kakac, S.: Review of two-phase flow instabilities. Metu. J. Pure and Applied Sciences 18, 171–252 (1985)
Kolev, N.I.: Simple analytical U-tube benchmarks appropriate for testing of pipe-network computer codes. In: Proceedings of ICONE10: Tenth International Conference on Nuclear Engineering, Arlington, Virginia (Washington, D.C.), USA ICONE10-22711, April 14-18 (2002)
Kolev, N.I.: Verification of IVA computer code for flow boiling stability analysis. In: Proceedings of ICONE-14 2006, Miami, Florida, USA, Paper ICONE14-89034 (July 17-20, 2006)
Lahey Jr., R.T., Moody, F.J.: The thermal-hydraulics of a boiling water nuclear reactor. ANS (1993)
Lee, S.Y., Lee, D.W.: Linear analysis of flow instabilities in an open two-phase natural circulation loop. Nucl. Eng. Des. 128, 317–330 (1991)
Maschek, W., Roth, A., Kirstahler, M., Meyer, L.: Simulation experiments for centralized liquid sloshing motions, Kernforschungszentrum Karlsruhe, KfK 5090 (December 1992)
Nakanishi, S.: Recent Japanese research on two-phase flow instabilities. In: Proceedings of Japan–US Seminar on Two-phase Flow Dynamics, Hemisphere Publishing Corporation, Washington, D.C (1981)
Nayak, A.K., Vijayan, P.K., Saha, D., Venkat Raj, V., Aritomi, M.: Linear analysis of thermohydraulic instabilities of the advanced heavy water reactor (AHWR). J. Nucl. Sci. Technol. 35, 768–778 (1998)
Park, G.C., Podowski, M., Becker, M., Lahey Jr., R.T.: The modeling of density-wave oscillations in boiling water nuclear reactors. In: Kakac, S., Ishii, M. (eds.) Advances in Two-Phase flow and Heat Transfer. Martinus Nijhoff Publishers, Boston (1983a)
Park, G.C., Podowski, M., Becker, M., Lahey Jr., R.T.: The development of NUFREQ-N, An analytical model for the stability analysis of nuclear coupled density-wave oscillations in boiling water nuclear reactors. NLIREC/CR3375 (1983b)

Park, G.C., Becker, M., Park, G.C.: Nodal analysis for reactor kinetics and stability. NUREG/CR-3377 (1983c)

Park, G.C., Podowski, M., Becker, M., Lahey, R.T., Peng, S.J.: The development of a closed-form analytical model for the stability analysis of nuclear-coupled density-wave oscillations in Boiling Water Reactors. Nucl. Eng. Des. 92, 253–281 (1986)

Peng, S.J., Podowski, M., Beker, M., Lahey Jr., R.T.: NUFREQNP computer code for the linear stability analysis of boiling water reactors. NUREG/CR-4116 (1985)

Peng, S.J., Podowski, M., Lahey Jr., R.T.: BWR linear stability analysis (NUFREQ-NP). Nucl. Eng. Des. 93, 25–37 (1986)

Taleyarkhan, R.P.: An analysis of density wave oscillations in ventilated boiling channels. PhD Thesis, RPI, Tray, NY (1982)

Taleyarkhan, R.P., Podowski, M.Z., Lahey Jr, R.T.: Ventilated channel instability analysis. 1. Heat Transfer 107, 175–181 (1985)

Taleyarkhan, R.P., McFarlane, A.F., Lahey Jr., R.T., Podowski, M.Z.: Benchmarking and qualification of the NUFREQ-NPW code for best-estimate prediction of multi-channel stability margins. Nucl. Eng. Des. 151, 151–171 (1994)

Rizwan-Uddin, D.J.: Some nonlinear dynamics of a heated channel. Nucl. Eng. Design 93, 1–14 (1986)

Saha, P.: Thermally induced two-phase flow instabilities, including the effect of thermal non-equilibrium between the phases. Ph.D. Thesis, School of Mechanical Engineering, Georgia Institute of Technology (1974)

Saha, P., Ishii, M., Zuber, N.: An experimental investigation of the thermally induced flow oscillations in two-phase systems. J. Heat Transfer, Trans. ASME 98, 616–622 (1976)

Saha, P., Zuber, N.: An analytical study of the thermally induced two-phase flow instabilities including the effects of thermal non-equilibrium. Int. J. Heat Mass Transfer 21, 415–426 (1978)

Schäfer, H., Beisiegel, A.: Feasibility of stability tests under natural circulation conditions in the Karlstein test loop, Siemens technical report E32/92/e14a, proprietary (March 17, 1992)

Schaffrath, A.: Experimentelle and analytische Untersuchungen zur Wirksamkeit des Notkondensators des SWR600/1000, Institut für Sicherheitsforschung und Reaktortechnik, Forschungszentrum Jülich GmbH, 3326, D294 (Diss. Universität Bochum 1996) (1996) ISSN 0944-2952

Stenning, A.H., Veziroglu, T.N.: Flow oscillation modes in forced convection boiling. In: Proceedings of the 1965 Heat Transfer and Fluid Mechanics Institute, pp. 301–316. Stanford University Press, Palo Alto (1965)

Stenning, A.H., Veziroglu, T.N.: Oscillations in two component two phase flow, vol. 1, NASA CR-72121; Flow oscillations in forced convection boiling, vol. 2, NASA CR72122 (1967)

van Bragt, D.D.B., Rizwan-Uddin, van der Hagen, T.H.J.J.: Effect of void distribution parameter and axial power profile on boiling water bifurcation characteristics. Nucl. Sci. Eng. 134, 227–235 (2000)

Wallis, G.B., Heasley, J.H.: Oscillations in two-phase flow systems. J. Heat Transfer, Trans. ASME 83, 363 (1961)

Wang, F., Hu, L., Chin, P.: Thermal and stability analysis of a two-phase natural circulation loop. Nucl. Sci. Eng. 117, 33–46 (1994)

Wolf, S., Chan, K.C., Chen, K., Yadigaroglu, G.: Dynamic instabilities in radiation-heated boiler tubes for solar central receivers. ASME Paper 82-WA/HT 8 (1982)

Yadigaroglu, G.: Two-phase flow instabilities and propagation phenomena, in two-phase flows in nuclear reactors. Von Karman Institute of Fluid Dynamics Lecture Series (1978)

Yadigaroglu, G., Lahey Jr., R.T.: A Lagrangian analysis of two-phase hydrodynamic and nuclear-coupled density wave oscillations. In: Proceeding of the Fifth International Heat Transfer Conference, Tokyo, September 3-7, vol. 4 (1975)

Yadigaroglu, G.: Boiling water reactor stability revised: The effect of flashing. Nucl. Eng. Des. 235, 1093–1105 (2005)

Zhou, Z.W., Yadigaroglu, G.: Stability of natural circulation with gravity-induced flashing. Trans. ANS 57, 379–381 (1988)

16.7 Steady-state single-phase nozzle flow

The prediction of steady-state pressure variation in a nozzle with variable cross-section is used in the test presented in this section. Tables 16.7.1 and 16.7.2 give the geometry, the initial and the boundary conditions for the tests performed in *Abuaf* et al. (1981). Subcooled water is used in these tests. The parameters for the particular experiment defined in Table 16.7.1 are chosen in order to have no flashing in the nozzle.

Table 16.7.1 Geometry of the test sections

Author	Geometry	Length	Hydraulic diameter	
		m	z / m	$D_{hyd,\max} / m$
BNL Abuaf	Circular *nozzle*	0.6	0	0.051
et al. (1981			0.02	0.051
			0.028	0.02546
			0.58	0.051
			0.6	0.051

Table 16.7.2 Initial conditions

Author	Test Nr.	Pressures	Mass flow rate (comp.)	Tempera-ture
		$p_{in}, p_{out} / bar$	$G / kg /(m^2 s)$	T_{in} / K
BNL Abuaf et al. (1981)	006	682 000 620 000	7010 (6319)	300.85
	008	695 000 661 000	4710 (4362)	300.05
	009	709 000 688 000	3130 (2991)	300.25

Equidistant discretization is used in this case. The computed results are compared in Fig. 16.7.1 with those obtained experimentally. The agreement is good. The mass flow rate as a function of the pressure difference is computed within a 10% error band. For practical analysis we recommend nonequidistant discretization with a finer grid around the throat, as discussed in *Kolev* (1986). It is also known that for accelerating flows the friction coefficient is larger than the steady-state value, which was not taken into account here.

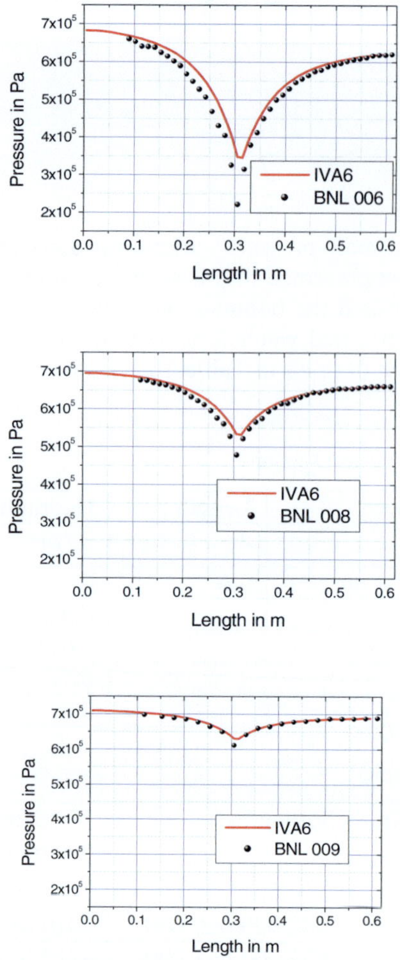

Fig. 16.7.1 Comparison between the IVA5 predictions of the pressure as a function of the axial coordinate with the measured pressure

References

Abuaf, N., Wu, B.J.C., Zimmer, G.A., Saha, P.: A study of non equilibrium flashing of water in a converging diverging nozzle, vol. 1 Experimental, vol. 2 Modeling, NUREG/CR-1864, BNL-NUREG-51317 (June 1981)

Kolev, N.I.: Transiente Zweiphasenströmung (Transient Two-Phase Flow). Springer, Heidelberg (1986)

16.8 Pressure waves – single phase

Pressure waves are common in piping systems like those presented in Fig. 16.8.1. In his remarkable work from 1898 *Joukowsky* responded to the need of the St. Petersburg administration to analyze the "strange strokes" in the urban city supply system if somebody close a valve to fast. *Joukowsky* devises an experimental facility and an ingenious method to record the pressure waves and discovered the first solution waves in the pipe networks. Moreover, systematically changing the pipe velocity before closure he found the relation for computing the first peak of the pressure that is used by present engineers. Then, not having a computer, he devised a graphical solution method based on the *Rieman*'s work from 1858–1859 known currently as a method of characteristics. So the pipe-network fluid dynamics was born.

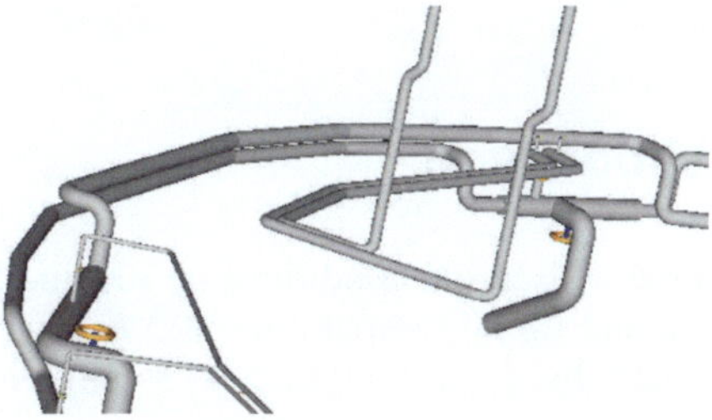

Fig. 16.8.1 Pipe network

In the following 100 years engineers wrote more than 100 single-phase computer codes based on the method of characteristics, performed many experiments in different geometries, and finally understood that there is no pure water in technical systems but water with dissolved gases and impurities that influence the inception of cavitation and many other processes, and that the pressure shocks caused by the collapsing bubbles in water, called water hammer, are much stronger then those in single-phase liquids.

Void in liquids causes nonlinear response to pressure waves depending on the frequency, a fact that is unfortunately used by the military engineers for amplification of the submarine pressure wave impact. Collapsing bubbles create a temperature increase at the center of the bubble up to a sun temperatures opening a new way for nuclear fusion in small laboratory scale *Lahey*, et al. (2005), *Nigmatulin* (2005). Another fascinating outcome of the cavitation is the reduction of skin friction on immersed bodies opening a new field for producing high-speed marine vehicles.

The kinetics of the void origination and disappearance can dramatically change the character of the processes. Last but not least, the waves cause movements of the structures as well as elastic or plastic change of the pipe cross-sections. Both phenomena have a feedback to the flow that may be very strong, depending on the pressure/elasticity ratio. As long as the methods do not cause cavitation, a single-phase method coupled with fluid–structure interaction does a good job but this is rather the exception in real systems than the rule. Moreover, the damage caused by water hammer still lead to large losses in industry.

It turned out that there is no single-phase method for an adequate description pressure waves in liquid systems in general.

Therefore, the sophistication of the methods for description of this complex of phenomena leads to the multiphase flow dynamics. Only the methods of multiphase flow dynamics allow detection in the analyses of the conditions for void inception, the kinetics of the void production and disappearance, and therefore provide the appropriate mathematical framework for general pressure-wave analyses in industrial pipe network systems.

The purpose of the present section is to document part of the verification basis for the multiphase computer code IVA for pressure waves in systems initially filled with gas or liquid.

16.8.1 Gas in a shock tube

Consider air in a 1-m pipe with initial conditions of 100 and 50 bar, respectively in the two separate halves, and initial temperature 600 K. The analytical solution to this problem is presented by *Oertel* (1966). We will compare the analytical solution for the problem defined at $\tau = 250 \mu s$ with the numerical solution predicted by IVA5. A detailed discussion to this problem is available in Chapter 4 of Vol. 1 of this monograph or in *Kolev* (1998).

Figure 16.8.2 shows the pressure as a function of time computed with the computer code IVA5 that exploits the first-order donor-cell method, as discussed in Chapter 12 of Volume 1 or in *Kolev* (1996). The code uses equations of state for air as a real gas. Surprisingly, the IVA5 donor cell method is more accurate than the first-order method of characteristics and the method of characteristics using third-order spatial interpolation *Kolev* (1998). The pressure at the two different ends of the pipe as a function of time is presented in Fig. 16.8.3. As discussed in Chapter 4 of Vol. 1 or in *Kolev* (1998), comparing Fig. 16.8.2 with the solutions obtained by using the internal energy form instead of the energy conservation in entropy form we obtained almost indistinguishable solutions. The total mass and energy conservation error is shown in Fig. 16.8.4. Again, we observe from Fig. 16.8.4 that the IVA5 numerical method is better than the first- and third-order method of characteristics.

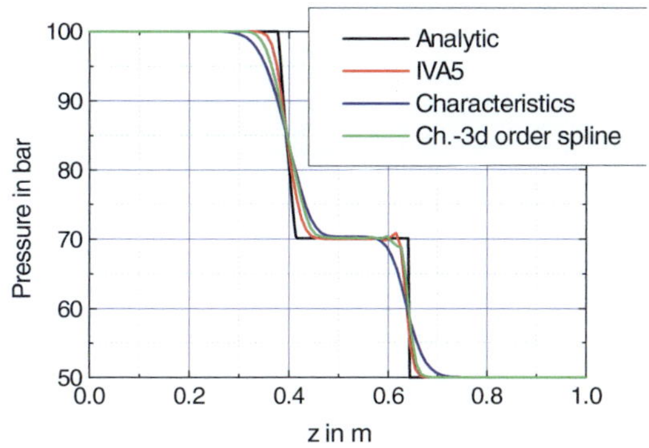

Fig. 16.8.2 Pressure as a function of the distance from the left closed end of the pipe at $\tau = 250 \mu s$

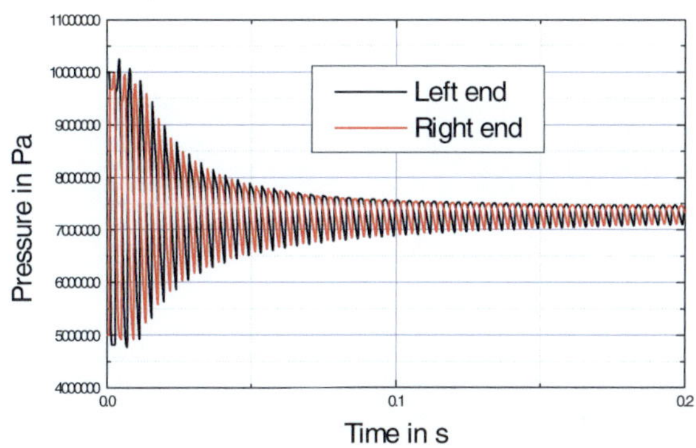

Fig. 16.8.3 Pressure at the two ends of the pipe as a function of time (air). The energy conservation is used in terms of specific entropy in IVA5

Now we perform a similar test with IVA5, replacing the air with a molten oxide mixture called corium at 3300 K initial temperature. The initial pressure distribution is the same as in the previous examples. The velocity of sound of the system is around 1500 m/s. For this test we developed a special set of equations of state and their derivatives, which are strictly consistent with each other following the instructions given in Chapter 3 of Volume 1. The pressure at the two ends as a function of time is shown in Fig. 16.8.5 and the corresponding cumulative relative mass and energy conservation error is given in Fig. 16.8.6. We see that the order of magnitude is 10^{-7}, which is much better than for the case of the strongly compressible gas. The slight difference in steady-state pressures is due to the geodetic pressure difference because the pipe is considered to be vertical and the fluid is very heavy.

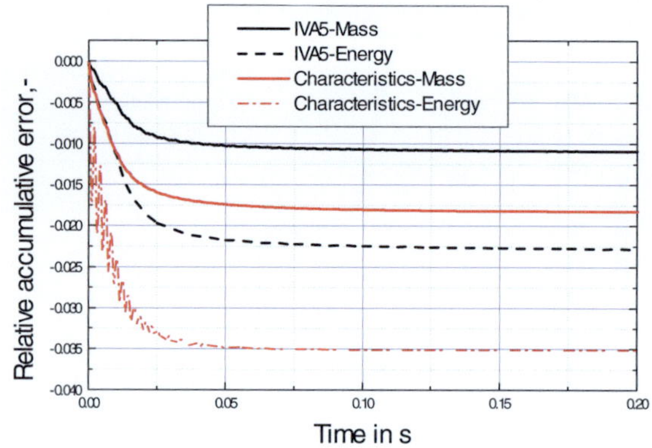

Fig. 16.8.4 Comparison between the overall mass and energy conservation error of the IVA5 numerical method and the first-order method of characteristics

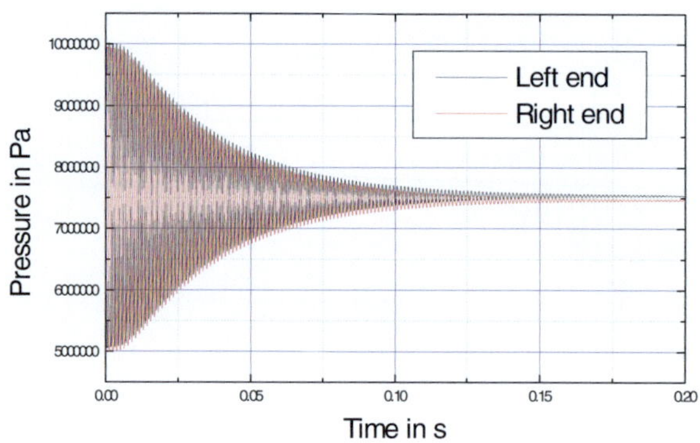

Fig. 16.8.5 Pressure at the two ends of the pipe as a function of time (corium melt 3300 K). The energy conservation is used in terms of specific entropy in IVA5

Conclusions

a) IVA5 exhibits the expected behavior for the first-order donor-cell solution method. The accuracy is better than the classical first-order method of characteristics and the method of characteristics using bicubic spatial spline interpolation.
b) For practical applications in large-scale facilities a spatial resolution of 1 cm is hardly achievable and therefore the expected errors for single-phase flows will be higher, corresponding to the size of the cells.

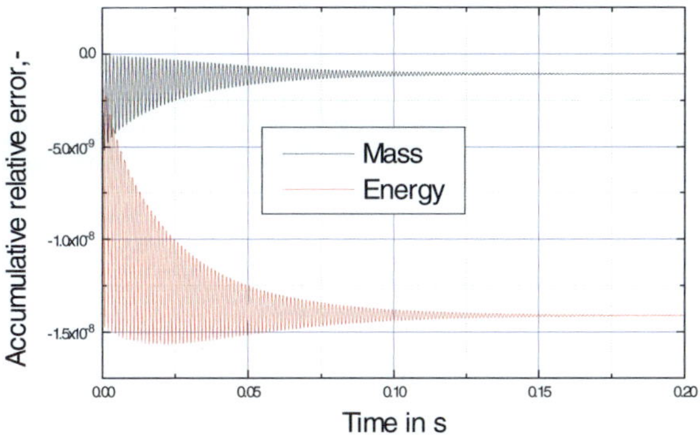

Fig. 16.8.6 Relative overall mass and energy conservation error as a function of time (corium melt 3300 K). The energy conservation is used in terms of specific entropy in IVA5

16.8.2 Water in a shock tube

The *Kellner* and *Gissler* (1984) experiment was used for this comparison. Pressure-wave propagation in a single pipe with five bends and one dead end was simulated by using a 1D representation – Fig. 16.8.8. The experiment was initiated by igniting an oxygen–hydrogen gas mixture. The imposed pressure function at the pipe inlet is shown in Fig. 16.8.7a. A comparison of predicted and measured responses as shown in Figs. 16.8.7b–d shows good agreement.

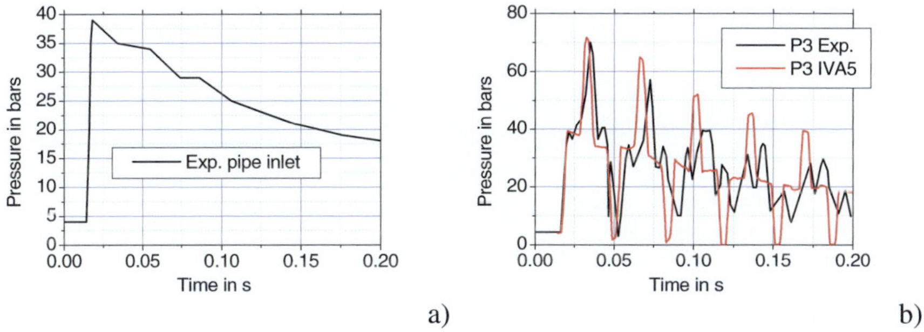

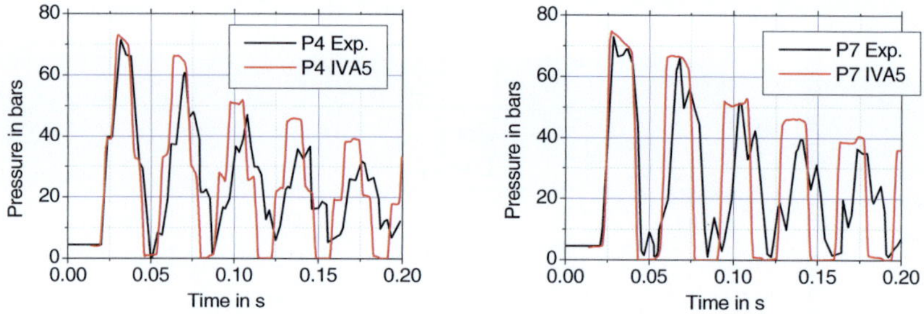

c) d)

Fig. 16.8.7 a Comparison of IVA5 prediction with the experimental data of *Kellner* and *Gissler* (1984). Pipe length 13.12 m, diameter 0.1 m, 4 bends at 4.8, 10.2, 12.6, 13.64 m from the inlet. Initial conditions: water at atmospheric temperature and pressure. a) Pressure at the pipe inlet as a function of time as measured by *Kellner* and *Gissler* (1984). b) At 4.08 m from the pipe inlet. c) At 8.18 m from the pipe inlet. d) At 12.5 m from the pipe inlet

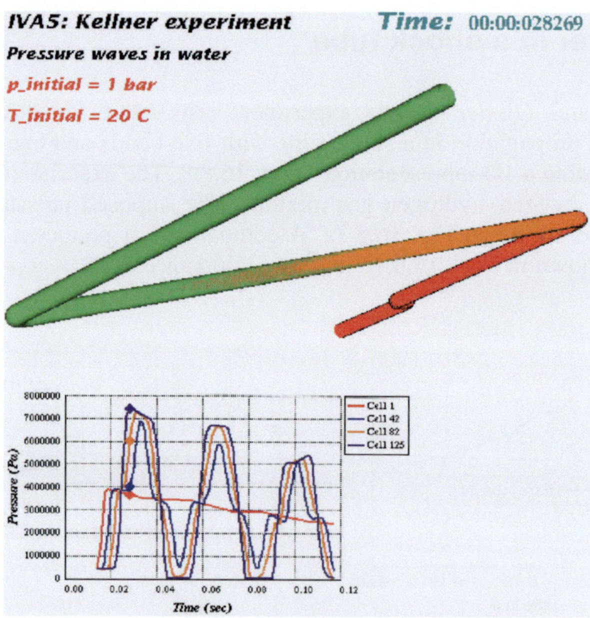

Fig. 16.8.8 A SONJA visualization of the pressure wave. Excitation pressure jump 40 bar results in a maximum of the pressure at the opposite end of about 80 bar

Conclusion. From comparison with experiments that feature pressure wave propagation in pipes initially filled with water, we conclude that the code is able to reproduce wave dynamics in single-phase water flow.

16.8.3 The 1983 Interatome experiments

In the framework of the research activities devoted to the German fast breeder nuclear reactor *Dumm* et al. (1983) performed in 1983 a systemic experimental study of pressure-wave propagation from simple to complicated pipe network systems with variety of modes of fluid–structure interactions: The report is still proprietary, hence I do not disclose here materials, geometrical sizes, fixed points, etc. The water quality was not quantitatively estimated in one group of the experiments. Degassed water is used in some of the experiments. Pressure waves are generated by hydrogen detonation and recorded as boundary conditions for successive code validations. Pressures at different position are measured together, in some experiments, with the characteristics of the structure motion. We confine our attention to those of the experiments that did not manifest plastic deformations and those that did not manifest strong pipe relocations. The reason is that IVA computer code takes into account only the elastic deformation of local volume and its feedback on the flow.

16.8.3.1 Experiment 1.2

Given a pipe with known material, geometry, wall thickness and fixed points as schematically presented in Fig. 16.8.9. The computation with IVA is performed with a spatial resolution of about 10 cm.

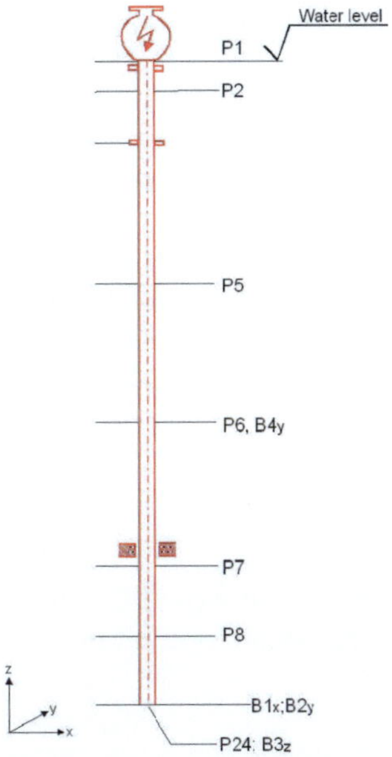

Fig. 16.8.9 Schematic pipe geometry for exp. 1.2 and 1.3

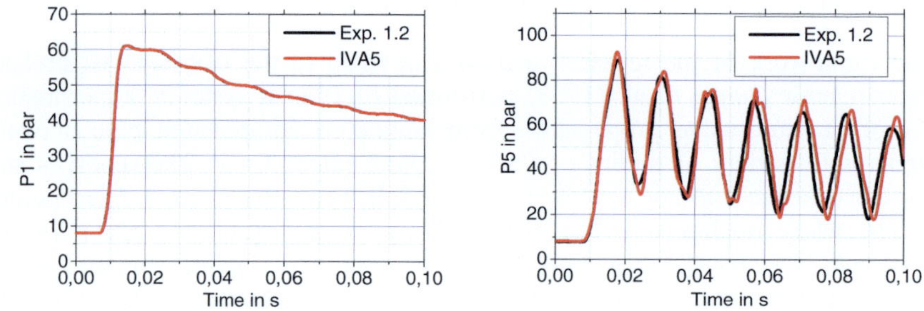

Fig. 16.8.10 Experiment 1.2: Pressure as a function of time. a) P1 pressure - boundary condition; b) Comparison with IVA prediction for P5 pressure.

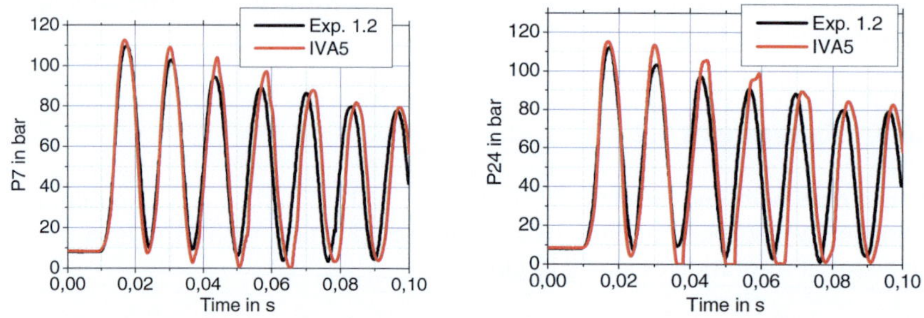

Fig. 16.8.11 Experiment 1.2: Pressure as a function of time. Comparison with IVA predictions for: a) P7; b) P24

The results of the computation are compared with data on Figs. 16.8.10 and 16.8.11 Note that without taking into account the elasticity of the pipe the comparison was poor. Only after taking into account the elastic change of the cross-section due to the variable pressure difference do we arrive at the results given in the figures. We realize that for a first-order donor-cell method with 10 cm spatial step the frequency and the magnitudes are well resolved. This result serves also as a recommendation of how to use the code for such pressure magnitudes/elasticity ratios.

16.8.3.2 Experiment 1.3

Given is the same geometry as in exp. 1.2. The important difference is in the stiffness of the pressure-jump boundary condition as given in Fig. 16.8.12.

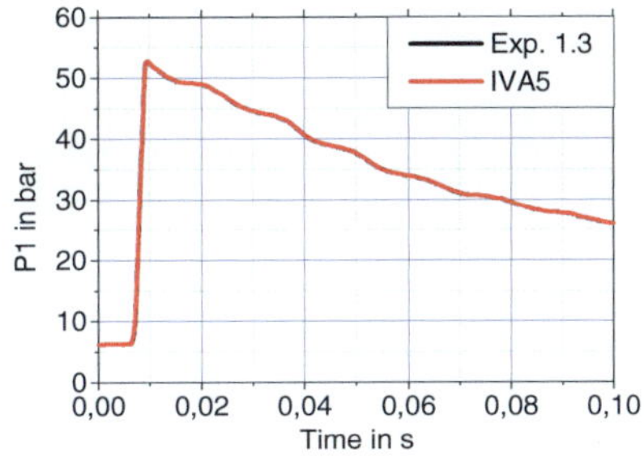

Fig. 16.8.12 Experiment 1.3: Pressure as a function of time. a) P1 pressure – boundary condition;

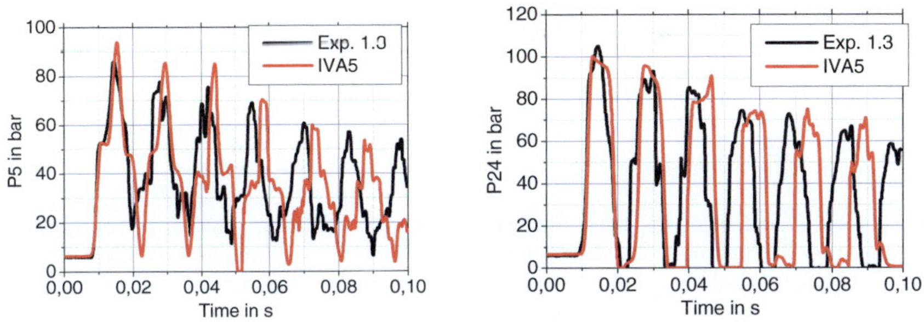

Fig. 16.8.13 Experiment 1.3: Pressure as a function of time. Comparison with IVA predictions for: a) P5; b) P24

The magnitude of the first wave is strongly connected with the stiffness of the pressure rump. The larger the stiffness, the larger the first magnitude. The reflection of the wave causes a tension state of the liquid and delayed nucleation and vapor generation. So, the smooth wave characteristics from the single-phase case change significantly. The results of the predictions are given in Fig. 16.8.13. The wave signature that depends on the void production is properly predicted. We realize that for increasing time the frequency shifts to a smaller frequency. We will analyze in more detail this phenomenon. First, we allow for the tension state of the water also for pressures less than 650 Pa. The following results are obtained with this option. Figure 16.8.14 presents the results with different bubble growth models. We realize that the differences are not substantial.

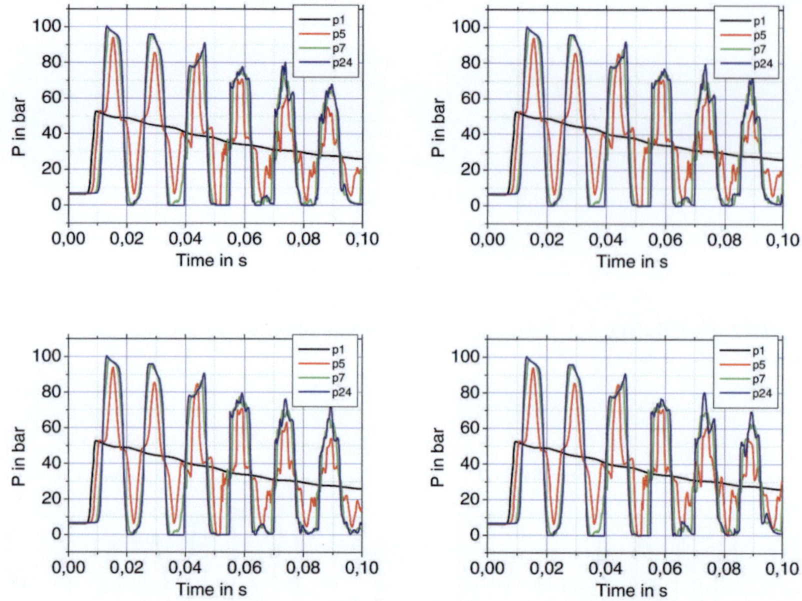

Fig. 16.8.14 Influence of the used flushing models: a) Implicit, internal circulation *Hunt* and *Issenberg*, thermal boundary layer in liquid; b) *Mikic* et al. model 1970 ; c) *Labunzov* – thermal-controlled bubble-growth model 1964; d) *Jones* and *Zuber*, variable pressure, thermal-controlled bubble-growth model 1978

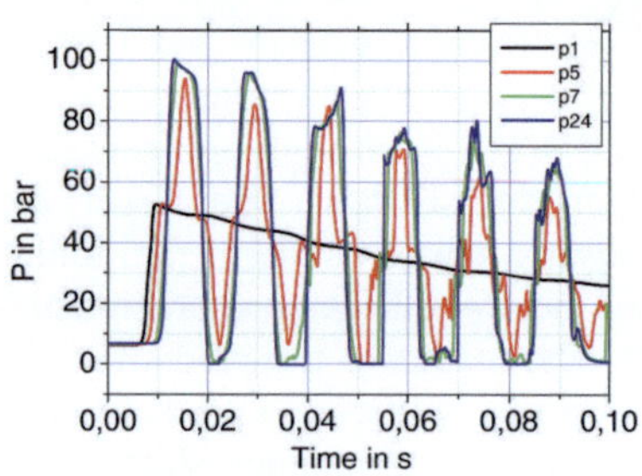

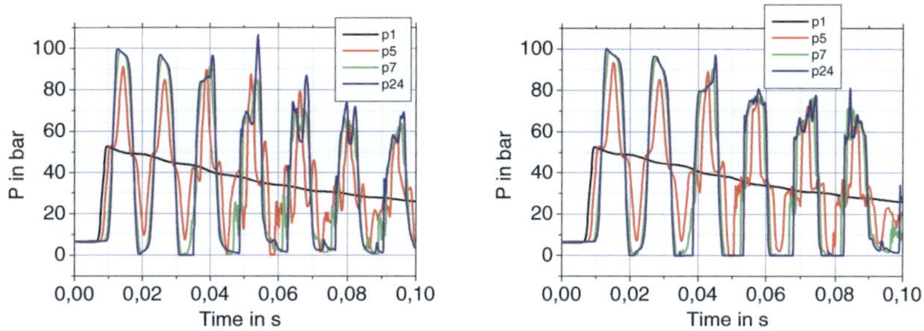

Fig. 16.8.15 Influence of the fluid structure interaction modeling on the frequency: a) Implicit, internal circulation *Hunt* and *Issenberg*, thermal boundary layer in liquid; e) as a) without fluid–structure interaction, elasticity modulus E = 2.0×10^{11} Pa; f) as a) without fluid–structure interaction, elasticity modulus E = 2.5×10^{11} Pa.

Figures 16.8.15a and b show computations with and without fluid–structure interaction. Obviously not taking into account the fluid–structure interactions, results in prediction of frequencies that are larger than the observed. Because our model does not take into account the increased stiffness of the flange and of the dead end we simply increase the elasticity modulus from 2 to 2.5×10^{11} Pa to see the tendency. As expected it increase slightly the frequency as seen in Fig. 16.8.15c compared to 16.8.15a. Therefore, we attribute the slight frequency shift not to flushing modeling but to the fluid–structure interaction modeling. More accurate reproduction of the frequencies requires more accurate structure models than those we use here.

16.8.3.3 Experiment 10.6

Given a pipe with known material, geometry, wall thickness and fixed points as schematically presented in Fig. 16.8.16. The type of the boundary condition is different from the proceeding experiments – Fig. 16.8.17a. The computation with IVA is performed with spatial resolution of about 10 cm.

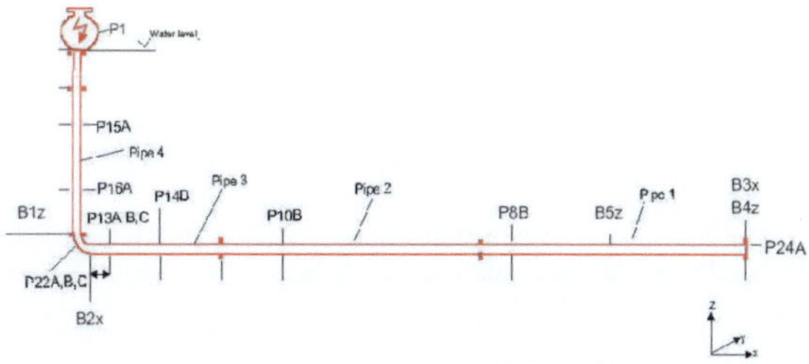

Fig. 16.8.16 Schematic pipe geometry for exp. 10.6

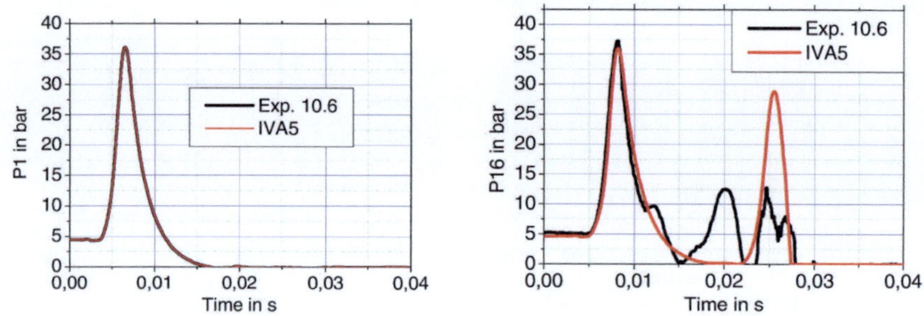

Fig. 16.8.17 Experiment 10.6: a) P1 pressure – boundary condition; b) Comparison with IVA prediction for P16 pressure.

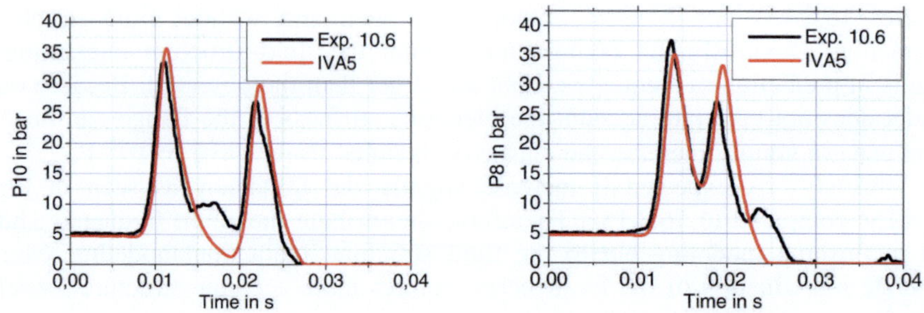

Fig. 16.8.18 Experiment 10.6: Pressure as a function of time. Comparison with IVA predictions for: a) P10; b) P8

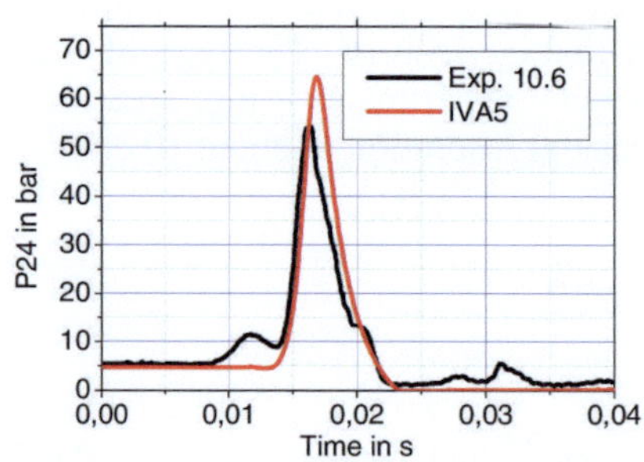

Fig. 16.8.19 Experiment 10.6: Pressure as a function of time. Comparison with IVA predictions for P24

The predictions of IVA are presented in Figs. 16.8.17, 16.8.18 and 16.8.19. The arrival of the wave at the end of the test section and its form is properly predicted.

Note that this is not the case if the elastic deformation of the pipe were not taken into account. The timing and the double-peak structure of the wave in the horizontal-pipe segment, see Fig. 16.8.18, is again properly predicted, indicating that the basics physics is reflected by the IVA model. The measurements in Fig. 16.8.17b indicate that there is a kind of earlier reflection either from the bend or from the stiff portion of the pipe. We do not model those places of the pipe that are stiff but consider the entire pipe in this simulation as elastic everywhere. It is obvious that this practice has to be improved in the future.

16.8.3.4 Experiment 11.3

This experiment differs slightly from 10.6 in the replacement of the bend with the T-junction, as given in Fig. 16.8.20.

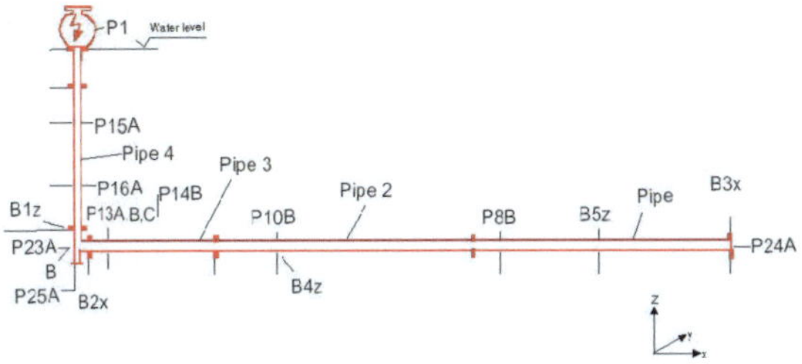

Fig. 16.8.20 Schematic pipe geometry for exp. 11.3

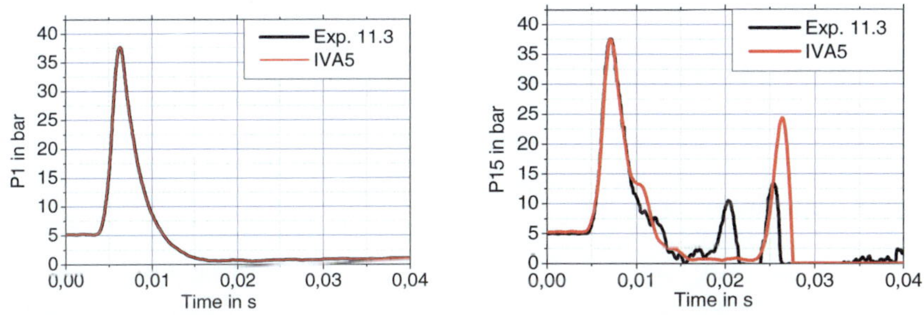

Fig. 16.8.21 Experiment 11.3: a) P1 pressure – boundary condition; b) Comparison with IVA prediction for P15 pressure.

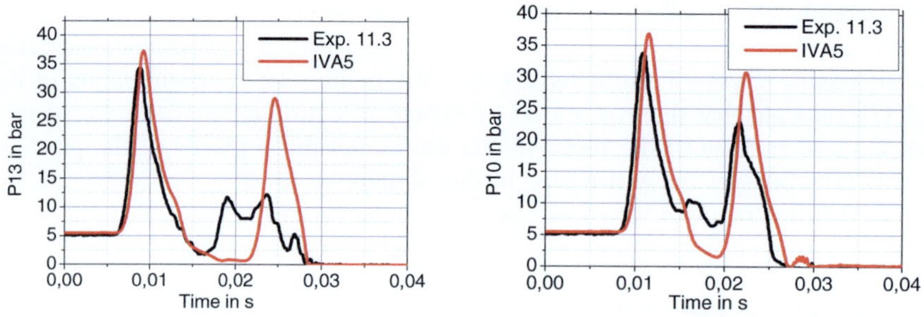

Fig. 16.8.22 Experiment 11.3: Pressure as a function of time. Comparison with IVA predictions for: a) P13; b) P10

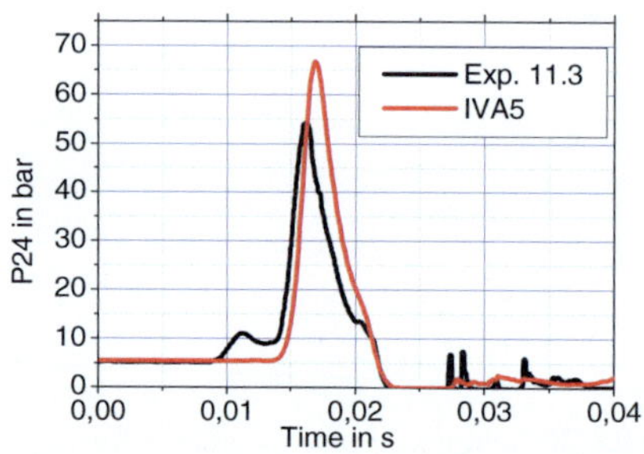

Fig. 16.8.23 Experiment 11.3: Pressure as a function of time. Comparison with IVA predictions for P24

The results are presented in Figs. 16.8.21, 16.8.22 and 16.8.23. The findings are very similar to those already discussed for experiment 10.6.

16.8.3.5 Experiment 21

Given a pipe network with known material, geometry, wall thickness and fixed points as schematically presented in Fig. 16.8.24. The pressure boundary condition is defined by Fig. 16.8.25a. The computation with IVA is performed with a spatial resolution of about 10 cm.

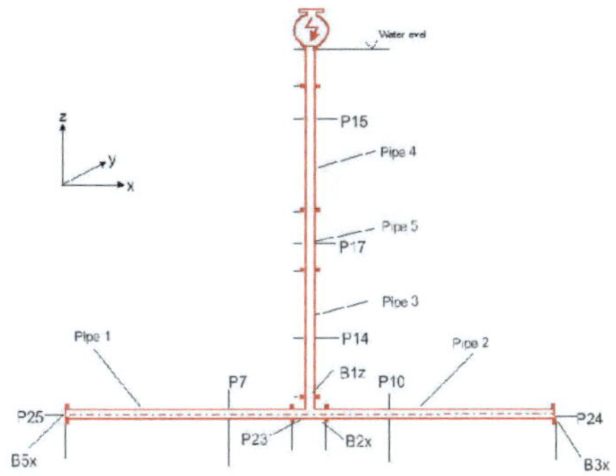

Fig. 16.8.24 Schematic pipe geometry for exp. 21

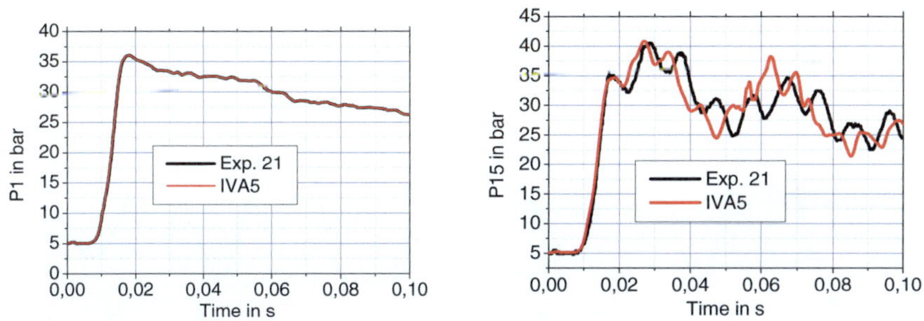

Fig. 16.8.25 Experiment 21: a) P1 pressure – boundary condition; b) Comparison with IVA prediction for P15 pressure.

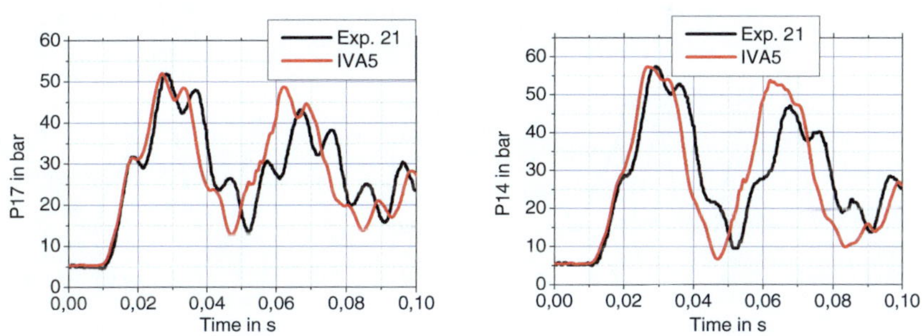

Fig. 16.8.26 Experiment 21: Pressure as a function of time. Comparison with IVA predictions for: a) P17; b) P14

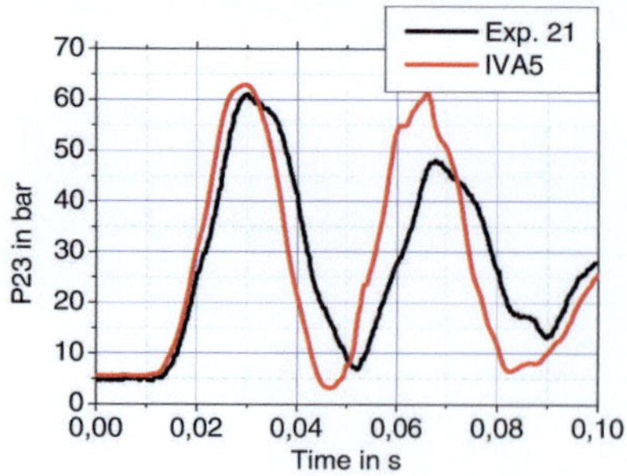

Fig. 16.8.27 Experiment 21: Pressure as a function of time. Comparison with IVA predictions for a) P23

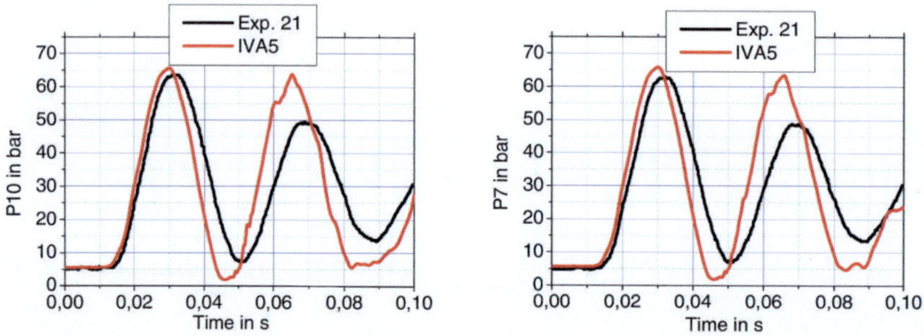

Fig. 16.8.28 Experiment 21: Pressure as a function of time. Comparison with IVA predictions for: a) P10; b) P7

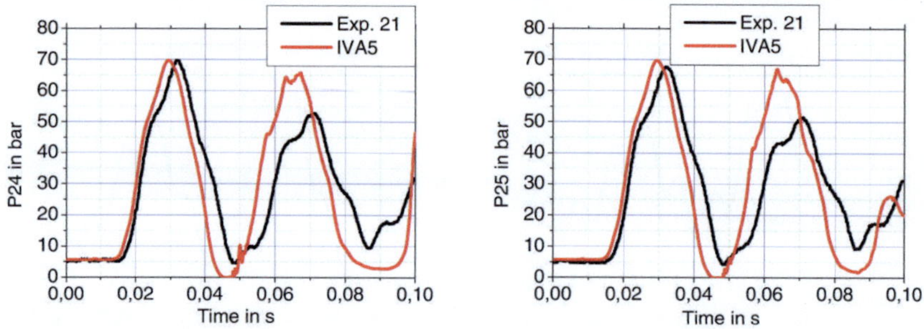

Fig. 16.8.29 Experiment 21: Pressure as a function of time. Comparison with IVA predictions for: a) P24; b) P25

First, we see that the signature of the waves is well reproduced by IVA. The observed symmetry presented in Figs. 16.8.28 and 16.8.29 is satisfactorily reproduced. The second wave magnitudes are more damped in reality than in the computation. We attribute this to the approximate fluid–structure interaction we use.

16.8.3.6 Experiment 5

Given a pipe network with known material, geometry, wall thickness and fixed points as schematically presented in Fig. 16.8.30.

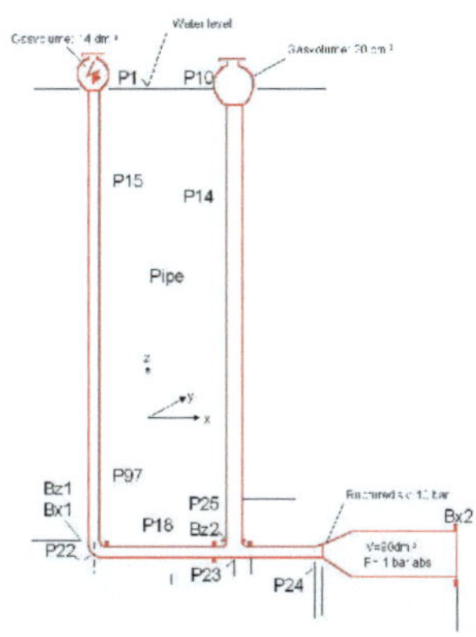

Fig. 16.8.30 Schematic pipe geometry for exp. 21

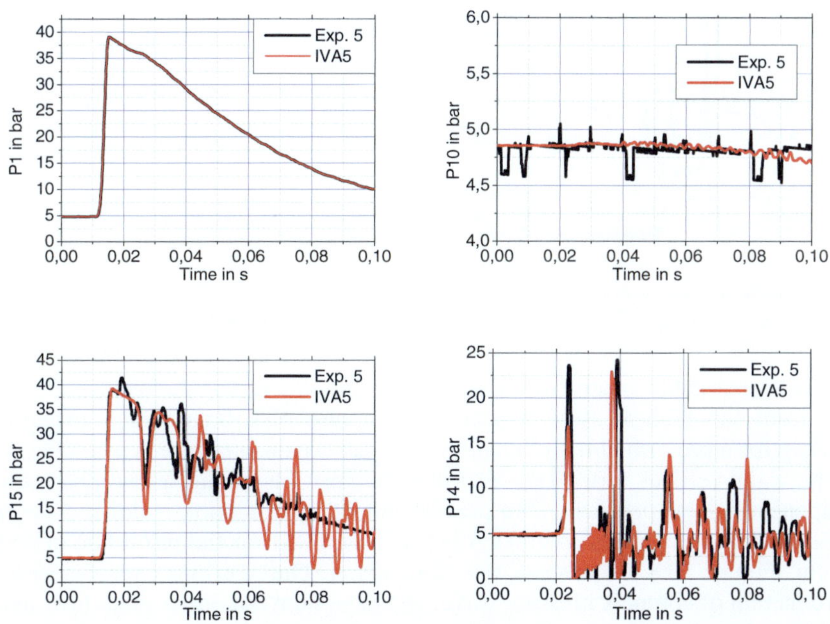

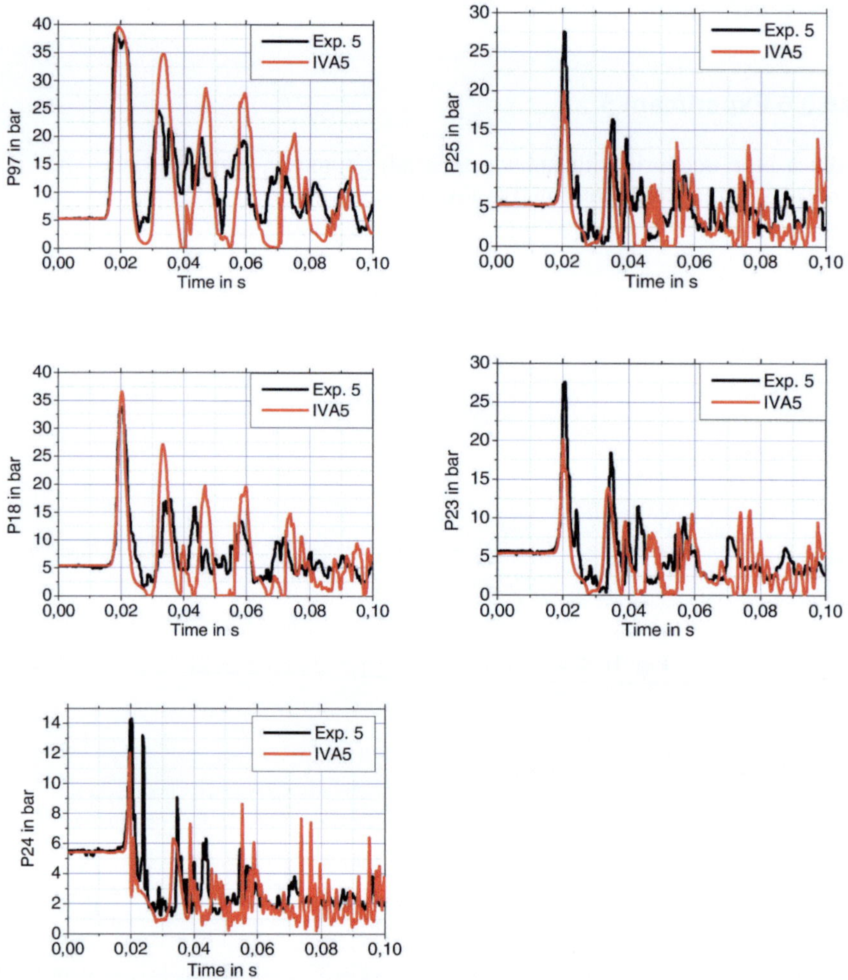

Fig. 16.8.31 Experiment 5: Pressure as a function of time. Comparison with IVA predictions. -1) though -9) from left to right and from top to the bottom.

This is a remarkable experiment having a low-pressure region divided by a rupture disk from the high-pressure region. The rupture disk breaks at a prescribed pressure for a prescribed time. In addition, a gas expansion space is available atop of the right vertical pipe. This is from the very beginning a two-phase two-component flow experiment. The pressure boundary condition is defined by Fig. 16.8.31a. The computation with IVA is performed with spatial resolution of about 10 cm. Figure 16.8.31 presents the comparison. One can not expect a perfect reproduction of complex pressure waves in this system but one realizes how many important features of this complex process are reproduced by IVA.

16.8.3.7 Experiment 15

Given a pipe network with known material, geometry, wall thickness and fixed points as schematically presented in Fig. 16.8.32. This is a experiment similar to 5 but not having a low-pressure region and rupture disk. In addition, a gas expansion space is available atop of the right vertical pipe having again different geometry compared to experiment No. 5. This again is from the very beginning a two-phase two-component flow experiment. The pressure boundary condition is defined by Fig. 16.8.31a. The computation with IVA is performed with a spatial resolution of about 10 cm.

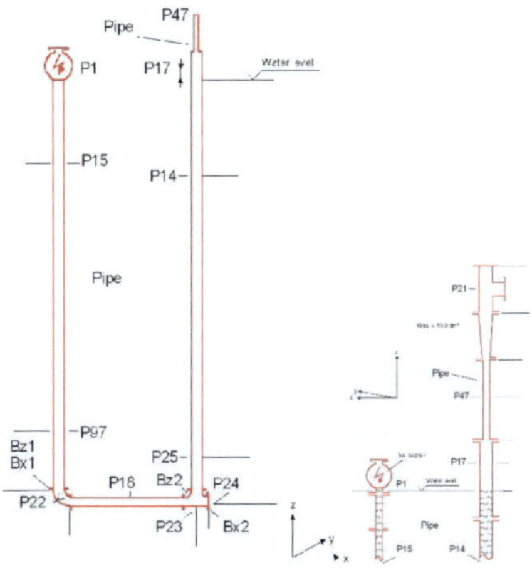

Fig. 16.8.32 Schematic pipe geometry for exp. 15

Figure 16.8.33 presents the comparison. Again, as in experiment 5 one can not expect a perfect reproduction of complex pressure waves in this system but one realize again how many important features of this complex process are reproduced by IVA.

Conclusions: The comparison with the above-discussed experiments demonstrates the capability of the IVA computational model to reproduce the most important characteristics of the process like cavitation due to the reflection of the strong pressure waves, collective bubble collapse leading to strong pressure wave generation, pressure wave propagation in pipe networks with rupture discs etc. Therefore, the use of the code in such fields has to be associated with uncertainties regarding magnitudes and frequencies documented here.

Acknowledgment: The data processing done by Iris Roloff-Bock from Framatome ANP and Eva Ziegerer from the University of Erlangen is highly appreciated.

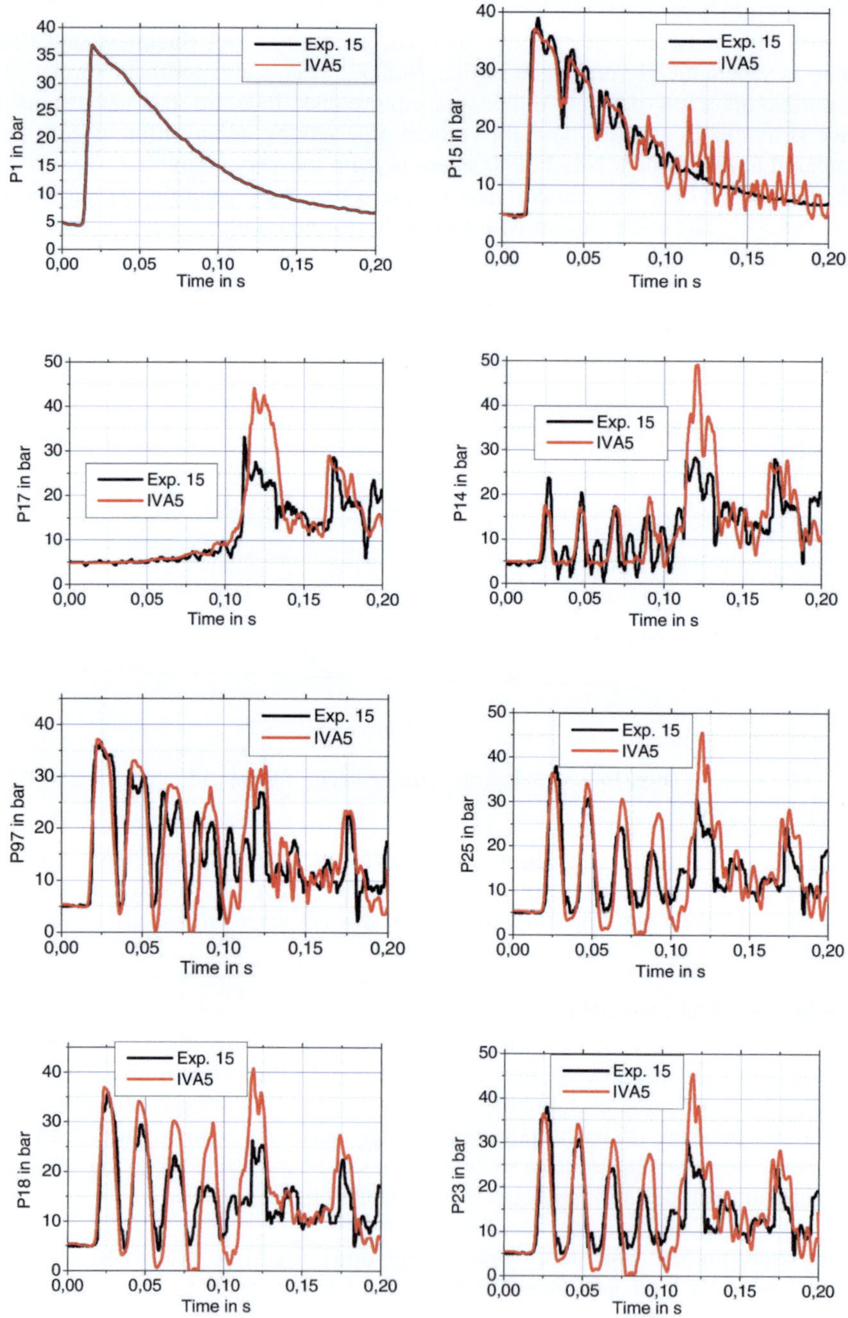
P1 in bar
P15 in bar
P17 in bar
P14 in bar
P97 in bar
P25 in bar
P18 in bar
P23 in bar
Time in s
Exp. 15
IVA5
0,00
0,05
0,10
0,15
0,20

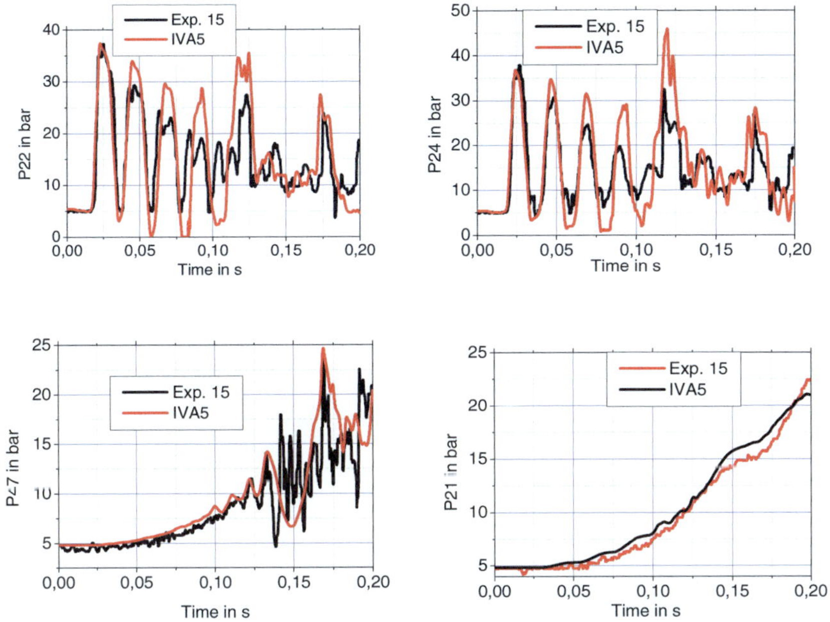

Fig. 16.8.33 Experiment 15: Pressure as a function of time. Comparison with IVA predictions: 1) though -12) from left to right and from top to the bottom.

16.8.4 Pressure-wave propagation in a cylindrical vessel with free surface (2D)

In order to study triggered melt–water interactions high gas pressure sources are usually used as a trigger. In order to model adequately the triggered melt-water interaction the trigger has first to be well defined and appropriately reproduced by the computer code. The experiment considered here was performed by *Huhtiniemi* et al. (1996) in the vessel presented in Fig. 16.8.34. At the bottom of the internal cylinder there is a 15 cm^3 gas capsule charged at 138.9 bar. Breaking the membrane releases the pressure wave into the water above. The water has a free surface. The results of the compared computed and measured pressure traces are presented in Figures 16.8.35(1)–(6). We see that the propagation of the first pressure pulse in the two-dimensional space regarding propagation speed, magnitude, and duration is appropriately predicted by the IVA computer code, except for the K5 magnitude, which is predicted to be smaller than the measured value by a factor of two.

Conclusion. Therefore, the method and the resolution used can be applied for trigger simulation in a melt–water interaction analysis.

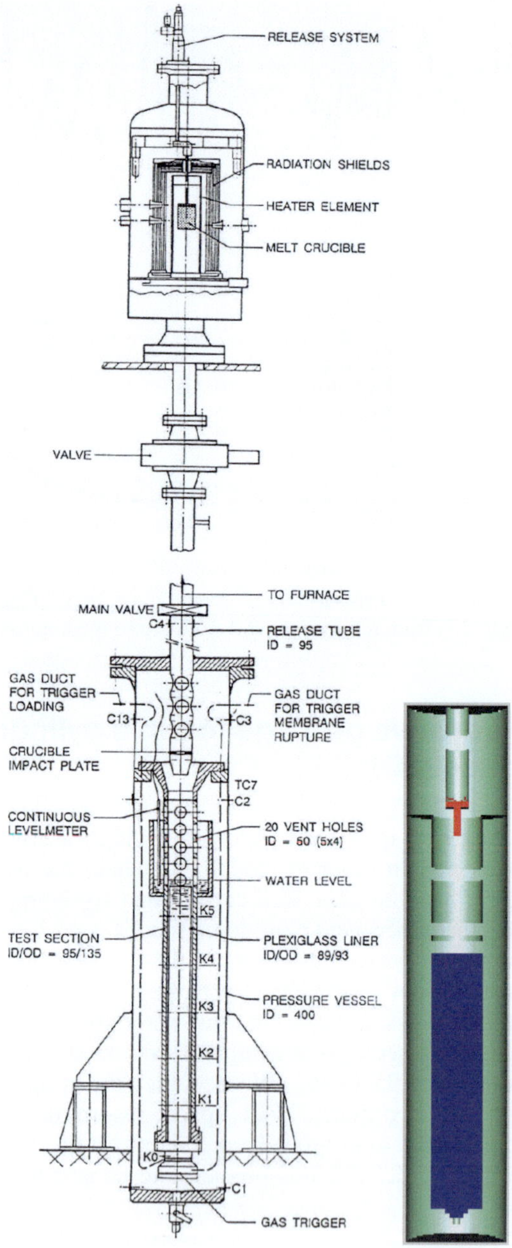

Fig. 16.8.34 KROTOS test facility *Huhtiniemi* et al. (1996) and IVA_5M computer model

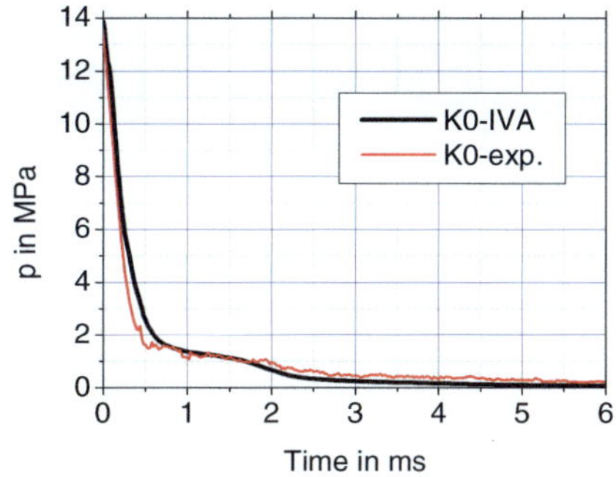
14
12
10
8
6
4
2
0
p in MPa
0 1 2 3 4 5 6
Time in ms
K0-IVA
K0-exp.

(1)

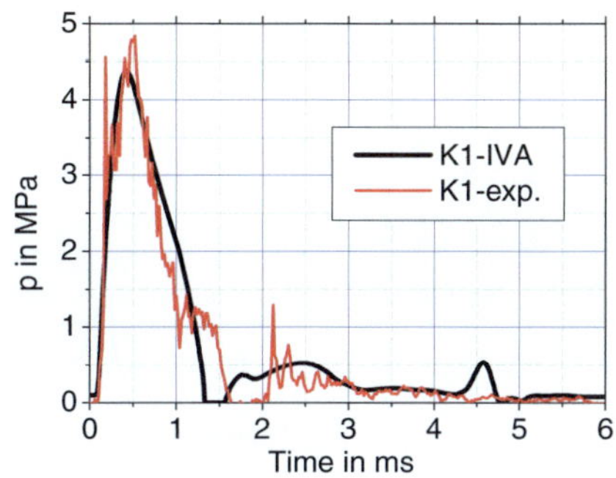
5
4
3
2
1
0
p in MPa
0 1 2 3 4 5 6
Time in ms
K1-IVA
K1-exp.

(2)

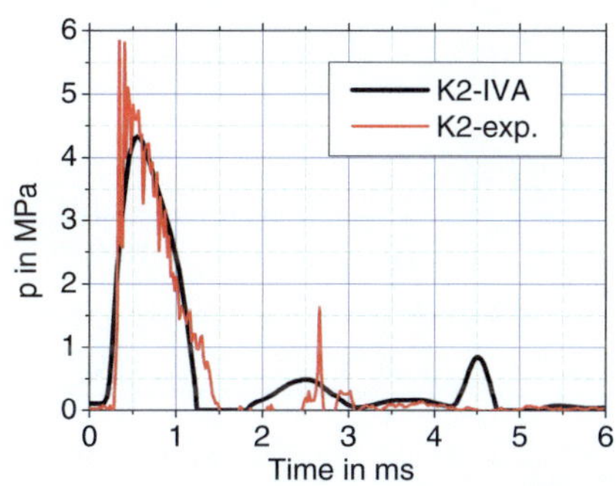
6
5
4
3
2
1
0
p in MPa
0 1 2 3 4 5 6
Time in ms
K2-IVA
K2-exp.

(3)

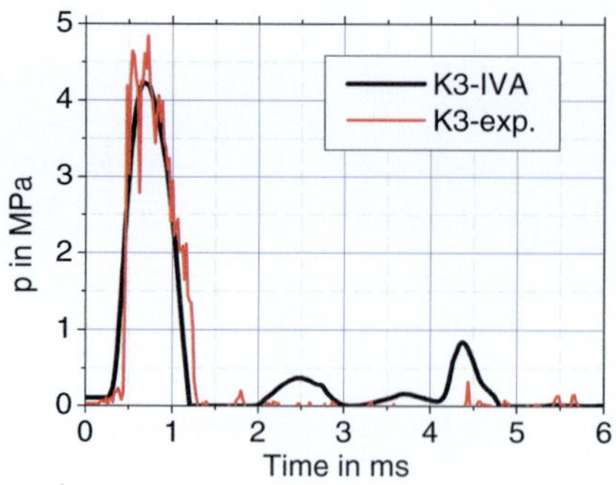

(4)

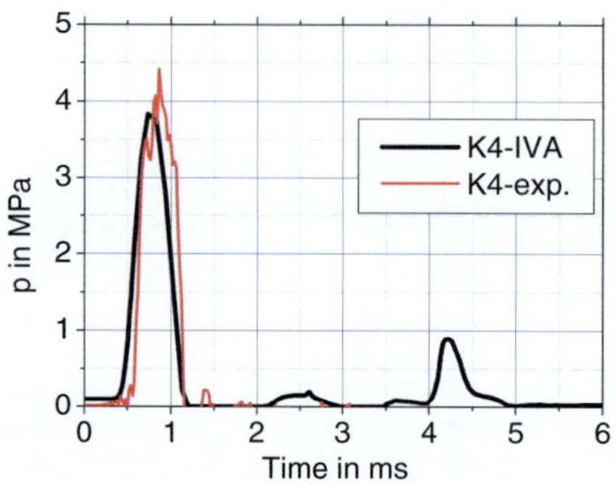

(5)

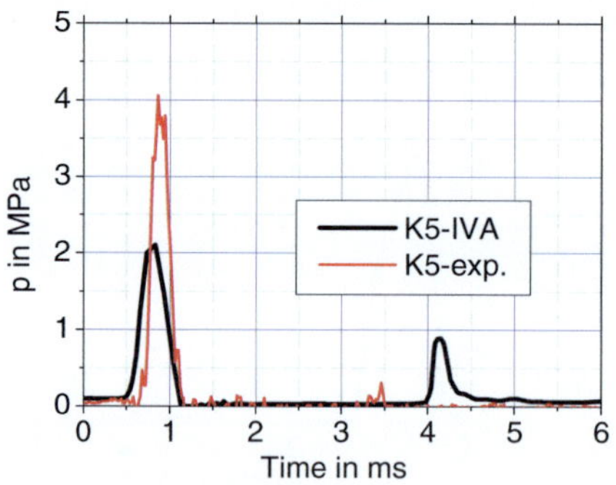

(6)

Fig. 16.8.35 Trigger pulse in pure water. The pulse was produced by a gas capsule charged at 138.9 bar, volume 15 cm^3. Pressure at the wall as a function of time. Distance from the bottom plate: (1) 0 m; (2) 0.15 m; (3) 0.35 m; (4) 0.55 m; (5) 0.75 m; (6) 0.95 m.

References

Dumm, K., Voss, J., Kellner, A., Dörner, G., Lang, H., Joist, G.: Untersuchungen zur Wechselwirkung zwischen Druckwellen und Bauteilen in flüssigkeitsgefüllten Systemen, INTAT 5204706, 62.04706.9, Interatom Internationale Atomreaktorbau GmbH (April 1983)

Huhtiniemi, I., Hohmann, H., Faraoni, R., Flied, M., Gambaretti, G., Klein, K.: KROTOS 38 to KROTOS 44: Data report, Technical Note No.I.96.37, Institute for Safety Technology, Reactor Safety Program, Joint Research Center, Ispra (March 1996)

Joukowsky, N.: Ueber den hydraulischen Stoss in Wasserleitungesröhren, Voss, Petersburg und Leipzig, 1990, pp. 1–72 (April 24, 1898)

Kellner, A., Gissler, D.: Programsystem SAPHYR: Anwendungsbeispiel II, Notiz Nr.70.02748.4, Interatom GmbH (February 7, 1984)

Kolev, N.I.: Three fluid modeling with dynamic fragmentation and coalescence fiction or daily practice? 7th FARO Experts Group Meeting Ispra (October 15-16, 1996); Proceedings of OECD/CSNI Workshop on Transient thermal-hydraulic and neutronic codes requirements, Annapolis, MD, USA (November 5-8, 1996); 4th World Conference on Experimental Heat Transfer, Fluid Mechanics and Thermodynamics, ExHFT 4, Brussels (June 2-6, 1997); ASME Fluids Engineering Conference & Exhibition, The Hyatt Regency Vancouver, Vancouver, British Columbia, Canada Invited Paper; Proceedings of 1997 International Seminar on Vapor Explosions and Explosive Eruptions (AMIGO-IMI), Aoba Kinen Kaikan of Tohoku University, Sendai-City, Japan (May 22-24, 1997)

Kolev, N.I.: On the variety of notation of the energy conservation principle for single phase flow. Kerntechnik 63(3), 145–156 (1998)

Lahey Jr., R.T., Taleyarkhan, R.P., Nigmatulin, R.I.: Sonofusion – fact or fiction? In: The 11th International Topical Meeting on Nuclear Reactor Thermal-Hydraulics (NURETH-11) Paper: 517, Popes' Palace Conference Center, Avignon, France, October 2-6 (2005)

Nigmatulin, R.: Nano-scale thermonuclear fusion in imploding vapour bubbles in acoustically cavitated deuterated liquid. In: The 11th International Topical Meeting on Nuclear Reactor Thermal-Hydraulics (NURETH-11) Paper: 517, Popes' Palace Conference Center, Avignon, France (October 2-6, 2005)

Oertel, H.: Stossrohre. Springer, Wien, New York (1966)

Riemann, B.: Über die Fortpflanzung ebener Luftwellen von endlicher Schwingungsweite, Abhandlungen der Königlichen Gesellschaft der Wissenschaften zu Göttingen. Band 8, S43–S65 (1858-1859)

16.9 2D: N_2 explosion in space filled previously with air

A full description of the experimental facility, instrumentation, and experimental procedure is given in *Wilchelm* and *Kirstahler* (1987), *Meyer* and *Kirstahler* (1988). An axially symmetric vessel was used – Fig. 16.9.1. The vessel was divided into high- and low-pressure regions by sliding doors. Above the sliding doors an aluminum rupture foil separated the low-pressure region from the rest of the facility. The high-pressure region consisted of a 0.6 MPa nitrogen source. The low-pressure region comprised 0.1 MPa air at room temperature. The experiment was initiated by igniting an oxygen-hydrogen gas mixture in the sliding door driving mechanism. This caused the sliding doors to accelerate, opening the cross-section, starting from the center within 0.4 ms. The pressure was measured at different points.

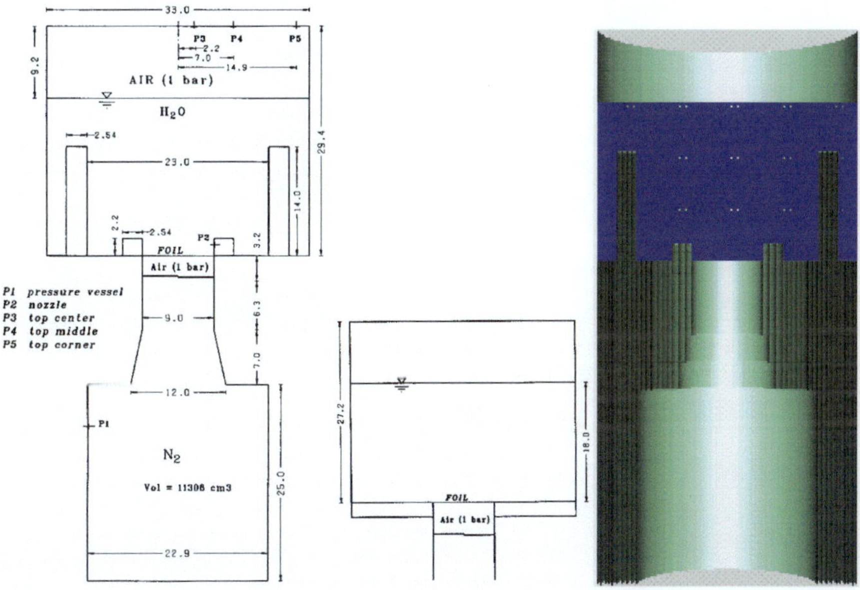

a) b) c)

Fig. 16.9.1 a) Geometry of the experimental test section with internals. b) Geometry of the experimental test section without internals. c) IVA_5M finite-volume grid for computational analysis

In the computational simulation a symmetrical process is assumed, neglecting the asymmetries observed in *Meyer* and *Kirstahler* (1988). The geometry was represented by 1144 cells (26×44) in only one angular sector with cylindrical coordinates – see Fig. 16.9.1c. Variable surface permeabilities were used to simulate the opening process for the sliding doors. The *Poisson*-type equation was solved for the whole (r, z) plane directly.

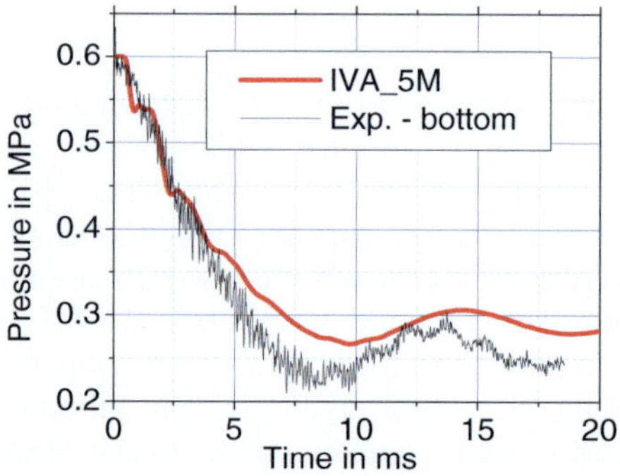

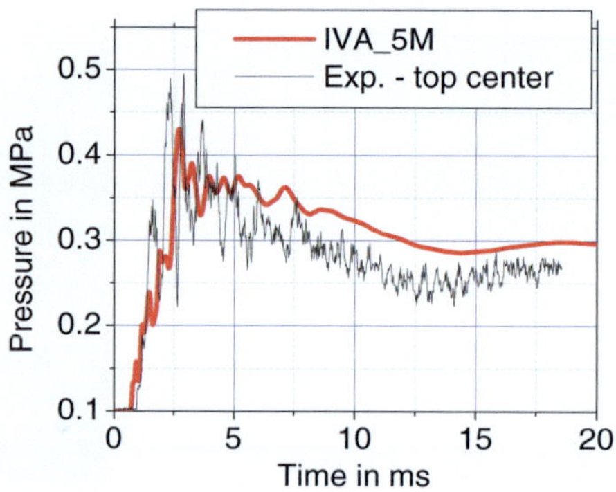

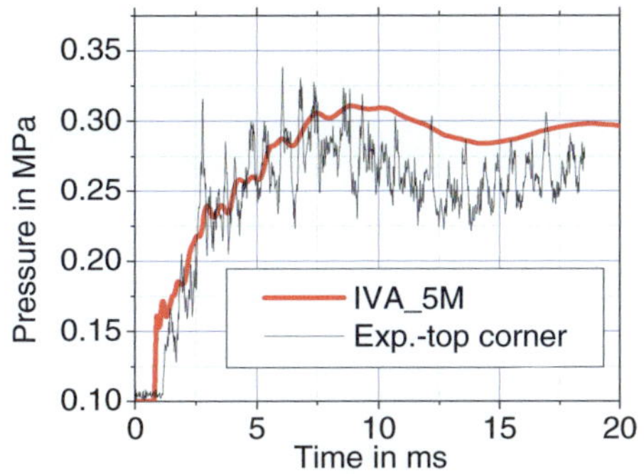

Fig. 16.9.2 Pressure as a function of time at different positions. An IVA_5M prediction for the *Meyer* and *Kirstahler* gas-in-gas explosion experiment

In this simulation it is the *multiblock boundary fitted* code architecture and the code integrator that are addressed, without addressing any empirical correlation, except for the gas state and transport properties.

The comparison between the predicted and the measured pressures is shown in Figs. 16.9.2. IVA_5M essentially provides a good prediction of the trends of the pressures at the *r*–*z* locations (0.022, 0.693), (0.07, 0.693) and (0.148, 0.693). We learn that especially in regions in which pressure spikes are expected the spatial resolution has to be selected carefully. This was already demonstrated in 1987 in *Kolev* (1987) by using IVA2 to the same test problem.

Conclusion. From comparison with experiments featuring gas explosions in a 2D space filled previously with gas we conclude that the code essentially provides a good prediction of the trends of the pressures. Especially in regions in which pressure spikes are expected the spatial resolution has to be selected carefully.

References

Kolev, N.I.: A three field-diffusion model of three-phase, three-component flow for the transient 3D-computer code IVA2/01. Nucl. Technol. 78, 95–131 (1987)

Meyer, L., Kirstahler, M.: Experimental investigation of the fluid dynamic of gas jet expansion in a liquid pool. In: Shah, K., Ganic, E.N., Yang, K.T. (eds.) Proc. 1st World Conf. Experimental Heat Transfer, Fluid Mechanics, and Thermodynamics, Dubrovnic, Yugoslavia. Elsevier, Amsterdam (1988)

Wilchelm, D., Kirstahler, M.: Versuchsanlage zur schnellen Gasinjektion in einen wassergefüllten Behälter, KfK-4321, Kernforschungszentrum Karlsruhe (1987)

16.10 2D: N_2 explosion in space with internals filled previously with water

Consider the very practical case of failure of the melt confined by a crust in in-vessel melt–water interaction analysis, see Fig. 16.10.1 Assume that the crust failure is restricted to a limited region. The melt penetrates this region, driven by gravity. Fragmentation and subsequent steam generation create a high-pressure zone. The vapor of this zone removes all the surrounding liquids (water and melt) towards the available venting cross-sections. The steam also penetrates the surrounding liquids by creating *Taylor* instability entrainment at the surface as discussed in *Corradini*, et al. (1980). Complicated pressure wave dynamics arises in this case.

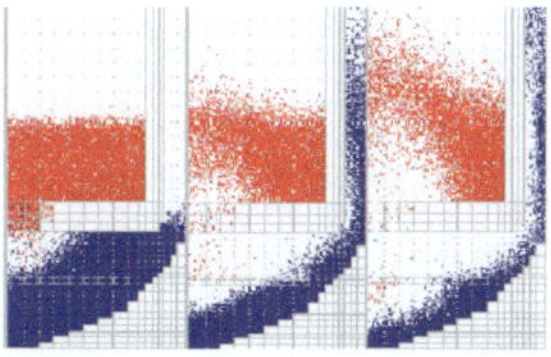

Fig. 16.10.1 Illustration showing steam generation after premixing and its action on the surrounding water and melt for in-vessel melt–water interaction

We analyze the capability of IVA5 to simulate the gas-driven explosions into liquid without strong evaporation, *Kolev* (1987), by comparing with the experimental data for fast acoustic processes in two-phase mixtures in a complex geometry obtained by *Wilchelm* and *Kirstahler* (1987), *Meyer* and *Kirstahler* (1988). The calculation is repeated for the same geometry as in the previous section with the same discretization and the same initial and boundary conditions, but with the following changes. The low-pressure region consists here of degassed water with an air volumetric fraction of 0.002 to 0.005 *Meyer* and *Wilhelm* (1994), and with air above the water at $p = 0.1$ MPa and room temperature, Fig. 16.10.1c. The walls containing the low-pressure region are transparent. Two high-speed cameras (90° apart, 7 frames/ms) were used to record the "bubble growth" from the region at the nozzle entrance. The photographs show bubble growth, quantitatively measured in *Meyer* and *Kirstahler* (1988) and entrainment of droplets from the "bubble surface" due to instability.

Fig. 16.10.2 An IVA6 prediction of the *Meyer* and *Kirstahler* gas-in-water explosion experiment. 1st row: water volume fractions as a function of space. 2nd row: pressure as a function of space. Parameter-time: 0, 0.0029, 0.0052, 0.0068, 0.0115, 0.02 s

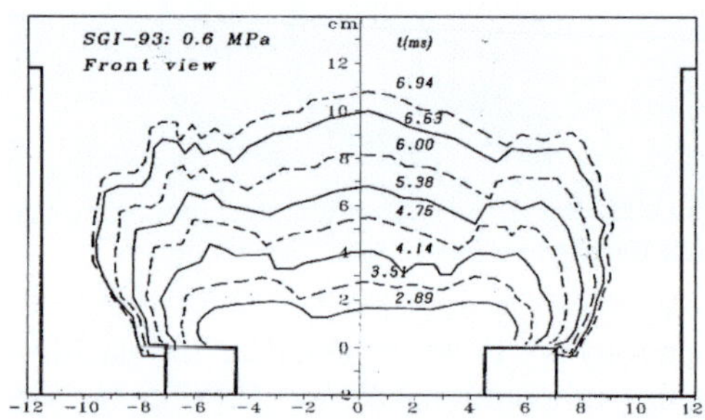

Fig. 16.10.3 Observed (bottom) gas-water "interface" as a function of space. Parameter – time

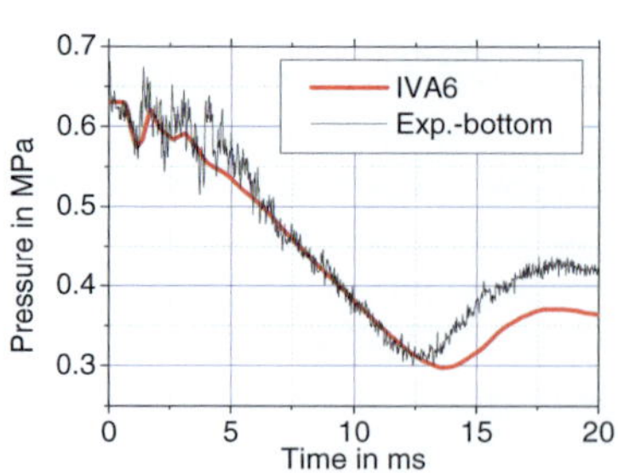

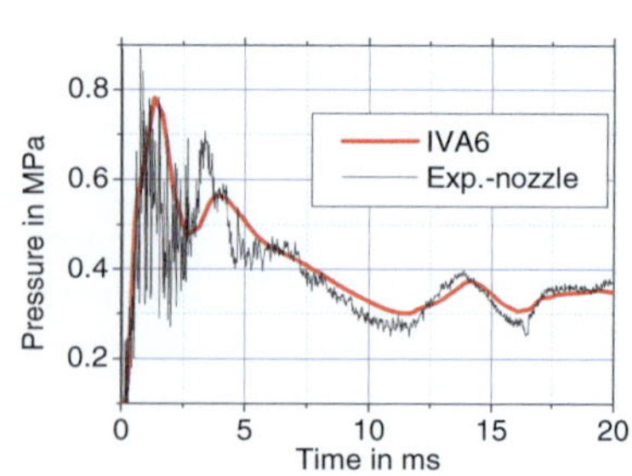

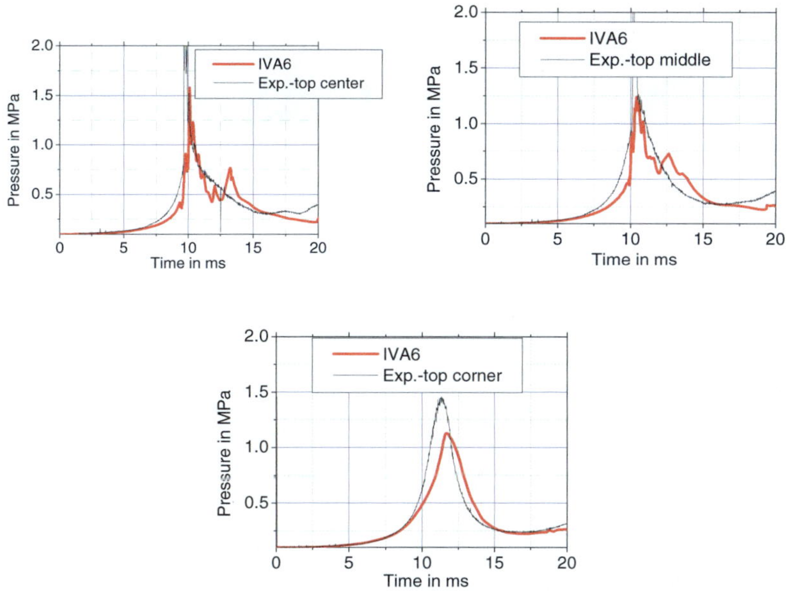

Fig. 16.10.4 An IVA6 prediction of the *Meyer* and *Kirstahler* gas-in-water explosion experiment. Pressure as a function of time at different positions

Figure 16.10.2 shows the predicted water volume fraction as a function of space. The dynamic displacement is evident from the time sequences. A comparison between the predicted and the measured lower gas/water "interface" is possible by comparison of Figs. 16.10.2 and 16.10.3. Figures 16.10.4a–e give a comparison of the computed and measured pressures at the bottom, nozzle, top-center, top-middle and top-corner, respectively. There is reasonable agreement between prediction and experiment.

Conclusion. From comparison with experiments on gas explosions in a 2D space with internals previously filled with liquid we conclude that there is reasonable agreement for fast material relocation. The code essentially provides a good prediction of the pressure impulses, which is essential for fluid structure interaction analysis. Again, just as in the single-phase case, the spatial resolution has to be selected carefully particularly for regions in which pressure spikes are expected.

Now, we simulate the same case with the multiblock boundary fitted computer code IVA_5M, *Kolev* (2004). The results are presented in Fig. 16.10.5.

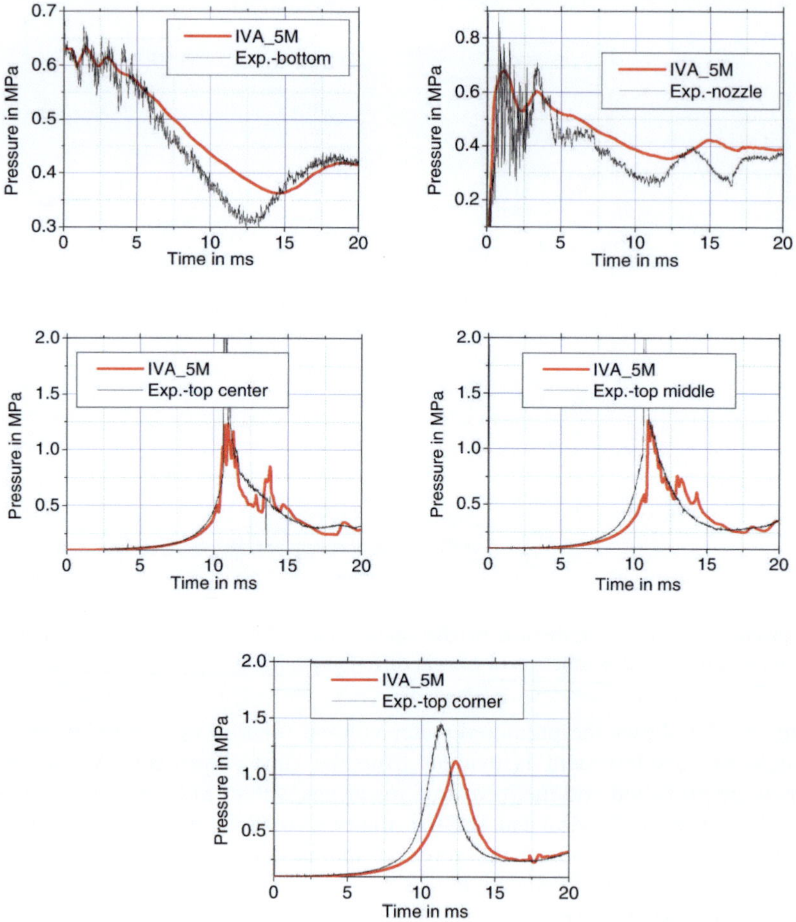

Fig. 16.10.5 An IVA_5M prediction of the *Meyer* and *Kirstahler* gas-in-water explosion experiment. Pressure as function of time at different positions

Conclusion. From comparison with experiments on gas explosions in a 2D space with internals previously filled with liquid we conclude that there is reasonable agreement for fast material relocation IVA_5M essentially provides a good prediction of the pressure impulses, which is essential for fluid structure interaction analysis. We see the expected adequate performance of the IVA_5M method under the new programming architecture and data management compared with the old architecture of single block in IVA6.

References

Corradini, M.L., Rohsenow, W.M., Todreas, N.E.: The effect of sodium entrainment and heat transfer with two-phase UO2 during a hypothetical core disruptive accident. NSE 73, 242–258 (1980)

Kolev, N.I.: IVA2 verification expansion phase experiment in SNR geometry, KfK 4310 (September 1987)

Kolev, N.I.: IVA_5M numerical method for analysis of three-fluid multi-component flows in boundary-fitted multi-blocks. Presented in Second M.I.T. Conference on Computational Fluid and Solid Mechanics, Boston, June 17-20. Computers & Structures, vol. 83, pp. 499–523 (2004)

Meyer, L., Kirstahler, M.: Experimental investigation of the fluid dynamic of gas jet expansion in a liquid pool. In: Shah, K., Ganic, E.N., Yang, K.T. (eds.) Proc. 1st World Conf. Experimental Heat Transfer, Fluid Mechanics, and Thermodynamics, Dubrovnic, Yugoslavia, September 4-9, Elsevier, New York (1988)

Meyer, L., Wilhelm, D.: Investigation of the fluid dynamics of gas jet expansion in a liquid pool, Kernforschungszentrum Karlsruhe, KfK 5307 (March 1994)

Wilchelm, D., Kirstahler, M.: Versuchsanlage zur schnellen Gasinjektion in einen wassergefüllten Behälter, KfK-4321, Kernforschungszentrum Karlsruhe (1987)

16.11 Film entrainment in pipe flow

Entrainment of continuous liquid is an important fragmentation mechanism. The UPTF test Nr.26, *Kolev*, et al. (2000) is appropriate to test the physical modeling in this field. Details of this study are provided in *Kolev* et al. (2000). We provide here a brief summary of the 1D result. In a 3D vessel, Fig. 16.11.1, steam is injected as presented in Fig. 16.11.2. The steam may leave the vessel only through the horizontal pipe presented in Fig. 16.11.1 and in more detail in Fig. 16.11.2. The initial pressure and the pressure of the exit of the pipe are specified. The water and the steam temperatures are 141–142 °C, and 144 through 148 °C, respectively. The pressure level of this particular experiment was 0.39–0.43 MPa.

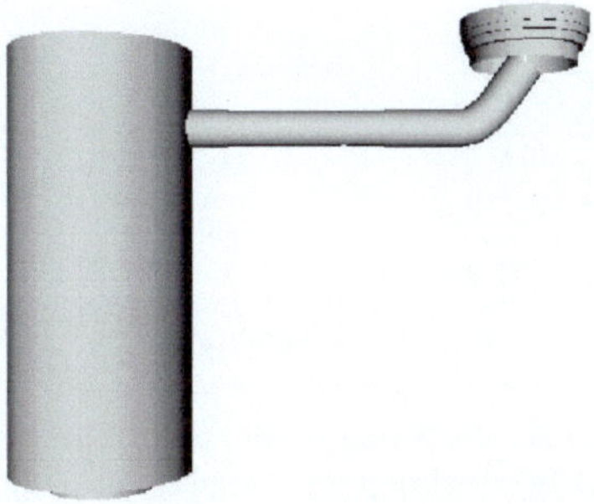

Fig. 16.11.1 IVA6 model for 1D network simulation including the steam generator inlet plenum

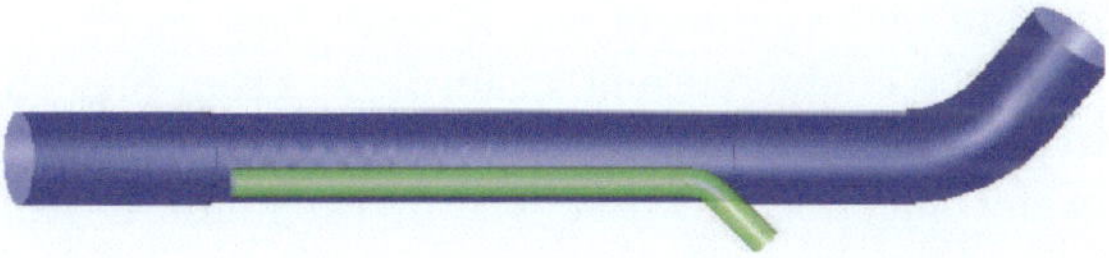

Fig. 16.11.2 IVA6 representation of the geometry – two-pipe system

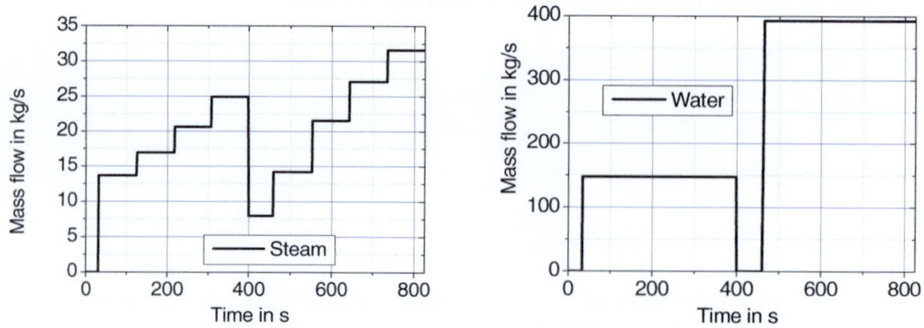

Fig. 16.11.3 a) Steam mass flow injected below the core simulator as a function of time. b) Water mass flow injected through the emergency core cooling pipe coaxial into the main coolant line as a function of time

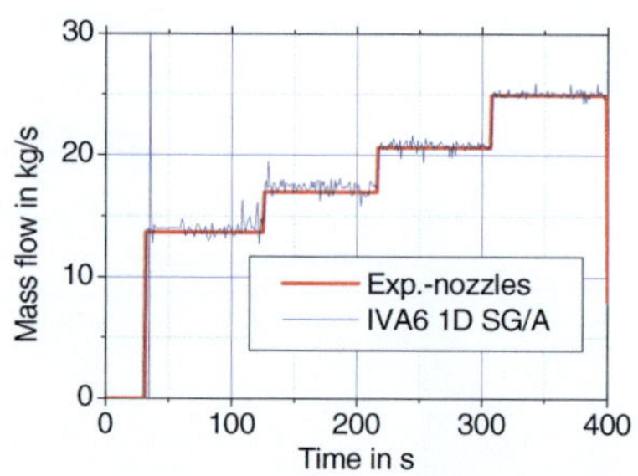

Fig. 16.11.4 Injected steam mass flow (exp.) below the core simulator as a function of time (red). Computed steam mass flow leaving the vessel as a function of time. 1D IVA6 simulation with modeling the steam generator simulators

Through the small pipe visible in Fig. 16.11.2 water is injected as shown in Fig. 16.11.3b. The target of the experiment was to study how much of the injected water is entrained from the countercurrent steam flow. From Fig. 16.11.3 we realize that there are two phases of this experiment – phase A and phase B corresponding to the stepwise water-injection function. The investigation in *Kolev* et al. (2000) indicated that in these two stages the entrainment phenomena happen at different places: in phase A in the pipe, and in phase B essentially inside the 3D vessel. In this section we consider only phase A. In this case we use a 5-pipe network model as presented in Fig. 16.11.1. Pipes 1 and 2 are shown in Fig. 16.11.2. Pipe 1 simulating the hot leg is discretized using 42 cells. Pipe 2 simulating the water injection is discretized using 5 equidistant cells. The main coolant pipe cross-section is correspondingly reduced with the injection pipe cross-section over the common distance. Pipe 3 is simulating the vessel, Fig. 16.11.2, and is discretized into 3 cells. The steam is injected through Pipe 4 in the uppermost cell of Pipe 3. The main coolant Pipe 1 is connected to the middle cell of the Pipe 3. The lower vessel cell (Pipe 3) is large and serves to collect the gravitationally separated water.

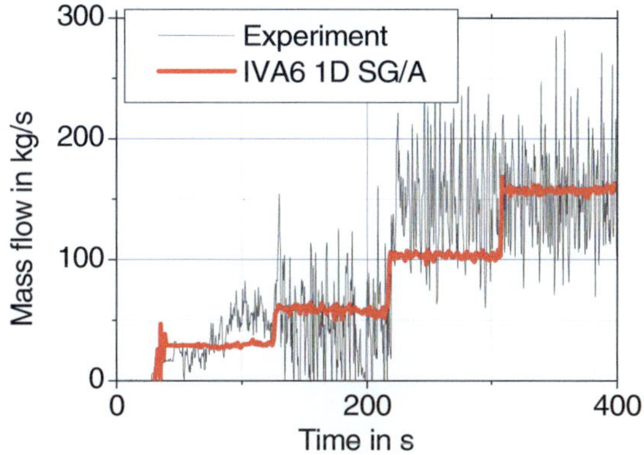

Fig. 16.11.5 Comparison between the measured and the computed total mass flow as a function of time at the gamma-densitometer position. The SG simulator considered in the 1D geometrical model

The results of the computation are presented in Figs. 16.11.4 and 16.11.5 and compared with the measurements.

Conclusion: Figure 16.11.5 presents the essence of this analysis showing that three velocity fields are a proper skeleton for the mathematical description of entrainment processes in pipe flows. One should be careful with replacing 3D flows by 1D cross-section-averaged flows and with the applicability of entrainment correlations for a particular problem – see the discussion in *Kolev* et al. (2000). As already stated in *Kolev* et al. (2000) the particular data bank for developing better entrainment correlations has to be extended for liquid inertia-dominated flows. Experimental activities on this topic are in progress in the Research Center Karlsruhe, Germany.

Phase B of this experiment is discussed in one of the following sections.

Reference

Kolev, N.I., Seitz, H., Roloff-Bock, I.: Hot-leg injection: 3D versus 1D three velocity fields modeling and comparison with UPTF experiment, Siemens report KWU NA-T/00/E027, R&D Project 903060, Verification IVA6 (2000)

16.12 Water flashing in nozzle flow

In order to verify the capability of IVA5 to model flashing phenomena we use the steady-state flashing experiment of *Abuaf* et al. (1981) and the becoming classical *Edwards* and *O'Brien* pipe experiment documented in *Edwards* and O'*Brien* (1970) for transient pipe blow-down. The second experiment will be discussed in the next section.

The geometry of the nozzle is given in Table 16.12.1. In Table 16.12.2 we summarize the initial and the boundary condition used in the test comparison here. Fig. 16.12.1 presents the result of the comparison.

Table 16.12.1 Geometry of the test sections

Author	Geometry	Length	Hydraulic diameter	
		m	z/m	$D_{hyd,\max}/m$
BNL Abuaf	Circular *nozzle*	0.6	0	0.051
et al. (1981)			0.02	0.051
			0.028	0.02546
			0.58	0.051
			0.6	0.051

Table 16.12.2 Initial conditions

Author	Test Nr.	Pressures		Mass flow rate (comp.)	Temperature
Abuaf et al. (1981)		p_{in}/bar	p_{out}/bar	$G/kg/(m^2s)$	T_{in}/K
	133	357 485	185 030	$w_{2in} = 4.55\ m/s$ (4190)	394.35
	224	341 200	196 000	4410 (4044)	394.05
	275	573 500	441 000	4290 (4193)	421.85
	281	688 700	431 000	5730 (5533)	421.95
	286	530 800	459 000	3580 (2706)	422.25
	292	504 700	473 000	2510 (2197)	422.05
	315	341 200	193 000	4410 (4202)	392.25
	320	323 700	169 000	4430 (4352)	387.55
	324	341 000	198 000	4410 (4004)	394.15
	340	321 000	249 000	4430 (3768)	394.45

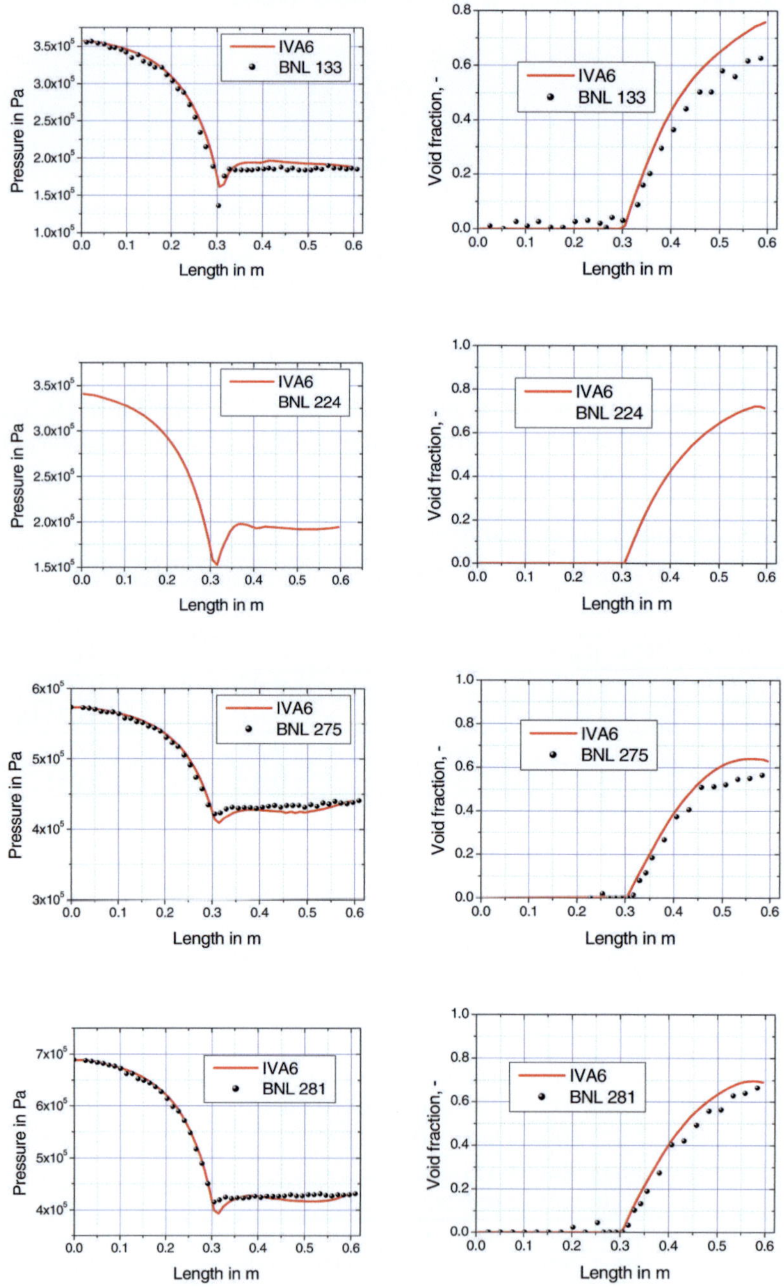

IVA6
BNL 133
Pressure in Pa
Length in m
Void fraction, -
IVA6
BNL 224
IVA6
BNL 275
IVA6
BNL 281

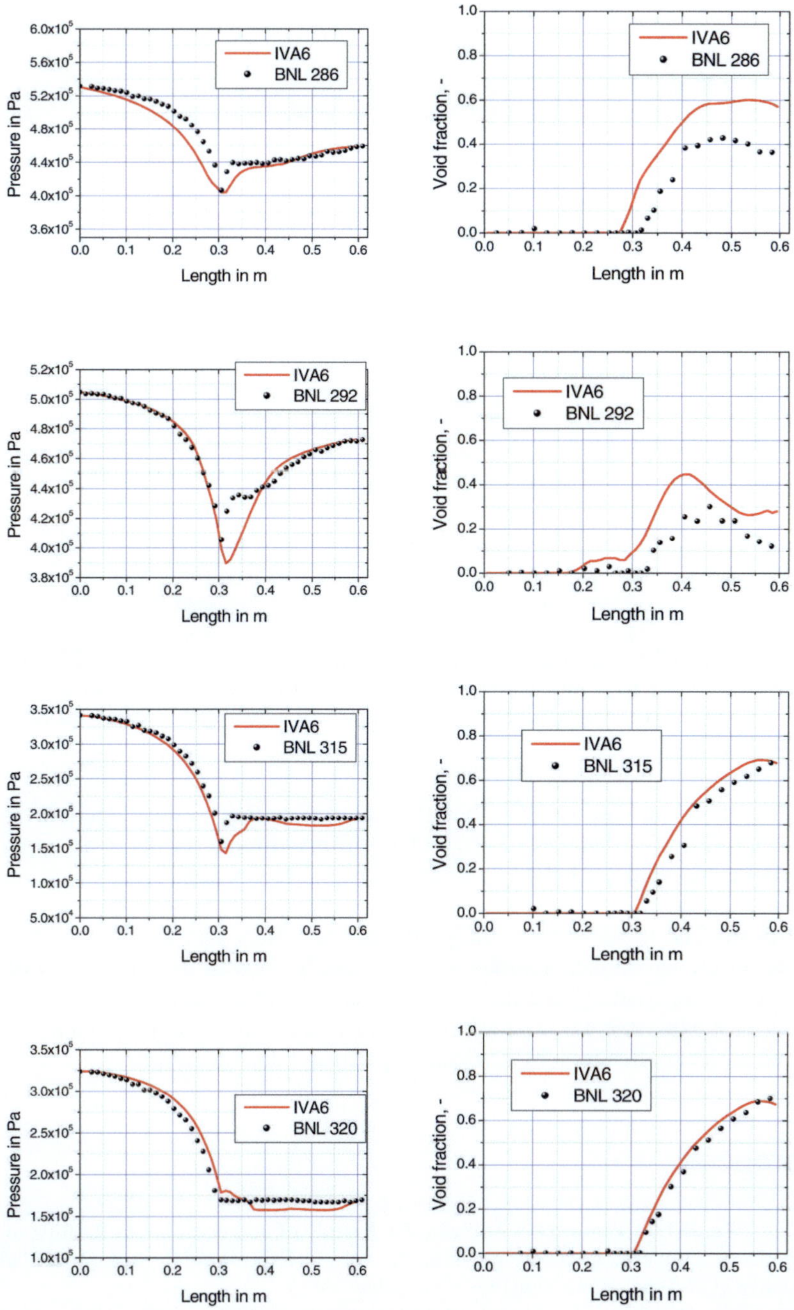
IVA6
BNL 286
Pressure in Pa
Length in m
Void fraction, -
IVA6
BNL 292
IVA6
BNL 315
IVA6
BNL 320

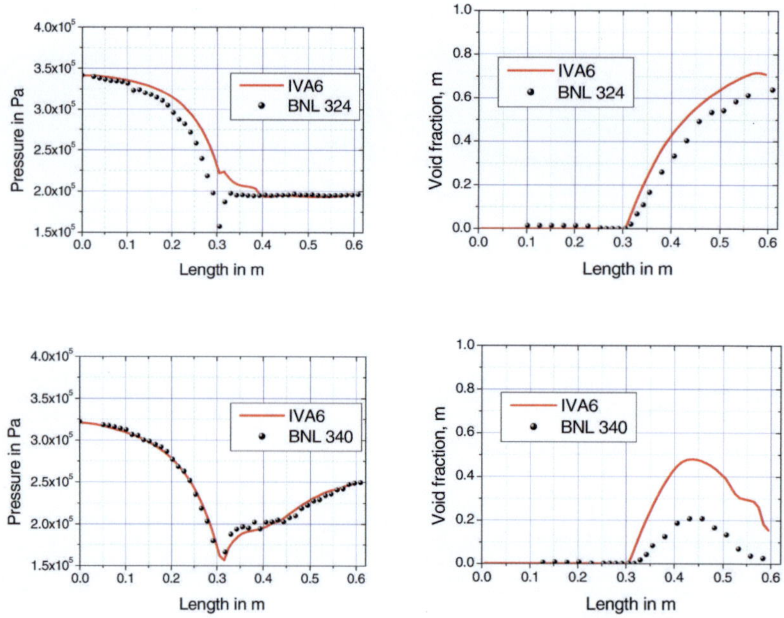

Fig. 16.12.1 An IVA5 prediction of the flashing flow in nozzle with variable cross-section. Comparison with the BNL measurements. In each pair of figures: left – pressure as a function of the spatial coordinate, right – local cross-section-averaged volume fraction as a function of the spatial coordinate

It is evident from Fig. 16.12.1 that the nucleation theory developed in *Kolev* (1995) and given in Chapters 15, 16 and 17 of this volume results in an appropriate pressure and void prediction.

Conclusion: The mean error of the predicted mass flow rate as a function of the pressure difference on the two ends is within 9%. In view of the complexity of this process the comparison of the pressure and the local cross-section-averaged vapor fraction with experimental data are classified as very good. Improvement can be achieved in the applications by using nonequidistant discretization.

References

Abuaf, N., Wu, B.J.C., Zimmer, G.A., Saha, P.: A study of non equilibrium flashing of water in a converging diverging nozzle, vol 1 Experimental, vol 2 Modeling, NUREG/CR-1864, BNL-NUREG-51317 (June 1981)

Edwards, A.R., O'Brien, T.P.: Studies of phenomena connected with the depressurization of water reactors. J. Brit. Nucl. Energy Soc. 9(1-4), 125–135 (1970)

Kolev, N.I.: The code IVA4: Nucleation and flashing model. Kerntechnik 60(6), 157–164 (1995); Also in: Proc. Second Int. Conf. On Multiphase Flow, Kyoto (April 3-7, 1995); ASME & JSME Fluid Engineering Conference International Symposium on Validation of System Transient Analysis Codes - Hilton Head (SC) USA (August 13-18, 1995); Int. Symposium on Two-Phase Flow Modeling and Experimentation, ERGIFE Place Hotel, Rome, Italy (October 9-11, 1995)

16.13 Pipe blow-down with flashing

16.13.1 Single pipe

As already mentioned, in order to verify the capability of IVA5 to model transient flashing phenomena we use the classical *Edwards* and *O'Brien* pipe experiment from 1970. In this experiment one of the ends of a horizontal pipe filled with subcooled water at high pressure is abruptly opened, Fig. 16.13.1.

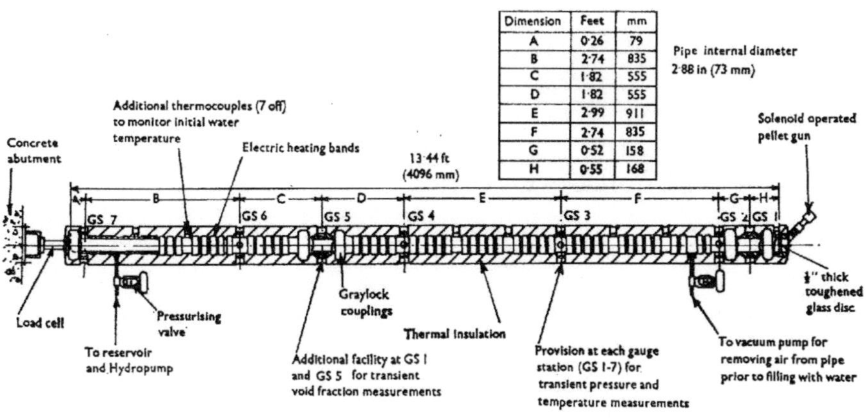

Dimension	Feet	mm
A	0·26	79
B	2·74	835
C	1·82	555
D	1·82	555
E	2·99	911
F	2·74	835
G	0·52	158
H	0·55	168

Fig. 16.13.1 General arrangement of the blow-down pipe of Edwards and O'Brien (1970)

The pressure reduction brings the water into a superheated state. The delayed nucleation and bubble growth cause the so-called pressure undershoot and then the relaxation to the saturation pressure corresponding to the initial water temperature. The results are given in Figs. 16.13.2, 16.13.3 and 16.13.4. It is evident from Figs. 16.13.2–16.13.4 that the nucleation theory developed in *Kolev* (1995) and presented in Chapters 15 to 17 of Vol. 2 results in an appropriate pressure prediction, Fig. 16.13.2. The somewhat faster emptying of the pipe in the computation for the later stage of the process, Figs. 16.13.3a and b is caused by several interactions in the computer code such as drag forces, flow pattern recognition, etc. which will not be discussed here. It can also be seen that the computed water temperature, Fig. 16.13.4a decreases faster than that determined experimentally. This is evidence of the release of more energy from the liquid for evaporation in the model than indicated by experimental data. This is consistent with the predicted void fraction at a given position, Fig. 16.13.4c. The large void fraction increase in the first 0.1 s is not predicted by the model.

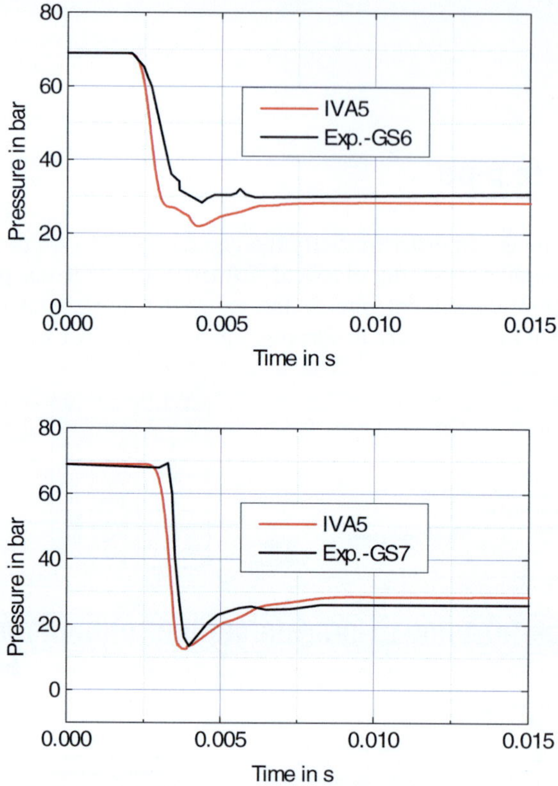

Fig. 16.13.2 Pressure at 0.072 and 0.914 m from the dead end of a pipe as a function of time. Comparison of IVA5 prediction with the experimental data of *Edwards* and *O'Brien*. Pipe diameter 0.0732 m, length 4.096 m

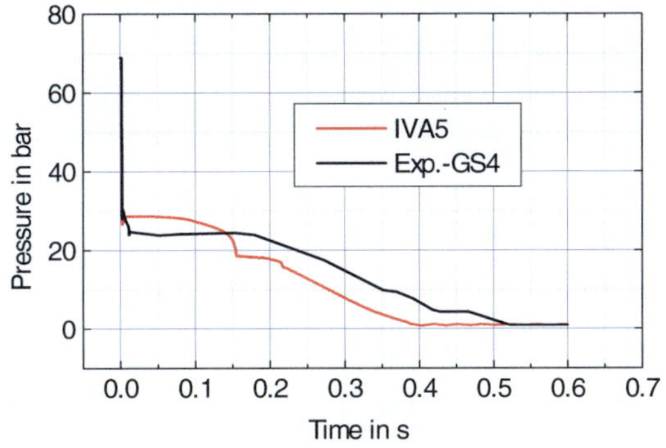

(a)

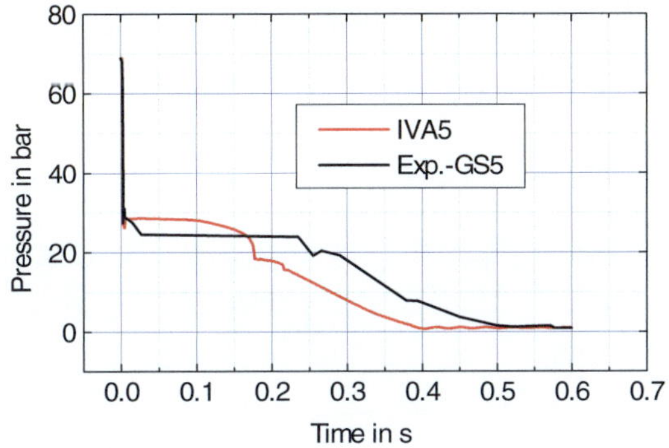

(b)

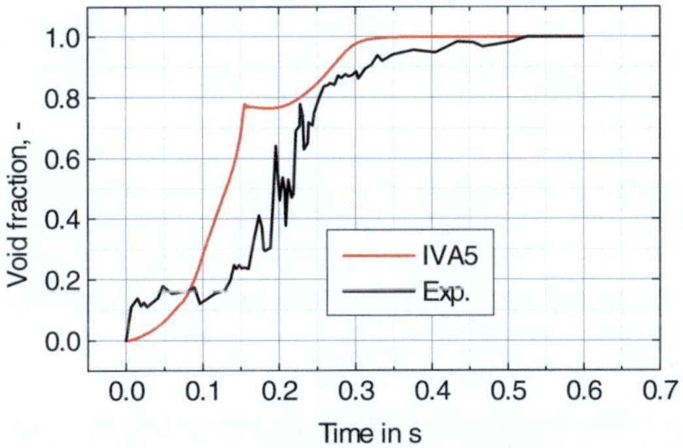

(c)

Fig. 16.13.3 a) b) Pressure at 2.024 and 2.469 m from the dead end of a pipe as a function of time. c) Steam volume fraction at 1.469 m from the dead end of the pipe as a function of time. Comparison of IVA5 prediction with the experimental data of *Edwards* and *O'Brien*. Pipe diameter 0.0732 m, length 4.096 m

The IVA5 force processor computes pipe forces related to hydrodynamics only in accordance with Chapter 8 of Volume 1. The results are presented in Fig. 16.13.4b and correspond to the predicted pressure history. The oscillating character of the measurements, Fig. 16.13.4b, is a result of the fluid–structure interactions. These interactions are not simulated in the computation.

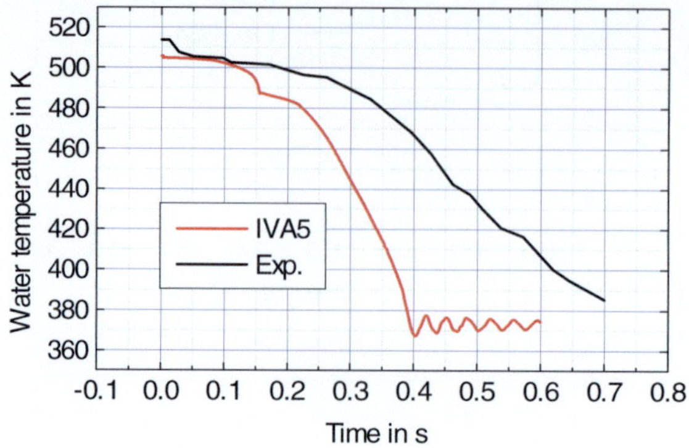

(a)

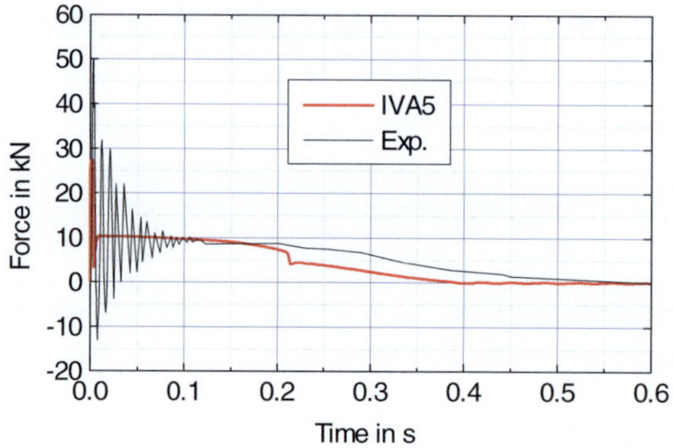

(b)

Fig. 16.13.4 a) Water temperature at 1.469 m from the dead end of the pipe as a function of time. Force as a function of time: b) IVA5 prediction without fluid-structure interactions compared with the *Edwards–O'Brien* measurements reflecting fluid structure interactions. Comparison of IVA5 prediction with the experimental data of *Edwards* and *O'Brien*. Pipe diameter 0.0732 m, length 4.096 m

Conclusion: From the comparison with experiments on pipe blow-down and water flashing we conclude that the nucleation theory presented in this monograph and used in the code results in an appropriate pressure prediction. The fast discharge pressure waves are well predicted. The evaporation model gives somewhat faster energy discharge from the liquid. In general, the blow-down process is properly predicted. The force processor works properly and provides the forces acting on a pipe network.

16.13.2 Complex pipe network

Figure 16.13.5 shows a complicated network consisting of a heat exchanger and pipes, valves etc. The heat exchanger has as primary medium water at 320 °C and about 172 bar pressure and as secondary medium water at 50 °C and about 9 bar. A break inside the heat exchanger is simulated. A flashing water–steam mixture enters the secondary side that is filled with subcooled water and creates a nonstable pressure increase and dramatic condensation shocks. As seen from Fig. 16.13.4 the first 0.2 s are characterized by strong condensation oscillation in the secondary side. Complicated interactions between the valves and the break are extremely challenging for the solution method. Analysis of processes in industrial network is becoming a routine activity in large companies.

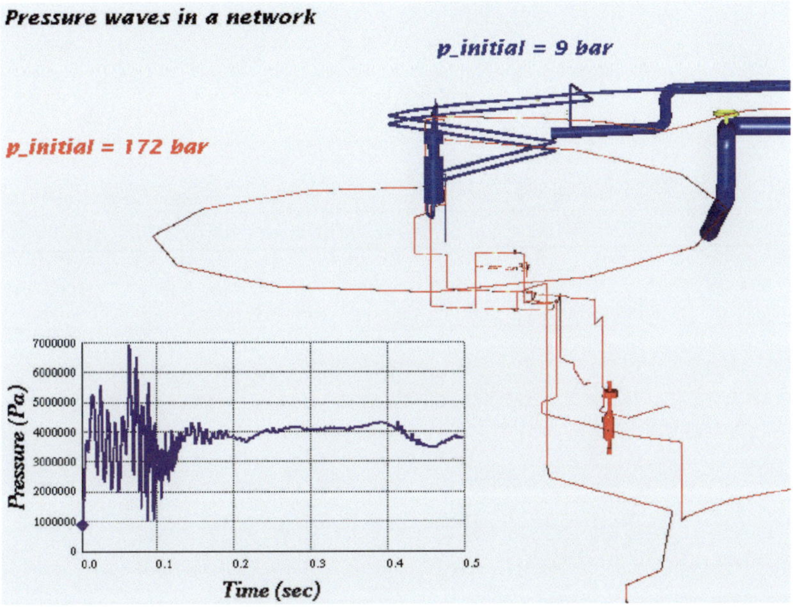

Fig. 16.13.5 Pipe break in a high-pressure heat exchanger. Pressure as a function of time. The visualization is performed using SONJA

References

Edwards, A.R., O'Brien, T.P.: Studies of phenomena connected with the depressurization of water reactors. J. Brit. Nucl. Energy Soc. 9(1-4), 125–135 (1970)

Kolev, N.I.: The code IVA4: Nucleation and flashing model. Kerntechnik 60(6), 157–164 (1995); Also in: Proc. Second Int. Conf. On Multiphase Flow, Kyoto (April 3-7, 1995); ASME & JSME Fluid Engineering Conference International Symposium on Validation of System Transient Analysis Codes - Hilton Head (SC) USA (August 13-18, 1995); Int. Symposium on Two-Phase Flow Modeling and Experimentation, ERGIFE Place Hotel, Rome, Italy (October 9-11, 1995)

16.14 1D boiling, critical heat flux, postcritical heat flux heat transfer

Boiling on a hot solid surface is an essential phenomenon in many technological processes. The purpose of this section is to compare the prediction of IVA6 with a variety of data from different authors for subcooled and saturated boiling in tubes and rod bundles.

Several authors measured the volume fraction of steam in heated channels. In this section we will use the set of data reported by *Sabotinov* (1974), *Pierre* (1965), *Egen* et al. (1957) and *Nylund* et al. (1968). The data vary over a relative wide range of parameters

$$38 \le G \le 2000 \ kg/(m^2 s),$$

$$1 \le p \le 140 \ bar,$$

$$\Delta T_{sub} \le 132 \ K,$$

$$20.3 \le \dot{q}''_w \le 1723 \ kW/m^2.$$

The geometry of the test sections is summarized in the Table 16.14.1 and the boundary conditions in Table 16.14.2.

Table 16.14.1 Geometry of the test sections

Author	Length	Hydraulic diameter	Heated diameter	Cross section
	m	D_{hyd}/m	D_{heat}/m	F/m^2
Sabotinov (1974), circular tube	1.5	0.0117	0.0117	0.0001075
Pierre (1965), rect. tube	1.2576	0.01777	0.01777	0.0004936
Egen et al. (1957), rect. tube	0.686	0.004756	0.004756	0.0000665
Nylund et al. (1968) 36 fuel rods, $d_{rod} = 0.0138 \ m$	4.375	0.0269	0.0366	0.01428
Bennett (1967), circular tube	3.6576 5.5626	0.01262	0.01262	0.000125448

Table 16.14.2 Boundary conditions

Author	Test No.	Pressure	Mass flow rate	Heat flux	Temperature
		p/bar	$G/kg/(m^2s)$	$\dot{q}''_w/\left(kW/m^2\right)$	T_{in}/K
Sabotinov (1974)					
	1	68.0585	2037	1137	504.15
	2	67.96043	1980	1130	523.15
	3	68.05849	1975	1138	539.15
	4	69.92177	1530	1360	476.15
	5	68.35269	1547	1367	504.15
	6	67.86236	2023	785	519.15
	7	68.05849	1979	786	536.15
	8	68.05849	2089	793	544.15
	9	67.86236	962	1688	461.15
	10	106.89303	1015	1723	456.15
	11	106.89303	2129	1723	514.15
	12	107.97176	2109	1718	539.15
					Subcooling
					$\Delta T_{sub}/K$
Pierre (1965)					
	1	13.79	1005	71.9	0.28
	9	41.38	921	287.7	2.3
	10	41.38	929	287.7	7
	13	55.17	890	287.7	2.6
Egen et al. (1957)					
	7	137.9	949	630.9	31
	8	137.9	895	946.4	34.3
	13	137.9	922	630.9	3.2
	16	137.9	1153	1577.3	74.3
Nylund et al. (1968)					
	313009	50	1107	436	4.4
	313018	49.7	1124	643	3.7
	313024	49.7	858	216	4.2
	313016	49.6	1208	426	19.3
	313014	49.7	1163	429	11
Bennett (1967), $L = 3.6576$ m					
	5442	68.93	4 813.8	2 082	15.95
	5407	68.93	1 939.1	1 496	14.65
	5424	68.93	2 562.8	1 655	10.34
	5456	68.93	1 328.9	1 317	24.13

Bennett (1967), $L = 5.5626$ m					
	5253	68.93	1 356.0	902	18.96
	5293	68.93	1 979.8	1 074	18.96
	5332	68.93	664.4	648	25.43
	5280	68.93	3851.	1 691	11.64

Some of the specific models used for mass transfer are indicated in the text of the figures. The comparison for the pipes is given in Figs. 16.14.1–16.14.3. The comparison for the heated rod bundle is given in Fig. 16.14.4.

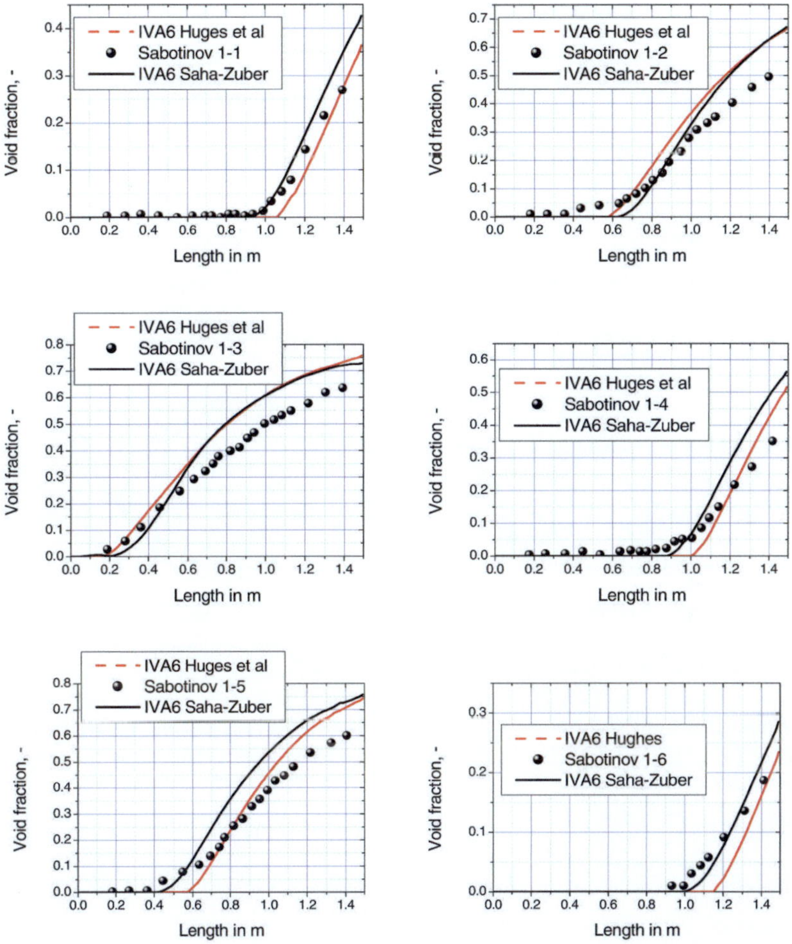

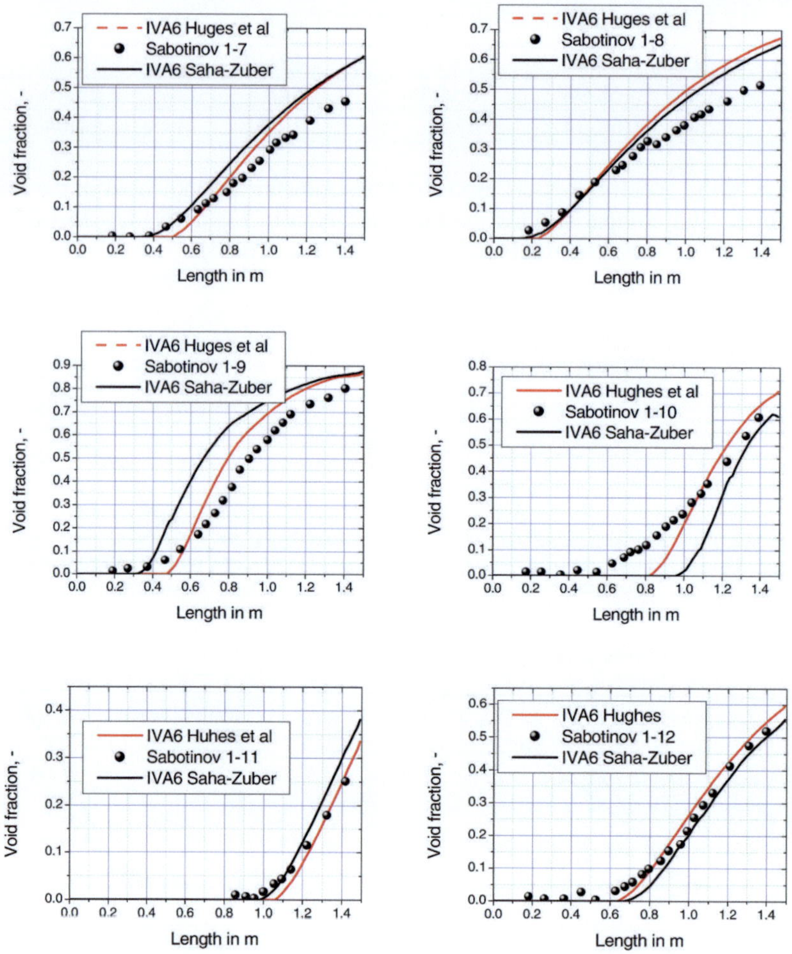

Fig. 16.14.1 a-l Steam volume fraction as a function of length measured from the bottom. Comparison between the IVA prediction and the data of *Sabotinov* (1974). The mass transfer is modeled using the proposal by *Hughes* et al. (1981) or using the approach by *Saha* and *Zuber* (1974) and *Levy* (1967). Saturated forced convection boiling is modeled using the *Chen* (1963) correlation

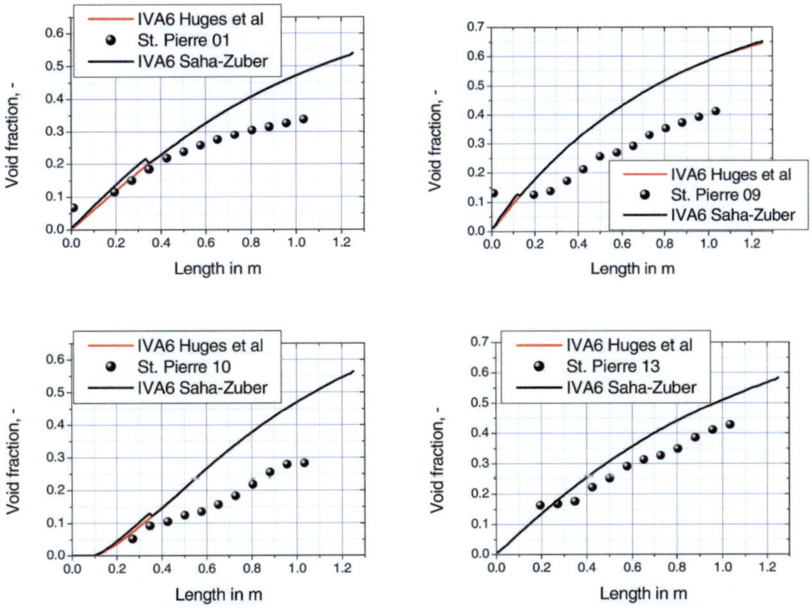

Fig. 16.14.2 Steam volume fraction as a function of length measured from the bottom. Comparison between the IVA6 prediction and the data of *Pierre* (1965). The mass transfer is modeled using the proposal by *Hughes* et al. (1981) or using the approach of *Saha* and *Zuber* (1974) and *Levy* (1967). Saturated forced convection boiling is modeled using the *Chen* (1963) correlation

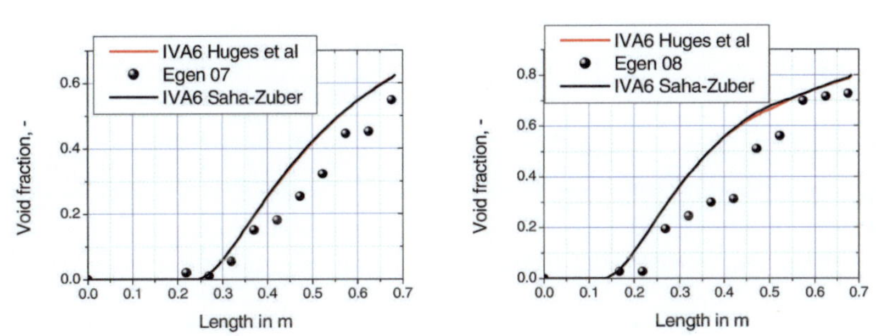

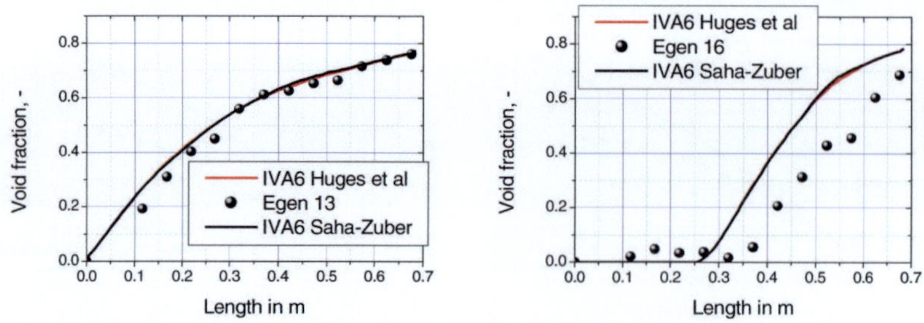

Fig. 16.14.3 Steam volume fraction as a function of length measured from the bottom. Comparison between the IVA6 prediction and the data of *Egen* et al. (1957). The mass transfer is modeled using the proposal by *Hughes* et al. (1981) or using the approach of *Saha* and *Zuber* (1974) and *Levy* (1967). Saturated forced convection boiling is modeled using the *Chen* (1963) correlation

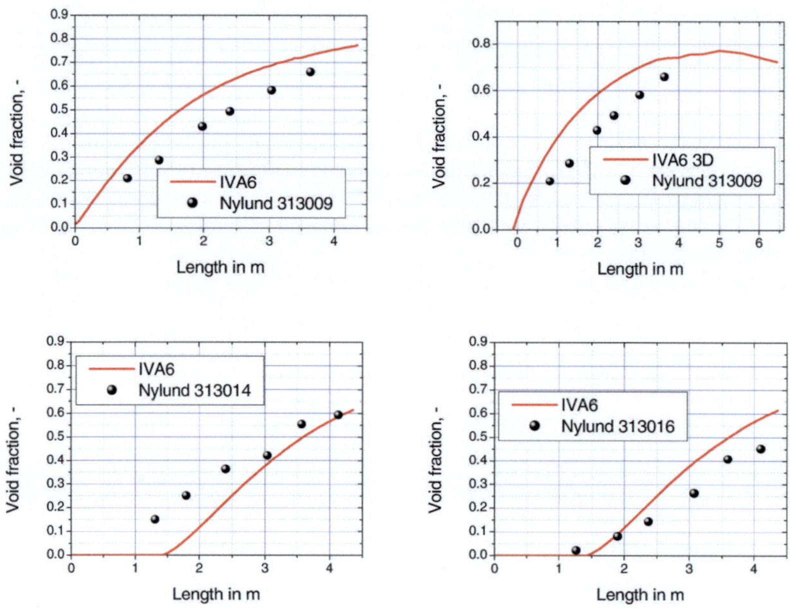

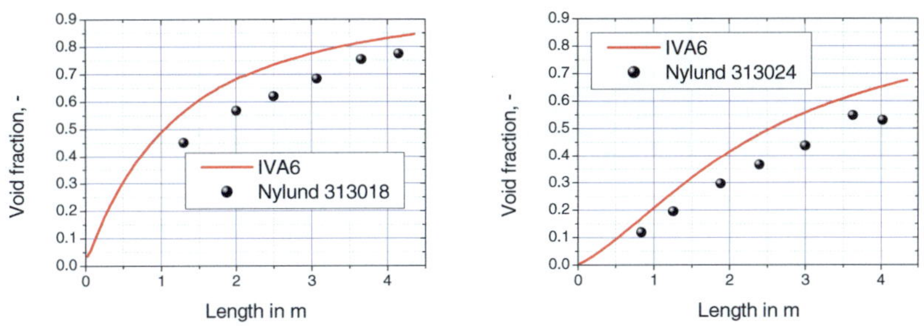

Fig. 16.14.4 Steam volume fraction as a function of length measured from the bottom. Comparison between the IVA prediction and the data of *Nylund* et al. (1968). The mass transfer is modeled using the proposal by *Hughes* et al. (1981). Saturated forced convection boiling is modeled using the *Chen* (1963) correlation

Valuable data set with systematic variation of one boundary condition by keeping the others constant for boiling in vertical pipes were reported by *Bartolomei* et al. (1982). 12-mm diameter uniformly heated pipe of 1.5 m length is used. Fixing the boundary conditions given in Table 16.14.3 the void fraction was measured by a gamma-ray densitometer within an error band of ± 0.04. The accuracy of the measurements of the boundary conditions was as follows: pressure 1%, mass flow rate 2%, heat flux 3% and temperature $\pm$ 1 K. All data except the points 11 and 15 has *Peclet* number larger then 70 000. The experimental data are presented in Fig. 16.14.1 as real void fraction as a function of the equilibrium vapor mass flow ratio, $\alpha_1 = \alpha_1\left(X_{1,eq}\right)$. At constant heat flux the equilibrium vapor mass flow ratio is proportional to the distance from the entrance of the pipe.

Table 16.14.3 Boundary conditions for the experiments performing by *Bartolomei* et al. (1982)

No.	p in Pa	ρw in kg/(m² s)	$\dot{q}''_{w2}$ in W/m²	T_2 in K
$p \approx const,\ \rho w \approx const,\ T_2 \approx const,\ \dot{q}''_{w2} \rightarrow variable$				
1	6 810 000	998	440 000	521
2	6 890 000	965	780 000	493
3	6 840 000	961	1 130 000	466
4	6 740 000	988	1 700 000	416
5	7 010 000	996	1 980 000	434
6	14 790 000	1878	420 000	603
7	14 740 000	1847	770 000	598
8	14 750 000	2123	1 130 000	583
9	14 700 000	2014	1 720 000	545
10	14 890 000	2012	2 210 000	563

$p \approx const,\ \rho w \to variable,\ T_2 \approx const,\ \dot{q}''_{w2} \approx const$				
11	6890 000	405	790 000	421
12	6890 000	986	780 000	493
13	6890 000	1467	770 000	519
14	6790 000	2024	780 000	520
15	11 020 000	503	990 000	494
15	10 810 000	966	1 130 000	502
17	10 810 000	1554	1 160 000	563
18	10 840 000	1959	1 130 000	563
$p \to variable,\ \rho w \approx const,\ T_2 \approx const,\ \dot{q}''_{w2} \approx const$				
19	3010 000	990	980 000	445
20	4410 000	994	900 000	463
21	6840 000	961	1 130 000	466
22	10 810 000	966	1 130 000	502
23	14 580 000	1000	1 130 000	533
24	6810 000	2037	1 130 000	504
25	10 840 000	1959	1 130 000	563
26	14 750 000	2123	1 130 000	583

The prediction of the *Bartolomei*'s experiments is presented in Fig. 16.14.5.

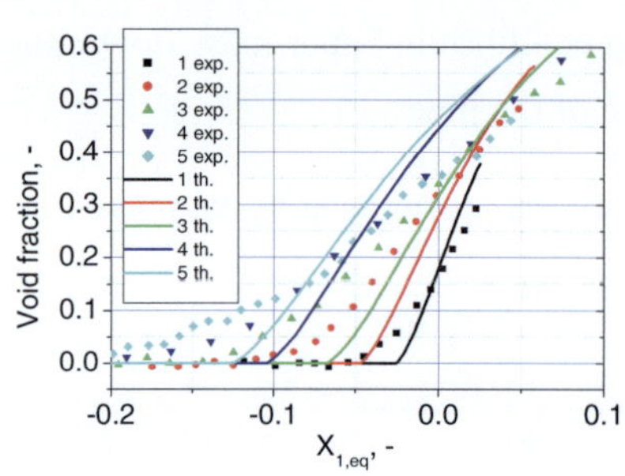

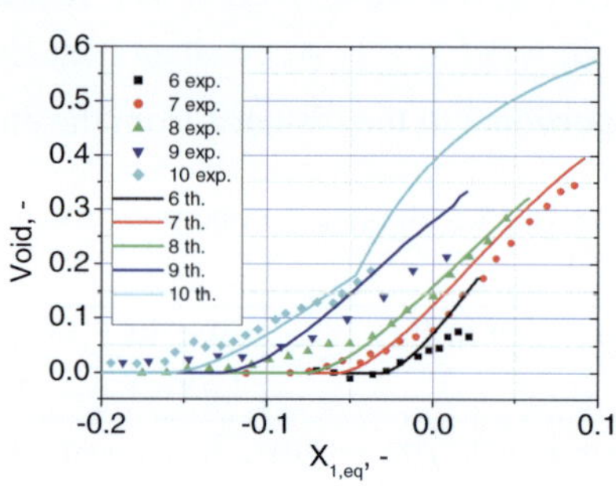

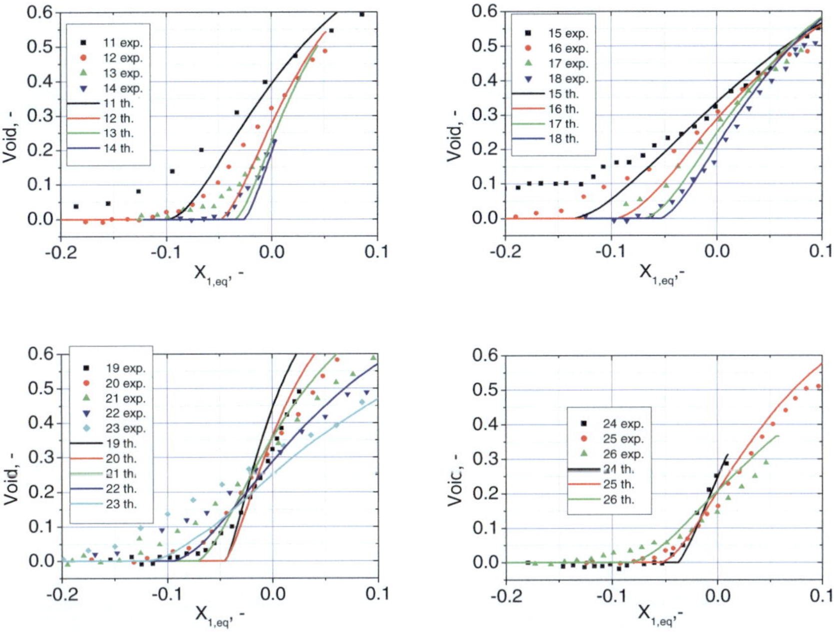

Fig. 16.14.5 Void fraction as a function of the homogeneous equilibrium mass fraction. Experimental data by *Bartolomei* et al. (1982). The mass transfer is modeled using the proposal by *Hughes* et al. (1981)

Conclusion. In general, we found slight overprediction of the volume fraction. This is attributed to the fact that the drag coefficient used in IVA6 particularly for channel flows is *not cross-section averaged.* It is well known that the void profile and the relative velocity profile have to be taken into account in the description of the mechanical interaction. This was the reason for the success of the drift flux model of *Zuber* and *Findlay*. Improvement of the model in this direction is possible if one uses the results of Section 4.2.

Bartolomei et al. (1980) performed experiments in a vertical 1.5 m pipe with 0.01203 m internal diameter and 2 mm wall thickness. The pipe was heated over the first 1 m. The flow was upwards directed. A gamma densitometer was used to measure the void fraction over the high of the pipe. The mean measurements errors reported are: p 1%, ρw 2%, $\dot{q}''_w$ 3%, $T_{2,in}$ 1 K, α_1 1%. The boundary conditions for the experiments considered here a given in Table 16.14.4.

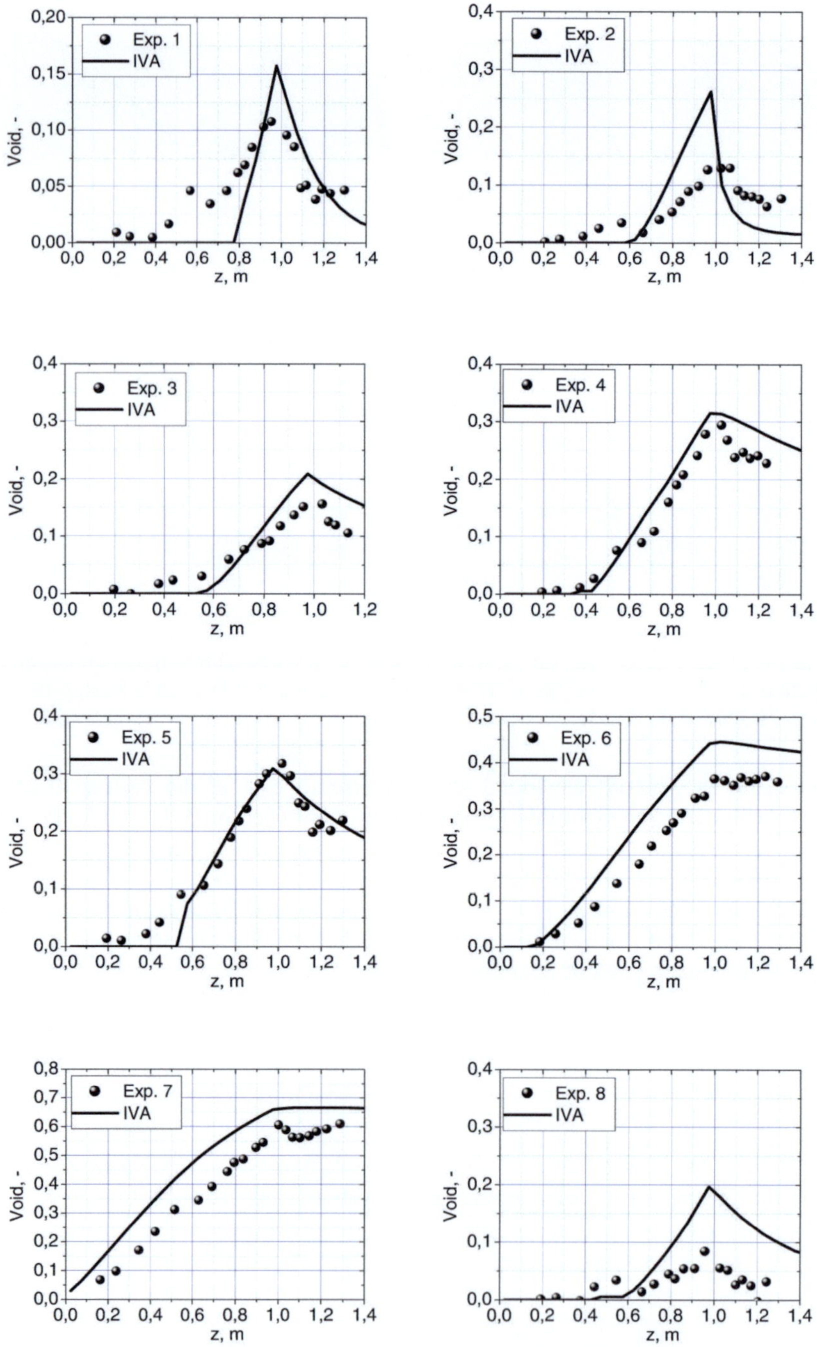
Exp. 1
IVA
Void, -
z, m
Exp. 2
IVA
Void, -
z, m
Exp. 3
IVA
Void, -
z, m
Exp. 4
IVA
Void, -
z, m
Exp. 5
IVA
Void, -
z, m
Exp. 6
IVA
Void, -
z, m
Exp. 7
IVA
Void, -
z, m
Exp. 8
IVA
Void, -
z, m

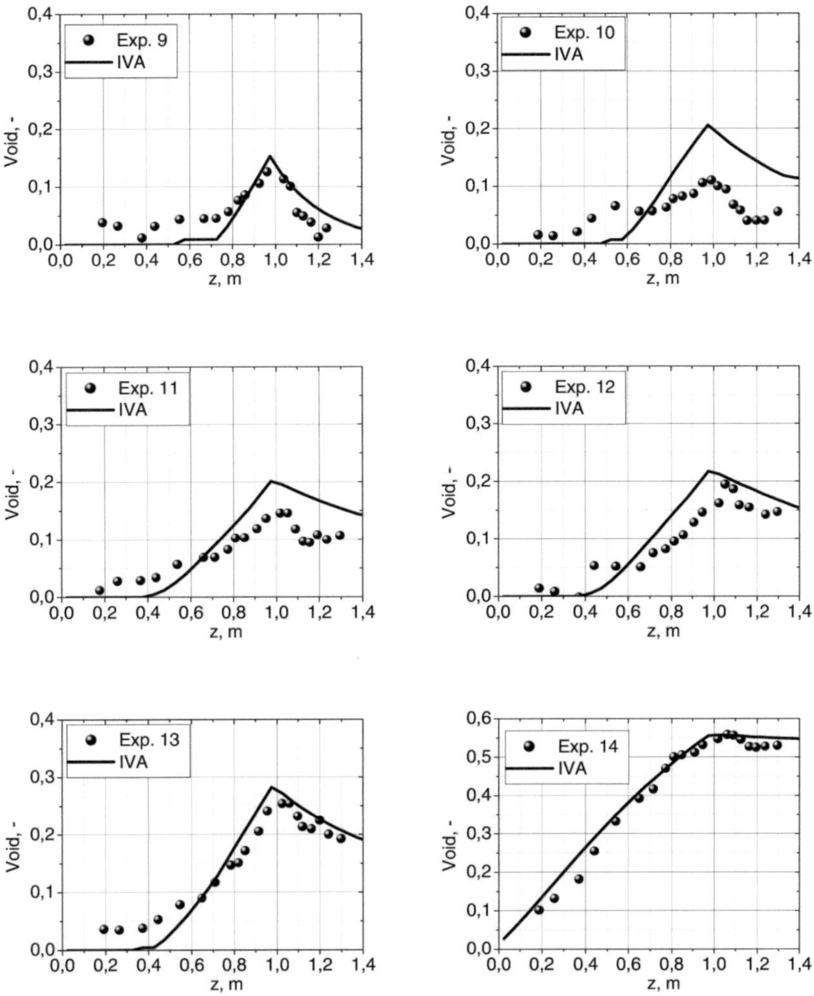

Fig. 16.14.6 Steam volume fraction as a function of length measured from the bottom. Comparison between the IVA prediction and the data of *Bartolomei* et al. (1980). The mass transfer is modeled using the proposal by *Hughes* et al. (1981). Modified *Hancox* and *Nicoll* multiplier: 0.2 instead 0.4. Saturated forced convection boiling is modeled using the *Chen* (1963) correlation

Table 16.14.4 Initial conditions for the *Bartolomei* et al. (1980) boiling pipe experiments

No	p	ρw	$\dot{q}''_w$	$T_{2,in}$
	MPa	kg/(m² s)	MW/m²	K
1	6.89	1000	1.2	452
2	6.89	1500	1.2	495

3	6.89	1500	0.8	519
4	6.89	2000	1.2	519
5	6.89	1000	0.8	503
6	6.89	1500	0.8	535
7	6.89	1000	0.8	534
8	10.8	1000	0.8	529
9	10.8	1000	1.2	493
10	10.8	1500	1.2	534
11	10.8	1500	0.8	558
12	10.8	2000	1.2	552
13	10.8	1000	0.8	544
14	10.8	1000	0.8	572

Here, the particular physical process that is tested in addition to the previous tests is the condensation in the nonheated part of the tube. The results are presented in Figures 16.14.6 a–j. We realize from this comparison the condensation coefficient in the *Hancox* and *Nicoll* model, see *Hughes* et al. (1981), of 0.4 has to be reduced to 0.2. Note that this type of model excludes the tracing of the interfacial area density of the bubble/liquid interface.

Figure 16.14.7 gives the comparison with the *Bennet* data. In these experiments critical heat flux occurs and the characteristic jump of the wall temperature is very well expressed.

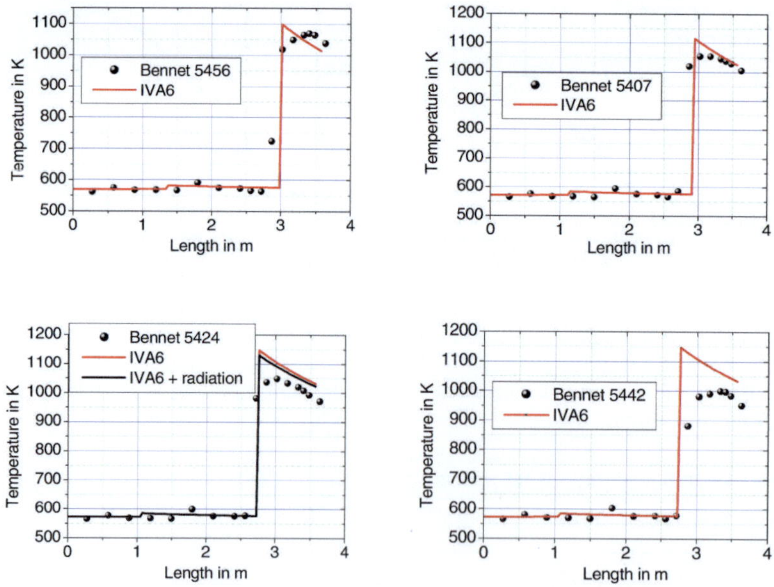

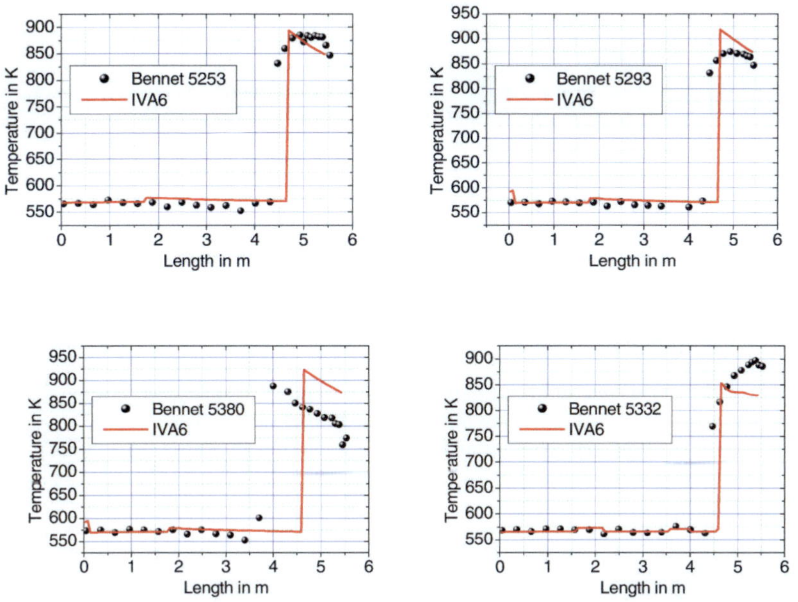

Fig. 16.14.7 a–h Wall temperature as a function of length measured from the bottom. Comparison between the IVA prediction and the data of *Bennet* (1967). The mass transfer in the sub-critical region is modeled using the proposal by *Hughes* et al. (1981). Saturated forced convection boiling is modeled using the *Chen* (1963) correlation. The critical heat flux is predicted by the 1995 look-up table (1996). The postcritical heat flux is computed using the correlation by *Miropolski* (1963) with a coefficient of 0.011

We see that the *Groeneveld* et al. (1996) look-up table predicts very well the onset of the critical heat flux. The forced convection nuclide boiling is also well predicted. The postcritical heat and mass-transfer modeling needs further sophistication especially for very long pipes. The effect of the radiative heat transfer is demonstrated for case 5424.

References

Bennett, A.W., et al.: Heat transfer to steam-water mixtures flowing in uniformly heated tubes in which the critical heat flux has been exceeded, AERE-R5373 (1967)

Bartolomei, G.G., Batashova, G.N., Brantov, V.G., et al.: Heat and Mass Transfer IV. Izd. ITMO AN BSSR, Minsk 5, 38 (1980) (in Russian)

Bartolomei, G.G., Brantov, V.G., Molochnikov, Y.S., Kharitonov, Y.V., Solodkii, V.A., Batashova, G.N., Mikjailov, V.N.: An experimental investigation of true volumetric vapor content with subcooled boiling in tubes. Thermal Eng. 29(3), 132–135 (1982)

Chen, J.C.: A correlation for film boiling heat transfer to saturated fluids in convective flow. ASME Publication-63-HT 34, 2–6 (1963)

Egen, R.A., Dingee, D.A., Chastain, J.W.: Vapor formation and behavior in boiling heat transfer, AEC Report BMI – 1167 (1957)
Groeneveld, D.C., et al.: The 1995 look-up table for critical heat flux in tubes. Nucl. Eng. Des. 163, 1–23 (1996)
Hughes, E.D., Paulsen, M.P., Agee, L.J.: A drift-flux model of two phase flow for RETRAN. Nucl. Technol. 54, 410 (1981)
Levy, S.: Int. J. Heat Transfer 10, 351–365 (1967)
Miropolskij, Z.L.: Heat transfer in film boiling of steam-water mixture in steam generating tubes. Teploenergetika 10(5), 49–53 (1963)
Nylund, D., et al.: Hydrodynamic and heat transfer measurements on a full-scale simulated 36-rod Marviken fuel element with uniform heat flux distribution, FRIG-2, Danish Atomic Energy Commission (1968)
St. Pierre, C.C.: Frequency-response analysis of steam voids to sinusoidal power modulation in a thin-walled boiling water coolant channel, Argon National Lab. Report, ANL-7041 (1965)
Sabotinov, L.S.: Experimental investigation of void fraction in subcooled boiling for different power distribution laws along the channel. Moskva PhD Thesis (1974) (in Russian)
Saha, P., Zuber, N.: Proc. Int. Heat Transfer Conference, Tokyo, Paper B4.7 (1974)

16.15 Flow boiling in bundles: 333 experiments for variety of bundles, flow regimes including dry out, steady state and transients

The computational analysis presented in this section is performed with IVA computer code. It serves (a) as a verification base for IVA for simulation of steady-state and transient boiling processes in rod bundles and (b) as a source of inspiration for improving the accuracy of the mathematical description of such processes in the future.

Regarding (a): 333 3D tests in bundles with 1, 16, 25, 36 heated rods, 7 different bundles with 64 heated rods from different laboratories are simulated: : 273 3D experiments on 6 bundles for CHF, 54 3D experiments on 7 bundles for void fraction, 2 3D experiments in a bundles for transients and 4 1D experiments in a subchannel for transient.

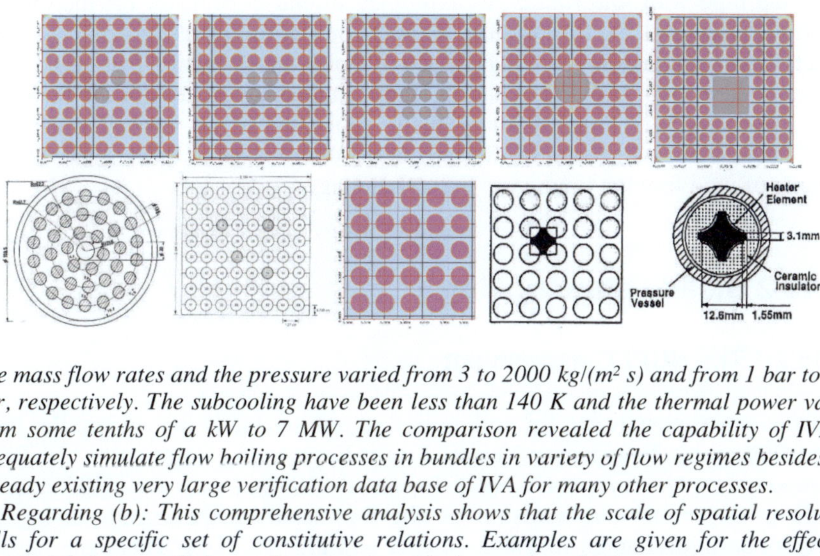

The mass flow rates and the pressure varied from 3 to 2000 kg/(m² s) and from 1 bar to 200 bar, respectively. The subcooling have been less than 140 K and the thermal power varies from some tenths of a kW to 7 MW. The comparison revealed the capability of IVA to adequately simulate flow boiling processes in bundles in variety of flow regimes besides the already existing very large verification data base of IVA for many other processes.

Regarding (b): This comprehensive analysis shows that the scale of spatial resolution calls for a specific set of constitutive relations. Examples are given for the effective interfacial drag coefficients. Comparison with the results of other authors is made and discussion is provided. In addition, by using different spatial resolution it was demonstrated that the turbulent void and droplets transport have to be described much more accurately in distributed parameters to provide a universal way of prediction of void and droplet dispersion with such important consequences like accurate void and dry-out prediction.

16.15.1 Introduction

The subject of this work will be the application of IVA to the class of processes generally known as two-phase boiling flow in rod bundles in order to (a) establish a verification base for IVA for simulation of steady-state and transient boiling processes in rod bundles and (b) identify problems that have be resolved in the future for improving the accuracy of the mathematical description of such processes. The results of this analysis are first reported in *Kolev* (2005a, b, c).

The problem with appropriate constitutive relations for a particular spatial scale resolution can be formulated as follows: Single-phase flow in bundles exhibits very heterogeneous spatial velocity profiles, *Rehme* (1978). In boiling two-phase flow the void fraction is also heterogeneous, *Hori* et al. (1994), *Hori* et al. (1995). Therefore, the cross-section-averaged properties differ significantly from the local flow properties. The constitutive relations for interfacial heat, mass, and momentum transfer are derived by procedures relying on averaging. So for instance, the bubble-drag coefficient derived in pool flow will not be appropriate for description of the mechanical interaction in gross spatial resolution analyses in rod bundles. Even bubble-drag coefficients derived for pipe flow and based on cross-section-averaged pipe flow will not be appropriate to describe the processes inside the pipe with *fine radial resolution*. A similar conclusion for the thermal profiles is also valid. So for instance, using a correlation for recognizing the point of net vapor generation, which is based on pipe cross-section-averaged water temperature, will not be appropriate to identify the same process in the boundary layer. Therefore one has to consciously apply the right set of constitutive equations to the desired type of discretization. We will demonstrate this with the example of void fraction prediction in rod bundles.

16.15.2 Steady-state boiling

16.15.2.1 The NUPEC experiment

Valuable experimental data for boiling in 8×8 rod unequally heated bundles have been collected by the Japanese Nuclear Power Engineering Corporation (NUPEC) and reported by *Morooka* et al (1991), *Yagi* et al. (1992), *Inoue* et al. (1995a, b, c). We first concentrate our attention on the 8×8 bundle experiments having the geometry definition given in Table 16.15.1 and Fig. 16.15.1a. This experimental arrangement is referred to as a low-burn-up configuration. In addition the high-burn-up configuration as given in Fig. 16.15.1b is also used. The central rod in the last case has a diameter of 34 mm. ï

Table 16.15.1 Dimensions of BWR bundle *Utsuno* et al. (2004)

Number of fuel rods	62
Outer diameter, mm	12.3
Heated length, m	3.7
Number of water rod, mm	2
Outer diameter of water rod, mm	15.0
Rod pitch, mm	16.2
Width of channel box, mm	132.5
Number of spacers	7
Spacer type	Grid

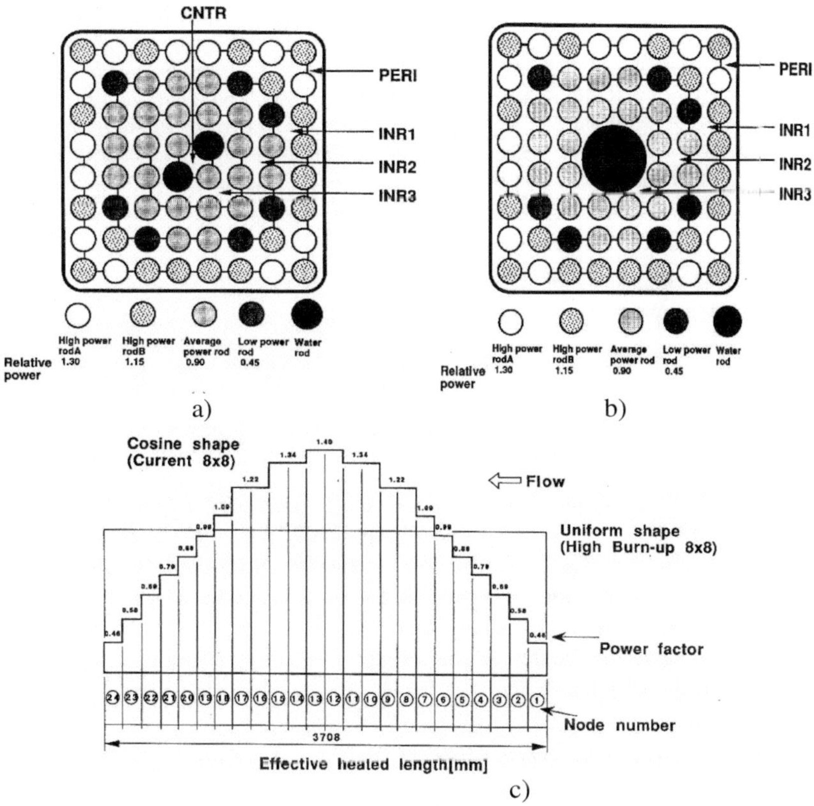

Fig. 16.15.1 Nonhomogeneous power release in the bundle, *Inoue* et al. (1995)

The boundary conditions for the considered experiments are given in Tables 16.15.2 and 16.15.3.

Table 16.15.2 Boundary conditions, nonuniform power release, NUPEC *Yagi* et al. (1992) p. 163.

Exit. eq. quality	p in MPa	Power in MW	G in kg/(m²s)	Subc. in kJ/kg (K)
0.2415	7.2	1.143	284	50.2 (9.42)
0.2447	7.2	2.313	568	50.2 (9.42)
0.2479	7.2	3.509	852	50.2 (9.42)
0.2410	7.2	6.458	1562	50.2 (9.42)
0.2489	7.2	8.219	1988	50.2 (9.42)

The subcooling is taken from *Inoue* et al. (1995) p. 393. The exit equilibrium cross-section-averaged quality as well as the experimental cross-section void fractions are taken from *Yagi* et al. (1992) p. 163. The power is then computed by using the internal subcooling, the exit quality and the mass flow rate. The axial power distribution as presented in Fig. 16.15.1c is taken from Fig. 3 in *Inoue* et al. (1995) p. 391. The lateral power distribution is taken from Fig. 4a *Inoue* et al. (1995) p. 391. The later distribution required slight renormalization to achieve strict energy conservation of the total energy release as a boundary condition. The spacers receive an irreversible pressure drop coefficient equal to 1 based on the bundle velocity due to lack of better knowledge.

Table 16.15.3 Boundary conditions, nonuniform power release *Inoue* et al. (1995) p. 394, high-burn-up geometry

Exit. eq. quality	p in MPa	Power in MW	G in kg/(m² s)	Subc. in kJ/kg (K)
0.2479	7.2	3.509	852	50.2 (9.43)
0.2410	7.2	2.313	1562	50.2 (9.43)

The cross-section-averaged void fractions are presented in the original sources *Yagi* et al. (1992) p.163 and *Inoue* et al. (1995) p. 394 as a function of the equilibrium quality. The equilibrium quality is computed be using the inlet subcooling and integrating over the height of the bundle using the axial energy distribution. The type of the channels for the low burn-up bundle, their number, the hydraulic and the heated diameter and the corresponding cross-section are given in Table 16.15.4.

Table 16.15.4 Subchannel characteristics for the NUPEC 8×8 low-burn-up bundle

Typ	nr.	γ_z	D_{hyd}	D_{heat}	F
1, internal	22	0.5472	1.4867E–02	1.4867E–02	1.4362E–04
2, corner	4	0.5930	9.9153E–03	1.7921E–02	1.7313E–04
3, periphery	36	0.5844	1.2188E–02	1.7298E–02	1.6711E–04
4, unheated	2	0.3266	7.2766E–03	0.	8.5725E–05
Total	64	0.5583	1.2994E–02	1.6291E–02	9.7575E–03

Note that *Aounalla* and *Coddington* (1999) used part of this data for verification of the two-fluid subchannel code VIPRE-02 in which the cross-flow is modeled by simplified momentum equations. *Naitoh* et al. (1999) and *Utsuno* et al. (2004)

used part of the NUPEC data for verification of drift flux subchannel code CAPE and TCAPE, respectively, in which the crossflow is modeled by mixing models defining crossflows, but not by momentum conservation.

Our computation in Cartesian coordinates uses 8×8×24 cells that form 64 rod-centered subchannels, see Fig. 16.15.4. We start the simulation with an arbitrary but meaningful initial state and continue the simulation until the steady state was established, controlling this by plotting the relative difference between inlet and outlet flows in %. An example is given in Fig. 16.15.2 for the low-burn-up geometry with mass flow rate 1988 kg/(m² s). For reaching a steady state solution through a transient analysis achieving a good accuracy of the time integration is crucial in order not to produce artificial density waves. In this case time steps of 1×10^{-3}s or less are appropriate.

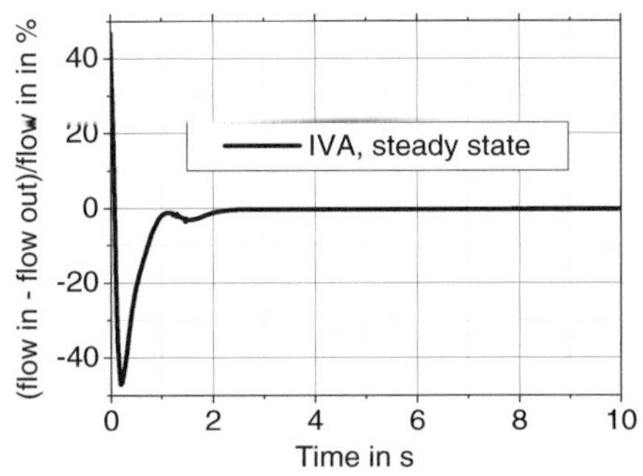

Fig. 16.15.2 Nonoscillating steady state for the NUPEC experiment with 1988 kg/(m² s)

Only the case with 284 kg/(m² s) resulted in an oscillating but stable steady-state solution. This will be discussed in a separate section. All other cases manifest nonoscillating steady-state solutions, as those characterized by Fig. 16.15.2.

16.15.2.1.1 Comparison of the IVA predictions with the NUPEC data

16.15.2.1.1.1 Cross-section-averaged axial profiles

The results of the computational analyses for the low-burn-up cases are presented in Fig. 16.15.3. The upper curve present the unchanged IVA set of drag coefficients that are applicable to fine resolution analyses.

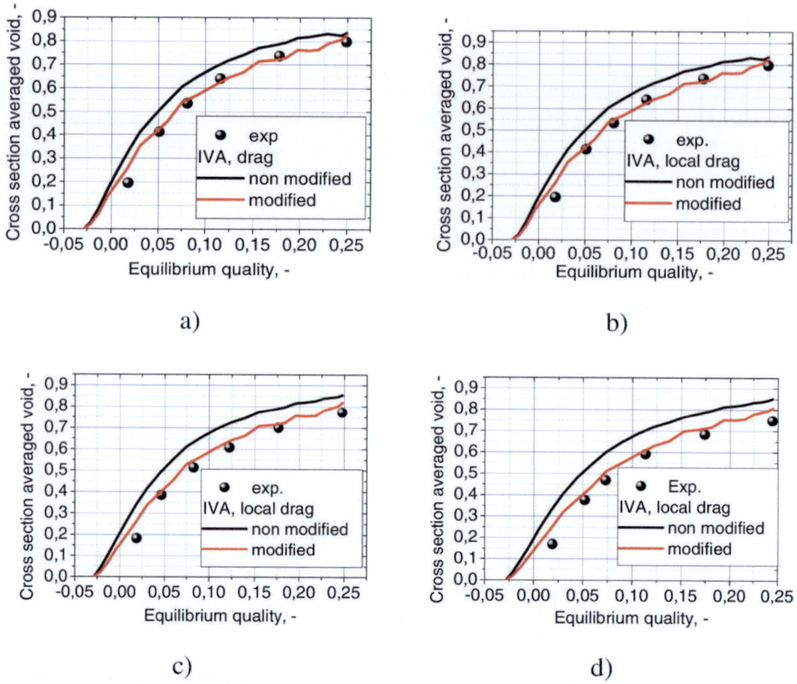

Fig. 16.15.3 NUPEC experiment with mass flow rate a) 1998 kg/(m² s); b) 1562 kg/(m² s); c) 852 kg/(m² s); d) 568 kg/(m² s). Void fraction as a function of the height. Parameter: local interfacial drag

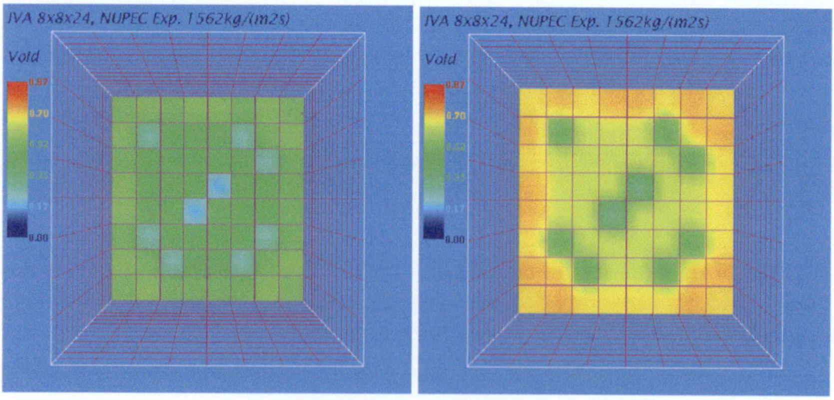

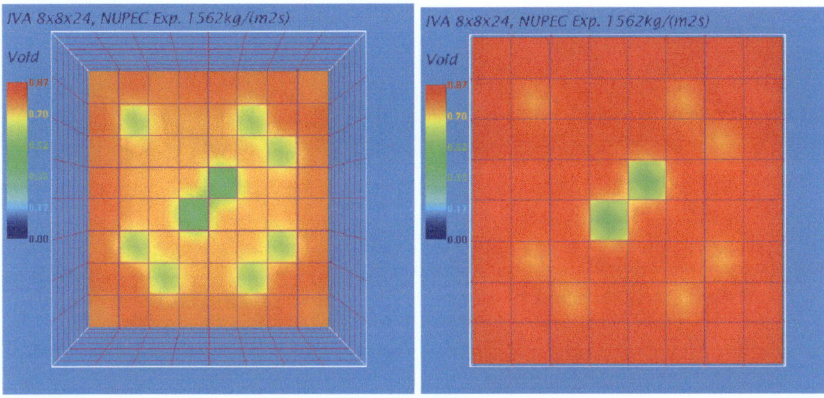

Fig. 16.15.4 NUPEC experiment with mass flow rate 1562 kg/(m²s): a) z = 1.3133 m, X_1 = 0.0596; b) z = 1.9313 m, X_1 = 0.124; c) z = 2.5492 m, X_1 = 0.184; d) z_{exit} = 3.6307 m, $X_{1,exit}$= 0.2489

The used rod-centered subchannel discretization is by far not a fine resolution. Unless otherwise noted we use the *Saha* and *Zuber* (1974) correlation for the initiation of the subcooled boiling and the *Levy* (1967) asymptotic model for the splitting of the heat into convection evaporation and condensation. As already mentioned in the introduction, there are velocity and void profiles that make the application of the local drag coefficient based on local parameters not appropriate. An ad hoc modification on the drag coefficients is made as presented in Appendix 16.15.1. The results with this modification are presented in each figure as a lower curve. The comparison with the data is then favorable. Figure 16.15.4 illustrates some details of the case with a mass flow rate 1562 kg/(m² s). The influence of the nonheated rods and of the low-power rods is clearly seen.

Now, let us compare the predictions with the measurements for the two high-burn-up cases using the modified drag coefficients. The solutions are presented in Figs. 16.15.5a and b.

In summary, the effect of the nonuniformity of the flow profiles inside the subchannels is manifested in effectively lower drag between the liquid and the vapor. The comparison with the data using the reduced effective drag is favorable. However, it can not be expected a priori to have a unique interfacial drag-reduction algorithm for all types of subchannels. To check this, we analyze in the following sections void fraction data collected in bundles of completely different types.

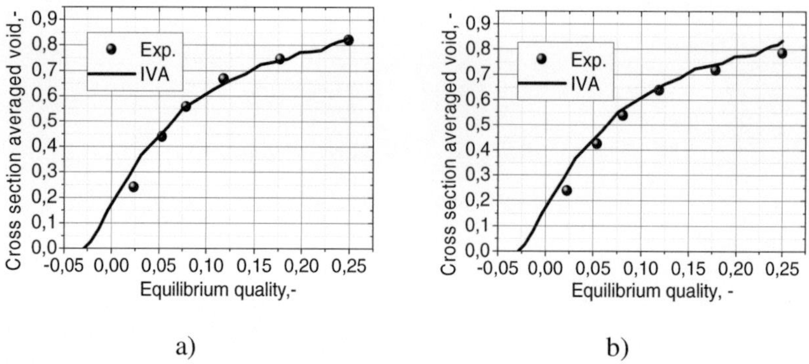

Fig. 16.15.5 NUPEC experiment with mass flow rate a) 1562 kg/(m² s); b) 852kg/(m² s). Void fraction as a function of the height

16.15.2.1.1.2 Regions averaged exit void fractions at different power level

Inoue et al. (1995) reported region-averaged void fractions at the exit of the *low-burn-up bundle* for the case with 1562 kg/(m² s). The definitions of the five regions are given in Fig. 16.15.6a. The measured region-averaged void fractions at the exit of the bundle are presented in Fig. 16.15.6b for different bundle power defined by equilibrium exit quality 0.05, 0.12 and 0.18, respectively. The computed void fractions are also presented for comparison in Fig. 16.15.6b.

Inoue et al. (1995) also reported a region-averaged void fractions at the exit of the *high-burn-up bundle* for the case with 1562 kg/(m² s). The definitions of the four regions are given in Fig. 16.15.7a. The measured region-averaged void fractions at the exit of the bundle are presented in Fig. 16.15.7b for different bundle power defined by equilibrium exit quality 0.05, 0.12 and 0.18, respectively. The computed void fractions are also presented for comparison in Fig. 16.15.7b.

For both bundles the prediction accuracy is very similar. Low-burn-up bundle (Fig. 16.15.6b): We see "good" agreement for the three external regions. For the central region the prediction with such gross discretization underpredicts the void fraction. Comparing the cases with smaller and larger resolution, b) and c) in Fig. 16.15.6, we realize that the comparison with the data is "better" for the gross resolution. This is an indication that the void mixing computed based only on the transport equation without turbulence modeling is not enough to describe appropriately this process. High-burn-up bundle: For low power the computed results are higher than the measured. To our view, the reason for the discrepancy is partially in the low resolution of the subchannel analysis. As already mentioned increasing the resolution alone without turbulence modeling will not improve the result. As we will see later, the use of the *Hughes* et al. (1981) correlation for splitting of the heat fluxes instead of *Saha-Zuber* (1974), *Levy* (1967) improves the prediction accuracy in the low-void region.

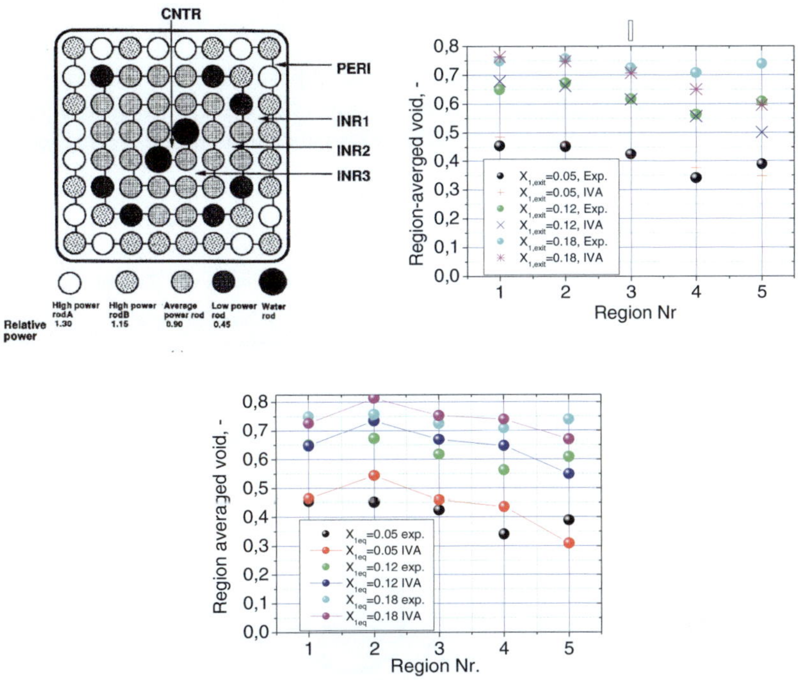

Fig. 16.15.6 a) Definition of the averaging regions, *Inoue* et al. (1995): PERI = 1, INR1 = 2, INR2 = 3, INR3 = 4, CNTR = 5; Region-averaged void fractions at the exit of the bundle for different bundle power corresponding to constant conditions but specified exit equilibrium quality: Resolution: b) 8×8×24; c) 20×20×24

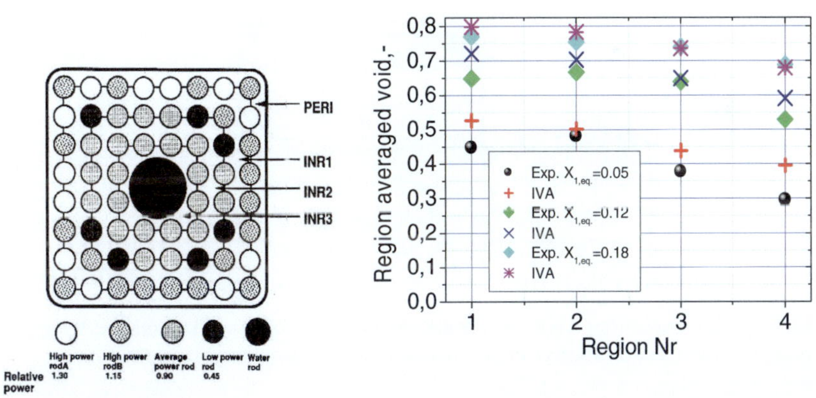

Fig. 16.15.7 a) Definition of the averaging regions, *Inoue* et al. (1995): PERI = 1, INR1 = 2, INR2 = 3, INR3 = 4; b) Region-averaged void fractions at the exit of the bundle for different bundle power corresponding to constant conditions but specified exit equilibrium quality

16.15.2.1.1.3 Local exit void fractions at different power level

Prediction for 15 additional proprietary NUPEC experiments in a framework of an OECD/USNRC benchmark, *Kolev* (2005a), are compared with the outlet void fractions based on subchannels. The 1365 experimental data are available at the exit of the bundle as a cross-section-averaged void fraction and as a 9×9 local subchannel data, see Figs. 16.15.8 and 16.15.9.

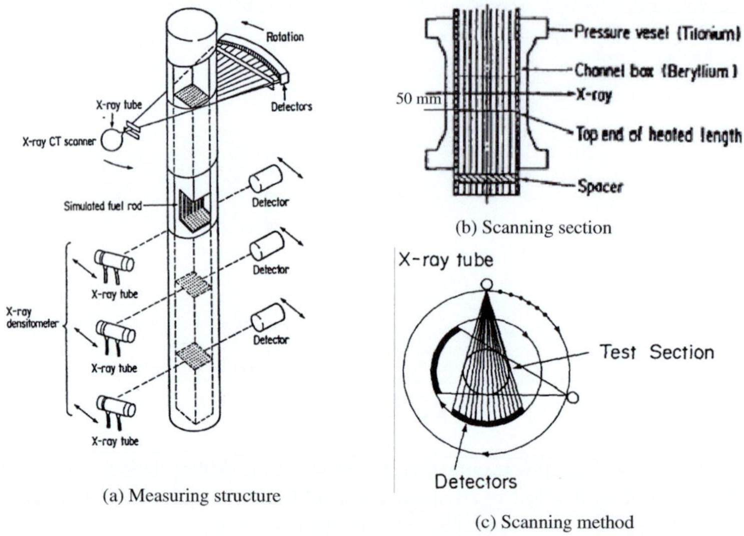

Fig. 16.15.8 Void-fraction measurement system

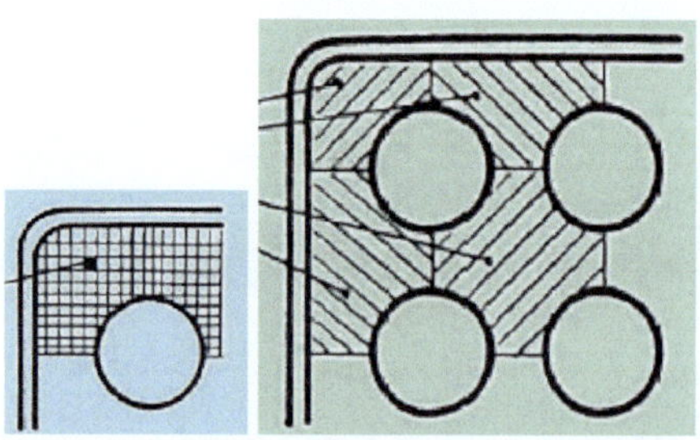

a) Pixel void fraction b) Subchannel void fraction

Fig. 16.15.9 Spatial resolution: a) of the experimentally measured void fractions; b) of the region averaged void fraction subject of comparison with the predicted void fractions

The geometry of the bundles is presented on Fig. 16.15.10. Comparison between the computed and measured exit void fractions are presented in Tables 16.15.5–16.15.12 for two different models especially for the subcooled boiling. Tables 16.15.5–16.15.8 contain the comparison for data predicted using the *Saha* and *Zuber* (1974) correlation for the initiation of the subcooled boiling and the *Levy* (1967) asymptotic model for the splitting of the heat into convection evaporation and condensation. Tables 16.15.9–16.15.12 contain the comparison for data predicted using the *Hughes* et al. (1981) approach for the initiation of the subcooled boiling and for the splitting of the heat into convection evaporation and recondensation.

Table 16.15.5 Cross-section-averaged void fractions computed by IVA and measured by NUPEC. IVA heat partitioning models: *Saha-Zuber* (1974) and *Levy* (1967)

Test No.		0011, assembly 0-1			0021, assembly 0-2			0031, assembly 0-3		
z in m		55	58	61	16	18	21	16	18	21
	Exp.	0.409	0.630	0.794	0.394	0.626	0.783	0.393	0.623	0.783
3.758	IVA	0.461	0.667	0.797	0.450	0.655	0.791	0.447	0.638	0.778
	Dev.%	+12.7	+5.9	+0.38	+14.2	+4.6	+1.0	+13.7	+2.4	+0.6

Table 16.15.6 Cross-section-averaged void fractions computed by IVA and measured by NUPEC. IVA heat partitioning models: *Saha-Zuber* (1974) and *Levy* (1967)

Test No.		1071, assembly 1			4101, assembly 4		
z in m		55	58	61	55	58	61
	Exp.	0.430	0.637	0.791	0.438	0.645	0.807
3.758	IVA	0.453	0.654	0.799	0.467	0.670	0.803
	Dev.%	+5.3	+2.7	+1.0	+6.6	+3.9	-0.5

Table 16.15.7 Local void-fraction mean error at the exit of the bundle. IVA heat partitioning models: *Saha-Zuber* (1974) and *Levy* (1967)

Test No.	0011, assembly 0-1			0021, assembly 0-2			0031, assembly 0-3		
	55	58	61	16	18	21	16	18	21
Mean error%	6.43	5.48	4.61	7.22	6.31	5.00	7.20	7.22	6.66

Table 16.15.8 Local void-fraction mean error at the exit of the bundle. IVA heat partitioning models: *Saha-Zuber* (1974) and *Levy* (1967)

Test No.	1071, assembly 1			4101, assembly 4		
	55	58	61	55	58	61
Mean error%	5.38	4.57	4.22	4.77	4.71	4.49

Tables 16.15.7 and 16.15.8 contain the square root mean error for the 9×9 subchannel data. The total mean error is 5.62%. The data structure is presented on Fig. 16.15.11 where the measured void local subchannel fractions are presented versus the computed for all 1365 data points. The spread is within +20 and –25% with the main bulk of the data grouped around the diagonal.

Table 16.15.9 Cross-section-averaged void fractions computed by IVA and measured by NUPEC. IVA heat partitioning models: *Hughes* et al. (1981)

Test No.		0011, assembly 0-1			0021, assembly 0-2			0031, assembly 0-3		
z in m		55	58	61	16	18	21	16	18	21
	Exp.	0.409	0.630	0.794	0.394	0.626	0.783	0.393	0.623	0.783
3.758	IVA	0.426	0.637	0.782	0.415	0.627	0.777	0.414	0.610	0.765
	Dev.%	4.16	1.11	–1.51	5.33	0.16	–0.77	5.34	–2.09	–2.30

Table 16.15.10 Cross-section-averaged void fractions computed by IVA and measured by NUPEC. IVA heat partitioning models: *Hughes* et al. (1981)

Test No.		1071, assembly 1			4101, assembly 4		
z in m		55	58	61	55	58	61
	Exp.	0.430	0.637	0.791	0.438	0.645	0.807
3.758	IVA	0.415	0.629	0.786	0.428	0.639	0.788
	Dev.%	–3.49	–1.26	–0.63	–2.28	–0.93	–2.35

Table 16.15.11 Local void-fractions mean error at the exit of the bundle. IVA heat partitioning models: *Hughes* et al. (1981)

Test No.	0011, assembly 0-1			0021, assembly 0-2			0031, assembly 0-3		
	55	58	61	16	18	21	16	18	21
Mean error%	4.97	4.70	4.53	5.33	5.82	4.81	5.66	6.87	6.47

Table 16.15.12 Local void-fractions mean error at the exit of the bundle. IVA heat partitioning models: *Hughes* et al. (1981)

Test No.	1071, assembly 1			4101, assembly 4		
	55	58	61	55	58	61
Mean error%	5.36	4.45	3.99	4.55	4.57	5.21

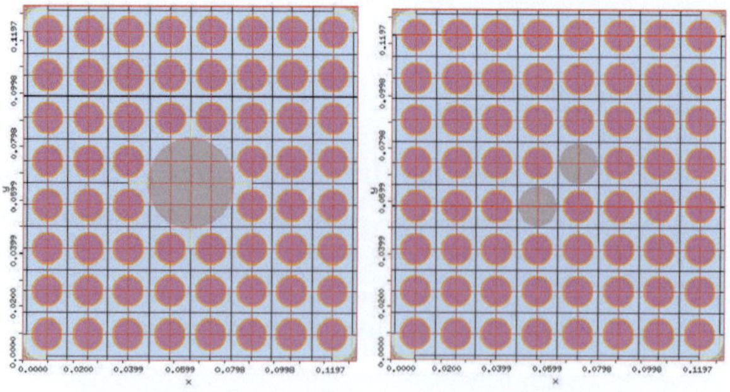

Assembly: 0-1, 1, 3 0-2

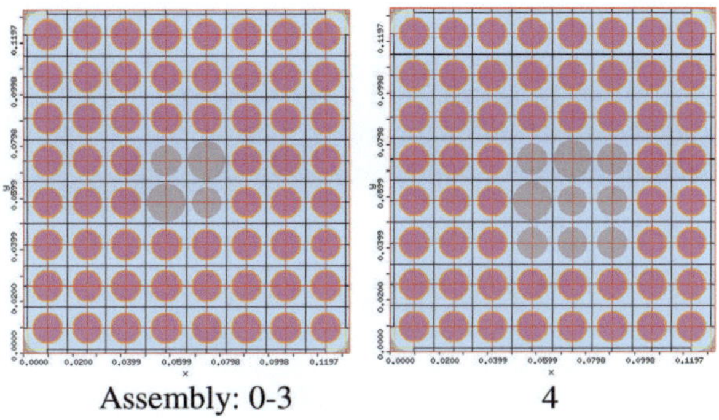

Fig. 16.15.10 IVA discretization models of the four bundles

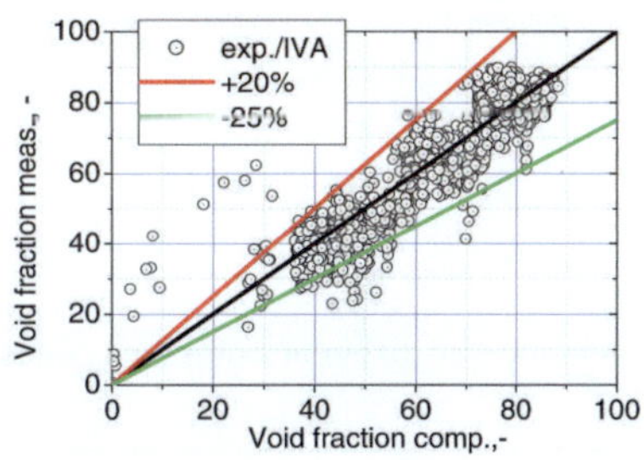

Fig. 16.15.11 Computed versus measured void fraction at the 91 subchannels at the exit of the bundle. Models for subcooled boiling: *Saha* and *Zuber* (1974) correlation for the initiation of the subcooled boiling and the *Levy* (1967) asymptotic model for the splitting of the heat into convection evaporation and recondensation. The total mean error is 5.62%

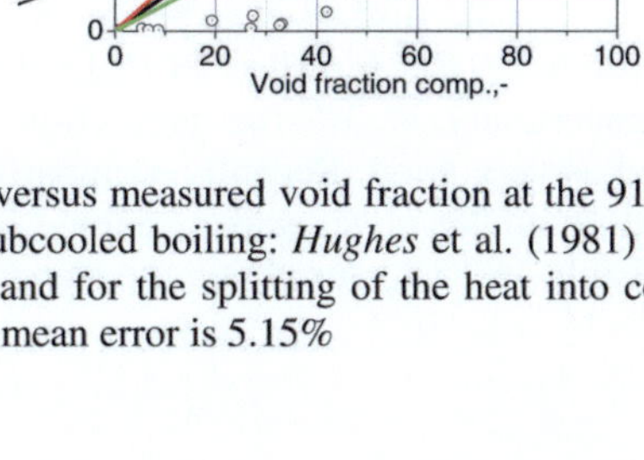

Fig. 16.15.12 Computed versus measured void fraction at the 91 subchannels at the exit of the bundle. Models for subcooled boiling: *Hughes* et al. (1981) approach for the initiation of the subcooled boiling and for the splitting of the heat into convection evaporation and recondensation. The total mean error is 5.15%

Tables 16.15.9 and 16.15.10 contain the square root mean error for the 9×9 subchannel data. The total mean error is 5.15%. The data structure is presented in Fig. 16.15.12 where the measured void local subchannel fractions are presented versus the computed for all 1365 data points. The spread is within +20 and –25% with the main bulk of the data grouped around the diagonal.

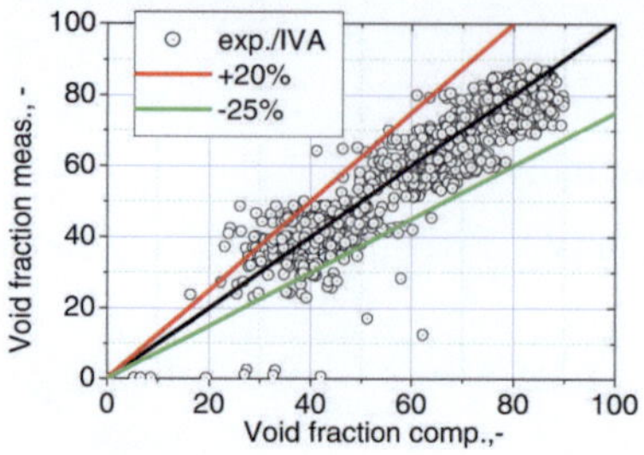

Fig. 16.15.13 As in Fig. 16.15.12: latest version of IVA. The total mean error is 5.07%

Table 16.15.13 Standard deviation for different computer codes reported in *Neykov* et al. (2006)

	1	2	3	4	5	6	7	8	9	10
01	0.18	1.7	1.2	1.44	1.55	2.13	0.94	0.63	2.36	1.86
02	0.09	1.67	0.89	1.44	1.98	4.98	4.21	0.12	2.15	1.14
03	0.39	2.01	1.25	1.24			4.49	0.3	2.13	1.4
1	1.08	0.76	0.28	0.0			3.18	1.76	1.1	1
4	2.11	1.02	1.09	0.88	4.45		2.01	1.58	0.37	0.84

1. MATRA (KAERI)
2. MARS (KAERI)
3. MONA (KTH)
4. CAPE Mod 1.0 (NUPEC)
5. TwoPorFlow (FZK)
6. COBRA-TF (PSU)
7. F-COBRA-TF (AREVA)
8. IVA (AREVA)
9. NEPTUNE -FLICA4 (CEA)
10. NASCA (TEPCO)

We recently recomputed again the discussed experiment with the last version of IVA. The total mean error is 5.07%, see Fig. 16.15.13. Having in mind that the reported experimental uncertainty is 3% the prediction without any turbulent mixing can be considered as very good. Recently, comparison of the prediction of this data set with several computer codes was presented by *Neykov* et al. (2006) with the result given in Table 16.15.13 for which the standard deviation computed as follows is used

$$\sigma = \sqrt{\frac{\sum_{k=1}^{K} \left(\alpha_{\exp,k} - \alpha_{pred,k} \right)^2}{(K-1)}}.$$

Conclusions:

- The cross-section-averaged void fractions predicted at the exit of the bundles with IVA depends on the used modeling approach for the subcooled boiling available in IVA: a) *Saha* and *Zuber* (1974) correlation for the initiation of the subcooled boiling and the *Levy* (1967) asymptotic model for the splitting of the heat into convection evaporation and recondensation; b) *Hughes* et al. (1981) approach for the initiation of the subcooled boiling and for the splitting of the heat into convection evaporation and recondensation.
- By using the model a) the predicted data agree excellent at high power (1%), agree well at averaged power (5.9%) and are up to 14.2% lower at 5% of the nominal power.
- By using the model b) the predicted data agree excellent at high power (2.3%) and well at averaged power (2.9%) and at 5% of the nominal power (5.34%). Obviously model b) is superior at low void fractions.
- The local void fractions at the exit of the bundle on a 9×9 subchannel basis are predicted with an error of 5.62% using model a) and of 5.15% using model b). The local accuracy increases with the increasing power. Again model b) contributes to better performance for all void fractions.
- As already reported in *Kolev* (2005b, 2005c) comparing the cases with smaller (10×10×24) and larger (18×18×24) resolution we obtain "better" agreement with the data for the gross resolution. This is an indication that the void mixing computed based only on the transport equation is not enough to describe appropriately this process. This is confirmed by this study too. In all cases the "void diffusion" from regions with higher void to regions with lower void is underestimated.
- Without appropriate turbulence modeling the accuracy of this method regarding predicting the local void fraction can not be increased.
- Fine resolution is required in the future and will be accomplished with appropriate constitutive relationships specially developed for fine resolution.

16.15.2.1.1.4 Comparison with *Utsuno* et al. (2004) analyses

Ursuno et al. (2004) reported analyses of the steady-state void data of NUPEC. The authors used the code TCAPE. The TCAPE code uses the drift-flux formulation of all axial equations. The crossflow is modeled by algebraic models that do not satisfy lateral momentum conservation. The simplified energy-conservation equation used is appropriate for slow transients only. In the case of the film flow an additional one-dimensional mass balance is introduced using entrainment and deposition sources. Therefore, convective transport of film and droplets across the subchannel is not permitted in the model. The TCAPE code does not make use of dynamic fragmentation and coalescence. Even with this simplification the code predicts "reasonably" a gross scale void distribution by using flow-channel centered subchannels discretization. As already mentioned, the apparently demonstrated good void intermixing is due to numerical diffusion and not due to appropriate turbulence description.

16.15.2.1.1.5 3D versus 1D analysis

Consider the moisture analyses of a complete boiling water reactor as reported by *Kolev* (2002b) for the geometry given in Fig. 16.15.14. The question whether the core has to be simulated in a pin-by-pin or bundle-by-bundle approach is important because the pin-by-pin resolution is still expensive. To illustrate that bundle-by-bundles analysis is accurate enough for the moisture optimization analyses we simulate one of the NUPEC experiments as presented in Fig. 16.15.15 by using the characteristic thermohydraulic parameters for the overall rod bundle and compare the results with the 3D representation. As seen from Fig. 16.15.15 the cross-section-averaged void profiles are almost not distinguishable.

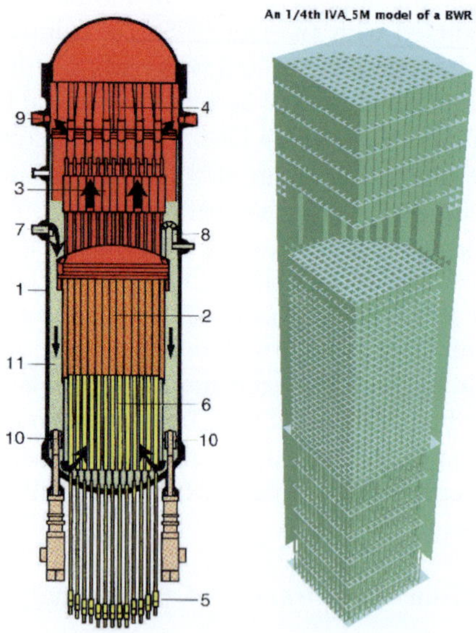

Fig. 16.15.14 a) Typical boiling water reactor: 1 reactor pressure vessel; 2 reactor core; 3 steam-water separators (cyclones); 4 steam dryer; 5 control rod drives; 6 control assemblies; 7 feed water inlet nozzles; 8 core spray line; 9 main steam outlet nozzle; 10 forced circulation pumps; 11 annular down comer. b) IVA_5M 1/4th geometry model of the control rod space, core, upper plenum and stand pipes.

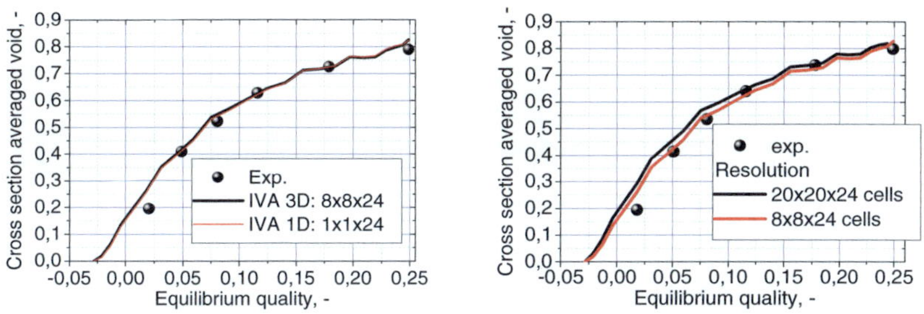

a) b)

Fig. 16.15.15 NUPEC experiment with mass flow rate 1562 kg/(m² s): Comparison between cross-section-averaged void fraction predictions using a) 3D (rod-centered channels) and 1D analysis; b) 3D rod centered channels 8×8×24 cells and 3D separate periphery layers, one rod belongs to 4 cells, 20×20×24 cells

Increasing the resolution to 20×20 horizontal cells gives slightly higher cross-section-averaged void fraction, as demonstrated in Fig. 16.15.15b. This allows the conclusion: For analyses of the moisture in BWRs the core representation in a bundle-by-bundle approach provide the appropriate large-scale three-dimensional void distributions at the exit of the core.

For approximating the bundle as a single channel for moisture analysis the averaged pressure drop coefficient for the bundle can be computed using the method proposed and verified on large data base by *Rehme* (1973, 1971, 1972). For the laminar regime the friction coefficient is $\lambda_{fr}\,\mathrm{Re} = Rm$, where the *Rehme*'s number (*Rehme*, 1973) is $Rm =$ 63.172, For the turbulent regime the friction factor in accordance with *Rehme* (1972) is $\sqrt{8/\lambda_{fr}} = 2.55\ln\left(\mathrm{Re}\sqrt{8/\lambda_{fr}}\right) - 0.255$. Here the bundle *Reynolds* number $\mathrm{Re} = \dot{m}D_h/(F\eta)$ is based on the total mass flow, total flow cross-section and on the effective bundle hydraulic diameter.

16.15.2.1.1.6 The oscillating steady-state for low mass fluxes

For large mass flows through the boiling bundle channels the steady state is nonoscillating. For low mass flows, however, the steady state is oscillating. As reported by *Yagi* et al. (1992) and *Inoue* et al. (1995c), oscillating void measurements are smoothed so that the data points are not only cross-section averages but also time averages. An example for a stable but oscillating steady-state solution is given in Figs. 16.15.16a and 16.16.16b for the NUPEC experiment with mass flow rate 284 kg/(m² s). Figure 16.15.17 presents the cross-section-averaged void fractions at different times within one cycle of oscillation. As we see they oscillate around the reported time-averaged values. This is an important advantage of looking for steady-state solutions as asymptotic solutions of transients. Important oscillating characteristics can be then recognized, something that is impossible by solving only the steady-state part of the systems of PDEs.

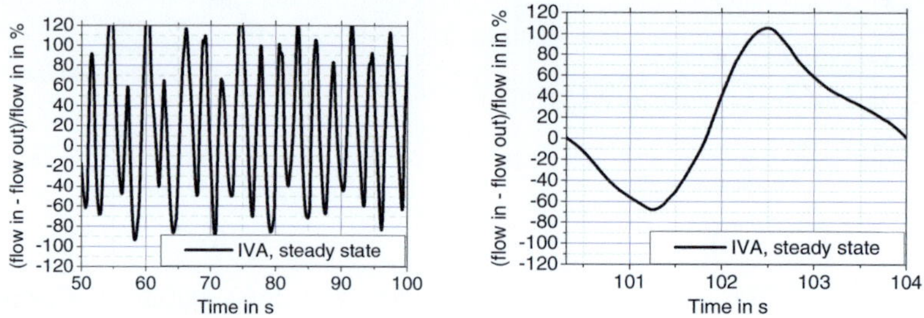

Fig. 16.15.16 Oscillating steady state for 284 kg/(m² s) case: a) long-term steady-state oscillations; b) one oscillation cycle

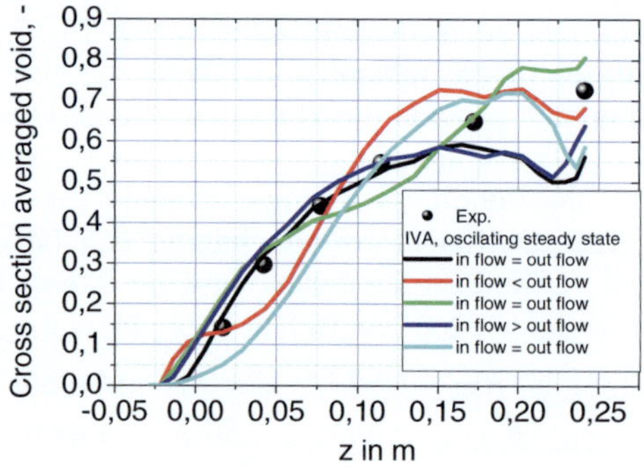

Fig. 16.15.17 NUPEC experiment with mass flow rate 284 kg/(m² s). Void fraction as a function of the height. Parameter – time within one oscillation period

16.15.2.2 The SIEMENS void data for the ATRIUM 10 fuel bundle

A set of proprietary data for void fraction in ATRIUM 10XP bundles at 4 peaking patterns was obtained by *Spierling* (2002) at the Karlstein Thermal Hydraulic Loop in AREVA NP, Germany. At about 70 bar 12 experiments with 2.2 m/s inlet velocity, 58 kJ/kg subcooling and axial choked cosine power profile chordial void profiles 3.24 m from the inlet are collected and then averaged over the cross-section *except* the water channel. 4 of the experiments are at low power and 8 at high power. All the 12 test sets are computed using IVA computer code. 10×10×37 3D-discretization (10×10 fuel centered subchannels with 37 axial cells) is applied. The structure of the errors is documented in Table 16.15.14. The maximum deviation of the predicted from the measured void was less than 8.7 and 7.55% for set a) and b), respectively. The averaged deviation is about 6.12 and 4.19% for set a) and b), respectively. Obviously, again set b) gives better performance.

In summary, the void prediction for AREVA NP bundle test is with uncertainty of 4.19% is even better the prediction of the previously reported NUPEC data.

Table 16.15.14 Measured and predicted cross-section-averaged void fractions at 3.24m from the inlet

Exp no.	Lateral peaking pattern	Void exp.	Void IVA, a)*	Deviation in %	Void IVA, b)*	Deviation in %
257	STS-66.1	0.519	0.524	0.96	0.482	–7.13
258		0.813	0.842	3.57	0.827	1.72
259		0.819	0.843	2.93	0.828	1.10
270	STS-68.3	0.468	0.519	10.2	0.476	1.71
271		0.784	0.840	7.14	0.827	5.48
272		0.779	0.840	7.83	0.828	6.29
328	STS-52.1	0.543	0.499	-8.10	0.502	–7.55
332		0.804	0.816	1.49	0.798	–0.75
339		0.839	0.861	2.62	0.843	0.48
261	STS-68.5	0.479	0.529	10.6	0.500	4.38
262		0.774	0.839	8.4	0.826	6.72
263		0.771	0.838	8.7	0.825	7.00

* Models for subcooled boiling: a) *Saha* and *Zuber* (1974) correlation for the initiation of the subcooled boiling and the *Levy* (1967) asymptotic model for the splitting of the heat into convection evaporation and recondensation; b) *Hughes* et al. approach (1981) for the initiation of the subcooled boiling and for the splitting of the heat into convection evaporation and recondensation.

16.15.2.3 The FRIGG experiment

The FRIGG experiments are reported by *Nylund* et al. (1968, 1970). The geometry given in Fig. 16.15.18b is defined as follows: 36 rods with 13.8 mm diameter, heated, 1 central rod with 20 mm diameter, unheated. The heated length of the rods is 4.375 m. The rods are placed within a cylinder with internal diameter 159.5 mm and positioned in three circles with 21.6, 41.7, and 62.2 mm radius, respectively. The circles contain 6, 12 and 18 rods, respectively, at equal arc distance from axis to axis along the circle. The bundle is 1/6-symmetric so that 1/6 of the circular cross-section is simulated. 8 spacers keep the rods parallel to each other. The spacers receive an irreversible pressure drop coefficient equal to 0.6 based on the bundle velocity. Discretization used for the IVA computation: The flow in the bundle is considered as 3D axis symmetric flow in cylindrical coordinates. 1/6 segment is simulated with three radial annuli defined by the radii 0.01, 0.03165, 0.05195 and 0.07975 m, respectively. The height is discretized using 36 cells. 8 spacers are modeled by a local irreversible pressure-loss coefficient.

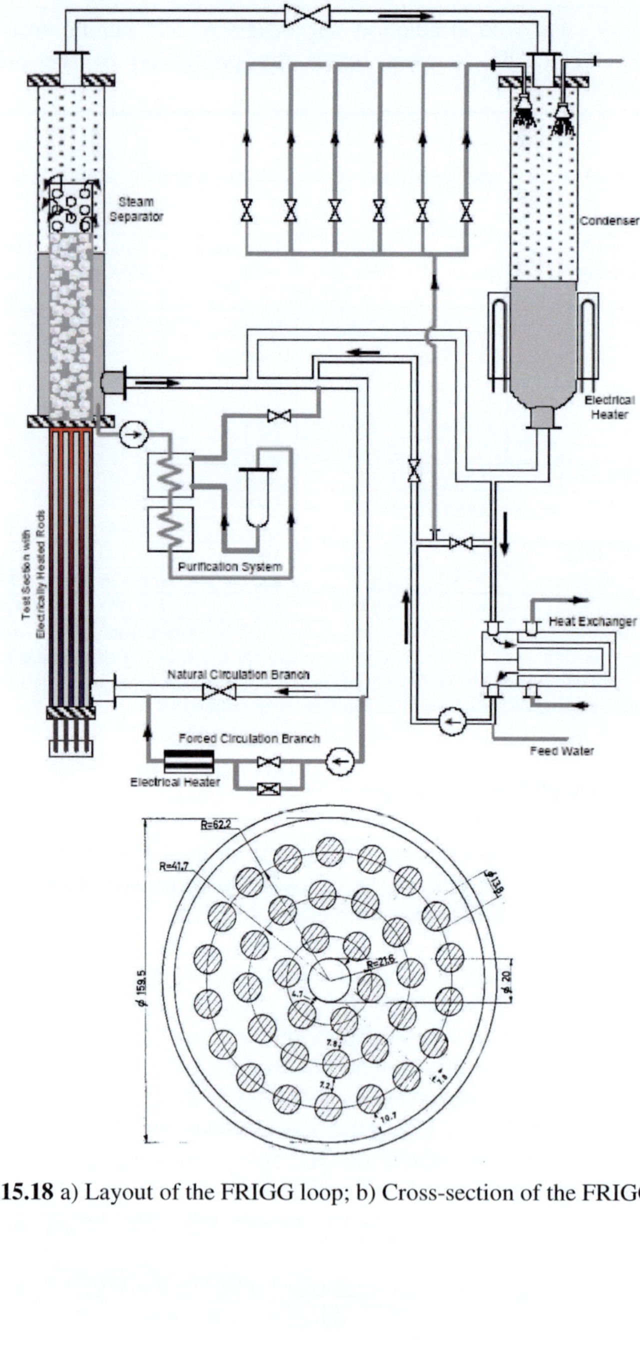

Fig. 16.15.18 a) Layout of the FRIGG loop; b) Cross-section of the FRIGG bundle

16.15.2.3.1 Uniform power distribution

We proceed with simulation of 5 experiments for which boundary conditions are defined in Table 16.15.15. The lateral and axial power distribution is uniform. The results are presented in Figs. 16.15.19 and 16.15.20.

Table 16.15.15 Boundary conditions, uniform power release FRIGG Nylund et al. (1968)

No.	p in MPa	Power in MW	G in kg/(m² s)	Subc. in K
313009	5	2.98	1107	4.4
313016	4.96	2.91	1208	19.3
313018	4.97	4.39	1124	3.7
313024	4.97	1.475	858	4.2

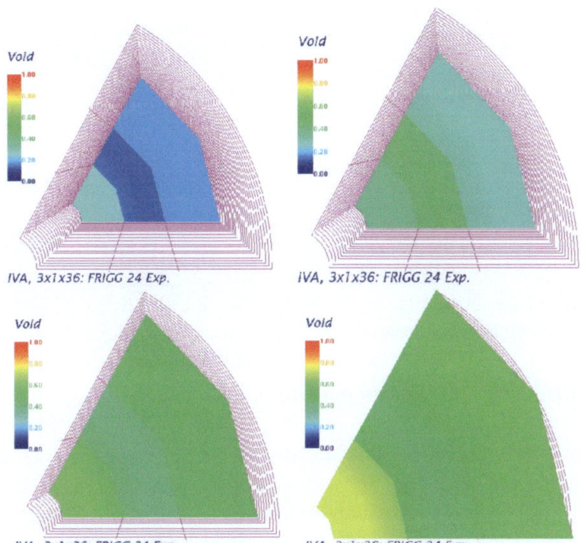

Fig. 16.15.19 1/6 of the FRIGG test section simulated by IVA (3×1×26cells). Void fraction at different levels (1, 2, 3, 4.3142 m) as a function of the radius

As for the NUPECC data the effect of the nonuniformity of the flow profiles inside the subchannels is manifested in effectively lower drag between the liquid and the vapor. The comparison at the predictions with the data using the reduced effective drag as in the previous section is favorable.

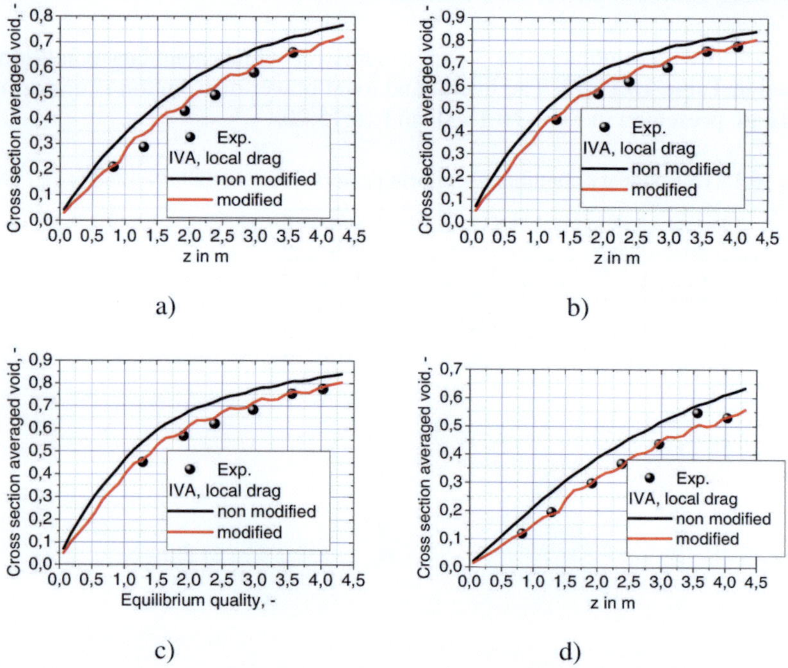

Fig. 16.15.20 FRIGG experiments: a) 3130009; b) 3130016; c) 3130018; d) 3130024: Void fraction as a function of the height. Parameter: local interfacial drag

16.15.2.3.2 Nonlinear axial power profile

We continue our analysis with simulation of other 4 experiments with boundary conditions defined in Table 16.15.16. The lateral power distribution is uniform. The smooth symmetric axial power distribution is defined by Table 16.15.17. We approximate the axial profile with 3% error with the Gauss function

$$f_z = 0.62914 + 1.74672 \Big/ \left\{ 2.47575\sqrt{\pi/2}\exp\left[\frac{2(z-2.182)^2}{2.47575^2}\right]\right\}.$$

After digitizing the profile a renormalization was necessary to guarantee the energy conservation.

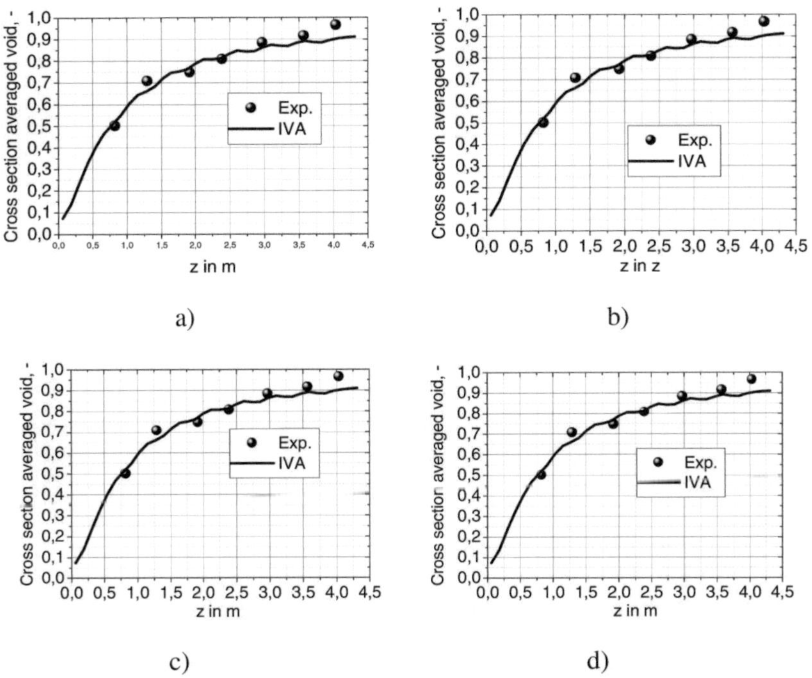

Fig. 16.15.21 FRIGG experiments: a) 1; b) 10; c) 13; d) 19: Void fraction as a function of the height. The modified drag coefficient for gas in flow is used

Table 16.15.16 Boundary conditions, nonuniform power release, FRIGG *Nylund* et al. (1970)

No.	p in MPa	Power in MW	G in kg/(m² s)	Subc. in K
1	4.88	1.66	703	1.5
10	4.87	5.55	645	2
13	4.99	5.55	688	22.89
19	5.03	4.52	681	26.11

Table 16.15.17 Axial power distribution

z in m	Ax. distr.
0.032	0.76
0.462	0.83
0.892	0.96
1.322	1.08
1.752	1.16
2.182	1.18
2.612	1.16
3.042	1.08
3.472	0.96
3.902	0.83
4.332	0.76

The results are presented in Figs. 16.15.21a–d. Again we see that the predictions coincide well with the measurements. Although the comparison with the data is favorable, a physically based model for the mechanical interaction for subchannel analyses is needed.

16.15.2.4 The THTF experiments: high pressure and low mass flow

The experiments performed in the Thermal Hydraulic Test Facility (THTF) by *Anklama* and *Miller* (1982) are characterized with such a low mass flows that in combination with appropriate low heat fluxes results in steady states in which the bundle is partially uncovered. The two-phase mixture level given in Table 16.15.18 is the essential parameter measured in these experiments. This is a very challenging problem for transient mathematical analysis because it possesses an oscillating steady state. The 8×8 rod bundle with quadratic arrangement is mounted in a 0.104×0.104 m shroud as shown in Fig. 16.15.23. The heated length was 3.66 m, the rod diameter 0.095 m and the axis to axis distance is 0.0127 m. 4 unheated rods with diameter 0.0102 m as indicated in Fig. 16.15.23 are used. The boundary conditions for the experiments analyzed are given in Table 16.15.18. The void fraction was intended to be measured by pressure difference along a given number of segments. This method works for real steady-state systems but is questionable for oscillating steady-state systems because the acceleration pressure drop components are not taken into account.

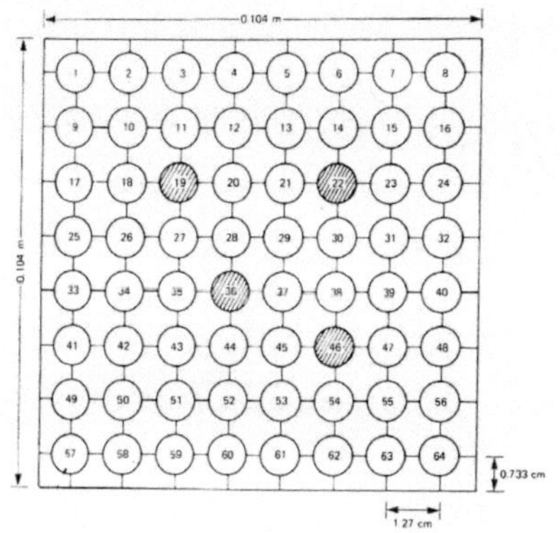

Fig. 16.15.22 Thermal Hydraulic Test Facility rod bundle cross-section, *Anklama* and *Miller* (1982)

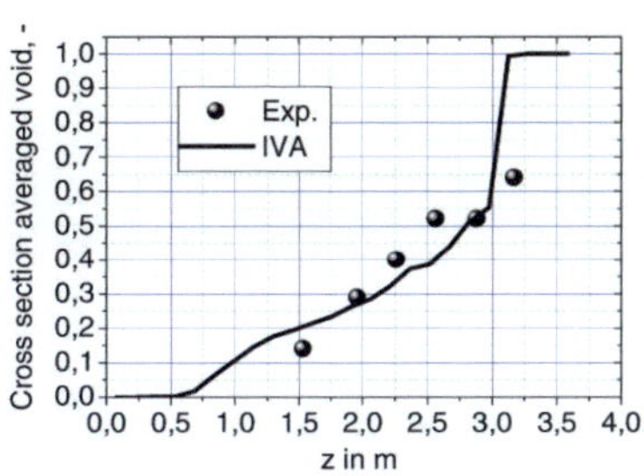

3.09.10DD

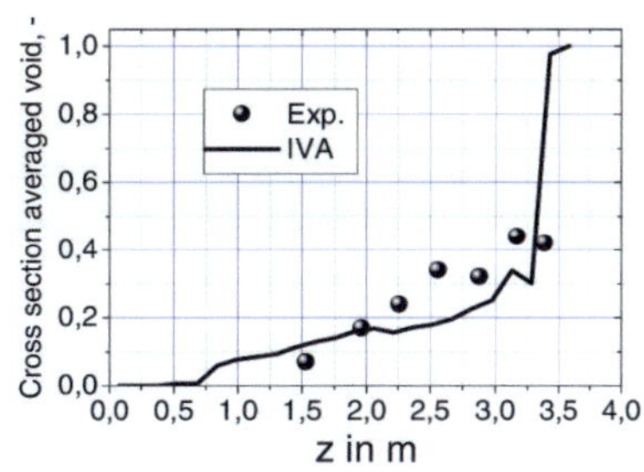

3.09.10EE

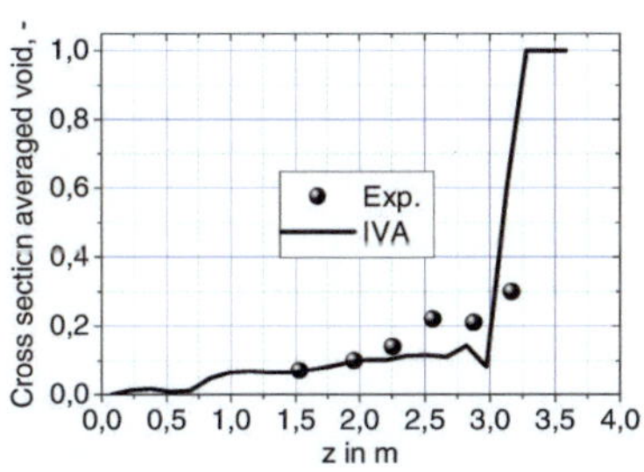

3.09.10FF

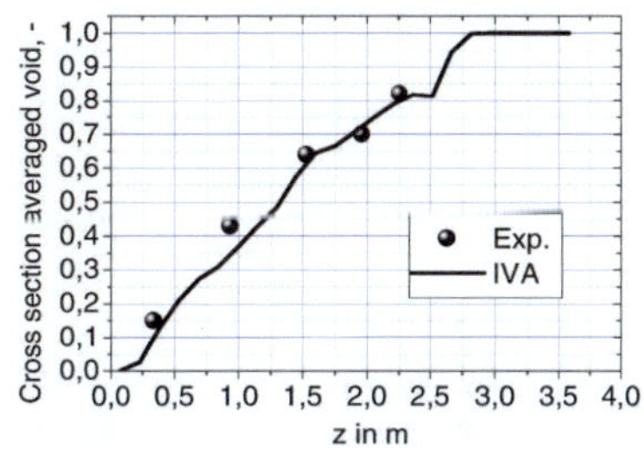

3.09.10I

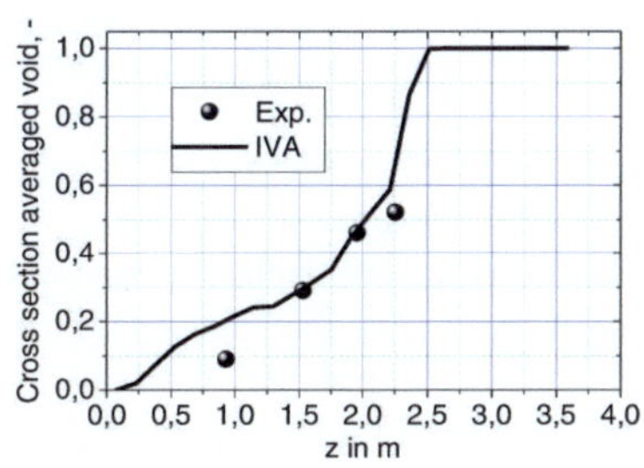

3.09.10J

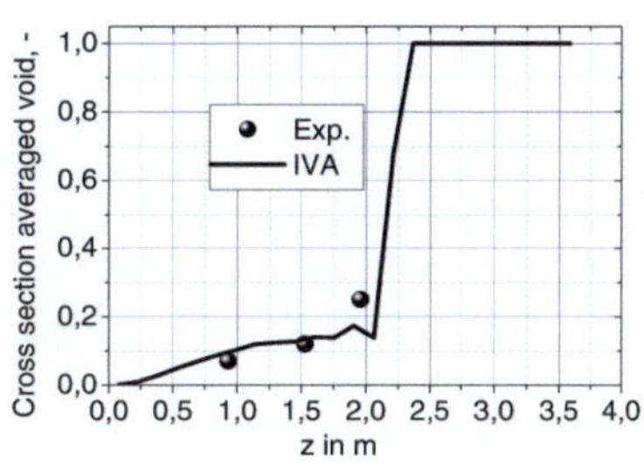

3.09.10K

Fig. 16.15.23 Snapshots of oscillating cross-section-averaged void fraction profiles in a bundle with partial uncovering of the bundle

Comparing the computed with the measured two-phase mixture level presented on Fig. 16.15.23 we see an acceptable agreement. This demonstrates also the IVA constitutive set to crossvariety of flow and heat-transfer regimes with their peculiarities.

Table 16.15.18 Boundary conditions, uniform power release, THTF *Anklama* and *Miller* (1982)

No.	p in MPa	Power in MW	G in kg/(m² s)	Subc. in K	Mixture level in m
3.09.10DD	8.1	0.28329	19.7	129.7	3.23 ± 0.04
3.09.10EE	7.7	0.14055	10.9	102.4	3.47 ± 0.03
3.09.10FF	7.5	0.07027	4.8	116.13	3.23 ± 0.04
3.09.10I	4.5	0.48753	29.3	56	2.62 ± 0.04
3.09.10J	4.2	0.23498	12.9	45.9	2.47 ± 0.4
3.09.10K	4.0	0. 7027	3.1	60.1	2.13 ± 0.3

16.15.2.5 Conclusions

Without special void-intermixing models the prediction of the void fraction of boiling flows in bundles with IVA computer code is possible with a square mean error of about 5%. The measurement error for such experiments is 3%. Future improvements are probably possible by introducing turbulence in the multiphase flow based on well-verified simulation techniques.

The production of turbulence due to increased skin friction is much smaller than the irreversible form-induced turbulence. However, the dissipation due to skin friction is substantial. The decay is well within the 30 hydraulic diameters as expected.

The recommended approach for large-scale analyses is to combine both effects: Changing of the hydraulic diameter and the cross-section over the grid length and setting the irreversible friction coefficient at the end of the grid reduced by the additional grid skin friction component. In this way the effect of the increased velocity on droplet fragmentation can be taken into account.

In general, using correlation for the interfacial drag coefficients derived from averaged pipe flows predicts strong cohesion between the phases. This was confirmed by all analysts using separated momentum equations worldwide. The ad hoc introduced reduction here has to be further investigated in the future.

The method proposed by *Hughes* et al. (1981) is recommended for description of the initiation of nucleate boiling in subcooled liquid and for the modeling of the splitting of the heat flux coming from the wall.

The so-called subchannel analysis or gross discretization analysis is discretization dependent. Therefore, fine-resolution analysis is required in the future to avoid this deficiency. For this purpose appropriate constitutive relationships have to be specially developed.

16.15.3 Transient boiling

16.15.3.1 The NUPEC transients in a channel simulating one subchannel of a PWR fuel assembly

Hori et al. (1994, 1995) reported a series of transient experiments on a specially designed 1.5-m long heated channel – Fig. 16.15.24a. The channel, Fig. 2 in Hori et al. (1995), simulates a single internal subchannel of a PWR fuel assembly. The initial conditions for the transients are defined in Table 16.15.19. Only one of the four parameters in Table 16.19.19 is varied in each experiment as shown in the left-hand side of Figs. 16.15.25–16.15.28. The others are kept constant.

Table 16.15.19 Initial conditions for the four transients, taken from Table 1 of *Hori* et al. (1994)

No.	Power in kW	Inlet temperature in °C	Pressure in MPa	Mass flow rate in $\times 10^6$ kg/(m² h)
Case 1	55	315	15.5	12
Case 2	55	315	15.5	12
Case 3	55	315	15.5	12
Case 4	75	305	15.5	12

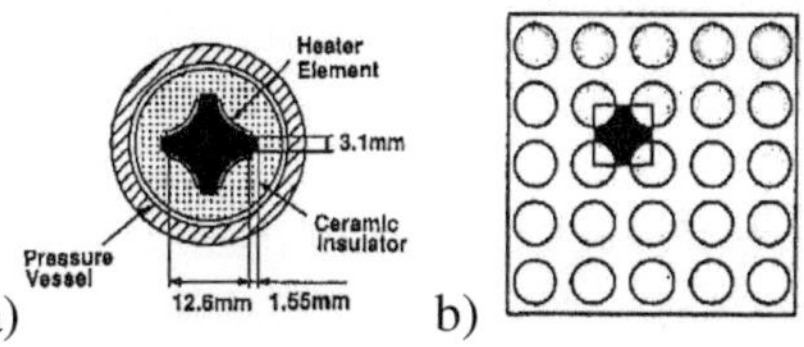

a) b)

Fig. 16.15.24 a) Cross-section of the channel simulating one subchannel of a PWR fuel assembly; b) 5×5 experimental fuel assembly

On the right of the corresponding figures the computed and the measured mixture densities are compared as they evolve during the transients. The set of constitutive relations used is: high-fidelity EOS, models for subcooled boiling: *Hughes* et al. approach (1981) for the initiation of the subcooled boiling and for the splitting of the heat into convection evaporation and recondensation. Note that only the temporal part of the virtual mass force is used in this code version. Changing the virtual mass coefficient from ½ for bubbly flow to the *Zuber*'s solution does not change this picture. The figures indicate that probably the spatial part is also important for transients like these.

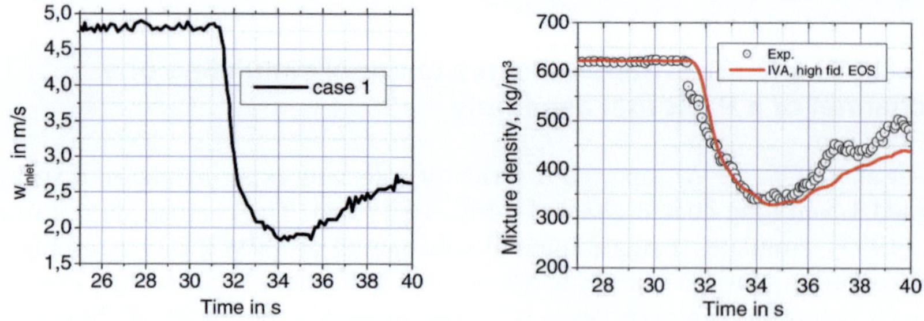

Fig. 16.5.25 Case 1: a) Velocity at the channel inlet as a function of time; Measured and computed mixture density at the exit of the channel as a function of time.

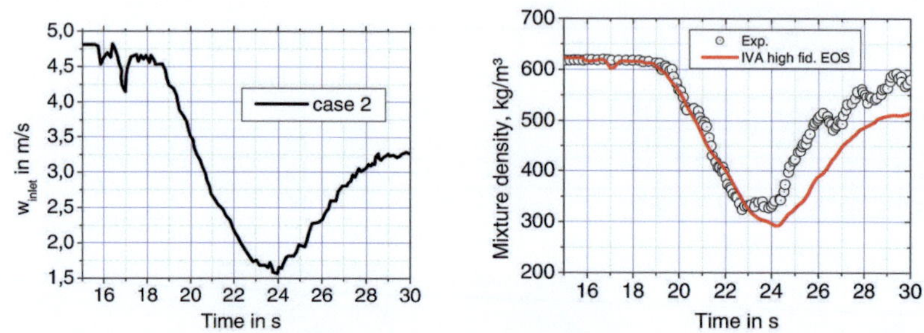

Fig. 16.15.26 Case 2: a) Velocity at the channel inlet as a function of time; Measured and computed mixture density at the exit of the channel as a function of time.

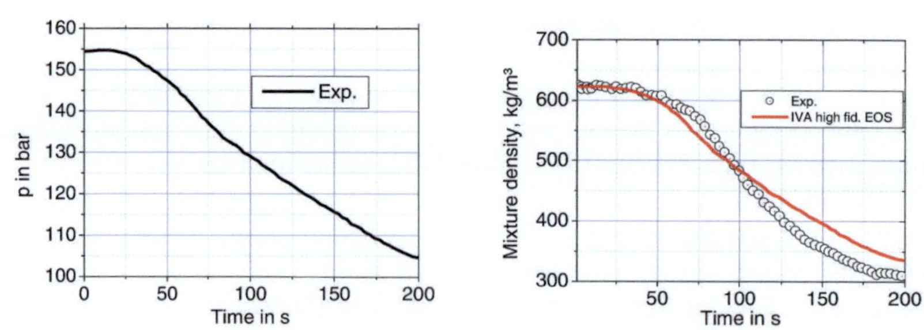

Fig. 16.15.27 Case 3: a) Averaged pressure as a function of time; Measured and computed mixture density at the exit of the channel as a function of time.

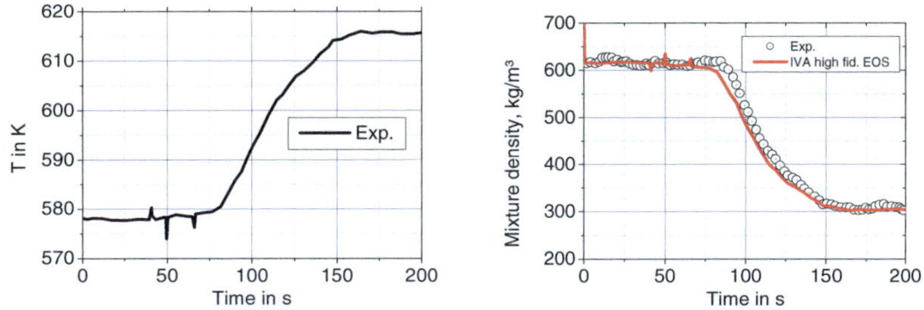

Fig. 16.15.28 Case 4: a) Inlet water temperature as a function of time; Measured and computed mixture density at the exit of the channel as a function of time.

The pressure reduction in Fig. 16.15.27 causes evaporation and decrease of the mixture density. In this case we have some uncertainties because we actually need the pressure histories at the both ends. Figure 16.15.27b gives the impression that the experimental pressure and density curves are not synchronized. The impression gained from the inlet-temperature-transient presented in Fig. 16.15.28b is similar. In any case, the agreement can be considered as good having in mind that we do not resolve the fine structure of the two-phase flow in the channel. The modification of the drag coefficients that has brought the improvement by the simulation of the steady-state experiments in rod bundles has brought also an improvement for prediction of the transient processes. In any case, we see again that the effective gas–liquid drag coefficients for rod-bundle-subchannel analyses are smaller than those correlated for averaged parameters on pipe geometry.

16.15.3.2 The NUPEC transients in PWR 5x5 fuel assembly

16.15.3.2.1 The IVA prediction

Hori et al. (1996, 1993) and *Akiyama* et al. (1995, 2005) reported series of transient experiments on a 3.66-m long heated 5×5 rod bundle, Fig. 16.15.24b, simulating a Japanese 17×17 PWR fuel assembly. The initial conditions for the transients are defined in Table 16.15.20. The axial power distribution is uniform. As reported by *Hori* et al. (1996) p. 802, the peripheral rods have 85% of the power of the internal rods. Averaged hydraulic characteristics of the bundle: for the laminar regime the friction coefficient is $\lambda_{fr}\,\mathrm{Re} = 61$ in accordance with *Rehme*'s (1973, 1971), for the turbulent regime the friction factor in accordance with *Reheme* (1972) is $\sqrt{8/\lambda_{fr}} = 2.5\ln\left(\mathrm{Re}\sqrt{8/\lambda_{fr}}\right) - 0.17$. We selected from the 4 experiments only two because the variable boundary conditions are specified only for them, *Akiyama* et al. (1995) Fig. 8, and Fig. 9. One of the four parameters in

Table 16.15.20 is varied in every experiment, as shown in the left-hand side of the figures 16.15.29 and 16.15.30. The others are kept constant.

Table 16.15.20 Initial conditions for the NUPEC 5x5 bundle transients, taken from Appendix 1 of *Hori* et al. (1996), p. 810.

Variable	Power in KW	Inlet temperature in °C	Pressure in MPa	Mass flow rate in $\times 10^6$ kg/(m² h)
power	2250	300	15.5	12
flow rate	2250	300	15.5	12

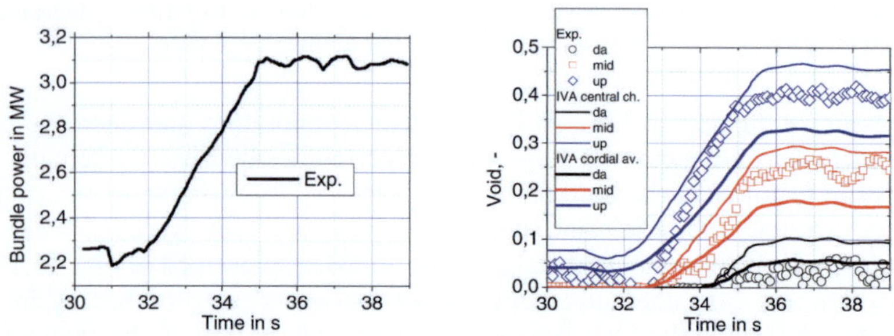

a) b)

Fig. 16.15.29 a) Bundle power as a function of time; b) NUPEC measurements and IVA results (central subchannel and central chordial average) for the void fraction at three different elevations as a function of time. Initial conditions: T_{in} = 300 °C, p = 155 bar, w_{in} = 4.588 m/s, bundle power 2250 kW

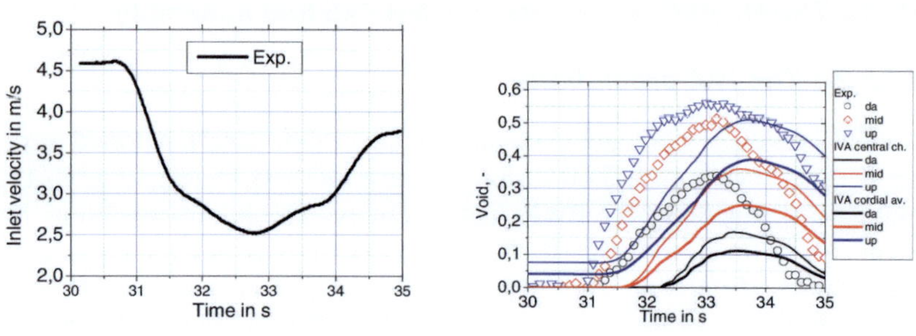

a) b)

Fig. 16.15.30 a) Inlet velocity as a function of time; b) NUPEC measurements and IVA results (central subchannel and central chordial average) for the void fraction at three different elevations as a function of time. Initial conditions: T_{in} = 300 °C, p = 155bar, w_{in} = 4.588 m/s, bundle power 2250 kW

Measured are the void fractions at three different geodetic levels. The distance between the levels is given by *Hori* et al. (1996) p. 811. The position of the first level is approximately taken from Fig. 1 of *Akiyama* et al. (2005). The type of void averaging presented on the figures is confusing. In *Hori* et al. (1993) p. 74 the authors presented for the pressure transient in Fig. 12 a line-averaged void fraction, whereas in *Akiyama* et al. (1995) for the power- and mass-flow transients the authors compared their data with predictions of the void in the central channel of the bundle. This nonprecise information forced us to put on Figs. 16.15.27b and 16.15.28b both the central channel void and the line averaged void along the *x*-axis. As expected the central channel void is larger than the averaged void along the beam length.

Regarding the power transient: The experimental data for the upper two levels are between the predicted both void fractions. At the lowest measured position the void is overpredicted, indicating probably that the subcooled boiling model needs more attention. The gradient of the change reflects the experimental observations.

Regarding the flow transients: There is considerable discrepancy between prediction and measurements. The moment of the intensive void formation is predicted later, the inclination of the void increase is predicted to be smaller, saying that the heat partitioning between evaporation and condensation is different. The predicted maxima of the curves are lower than the measured.

16.15.3.2.2 Comparison with Aounallah et al. (1999), Macian et al. (2000) and Wang et al. (2005) analyses

Analyses with the two-fluid computer code VIPRE-02 (*Kelly* et al. 1992) of the openly published NUPEC data that are used also in this study are already reported by *Aounallah* and *Coddington* (1999) and by *Macian* et al. (2000). VIPRE-02 is a two-fluid computer code using six equations for describing predominant axial flow. The momentum equations in the other two lateral directions are simplified for a normal velocity component from or to the neighboring cannels using 1 instead of 9 nonhomogeneous convection terms in a transformed coordinate system. IVA possesses a three-fluid model with conservation equations that do not neglect any terms. Regarding selection of the drag coefficient correlations and boiling mechanisms for this particular class of processes as long as two of the fluids in IVA are addressed, the codes are similar except for the use of dynamic fragmentation and coalescence in IVA but not in VIPRE-02. The observation made by *Aounallah* and *Coddington* (1999) and by *Macian* et al. (2000) are confirmed by our study too.

> In general, effectively less drag between void and liquid is observed in the experiments than in the separated-momentum equation predictions with correlation gained on flows with other geometry.

Macian et al. interpret the underprediction of the void as a possible deficiency of the subcooled boiling instead of the effectively lower drag. This difference in the interpretation can be clarified in the future only if fine-scale analyses are done.

The discrepancy with the data in Fig. 16.15.30 is also observed by *Macian* et al. The reasons remain to be clarified. *Wang* et al. (2005) performed a comparison between VIPRE two-fluid code and 6 FRIGG experiments comparing also void fraction in lateral bundle zones. The authors come to the conclusion that additional modeling of the radial redistribution is required to better predict void in neighboring subchannels with quite different power load. Over prediction of the axial voids in the low-power regions is also reported, which tells us that the subcooled boiling model in VIPRE needs further sophistication.

16.15.3.2.3 Comparison with Hori et al. (1994)

Hori et al. (1994) reported that the comparison with their transient measurements can be better reproduced in the acceleration phase by homogeneous models and in the deceleration phase by drift-flux models. This is in line with our observation here and with the observation made by *Aounallah* and *Coddington* (1999) and by *Macian* et al. (2000).

16.15.3.2.4 Conclusion

In general, averaged measurements over a beam line have to be compared with averages of rows of fine mesh cells extracted from computational results. The larger the size of the cells the less informative is the comparison.

16.15.4 Steady-state critical heat flux

Nucleate and flow boiling is a very good cooling mechanism. Increasing heat fluxes in the wall may lead to heat-transfer regimes that have considerably lower heat transfer. In such cases the frequency of the bubble production and their bubble departure size allow them to touch each other. Then, a film forms, reducing the heat-transfer coefficient. This regime is calling departure from nucleate boiling and designated by DNB. Another important case is the boiling of a flowing film which is a good heat-transfer mechanism. If the film dries out, the heat transfer is controlled by the steam flow, which again makes the cooling much worst. This regime is called dry out and designated by DO. Both mechanism may happen in technical facilities with the first one causing strong temperature jump at the wall the second. If they last long enough both regimes can destroy the facility.

16.15.4.1 Initial 0D guess

Problem: Check whether it is possible to use the 2005 look-up table primary designed for heated pipes also for heated rod bundles.

Solution: I exploit deliberately a set of simplifying assumptions usually used in the 1960s:

1. Fictive subchannels corresponding to the number of rods are so defined as to be equivalent in geometry.
2. Equal mass flow rate to each subchannel.
3. The critical heat flux occurs at the exit of the hot channel as a function of the exit parameters. The phenomenon is similar to pipe flow. Therefore, the *Groeneveld*'s look-up table is expected to give the right dependence on mass flow rate, quality and subcooling.
4. Uniform boundary conditions for all channels.
5. Grid turbulization as well as the axial power distribution is not specifically treated.

The procedure I use is: Unless $\left|\frac{q''_{crit,i}\left(z_{\max}\right)-q''_i\left(z_{\max}\right)}{q''_{crit,i}\left(z_{\max}\right)}\right|<\varepsilon$ increase q with a Δq starting with q_0. Here the power of the hot channel is $q_i = f_{lat,i}\, q/n_{rods}$, where, as per the definition, the bundles power is the sum of the subchannel powers $q=\sum_{i=1}^{n_{rods}} q_i$ and the lateral distribution factor obeys $\frac{1}{n_{rods}}\sum_{i=1}^{n_{rods}} f_{lat,i}=1$. The exit equilibrium mass flow quality is $X_{1,i}=\frac{1}{\Delta h}\left(\frac{q_i}{\dot{m}_i}-\Delta h_{sub}\right)=\frac{1}{\Delta h}\left(f_{lat,i}\frac{q}{\dot{m}}-\Delta h_{sub}\right)$. Here, by virtue of assumption 2 I use $\dot{m}/\dot{m}_i=n_{rods}$. The local heat flux at the exit of the hot channel i, $q''_i\left(z_{\max}\right)=f_{z_{\max},i}\frac{f_{lat,i}\, q/n_{rods}}{\pi D_{rod}L_{rod}}$, is compared with the local critical heat flux. The critical heat flux at the exit of channel i is defined by the 2005 look-up table $q''_{crit,i}\left(z_{\max}\right)=f\left(p,G,X_{1,i},D_{hyd}\right)$. Dividing this value by the lateral hot channel factor and recomputing the averaged flux into bundle power we obtain the final result $q=\frac{q''_{crit,i}\left(z_{\max}\right)}{f_{lat,i}}\pi D_{rod}L_{rod}n_{rods}$. Note that the last expression is based on the intuitive assumption that this value may represent the averaged heat flux in the hot channel. It is only a *hypothesis*. It is very interesting to compare the prediction of this "primitive" approach with experimental data and to see how it works.

8×8 NUPEC BWR bundle tests: First I simulate three data sets for experiments with the 8×8 NUPEC BWR bundle tests (Fig. 16.15.31) as proposed in the international benchmark by *Neykov* et al. (2005). The *first* and the *second* tests are performed with the same axial power distribution but with different lateral power-peaking pattern. They serve for analysis of the influence of the lateral power distribution. The *first* and the *third* tests are performed with the same lateral power-peaking pattern but with different axial power distribution. They serve for analysis of the influence of the axial power distribution.

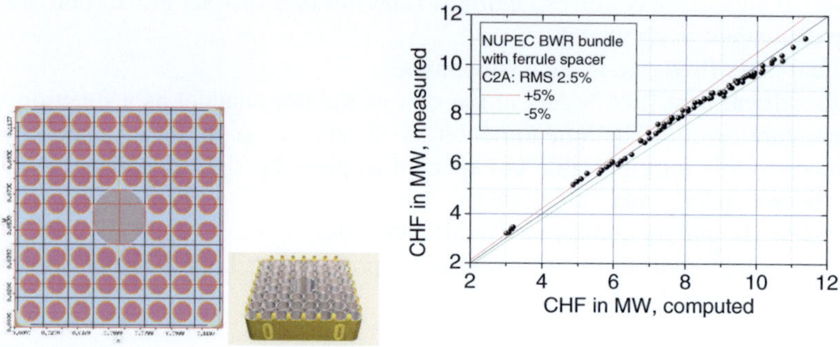

Fig. 16.15.31 Geometry of the 8x8 NUPEC CHF bundle tests: a) cross-section; b) ferrule spacer; c) Measured versus computed critical bundle power for *cosine* profile with $f_{z,\max}$ = 1.4 and lateral peaking pattern characterized by f_{xy} = 0.89 to 1.3. Data by NUPEC: p = 5.48 to 8.67 MPa, $\Delta h_{2,sub}$ = 22.61 to 137.26 kJ/kg, $\dot{m}$ = 9.98 to 65.52 t/h. C2A tests, 79 points

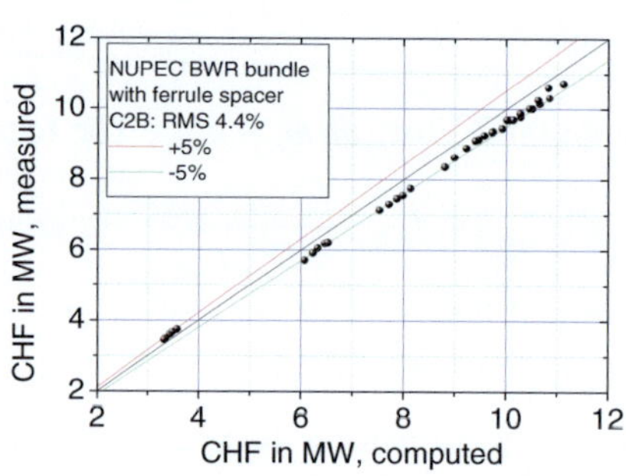

Fig. 16.15.32 Measured versus computed critical bundle power for *cosine* profile with $f_{z,\max}$ = 1.4 and lateral peaking pattern characterized by f_{xy} = 0.99 to 1.18. Data by NUPEC: p = 7.14 to 7.20 MPa, $\Delta h_{2,sub}$ = 21.06 to 128.57 kJ/kg, $\dot{m}$ = 9.93 to 64.85 t/h. C2B test, 36 points

The measured versus the predicted bundle power is presented in Figs. 16.15.31c, 16.15.32 and 16.15.33.

Comparing Figs. 16.15.31c and 16.15.2 we see that for flatter lateral profiles this method slightly systematically underpredicts the measurements.

Comparing Figs. 16.15.32 and 16.15.33 we see that for the two different axial power profiles this method predicts similar measurements.

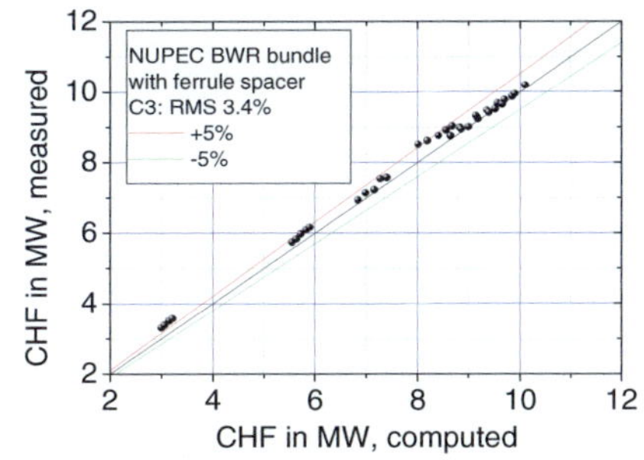

Fig. 16.15.33 Measured versus computed critical bundle power for *inlet peak* profile with $f_{z,\max} = 1.4$ and lateral peaking pattern characterized by $f_{xy} = 0.89$ to 1.3. Data by NUPEC: $p = 7.10$ to 7.19 MPa, $\Delta h_{2,sub} = 21.7$ to 125.74 kJ/kg, $\dot{m} = 9.93$ to 65.02 t/h. C3 test, 36 points

> Having in mind the 2005 look-up table possesses mean error of 7.1% the above predictions with mean error of 2.5, 4.4 and 3.4% can hardly be made better.

AREVA ATRIUM10 BWR tests: Figures 16.15.34 and 16.15.35 present the measured versus the predicted bundle power for a specific tests for AREVA BWR bundles ATRIUM 10.

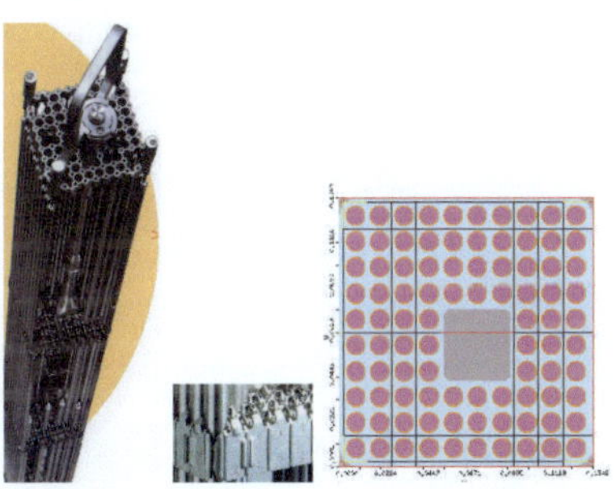

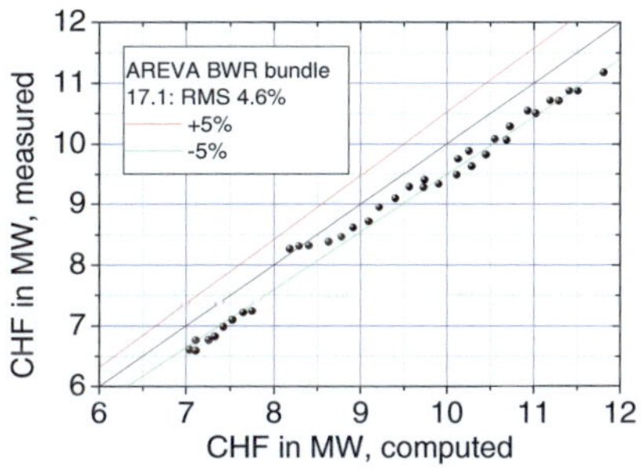

Fig. 16.15.34 a) BWR ATRIUM™ 10 bundle with ULTRAFLOW™ spacer; b) Measured versus computed critical bundle power for *cosine* profile with $f_{z,\max} = 1.4$ and lateral peaking pattern characterized by $f_{xy} = 0.751$ to 1.255: mean error 4.6%, slight systematic under prediction. Data by *Fleiss* et al. (1992): $p = 69$ bar, $\Delta h_{2,sub} = 23$ to 184 kJ/kg, $\dot{m} = 6.3$ to 18.90 kg/s. 40 points

There are slight differences in the geometry. For both cases we see mean errors of 4.6 and 3.95%, respectively, which again lie inside the accuracy of the 2005 look-up table. Again, prediction with better accuracy can hardly be achieved.

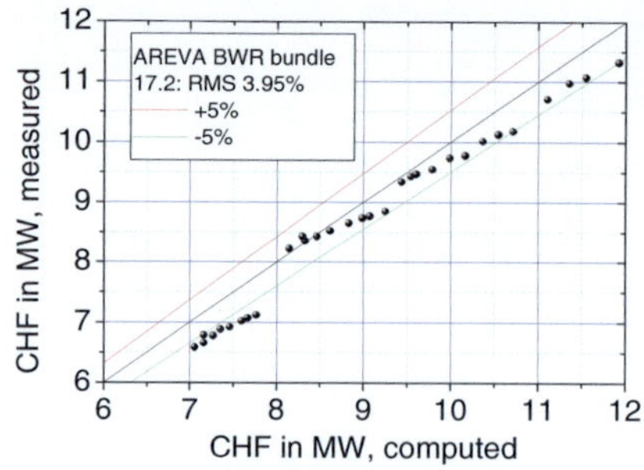

Fig. 16.15.35 Measured versus computed critical bundle power for *cosine* profile with $f_{z,\max}$ = 1.4 and lateral peaking pattern characterized by f_{xy} = 0.731 to 1.251: mean error 3.95%, slight systematic under prediction. Data by *Fleiss* et al. (1992): p = 68.8 to 69.5 bar, $\Delta h_{2,sub}$ = 23 to 188 kJ/kg, $\dot{m}$ = 6.33 to 15.7 kg/s. 37points

AREVA tests for PWR bundles with FOCUS grids: Unlike the previous four data sets this one is for pressures up to 165 bar. The bundle contains 5x5 rods and 5 grids. Again, the simple method predicts the CHF with mean error of 7.24 that is within the accuracy of the 2005 look-up table, see Fig. 16.15.36.

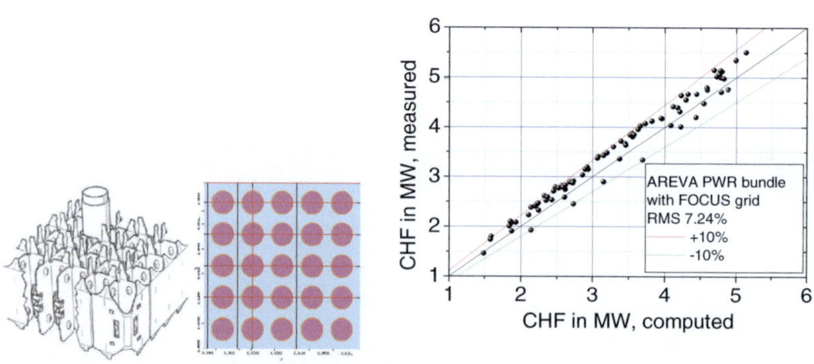

Fig. 16.15.36 AREVA PWR bundle with FOCUS grid: Computed versus measured bundle power: mean error 7.5%, slight systematic under prediction. Data by *Vogel* et al. [34] (1991): p = 69.3 to 166.6 bar, $T_{2,\text{in}}$ = 179.4 to 329.5 °C, G = 1186.6 to 3629 kg/(m² s), q = 1.4643 to 5.5015 MW. 82 points

Conclusion

The check whether it is possible to use the 2005 look-up table primary designed for heated pipes also for heated rod bundles gives the surprising result that the bundle critical power for 5 data sets of three different bundles and different power distributions all together 273 experiments are predicted by a simple method described above using the 2005 look-up table within the accuracy reported by the authors of this table. Therefore, unless a better prediction method is developed it is a good idea to use the above method as the best predictive instrument. The influence of the grid design can in addition be taken into account by an empirical coefficient derived from experiment.

16.15.4.2 Pressure drop for boiling flow in bundles

Finally the check of the pressure drop in boiling bundles will be briefly discussed. First, I use the data sets for experiments with the 8×8 NUPEC BWR bundle tests (Fig. 16.15.31) as proposed in the international benchmark by *Neykov* et al. (2005). The single-phase friction coefficient for the specific ferule grid spacer is fitted to the data with a mean error of 1%. The predicted versus the computed friction pressure drop only is given in Fig. 16.15.37.

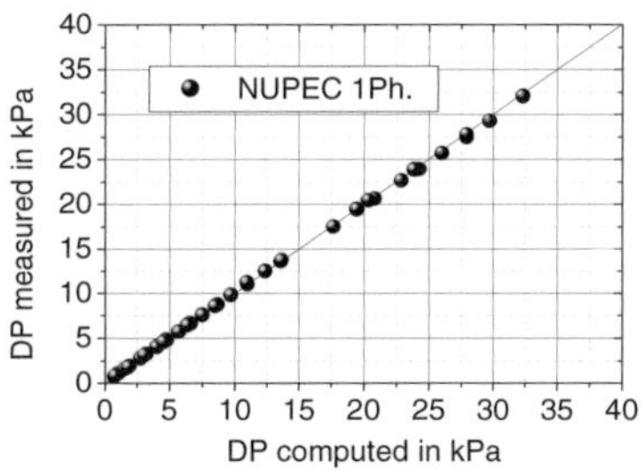

Fig. 16.15.37 Measured versus computed friction pressure drop (wall friction+7 grids). mean error 1%.

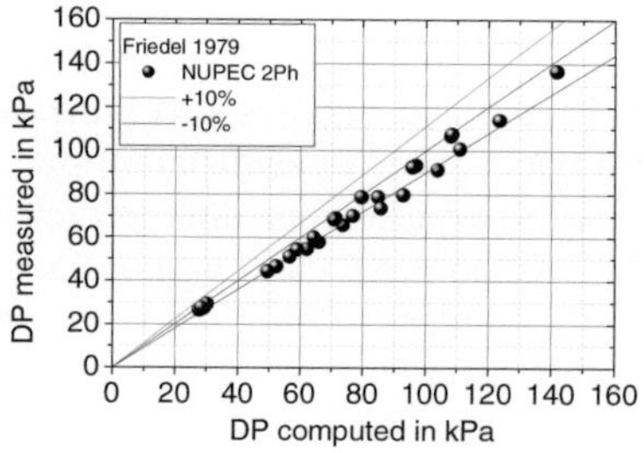

Fig. 16.15.38 Measured versus computed pressure drop for boiling channels. Two-phase friction multiplier by *Friedel* (1979), mean error 7.38%.

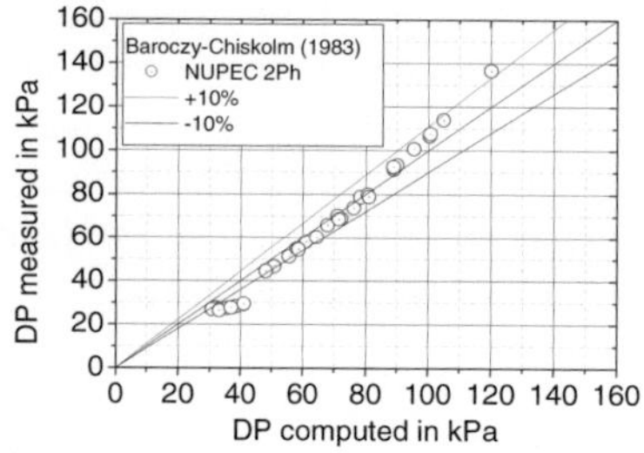

Fig. 16.15.39 Measured versus computed pressure drop for boiling channels, 33 experiments. Two-phase friction multiplier by *Baroczy-Chiskolm* modified by *Chisholm* (1983), mean error 11.8%

In the next step, I analyze the two-phase flow data. I use the already-obtained correlation for the irreversible pressure loss coefficient at the grid and two phase flow multiplier proposed by *Friedel* (1979). We see from Fig. 16.15.38 that *Friedel*'s correlation predicts the data with mean error 7.38% and from Fig. 16.15.39 that the *Baroczy* correlation from 1965 modified by *Chisholm* (1983) predicts the data with mean error 11.8%.

Next, we analyze 96 data points for the AREVA bundles with the FOCUS grid spacers. Again the single-phase irreversible pressure loss coefficients of the spacers are fitted with an accuracy of 1%. Then, the total pressure loss for boiling flow is computed. The *Friedel* 1979 correlation for the two-phase multiplier is used. A mean error of 9% is found.

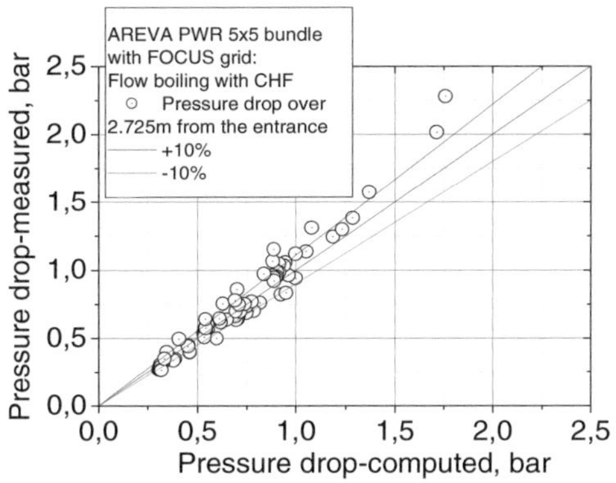

Fig. 16.15.40 Measured versus computed total pressure drop in a 5×5 bundle with FOCUS grid. 96 experiments

Conclusion: Note that *Friedel*'s correlation approximates 25 000 experimental points for vertical upwards cocurrent flow and for horizontal flow with 30 to 40% standard deviation for one- and two-component flow. Therefore, better accuracy can not be expected for this particular comparison. The two-phase total pressure drop in bundles with boiling flows is best predicted by *Friedel*'s correlation with a mean error between 7.38 and 9%. So we recommend the use of *Friedel*'s correlation for such cases.

16.15.4.3 3D analysis

First, I design a preprocessor and function for IVA that for given mass flow, pressure, subcooling and geometry of the bundles that predicts the critical bundle power and parameter by the simplified procedure presented in the previous section. Then I perform 3 computations: (a) with the so estimated power; in (b) and (c) we vary this power by ± 5%, respectively, and for each of them compute the portion of the surface that is dry. Dry is defined as either film thickness less the 10 μm in the cell or the 2005 look-up table predicts conditions fro CHF (DO or DNB). In most of the BWR cases I have analyzed, dry out is observed. However there are cases in which the film thicknesses are large and the look-up table identifies DNB. How important the grid modeling is, is demonstrated in the following example: For the case 505500 of the NUPEC experiments I consider the grid in Fig. 16.15.31b without and with small vanes at the periphery. Figure 16.15.41 shows the differences.

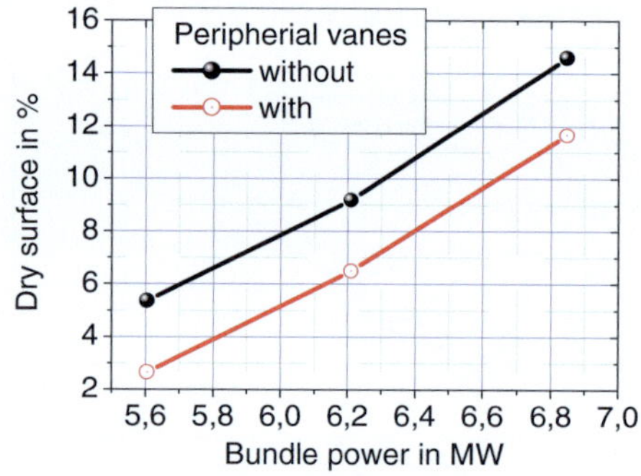

Fig. 16.15.41 Portion of the dry surface as a function of the power without and with blades

It is obvious that the vanes redirect unused liquid from the periphery to the fuels and improve the wetting process. It is remarkable that such an effect is possible to be taken into account in large-scale discretization as those shown in Fig. 16.15.31a. Therefore, in all future computation I do consider the peripheral blades. One example of the results obtained with this procedure is given in Table 16.15.21. The computations in this subject are still going on and the final analysis will be provided later.

Table 16.15.21 Computed dry out heat flux and the corresponding dry portion of the surface of the fuels in the bundle.

Exp. ID	$\dot{m}$ kg/s	P bar	ΔT_{sub} K	$\dot{Q}_{comp}$ MW	$\dot{Q}_{\exp}$ MW	F_{dry} / F_{heated} % for		
						$\dot{Q}_{comp}$ -5%	$\dot{Q}_{comp}$	$\dot{Q}_{comp}$ +5%
SA505500	5.60	54.9	−10.07	6.216	6.13	2.65	6.49	11.66
SA505501	5.58	54.9	−10.15	6.203	6.13	2.37	4.87	8.56
SA505600	5.59	55.1	−16.92	6.323	6.23	0.63	1.86	3.15
SA505800	5.61	55.0	−26.14	6.504	6.39	0.57	1.83	3.14
SA505900	5.59	54.9	−5.11	6.121	5.98	0.63	1.72	3.48
SA510500	15.29	54.8	−11.17	10.129	9.72	0.61	1.64	3.41
SA510501	15.31	55.1	−12.39	10.159	9.81	0.66	1.69	3.54
SA510600	15.19	55.1	−19.25	10.413	10.09	0.54	1.38	2.84
SA510601	15.37	55.2	−19.37	10.431	10.19	0.61	1.46	2.89
SA510800	15.23	55.1	−27.30	10.720	10.20	0.56	1.39	3.03
SA510900	15.19	55.2	−6.94	9.895	9.560	0.54	1.38	2.84
SA510901	15.29	55.1	−6.88	9.920	9.660	0.61	1.46	2.89
SA512500	18.19	55.4	−12.75	10.702	10.410	0.56	1. 39	3.03
SA605500	5.58	71.6	−9.46	5.667	5.770	0.21	2.39	5.40
SA605502	5.58	71.7	−9.63	5.661	5.730	0.19	2.46	6.76
SA610503	15.33	71.7	−11.15	8.956	8.850	0.00	0.38	1.01
SA610504	15.41	71.7	−10.90	8.966	8.910	0.00	0.38	0.99
SA610600	15.29	71.8	−16.97	9.217	9.200	0.02	0.35	1.08
SA610700	15.33	71.3	−20.56	9.445	9.370	0.00	0.49	1.34
SA610701	15.24	72.1	−21.63	9.397	9.380	0.00	0.37	1.15
SA610800	15.36	72.4	−26.39	9.634	9.520	0.00	0.54	1.15
SA610900	15.31	72.7	−7.00	8.654	8.660	0.00	0.10	0.73

The portion of the fuel surface that is dry is an inherent property of the process. It is detected by measuring with a finite number of thermocouples. It is simulated with a finite number of volumes containing again a finite number of portions of the fuel bundle-surface.

> Therefore, there is a specific definition of CHF associated with the density of the thermocouples and their distribution in each experiment. This important source of uncertainty was never considered in the literature as far I know but it has to be considered in the future by transferring experimental observation to real plant behavior.

As will be demonstrated in the next section the mean error of the 1D prediction of the location of the dry out is about 40 cm due to the uncertainties in the involved constitutive models that are going to be next discussed. They remain also in 3D. Therefore, more effort is necessary to increase the accuracy of the involved correlation, to derive a complete set of source terms for generation of turbulence in all flow patterns, to accomplish the right coupling between deposition and local degree of turbulence, and finally to derive an appropriate mechanistic criterion for identification of dry out. I strongly recommend continuation of this line of research theoretically and experimentally.

16.15.4.4 Uncertainties

Trying to compute dry-out heat flux by using three fluid models with constitutive relations gained from adiabatic flow I found in *Kolev* 1985a, b that the uncertainty of the entrainment and deposition models does not allow very accurate prediction of the location of the dry out. Let us analyze some reasons for this situation, which still has not changed very much since those days.

16.15.4.4.1 Interfacial drag

The interfacial drag between gas and film in annular two-phase flow controls the relative velocity between bulk flow and film. This relative velocity controls the entrainment of droplets. For the computation of the drag coefficients different approximations exist, all of them reported by their authors to be based on experiments. For comparison between the predictions of different correlations, see the review in *Kolev* 2004b, as a function of the gas *Reynolds* number is given in Fig. 16.15.42.

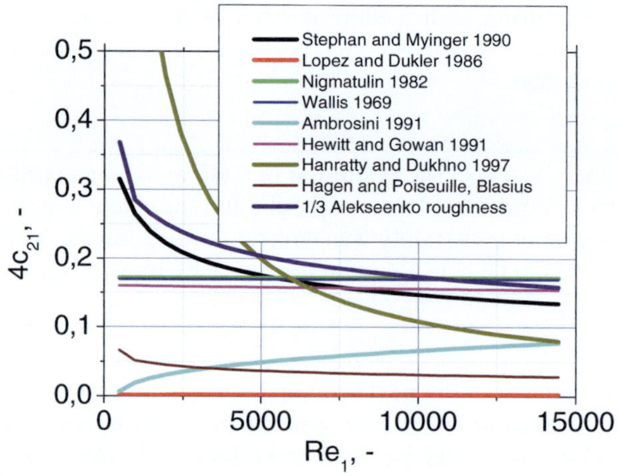

Fig. 16.15.42 Gas–film friction coefficient as a function of the gas *Reynolds* number

The correlations by *Wallis*, et al. and *Nigmatuling* do not differ from each other much. The prediction of the correlation by *Stephan* and *Myinger* converges to the prediction of the above-mentioned correlations for higher *Reynolds* number, but differs greatly for low *Reynolds* numbers. If one uses the formula of *Alekseenko* and takes 1/3th of the film amplitude as the effective roughness the *Nikuradze* formula predict results close to those predicted by *Stephan* and *Myinger*. For comparison the prediction by the *Blasius* correlation for *Reynolds* number larger then 1187 and by the *Hagen* and *Poiseuille* correlation for lower *Reynolds* numbers is also given in Fig. 16.15.42. It is obvious that the final state of the knowledge in this field is not achieved.

16.15.4.4.2 Entrainment

The droplet entrainment in annular two-phase flow influences the film thickness. For the computation of the entrainment different approximations exists, all of them reported by their authors to be based on adiabatic experiments.

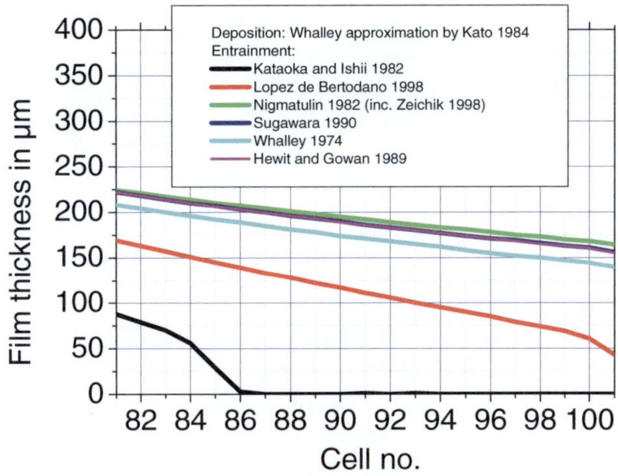

Fig. 16.15.43 Film thickness as a function of the distance from the entrance. Boiling water in vertical circular pipe, *Bennett* et al. 1967: Test nr. 5253, vertical pipe, 0.01262 m inner diameter, 5.5626 m length, uniformly heated with 199 kW, inlet water flow from the bottom: 68.93 bar and 538.90 K

Kawara et al. (1998) compared 9 correlations for entrainment obtained by different authors or teams. They found differences of six orders of magnitude at low entrainment rations and three orders of magnitude at high entrainment ratios. We give here another example. We consider here one of the many vertical boiling pipe experiment by *Bennett* et al. 1967 experiment. Simulating the flow with a three-fluid model (IVA) using different entrainment correlations, see the review in *Kolev* 2004b, we obtain the film thickness as a function of the axial coordinate as presented in Fig. 16.15.43. We know from the measurements that the film disappear in position 8/10 of the total pipe length.

The largest entrainment, and therefore the closest to the observed result, is predicted by *Kataoka* and *Ishii*'s correlation. None of the models takes the change of the entrainment due to the wall boiling into account. Therefore, a final state of this research field is still not reached.

16.15.4.4.3 Deposition

Kawara et al. 1998 compared 11 deposition coefficients obtained by different authors or teams. They found differences up to two orders of magnitude, depending on the local droplet concentration.

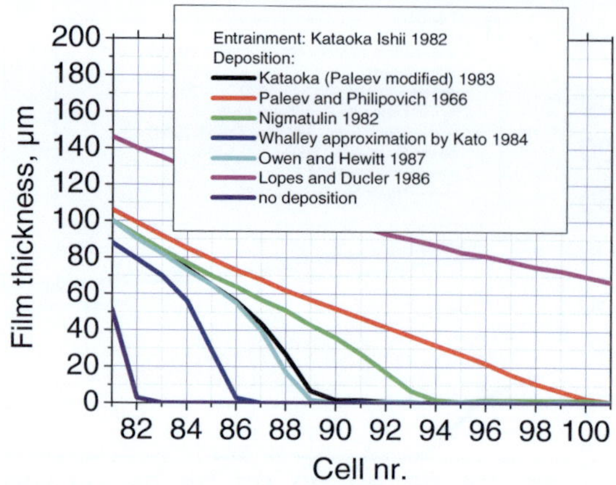

Fig. 16.15.44 As in Fig. 16.15.39.

I make 7 computations changing only the deposition models keeping the entrainment model unchanged. The results regarding the film thickness are presented in Fig. 16.15.44. The smallest deposition, and therefore the closest to the observed result, is predicted by the *Kato*'s correlation. Note that we do not take into account the deposition suppression due to boiling of the film, which is obviously important. Repeating the computation with virtually no deposition gives the best results. Note that *Hoyer* and *Stepniewski* reported already in 1999 that deposition suppression due to boiling is important for dry-out prediction at high void fractions in boiling flows.

Therefore, a final state of this research field is still not reached.

The influence of the boiling on the deposition will be discussed next.

16.15.4.4.4 Deposition and entrainment changes due to nucleate boiling

Boiling films manifest a few additional phenomena controlling the film dynamics that are not present in adiabatic films.

Modified deposition: Vapor produced through the film blows into the gas core and dumps the penetration of the gas pulsation into the boundary layer. Therefore, there is an impact on the gas boundary layer close to the interface resembling blowing of gas from the film/gas interface. This phenomenon hinders deposition. Assuming that all generated vapor enters the bulk flow perpendicular to the wall we have for the effective gas velocity

$$u_{1\_blow} = \frac{D_{heat}}{4} \frac{\mu_{21}}{\rho_1}. \qquad (16.15.6)$$

If $u_{1\_blow} > V_1'$ the deposition is effectively nonexisting. If $u_{1\_blow} > V_1'$ the effective gas fluctuation velocity is reduced to $V_1' - u_{1\_blow}$. *Doroschuk* and *Levitan* (1971), *Guguchkin* et al. (1985) reported data that prove that the vapor blow from the film into the gas bulk reduces deposition. *Milashenko* et al. 1989 recommended ignoring the deposition in boiling flows.

Modified entrainment: The steam mass flow generated at the wall surface contributes substantially to the fragmentation of the liquid by two mechanisms: At low film velocity if the bubble departures diameter is larger than the film thickness the bubble burst causes additional entrainment. At high film velocities the bubble departure diameter is small but there is a vapor net flow perpendicular to the wall crossing the film. The vapor creates two-phase unstable structure that eventually break up and release the vapor into the gas core flow by entraining additional droplets.

Milashenko et al. collected in 1989 data for boiling flow in a 0.0131-m inside-diameter pipe of 0.15 and 1 m heated lengths. Setting the deposition rate μ_{32} equal to zero the authors correlated their data for an effective entrainment by the following correlation

$$\mu_{23\_boiling\_film} + \mu_{23\_adiabat} - \mu_{32} = \frac{1.75}{\pi D_h} \alpha_2 \rho_2 w_2 \left(\dot{q}''_{w2} 10^{-6} \frac{\rho''}{\rho'} \right)^{1.3}, \qquad (16.15.7)$$

indicating a strong dependence on the wall heat flux. Note that the asymptotic value for zero heat flux will produce zero effective entrainment, which can not be true. Nevertheless, this is the ultimate prove that boiling process influence the effective entrainment.

Probably a better scaling velocity of the additional to the adiabatic entrainment is given by Eq. (16.15.6), which will be linearly dependent on the heat flux, $w_{1\_blow} = \dot{q}''_{w2} / (\rho_1 \Delta h)$, for saturated film. This explains why *Milashenko* et al. are forced to introduce the constant 10^{-6} being of order of $1/\Delta h$. Therefore

$$\mu_{23\_boiling\_film} = f \rho_2 \frac{\mu_{21}}{\rho_1} \frac{D_{heat}}{4} \qquad (16.15.8)$$

with function $f < 1$ that remains to be found is probably the better physical basis for data correlation. *Kodama* and *Kataoka* reported in 2002 a dimensional correlation for the net entrainment rate due to the bubble break up with accuracy up to a constant that has to be derived from experiments

$$(\rho w)_{23\_boiling\_film} + (\rho w)_{23\_adiabat} - (\rho w)_{32}$$
$$= const \frac{\dot{q}''_{w2}}{\Delta h \rho''} \exp\left[-\frac{\delta_2 \sqrt{\tau_{2w}/\rho_2}}{158.7 (\eta''/\eta')^{2.66} 30 w_2} \right]. \qquad (16.15.9)$$

We see here that the group $w_{1\_blow} = \dot{q}''_{w2} / (\rho_1 \Delta h)$ is used correctly but again for adiabatic flow the asymptotic is not correct.

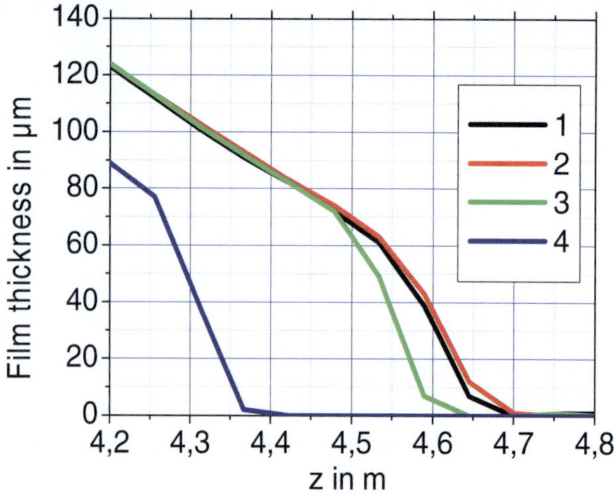

Fig. 16.15.45 Film thickness as a function of the axial coordinate for conditions of Bennett 5253 experiment. (1) Entrainment, deposition like in adiabatic flow; (2) Like (1) with deposition multiplied by ratio of the gas turbulent kinetic energy to the equilibrium turbulent kinetic energy; (3) Like (1) + (2) with suppression of the deposition due to the vapor blow from the film; Like (1) + (2) + (3) with entrainment enhancement due to the vapor blow through the film; Experimental position of the dry out about 4.4 m.

Now, we demonstrate the effect of the above-discussed phenomena again on the *Bennett* experiment as given in Fig. 16.15.45. Using f of the order of 0.02 and repeating the computations for other experiments as given in Table 16.15.22 we see that the position of the dry out of the film can be predicted with mean error of 0.41 m.

Table 16.15.22 Position of predicted and computed film dry out for some *Bennett* heated-pipe experiments (Mean error 0.41 m)

Bennett exp. ID	z_{DO} exp. ± 0.076	z_{DO} IVA	Δ
5253	3.89	4.31	0.42
5293	3.89	4.42	0.53
5332	4.39	3.37	–1.02
5380	3.81	3.7	–0.11
5407	2.79	3.02	0.23
5424	2.64	2.94	0.30
5442	2.79	3.27	0.48
5456	2.79	2.94	0.15

Conclusion: Important elements of the three-fluid models are developed based on adiabatic flows and are associated with considerable uncertainty. One can not expect to reach a better description of the dry-out performance only improving the grid models without substation reduction of those uncertainties for boiling flows.

The mean error of the 1D prediction of the location of the dry out is about 40 cm due to the uncertainties in the involved constitutive models. These uncertainties remain also in 3D predictions. Therefore, more effort is necessary to increase the accuracy of the involved correlation, to derive a complete set of source terms for generation of turbulence in all flow pattern, to accomplish the right coupling between deposition and local degree of turbulence, and finally to derive appropriate mechanistic criterion for identification of dry out.

16.15.4.4.5 Residual film thickness at DO?

Using correlation for entrainment and deposition gained from adiabatic experiments results obviously to overprediction of the deposition and underprediction of the entrainment. The results of such computations are always predicting some film thickness at places where the wall has to be dry. This led some authors to introduce some critical film thickness depending on the local parameter as already mentioned in the introduction. Regarding the experimental evidence: At the dry out *Milashenko* et al. 1989 experimentally observed a residual film mass flow rate of 0.02 kg/s. *Shiralkar* and *Lahey* (1973) reported finite film flow rates at the location of the CHF appearance. Regarding the dry-out film thickness *Groeneveld* 2001 reported that in all of his dry-out experiments (more then 30 years experimental research) a complete drying of the film was observed that contradicts *Milashenko* et al. 1989.

Conclusion: The influence of the boiling film on enhancing of the entrainment and suppression of the deposition has to be taken in any case into account. Then, the film nucleate boiling has to be exanimate for departure from nucleate boiling. If there are no local conditions for DNB then the DO requires dry film. This logic allows finite film thickness and boiling crisis simultaneously in cases of DNB.

16.15.4.4.6 Amount of the liquid in the core at the onset of film flow

Some authors reported that at the onset of the film flow there is an immediate amount of entrained liquid inside the gas core; see *Barbosa* et al. (2002). This is still source of uncertainty because the dry out point depends on such information too.

16.15.5 Outlook – towards the fine-resolution analysis

Associating large flow cross-sections in one subchannel as still practiced worldwide naturally limits the prediction accuracy. Therefore, future analyses have to concentrate on much finer resolution of the processes as demonstrated in Fig. 16.15.46.

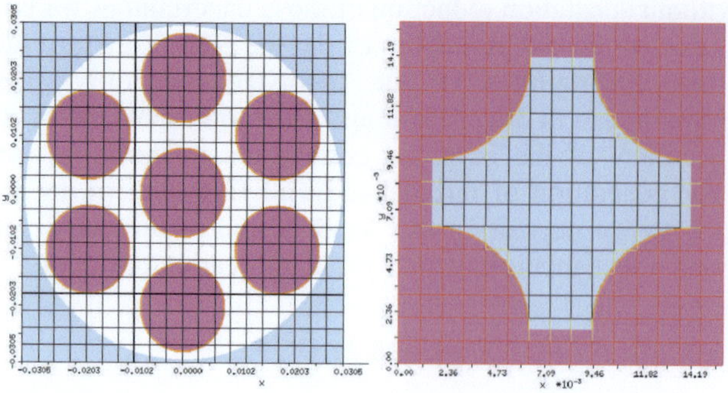

Fig. 16.15.46 Abandoning of the subchannel analyses – a challenge for the near future

The challenges in this area are associated with the constitutive relationships. Empirical correlations that are gained by averaging values across pipe cross-section are not valid for the subscale and have to be systematically replaced by new ones.

16.15.6 Conclusions

This work is a demonstration of the achievable accuracy that can be obtained by simulating steady-state and transient boiling in rod bundles by IVA computer code crossing all known three-fluid flow pattern and heat-transfer mechanisms and using large-scale discretization. 333 three-dimensional tests in bundles with 1, 16, 25, 36 heated rods, 7 different bundles with 64 heated rods from different laboratories are simulated: 273 3D experiments on 6 bundles for CHF, 54 3D experiments on 7 bundles for void fraction, 2 3D experiments in a bundles for transients and 4 1D experiments in a subchannel for transient. The mass flow rates and the pressure varied from 3 to 2000 kg/(m² s) and from 1 bar to 200 bar, respectively. The subcoolings were less than 140 K and the thermal power varies from some tenths of a kW to 7 MW. The comparison revealed the capability of IVA to adequately simulate flow boiling processes in bundles in variety of flow regimes besides the already existing very large verification data base of IVA for many other processes.

Simultaneously, this work was also a source of inspiration for improving the accuracy of the mathematical description of such processes in the future. This comprehensive analysis shows that the scale of spatial resolution calls for a specific set of constitutive relations. Examples are given for the effective interfacial drag coefficients. Comparison with the results of other authors was made and discussion was provided. In addition, by using different spatial resolution was demonstrated that the turbulent void and droplets transport have to be described much more accurately in distributed parameters to provide a universal way of prediction of void and droplet dispersion with such important consequences like accurate void and dry-out prediction.

Appendix 16.15.1 Some relevant constitutive relationship addressed in this analysis

IVA computer code has absorbed in the last 25 years hundreds of constitutive relationships for variety of processes as described by *Kolev* (2004). We address here only those that are subject of modification influencing the void fraction prediction in bundles. Note that for the computation of the drag coefficients the procedure described in *Kolev* (vol. 1, 2004). Chapter 2 was initially used.

1. Transition between bubbly and slug flow: If $D_1 > 6\lambda_{RT}$ $\alpha_{1,B-Ch} = 0.25$, else if $D_1 < \lambda_{RT}$ $\alpha_{1,B-Ch} = 0.3$, else $\alpha_{1,B-Ch} = 0.3 - 0.01(D_1/\lambda_{RT} - 1)$, where λ_{RT} is the *Raleigh–Taylor* wavelength defined as follows $\lambda_{RT} = \sqrt{\sigma_2/(g\Delta\rho_{21})}$. For the cases considered here the *Mishima* and *Ishii* (1984) correlation for transition from slug into churn turbulent flow $\alpha_{1,slug\ to\ churn} = 1 - 0.813\left\{\frac{(C_0 - 1)\ |j| + 0.35\ V_{TB}}{|j| + 0.75\ V_{TB} b_1}\right\}^{0.75}$ does predict transition to from bubble to slug but not to churn turbulent flow. Here the drift flux distribution coefficient for slug flow is $C_0 = 1.2$, the slug (*Taylor* bubble) raising velocity is $V_{TB} = \sqrt{\frac{\rho_2 - \rho_1}{\rho_2} g D_h}$, the mixture volumetric flux is $j = \alpha_1 V_1 + (1-\alpha_1) V_2$, and $b_1 = \left(\frac{\rho_2 - \rho_1}{\eta_2^2/\rho_2} g D_h^3\right)^{1/18}$.

2. Drag force for distorted bubbles: Nonmodified *Ishii* and *Chawla* (1979)

$$c_{21}^d = \frac{2}{3}(D_1/\lambda_{RT})\left(\frac{1+17.67 f^{6/7}}{18.67 f}\right)^2; \; f = (1-\alpha_d)^{1.5}.$$

3. Drag force for slug flow: None modified *Ishii* and *Chawla* (1979)

$$c_{21}^d = 9.8(1-\alpha_1)^3.$$

4. Drag force for distorted bubbles: Modified

$$c_{21}^d = 0.01\frac{2}{3}(D_1/\lambda_{RT})\left(\frac{1+17.67 f^{6/7}}{18.67 f}\right)^2.$$

5. Drag force for slug flow modified: Here we use the expression derived for cup bubbles instead for slugs multiplied by 0.1,

$$c_{21}^{d} = 0.1\frac{8}{3}(1-\alpha_1)^2 .$$

6. Instead of the primarily used model by *Kataoka* et al. (1983), *Bertodano* et al. (1998) model for entrainment was used here. It was verified for low pressure p = 1.2 bar and 10 < $\rho_2 w_2 4\delta_{2F} / \eta_2$ < 9700, 2.5 < $\alpha_1 \rho_1 V_1 D_h / \eta_1$ < 17.e4, 3 < $\rho_1 \left(\alpha_1 V_1\right)^2 D_{3E} / \sigma$ < 20. The *Bertodano* et al. correlation gives the same results as the *Kataoka* et al. correlation for low pressure and relative low mass flow rates but better agreement with the data for high pressure and large gas mass flow rates.

7. The diffusion droplet deposition rate was computed using the correlation by *Paleev*'s *and Filipovich* (1966) slightly modified by *Kataoka* and *Ishii* (1983). *Kataoka* and *Ishii* provided verification based on data in the following region 277 < $\rho_2 (1-\alpha_1)\left|w_{23}\right| D_h / \eta_2$ < 5041, 1414 < $\frac{\rho_1 (\alpha_1 w_1)^2 D_h}{\sigma_2}\left(\frac{\rho_2-\rho_1}{\rho_1}\right)^{1/3}$ < 9602, 0.0095 < D_h < 0.032.

Nomenclature

c_{21}^{d}	bubble drag coefficient due to liquid action, dimensionless
D_1	bubble diameter, m
g	gravitational acceleration, m/s²
V_l	velocity of field *l*, m/s
α_l	local volume fractions of the fields *l*, dimensionless
λ_{RT}	*Rayleigh-Taylor* wavelength, m
ρ_1	gas density, kg/m³
ρ_2	liquid density, kg/m³
ρw	mixture mass flow rate, kg/(m² s)
σ_2	viscous tension, N/m
η_1	dynamic gas viscosity, kg/(m s)
η_2	dynamic liquid viscosity, kg/(m s)

References

Akiyama, Y., Hori, K., Miyazaki, K., Nishioka, H., Sugiyama, S.: Evaluation of the bundle void fraction measurement data of PWR fuel assembly. In: The 3rd JSME/ASME Joint International Conference on Nuclear Engineering, Kyoto International Conference Hall, Kyoto, Japan, April 23-27, vol. 1, pp. 211–216 (1995)

Akiyama, Y., Hori, K., Miyazaki, K., Nishioka, H., Sugiyama, S.: Pressurized water reactor fuel assembly subchannel void measurement. Nucl. Technol. 112, 412–421 (2005)

Anglart, H., Nylund, O.: CFD application to prediction of void distribution in two pahas bubbly flow in rod bundles. Nucl. Eng. Des. 163, 81–98 (1996)

Anklama, T.M., Miller, R.F.: Void fraction under high pressure, low flow conditions in rod bundle geometry. Nucl. Eng. Des. 75, 99–108 (1982)

Aounalla, Y., Coddington, P.: Assessment of VIPRE-02 void fraction prediction against NUPEC experimental BWR data. In: 9th International Topical Meeting on Nuclear Thermal Hydraulics (NUTETH-9) San Francisco, California (October 3-8 , 1999)

Barbosa, J.R., Hewitt, G.F., König, G., Richardson, S.M.: Liquid entrainment, droplet concentration and pressure gradient at the onset of annular flow in vertical pipe. Int. J. Multiphase Flow 28, 943–961 (2002)

Bennett, A.W., et al.: Heat transfer to steam-water mixtures flowing in uniformly heated tubes in which the critical heat flux has been exceeded, AERE-R5373 (1967b)

Chisholm, D.: Two-phase flow in pipelines and heat exchanger, p. 110. George Godwin, London and New York (1983)

Doroschuk, V.E., Levitan, L.L.: Investigation of droplet deposition from annular dispersed steam-water flow core and wall liquid film. Therm. Phys. High Temp. 9, 591–596 (1971)

Fuel Economic and Reliable, AREVA, France (2004), see also in Adv. Nucl. Power No 3, The Magazine of Framatome ANP (January 2002)

Fleiss, Knapp, Which: STS-17.2, STS-17.1: Messung der Siedeübergangsleistungen und Druckverluste am ATRIUM-10-9Q-Testbündel, e321/92/43, e321/92/44, Karlstein, proprietary (1992)

Friedel, L.: New friction pressure drop correlations for upward, horizontal, and downward two-phase pipe flow. Presented at the HTFS Symposium, Oxford, September 1979 (Hoechst AG Reference No. 372217/24 698) (1979)

Groeneveld, D.C.: Private communication (2001)

Guguchkin, V.V., Nigmatulin, B.I., Markovich, E.E., Vasilev, N.I., Arestenko, Y.P., Ivanovskaya, V.I.: Peculiarities in droplet movement and bubbles in wall region. Theses of a report Presented at VII All-Union Conf. an Two Phase Flow in Power Machines and Apparatus, Leningrad, vol. 3, pp. 316–317 (1995)

Hori, K., Miyazaki, K., Kurosu, T., Sugiyama, S., Matsumoto, J., Akiyama, Y.: In bundle void fraction measurement of PWR fuel assembly. In: Proceedings of ICONE 2, vol. 1, pp. 69–76 (1993)

Hori, K., Akiyama, Y., Miyazaki, K., Kurosu, T., Sugiyama, S.: Transient void fraction measurement in a single Channel simulating one channel for a PWR Fuel Assembly. In: 10th Proceeding of Nuclear Thermal-Hydraulics, ANS Winter Meeting, pp. 56–68 (1994)

Hori, K., Akiyama, Y., Miyazaki, K., Kurosu, T., Sugiyama, S.: Void fraction in a single channel simulating one subchannel of a PWR fuel assembly. In: Proceedings of the First International Symposium an Two-Phase Flow Modelling and Experimentation (1995)

Hori, K., Akiyama, Y., Miyazaki, K., Nishioka, H., Takeda, N.: Total evaluation of in bundle void fraction measurement test of PWR fuel assembly. In: Proceedings of ICONE-4, vol 1, Part B, pp. 801–811 (1996)

Hoyer, N., Stepniewski, M.: Dry out prediction in bundle geometry. In: 9th Int. Top. Meeting on Nuclear Reactor Thermal Hydraulics (NURETH-9), San Francisco, California, October 3-8 (1999)

Hughes, E.D., Paulsen, M.P., Agee, L.: A drift-flux model of two-phase flow for RETRAN. Nucl. Technol. 54, 410–420 (1981)

Inoue, A., Futakuchi, M., Yagi, M., Kurosu, T., Mitsutake, T., Morooka, S.: Void fraction distribution in a boiling water reactor fuel assembly and the evaluation of subchannel analysis codes. Nucl. Technol. 112(3), 388–400 (1995)

Inoue, A., Kurosu, T., Oaki, T., Futaguchi, M., Yagi, M., Morooka, S., Hoshide, A., Ishizuka, T., Yoshimura, K., Yoshida, H., Mitsutake, T., Abe, N., Kozima, A., Nakajima, T., Nakamura, S., Sawa, T.: Void measurement of BWR fuel assembly. J. At. Energy Soc. Japan 37(8), 710–720 (1995) (in Japanese)

Inoue, A., Kurosu, T., Oaki, T., Yagi, M., Mitsutake, T., Morooka, S.: Void fraction distribution in boiling water reactor fuel assembly and evaluation of subchannel code. J. Nucl. Sci. Technol. 32(7), 629–640 (1995)

Ishii, M., Chawla, T.C.: Local drag laws in dispersed two-phase flow, NUREG/CR-1230, ANL-79-105 (December 1979)

Kataoka, I., Ishii, M., Mishima, K.: Trans. ASME 5, 230–238 (1983)

Kataoka, I., Ishii, M.: Entrainment and deposition rates of droplets in annular two-phase flow. In: Mori, Y., Yang, W.-J. (eds.) ASME-JSME Thermal Engineering Joint Conference Proceedings, Honolulu, Hawaii, vol. 1, pp. 69–80 (1983)

Kawara, Z., Kataoka, I., Serizawa, A., Ko, Y.J., Takahashi, O.: Analysis of forced convective CHF based on two-fluid and three-fluid model. In: Heat Transfer 1998, Proc. of the 11th IHTC, Kyongju, Korea, August 23-28, vol. 2, pp. 103–108 (1998)

Kelly, J.M., Stewart, C.W., Cuta, J.M.: VIPRE-02 – A two-fluid thermal-hydraulics code for reactor core and vessel analysis: Mathematical modelling and solution methods. Nucl. Technol. 100, 246–259 (1992)

Kodama, S., Kataoka, I.: Study on analytical prediction of forced convective CHF in the wide range of quality. In: Proceedings of ICONE10, 10TH International Conference on Nuclear Engineering Arlington, VA, USA, Paper nr ICONE10-22128 (April 14-18, 2002)

Kolev, N.I.: Transient three-phase three-component non-equilibrium non-homogeneous flow described by 1D-three-velocity fields slip model. In: Proc. Specialists Meeting on Small Break LOCA Analysis in LWRs, Pisa, vol 3, June 23-27, pp 363–378 (1985a)

Kolev, N.I.: Transiente Dreiphasen Dreikomponenten Strömung, Teil 2: Eindimensionales Schlupfmodell Vergleich Theorie experiment. KfK 3926 (August 1985b)

Kolev, N.I.: Simple analytical U-tube benchmarks appropriate for testing of pipe-network computer codes. In: Proceedings of ICONE10: Tenth International Conference on Nuclear Engineering – Arlington, Virginia, Washington, D.C., USA ICONE10-22711 (April 14-18, 2002a)

Kolev, N.I.: 3D-Multiphase-Fluid-Dynamik: Reduzierung der Dampffeuchte am RDB-Austritt der Siedewasserreaktoren, Betriebsleitertagung, Geneve (November 27-29, 2002b); in English: Controlling the moisture content in the steam by using nozzles in BWR's. In: Elevent International Conference on Nuclear Engineering, ICONE 11, Keio Plaza Inter-Continental, Shinjuku, Tokyo, Japan, ICONE11-36306 (April 20-23, 2003)

Kolev, N.I.: Multiphase Flow Dynamics, 2nd edn., with 114 Figures and CD-ROM. Fundamentals, vol. 1. Springer, Berlin (2002, 2004), `http://www.springeronline.com/east/3-540-22106-9`, ISBN 3-540-2206-0

Kolev, N.I.: Multiphase Flow Dynamics, 2nd edn., with 81 Figures. Thermal and mechanical interactions, vol. 2. Springer, Berlin, `http://www.springeronline.com/east/3-540-22107-7`, ISBN 3-540-22107-7

Kolev, N.I.: 25-Years three-fluid modeling-experience: successes and limits. In: Proceedings of ICONE-12 2004, Arlington VA, USA, Paper ICONE12-49054 (April 25-29, 2004); Also in 4th ECCOMAS (European Congress on Computational Methods in Applied Sciences and Engineering, Finland (July 24-28, 2004); Also in 6th WCCM (World Congress of Computational Mechanics), Beijin (September 5-10, 2004); In extended form in 6th International Topical Meeting on Nuclear Reactor Thermal Hydraulics, Operations and Safety, Nara-Ken New Public Hall, Nara, Japan, NUTHOS6-N6P380 (October 4-8, 2004)

Kolev, N.I.: IVA Simulations of exercises 1 of phase 1 and 2 of the OECD/NRC benchmark based on NUPEC BWR full-size fine-mesh bundle tests (preliminary). In: Second Workshop on OECD/NRC Benchmark based on NUPEC BWR Full-size Fine-mesh Bundle Tests (BFBT), State College, PA, USA (June 27-29, 2005a)

Kolev, N.I.: Do we have appropriate constitutive sets for sub-channel and fine-resolution 3D-analyses of two-phase flows in rod bundles? In: Mathematics and Computation, Supercomputing, Reactor Physics and Nuclear and Biological Applications Palais des Papes, Avignon, France, September 12-15, on CD-ROM. American Nuclear Society, LaGrange Park (2005b)

Kolev, N.I.: Flow boiling in bundles: Comparison of IVA predictions for 153 experiments for variety of bundles, flow regimes including DO, steady state and transients. In: The 11th International Topical Meeting on Nuclear Reactor Thermal-Hydraulics (NURETH-11) Log Number: 547, Popes Palace Conference Center, Avignon, France, October 2-6 (2005c)

Levy, S.: Int. J. Heat Mass. Transfer 10, 351–365 (1967)

Lopez de Bertodano, M.A., Assad, A., Beus, S.: Entrainment rate of droplets in the ripple-annular regime for small vertical ducts. In: CD Proceedings of the Third International Conference on Multiphase Flow, ICMF 1998, Lyon, France (June 8-12, 1988)

Lucchini, F., Marinelli, V.: Experimental data on burn-out in simulated BWR fuel bundle. Nucl. Eng. Des. 31, 371–378 (1974)

Macian, R., Aounallah, Y., Coddington, P., Stangroom, P.: Assessment of RETRAN-3D and VIPRE-02 void prediction against experimental transient data. In: Proc. of ICONE 8, 8th Int. Conf. on Nuclear Engineering, Baltimore, MD, USA, April 2-6 (2000)

Milashenko, V.I., Nigmatulin, B.I., Petukhov, V.V., Trubkin, N.I.: Burnout and distribution of liquid in evaporative channels of various lengths. Int. J. Multiphase Flow 15(3), 393–402 (1989)

Mishima, K., Ishii, M.: Flow regime transition criteria for upward two-phase flow in vertical tubes. Int. J. Heat Mass Transfer 27(5), 723–737 (1984)

Morooka, S., Inoue, A., Oishi, M., Aoki, T., Nagaoka, K., Yoshida, H.: In-bundle void measurement of BWR fuel assembly by X-ray CT Scanner. In: The 1st JSME/ASME Joint International Conference on Nuclear Engineering, pp. 237–243 (1991)

Naitoh, M., Iwashita, T., Ando, Y.: Boiling transition analysis code for LWR fuel bundles in the IMPACT project. In: 9th International Topical Meeting on Nuclear Thermal Hydraulics (NUTETH-9) San Francisco, California (October 3-8, 1999)

Neykov, B., Aydogan, F., Hochreiter, L., Ivanov, K., Utsuno, H., Fumio, K., Sartori, E., Martin, M.: NUPEC BWR full-size fine-mesh bundle test (BFBT) nenchmark, vol 1: Specifications, US NRC, OECD Nuclear Energy Agency, NEA/NSC/DOC, p. 5 (November 2005)

Neykov, B., Ivanov, K., Hochreiter, L.E., Avramova, M.: Summary of comparison and analysis of submitted resilts for exercise I-1. In: 3rd Workshop on OECD/NRC BFBT Benchmark, Pisa, Italy, April 27-29 (2006)

Nylund, O., et al.: Hydrodynamic and heat transfer measurements on a full scale simulated 36-rod Marviken fuel element with uniform heat flux distribution. In: FRIGG-2, AB Atomenergi, Stockholm, Sweden, ASEA, VÄSTERAS, R-4-447/RTL-1007 (1968)

Nylund, O., et al.: Hydrodynamic and heat transfer measurements on a full scale simulated 36-rod BHWR fuel element with non-uniform axial and radial heat flux distribution. In: FRIGG-4, AB Atomenergi, Stockholm, Sweden, ASEA, VÄSTERAS,R4-502/RL-1253 (1970)

Paleev II, Filipovich, B.S.: Phenomena of liquid transfer in two-phase dispersed annular flow. Int. J. Heat Mass Transfer 9, 1089 (1966)

Rehme, K.: Laminar Strömung in Stabbündeln. Chemie-Ing-Tech 43, 962–966 (1971)

Rehme, K.: Pressure drop performance of rod bundles in hexagonal arrangements. Int. J Heat Mass Transfer 15, 2499–2517 (1972)

Rehme, K.: Simple method of predicting friction factors of turbulent flow in non-circular channels. Int. J. Heat Mass Transfer 16, 933–950 (1973)

Rehme, K.: The structure of turbulent flow trough a wall subchannel of rods bundle. Nucl. Eng. Des. 45, 311–323 (1978)

Roloff-Bock, I.: 2D-grid generator for heterogeneous porous structures in structured Cartesian coordinates, Framatome ANP, proprietary (2005)

Saha, P., Zuber, N.: Proc. Int. Heat Transfer Conf., Tokyo, Paper 134.7 (1974)

Schäfer, H., Beisiegel, A.: Feasibility of stability tests under natural circulation conditions in the Karlstein test loop, Siemens technical report E32/92/e14a, propriatary (March 17, 1992)

Shiralkar, B.S., Lahey, R.T.: The effect of obstacles on a liquid film. J. Heat Tansfer, Trans. ASME 95, 528–533 (1973)

Spierling, H.: Regional void fraction in ATRIUM 10XP at various peaking patterns, Framatome-ANP report, FGT, A1C-1311669-1, proprietary (September 24, 2002)

Utsuno, H., Ishida, N., Masuhara, Y., Kasahara, F.: Assessment of boiling transition analysis code against data from NUPEC BWR full-size fine-mesh bundle tests. In: The 6th International Conference on Nuclear Thermal Hydraulics, Operations and Safety (NUTHOS-6) Nara, Japan, Paper ID. N6P041 (October 4-8, 2004)

Wang, G., Hochreiter, L.E., Sung, Y., Karoutas, Z.E.: VIPRE code void fraction prediction assessment using FRIGG loop data. In: 13th Int. Conf. On Nuclear Engineering, Beijing, China, May 16-20 (2005)

Windecker, G., Anglart, H.: Phase distribution in BWR assembly and evaluation of multidimensional multi-field model. In: 9th Int. Top. Meeting on Nuclear Reactor Thermohydraulics, NURETH-9, San Francisco, California (October 3-8, 1999)

Vogel, Bruch, Wang: SIEMENS Test Section 52 (DTS52) Description of experiments, KWZ BT23 1991 e 244, Erlangen, proprietary (August 19, 1991)

Yagi, M., Mitsutake, T., Morooka, S., Inoue, A.: Void fraction distribution in BWR fuel assembly and the evaluation of subchannel code. In: Ninokata, H., Aritomi, M. (eds.) Subchannel Analysis in Nuclear Reactors, Tokyo, Japan, pp. 141–167 (1992)

16.16 Film boiling

Film boiling on spheres, Fig. 16.16.1, is a basic heat-transfer mechanism in melt–water interaction processes. For this reason, we will address this topic in depth.

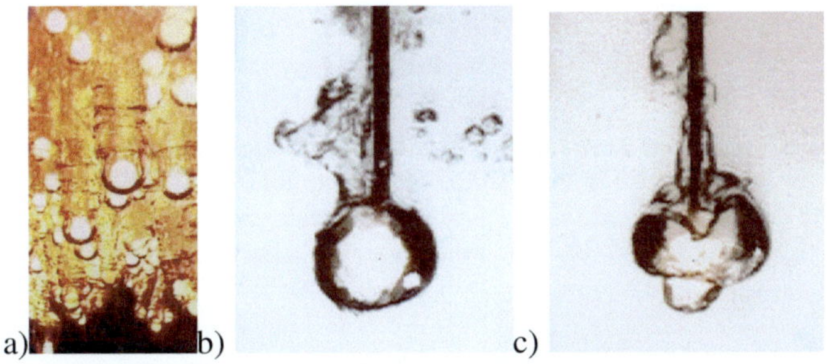

Fig. 16.16.1 a) Film boiling in a family of 2500 *K* spheres dropped into water. b-c) Film boiling on a single sphere in natural convection. With the courtesy of *L. Meyer*, FzK Karlsruhe, Germany

First we derived in Chapter 21 of Vol. 2 an analytical solution for film boiling heat transfer at a heated vertical surface, *Kolev* (1997a, c, 1998) and verified it with experimental data obtained at the Royal Institute of Technology in Sweden by *Okkonen* et al. (1996). The data are for vertical heated walls with $\Delta z = 1.5$ m. Water at atmospheric pressure and subcooling of $\Delta T_{sc} = 3$ to 42 K was used as a coolant. The wall superheating was considerable: $\Delta T_{sp} = 487$ to 1236 K. The heat-transfer coefficient was defined by *Okkonen* with respect to the wall–liquid temperature difference. This is recomputed here with respect to the wall-saturation temperature difference for comparison. Predicted values were obtained the theoretical result from Section 22.1 of this volume 2, see also in *Kolev* (1997b). A comparison of the predicted heat-transfer coefficients with 43 experimental points is given in Fig. 16.16.2. It is evident that all data predicted are inside the error band of $\pm 10\%$, with a mean-averaged error of less than $\pm 4\%$ when the modified theory expressed by Eq. (22.86) in Vol. 2 is used.

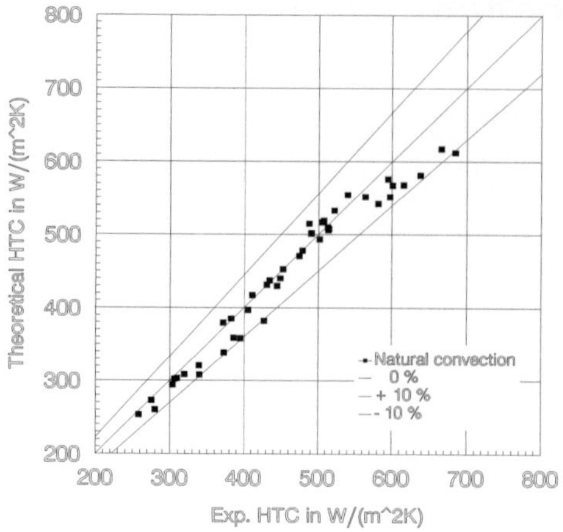

Fig. 16.16.2 Comparison between the film-boiling model and the data base of the Royal Institute of Technology (KTH Sweden), *Okkonen* et al. (1996). The *Lanzenberger* radiation model (1998) is incorporated. $\Delta z = 1.5m$ vertical wall. Water at atmospheric pressure and subcooling ΔT_{sc} = 3 to 42 K. Wall superheating ΔT_{sp} = 487 to 1236 K

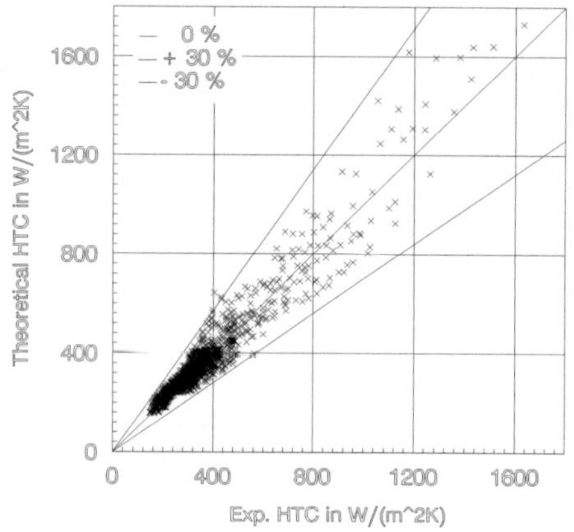

Fig. 16.16.3 Comparison between predicted and calculated heat-transfer coefficients for all data for spheres. Error band: ±30%

We then modify the plane solution to one appropriate for spherical geometry in Chapter 22 Vol. 2 and verify it with a large number of experimental data provided by *Liu* and *Theofanous* in 1995. The data consist of single- and two-phase flow cases. The single-phase data cover subcooling from 0 to 40 K, liquid velocity from 0 to 2.3 m/s, sphere superheating from 200 to 900 K and sphere diameter from 6 to 19 mm. The two-phase data are obtained for

(a) upward flow with a void fraction from 0.2 to 0.65, water velocity from 0.6 to 3.2 m/s, and steam velocity from 3 to 9 m/s, and
(b) downward flow with a void fraction from 0.7 to 0.95, water velocity from 1.9 to 6.5 m/s, and steam velocity from 1.1 to 9 m/s.

Figure 16.16.3 presents the comparison between the predicted and calculated heat-transfer coefficients for all data. The error band is found to be $\pm 30\%$.

References

Kolev, N.I.: Verification of the IVA4 film boiling model with the data base of Liu and Theofanous. In: Proceedings of OECD/CSNI Specialists Meeting on Fuel-Coolant Interactions (FCI), JAERI-Tokai Research Establishment, Japan (May 19-21 1997a)

Kolev, N.I.: IVA4 analysis of the FARO L14 experiment (ISP 39). In: International Seminar on Vapor Explosions and Explosive Eruptions (AMIGO-IMI), Aoba Kinen Kaikan of Tohoku University, Sendai-City, Japan (May 22-24, 1997b)

Kolev, N.I.: Film boiling: vertical plates. In: Proceedings of 4th World Conference on Experimental Heat Transfer, Fluid Mechanics and Thermodynamics EXHFT 4, Brussels, Belgium (June 2–6, 1997c)

Lanzenberger, K.: Thermal radiation in multiphase flow, KWU NA-T/98/E011. 1998 Karl Wirtz Award for Young Scientists: K. Lanzenberger, Thermal radiation in multiphase flow - Application to the severe accident scenario of molten fuel coolant interaction (MFCI), Siemens AG, Power Generation (KWU) (1998)

Liu, C., Theofanous, T.G.: Film boiling on spheres in single- and two-phase flows part 1: Experimental studies ANS Proceedings, Part 2: A Theoretical Study, National Heat Transfer Conference, Portland (August 1995)

Okkonen, T., et al.: Film boiling on a long vertical surface under high heat flux and water subcooling conditions. In: Proc. of the 31st Nat. Heat Transfer Conference, Houston, Texas (August 3-6, 1996)

16.17 Behavior of clouds of cold and very hot spheres in water

We now verify the capability of IVA5 to describe multiphase flows consisting of cold or hot spheres dropped into water. In this way, the mechanical and thermal interaction is checked without having the effect of the melt fragmentation. We use for this purpose QUEOS experiments performed at Forschungszentrum Karlsruhe, Germany. In the QUEOS facility, see Fig. 16.17.1 (left), 62 experiments have been performed. The first 12 are documented by *Meyer* and *Schumacher* (1996). Experiments with numbers 13 to 39 are presented by *Meyer* (1997). A comparison between experiments with cold and hot spheres, employing three different sphere types in each case in a wide (180 mm) and short jet was presented by *Meyer* (1996). The data of an experimental series with a smaller jet diameter (100 mm) and longer pours were published by *Meyer* (1998). Important measurements of void fraction in hot sphere three-phase flow are reported by *Meyer* and *Kuhn* (1999).

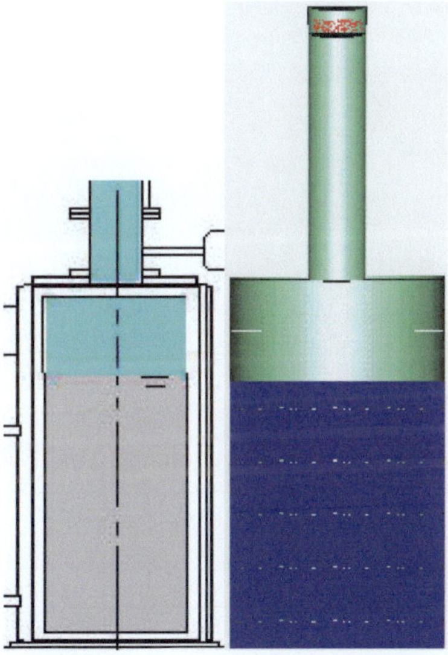

Fig. 16.17.1 Modeling of the QUEOS facility (left) by the IVA5 computer code (right)

We select from the 62 experiments three characteristic ones as listed in Table 16.17.1, and compare the predictions of IVA5 with the reported observations. The geometrical model for IVA5 is presented in Fig. 16.17.1 (right).

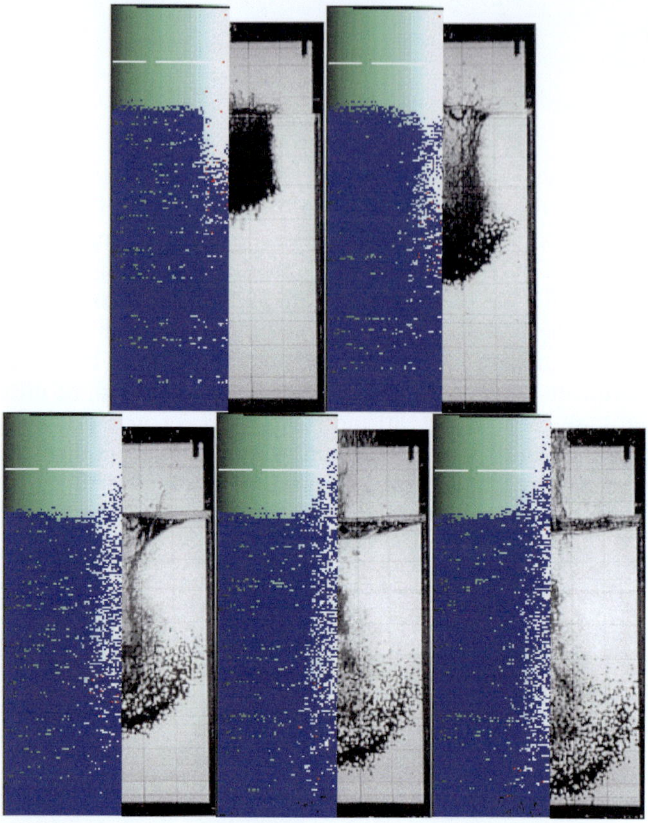

Fig. 16.17.2 Comparison between IVA5 predictions and the Q06 experiment for a cloud of cold solid zirconia spheres dropped into water: 0.62, 0.70, 0.78, 0.86, 0.94 s

Table 16.17.1. Summary of the QUEOS experimental conditions

No.	Material of spheres	D_3	Mass	Volume	Number	T_3	Pipe ID	T_2
		mm	kg	cm^3	-	K	mm	
06	ZrO_2	10	7	1900	2340	300	207 open	Sat
12	Mo	4.2	6.9	1183	16600	2300	207 open	Sat
47	ZrO_2	4.95	10	2600	25900	1833	100 open	Sat

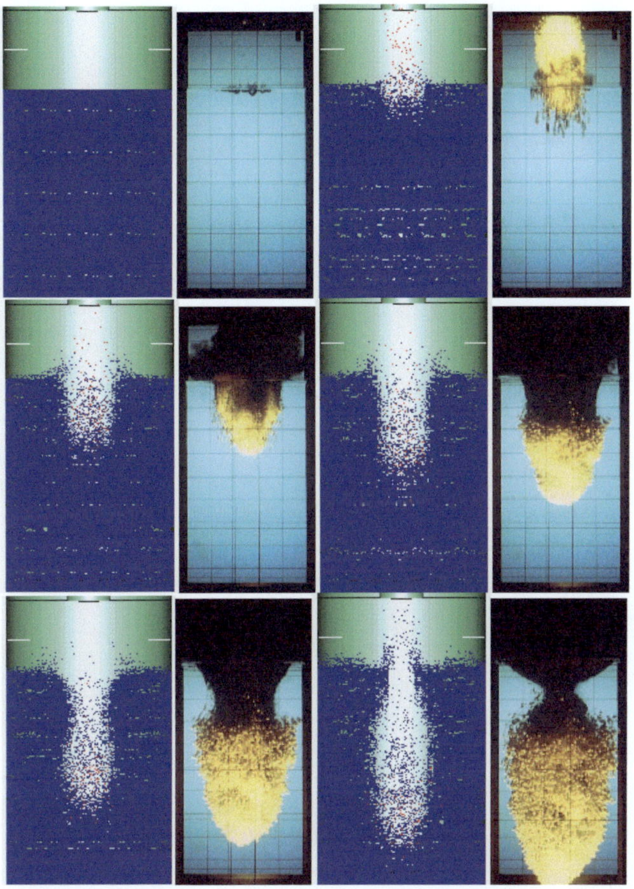

Fig. 16.17.3 An IVA5 prediction of the *Meyer* and *Schumacher* hot spheres-in-water experiment. Water volume fraction as a function of time: 0, 0.0238, 0.0734, 0.13, 0.1924, 0.2452 s. Experiment Q12: cloud of hot solid molybdenum spheres dropped into water

The particle penetration into the water and the corresponding water relocation are best illustrated using the experiment with No. 06 characterized by cold particles. Only the mechanical interaction is therefore modeled. The results are presented in Fig. 16.17.2. The time elapsing between the first particles touching the water surface and their touching the bottom is measured to be 0.4 s. The computed time was 0.397 s.

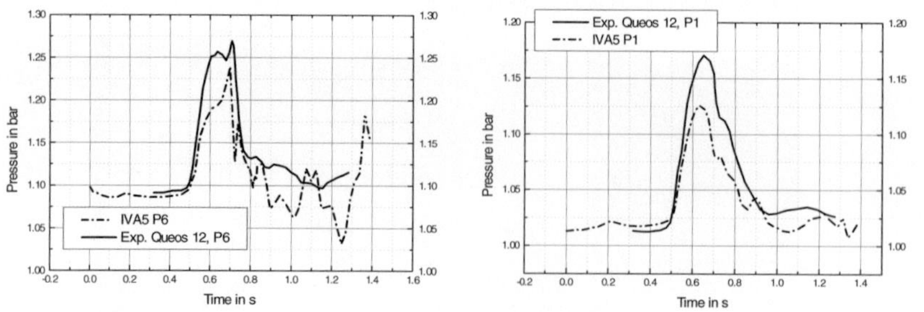

a) b)

Fig. 16.17.4 Pressure as a function of time at a) P6 position – underwater space; b) P1 position – gas space

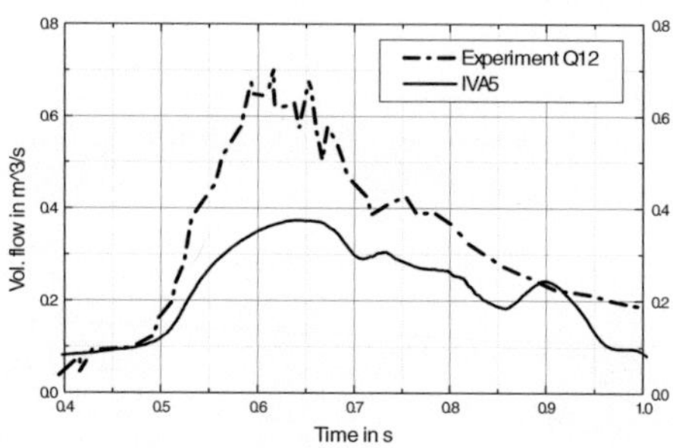

Fig. 16.17.5 Outlet volumetric flow as a function of time

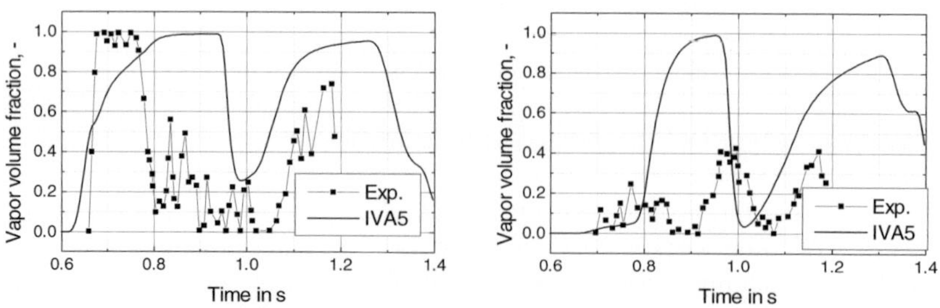

a) b)

Fig. 16.17.6 Void fraction as a function of time at a) the centerline 40 cm from the bottom; b) 15 cm from the centerline, 40 cm from the bottom as a function of time

Experiment No. 12 was performed with the same geometry but the spheres consisted of a very good heat conductor heated up to 2300 K, the highest temperature ever reached in this kind of experiment. The predicted water volume fractions are presented in Fig. 16.17.3. The photographs associated with this experiment are also shown in Fig. 16.17.3. A good agreement between the predicted and the experimental propagation velocity and level swell is obtained. Probably the lack of turbulence description makes the predicted radial expansion of the particles less than that measured.

The generated pressure as a function of time at two different positions is presented in Figs. 16.17.4a and b. The corresponding outlet steam flow is given in Fig. 16.17.5. IVA5 underpredicts the vapor generation in this case, which is consistent with the pressure under prediction in Figs. 16.17.4a and b. For this simulation the steady-state friction loss coefficient of 4.54 is used *Meyer* and *Schumacher* (1996). The appropriate modeling of this coefficient is very important and can be considerably improved for this case.

Void fraction measurements are extremely difficult in multiphase systems. In the QUEOS No. 47 experiment void fraction measurements are provided for points located 40 cm from the bottom at the symmetry axis and 15 cm from the symmetry axis without specifying the uncertainty band. The predicted and the measured results are presented in Figs. 16.17.6a and b. The secondary bubble release as observed by the experiment is predicted by IVA5. The gas front in the measurements on the centerline is very sharp. IVA5 predicts smooth origination of the gas front. The deviation between prediction and measurements for the secondary cycle increases.

Conclusions. The comparison presented in this chapter demonstrates that the main features of the mixing of cold and hot solid spheres with water are well predicted by IVA5.

One should keep in mind that *Lagrangian* presentation of the motion of each single particle would considerably improve the position prediction of the clouds of particles. For small-scale experiments this has already been implemented by several authors. However, with the computational technology currently available it is not yet possible to apply this technique to large industrial systems having hundreds of m^3 with 10^{10} particles per m^3. Instead, master particles are used to represent a number of real particles, and this procedure has been used since the original PIC codes *Crowe* et al. Replacing large number of particles with a single master particle also results in numerical diffusion.

References

Crowe, C.T., Stock, D.E., Sharma, M.P.: The particle-in-cell (PSI-cell) model for gas droplets flows. ASME, Trans., Series I-J. Fluids Eng. 99, 325–332 (1977)

Meyer, L., Schumacher, G.: QUEOS a simulation-experiment of the premixing phase of steam explosion with hot spheres in water base case experiments, Wissenschaftliche Berichte FZKA 5612, Forschungszentrum Karlsruhe Technik und Umwelt, Karlsruhe (April 1996)

Meyer, L.: The interaction of falling mass of hot spheres with water. In: ANS Proceedings 1996 National Heat Transfer Conference, Houston, Tx, August 3-6, vol. 9, pp. 105–114. American Nuclear Society, La Grange Park (1996)

Meyer, L.: QUEOS a simulation-experiment of the premixing phase of steam explosion with hot spheres in water, Results of the Second Test Series, Internal Report 32.21.02/INR1962/PSF3267, Forschungszentrum Karlsruhe Technik und Umwelt, Karlsruhe (July 1997)

Meyer, L.: QUEOS, an experimental investigation of premixing phase with hot spheres. Nucl. Eng. Des. 18 (1998)

Meyer, L., Kuhn, D.: The interaction of very hot particles falling into water. In: 2nd Int. Symp. on Two-Phase Flow Modeling and Experimentation, Pisa, Italy, May 23-25 (1999)

16.18 Experiments with dynamic fragmentation and coalescence

Experiments with real molten material that are prototypic nuclear reactor materials are very important in the course of the code validation for melt–water interactions. Now, we step forward to model four of the FARO tests performed by JRC Ispra in Italy. The purpose of this section is to verify the code for a class of processes called nonexplosive melt–water interaction. All elements of the code are addressed in such a demanding simulation, e.g., code architecture, numerical methods, constitutive models, etc. The simulation of the FARO test will reveal the capability of the code to handle such flows with simultaneous evaporation condensation, and fragmentation of all participating velocity fields, etc. Among these, there are some important constitutive models, which are addressed during the computation, and their performance within the overall system will be also tested.

16.18.1 L14 experiment

Geometrical model, initial and boundary conditions. Detailed information about this test is available in *Magallon* and *Leva* (1996), *Hohmann* et al. (1994), *Annunziato* et al. (1996), *Benuzzi* and *Magalon* (1994). The geometry of the TERMOS vessel used is reproduced from *Benuzzi* and *Magalon* (1994) in Fig. 16.18.1a. The initial and boundary conditions, and an analysis of the observed phenomena are presented in *Hohmann* et al. (1994), *Annunziato* et al. (1996). The water volume was 0.798 m^3 and the gas volume 0.761 (Fig. 16.18.1) + 0.5 (separator, connection lines) m^3 including the expansion volume. This means that the initial gas volume fraction was of the order of 0.612. 125 kg melt (80/20 w% UO_2/ w% ZrO_2) with initial temperature ~3073 K are dropped in 623 kg saturated water having 2.05 m depth. The gas (77 w% steam + 23 w% argon) was under 5.1 MPa initial pressure and 536 K initial temperature. The simulation was performed assuming an axisymmetric process in 2D geometry with ($12\times 64=768$, 648 in the vessel) cells.

Results. Figure 16.18.2a shows the predicted melt release as a function of time and space. The computed melt outlet velocity multiplied by the local melt volume fraction in the axis of the outlet nozzle is presented in Fig. 16.18.2b. In *Hohmann* et al. (1994) the outlet velocity was estimated approximately without feedback from pressure buildup and assuming a one-dimensional process and no flow separation in the nozzle. On the basis of such assumptions, the maximum of the predicted velocity in *Hohmann* et al. (1994) was 3 m/s. IVA5 predicts a maximum of about 3m/s. Figure 16.18.2b demonstrates the feedback effects of the pressure in the vessel on the outflow and the flow separation at the end of the discharge process.

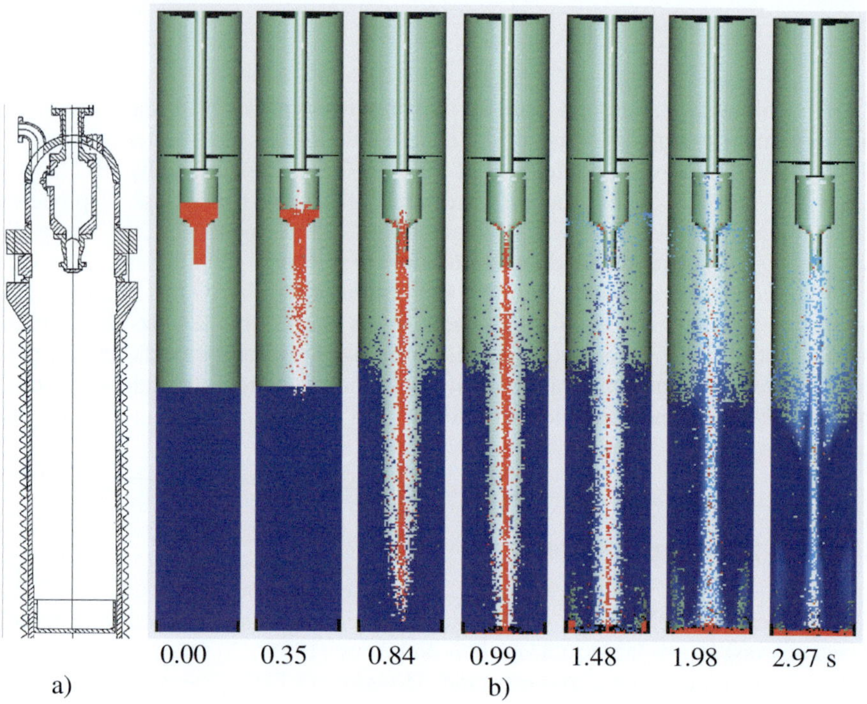

Fig. 16.18.1. a) IVA5 geometry model of the FARO L14 experiment. Number of cells: 768. Number of cells in the FARO vessel 648. b) Gas volume fraction as a function of space. Parameter: time

Melt falls down, interacts with water, and reaches the bottom. The processes of melt sloshing at the vessel bottom are presented in Fig. 16.18.3.

The experimentally observed leading edge position as a function of time is given in Fig. 16.18.4a. It is indicated by thermocouples placed at discrete positions in the vessel. The computed leading edge position is diffuse due to (a) fragmentation and (b) numerical diffusion. We take as a measure the easily recognizable interface between water and gas as seen in Fig. 16.18.1b. From Fig. 16.18.4a we see a good agreement between the IVA5 prediction and the measurements. Nevertheless, numerical diffusion associated with the first-order donor-cell method has to be reduced in future refinement of the IVA5 method.

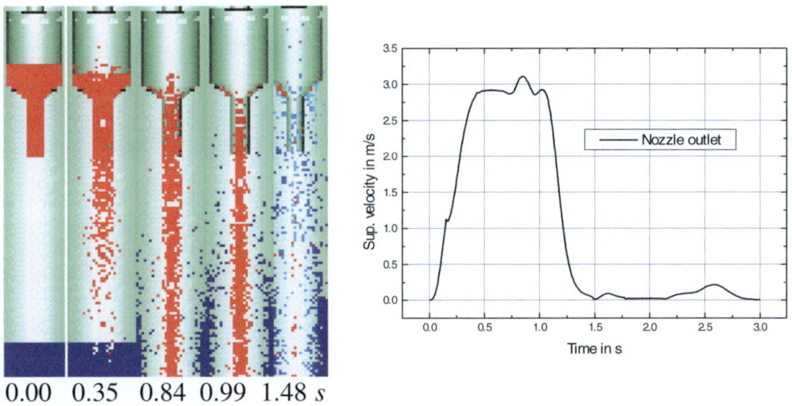

Fig. 16.18.2 a) Melt discharge as a function of time and space. b) Superficial velocity in the outlet nozzle as a function of time

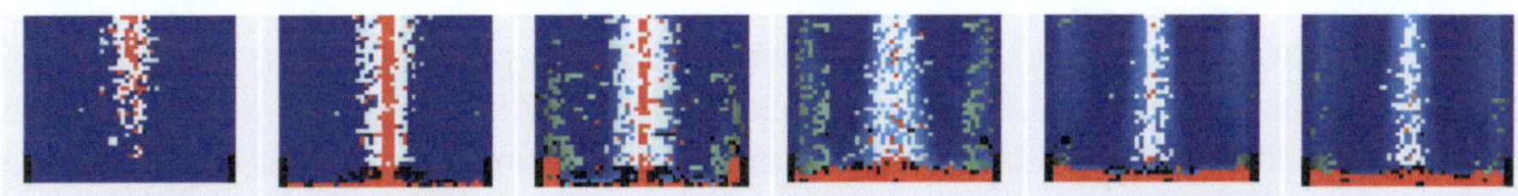

Fig. 16.18.3 Temperature of the melt presented only at places occupied by melt at the bottom as a function of space. Parameter – time: 0.84, 0.99, 1.48, 1.98, 2.51, 2.97 s

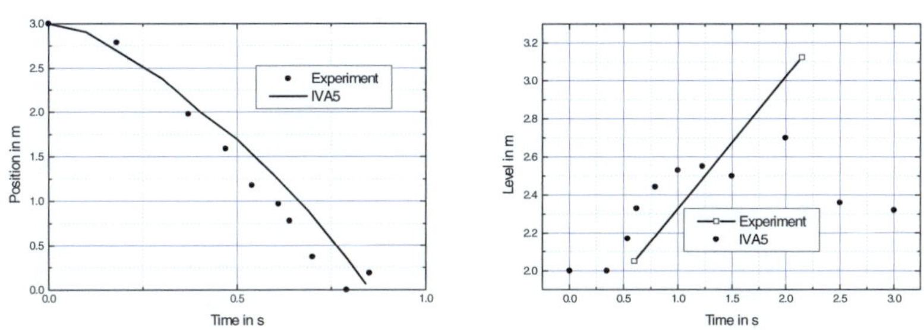

Fig. 16.18.4 a) Jet leading edge position in *m* versus time in s. b) Mixture level (IVA5: r = 0.18 m, volume fraction of water between 5 and 10 %) as a function of time

Figure 16.18.4b gives the water displacement in the course of the melt–water interaction. Discrete signals indicating the approximate position of the water level are available. The level in the computation was defined as the boundary with cells with less than or more than 5% water volume concentration. As seen from Fig. 16.18.4b the level is a very strong function of the radius and one signal curve is not representative. The estimated computed position for the level in accordance with the above criterion is given in Fig. 16.18.4b. In view of the many uncertainties associated with this comparison this prediction seems reasonable.

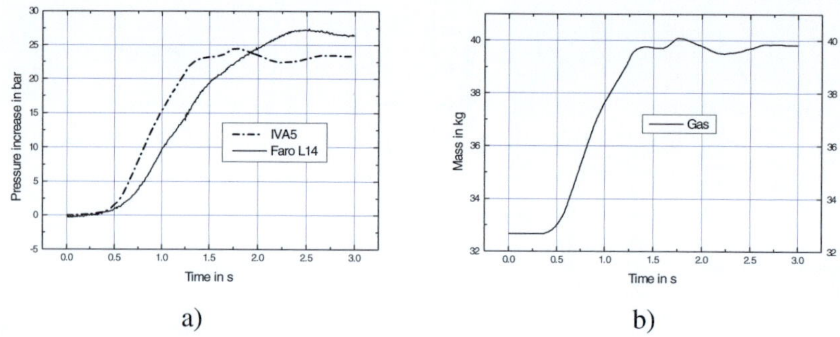

Fig. 16.18.5 a) Pressure increase as a function of time; b) Gas mass as a function of time

Figure 16.18.5a gives the most important information – the comparison between the predicted and the measured pressure as a function of time. The agreement is good. Comparing Figs. 16.18.5a and b we realize the proportionality between the evaporated mass and the pressure increase of the system. Presenting the pressure change with respect to the initial pressure as a function of the vapor mass change with respect to the initial vapor mass results in Fig. 16.4.1 in Section 16.4. It is a very interesting effect, as discussed in the introduction. The final turn off is due to the condensation at the end of the process.

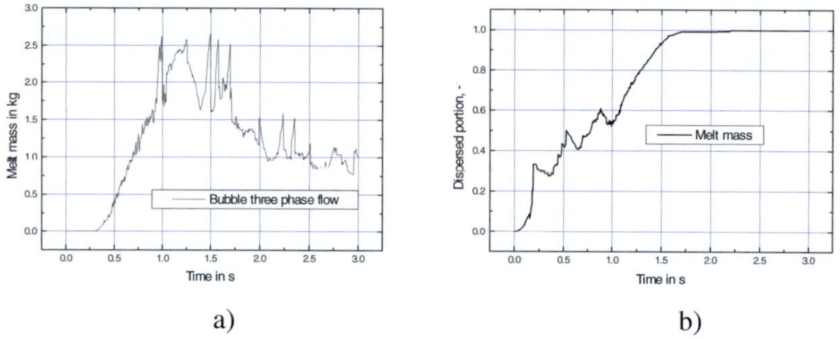

Fig. 16.18.6 a) Premixing melt mass in kg-melt mass being in dispersed melt-continuous water-gas bubble flow as a function of time in s; b) Part of the total melt being fragmented (dimensionless) as a function of time in s

Figure 16.18.6a shows the predicted mass of melt in the film boiling bubble flows. This is the most effective vapor production mechanism. From the comparison between Figs. 16.18.5a, 16.18.5b and 16.18.6a we see that the pressure increase in the first second is caused mainly by film boiling in three-phase flow. IVA5 predicts complete melt fragmentation after 1.7 s in this case, as indicated on Fig. 16.18.6b. The debris bed analysis after the experiment shows 16% nonfragmented melt. The reason for this disagreement is simple. IVA5 does not predict the freezing interaction with the cold bottom and allows the melt to slosh further in the lower head as presented in Fig. 16.18.3. In the computation this causes freezing of the fragments before building a cake at the bottom. The computed arrival time for the melt at the water surface is about 0.38 s, which is some 17% faster than the measured 0.46 s. The computed arrival of the melt at the bottom is about 0.862 s, which is some 4.2% faster than the measured 0.9 s. The maximum of the cumulative energy conservation error in this computation was below 0.4%. For more discussion on this issue see *Kolev* (1997a).

16.18.2 L20 and L24 experiments

The effect of the pressure on the nonexplosive melt–water interaction was studied in experiments L20 and L24 *Annununziato*, et al. (1996, 1997). They were performed in the same geometry with some minor changes of the initial conditions except the initial pressure. The initial pressure was 20 and 5 bar, respectively. Figure 16.18.7 shows the material relocation as a function of space 1 s after the beginning of the experiments. We see that the lower the pressure the larger the produced volume due to evaporation.

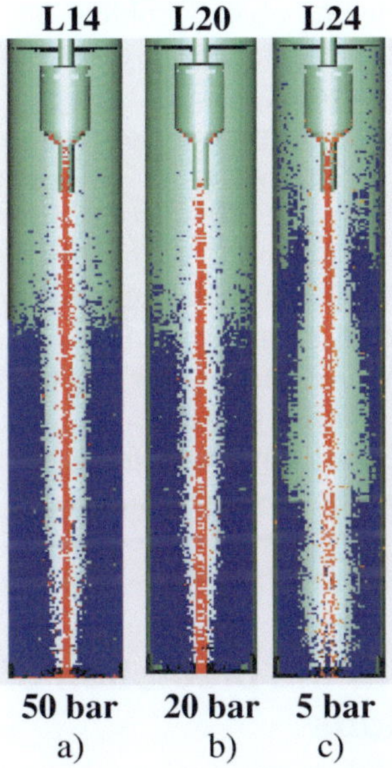

Fig. 16.18.7 The influence of the initial pressure on FCI. Material relocation after 1 s in the FARO experiments within the same geometry under virtually the same initial and boundary conditions except the initial pressure. Initial pressure: a) 50 bar; b) 20 bar; c) 5 bar

This causes much stronger material relocation if a gas volume is available for water depletion. This is an extremely important observation for the low-pressure melt–water interaction in real nuclear systems. Figures 16.18.5a, 16.18.8a and 16.18.8b present the comparison between the predicted and measured pressures as a function of time. We see an interesting tendency of IVA5 to slightly over predict the initial pressure change with time at 50bar, to predict a very good pressure change at 20 bar and to underpredict the pressure change at low pressure.

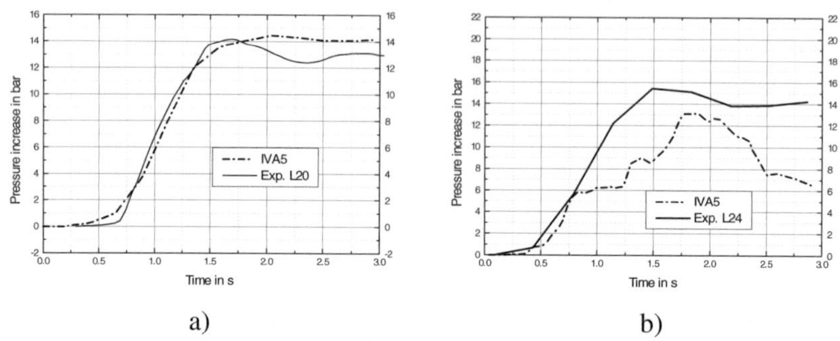

Fig. 16.18.8. a) Pressure as a function of time for the FARO L20 experiment. Initial pressure: 20 bar. b) Pressure as a function of time for the FARO L24 experiment. Initial pressure: 5 bar

Bearing in mind the complexity of the problem and the fact that many of the experimentally obtained empirical models are subject to known uncertainty, this agreement can be classified as good.

Note that *Yamano* et al. (1992) performed experiments with 20 kg of molten thermite penetrating 0.78 m³ of water at atmospheric conditions and initial relocation velocity 8 m/s. Increasing the pressure from 1 to 16 bar suppresses the explosion. *Mitchel* et al. (1981) reported a similar experiment with 54 kg of melt. Increasing the pressure to 10.9 bar suppresses the steam explosion. *Nelson* and *Duda* (1981, 1985) performed a small-scale droplet test in which increasing the external trigger pressure (impulse and energy) was required in order to induce an explosion by increasing the ambient pressure.

Thus, the increase in pressure has two effects on melt–water interaction: a) it decreases the intensity of the macroscopic material relocation and b) it increases the vapor film stability.

16.18.3 Uncertainty in the prediction of nonexplosive melt–water interactions

Results from applications of computer models are subject to uncertainty. This is basically due to model and parameter uncertainties. Model uncertainty arises in areas such as phenomenology, model formulation and numerical algorithms and is often partly due to intentional simplifications.

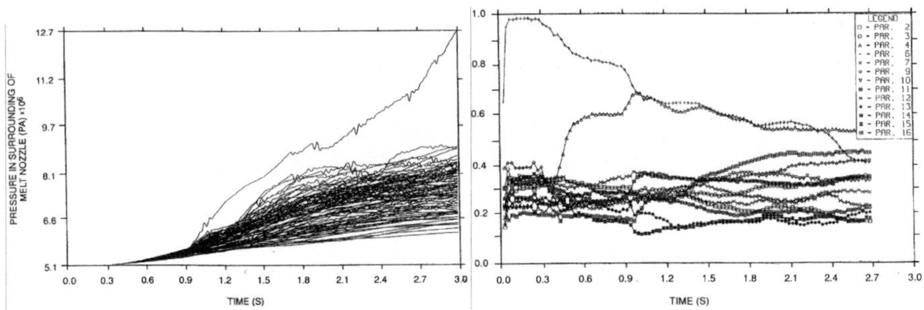

a) b)

Fig. 16.18.9 a) Pressure in the vessel as a function of time. The pressure time history is the most important information if the code is used for practical design purposes. 96 alternative time histories were obtained with 96 randomly selected vectors of parameter values. b) Sensitivity measures of the pressure with respect to each of the 13 considered uncertainties. The main contributions to the uncertainty of the pressure as a function of time come from the emissivity coefficient (No.6) and from the stable particle size (No. 4). The initial temperature (No. 16) exhibits growing influence over time. The influence of the latter two is in the opposite direction to the emissivity coefficient

An uncertainty analysis is to provide a quantitative statement about the combined influence of potentially important uncertainties on the results. Additionally, and often most important, it provides quantitative sensitivity statements that rank the uncertainties with respect to their contribution to model output uncertainty. In this way it helps to guide research, model development and measurement efforts in order to reduce model output uncertainty most effectively *Kolev* and *Hofer* (1996).

Uncertainty analyses for nonexplosive interactions using IVA4 computer code are reported by *Kolev* and *Hofer* (1996). The FARO L14 experiment was simulated at that time and the two main results are given in Figs. 16.18.9a and b. The lessons from this analysis are also applicable for IVA5 computer code. They are summarized in the conclusion section.

16.18.4 Conclusions

1. The local volume- and time-averaged multifield conservation equations in porous media presented in Volume 1, Chapters 1 and 5, as well as in *Kolev* (1995b, 1998, 1994a, 1994b, 1997b) and the concept of dynamic fragmentation and coalescence for all of the velocity fields described in Vol. 2 *Kolev* (1993) and implemented in IVA5 computer code *Kolev* (1995a) are powerful tools for simulation of such complicated multiphase flows with strongly nonlinear interfield interactions.
2. Due to the combined influence of the uncertainty of the initial melt temperature and of the model uncertainties for the analyzed FARO L14 nonexplosive melt–water interactions, prediction is only possible for the following:

(a) pressure increases within a (90%, 95%) tolerance interval of 24 bar width;
(b) gas temperatures within a (90%, 95%) tolerance interval of 50 K width;
(c) water temperatures within a (90%, 95%) tolerance interval of 30 K width;
(d) melt temperatures within a (90%, 95%) tolerance interval of 110 K width.

The uncertainties of the

(a) radiation emissivity coefficient of the melt surface;
(b) final particle size after fragmentation in the separate effect fragmentation models (attention: not included in the FARO experiment); and
(c) initial melt temperature;

are responsible for most of the uncertainty of the final results over most of the time.

3. The IVA5 computational model essentially gives a good prediction of the main characteristics of the FARO L14, L20 and L24 experiments.
4. The fragmentation and film boiling models, as described in this volume form a sound basis for nonexplosive melt–water interaction modeling.
5. Numerical diffusion associated with the first-order donor-cell method has to be reduced in future refinement of the IVA5 method – see Chapter 12 in Volume 1 for ideas for high-order numerical methods.
6. The modeling of the stagnation of packed solid particles at horizontal surfaces requires further attention.
7. The influence of the errors inevitably associated with the numerical integration of the 21 nonlinear conservation equations with strongly nonlinear source terms has to be monitored in practical applications.

16.18.5 L28, L31 experiment

The last experiment of the FARO research team, L28, was performed in the modified geometry given in Fig. 16.18.10a *Silverii* and *Annunziato* (1999). The initial conditions and the geometry for the simulations performed with IVA5 are given in Fig. 16.18.10b, first picture. The melt mass is 174.9 kg. The initial pressure 5 bar, the water temperature 151.4 °C, the steam temperature 181 °C and the melt temperature 2973 °C. Figure 16.18.10 b) shows the volume fractions of melt and water as a function of time and space.

Figure 16.18.11a shows the pressure as a function of time. Compared with the material relocation we see that the pressure in the computation is controlled by the depletion dynamics. IVA5 predicts an increase in pressure about 4bar larger at the end of the process compared to the measurements. Figure 16.18.11b shows the accumulative melt mass transported into the external region. We see that IVA5 also predicts this phenomenon, 6.8 kg are measured and about 12 kg are computed. The maximum depleted water in the external region is computed to be about 290 kg. The measured value for this is about 180 kg. The computed water expulsion after 3.5 s does not take place in the experiment.

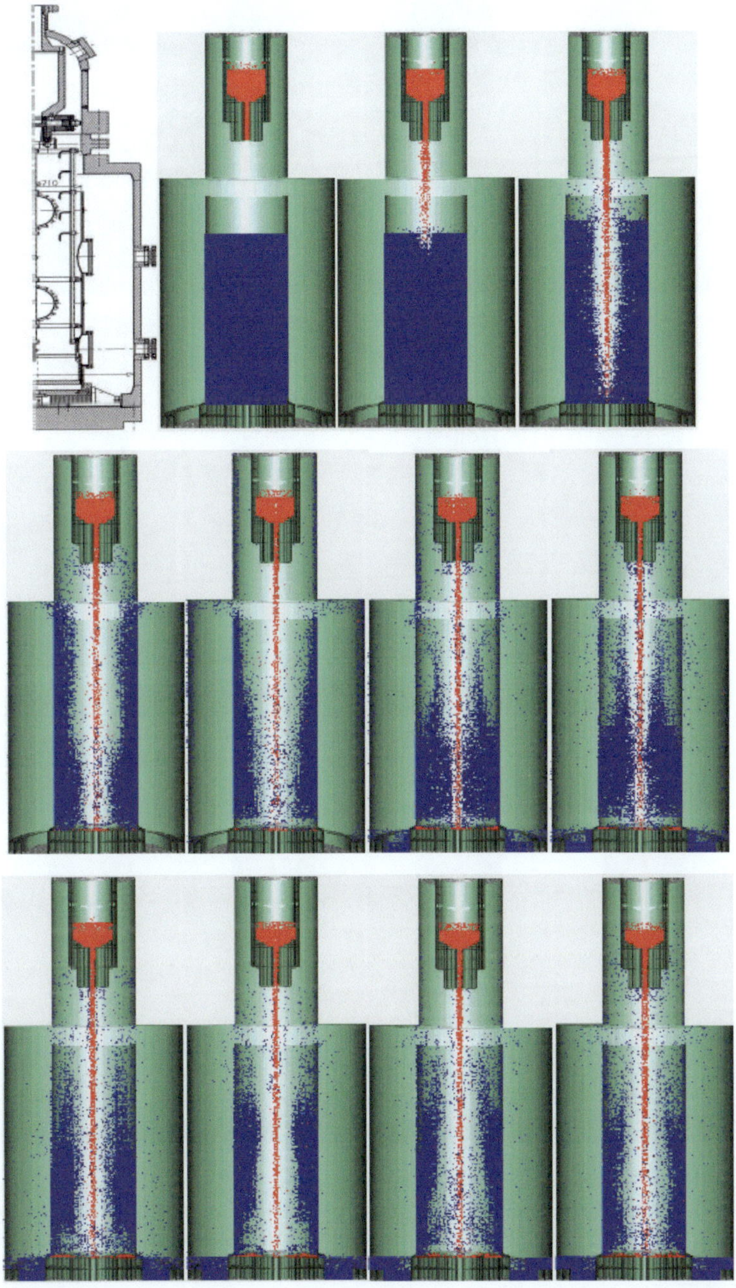

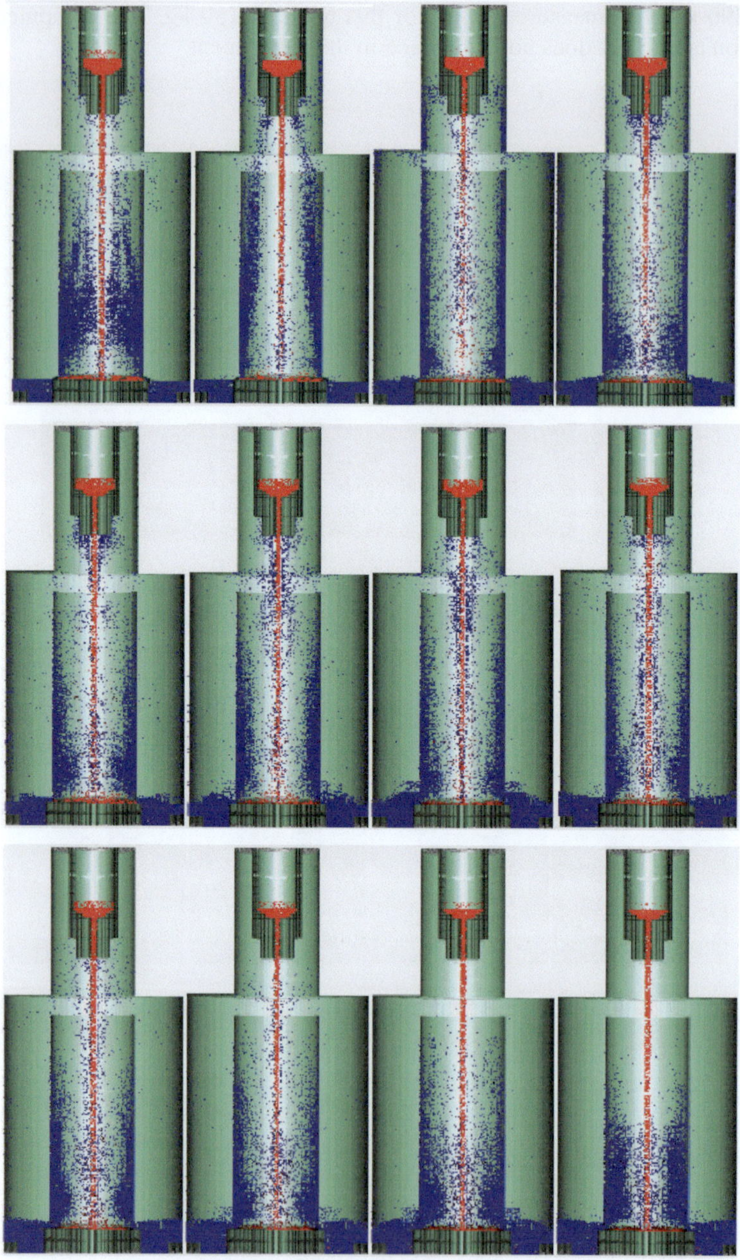

Fig. 16.18.10 a) Geometry of the FARO FAT vessel. b) Geometry representation of the IVA5 computation. Initial conditions. An IVA5 pretest simulation of FARO-FAT L28 experiment. 179.4 kg corium dropped into water. Volumetric fractions of melt and water as a function of time: 0, 0.34, 0.74, 1, 1.26, 1.76, 2.00, 2.26, 2.5, 2.76, 3, 3.26, 3.5, 3.76, 4, 4.26, 4.5, 4.76, 5, 5.26, 5.5, 5.75, 6 s.

Conclusions. a) The important two-dimensional depletion phenomena are predicted by the code. b) Keeping in mind the uncertainty inherent in the mathematical modeling of melt–coolant interactions the general pressure level is predicted well by the code. c) The pressure signature as a function of time indicated that there is still room for improvements to the model.

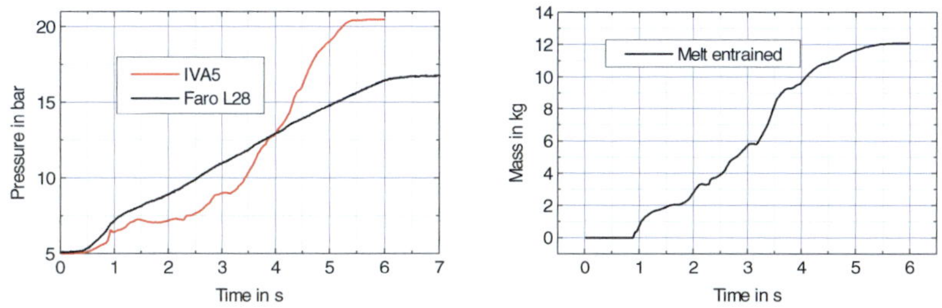

Fig. 16.18.11 a) An IVA5 pretest simulation of FARO-FAT L28 experiment. 174.9 kg corium dropped into water. Pressure as a function of time. b) Melt carried by the water-steam mixture into the external region. Measured carryover 6.8 kg, computed 12 kg

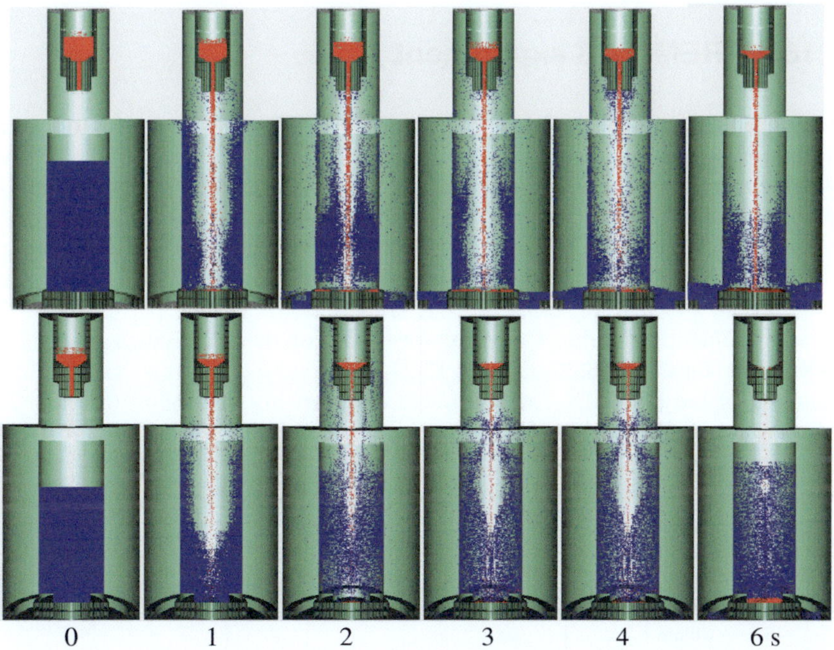

Fig. 16.18.12 Comparison between the material relocation predicted by IVA5 for the L28 (top) and L31 (bottom) tests

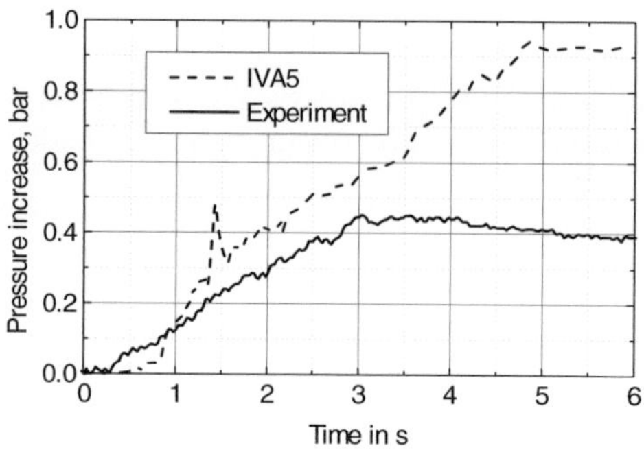

Fig. 16.18.13 L31 pressure increase as a function of time

The counterpart experiment to L28 having subcooled water at the beginning of the experiment was L31 *Silverii* and *Magallon* (1999). The simulation with IVA5 as shown in Fig. 16.18.12 predicts properly the strongly reduced vapor production due to the water subcooling, the resulting reduced material relocation and reduced increase in pressure, Fig. 16.18.13, compared to L28.

16.18.6 PREMIX-13 experiment

Introduction. Over the last 30 years experiments with melt dropped into water in different modes have shown that there are considerable differences in the behavior of different materials. One prominent example is the different behavior of UO_2 and alumina. The explanation we have for these differences and the arguments are presented in Chapter 9 of Volume 1. In order to investigate the physical nature of this difference experiment activities are performed. The PREMIX experiments, *Keiser*, et al. (1998a, b, 1999, 2001), *Huber* et al. (1996) performed at the Research Center Karlsruhe, Germany form part of these activities – see Table 16.18.1 and Figs. 16.18.20 and 16.18.21. The subject of this section is the simulation of the PREMIX 13 experiment and drawing of conclusions about the quality of the modeling technique.

Table 16.18.1 Conditions for the PREMIX tests PM12 through PM18

No	Melt						Water			System
	Mass released kg	Nozzle diam mm	Init. Driv. Press. MPa	Speed[1] m/s	Falling height mm	Release time s	Depth m	Temp. K	ΔT_{sub} K	Press. MPa
12	29.0	60	0.046	3.8–4.2	193	0.95	1.36	372	1	0.1
13	23.8	60	0.052	4.0–4.2	213	1.05	1.34	372	1	0.1

14	23.2	60	0.066	4.5–4.7	213	0.95	1.34	372	1	0.1
15	23.1	60	0.03	3.2–3.3	323	1.25	1.23	416	8	0.5
16	60.4	48	0.01[2]	1.9–2.5	223	5.00	1.33	419	5	0.5
17	16.0	48	0.01[2]	1.7–2.6	223	1.70	1.33	290	104	0.22
18	14.6	48	0.015[2]	1.9–2.9	223	1.50	1.33	370–263[3]	26–34	0.22

[1] on first melt–water contact-maximum; both values are calculated by the experimentalists.
[2] averaged value; pressure was controlled during the melt release.
[3] the scatter was due to internal convection prior to the start of the melt release.

The PREMIX-13 experiment. Briefly, the PREMIX 13 experiment consisted of the injection of about 23.8 kg of alumina at 2600 K through a nozzle of inner diameter 6 cm and length 21 cm into a water pool 1.34 m deep with initial temperature 372 K at atmospheric pressure. The geometry of the test section is given in Figs. 16.18.22 and 16.18.23. The dynamic viscosity of the melt was measured at 2217 K and reported to be less than 4 cp. The equivalent diameter of the cylinder containing the water was 0.684 m. The melt was injected by inert gas with a volume of 0.0761 m³ and an initial pressure of 0.152 MPa. The distance between the nozzle outlet and the water surface was 21.3 cm. The circumferential space around the melt supplier with an external diameter of 40 cm forms a release path for the expansion. Release pipes of diameter 10 cm were installed at four azimuthal positions on the top of the vessel. The variables measured were pressure at various positions, temperatures and collapsed water level. The form of the interaction was filmed with a high-speed camera through appropriate glass windows.

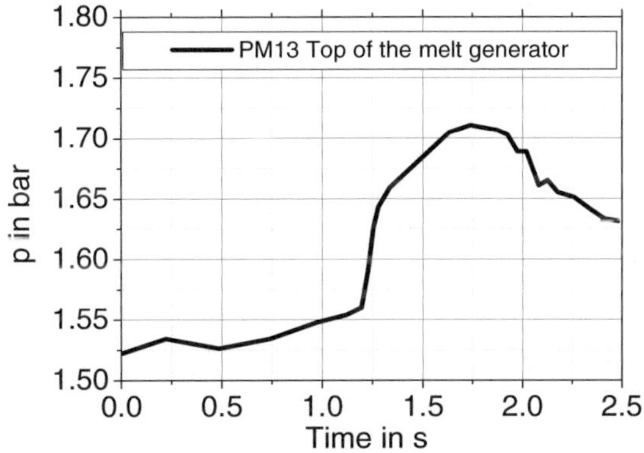

Fig. 16.18.14 Boundary condition: pressure at the crucible gas supply-and-venting pipe for PM13

IVA5 model. The initial conditions and the geometrical model used in IVA5 simulation are shown in Fig. 16.18.16. All parameters are kept as in the experiment. The exceptions are that a) the model cuts the space below the core catcher filled with water and b) the pressure at the gas venting pipe atop of the melt generator was not set as a boundary condition as given in Fig. 16.18.14. Instead, the venting pipe is considered to be isolated from the test section. As we will show in a moment, we will remove this assumption for the analysis of PREMIX 17 and 18. Use is made of the geometrical symmetry and the process is simulated as two dimensional inside the ¼ sector. 32×67 discretization cells are used for the r and z directions, respectively.

Results of the simulation. Figure 16.18.16 shows the material displacement and interaction. The participating phases are represented by different colors only at those positions occupied by them during the transient. This kind of presentation makes the figures self-explanatory. The experimentally observed jet dispersion is given in Fig. 16.18.17. The pressures at various different positions are shown in Fig. 16.18.18. The measured pressures are also shown in this figure.

PREMIX 15 differs from PREMIX 13 mainly in the initial pressure, which was 0.5 MPa.

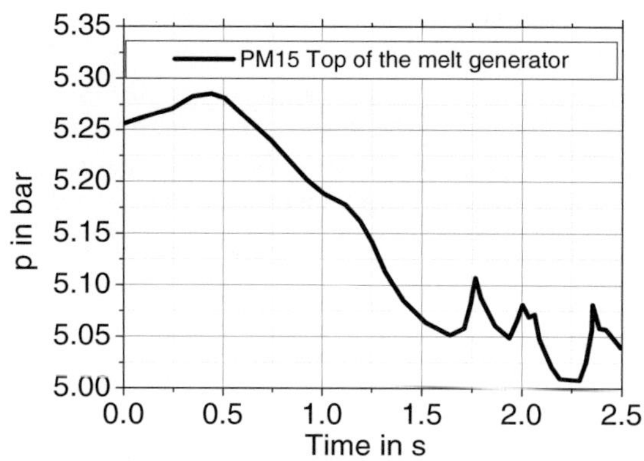

Fig. 16.18.15 Boundary condition: pressure at the crucible gas supply-and-venting pipe for PM15

Similarly to the previous case the venting pipe pressure history recorded in Fig. 16.18.15 was isolated. The IVA5 predictions of the PREMIX 15 are given in Figs. 16.18.19 and 16.18.21. The observed material relocation is presented in Fig. 16.18.20.

Conclusions

1. The overall level of predicted pressure is similar to those observed. At elevated pressure IVA5 over predicts the pressure. The trend of higher initial pressure

and a lower pressure increase is properly predicted by IVA5. Surprisingly, the fluctuations predicted in the computations were not observed experimentally. This is even more surprising, as such processes are well known to be very noisy.

2. The computed volumetric flow rate starts with a similar gradient to that observed, but reaches a plateau that is two orders lower than that measured, a result of not modeling the turbulent jet disintegration as will be discussed later. The drop in the gas volumetric flow rate for a period of 0.5 s is due to water accumulation in the region ahead of the discharge pipe. The representation of the discharge pipe 10 cm diameter by porosity in a 2D manner is not appropriate. Real 3D analysis is recommended.
3. One of the greatest discrepancies between the experiment and the computation is the form of the melt downstream of the nozzle. The observed form resembles very much the process after a spray nozzle. The observed diameter of the dispersed jet increases from 6 to 10 cm from the nozzle outlet to the initial water level. As a result, the melt reaches the water surface prefragmented. This influences greatly the later behavior of the process. The experiment gives the impression of cooling of a predispersed family of droplets that reaches the surface and descends slowly by "eroding" water with a large horizontal cross-section providing appropriate venting of the generated steam without substantial water entrainment. The code simulates jet behavior characterized by no fragmentation in the gas and intensive fragmentation during the interaction with the water. A kind of channeling produces large-scale instabilities in the water environment with a cyclic character, degraded in a later phase to complete intermixing behavior. To explain the differences a new simple jet fragmentation model was derived in Chapter 10, Eq. (10.43),

$$\Delta L_j / D_j = \frac{1}{2} V_j / \left\{ \left(c_{cj} / 2 \right)^{1/2} \left(\rho_c / \rho_j \right)^{1/2} \Delta V_{cj} + c V_j \right\},$$

where ΔL_j is the fragmentation length, D_j is the jet diameter, V_j is the jet velocity, c_{cj} is drag coefficient exerted by the jet submerged in continuum, ρ_c and ρ_j are the densities of the surrounding continuum and of the jet, ΔV_{cj} is the velocity difference between the continuum and the jet, and cV_j is the lateral fluctuation velocity with a constant c being around 0.1. The simple model reproduces well the existing limiting cases known from the literature. Simple jet-fragmentation analysis assuming a radial velocity fluctuation of 10% from the axial velocity magnitude results in complete fragmentation of 30 cm, which is exactly what was observed in the experiment. Thus, the following conclusion is made:

Turbulence modeling in multiphase flow is important for predicting fragmentation processes. In cases where the fragmentation is the main driving force of the process the turbulence modeling starts to be the crucial element of the model.

4. The pressure boundary conditions as specified in Figs. 16.18.14 and 16.18.15 have to be taken into account. This will be done in the next section by simulating PREMIX 17 and 18 where substantial improvement of the pressure prediction is obtained.

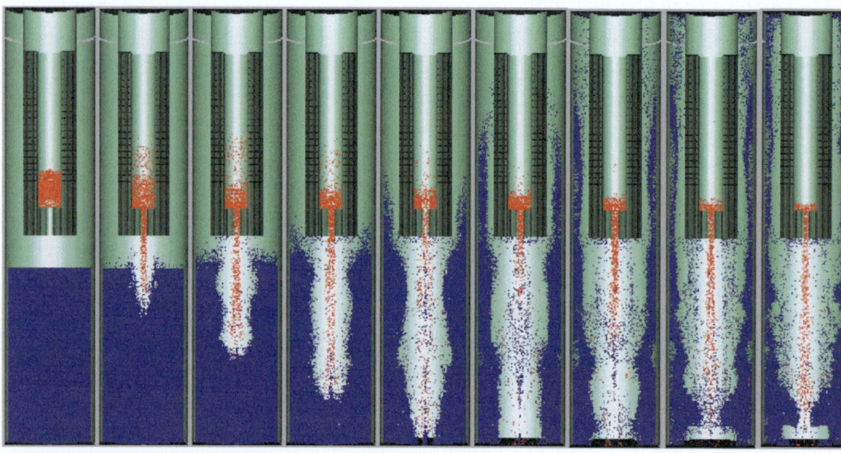

Fig. 16.18.16 An IVA5 simulation of PREMIX 13. Volumetric fractions of alumina and water as a function of space. Parameter: time; the time step between the pictures is 0.1 s

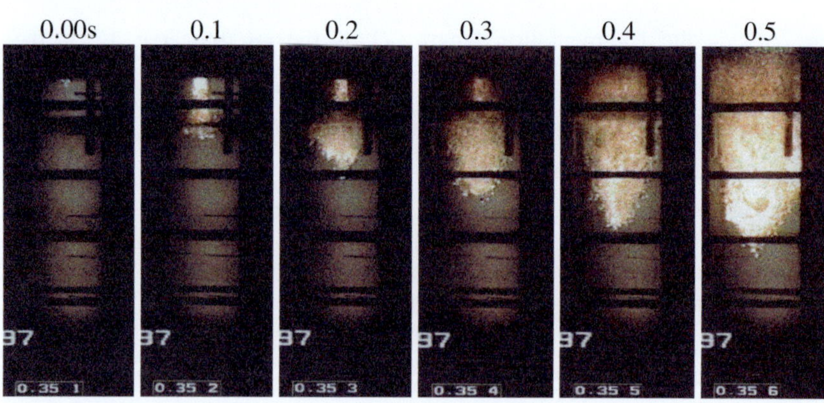

Fig. 16.18.17 Photographs of hot alumina in water. PREMIX 13 experiment, *Kaiser*, et al. (1998a)

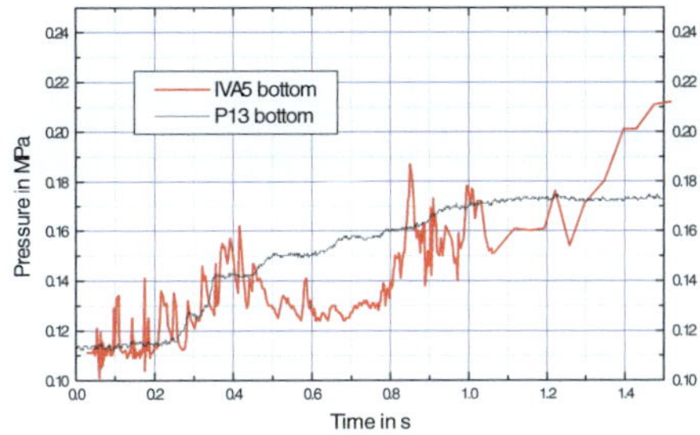

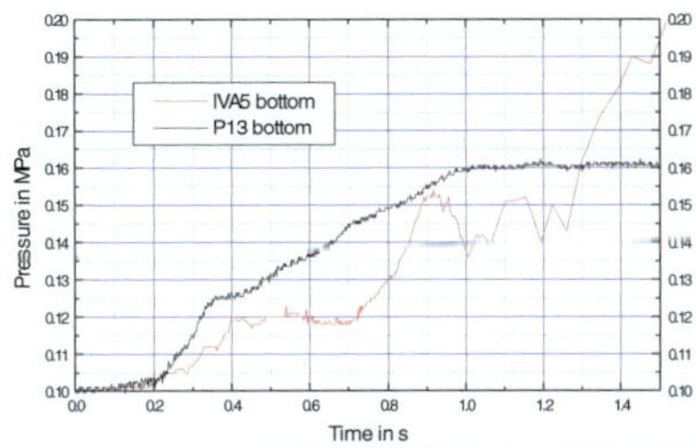

Fig. 16.18.18 Pressure as a function of time at different vertical positions

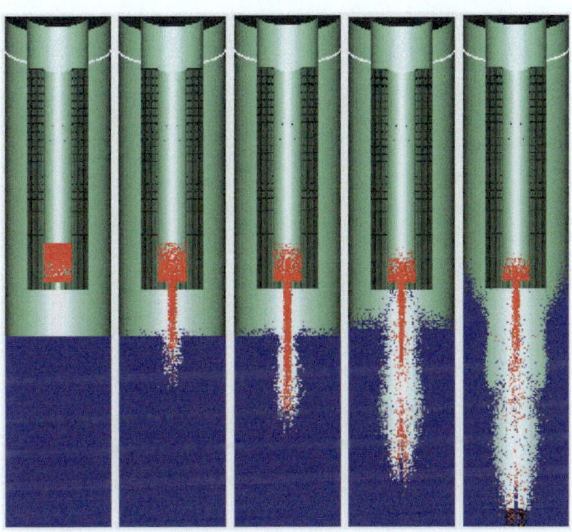

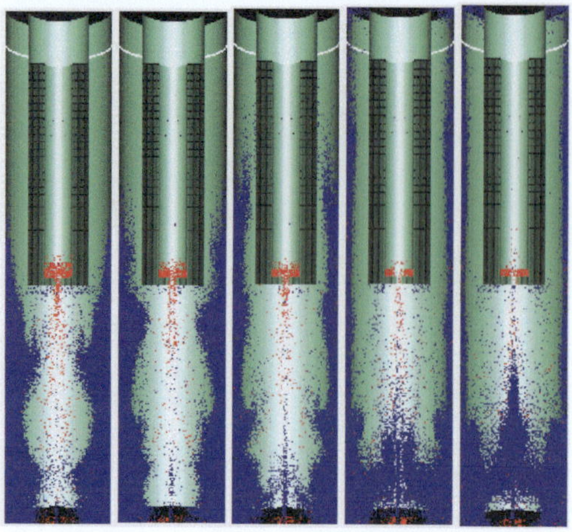

Fig. 16.18.19 PREMIX 15: Pressure as a function of time at different vertical positions. Parameter: time, the time step between the pictures is 0.1 s

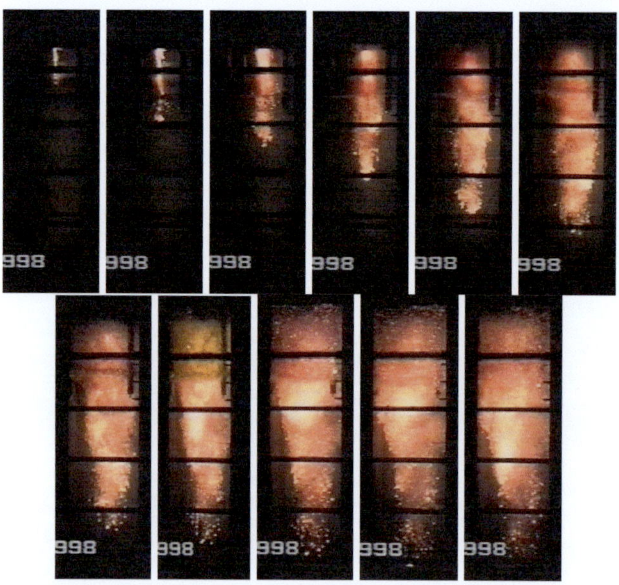

Fig. 16.18.20 Photographs of hot alumina in water. PREMIX 15 experiment *Kaiser*, et al. (1998b). First picture: water level reached by the melt jet. The time between two pictures is 0.1 s

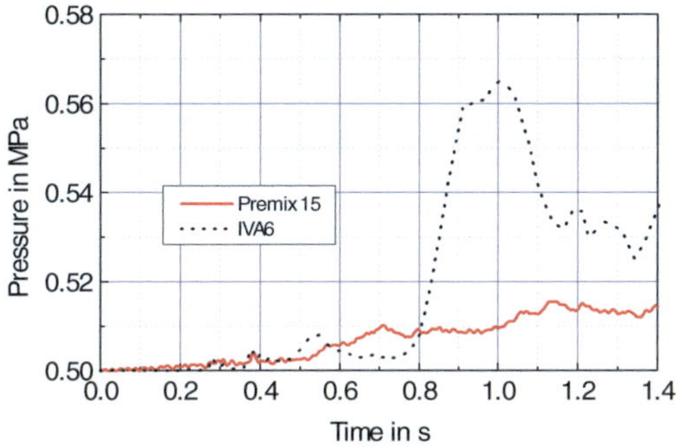

Fig. 16.18.21 IVA5 prediction of the PREMIX 15 experiment

16.18.7 PREMIX 17 and 18 experiments

Geometry: As mentioned before, the PREMIX experimental program was carried out over the years 1996 to 2001, *Kaiser*, et al. (1998a, b, 1999, 2001), *Huber* et al. (1996) at Research Center Karlsruhe (FzK) in Germany. The geometrical arrangement used serves (a) to produce alumina in a melt container by igniting thermite, (b) to separate the iron from the thermite by the melting of the first steel membrane that has a ring form, and (c) to release predominantly alumina, if possible, through the melting of the second steel membrane (0.2 mm thick). The melt container geometry is reproduced from *Kaiser*, et al. (2001) in Fig. 16.18.22. The container is connected to the gas supply and venting systems at the top. The melt is then released into a water container as shown in Fig. 16.18.23. Above the bottom there is a fragment catcher. The products of the melt–water interactions are convected in the internal space of the facility. In some of the experiments, but not in the particular PM17 and 18 that are of interest here, a venting system as shown in Fig. 16.18.22b was used.

Initial conditions: The initial conditions relevant for the computational analysis are given in Table 16.18.2. Note that not all the melted mass was found in the interaction volume probably due to crust formation. The melt-release temperature was given in *Kaiser*, et al. (2001) as 2600 K. The content of iron in the released alumina was estimated in *Kaiser*, et al. (2001) to be less than 10%. The distance between the nozzle outlet and the water surface was 223 mm.

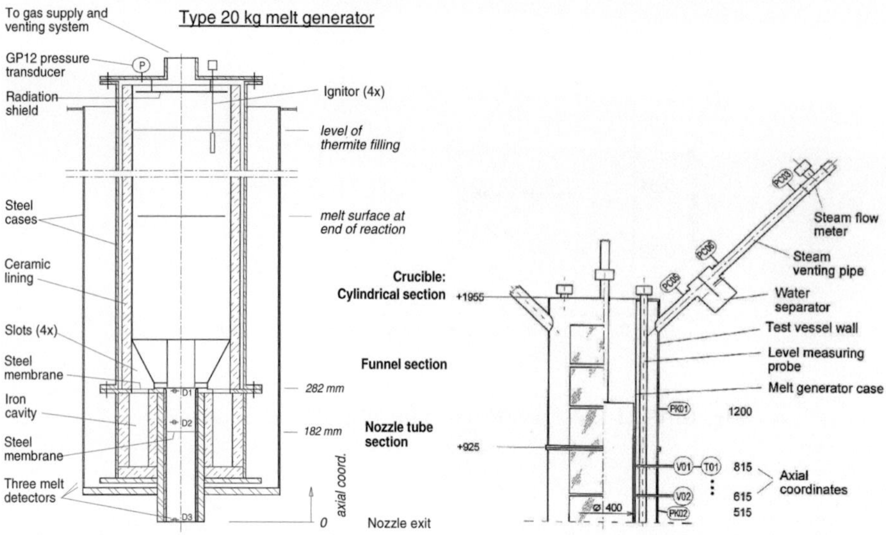

Fig. 16.18.22 a) Schematic diagram of the melt generator. b) Venting lines. Locations of the pressure transducers mounted in the annular gas space and in the venting line no. 3

Boundary conditions: The experimental space of interest is a closed space except for the so-called "gas supply-and-venting pipe". The only variable boundary condition was applied by the specially designed automatic procedure at the crucible gas supply-and-venting pipe.

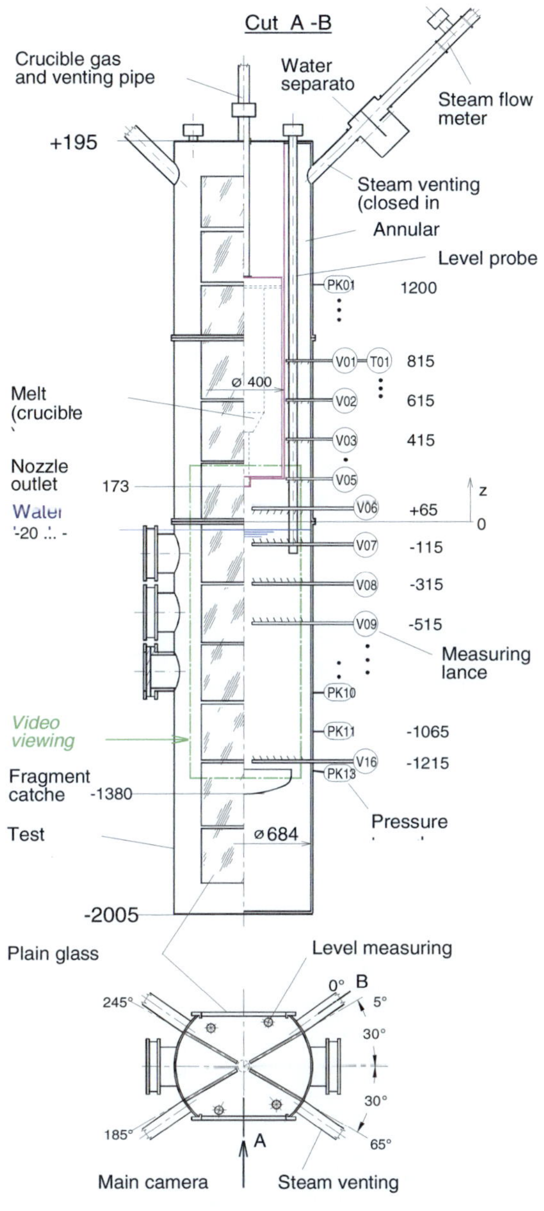

Fig. 16.18.23 The PREMIX test facility

We will not simulate this pipe but replace its action by a boundary condition prescribing pressure as a time function at the pipe cross-section just at the top of the interaction vessel. We reproduce for convenience the measured curves given in

Fig. 16.18.24 for PM17 and 18 with a much smaller number of characteristic points than that experimentally obtained.

Table 16.18.2 Initial conditions

No	Date	Melt mass/*kg* produced/ released	Melt mass/*kg* interacting*	*T*-water sub. /K	*p*/MPa	Duration of melt release/s
PM17	20.4.99	15/14.36	7.1+5.9+0.66 +0.76	104	0.2	2
PM18	07.7.99	15/14.56	5.48+5.68+3.4 4+0.	26	0.2	2

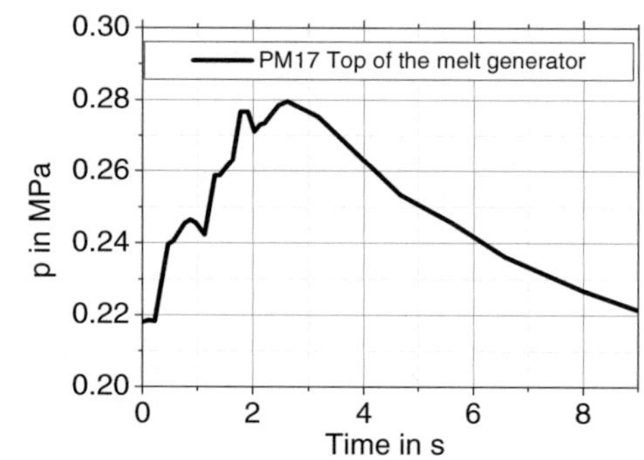

a)

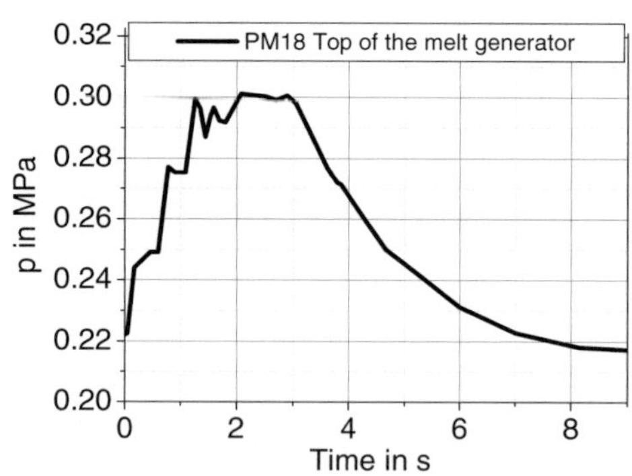

b)

Fig. 16.18.24 IVA_5M boundary condition: pressure at the crucible gas supply-and-venting pipe: a) for PM17; b) PM18

Important. Note that in the experiment all curves are related to a zero time point defined as the moment at which the jet touches the water surface.

What was measured? The experimentally measured pressures at different positions were acquired with a frequency of 10 kHz (some pressures have been redundantly measured with 50 kHz). The data made available from the experimentalists are then compressed to 1 kHz without smoothing. Another important part of the data acquisition system was the film production taken through specially designed glass windows as shown in Fig. 16.18.23. The installed mass flow measurement at the venting pipes was not active during PM17 and 18 because the pipes are isolated from the interaction vessel.

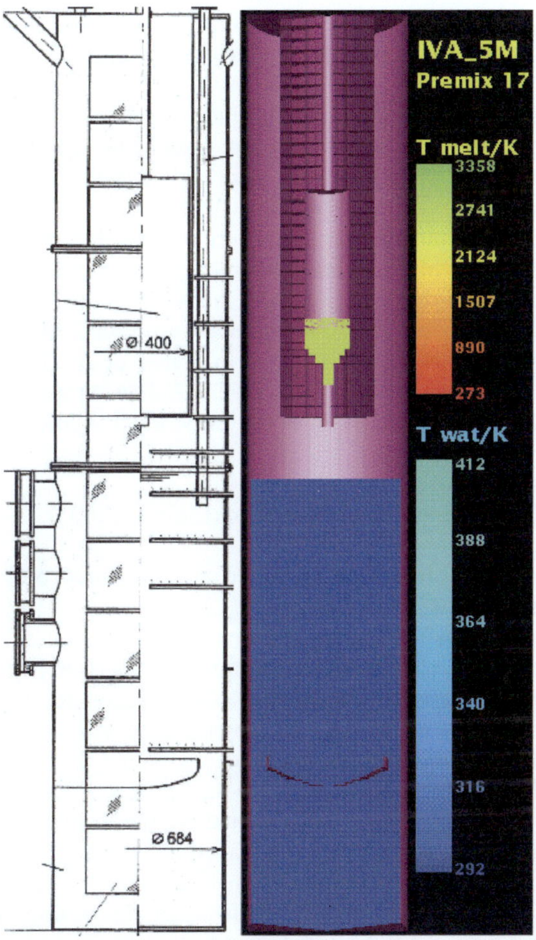

Fig. 16.18.25 Comparison between the real PREMIX vessel geometry and the IVA_5M geometry model

The IVA_5M model: Geometry: as will be demonstrated in a moment, many details of the geometrical representation have an important impact on the adequate description of the processes. We assume that the process is axisymmetric, and represent the process as a two-dimensional using 2268 finite-volume cells – 27 radial and 84 axial nonequidistant steps in cylindrical geometry as presented in Fig. 16.18.25. Realizing the impact of appropriate modeling of the form of the debris catcher we increase the resolution in this region.

Variable-geometry boundary conditions. We start with a closed lower steel membrane. Experiments in IKET-FzK demonstrated that steel membranes sustain oxide jets for longer before melting completely compared for instance with brass membranes. We do not know exactly the time delay of melting of the lower membrane with respect to the upper one and judge the opening characteristics from the diameter of the jet filmed during the experiment.

Initial conditions. The initial conditions for the water are posed exactly as reported from the experiment. We use air as an inert gas with the environmental temperature outside the melt generator and with a melt temperature above the melt.

Some natural uncertainties in the initial conditions still remain due to the nature of such experiments. During the processes in the melt crucible we do not take into account that the thermite reaction is usually inhomogeneous, and that the resulting mixture consists of the nonmiscible alumina and iron with 53.1 mass% iron – see the discussion in *Huber* et al. (1996) p. 3. In *Huber* et al. (1996) p. 4 only 81% alumina was found in the released debris and in *Kaiser*, et al. (2001) the figure of 90% is given. We use thermal and transport properties for pure alumina but not for the alumna–iron suspension.

Comparison between theory and experiment, *Kolev* (2003)

PM17. Figure 16.18.26 shows the predicted and measured pressures in the gas. The agreement is good.

Figure 16.18.27 shows the predicted and measured pressures in the water. We realize that the magnitude of the averaged pressure excursion is properly predicted by the code. The strong water subcooling is the reason for the high-frequency oscillation. The oscillations by IVA_5M are predicted with a numerical resolution of 500 kHz. The pressure oscillation frequency found in the computation was about 70% higher than those found in the experimental measurements. A parametric study using PM18 reveals that the stepwise modeling of the smooth core catcher introduces its own numerical frequencies, which overlay those generated by the condensation oscillations.

Figure 16.18.28 shows the observed melt–water interaction. The time zero is set to be about the time of the first melt–water contact. The simulated results are presented in Fig. 16.18.29. From the velocity vectors that are not given here, we realize that the process consists of two well-separated phases: (a) the release phase, and (b) the natural three-phase circulation phase of cooling the sloshing melt in the catcher. In the first phase the descending melt jet fragments without strong net vapor production because of the water subcooling. In the second, the melt sloshing from the center to the external boundary and back interacts with water, producing much more steam and building the pressure in this strong nonequilibrium system.

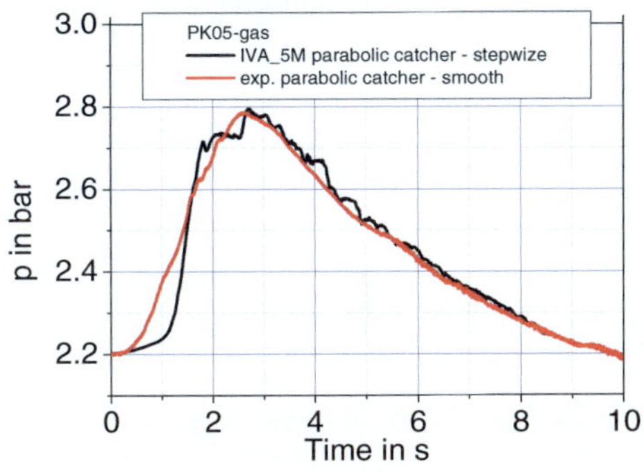

Fig. 16.18.26 Wall pressure in the gas PK05 ($z = 2.12$ m) as a function of time

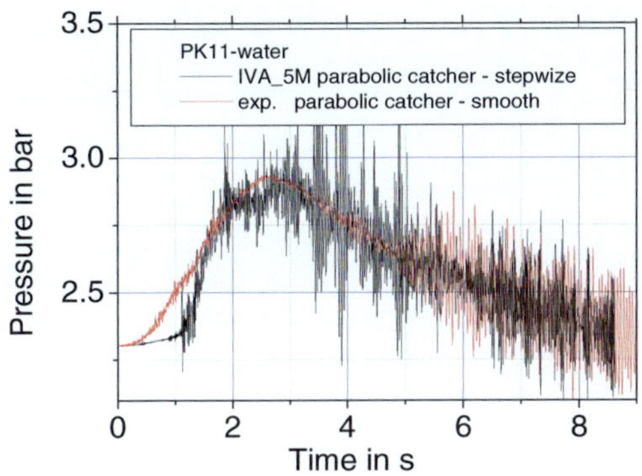

Fig. 16.18.27 Wall pressure in the water PK11 ($z = 0.94$ m) as a function of time

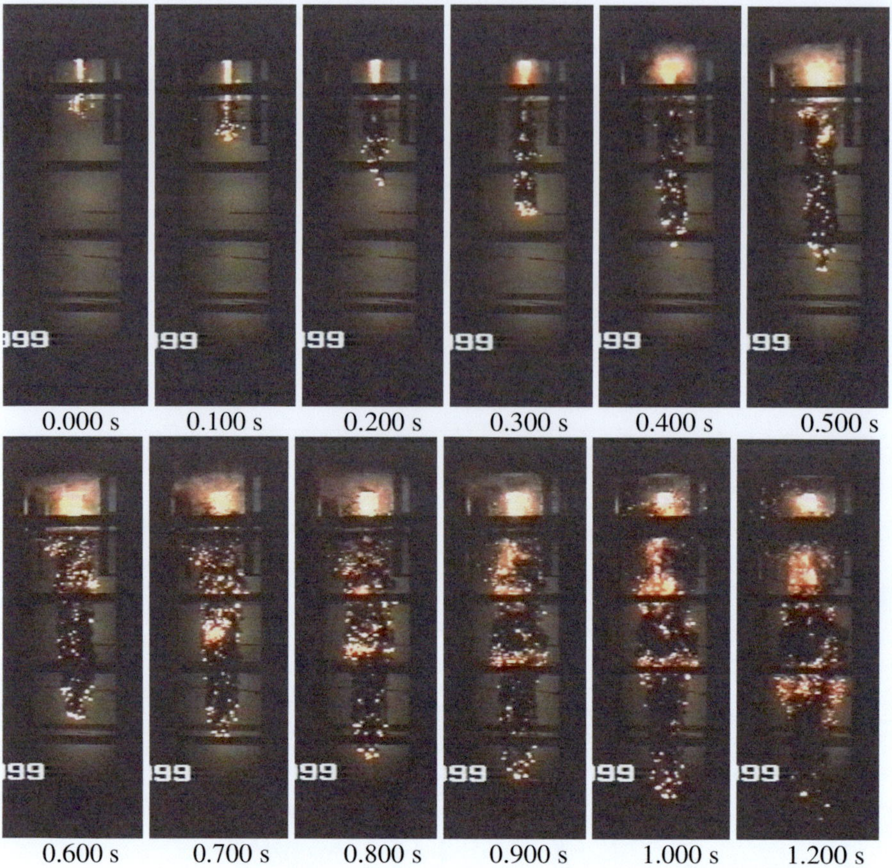
0.000 s
0.100 s
0.200 s
0.300 s
0.400 s
0.500 s
0.600 s
0.700 s
0.800 s
0.900 s
1.000 s
1.200 s

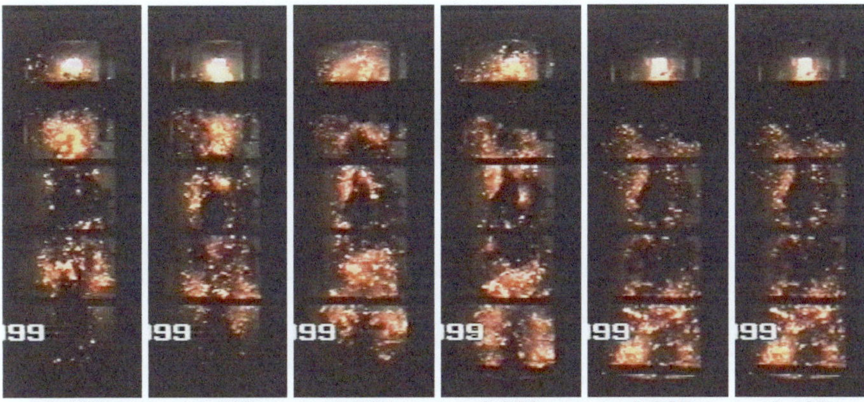

1.400 s 1.600 s 1.800 s 2.000 s 2.200 s 2.400 s

Fig. 16.18.28 PM17: Video pictures showing the three-phase flow in the interaction zone (larger time steps)

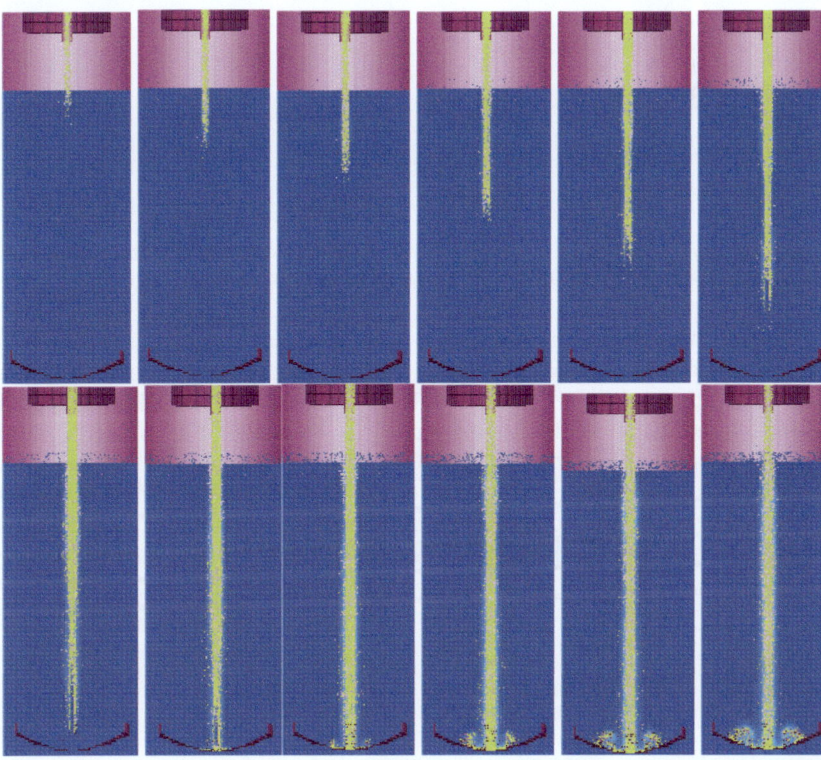

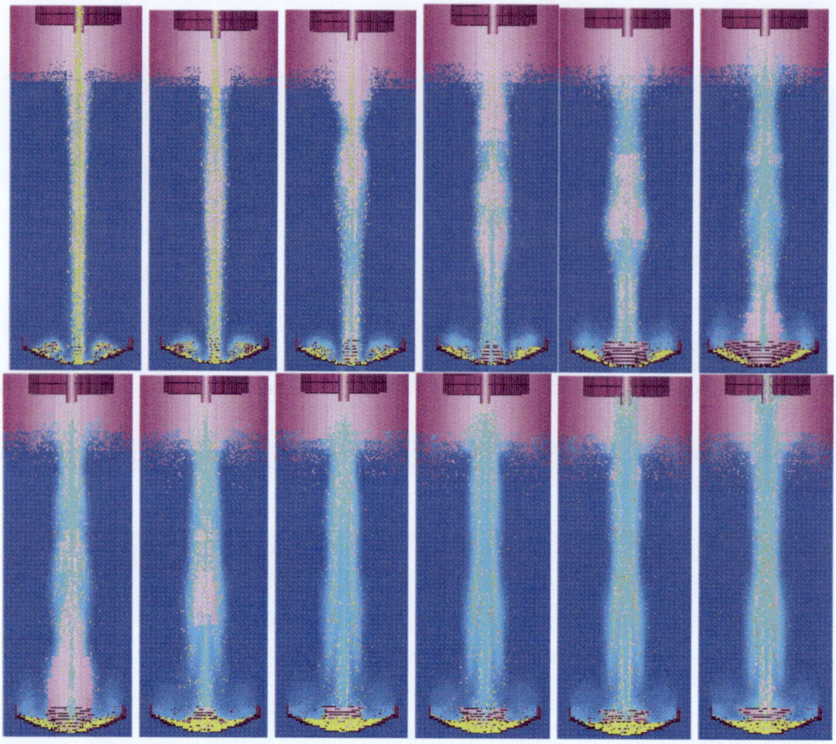

Fig. 16.18.29 IVA_5M simulation of PM17: video pictures showing the three-phase flow in the interaction zone (time step 0.1 s)

The resultant three-phase mixture moves upwards along the axis, and downwards in the external region. The suspension of the melt fragments is an observed fact in the experiment and is well predicted by the code. Strong radial temperature differences are predicted inside the water, as shown in Fig. 16.18.30, which explain the increasing net vapor production with the time. Then, the three-phase convection enforces mixing and the steam produced is predominantly condensed. The sloshing process in the catcher disappears with the time, thus reducing surface fragmentation and slowing down the interaction.

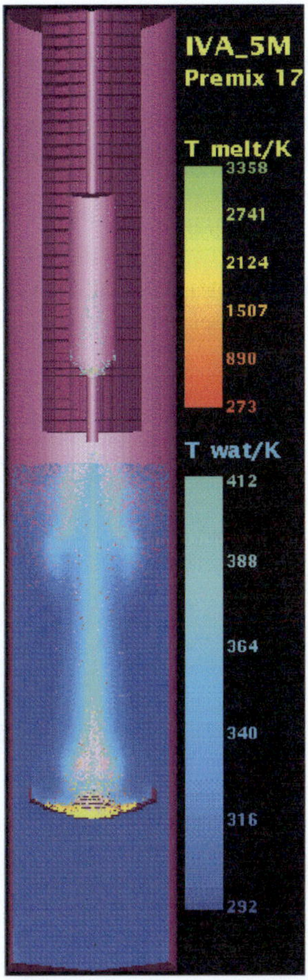

Fig. 16.18.30 Water temperature 5.55 s after the first melt–water contact

Not modeled in the simulation is the heat transfer from the melt through the catcher into the water, which is associated with additional melt freezing at the bottom.

PM18. In the next simulation we use the same initial conditions as in the case of PM17, except for the initial water temperature. The resulting pressures are presented in Figs. 16.18.31 and 16.18.32. As already mentioned in the case of PM17 the processes inside the catcher strongly influence the pressure build. The catcher used has a smooth surface. In the simulation the parabolic form of the catcher is represented by steps. In the next simulation we represent the catcher with a cylindrical container having a plane bottom. The results are presented in Figs. 16.18.33 and 16.18.34. We see that the flat bottom catcher does not generate high-frequency oscillations in the initial phase but does generate such oscillations in the second phase. The magnitude of the pressure is higher. In the case of the stepwise catcher the magnitude of the pressure increase is the right one, but the condensation oscillations are suppressed in the second phase. This analysis clearly indicates (a) that the essential physics of melt–water interaction is contained in the code and (b) that a further increase of the quality of the simulation of this process can be obtained only by using a boundary fitted representation of the form of the core catcher. Interpreting the simulation of PM17 in this light we see that the stepwise representation of the parabolic catcher really introduces numerical oscillation in the first part of the experiment and this is superimposed on the actual existing condensation oscillations that are also predicted by the code.

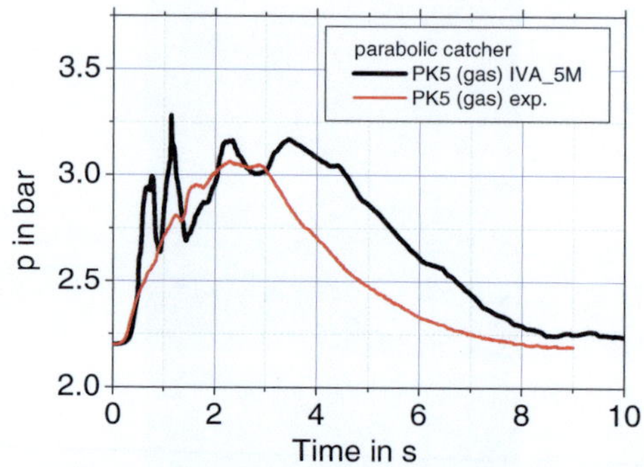

Fig. 16.18.31 Wall pressure in the gas PK05 as a function of time. Parabolic core catcher as presented in Figs. 16.18.25 and 16.18.30

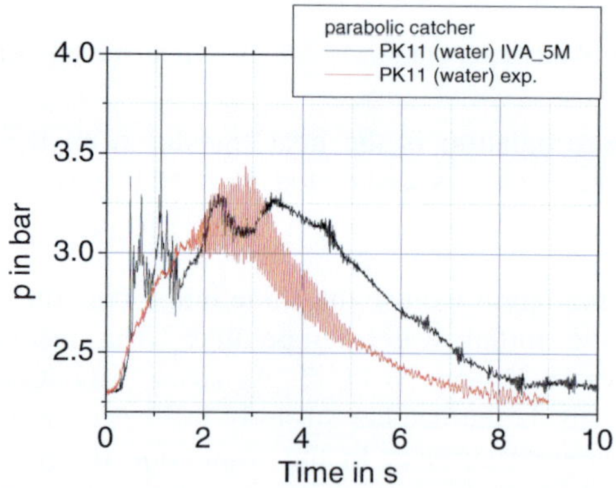

Fig. 16.18.32 Wall pressure in the water PK11 as a function of time. Parabolic core catcher as presented in Figs. 16.18.25 and 16.18.30

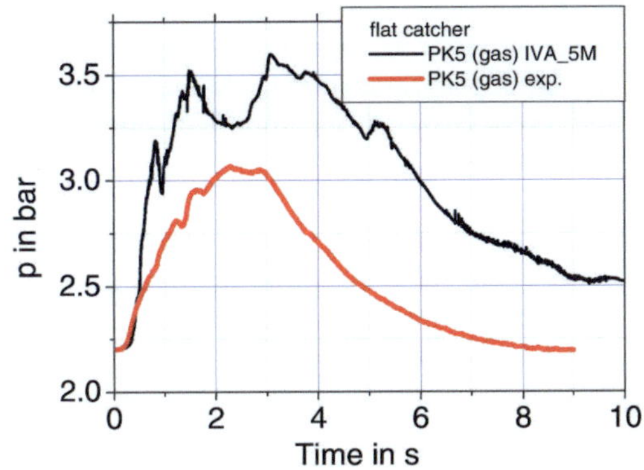

Fig. 16.18.33 Wall pressure in the gas PK05 as a function of time. Cylindrical catcher with flat bottom

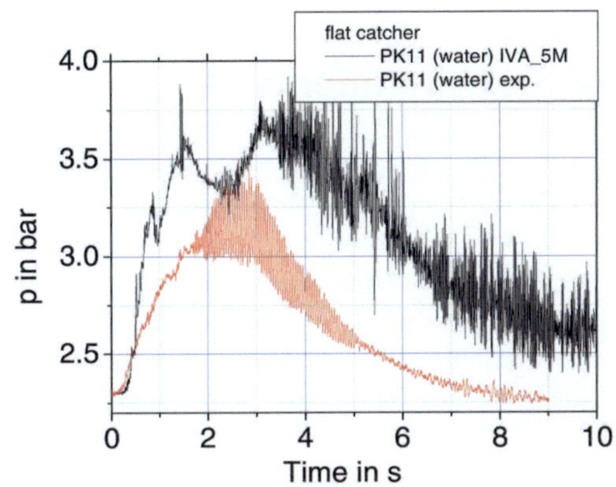

Fig. 16.18.34 Wall pressure in the water PK11 as a function of time. Cylindrical catcher with flat bottom

Conclusions. Using the PREMIX 17 and 18 experiments the capability of the IVA_5M computer code is tested to predict interaction of molten alumina with strongly subcooled water for release of melt close to the water surface. The following conclusions are drawn from this study:

(a) The experiment shows two very different characteristic features: the first is jet–water interaction without strong net vapor production, and the second is three-phase natural circulation that cools the sloshing melt inside the core catcher, the so-called three-phase chimney effect. The second phase is characterized by entrainment of melt into the buoyancy-driven upward three-phase mixture along the axis and cyclic removal of steam clouds. Both processes are well recognizable in the computer code simulation.

(b) The magnitude of the pressure increase is properly predicted.

(c) The importance of the appropriate representation of the geometry of the melt catcher is clearly demonstrated. Replacing the smooth parabolic surface of the catcher with a stepwise wall generates fluctuations that are not observed in the initial phase.

(d) The approximate representation of the catcher as a cylinder with a flat bottom demonstrates the lack of such oscillations in the initial phase, and the generation of such oscillations in the second phase in the computations, a phenomenon that was in fact observed;

(e) A boundary-fitted representation of the flow catcher is recommended for such types of analysis.

(f) The constitutive models addressed to control the processes in such interactions are functioning properly in the code.

As far as this author knows this is the first successful simulation of PREMIX 17 and 18 with a system computer code.

16.18.8 RIT and IKE experiments

The findings obtained by RIT experiments are systematically compared with the formalism used in IVA5, *Kolev* (1999). The most important conclusions are as follows. a) The void fraction in the region of continuous water does not influence the fragmentation characteristics. This simply says that it is permissible to use models for liquid–liquid fragmentation also in bubbly three-phase flows without boiling *Haraldsson* et al. (1999). b) The liquid-liquid jet fragmentation obeys the *Taylor* solution, *Batchelor* (1958). The *L* over *D* relation of *Taylor* is useful with a constant of about 10. It confirms *Bracco*'s result that this constant is about 7 *Bracco* (1985).

Finally, it became clear that the spectral distribution of particles, as observed in FARO, PREMIX and RIT, could not be obtained by computer models with a single group model for the dispersed field.

Bürger and coworkers, *Bürger*, et al. (1996), *Bürger* (1998), *Bürger*, et al. (1998), *Coste*, et al. (1999) found out that the hydrodynamic fragmentation is the most important fragmentation mechanism in high *Weber* number droplet fragmentation and that the thermal enhancement may be of the order of 15% at maximum. Comparing different fragmentation models in this region *Coste*, et al. (1999) find that the *Pilch* based hydrodynamic models represent best the findings of the RIT experiments, *Kolev* (1999). It confirms the decision to use in the IVA code series *Pilch*-based models for melt-droplet fragmentation description – see *Silverii* and *Magallon* (1999), *Kolev* (1999).

References

Annunziato, A., Addabbo, C., Leva, G.: OECD-CSNI International standard problem 39 on FARO test L-14, Reference specification, JRC Technical Note I.96.xx, Draft (1996)

Annununziato, A., Addabbo, C., Magallon, D.: FARO test L-20 quick look report, Technical Note No. I.96.163, JRC Ispra, Italy (October 1996)

Annununziato, A., Addabbo, C., Magallon, D.: FARO Test L-24 Quick Look Report, Technical Note No. I.97.185, JRC Ispra, Italy (1997)

Batchelor, G.K. (ed.): Collected works of G. I. Taylor. Cambridge Univ. Press, Cambridge (1958)

Bracco, F.V.: Modeling of engine sprays. In: Proc. International Congress & Exposition Detroit, Michigan, February 25-March 1, pp. 113–136 (1985)

Benuzzi, A., Magalon, D.: FARO-LWR programme L-14 test quick look report, JRC Technical Note I.94.171 (December 1994)

Bürger, M., Buck, M., Saied-Ahmad, A., Schatz, A.: Experimental and theoretical investigation on the fragmentation of melt drops in relative flow, Report INV-MFC (98)-D016, Institut für Kernenrgetik und Energiesysteme, Universität Stutgart, IKE 2–135 (December 1996)

Bürger, M.: Comparison and theoretical interpretation of the experiments on hydrodynamic drop fragmentation, Report INV-MFC (98)-D014, Institut für Kernenrgetik und Energiesysteme, Universität Stutgart, IKE 2-FB-16 (Juli 1998)

Bürger, M., Cho, S.H., von Berg, E., Schatz, A.: Modeling of drop fragmentation in thermal detonation waves and experimental verification, Report INV-MFC (98)-D016, Institut für Kernenrgetik und Energiesysteme, Universität Stutgart, IKE 2-FB-15 (November 1998)

Coste, P., Valette, M., Berthoud, G.: Pb-Bi droplets fragmentation in liquid water (Li et al. experiment, ICMF 98): comparison of the existing code correlations and calculations with MC3D. In: 6th MFCI Project Meeting, CEA Grenoble, France, June 23-24 (1999)

Haraldsson, H.O., Li, H.X., Dinh, T.N., Green, J.A., Sehgal, B.R.: Effect of solidification and coolant voiding on breakup of molten metal jet in water, Royal Institut of Technology (KTH), NPS, EU report INV-MFCI(00)-D048, Report EU-KTH/NPS/MFCI1-9907-SE, Stockholm (July 6, 1999)

Hohmann, H., Magallon, D., Huhtiniemi, I., Annunziato, A., Yerkess, A.: Advance in the FARO/KROTOS Melt Quenching Test Series. In: 22nd Water Reactor Safety Meeting, Bethesda (Maryland) (October 24-26, 1994)

Huber, F., Keiser, A., Steinbrück, M., Will, H.: PREMIX, Documentation of the results of experiments PM01 to PM06, FZKA 5756, Forschungszentrum Karlsruhe, Germany (März 1996)

Kaiser, A., Schütz, W., Will, H.: PREMIX experiment, EC 4th Framework Program on Reactor Safety Shared Cost Actions, Contract no FI4s-CT96-0037, MFCI PROJECT, Yearly Progress Report no 1 (1998a)

Kaiser, A., Schütz, W., Will, H.: PREMIX: Investigation of premixing behavior of a hot melt being poured into water. Test PM15 (December 17, 1998b); 5th MFCI Project Meeting, Forschungszentrum Karlsruhe, Germany (September 16, 1998)

Keiser, A., Schütz, W., Will, H.: PREMIX Tests PM12, PM13 and PM14, Documentation and evaluation of experimental data, FZKA 6370, Forschungszentrum Karlsruhe, Germany (November 1999)

Keiser, A., Schütz, W., Will, H.: PREMIX Experiments PM12-PM18 to investigate the mixing of hot melt with water, FZKA 6380, Forschungszentrum Karlsruhe, Germany (Juli (2001)

Kolev, N.I.: Fragmentation and coalescence dynamics in multi-phase flows. Exp. Thermal Fluid Sci. 6, 211–251 (1993)

Kolev, N.I.: The code IVA4: Modeling of mass conservation in multi-phase multi-component flows in heterogeneous porous media. Kerntechnik 59(4-5), 226–237 (1994a)

Kolev, N.I.: The code IVA4: Modeling of momentum conservation in multi-phase multi component flows in heterogeneous porous media. Kerntechnik 59(6), 249–258 (1994b)

Kolev, N.I.: IVA4 Computer code: Dynamic fragmentation model for liquid and its application to melt water interaction. In: Proc. ICONE-3, The Third International Conf. on Nucl. Engineering, Nuclear Power and Energy Future, Kyoto, Japan, April 23-27 (1995a); Presented at the Workshop zur Kühlmittel/ Schmelze - Wechselwirkung, Köln, Germany (November 14-15, 1994)

Kolev, N.I.: The code IVA4: Second law of thermodynamics for multi phase flows in heterogeneous porous media. Kerntechnik 60(1), 1–39 (1995b)

Kolev, N.I., Hofer, E.: Uncertainty and sensitivity analysis of post experimental IVA4 simulations of melt water interaction. Exp. Thermal Fluid Sci. 13, 98–116 (1996)

Kolev, N.I.: IVA4 analysis of the FARO L14 experiment (ISP 39). In: International Seminar on Vapor Explosions and Explosive Eruptions (AMIGO-IMI), Aoba Kinen Kaikan of Tohoku University, Sendai-City, Japan (May 22-24, 1997a)

Kolev, N.I.: Comments on the entropy concept. Kerntechnik 62(1), 67–70 (1997b)

Kolev, N.I.: On the variety of notation of the energy conservation principle for single phase flow. Kerntechnik 63(3), 145–156 (1998)

Kolev, N.I.: The IVA5 dynamic droplet fragmentation and coalescence models and their verification data base: Part 1 (INV-MFCI(99)-D027) and Part 2 (INV-MFCI(99)-D028) (1999)

Kolev, N.I., Seitz, H., Roloff-Bock, I.: Hot-leg injection: 3D versus 1D three velocity fields modeling and comparison with UPTF experiment, Siemens report KWU NA-T/00/E027, R&D Project 903060, Verification IVA6 (2000)

Kolev, N.I.: Analysis pf the PREMIX 17 and 18 experiments with IVA_5M computer code. In: The 10th International Topical Meeting on Nuclear Reactor Thermal Hydraulics, Seoul, Korea, E00106 (October 5-9, 2003)

Magallon, D., Leva, G.: FARO LWR programme test L-14 data report, JRC Technical Note I.96.25 (February 1996)

Mitchel, D.E., et al.: Intermediate scales steam explosion phenomena: Experimental analysis SAND8-0124 (1981)

Nelson, L.S., Duda, P.M.: Steam explosion experiments with single droplet of CO_2 laser melted with iron oxide. Report NUREG/CR-2295 (SAND 81-1346) (September 1981)

Nelson, L.S., Duda, P.M.: Steam explosion experiments with single droplet of CO_2 laser melted with iron oxide. Part 2. Parametric Studies. Report NUREG/CR-2718 (SAND 81-1105) (September 1985)

Silverii, R., Annunziato, A.: FARO LWR program, Test L-28 data report, Technical Note No.I.99.76, INV-MFCI(99)-D033, JRC Ispra, Italy (1999)

Silverii, R., Magallon, D.: FARO LWR Programme Test L-31 Data Report, Technical Note No. I.99.100, INV-MFCI(99)-D035, JRC, ISIS, Ispra (VA), Italy (June 1999)

Yamano, N., Sugimoto, J., Moryama, Y., Soda, K.: Studies of fuel coolant interactions during core melt accident of nuclear power plants. In: NURETH-6th Int. Meeting On Thermal-Hydraulics, pp. 271–281 (1992)

16.19 Assessment for detonation analysis

In order to asses the IVA5 capabilities for detonation analysis is underway. We first check the capability of the thermal properties to operate properly in such conditions. The thermal properties for the materials: uranium dioxide, nuclear reactor corium (mixture consisting of 76% UO_2, 24% ZrO_2), zirconium, zirconium dioxide, stainless steel, alumina, silicon dioxide, iron oxide, molybdenum, aluminum and boron oxide are obtained by using the recommendations of Chapter 3 of Volume 1.

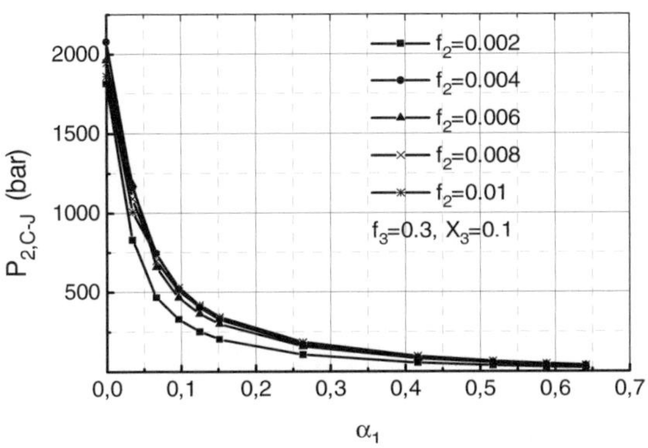

Fig. 16.19.1 Pressure behind the detonation discontinuity as a function of the void fraction before the discontinuity for UO_2. α_3 and X_3 are the void fraction and the melt mass concentration in front of the detonation wave, and f_2 and f_3 are the entrainment fractions of the coolant and of the melt after the detonation front

Then we perform detonation analysis with the solution proposed in Chapter 9 of Vol. 1. Figures 16.19.1, 16.19.2 and 16.19.3 show the pressure behind the detonation discontinuity as a function of the void fraction before the discontinuity for UO_2, Al_2O_3 and Fe for the same initial conditions. The entrainment of the melt into fine particles was set to 30%. The entrained water in thermal equilibrium behind the discontinuity is varied between 2 and 8%. We learn from this analysis that Al_2O_3 and Fe are able to produce a stronger detonation wave than UO_2. This phenomenon really happens in the experiments and seemed to be unexplained up to now in spite of the large number of theories in the literature.

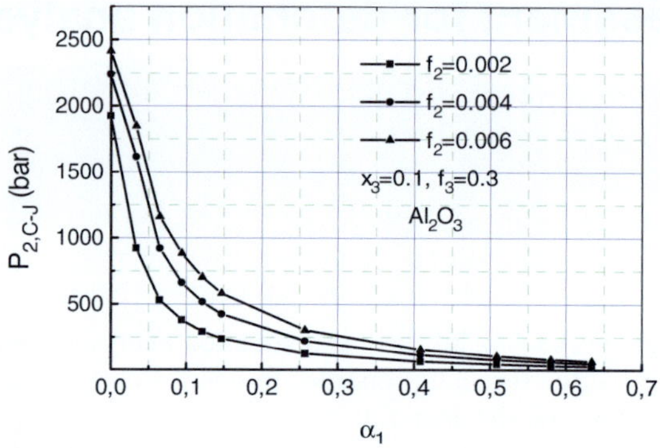

Fig. 16.19.2 Pressure behind the detonation discontinuity as a function of the void fraction before the discontinuity for Al_2O_3

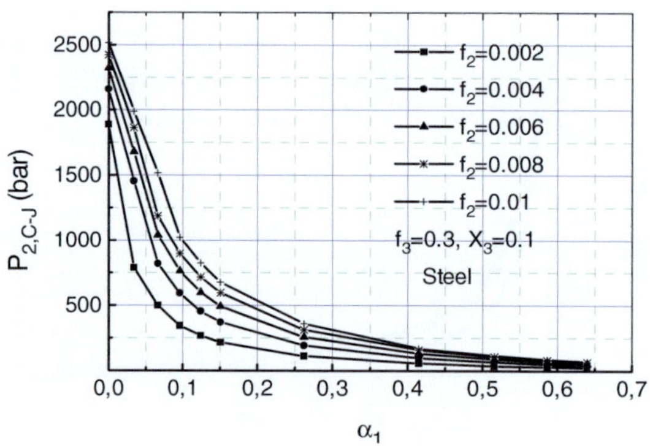

Fig. 16.19.3 Pressure behind the detonation discontinuity as a function of the void fraction before the discontinuity for Fe

Conclusion. The melt, water and gas properties work well in the detonation case. The substantial differences of the explosivity of the mixtures of different molten materials with water is explained with the differences of the caloric properties of the melt.

The reader will find additional information on this topic in *Kolev* and *Hulin* (1999), *Kolev* (2000).

References

Kolev, N.I., Hulin, H.: Detonation waves in melt-coolant interaction, Part.2: Applied analysis, MFCI Project. In: 6th Progress Meeting, CEA, Grenoble. EU Nr. (June 23-24, 1999)

Kolev, N.I.: Detonation waves in melt-coolant interaction. Part 1: Kernenergie 65, 254–260 (2000)

16.20 Other examples of 3D capabilities

16.20.1 Case 1. Rigid-body steady-rotation problem

This test problem reported in *Harison* and *Johnson* (1992) presents a hollow cylinder with symmetric flow in the azimuthal direction – Fig. 16.20.1. No axial and radial flow exists. The mass-conservation equation gives $\frac{\partial v}{\partial \theta} = 0$. The r direction momentum equation simplifies to $\rho \frac{v^2}{r} = \frac{\partial p}{\partial r}$, and the θ direction momentum equation gives $\frac{\partial p}{\partial \theta} = 0$. For constant rotational frequency ω, $\left(v(r) = r\omega\right)$, the analytical solution of the radial momentum equation is $p - p_0 = \frac{1}{2}\rho\omega^2\left(r^2 - r_0^2\right)$ or $p - p_0 = \frac{1}{2}\rho\left[v(r)\right]^2\left[1 - \left(\frac{r_0}{r}\right)^2\right]$.

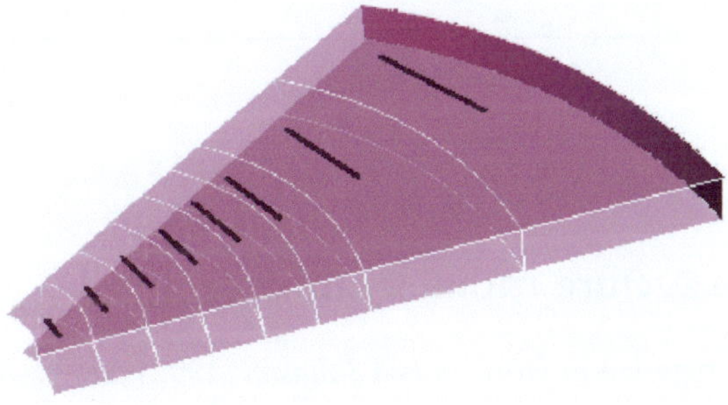

Fig. 16.20.1 Geometry of the test problem for rigid-body steady rotation

The discretization is presented in Fig. 16.20.1. The prescribed azimuthal velocity, the analytical pressure and the IVA5 prediction are presented in Table 16.20.1 and in Fig. 16.20.2. We see a very good agreement.

Table 16.20.1 Angular velocity and pressure as function of the radius

r [m]	v_θ [m/s]	$p_{analyt.}$ [10^5 Pa]	p_{IVA5} [10^5 Pa]
0.75	-	5.0	5.0
1.25	0.1923	5.00012	5.00009
1.75	0.2692	5.00030	5.00027
2.25	0.3462	5.00053	5.00051
2.75	0.4231	5.00083	5.00080
3.25	0.5	5.00118	5.00116
3.75	0.5769	5.00159	5.00157
4.75	0.7308	5.00260	5.00270
6.5	1.0	5.00492	5.00513

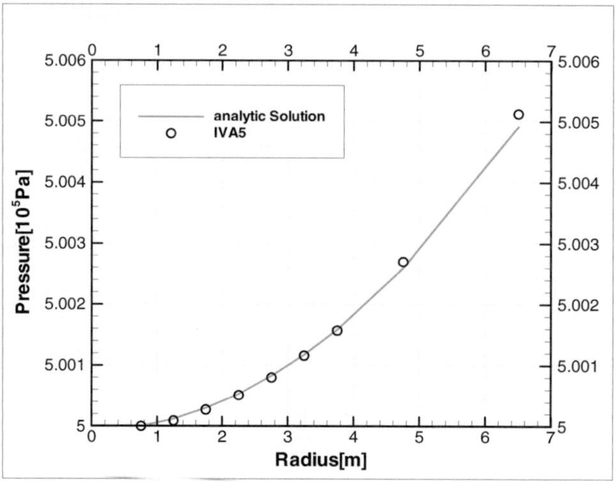

Fig. 16.20.2 Pressure as function of the radius

16.20.2 Case 2. Pure radial symmetric flow

This test problem reported in *Harison* and *Johnson* (1992) presents a hollow cylinder with symmetric flow in the radial direction. No axial and azimuthal flow exists. The mass conservation equation gives $\frac{\partial}{\partial r}(ru) = 0$. The r direction momentum equation simplifies to $\rho \frac{1}{2}\frac{\partial u^2}{\partial r} = -\frac{\partial p}{\partial r}$, and the θ direction momentum equation gives $\frac{\partial p}{\partial \theta} = 0$. From the mass conservation we have $u = u_0 \frac{r_0}{r}$. The analytical solution of the radial momentum equation is the well-known *Bernoulli* equation

$$p - p_0 = -\frac{1}{2}\rho\left(u^2 - u_0^2\right) \text{ or } p - p_0 = \frac{1}{2}\rho u_0^2\left[1 - \left(\frac{r_0}{r}\right)^2\right].$$

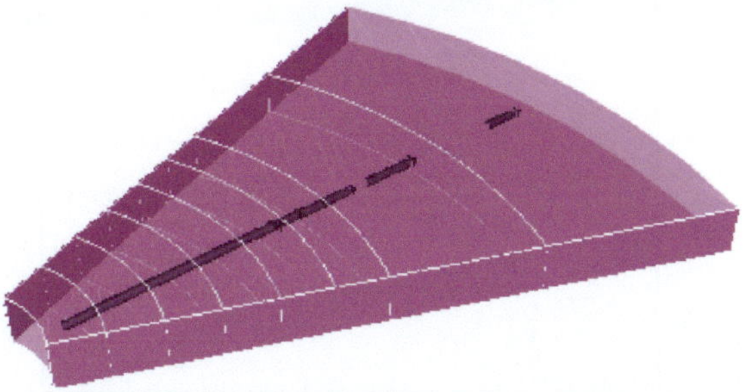

Fig. 16.20.3 Geometry of the test problem for pure radial symmetric flow

The discretization is presented in Fig. 16.20.3. The analytical pressure and the IVA5 prediction are presented in Table 16.20.2 and in Fig. 16.20.4 for radial velocities given in Table 16.20.3. Only the inlet radial velocity is prescribed. All others are computed by the code. We see a very good agreement.

Table 16.20.2 Pressure as function of the radius

r [m]	$p_{analyt.}$ [10^5 Pa]	p_{IVA5} [10^5 Pa]
1.25	5.0	5.0
1.75	5.0661	5.06615
2.25	5.0934	5.09337
2.75	5.1071	5.10715
3.25	5.1151	5.11508
3.75	5.1200	5.12005
4.75	5.1257	5.1257
6.5	5.1301	5.13006

Table 16.20.3 Radial velocity as a function of the radius

r [m]	v_r [m/s]
1.0	6.5
1.5	4.3333
2.0	3.25
2.5	2.6
3.0	2.1667
3.5	1.8571
4.0	1.625
5.5	1.1818
7.5	0.8667

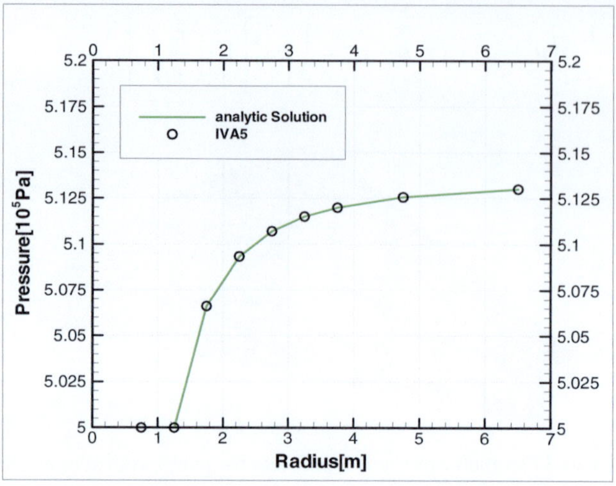

Fig. 16.20.4 Pressure as a function of the radius

16.20.3 Case 3. Radial-azimuthal symmetric flow

This test problem reported in *Harison* and *Johnson* (1992) presents a hollow cylinder with symmetric flow in the radial and azimuthal directions, Fig. 16.20.5, in fact a superposition of the previous two cases, rigid-body steady rotation and pure radial symmetric flow. No axial flow exists.

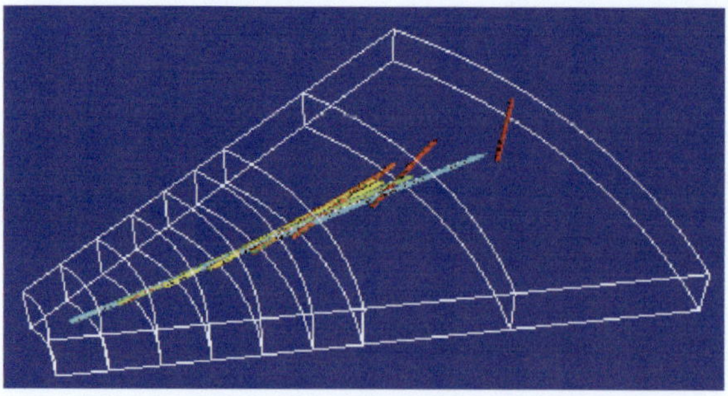

Fig. 16.20.5 Geometry of the test of the problem for radial-azimuthal symmetric flow

The mass conservation equation gives $\frac{\partial}{\partial r}(ru) = 0$. The r direction momentum equation simplifies to $\rho\left(\frac{1}{2}\frac{\partial u^2}{\partial r} - \frac{v^2}{r}\right) = -\frac{\partial p}{\partial r}$, and the θ direction momentum equation gives $\rho u\left(\frac{\partial v}{\partial r}+\frac{v}{r}\right)=-\frac{1}{r}\frac{\partial p}{\partial \theta}$. From the mass conservation we have $u=u_0\frac{r_0}{r}$. From the azimuthal symmetry, $\frac{\partial p}{\partial \theta}=0$, the θ direction momentum equation gives $\frac{\partial v}{\partial r}+\frac{v}{r}=0$ or $v=v_0\frac{r_0}{r}$. Taking into account the two solutions of the mass and of the θ momentum equation the radial momentum equations gives $\rho\frac{1}{r}\left(u^2+v^2\right)=\frac{\partial p}{\partial r}$ or $\rho\left(u_0^2+v_0^2\right)r_0^2\frac{1}{r^3}=\frac{\partial p}{\partial r}$ or

$$p-p_0=\rho\left(u_0^2+v_0^2\right)r_0^2\frac{1}{2}\left(\frac{1}{r_0^2}-\frac{1}{r^2}\right).$$

The analytical pressure and the IVA5 prediction are presented in Table 16.20.4 and in Fig. 16.20.6. We see a very good agreement.

Table 16.20.4 Pressure as function of the radius

r [m]	$p_{analyt.}$ [10^5 Pa]	p_{IVA5} [10^5 Pa]
0.75	5.0	5.0
1.25	5.0	5.00009
1.75	5.06624	5.06642
2.25	5.09350	5.09388
2.75	5.10729	5.10795
3.25	5.11523	5.11623
3.75	5.12021	5.12162
4.75	5.12587	5.12840
6.5	5.13023	5.13519

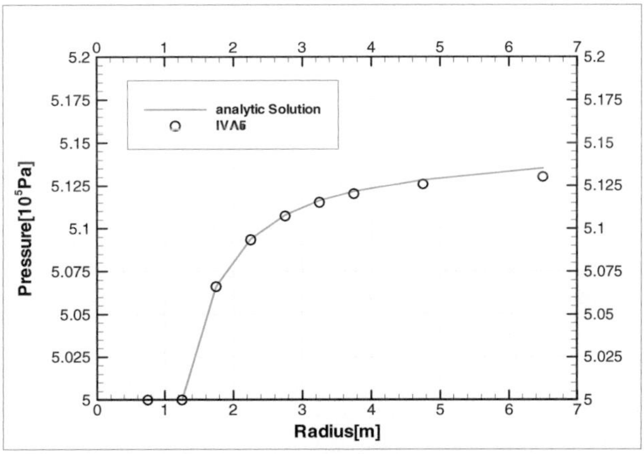

Fig. 16.20.6 Pressure as a function of the radius

16.20.4 Case 4. Small-break loss of coolant

The *Kevchishvili* and *Dementev* small-break loss-of-coolant experiment was performed at the Moscow Energetic Institute to study strong separation effects *Kevchishvili* and *Dementev* (1985). The 36 uniformly heated rods 16 mm in diameter and 1700 mm long are mounted within a shroud 159 mm in diameter in a pressure vessel 309 mm in diameter and 2100 mm high as shown in Fig. 16.20.7a. We simulate an experiment with a thermal power input of 67 kW and a cold leak from a break nozzle 22 mm in diameter. The initial state of the water inside the vessel was p = 123 bar, T = 558 K, and void fraction equal to zero. The void fractions in the rod bundle in the three different locations shown in Figs. 16.20.8a, b and c were recorded as a function of time.

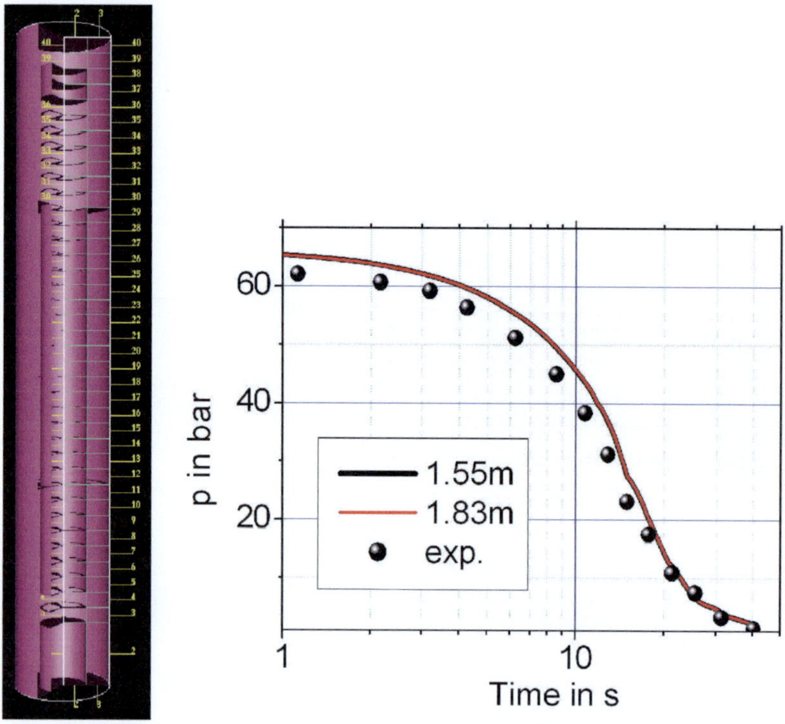

a) b)

Fig. 16.20.7 a) IVA6 geometry model for the *Kevchishvili* and *Dementev* small-break loss-of-coolant experiment. b) Pressure at level z = 1.83 m as a function of time. Comparison of the IVA6 prediction with the experiment of *Kevchishvili* and *Dementev* (1985). Heated bundle, 67 kW

I simulate the geometry with 2 radial, 3 angular, and 39 axial zones – 117 cells – as shown in Fig. 16.20.7a. I assume that the "vena contracta" coefficient at the nozzle is 0.8. The comparison with the experimental data is shown in Figs. 16.20.7b and 16.20.8. From this comparison the following conclusions can be drawn. The steam discharge between 7 and 11 s is stronger due to friction in the discharge pipe, which was not considered. The pressure compares well with the measurement. The end of the stratification at the three different positions of the test section is well predicted. In the upper part the stratification is well predicted. In the mid and lower parts the stratification is well predicted except for the region between 7 and 10 s where it happens faster in the simulation.

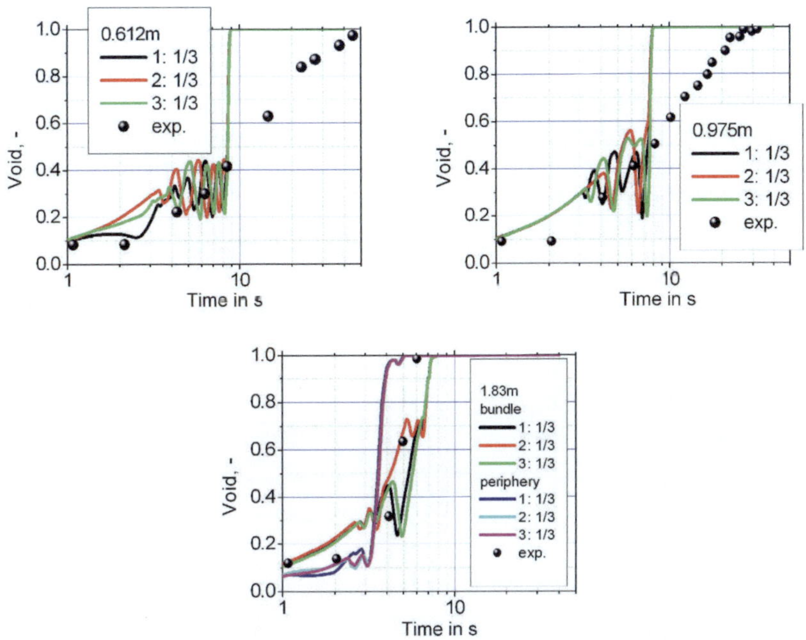

Fig. 16.20.8 The void fraction in a rod bundle on level z = 1.83, 0.975, 0.612 m as a function of time. Comparison of the IVA6 prediction with the experiment of *Kevchishvili* and *Dementev* (1985). Heated bundle, 67 kW

16.20.5 Case 5. Asymmetric steam–water interaction in a vessel

Now we turn back to the UPTF test No. 26 already introduced in Section 16.7. The phase B is of particular interest in this section. As already found in *Kolev* (1999) through a 3D simulation, in phase B the main steam–water interaction processes leading to entrainment happen in the vessel, whereas in phase A the 3D analysis results in no carryover from the vessel, supporting the correctness of the 1D analysis in Section 16.7. We present here details of the 3D analysis. The geometry of the vessel is presented in Fig. 16.20.1. We take advantage of the symmetry and simulate half of the vessel using 1200 ($12\times 5\times 20$) computational cells as shown in Fig. 16.20.2. The resolution of the discretization is higher in the upper plenum of the vessel. We simulate the asymmetric injection inside the integration volume by defining distributed mass sources at the corresponding level. The water injection is simulated as if directly injected into the vessel at the vessel wall and not as in the real experiment 1.5 m apart from the vessel inside the main coolant line. The steam mass conservation on average is clearly demonstrated in Fig. 16.20.3 where the steam injection nozzle mass flow is also presented.

As we realize from Fig. 16.20.3, there is a strong feedback of the water mass flow rate on the steam mass flow causing strong oscillations. Figure 16.20.3 presents the main result of this analysis. It shows no in-vessel water carryover during phase A and almost complete carryover after 600 s. There was water suction from the bottom of the test vessel as presented in Fig. 16.20.4a. This suction was not considered in our computation, which explains the overprediction of the entrainment in phase B.

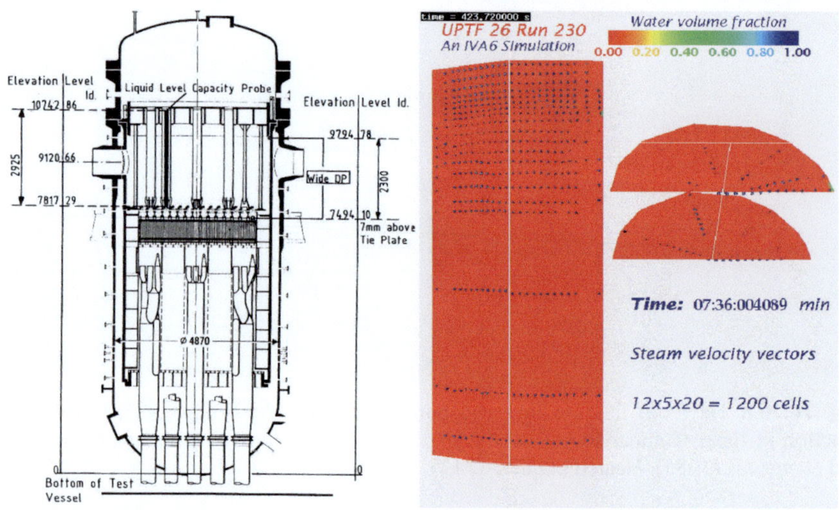

Fig. 16.20.9 3D test section simulating a full scale PWR for hot-leg water injection in a steaming core. Initial conditions for Phase A and intermediate state between Phase A and Phase B

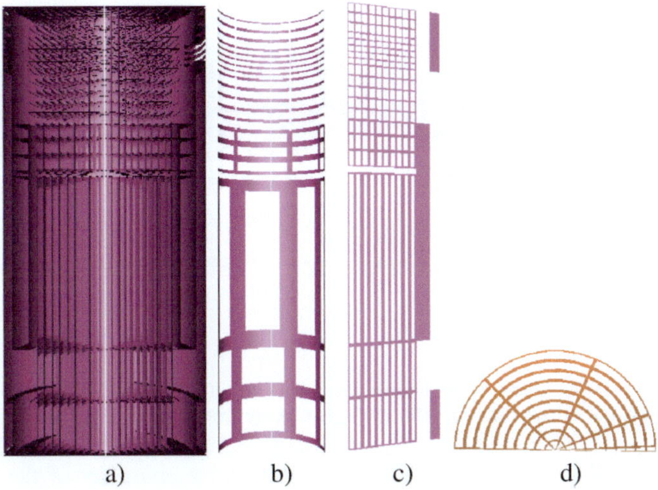

Fig. 16.20.10 a) IVA6 reactor pressure vessel geometry model for UPTF-Test 26 Run 230 simulation. b, c, d) Illustration of how the complicated technical structures are modeled by means of radial, angular, and axial surface permeabilities, respectively. The colored surface in the last three pictures is nonpermeable for flow in the corresponding plane

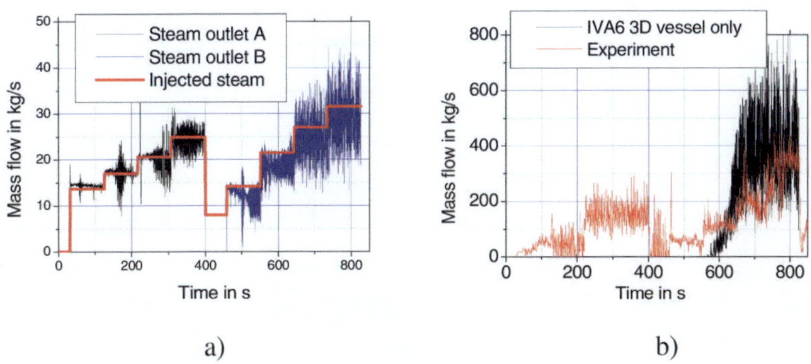

Fig. 16.20.11 3D case. a) Mass flow of injected steam and of the steam leaving the vessel as a function of time. b) Total mass flow of steam + water at the densitometer position as a function of time

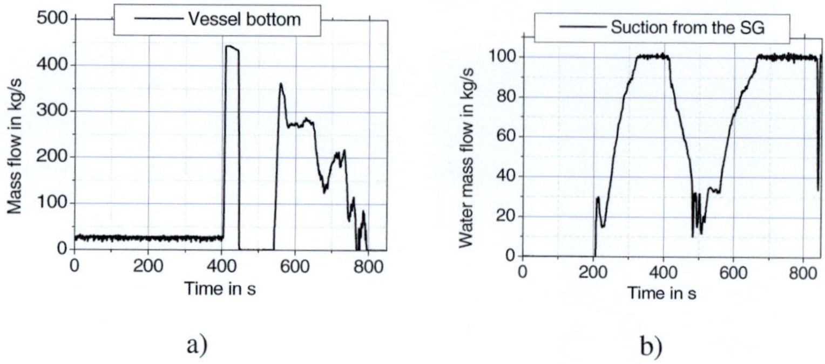

Fig. 16.20.12 a) Water mass flow suction through the bottom of the vessel as a function of time. b) Water mass flow suction from the steam generator separator as a function of time

A time-averaged Fig. 16.20.11 indicates the stepwise character of the total mass flow rate reflecting in a way the stepwise character of the steam injection. This is a clear indication that in both cases entrainment is happening inside the pipe. While, as already mentioned, in phase A this entrainment is dominant, in phase B we have superposition of entrainment inside the vessel and inside the pipe. The next 3D representation of the water content and the steam velocity vectors additionally clarifies this point – Figs. 16.20.5 and 16.20.6.

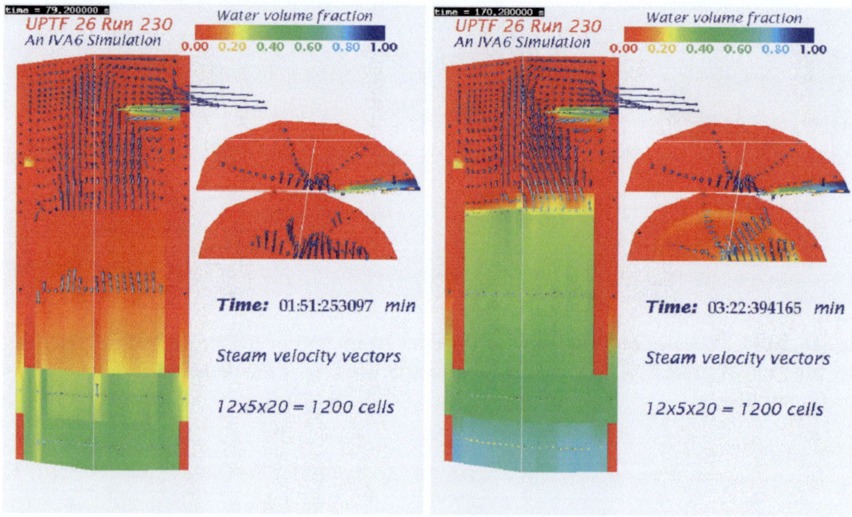

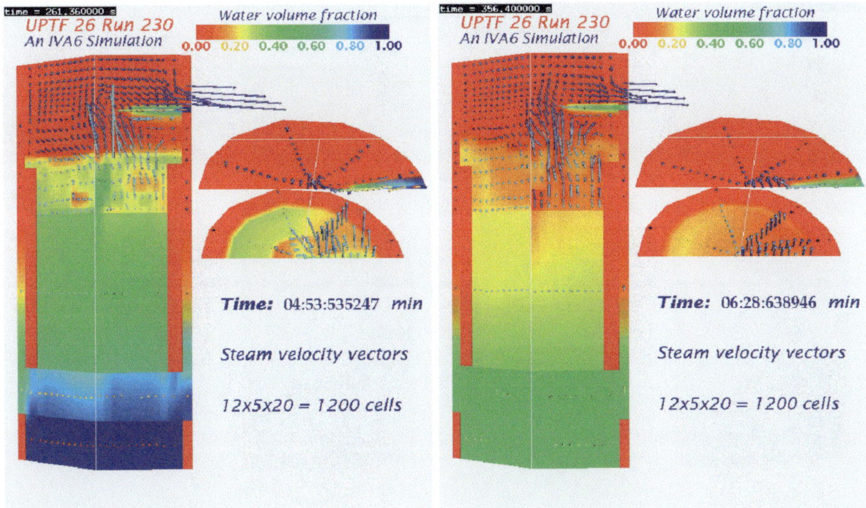

Fig. 16.20.13 Phase A-1, 2, 3, 4 Water volume fraction and steam velocity vectors as a function of space. The two horizontal cuts (left) are at the level above the steam injection and at the water injection

Conclusions. We observe a reasonable response of IVA6 during the simulation of UPTF experiment No. 26 run 230. We learn from this analysis that the combination of 3D representation of the vessel with 1D representation of the pipe will improve the result. Furthermore, complete multiblock simulation of the vessel and the large (0.75 m diameter) pipe is the desired approach for the future sophistication of the modeling technology.

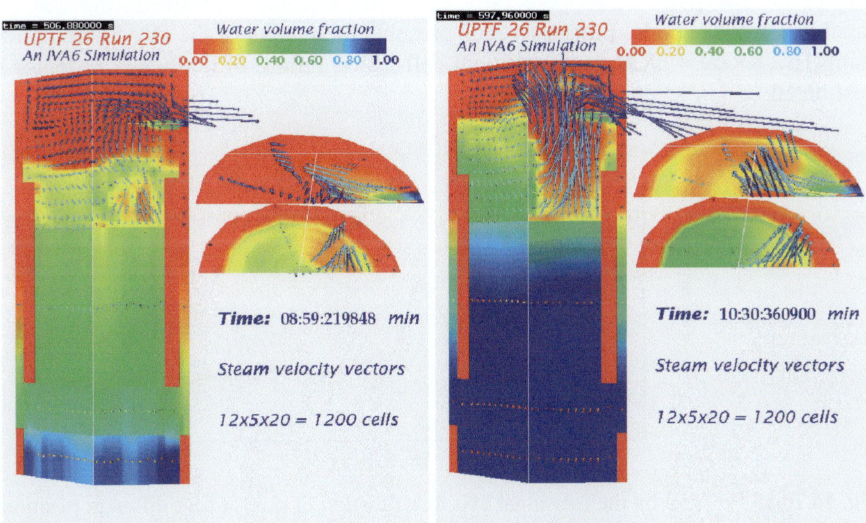

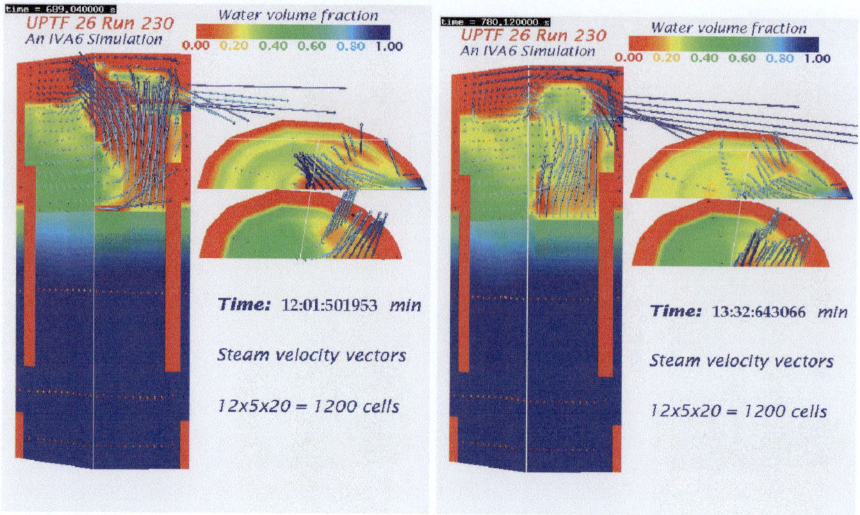

Fig. 16.20.14 Phase B-1, 2, 3, 4 Water volume fraction and steam velocity vectors as a function of space. The two horizontal cuts (left) are at the level above the steam injection and at the water injection

16.20.6 Case 6. Melt relocation in a pressure vessel

Between 1995 and 1998 we performed hundreds of 2D and 3D simulations mainly applied to the typical pressurized-water reactors (PWRs). Some of them are presented in Chapter 13 of Volume 1. The PWRs Steam Explosion Study is not a subject of this publication. We only give one example demonstrating the 3D capabilities of the code. Figure 16.20.7 presents the results of the IVA5-simulation using $16\times 8\times 69 = 8832$ computational cells. Side failure of the heavy reflector is postulated.

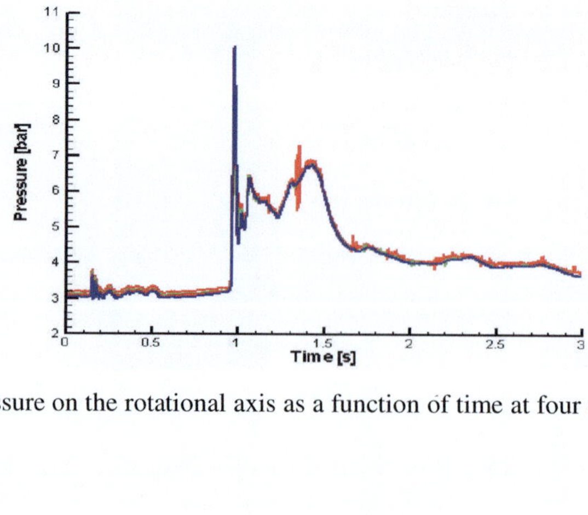

Fig. 16.20.15 Pressure on the rotational axis as a function of time at four different positions in the lower head

The material relocation is presented in Figs. 16.20.7a–c. The pressure on the axis as a function of time is given in Fig. 16.20.7. We see the full capability of IVA5 to model transient multiphase 3D flows in complicated geometry.

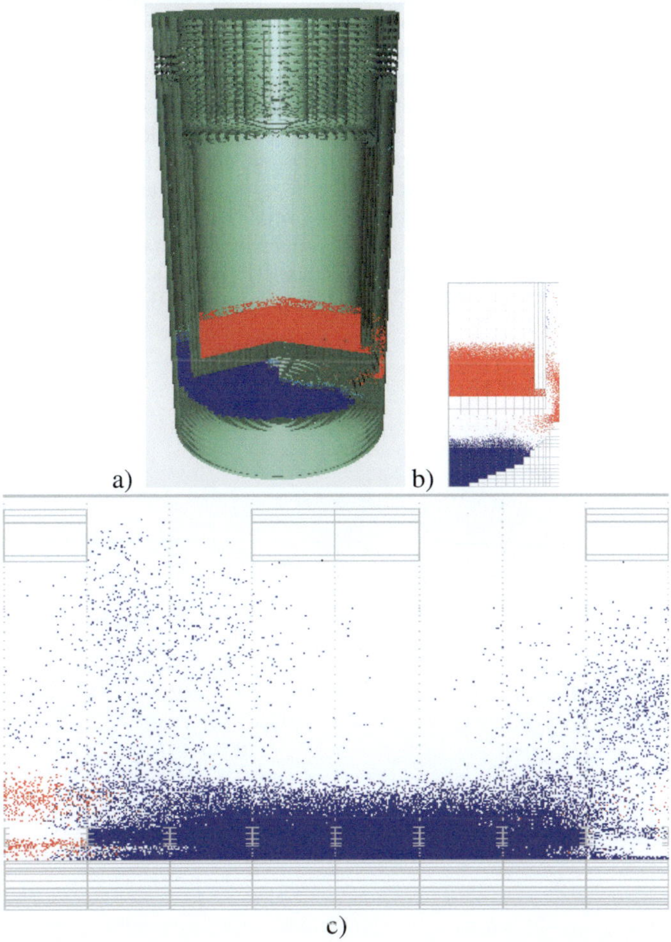

Fig. 16.20.16 a) Melt and water relocation in two different vertical planes 1 s after the begging of the failure of the heavy reflector. b) Melt and water inverter in the vertical plane crossing the failure of the heavy reflector. c) Melt and water in the downcomer, cylindrical surface close to the RPV wall

References

Harison, R.G., Johnson, E.C. (eds.): Developmental assessment of the multidimensional component in RELAP5 for Savannah River Site thermal hydraulic analysis, EGG-EAST-9803, DE93 001942. EG&G Idaho, Inc., Idaho Falls (1992)

Kevchishvili, N.A., Dementev, B.S.: Investigation of the influence of the decay heat on the blow down characteristics of steam-water mixtures. Teploenergetika 7, 67 (1985)

Kolev, N.I.: Verification of IVA5 computer code for melt-water interaction analysis, Part 1: Single-phase flow, Part 2: Two-phase flow, three-phase flow with cold and hot solid spheres, Part 3: Three-phase flow with dynamic fragmentation and coalescence, Part 4: Three-phase flow with dynamic fragmentation and coalescence – alumina experiments. In: Proc of the Ninth International Topical Meeting on Nuclear Reactor Thermal Hydraulics (NURETH-9), San Francisco, California, October 3-8 (1999)

Kolev, N.I., Seitz, H., Roloff-Bock, I.: Hot-leg injection: 3D versus 1D three velocity fields modeling and comparison with UPTF experiment, Siemens report KWU NA-T/00/E027, R&D Project 903060, Verification IVA6 (2000)

16.21 General conclusions: Is it possible to design a universal multiphase flow analyzer?

Comparison with a variety of experiments demonstrated the power and the weaknesses of the methods described in this monograph and qualified IVA computer code as a code representing the state-of-the-art in the field of multiphase flows. The code is capable of predicting multiphase flow behavior in complicated 3D geometries and industrial networks. The code is able to predict melt–water interaction with a well-quantified uncertainty band. Reducing this uncertainty band will demand future sophistication in the directions indicated in many places in this work.

IVA has already been successfully applied for the melt–water interaction study of the Framatome European Pressurized Water Reactor (EPR), for the Boiling Water Reactor SWR1000–for more details see *Kolev* (2002, 2003, 2004), for the research reactor FRM II in Munich and many other studies of multiphase processes in the nuclear industry.

After years of research in this field I ask myself the following question: Is it possible to design a universal multiphase flow analyzer? In what follows I will try to give the answer.

Engineers and scientists created hundreds of computer codes for description of more or less specific realizations of multiphase flows. If one compares the structure of these codes one is astonished by the waste of human resources for programming repeating model elements like equations of state, friction laws in a variety of geometry, heat-transfer coefficients, algebraic equation solvers, data-handling procedures, graphical environment, etc. It is hard to believe that the best solution for the specific subphenomenon is available in all codes. Looking in other branches of technology like computer-chips production we realize that the revolutionary idea of having common "chips" within complex applications is very far from its practical realization in computational multiphase flow dynamics. Following this line of arguments I expressed several times in my publications explicitly or implicitly (see *Kolev* 1990, 2000) the idea that it is possible to create a universal multiphase flow analyzer in the sense of computer architecture, that is capable of absorbing adequate multiphase knowledge data base specified in Appendix 1. The subject of this chapter is to summarize some of the main ideas, some of them already realized by this author, on the way of creating such a computer-code architecture, to illustrate how they work, and to propose an outlook regarding what are the challenges in the future developments.

We confine our attention deliberately to the solution of the so-called local volume and time-averaged system of PDEs for a simple reason: Direct numerical resolution of interacting fluids is possible as demonstrated for small scales by many researchers, but for real large-scale engineering problems it is still very expensive. I believe that this will remain so for decades.

16.21.1 The idea of multiple velocity fields

A field is defined as a part of the flow having its own well-defined spatially and temporally evolving boundary to its environment: another field or a wall. *Zimansky* (1968) uses the same definition for phase but we understand phase as a synonym for state of aggregate. A field possesses many attributes. One flow attribute is for instance the velocity. Thermodynamic attributes are the density specific enthalpy, entropy, etc. Transport attributes are, e.g., viscosities, conductivities, etc. Constituent attributes are for instance the number of chemical components inside the field, their mass concentration etc. The field is in itself a heterogeneous structure, e.g., internal circulation of a drop, boundary layers for thermal and mechanical interactions, etc. The field can be represented with its internal averaged properties and boundary-layer properties. A complex flow model my then be constructing as consisting of many fields inside the mixture.

The mixture is characterized by the number of the fields and their volumetric concentrations inside the field. The scale at which the mixture is modeled may be large enough to contain several processes that happen in subscales. This is valid for the time scale as well for the spatial resolution scale. Therefore, local volume and time averaging is necessary to obtain the conservation laws valid on the macroscale. During the averaging a record is kept for the subscale especially how formally the processes are incorporated in the large-scale method. Mathematical consistency is crucial for creating functioning general architecture.

16.21.2 Do we have adequate mathematical reflection of the conservation laws based on averaging?

The answer is yes. The system of partial differential equations reflecting the instantaneous and local conservation of mass, momentum und energy has existed since the 19th century. The 20th century generated many ideas about averaging this equation over large-scale volume, e.g., assembly averaging, statistical averaging etc. Among the mathematical tools for averaging the spatial averaging theorem by *Slattery* (1967), *Whitaker* (1967, 1969, 1977, 1985), *Anderson* and *Jackson* (1967), together with the *Leibnitz* rule allows application of averaging of the instant conservation equations over a volume. Considering as one special field the structure defining the channel, results in the so-called porous body concept *Sha*, et al. (1984). In *Kolev* (1990) and in this book the fields are considered as consisting of many chemical components allowing molecular diffusion inside them if they are continuous. The next step is time averaging also called Reynolds averaging of the resulting systems. The resulting system is then what we need to integrate in space and time to describe in averaged terms the behavior of the field. During the averaging processes a large number of terms originate having particular meaning. The elaboration of these terms is a long-lived task in the science. A useful representation of the sate-of-the-art in this field is given in Vol. 2 of this monograph.

16.21.3 Coexisting fields

Now, we can define the number of the abstract fields to which we would like to associate physical properties. The maximum number of the fields have to be specified and then the place holders for the interaction terms. The general approach requires clear physical and mathematical understanding of the structure of these terms. Usually they are strongly nonlinear.

16.21.4 Geometry definition

We adopt the following strategy: We consider the technical systems as a combination of pipe-networks with valves, pumps and other components that may be connected with 3D volumes. That is why our geometry definition contains three elements: (1) pipe network; (2) multiple blocks in boundary fitted coordinates interconnected to each other. In many daily applications (3) a single 3D-block in Cartesian or cylindrical geometry is sufficient and very useful. It is useful also for development purposes. The set of conservation equations is then defined in this region. The separation of the computational region from the real wall requires boundary conditions defining the action of the environment on the specified volume.

16.21.5 Solvers

The system of partial differential equations is usually transferred in a discrete form with a priory specified accuracy of discretization. Designing working solvers for the resulting nonhomogeneous algebraic system with variable coefficients is an art. The challenge is to design as strong as possible coupling within the different processes within a single step. A powerful tool for this is the entropy concept presented in this book. Some examples are given in this work. This method will surely be improved in the future.

16.21.6 Summary of the ideas

I deliberately do not go into details in this presentation in order to make the main structure of the universal flow analyzer clear:

(1) The flow is represented by coexisting fields inside a control volume.
(2) The control volume having simple or complex geometry is separated from the environment by surfaces at which the action of the environment is replaced by boundary conditions.
(3) Each field possesses a variety of attributes originating from the interaction with the environment and from the mathematical procedure of the averaging.
(4) The coexisting fields are described by a system of well-founded partial differential equations that are local volume and time averaged.
(5) The geometry is a combination of pipe networks, multiple interconnected volumes that are boundary fitted.
(6) The solution of discretized form is based on strong coupling of the interactions between the fields. The entropy concept is a powerful tool for simplifying the mathematical description without losing generality.

16.21.7 Is there any chance to overcome the overwhelming complexity and design algorithms that are of practical use?

Yes. We will give here one example based on three multicomponent velocity fields in porous structure that may change its geometry in time. The foundations of the computer code IVA described in this book allows applications of this multiphase flow analyzer for mathematical simulation of variety of processes. From two-phase gas-plasma multicomponent hydrogen detonation in pipe-network with dissociation of the gases, Fig. 16.1.3, through condensation water-steam shock waves in complex pipe networks, Fig. 16.1.4, gas solution and dissolution in liquids, dissolved gas release from water in pipe network and gas-slug formation and transport, Fig. 16.1.5, diesel injection problems, Fig. 16.1.6, particles sedimentation in water, Fig. 16.1.7, turbulent mixing and transport in a NPP sump, Fig. 16.1.8, thermite injection by a high-pressure steam–hydrogen mixture into an air environment, melt–water interaction in postulated SWR 1000 severe accidents, alumina melt jet dropped into a subcooled water, urania melt jet dropped into water, Fig. 16.1.9, void formation in existing, Fig. 16.1.10, or future boiling water reactors, Fig. 16.1.11, all this demonstrates the capability of the single model architecture to handle different material systems, different intensities of interactions, large variety of the spatial and temporal scales of the simulated processes.

However, if moving to new application, validation with a set of particular experimental data is mandatory, for elaboration of the achievable prediction accuracy, experience for the integration of the conservation equations under the particular initial and boundary conditions, etc.

Conclusions

1) Computational analysis of multiphase flows in practice requires special computer code architecture that is not trivially derivable from the single-phase fluid mechanics and thermodynamics.
2) Universal code architecture follows rigorously the analytical procedure of gaining the local volume- and time-averaged conservation equations reflecting elements like fields, their attributes, their interfacial exchange terms, etc.
3) The coupling between the nonlinear equations with variable coefficients in the system is designed as close as possible within a single analytical step. The entropy concept facilitates the coupling most efficiently.
4) Such an approach allows the best single model of a particular process or mathematical manipulation procedure to be multiply used by all different applications.
5) Such an approach unifies the efforts of teams in using components with prescribed interfaces derived from the basic derivation of the conservation equations. So, with time the number of the model elements will grow and become exchangeable. The data base behind the analyzer will also grow.
6) Graphical visualization of the processes is crucial not only for the development but later also for understanding better the simulated processes.

Appendix 1 Knowledge data base required to design a universal multiphase flow analyzer

1. technical thermodynamics; simple and multicomponent systems, equations of state;
2. behavior of metastable liquids;
3. heat and mass transfer in single- and complex multiphase systems;
4. fluid mechanics;
5. review of the world literature in particular fields of multiphase flows and processes associated with them;
6. multiphase flows; transient; one-, two- and three-dimensional; in simple geometry; in complicated geometry;
7. interaction of multiphase flows with technical structures; evaporation; condensation; spontaneous evaporation; spontaneous condensation; cavitation; flow instabilities; flow-induced vibrations; starting of steam generators; computation of forces caused by fluid transients;

8. design of components, apparatus and facilities based on mathematical modeling of complex thermal processes including complicated multiphase flows;
9. pressure and concentration waves in pipes, complex pipe systems, complex three-dimensional equipment;
10. analysis of explosive fluid–fluid interactions, melt–water interactions; steam explosion in nuclear facilities, in metallurgical facilities, in manufacturing, in space technology, in combustion motors; hydrogen deflagration and detonation, powder combustion and powder detonation; detonation of metastable fluids;
11. numerical methods for analysis of complex flow processes;
12. design of complex computer codes for simulation of the system behavior of coupled components in single-, two-, and multiphase flow region;
13. computing: numerical methods for integration of systems of ordinary and partial differential equations;
14. uncertainty analysis of computational methods;
15. nuclear technology: developing and applying mathematical methods for description and simulation of the physics of thermohydraulic processes in the reactor, primary-, secondary, and tertiary circuit, in the containment in normal operation, in transients and by severe accidents with catastrophic consequences;
16. quality assurance of complicated thermohydraulic computations;
17. quality assurance of developments of code systems for modeling of complex thermodynamic processes;
18. education of specialists in all of the above fields, courses, consultations

References

Anderson, T.B., Jackson, R.: A fluid mechanical description of fluidized beds. Ind. Eng. Fundam. 6, 527 (1967)

Kolev, N.I.: Derivatives for the state equations of multi-component mixtures for universal multi-component flow models. Nuclear Science and Engineering 108, 74–87 (1990)

Kolev, N.I.: Needs of industrial fluid dynamics applications. Invited lecture, 2000 ASME Fluids Engineering Division Summer Meeting (FEDSM), Industry Exchange Program, Sheraton Boston Hotel, Boston, Massachusetts, June 11-15 (2000)

Kolev, N.I.: 3D-Multiphase-Fluid-Dynamik: Reduzierung der Dampffeuchte am RDB-Austritt der Siedewasserreaktoren, Betriebsleitertagung, Geneve (November 27-29, 2002); in English: Controlling the moisture content in the steam by using nozzles in BWR's, Eleventh International Conference on Nuclear Engineering, ICONE 11, Keio Plaza Inter-Continental, Shinjuku, Tokyo, Japan (April 20-23, 2003); ICONE11-36306, The 10th International Topical Meeting on Nuclear Reactor Thermal Hydraulics, Seoul, Korea, p. E00301 (October 5-9, 2003)

Kolev, N.I.: Controlling the moisture content in the steam by using nozzles in BWR'S. In: Eleventh International Conference on Nuclear Engineering, ICONE 11, Keio Plaza Inter-Continental, Shinjuku, Tokyo, Japan, ICONE11-36306; ETPFG, European Two-Phase Flow Group Meeting 2003 Norway (April 20-23, 2003); The 10th International Topical Meeting on Nuclear Reactor Thermal Hydraulics (NURETH-10) Seoul, Korea (October 5-9, 2003)

Kolev, N.I.: External cooling - the SWR 1000 severe accident management strategy Part 1: Motivation, strategy, analysis: melt phase, vessel integrity during melt-water interaction. In: Proceedings of 6th International Topical Meeting on Nuclear Reactor Thermal Hydraulics, Operations and Safety, Nara-Ken New Public Hall, Nara, Japan, NUTHOS6-N6P381 (October 4-8, 2004)

Sha, T., Chao, B.T., Soo, S.L.: Porous-media formulation for multi-phase flow with heat transfer. Nucl. Eng. Des. 82, 93–106 (1984)

Slattery, J.C.: Flow of viscoelastic fluids through porous media. AI. Ch. E. J. 13, 1066 (1967)

Whitaker, S.: Diffusion and dispersion in porous media. AI. Ch. E. J. 13, 420 (1967)

Whitaker, S.: Advances in theory of fluid motion in porous media. Ind. Eng. Chem. 61(12), 14–28 (1969)

Whitaker, S.: Experimental principles of heat transfer. Pergamon Press Inc., New York (1977)

Whitaker, S.: A simple geometrical derivation of the spatial averaging theorem. Chem. Eng. Educ., 18–21, 50–52 (1985)

Zemansky, M.W.: Heat and thermodynamics, 5th edn. McGraw-Hill Book Company, New York (1968)

Index

D

E

F

G

H

I

J

K

L

M

N

O

P

R

MIX
Papier aus verantwortungsvollen Quellen
Paper from responsible sources
FSC® C105338

If you have any concerns about our products,
you can contact us on
ProductSafety@springernature.com

In case Publisher is established outside the EU,
the EU authorized representative is:
Springer Nature Customer Service Center GmbH
Europaplatz 3, 69115 Heidelberg, Germany

Printed by Libri Plureos GmbH
in Hamburg, Germany